AF676730

Environmental effects on gut health in production animals

Environmental effects on gut health in production animals

Edited by

Michael H. Kogut
Mariano E. Fernández-Miyakawa

BRILL | WAGENINGEN ACADEMIC

The Library of Congress Cataloging-in-Publication Data is available online at https://catalog.loc.gov
LC record available at https://lccn.loc.gov/2024938400

Typeface for the Latin, Greek, and Cyrillic scripts: "Brill". See and download: brill.com/brill-typeface.

ISBN 978-90-04-69545-0 (hardback)
ISBN 978-90-04-69546-7 (e-book)
DOI 10.3920/9789004695467

Copyright 2024 by Michael H. Kogut and Mariano E. Fernández-Miyakawa. Published by Koninklijke Brill BV, Leiden, The Netherlands.
Koninklijke Brill BV incorporates the imprints Brill, Brill Nijhoff, Brill Schöningh, Brill Fink, Brill mentis, Brill Wageningen Academic, Vandenhoeck & Ruprecht, Böhlau and V&R unipress.
Koninklijke Brill BV reserves the right to protect this publication against unauthorized use. Requests for re-use and/or translations must be addressed to Koninklijke Brill BV via brill.com or copyright.com.

This book is printed on acid-free paper and produced in a sustainable manner.

Contents

Tables and figures

Tables

Figures

CHAPTER 1

Introduction

Michael H. Kogut and Mariano E. Fernández-Miyakawa

There is a direct relationship between animal health, welfare, and performance and a "healthy" gastrointestinal tract (GIT), with the GIT responsible for regulating physiological homeostasis that provides the animal the ability to withstand infectious and non-infectious stressors. Gut health encompasses a number of physiological and functional features, including nutrient digestion and absorption, host metabolism and energy generation, a stable microbiome, mucus layer development, barrier function, and mucosal immune responses all of which are required to interact together so that the animal is able to perform its physiological functions near 100% of its genetic potential. Comprehension of gut health requires the elucidation of the interactions between all of these components. Understanding the interactions between the environmental stressors and the diverse physiological feature of the GIT emphasizes the extent of areas encompassed by gut health and the ability to regulate animal production.

Optimal gut health is of vital importance to the performance of production animals (fish, poultry, swine, cattle). Production animals reared under conventional conditions of intensive production are subjected to various exogenous and endogenous sources of environmental factors that can impact gut health. Exogenous factors are environmental stressors derived from external sources connected with diet, infectious disease, mycotoxin exposure, climate (heat and cold), management practices, biosecurity level, housing, litter, feed access, quality, and components. Endogenous factors are host-related such as age, sex, genetics, and breed or are made within the host's body, such as hormones and neurotransmitters, in response to a stimulus, stressor, or trigger. Endogenous factors serve to communicate signals both locally and distantly in the body. Recent studies are beginning to understand the relationship between these different factors and a 'healthy' gastrointestinal tract (GIT) that is necessary to maximize production and improve animal health and welfare.

To date, the literature has focused on the development of alternatives to antibiotics for sustainable animal production advances in the development of functional feed additives such as pre-, pro-, and synbiotics, phytobiotics, nutritional antioxidants, essential oils, organic acids, anti-nutritional enzymes, and immune modulators for the regulation of gut health and functionality. However, exposure to microorganisms in the local environment influence the intestinal microbiota; this is especially poignant in poultry that hatch without the presence of the hen's microbiota and are colonized from environmental sources only (Rychik, 2020; Ducatelle *et al.*, 2023).

© MICHAEL H. KOGUT AND MARIANO E. FERNÁNDEZ-MIYAKAWA, 2024 | DOI:10.3920/9789004695467_002

Instead, this book will concentrate on the interactions between the exogenous and endogenous environmental stressors on the diverse physiological features of the GIT including nutrient digestion and absorption, host metabolism, the microbiome, barrier function, and mucosal immune responses and the ability of a healthy gut to regulate animal production.

References

Ducatelle, R., Goosens, E., Eeckhaut, V., Van Immerseel, F. (2023). Poultry gut health and beyond. *Animal Nutrition* 13: 240–248. https://doi.org/10.1016/j.aninu.2023.03.005.

Rychlik, I. (2020). Composition and function of chicken gut microbiota. *Animals* 10: 103. https://doi.org/10.3390/ani10010103.

CHAPTER 2

Management effects on poultry gut health

Edgar O. Oviedo-Rondón

Prestage Department of Poultry Science, North Carolina State University, Raleigh, NC 27695, USA; eooviedo@ncsu.edu

Abstract

Decisions and actions related to poultry management can determine birds' developmental, metabolic, physiological, endocrinological, and immune responses to maintain adequate intestinal health. The environment, husbandry, sanitary, and feeding management practices also affect the gut microbiota of avian species. This chapter will review the role of the primary management factors that may affect immediate and long-term gut health. The effects of breeder management and nutrition on pullet and hen microbiota development, egg properties, transfer of nutrients, microbiota, hormones, and growth factors to offspring and their gut development will be discussed. The role of the embryo and early post-hatch development will be emphasized. Perinatal topics include the impact of hatching egg management and disinfection, incubation conditions, transportation to the farm, placement in the chicken house, and brooding. These typical husbandry activities can influence fundamental mechanisms dictating the ability of birds to develop appropriate responses when facing gut health challenges. The early access to feed and water, contact with environmental microbiota, and antibiotic applications will be debated. Housing type and environmental conditions may be more prominent in gut health than nutrition and feeding. High stocking density and heat stress can be the primary triggers for gut health issues, and their implications will be described. Lighting is an excellent tool to control bird feeding behavior and mitigate stress responsible for many of the gut problems discussed. Good water quality should always be considered for poultry gut health. House ventilation and litter management can aid in gut microbiome development and intestinal health. Finally, feeding management and minimizing stress during feed withdrawal can contribute to maintaining healthy guts until the processing plant. Considering all these management factors, it is possible to establish the main pillars of gut health, and all other disease control strategies could be more efficient.

Keywords

housing – lighting – feeding – incubation – microflora – gut development

© EDGAR O. OVIEDO-RONDÓN, 2024 | DOI:10.3920/9789004695467_003

1 Introduction

Management decisions and daily husbandry practices affect the development of poultry's gastrointestinal tract (GIT), microbiome, and immunity (Wickramasuriya *et al.*, 2022). Environmental comfort, biosecurity, and sanitary conditions also depend on management. Consequently, management is pivotal in preventing gut health and food pathogens (Tsiouris, 2016; Sibanda *et al.*, 2018; Shynkaruk *et al.*, 2022). This premise has become more evident during the past decade with the growing trend of antibiotic-free production worldwide (Smith, 2011, 2019), where management has become the primary tool to control gut health issues (Oviedo-Rondón, 2019).

The intestinal health of poultry starts with hen health and proper incubation conditions. Hens transfer their genetic material, nutrients, hormones, and microbiota into the egg. The characteristics of the hatching egg will affect the embryo and offspring post-hatch development. The egg albumen and eggshell properties can also influence embryo GIT development. Egg properties can affect gas exchange and moisture loss during incubation, which is essential for proper embryo development (Olkowski *et al.*, 2015). Breeder age, management, and nutrition during rearing and laying phases affect nutrient and immunity transfer to the egg, hen gut microbiome, hormonal status, egg nutrient content, and eggshell traits. The sanitary conditions of breeder houses, vaccination programs, immunological challenges, and infections can also modify the egg properties and progeny responses to pathogens.

Egg disinfection, egg transport, and manipulation can affect the eggshell microbiome, which could be transferred to offspring at hatch. Egg storage can negatively affect the hatchability, chick quality, performance, and immunocompetence parameters of broiler chickens (Goliomytis *et al.*, 2015). In hatchlings, intestinal morphology and nutrient transporters can be affected for prolonged egg storage (Yalçin *et al.*, 2016, 2017).

Incubation conditions are known to affect embryo development. Temperature, humidity, egg turning, ventilation, air gas concentrations, and lighting can affect GIT development and immunity (Oznurlu *et al.*, 2010). Temperature profiles can affect the time of hatch and the length of the hatch window (Agyekum *et al.*, 2022). The delay in having contact with feed and water after hatching may slow down vitelline sac absorption (van der Wagt *et al.*, 2020), gut-associated lymphoid tissue (GALT) development (Shira *et al.*, 2005), and increase exposure to microorganisms present in the surrounding eggshells (Ganjali *et al.*, 2015; Proszkowiec-Weglarz *et al.*, 2020, 2022; de Jong *et al.*, 2022). Research studies have commonly compared the impacts of feed delay between immediate access to feed and water post-hatch and delays of 36 or 48 hours (Ganjali *et al.*, 2015;). However, any delay in feeding can also affect hatchling gut development if, additionally, the perinatal environment in the hatcher, hatchery processing room, or during transportation to the farm is not adequate in temperature, humidity, or with low oxygen concentration and high

CO_2 (Barri, 2008; Barri *et al.*, 2011; Shinde *et al.*, 2015). Post-hatch feed and water deprivation have long-lasting effects on GALT development (Shira *et al.*, 2005), performance, and welfare of chickens (de Jong *et al.*, 2017).

Transportation conditions can affect the development of the intestine and yolk sac utilization in chickens (Mikec *et al.*, 2006). Time to access feed and water is critical to maintaining gut mucosa development, morphology, and immune response (Barri, 2008; Barri *et al.*, 2011). Early brooding temperatures, ventilation, water temperature, and light can influence feed and water consumption. These in-house environmental factors will influence poultry's voluntary feed intake, stress, and gut physiology for life.

Litter quality depends mainly on management pre-chicken placement, ventilation, and water line management. Litter microbial load can greatly influence gut microbial succession and health (Schokker *et al.*, 2021; Bindari and Gerber, 2022). Litter quality, bedding material, frequency of cleaning, and management practices between flocks and during flocks can affect how microorganisms such as protozoa, bacteria, and viruses enter in contact with poultry and start colonizing their guts (Bucher *et al.*, 2020; de Toledo *et al.*, 2020; Díaz Carrasco *et al.*, 2022).

Water line pressure, temperature, and physicochemical properties are essential in water ingestion and excretion, gut pH, and function. Water characteristics vary in every house and farm. Consequently, the water treatment methodologies vary according to the source, region, and farms. Frequently, water is used to provide medications and nutritional supplements. Interventions in drinking water can affect microbiological, physical, and chemical properties.

At harvest, the length of the withdrawal period, environmental conditions, stress, and time of transport to the plant may change plasma hormone and metabolite concentrations (Nijdam *et al.*, 2005) and alter the gut microbiota. Under those conditions, pathobionts of food safety implications proliferate and affect their presence in the carcasses at the processing plant. This chapter will review these management factors studied and demonstrate implications on important gut health indicators. Nutritional aspects and the effects of feed additives will not be included in this discussion.

2 Biosecurity

Biosecurity programs and practices are the means to prevent the introduction, spread, and persistence of pathogens in livestock and poultry production systems. Gut health is easier to achieve when biosecurity is maintained in the production system (Sibanda *et al.*, 2018; Smith, 2019; El-Hack *et al.*, 2021). The structure of biosecurity programs, principles, and implementation have been extensively discussed (Van Meirhaeghe *et al.*, 2019). Potential infection with new pathogens that directly affect the intestine or other organs with systemic effects may activate responses that end up with dysbiosis or more complex intestinal health problems like necrotic enteritis in broilers (Tsiouris, 2016), and layers (Goossens *et al.*, 2020). Biosecurity measurements are partially effective for controlling common pathogens (Assefa, 2022). The breaches in daily biosecurity practices are frequent. However, constant training and new electronic technologies can help to improve compliance with biosecurity rules (Racicot *et al.*, 2012a, b). Generally, audits or constant retraining do not influence medium-term compliance with biosecurity. Visible cameras or tracking sensors can help to increase short-term compliance with biosecurity measures required when entering and exiting poultry barns. However, Racicot *et al.* (2012a) observed that compliance declined six months after implementing video cameras and electronic sensors. This indicates constant retraining is necessary, but human factors such as personality traits, experience, and education should be considered to enhance the efficacy of biosecurity (Racicot *et al.*, 2012c). Since it is difficult to achieve compliance with biosecurity, other management factors may play a more significant role in gut health.

3 Management of pullet and hen breeders on offspring gut health

Plenty of evidence indicates that an individual's gut health may be affected by the maternal background (Babacanoğlu and Yalçin, 2014; Akhlaghi *et al.*, 2013; Bowling *et al.*, 2018; Darras, 2019). Hens may transmit to progeny the genetics, nutrients, hormones, and immunity (Verwoolde *et al.*, 2022) in the egg and set up egg albumen and eggshell properties for embryo development. Together with the egg, microbiota inside and outside the egg are transferred to the offspring. The development of pullet microbiota may impact microbiomes in the hen, which can influence the egg microbiota and the initial development of hatchling microbiota.

3.1 *Development of maternal gut microbiota and egg microbiota*

The development of hen gut microbiota is influenced by the feeding programs, amounts fed, feed composition, and vaccination programs during rearing. Broiler breeder's fecal microbiota contributes to the development of microbiota inside the egg (Lee *et al.*, 2019; Wen *et al.*, 2021) and on the surface (Trudeau *et al.*, 2020) of laid eggs. Hen's oviduct microbiota was studied by Lee *et al.* (2019), and the results indicated that the microbial composition and relative abundance of bacterial species are stable throughout the entire reproductive tract. Wen *et al.* (2021) observed increased microbial diversity along the reproductive tract from the infundibulum to the cloaca. This microbiota is transferred to the embryo through the egg white, constituting most of the embryo's gut microbial populations (Lee *et al.*, 2019). Yolk microbiota also influences offspring gut microbiota by increasing the colonization of *Proteobacteria*, *Firmicutes*, and *Bacteroidetes* in the intestine (Ding *et al.*, 2022). The yolk microbiota is functionally related to the amino acid, carbohydrate, and lipid metabolisms during chicken embryogenesis (Ding *et al.*, 2022).

The eggshell microbiota also plays a vital role in developing chicken intestinal microbiota, especially in the jejunum and ileum (Maki *et al.*, 2020). The correlation of microbiotas at the family level between the hen and embryo has been estimated at 0.40 and between the embryo and the chick at 0.52 (Ding *et al.*, 2017). This data means that some microbial colonizers detected in the gut of 4- and 19-day embryos came directly from the hen through the reproductive tract and not only the eggshell. The microbial communities of these embryonic gut microbiomes vary depending on the breeder's genetic line and feeding pattern. Once developed in the embryo, at least half of that microbiota will be similar in the chick and influence the microbiota post-hatch in broilers.

The gut microbial abundance and diversity are affected by environmental factors, nutrition, and host (pullet) genetic variation during development. Shterzer *et al.* (2020) concluded that there is a significant overlap between the intestinal and reproductive tract microbiomes in broiler breeders. Su *et al.* (2021) observed a similar relationship between the digestive and reproductive tract microbiotas in laying hens. Gut microbiota may affect hen metabolism and egg quality. Dai *et al.* (2022) discussed the critical connections between intestinal microbiota, their metabolites and egg quality, eggshell quality, egg safety, hen bone metabolism, bile acids, and hen lipid metabolism. Consequently, broiler breeders' intestinal health management may influence the offspring's gut and general health (Bowling *et al.*, 2018; Ebeid *et al.*, 2022).

3.2 *Feed restriction programs on microbiota and pathobionts*

Several management factors during breeder pullet rearing may influence intestinal health. Feed restriction programs commonly practiced worldwide for growth control of broiler breeder pullets and cockerels induce changes in growth rates, metabolism, behavior, gut development, corticosterone, and many other hormones (De Beer *et al.*, 2007, 2008; Ekmay *et al.*, 2010) immune response (Bowling *et al.*, 2018), and gut microbial programming (Yan *et al.*, 2021; Ebeid *et al.*, 2022). The effects of pullet feed restriction in gut microflora have been observed in broiler breeders (Yan *et al.*, 2021) and growing breeding ducks (Li *et al.*, 2022). Feed restriction can change the diversity of intestinal microorganisms. Feed-restricted pullet breeders had a higher abundance of *Mucispirillum* but a lower abundance of *Cloacibacillus*, *Clostridium* XlVa, and *Clostridium* IV (Yan *et al.*, 2021). Feed restriction caused an increase in the abundance of *Firmicutes* and decreased the abundance of *Bacteroidetes* in breeding ducks (Li *et al.*, 2022). These results indicated feed restriction represents chronic stress during the rearing of poultry breeders that causes altered gut microbiota composition.

These changes in microbiota can increase gut susceptibility to being colonized by pathogens. The feed restriction program selected for breeder pullets can affect the colonization and litter prevalence of pathogens like *Salmonella* and *Campylobacter* (Wilson *et al.*, 2018). Wilson *et al.* (2018) evaluated the impact of three feeding programs: skip-a-day (SAD) in trough feeders, every-day feeding in trough feeders (EDT), and every-day on the pen litter (EDL), on colonization with *Salmonella typhimurium*. *Salmonella* was incorporated into each treatment group by including *Salmonella*-positive birds at five weeks of age. This management was an intelligent way to simulate commercial conditions of transversal infection. The *Salmonella* prevalence was evaluated several times during rearing with stepped-on drag swabs and samples of ceca and spleens. Results indicated that SAD had a higher *Salmonella* prevalence than the EDT and EDL pullets. By 16 and 20 weeks, only SAD had recovered *Salmonella* (20%). The challenge with *Campylobacter coli* was conducted at 21 wk via oral gavage. At 22 and 26 weeks, *Campylobacter* was isolated from ceca of 80 to 100% of the SAD pullets, while only 30 to 60% of EDL or 40 to 95% of EDT were positive. These data indicated that SAD could contribute to higher persistence of *Salmonella* and *Campylobacter* ceca colonization and litter prevalence. The feeding restriction program and diet composition may affect microbiota and response to gut pathogens. Nutrient density (protein level) of pullet breeders affected oocyst cycling and lesion development during coccidiosis vaccination in broiler breeders (Oden *et al.*, 2012).

Wilson's experiment (2018) also provided insight into breeder management practices that may contribute to the intestinal health of breeders and, consequently, the offspring. As done commercially with spin feeders, feeding breeders on the litter provided contact with litter microbiota and potentially additional microbial seeding with competitive exclusion properties to reduce the colonization of pathogenic bacteria. Feeding on the litter expanded the feeder space for all pullets in each flock, increasing feeding time and allowing less aggressive pullets to access feed. This practice minimizes the instances when some pullets do not have a chance to eat, causing uniformity, health, and welfare issues (de Jong *et al.*, 2005).

3.3 *Breeder vaccination*

The vaccination program of the breeder and the maternal immunity can also affect the response of progeny offspring to gut pathogens and carcass contamination. Dórea *et al.* (2010) reported that broiler breeder flocks vaccinated against Salmonella had broiler flocks with a lower prevalence of Salmonella at processing than non-vaccinated flocks. Breeders were vaccinated with a live-attenuated *Salmonella enterica* serovar *Typhimurium* vaccine (Megan VAC-1) followed by a killed *Salmonella* bacterin consisting of *S. enterica* serovar Berta and *S. enterica* serovar Kentucky. Hens had a lower prevalence of *Salmonella* in the ceca (38.3% versus 64.2%) and the reproductive tracts (14.22% versus 51.7%). Broiler chicks from vaccinated breeders had lower *Salmonella* isolations (18.1% versus 33.5%). Fewer environmental samples containing *Salmonella* (14.4% versus 30.1%) were observed in farms with offspring of vaccinated flocks. Finally, lower *Salmonella* prevalence was observed in broilers from vaccinated breeders at the processing plants (23.4% versus 33.5%).

Similar results were observed by Berghaus *et al.* (2011) after using a killed *Salmonella* vaccine containing *Salmonella typhimurium*, *Salmonella enteritidis*, and

Salmonella kentucky and evaluating 58 commercial broiler flock houses. Vaccinated breeder flocks had significantly higher Salmonella-specific antibody titers than the unvaccinated ones, although they did not differ significantly concerning environmental *Salmonella* prevalence or loads. Offspring broiler flocks of vaccinated breeders had significantly lower *Salmonella* prevalence (62%) and 50% lower loads than the progeny of unvaccinated breeders. These results indicated that vaccination of broiler breeder pullets increased humoral immunity in the breeders and reduced *Salmonella* prevalence and loads in their broiler offspring. However, it did not significantly decrease *Salmonella* in the breeder farm environment.

The exposure of hens to dust and environments similar to the ones that the progeny will be exposed to may help to prepare the offspring to cope with everyday challenges. Verwoolde *et al.* (2022) demonstrated that the activation of the maternal innate immune system might affect both innate and adaptive immune responses in the next generation. Hen contact with bacterial lipopolysaccharides leads to tolerance of innate cells to β-glucans in the chick offspring, resulting in a lack of specific IgY antibody production.

3.4 *Housing and feeding management on maternal hormones and metabolism*

Pullets are generally housed in dark or brown houses to control lighting. These housing conditions often mean more difficulties in providing appropriate house ventilation and cooling. House environmental stress conditions and feeding restriction programs can cause chronic stress during rearing (de Jong *et al.*, 2005). Stress can cause changes in pullet behavior, such as increased activity, more foraging, and feather pecking (Yan *et al.*, 2021; Ebeid *et al.*, 2022). The gut microbiota of pullets can be affected by mild heat stress. Wang *et al.* (2020) observed a lower abundance of cecum bacteria in layer-pullets at 30 °C than those maintained at 24 °C due to heat stress response. The abundance of *Firmicutes* bacteria in the lung and cecum flora was also lower for pullets at 30 °C than those exposed to 24 °C.

Feeding programs with severe quantitative restriction or with days off feed increase serum corticosterone levels up to two-fold during the rearing (De Beer *et al.*, 2008; Wilson *et al.*, 2018) and laying phases and alter lipogenic gene expression and lipid metabolism (De Beer *et al.*, 2007; Ekmay *et al.*, 2010). Stress hormones can be transferred to the egg, negatively affecting embryo immunity and gut development (Saino *et al.*, 2005; Haussmann *et al.*, 2012; Babacanoğlu and Yalçin, 2014; Bowling *et al.*, 2018). The egg deposition of corticosterone and the hepatic expression of glucocorticoid metabolic enzymes is higher in fast-growing chicken genetic lines (Ahmed *et al.*, 2013). As genetic selection advances, this factor should be considered because the stress related to breeder feeding management is constantly gaining more relevance for its implications in the offspring.

Maternal thyroid status should also be taken into consideration for offspring gut health. The thyroid hormone levels can be affected by genetic selection, environment, and nutrition (Decuypere *et al.*, 2005; Leko *et al.*, 2021). Maternal thyroid hormones deposited in the egg yolk can affect embryo intestinal and immune system development (Akhlaghi *et al.*, 2013; Darras, 2019). As the genetic selection for fast growth advances, all its consequences on the hormonal status of breeders should be considered to optimize progeny health.

The fatty acid profile of the yolk can be affected by breeder age (Yilmaz-Dikmen and Sahan, 2009; Şahan *et al.*, 2014), the feeding programs during rearing (Ekmay *et al.*, 2010), and diet (Yilmaz-Dikmen and Sahan, 2009). Embryos have selective absorption of some fatty acids from the vitelline sac during incubation (Şahan *et al.*, 2014; van der Wagt *et al.*, 2020). After 15 days of incubation, chicken embryos preferentially utilize long-chain omega-3 (n-3) polyunsaturated fatty acids (PUFA). The yolk fatty acid profile can affect embryo livability (Yilmaz-Dikmen and Sahan, 2009), immunity, and gastrointestinal development. Thanabalan and Kiarie (2021) concluded that feeding breeders n-3 PUFA decreases inflammatory states and enriches hatching eggs with n-3 PUFA and immunoglobulins. Enrichment of n-3 PUFA in hatchling tissues reduces proinflammatory eicosanoids and bolsters progeny immunocompetence by receiving higher antibody titers against routine vaccination. The n-3 PUFA also modulates breeder GIT microbiota with consequences of microbial colonization and succession of chicks.

4 Management of hatching eggs and incubation

Every aspect of hatching egg management may have implications on embryo development and the post-hatch health of hatchlings. Temperature, humidity, and vibrations during egg handling and storage may affect the embryo and extra-embryonic membranes. Modifications of microbiota due to handling and disinfection could alter microbiota for hatchlings.

4.1 *Egg disinfection*

Disinfection of hatching eggs is a common management practice to reduce bacterial loads and pathogenic bacteria in the incubator and hatchlings. Formaldehyde was the most common disinfectant used worldwide, but nowadays, many other alternative products are in use (Tebrün *et al.*, 2020; Oliveira *et al.*, 2022). Some egg disinfectants can be efficacious in reducing bacterial loads but have many other implications in hatchability and health post-hatch. The effects of disinfectants on the egg and eggshell microbiota should be considered for the system's sustainability.

Olsen *et al.* (2017) demonstrated the changes in the eggshell microbiome after disinfection protocols with a wash in chlorine, two fumigations, and a spray. Disinfection reduced the bacterial load from more than 10^4 CFU, the typical natural value in the initially visibly clean eggs, and 10^5 CFU in the initially visibly dirty eggs to less than 10 CFU per sample after successful disinfection. In the microbiome of both groups of eggs, the phyla *Firmicutes*, *Proteobacteria*, and *Bacteroidetes* had the highest abundances. Within the phyla, *Firmicutes*, the relative abundances of *Clostridiales* decreased while *Lactobacillus* increased from before to after final disinfection. This indicates that well-conducted disinfection processes can also provide a healthier eggshell microbiome. Consequently, Olsen *et al.* (2017) concluded that the evaluated disinfection procedure was adequate, and the microbiome of initially visibly clean and dirty eggs was highly similar after the final disinfection step. The visibly dirty eggs had a less diverse microbiome and were enriched with pathobiont bacteria compared to the microbiome of visibly clean eggs. Then, exclusion or segregation and disinfection of dirty eggs could be necessary.

Egg and hatchery disinfection procedures may reduce the egg's defense mechanisms against pathogenic bacteria and reduce microbial diversity in the environment and, consequently, for hatchlings (Maki *et al.*, 2020). Pathogen contamination of eggs and hatchlings in the hatchery due to egg and chick manipulation during the incubation process by contact with hands, equipment that was not well disinfected, or other contaminated eggs and chicks becomes a more significant risk when egg-defense mechanisms and microbial diversity are reduced (Wilkinson *et al.*, 2020).

4.2 *Incubation and early management of gut embryo and hatchling development*

Proper incubation and chick quality are essential in antibiotic-free production systems and gut health (Smith, 2019). Incubation conditions can modify embryo metabolism, tissue development, and yolk utilization (van der Wagt *et al.*, 2020). Subtle changes in incubation temperature can significantly impact embryo fatty acid and yolk utilization (Nangsuay *et al.*, 2017; Dayan *et al.*, 2020; van de Wagt *et al.*, 2020). Reducing temperature for six hours from 37.6 to 36.6 °C each day during embryonic age 10 to 18 can affect fatty acid profiles, catalase, and nitric oxide production of brain and liver tissues (Yalçin *et al.*, 2012). Dayan *et al.* (2020) observed that incubation temperatures 1.5 °C below or above the control (37.8 °C) during the 21 days of embryo development caused changes in the expression of yolk sac genes.

These genes are involved in lipid uptake and metabolism (LRP2, ApoA1), oligopeptides uptake (PepT1), gluconeogenesis (FBP1), glycogenesis (GYS2), and thyroid hormone regulation (DIO1, DIO2). The changes in metabolism caused the residual yolk at hatch was 18.18 and 29.99% for the cold and hot incubations, which differed significantly from the control, 11.12%. Proper calibration and periodic maintenance of incubator machines are necessary to obtain optimum uniform temperature, humidity, and ventilation conditions. Egg segregation by breeder age, egg size or weight helps to provide the best incubation temperature profiles and conditions for each group.

On the other hand, the thermal manipulation or periodic thermal stimulation of the embryos in single-stage machines can prepare the hatchling to cope with heat stress. This procedure has been called circadian incubation. Al-Zghoul *et al.* (2019)

used thermal manipulation, raising machine temperature to 39 °C and 65% relative humidity for 18 h daily from embryonic days 10 to 18. The hatchlings were raised under standard temperatures and challenged with chronic heat stress (35 °C) from days 28 to 35. The findings of this study suggested broilers subjected to thermal manipulation can have better nutrient digestion and absorption, thermal control, lower mortality rates, and better body weight gain during chronic stress than broilers from standard incubation.

One hatchery management practice that has been proposed for almost three decades but is still not widespread is *In-ovo* feeding. Multiple research studies have proven that *In Ovo* applications of amino acids, trace minerals, vitamins, glucose, antibiotics, prebiotics, probiotics, synbiotics, creatine, follistatin, L-carnitine, CpG oligodeoxynucleotide, growth hormone, polyclonal antimyostatin antibody, peptide YY, and insulin-like growth factor-1 have positive effects on hatchlings (Das *et al.*, 2021). Some of these products do not affect hatchability. Still, they may improve body weight at hatch, physiological functions, immune responses, gut morphology, and gut microbiome, and some improve post-hatch live performance and overall health of poultry (Das *et al.*, 2021). Despite all those benefits, commercial application is still uncommon since commercial tests have not been that successful, and priority has been given to vaccination that is applied more frequently between day 18 and day 19.5 of incubation. Depending on the interest, the compounds for *In Ovo* feeding are injected between day 12 or earlier and 17 to 18 of incubation via air sac or amnion. The difference in timing and other logistic issues have made the application of *In Ovo* feeding complex, but its potential to contribute to gut health is undeniable.

Gut health depends on proper early development. Elevated embryonic incubation temperature can negatively affect post-hatch relative nutrient transporter gene expression, intestinal epithelium integrity, organ development, and the performance of broiler chickens (Barri *et al.*, 2011). Incubation temperature also affects the time of hatch and hatch window (Bergoug *et al.*, 2013). Broad hatch windows indicate that more hatchlings may delay access to feed and water (Lamot *et al.*, 2014; Wijnen *et al.*, 2022). Delayed access to feed and water for 24 hours or more delays residual yolk utilization (van der Wagt *et al.*, 2020), increases stress (Wijnen *et al.*, 2022), reduces body weight, increases mortality, retarded development of gut segments during the first week of life (de Jong *et al.*, 2017). Delayed access to feed post-hatch can impair and delay the development of the GALT for up to two weeks, especially in the hindgut (Shira *et al.*, 2005) and general immune development (Panda *et al.*, 2015). Proszkowiec-Weglarz *et al.* (2020) reported that delayed access to feed post-hatch might affect tight junction structure and function, consequently, the barrier function and microbiota development (Proszkowiec-Weglarz *et al.*, 2022). Consequently, all factors related to gut health are implicated.

Wijnen *et al.* (2022) recently studied the effects of incubation temperatures and delayed feeding on resilience against colibacillosis infection. The incubation temperatures were evaluated as eggshell temperature (EST), 37.8 or 38.9 °C, during mid-incubation (7 to 14 days) in a factorial combination with early feeding post-hatch or a 48-hour delay. At day eight, post-hatch avian pathogenic *E. coli* were inoculated intratracheally. At higher EST, early feeding caused a higher incidence of systemic *E. coli* infection than delayed feeding. Whereas at control EST, systemic infection was not different between feeding strategies. Regardless of EST, early feeding, compared to delayed feeding, resulted in a lower incidence of local infection and higher growth until d 35. In conclusion, early feeding could be considered a strategy to enhance broiler resilience, but only when EST is not too high.

The time, environmental conditions, and access to feed during chick transportation from the hatchery to the farm can affect gut development (Barri, 2008) and yolk absorption (van der Wagt *et al.*, 2020). Early feeding can affect chickens' lymphoid organs and immune competence (Taha-Abdelaziz *et al.*, 2018). The on-farm hatching has been proposed to solve the potential issues that chick processing and transportation logistics cause in early feeding. In on-farm hatching systems, trays of eggs are brought to the broiler farm between 17 and 19 days of incubation. The system allows chicks to have interrupted access to feed and water after hatching. Evaluations of this management practice showed some potential benefits on welfare compared to traditional hatching at the hatchery (de Jong *et al.*, 2019; Giersberg *et al.*, 2021). Recently, Jerab *et al.* (2023) reported that on-farm hatched broilers had a lower antimicrobial treatment incidence and had more antimicrobial-free flocks. Using on-farm hatching resulted in a 5.6 times lower probability of antimicrobial use. Taking into consideration that the most common indications for antimicrobial use were respiratory (37.51%, n = 1,788/4,766) and gastrointestinal disorders (30.49%, n = 1,453/4,766), it is evident that on-farm hatching could be an essential management practice for antibiotic-free production and enhancing gut health in any system.

There are multiple implications on gut health and food safety of a simple management decision and practice as providing access to feed and water as soon as birds hatch or are ready to have it. However, achieving this goal in large flocks is not as easy as it sounds due to the significant variability in development that diverse groups of embryos may have.

4.3 *Interactions among breeder, incubation, and placement factors*

We have been reviewing management factors that affect gut health independently. However, several factors intermingle in practice to cause a gut health issue. Several recent studies have evaluated the combination of these factors to establish the degree of importance or understanding of mechanisms of interactions. Cardeal *et al.* (2021) assessed the effects of breeder age (29 or 55 weeks), placement time (24 or 48 h after pulling), and the use of a pre-placement feed (PPF) at 2.5 g/chicks inside transport boxes on yolk, sac absorption, duodenum villus height and crypt depth and performance of male broiler chicks. Early access to feed stimulated yolk sac absorption until four days of age, and this effect was more substantial for chicks from 29-week-old broiler breeders.

On the other hand, placement within 48 h negatively impaired the development of duodenum mucosa, villus, and crypt until seven days of age for chicks from young broiler breeders. Still, the use of PPF could reduce this adverse effect. These results suggested that PPF inside chick transport boxes to the farm is recommended for chicks from young broiler breeders who will be submitted to delayed feed access for more than 24 hours to improve GIT development.

Yalçin *et al.* (2016, 2017) evaluated the effects of egg storage and brooding temperature. This research group evaluated the effects of egg storage duration (ESD) and brooding temperature (BT) on body weight, intestine development, and nutrient transporters of broiler chicks. Eggs were stored at either 18 °C for three days (ESD3-18 °C) or 14 °C for 14 days (ESD14-14 °C) before incubation. Chickens were exposed to three BTs. Temperatures were initially set at 32 °C, 34 °C, and 30 °C for control (BT-Cont), high (BT-High), and low (BT-Low) BTs, respectively. The BTs were decreased by two °C each on days 2, 7, 14, and 21. BT-High resulted in the heaviest body weights at seven days, especially for ESD14-14 °C chicks. This result was consistent with longer villus and larger villus areas of ESD14-14 °C chicks at BT-High conditions. From 14 days to slaughter age, BT did not affect broiler weight. ESD3-18 °C chicks were heavier than ESD14-14 °C chicks for up to 28 days.

The gene expression of H^+-dependent peptide transporter PepT1 and Na-dependent glucose SGLT1 was significantly higher in ESD3-18 °C chicks than in ESD14-14 °C on the hatch day. These transporters are essential amino acid and glucose uptake mediators across the apical cell gut membranes. Reduction in their activity causes reduced nutrient absorption; consequently, bacteria could use these nutrients, leading to dysbiosis. There was significant egg storage by BT interaction for PepT1 and SGLT1 transporters on day 14. Chicks with ESD14-14 °C had significantly higher expressions of PepT1 and SGLT1 at BT-Low than those at BT-Cont. ESD14-14 °C chicks upregulated PepT1 gene expression 1.15 and 1.57-fold at BT-High and BT-Low, respectively, compared with BT-Cont, whereas PepT1 expression was downregulated 0.67 and 0.62-fold in ESD3-18 °C chicks at BT-High and BT-Low. These results indicated that pre-incubation egg storage conditions and BTs affected intestine morphology and the expression of PepT1 and SGLT1 nutrient transporters in broiler chicks. All these factors can play essential roles in gut health.

4.4 *Implications of early dysbiosis and use of antibiotics*

The early and fast development of microbiota is essential for future gut health. Contact with beneficial microbiota in early life is vital for life programming in chickens to enhance productivity, immunity, and health (Broom and Kogut, 2018; Rubio, 2019; de Jong *et al.*, 2022; Bindari and Gerber, 2022; Wickramasuriya *et al.*, 2022). Pre and early post-hatch stimulation of microbiota seeders from the maternal origin and posterior contact with an environment with favorable microbiota are pivotal to developing a diverse, rich, and resilient microbiota (Rubio, 2019; Bindari and Gerber, 2022). Another option is to provide feed additives that modulate or promote healthy microbiomes, such as specific enzymes, plant derivatives, sodium butyrate, β-glucans, pre-and probiotics, and symbiotics.

Reduced environmental bacterial load during early development and gut colonization has detrimental consequences (Wilkinson *et al.*, 2020). This could partly explain the success of poultry litter reuse worldwide when regulations allow it (Bindari and Gerber, 2022). Several reports indicate that the reuse of treated broiler litter is a safe practice and substantially decreases the bacterial load of *Salmonella* and other pathogens by a potential competitive exclusion (Roll *et al.*, 2011; Bucher *et al.*, 2020; Bindari and Gerber, 2022).

Disruption of intestinal microbiota composition during immune development negatively affects immune responses later in life. Simon *et al.* (2016) investigated the effects of antibiotic treatment during the first week of the life of pullet layers on the specific antibody response 14 weeks after cessation of antibiotic treatment. A heavy cocktail of antibiotics was provided during the first week. This treatment consisted of vancomycin, neomycin, metronidazole, and amphotericin-B provided by oral gavage every 12 hours, and ampicillin and colistin in drinking water. The fecal microbiota was similar to the control birds after two weeks post-antibiotic treatment. The antibiotic-treated pullets had lower immunity against an intra-tracheal challenge. The challenge with lipopolysaccharide and human serum albumin was provided 12 weeks post-treatment. Immune response against the two challenges was measured ten days post-challenge. The antibiotic-treated birds had lower T cell-dependent antibody titers than the control pullets. In broilers, Schokker *et al.* (2017) observed that antibiotic treatment (amoxicillin) via drinker water in day-old broiler chickens affected intestinal microbiota temporarily and reduced their intestinal innate immune competence (reduced $CD4^+$ T cells and macrophages) for two weeks. Consequently, management practices, including antibiotic treatments early in life, may negatively affect intestinal health weeks later. These findings partially explain how poultry growing programs without antibiotics can often achieve even better final live performance, welfare, and health results than those using antibiotics (Smith, 2019).

However, dysbiosis in young poultry can be caused by many environmental and management factors. One of them is the temperature during brooding. Liu *et al.* (2019) evaluated the effects of acute (0.5, 1, 3, 6, and 12 hours) and chronic cold stress (5, 10, and 20 days) during brooding after day 15 on Shaweimate male quails, compared to brooding at 28 °C. Both cold stress treatments increased ceca oxidative stress and inflammatory markers, especially chronic cold stress. The ceca microbiota was also shifted by cold stress during brooding. Early gut health can be a marker of future health status. Caekebeke *et al.* (2020) concluded that the prevalence of gut abnormalities at day ten could predict gut health scores later.

5 Housing, stocking density, heat stress, and lighting

5.1 *Housing*

The diet and nutritional interventions with processing or feed additives can modify the gut microbiota of broilers, but the effect of housing conditions on microbiota composition and functionality is more substantial (Santos *et al.*, 2012; Kers *et al.*, 2019; Hoffmann *et al.*, 2020; Aruwa *et al.*, 2021). Therefore, to understand gut microbiomes and gut health in poultry, it is essential to be aware of the significant impact of housing conditions on the obtained results.

Recent studies have found evidence of direct relationships between environmental stress and welfare status due to the housing and production system that broilers experience at farm with the shedding of foodborne pathogens, including *Campylobacter* and *Salmonella* (Santos *et al.*, 2012; Alpigiani *et al.*, 2017; Iannetti *et al.*, 2020). These relations are tightly connected with the gut microbiota composition. Poultry house environmental factors, such as litter, feed access, and climate, can affect the composition of intestinal microbiota in both layer- and meat-type birds (Kers *et al.*, 2018). Iannetti *et al.* (2020) reported differences in *Campylobacter* and *Salmonell*a fecal shedding among chicken flocks of variable animal welfare levels. Farm stress can increase susceptibility to these pathogens, and maintaining high animal welfare can improve the safety of poultry meat products.

Housing environments have become more variable in layers due to welfare concerns. Housing differs in group size, freedom of movement, and environmental complexity. Pullets reared in houses with a complex environment, including perches and litter materials, had less fearfulness, lower plasma corticosterone, improved gut microbial functions, lower relative mRNA expression of glucocorticoid receptors, and elevated mRNA expressions of stress-related genes CRH, BDNF, and NR2A in the hypothalamus, compared to pullets in groups housed in less complex environments. These metabolic changes affect the brain-gut axis and can help them to cope with future challenges (Du *et al.*, 2022).

Cage systems can be divided into conventional cages (CC) and enriched cages (EC). The housing system impacts hen immunity, but the findings are not consistent. However, most studies that included the heterophil: lymphocyte (H:L) ratios reported higher ratios in CC hens. These observations could indicate potential higher stress, probably immunosuppression, or lower resilience before a challenge (Hoffmann *et al.*, 2020).

Wiersema *et al.* (2021) evaluated the impact of the CC and cage-free housing (CF) on intestinal permeability, morphology, and microbial communities in commercial hens. In this study, several parameters favored gut health in hens housed in CC. Hens housed in the CF had increased jejunal villus height and crypt depth compared with hens from the CC. However, hens from the CC tended to have a greater villus height to crypt depth ratio in the jejunum and ileum than in the CF. The microbial community diversity and richness were higher in CC hens. These hens had an increased number of over-represented operational taxonomic units, OTUs (46/64), including *Romboutsia* sp. (30.80%), *Lactobacillus kitasatonis* (17.16%), and *Lactobacillus aviarius* (11.15%). The OTU is the basic unit used in numerical taxonomy. These units refer to an individual, species, genus, or class.

5.2 *Stocking density*

The effects of high stocking density on live performance, gut health, and welfare have been extensively documented and investigated in broilers (Tsiouris *et al.*, 2015; Heidari and Toghyani, 2018; Goo *et al.*, 2019; Jobe *et al.*, 2019; Nasr *et al.*, 2021), layers (Hofmann *et al.*, 2021), turkeys (Butterworth, 2019) and ducks (Xiong *et al.*, 2020). High stocking densities tend to have negative implications. They potentially can increase gut health issues such as coccidiosis (López-López *et al.*, 2022) and necrotic enteritis (Tsiouris *et al.*, 2015) and modify gut microflora suggesting subclinical dysbiosis (Kers *et al.*, 2018, 2019) especially when house ventilation is not sufficient to maintain proper environmental conditions (Green and Xin, 2009; Farhadi *et al.*, 2016). Broiler and layer stocking density is regulated, or different entities worldwide have given welfare recommendations. However, there are considerable variations in the densities recommended, probably due to the prevalent house environmental control capabilities and level of acceptability of welfare information in each country. In Europe, the recommended stocking density in the Netherlands is 45 to 54 kg/m^2; in the United Kingdom is 40 kg/m^2; and in Switzerland is 30 to 36 kg/m^2. In the United States, the welfare guidelines of the National Chicken Council (NCC, 2022) suggested 31.74 kg/m^2 for broilers with live weight below 2.04 kg and 41.5 kg/m^2 for broilers weighing up to 3.4 kg (Nasr *et al.*, 2021). Research and welfare recommendations indicate that optimum stocking density varies with birds' age, weight, and housing conditions.

There is some variability in the results mainly due to confounding factors between group size in the same area and variations in physical space, feeder, and drinker space allowed to the groups. However, in layers, the impacts of stocking density during pullet rearing on the immune system and welfare of hens have not been investigated. Hofmann *et al.* (2021) compared pullet flocks reared at either low (13 birds/m^2) or high (23 birds/m^2) stocking densities but in identical group sizes from weeks 7 to 17. Later, hens were kept at a similar stocking density of 2.4 birds/m^2 until 28 weeks. At the end of the rearing period, pullets kept under high stocking density had lower numbers of T lymphocytes, especially $\gamma\delta$ T cells in the blood, spleen, and cecal tonsils, and displayed a higher H:L ratio. These effects were persistent during the laying period, even though the stocking density in this phase was identical. Moreover, hens from the high stocking density pullet treatment were less active, had more pecking behavior, and had a worse physical appearance. Then, stocking density during rearing may have prolonged effects on hens' life and health.

5.3 *Heat stress*

Climate change and increasing variability in environmental temperatures and precipitation patterns are challenges for the poultry industry. Heat stress-associated gut health and food safety issues have gained particular importance due to public awareness and copious available scientific information (Tsiouris *et al.*, 2018; Rostagno, 2020; Brugaletta *et al.*, 2022). Even the best poultry housing available may have difficulties guaranteeing optimum temperatures and relative humidity to poultry depending on their growth rate and productive stage when external environmental conditions are adverse.

Increased air velocity is the most effective method to release convective heat, but it is strongly related to the temperature difference between the surfaces of animals and the surrounding air. Then, the efficacy of air velocity declines when air temperature approaches 40 °C (Akter *et al.*, 2022). Evaporative cooling pad systems are employed in tunnel ventilation to lower temperature and increase heat dissipation from birds. Additionally, sprinkler systems are used to partially wet the birds and allow water evaporation that absorbs heat directly from the body of the birds and surrounding air. In both cases, maintaining a large vapor pressure gradient is crucial, keeping relative humidity low (Moon *et al.*, 2023). Combining sprinklers with cool cells improves cooling efficacy and reduces cooling water use by 64%. However, it is necessary to operate the house slightly warmer (2.2 °C above) and drier by delaying the cool cell system is critical in maintaining litter quality when using a cool cell pad with sprinklers. However, when the prevalent environmental relative humidity is higher than 85%, and the temperature is higher than 35 °C, all these three methods could have lower efficacy. Then, current environmental control may not satisfy the unique needs of stressed poultry, and gut health issues are observed (Rostagno, 2020; Brugaletta *et al.*, 2022).

Nutritional interventions to mitigate heat stress include improving precision on nutrient composition, avoiding nutrient excess, increasing amino acid balance with crystalline sources, increasing vitamin levels (A, E), supplementation of vitamin C, supplementation of Se, Ca, Na, Cl, and K (Kumari and Nath, 2018). However, these nutritional strategies often only mitigate some effects, and their efficacy is low. Hence, there is a crucial need to explore effective strategies, including genetic selection, early programming during incubation, and post-hatch life, to enhance thermo-tolerance and productivity of poultry around the world (Nawab *et al.*, 2018; Goel, 2021).

Heat stress affects the gut-brain axis and feed intake and may affect the incidence and severity of gut health issues, including avian necrotic enteritis (Calefi *et al.*, 2016; Rostagno, 2020). The gut-brain axis refers to the bidirectional biochemical signaling between the central nervous system and the GIT, including the microbiota. There are evident relationships between the degree of intestinal lesions, the chickens' behavioral outcomes, brain activity, and serum levels of corticosterone. Heat-stressed animals frequently have damage to microstructures of the mucosal epithelia, increased oxidative insults, reduced immunity, and increased permeability of the gut to toxins and pathogens (Quinteiro-Filho *et al.*, 2010; Patra and Kar, 2021; Hosseindoust *et al.*, 2022). The disruptions in the gut mucosa barrier integrity make the animal more vulnerable to the invasion of pathogenic microbes and the translocation of antigens to the blood circulation, causing systematic inflammations and immune responses.

Moreover, digestion and absorption of nutrients are impaired due to reduced enzymatic activity in the digesta, reduced surface areas for absorption, lower nutrient transporter in the mucosa (Al-Zghoul *et al.*, 2019), liver dysfunction, and injury to the mucosal structure and altered expressions of the nutrient transport proteins and genes. The systematic hormonal changes due to heat stress and alterations in immune and inflammatory responses often cause reduced feed intake and worsen production performance in poultry (Patra and Kar, 2021). Al-Zghoul *et al.* (2019) observed that during chronic heat stress (35 °C between 28 and 35 days of age), the mRNA expression levels of mucosa glucose transporters (SGLT1, GLUT2), fatty acid transporter (FABP1), and trypsin were significantly decreased after one day of expose. This expression continued to decrease during the stress and after it had stopped. The disruption in the digestive function leads to microbiota changes, pathobiont proliferation, and potential gut health issues.

Exposure to high temperatures alters the microbial community structure and the functional features of the gut microbiota of broilers (Shi *et al.*, 2019), layers (Xing *et al.*, 2019; Zhu *et al.*, 2019; Zhou *et al.*, 2022), turkeys (Farghly *et al.*, 2018), ducks (Tian *et al.*, 2020). Gut microbial diversity is vital for nutrient metabolism, protection against pathogens, and immune system development. Microbial diversity is determined using sequencing data from 16S rRNA genes amplified from the DNA of digesta or gut samples. This data is clustered into OTUs, in which similar 16S rRNA sequences are considered to belong to the same microbial genetic taxa. Each sample may have many taxa, and to describe the differences in the composition of microorganisms, the diversity of the sample can be described by alpha, beta, and gamma diversity.

Alpha diversity measures the richness (number of taxa) or evenness (the relative abundances of those taxa) of an average sample of a treatment or condition. Beta diversity measures the similarity or dissimilarity of two communities among samples in a treatment. The gamma diversity describes the overall species diversity across communities for each condition. The alpha- and beta-diversity of gut microbiota composition are altered due to heat stress, with more significant colonization of pathogenic microbiota groups (Patra and Kar, 2021).

At the phylum level, the composition of *Firmicutes*, *Tenericutes*, and *Proteobacteria* in heat-stressed broilers increased compared to the control group, and *Bacteroidetes* and *Cyanobacteria* in heat-stressed broilers were reduced. In addition, the composition of *Anaeroplasma* and *Lactobacillus* phyla in the heat-stressed group was increased. In contrast, in the heat-stressed group, the *Bacteroides*,

Oscillospira, Faecalibacterium, and *Dorea* genera decreased more than the control group (Shi *et al.*, 2019). The alterations in microbiota are so evident that specific changes in poultry gut microbiota composition have already been considered a biomarker of heat stress in monogastric species (He *et al.*, 2021).

Zhu *et al.* (2019) also observed that heat stress induced a significant taxonomic perturbation in the gut microbiome of caged laying hens. The fecal microbial communities were dominated by *Firmicutes, Bacteroidetes*, and *Proteobacteria* phyla. The *Firmicutes* were significantly decreased, and *Bacteroidetes* were increased in the fecal microbiota under heat stress. The functional microbiota evaluation indicated that cysteine and methionine metabolism and benzoate degradation were more abundant in heat-stressed hens. At the same time, retinol metabolism and phenylpropanoid biosynthesis were decreased by heat stress.

5.4 *Lighting*

Light is a valuable environmental factor that affects poultry performance and health. The length and frequency of dark periods, the intensity of light, the wavelength (light color), and the total photoperiod can impact behavior, feed intake, body weight gain, and resilience to necrotic enteritis infection (Rodrigues *et al.*, 2018; Rodrigues and Choct, 2019; Hoffmann *et al.*, 2020; Abo-Al-Ela *et al.*, 2021; Du *et al.*, 2022). Lighting programs with extended dark periods encourage crop utilization, and enhancing the function of this organ contributes to the regulation of the whole digestive tract (Classen *et al.*, 2016; Rodrigues and Choct, 2018). Hieke *et al.* (2019) demonstrated that prolonged photoperiods (23/1 h Light/Dark) had a differential microbiota acquisition in layer chickens compared to those in a 12/12 h Light/Dark cycle during the first three weeks of life. Consequently, GIT physiology and gut microbiota can be affected by light, but immunity is also affected.

The photoperiod affects melatonin release, directly related to immunity and reducing stress. The peak level of melatonin is observed in chickens only after four hours of darkness (Hoffmann *et al.*, 2020). Consequently, dark periods should be longer than five hours in continuous or intermittent lighting programs. Considering the scientific evidence, it is understandable that lighting programs have been identified as an essential management tool in antibiotic-free broiler production (Smith, 2019; Shynkaruk *et al.*, 2022).

Intermittent lighting programs applied during the most critical period for necrotic enteritis risk (18 to 24 days of age) may help improve broilers' resilience to this gut health issue (Rodrigues *et al.*, 2018). The intermittent light program increases plasma triiodo thyroxine in heat-stressed broilers, modulating carbohydrate, fat, and protein metabolism. It can also decrease pro-inflammatory cytokines affecting the gut and growth (Abo-Al-Ela *et al.*, 2021).

The white light intensity can also cause immunosuppressive stress in layers and broilers. Low light intensities (around 10 lux) tend to have no detrimental effects on immune responses against viral or bacterial antigens. Prolonged exposure to 25 or

30 lux or higher light intensity tends to cause higher H:L ratios and lower humoral or cellular responses to vaccination or antigen stimulation (Hoffmann *et al.*, 2020).

Monochromatic lights (red, green, and blue) influence immune responses in poultry. Green and blue lights enhance T-lymphocyte proliferation and antibody production and reduce IL1β production. Green lights are more effective in the first three weeks of life and blue later on. Blue lights reduce the heat-induced oxidative stress by lowering MDA level, enhancing the antioxidant SOD and CAT gene expression and activities, and regulating the heat shock proteins HSP70, HSP90, HSF1, and HSF3 gene expression (Abdo *et al.* 2017). These effects on the immune system or reducing stress help to minimize gut metabolic costs, improve nutrient utilization and growth, and increase survivability during or after heat stress. Blue and green monochromatic lights improve gut health by improving mechanical and immunological barriers of the intestinal mucous membrane. These monochromatic lights increase the villus height of the small intestine and the numbers of goblet cells, intestinal intraepithelial lymphocytes, and IgA+ cells (Xie *et al.* 2011). Green light can also improve the development of broiler cecal tonsils during the first two weeks of life (Li *et al.*, 2014).

6 Water quality

Water is the nutrient with the greater flow in the GIT during the whole life of birds. The physicochemical properties may have a significant impact on GIT physiology. Water hardness and alkalinity can have a buffering effect on the crop and gizzard/proventriculus pH and consequently reduce the first phase of digestion (Kierończyk *et al.*, 2016). This buffering effect can negatively affect the efficacy of exogenous enzymes like phytase and carbohydrase. When the pH value of the crop is above 6.5, the enzyme activity can be reduced by 10–15% compared to the effects observed at pH 4.5–5.5 (Kierończyk *et al.*, 2016). Endogenous enzymes and nutrient digestibility can be affected by water pH. Bassi *et al.* (2022) demonstrated that acidified water resulted in better crude protein digestibility, and alkaline water (8.2 pH) improved Ca and P digestibility in turkeys. Phytase supplementation increased nutrient utilization at both water pH values, but the exogenous enzyme effect was attenuated in combination with alkaline water.

For this reason, water quality should be controlled and improved as much as possible to minimize the adverse effects of hardness and high carbonate concentrations. High salt and solid concentrations in the water and basic pH generally favor biofilm production and endotoxins in water lines and drinkers due to the proliferation of algae and microbes (Cox and Pavic, 2010). Biofilm and salt deposition in the water lines can also increase clogging issues in nipples, affecting water availability for birds. Reduced water intake, caused by any means, adversely affects gut physiology, digestion, and live performance (Oviedo-Rondón *et al.*, 2019).

The microbial quality of water is also fundamental in gut health and impacts food safety. Specific pathogens such as *Campylobacter* may have high survival rates in water. Consequently, drinking water can be a reservoir for birds. Generally, water chlorination and acidification can be effective against *Campylobacter*. The drinker system type may play a role in avoiding pathogens. Open-bell drinkers and nipples with cups tend to accumulate more pathogens than those without (Sibanda *et al.*, 2018).

7 House ventilation

House ventilation is necessary to evacuate moisture produced by birds, drips from water lines, gas brooders, and the cooling system. At the same time, ventilation helps to maintain air quality. The microbial degradation of excreta, litter, and feed wastage can produce gases like ammonia (NH_3) and hydrogen sulfide (H_2S). High concentrations of NH_3 (> 25 ppm) occur in broiler and turkey houses and aviaries with litter-based housing for layers. The H_2S concentrations (> 20 ppm) are more common in caged layers and broiler breeders where excreta accumulate in pits or under slats. Exposure to more than 25 ppm of NH_3 decreases antibodies against the Newcastle disease virus and immunoglobulins like Ig Y and IgM (Hoffmann *et al.*, 2020). The production of IgA in the duodenum was reduced when broilers were exposed to 70 ppm NH_3. The NH_3 exposure also reduces respiratory and intestinal tract interleukins and cytokines (Zhou *et al.*, 2020).

Han *et al.* (2021) confirmed that exposure to 25 ppm NH_3 for 14 days or 35 ppm for seven days reduced broiler growth performance but also caused changes in the structure of the cecal microbiota. At the phylum level, the relative abundance of *Proteobacteria* increased in the ceca after 35 ppm NH_3 exposure. At the genus level, 35 ppm NH_3 increased *Escherichia-Shigella*'s relative abundance and decreased *Butyricicoccus*, *Parasutterella*, *Lachnospiraceae*_UCG-010, *Ruminococcaceae*_UCG-013, and *Ruminococcaceae*_UCG-004. These changes in cecal microbiota composition were correlated with worse growth performance. The shift in microbiota could also be related to lower short-chain fatty acid production, affecting intestinal epithelial integrity, the proliferation of epithelial cells, anti-inflammatory pathways, and expression of tight junction proteins. Consequently, exposure to high NH_3 concentrations can have a direct negative impact on gut health. The pathological or clinical issues due to NH_3 exposure are more evident when there are concurrent stressors such as heat, coccidiosis, and bacterial dust endotoxins (Roque *et al.*, 2015; Hoffmann *et al.*, 2020).

Exposure to H_2S also affects the immune system. Broilers exposed to 20 ppm H_2S had activated inflammatory responses due to an increase in pro-inflammatory cytokines like TNF-α, IFN-γ, IL-6, IL-8, and IL-17 and a decrease in anti-inflammatory cytokines like IL-2, IL-4, and IL-10 in blood, spleen, and the bursa of Fabricius (Chi *et al.*, 2019). Moreover, Hu *et al.* (2018) observed oxidative stress, inflammation, apoptosis, imbalance in Th1/Th2 lymphocytes, and higher mRNA expressions of IgY, IgM, and IgA in the bursa of Fabricius causing immune injury in broilers.

8 Litter management and impacts on microflora, pathogens, and health

Most broilers, broiler breeders, turkeys, and cage-free layers are raised on litter in the Americas, Europe, the Middle East, and parts of Asia. Other countries use slats or cages, and this imposes different gut microbial development, immunity development, and intestinal health issues (Santos *et al.*, 2012; Farhadi *et al.*, 2016; Hofmann *et al.*, 2020; Wiersema *et al.*, 2021). Certain bedding materials like straw and rice hulls are associated with worse broiler live performance (de Toledo *et al.*, 2020). Litter moisture content is probably the main driver of litter microbiota composition. However, litter humidity can be influenced by the type of bedding materials used and the management given to the litter, which is highly variable (Díaz Carrasco *et al.*, 2022).

For example, cleaning frequency varies around the world. The reuse of poultry litter in several flocks is common in the Americas and parts of Asia but not in Europe, where litter is removed after each flock, and the house is cleaned, disinfected, and replaced with new bedding material. Litter reuse is a widespread management practice to minimize cost and environmental impact. Then, it is considered a more sustainable practice. Reusing litter affects the gut microbiome, and contrary to a more common belief, it seems to have positive effects on controlling pathogens like Salmonella (Roll *et al.*, 2011; Bucher *et al.*, 2020) and maintaining healthy antibiotic-free flocks (Smith, 2019).

In reused litter, the halotolerant/alkaliphilic bacteria tend to increase in litter while certain butyrate-producing and *Lactobacillus* species are augmented in the gut of young chicks (Wang *et al.* 2016). These chickens also tended to have a lower abundance of *Clostridium perfringens* than those reared in fresh litter (Wei *et al.*, 2013). Additionally, litter reuse during six cycles did not affect *Campylobacter* and *Escherichia coli* levels in litter or broiler cecal contents (Chinivasagam *et al.*, 2016). This research results partially explain reports indicating that broiler flocks raised on reused litter have a lower incidence of gut health issues (Smith, 2011, 2019).

It is necessary to conduct litter treatments to avoid issues with gases in reused litter-based systems. Failures in litter treatments can also negatively affect the house environment with more NH_3 emissions and concomitant poultry welfare issues. Litter treatments generally have positive effects (Toledo *et al.*, 2020; Bindari and Gerber, 2022), but it is crucial to observe results depending on the conditions. Acidifiers reduce pH, moisture, ammonia, and pathogenic bacteria, improving weight gain and decreasing mortality. Superphosphate and adsorbents reduce litter pH and moisture. Gypsum application reduces ammonia and improves feed conversion. However, alkalizing agents raise the pH but tend to worsen feed conversion, increase mortality despite decreasing moisture, and reduce pathogenic bacteria (Toledo *et al.*, 2020).

In layers, the different housing types, like deep litter, perchery, and free range systems, provide substrates that pullets or hens pick and exchange microbiota. The chemical and microbiological characteristics of litter in laying hen aviaries are different from broilers with lower loads of total coliform and *Staphylococcus* species (Díaz Carrasco *et al.*, 2022). However, due to the constant contact to develop natural behaviors like bathing with litter and nesting, litter is expected to impact the gut microbiome and health in layers.

9 Feeding practices and feed withdrawal

9.1 *Feeding practices*

Genetic selection of broilers, turkeys, and ducks (Nielsen, 2004; Howie *et al.*, 2010; Li *et al.*, 2020) for fast-growth and high-meat yield has increased their voluntary feed intake and changed their feed intake patterns. Due to disturbed satiety and hunger mechanisms, broilers eat to their maximum physical capacity (Bokkers and Koene, 2003). On a global basis, the common broiler management practices are *ad libitum* feeding and prolonged light photoperiods to stimulate feed intake and growth. However, these practices may have negative consequences on bird health (Rodrigues and Choct, 2019), including 1) overconsumption of feed which negatively impacts feed efficiency, nutrient digestibility, gut health, and ultimately financial returns; 2) induction of physiological growth-related problems, such as leg problems, acute metabolic (*i.e.*, sudden death syndrome) and chronic problems (*i.e.*, ascites); 3) impairment of the immune system of birds, and 4) promotion of a series of ocular-related problems. These issues are uncommon in laying hens since satiety and hunger mechanisms are equally involved in layer chickens (Bokkers and Koene, 2003).

Rodrigues and Choct (2018) proposed considering feed structure and composition, frequency of feeding, and light regimes as management tools to control the voluntary feed intake of broilers and their eating patterns. As discussed earlier, meal feeding or intermittent lighting can increase the residency time of feed in the crop when feed is unavailable and enhance foregut digestion and absorption to minimize undigested nutrients in the hindgut. The feed structure and form can mainly stimulate gizzard function and reduce its pH and retention time in this section of the GIT. Feed structure is related to the particle size and distribution post-grinding and even post-pelleting. The feed form (mash *vs.* pellets) can dictate the frequency of meals by the larger quantities a bird obtains in each visit to a feeder with pellet feed than with mash.

Finally, dietary materials containing fiber, like hulls and whole cereals or wood shavings, can obtain similar gizzard stimulus. In dietary fiber, it is necessary to understand the effects of level, source, and, in particular, the water solubility of fibers. The insoluble fiber can improve poultry health and performance, especially when coarsely ground, to stimulate gizzard activity and retention of digesta in this section. The soluble and partly soluble fibers constituted by arabinoxylans, β-glucans, and pectins mainly have anti-nutritive effects in poultry but can be prebiotic in small quantities. These soluble fibers have higher water holding and swelling capacity that promote digesta viscosity, negatively affecting digestibility and performance (Singh and Kim, 2021).

9.2 *Feed withdrawal and intestinal health*

Gut health is critical for poultry meat safety. However, many stressful practices are necessary before slaughtering, such as feed withdrawal, momentary water restriction, transportation and holding at the processing plant, and handling. Feed withdrawal is necessary to diminish carcass contamination with digesta and manure. However, feed withdrawal and water deprivation can cause gut perturbations during grow-out, cropping practices, and before processing (Cox and Pavic, 2010). After 5 or 6 h of feed withdrawal, gut mucosa traits shift, making chickens more susceptible to mucoid enteritis. Feed withdrawal stimulates mucin production by goblet cells in the intestinal mucosa, and extra mucin is avidly used by bacteria that irritate the mucosa, generating more mucus and inflammatory reactions. The disruption in mucus properties also facilitates the adherence of enteric pathogens (Thompson *et al.*, 2008).

Catching, handling, and transportation to the processing plant can cause significant stress to broilers and shift microbiota from non-pathogenic to pathogenic bacteria (Bello *et al.*, 2017). Dimming lighting during catching, training catchers for better handling and better ventilation, and reducing transportation and holding time pre-slaughter can minimize these adverse effects (Benincasa *et al.*, 2020).

10 Conclusions

Management decisions and practices are pivotal in gut health. Management interventions may start with implementing and practicing biosecurity at every level of the production chain. However, environmental control to guarantee comfort and minimize stress to birds starting in the previous generation plays a vital role in establishing healthy microbiota, maintaining proper development, and stimulating immunity. Hatching egg management, disinfection, and contact with all surfaces can affect the microbiota of hatchlings. Incubation conditions can impact embryo development, hatching window, and time to access feed and water. Management aspects such as housing system, stocking density, lighting, heat stress, house ventilation, bedding type, litter management, and water quality can affect the development of a bird and its microbiota. Environmental stress and contact with antibiotics can delay microbiota development, having long-term effects on the development of immunity. Deviations from optimum conditions can cause perturbation of gut functions, stress, and dysbiosis that concurrent factors can complicate. Lighting programs, light intensity, and color can be used to modulate feeding behavior and minimize stress for better gut health. Finally, stress before processing should be minimized to avoid shifting microbiota before arrival at the processing plant and putting food safety at risk as the last indicator of flock gut health.

References

Abdo, S.E., El-Kassas, S., El-Nahas, A.F., Mahmoud, S. 2017. Modulatory effect of monochromatic blue light on heat stress response in commercial broilers. Oxidative Med Cell Longev. 2017: 1351945–1351913. https://doi.10.1155/2017/1351945.

Abo-Al-Ela, H.G., El-Kassas, S., El-Naggar, K., Abdo, S.E., Jahejo, A.R., and Al Wakeel, R.A., 2021. Stress and immunity in poultry: light management and nanotechnology as effective immune enhancers to fight stress. Cell Stress and Chaperones 26: 457–472. https://doi.org/10.1007/s12192-021-01204-6.

Agyekum, G., Okai, M.A., Tona, J.K., Donkoh, A. and Hamidu, J.A., 2022. Impact of incubation temperature profile on chick quality, bone, and immune system during the late period of incubation of Cobb 500 broiler strain. Poultry Science: 101999. https://doi.org/10.1016/j.psj.2022.101999.

Ahmed, A.A., Ma, W., Guo, F., Ni, Y., Grossmann, R. and Zhao, R., 2013. Differences in egg deposition of corticosterone and embryonic expression of corticosterone metabolic enzymes between slow and fast-growing broiler chickens. Comparative Biochemistry and Physiology Part A: Molecular & Integrative Physiology 164(1): 200–206. https://doi.org/10.1016/j.cbpa.2012.09.004.

Akhlaghi, A., Zamiri, M.J., Ahangari, Y.J., Mollasalehi, M.R., Shojaie, H., Atashi, H., Navidshad, B., Akhlaghi, A.A. and Dadpasand, M., 2013. Growth performance and intestinal morphology in broiler chickens produced from hyperthyroid breeder hens. Animal production science 53(10): 1046–1051. https://doi.org/10.1071/AN12302.

Alpigiani, I., Abrahantes, J.C., Michel, V., Huneau-Salaün, A., Chemaly, M., Keeling, L.J., Gervelmeyer, A., Bacci, C., Brindani, F., Bonardi, S. and Berthe, F., 2017. Associations between animal welfare indicators and Campylobacter spp. in broiler chickens under commercial settings: A case study. Preventive Veterinary Medicine 147: 186–193. https://doi.org/10.1016/j.prevetmed.2017.09.005.

Akter, S., Cheng, B., West, D., Liu, Y., Qian, Y., Zou, X., Classen, J., Cordova, H., Oviedo, E. and Wang-Li, L., 2022. Impacts of Air Velocity Treatments under Summer Condition: Part I – Heavy broiler's surface temperature response. Animals 12(3): 328. https://doi.org/10.3390/ani12030328.

Al-Zghoul, M.B., Saleh, K.M.M. and Jaradat, Z.W., 2019. Expression of digestive enzyme and intestinal transporter genes during chronic heat stress in the thermally manipulated broiler chicken. Poultry Science 98(9): 4113–4122. https://doi.org/10.3382/ps/pez249.

Aruwa, C.E., Pillay, C., Nyaga, M.M., 2021. Poultry gut health – microbiome functions, environmental impacts, microbiome engineering and advancements in characterization technologies. Journal of Animal Science Biotechnology 12: 119. https://doi.org/10.1186/s40104-021-00640-9.

Assefa, B., 2022. Biosecurity: bioexclusion and biocontainment measures as effective poultry health management strategy. Preprints 2022010461 https://doi.org/10.20944/preprints202201.0461.v1.

Babacanoğlu, E., and Yalçin S.U., 2014. Hormone-mediated maternal stress affects embryonic development during incubation without adverse effect on chick weight and body composition. British poultry science. 2014 Mar 4; 55(2): 186–93. doi.org/10.1080/00071668.2013.878784.

Babić Leko, M., Gunjača, I., Pleić, N. and Zemunik, T., 2021. Environmental factors affecting thyroid-stimulating hormone and thyroid hormone levels. International Journal of Molecular Sciences 22(12): 6521. https://doi.org/10.3390/ijms22126521.

Barri, A., 2008. Effects of incubation temperature and transportation stress on yolk utilization, small intestine development, and post-hatch performance of high-yield broiler chicks (Doctoral dissertation, Virginia Tech).

Barri, A., Honaker, C.F., Sottosanti, J.R., Hulet, R.M. and McElroy, A.P., 2011. Effect of incubation temperature on nutrient transporters and small intestine morphology of broiler chickens. Poultry Science 90(1): 118–125. https://doi.org/10.3382/ps.2010-00908.

Bassi, L.S., Teixeira, L.V., Sens, R.F., Sonálio, K.C., Santos, M.C., Kuritza, L.N. and Maiorka, A., 2022. Effect of phytase supplementation and drinking water pH for turkey poults. Livestock Science, 265: 105069. https://doi.org/10.1016/j.livsci.2022.105069.

Benincasa, N.C., Sakamoto, K.S., da Silva, I.J.O. and Lobos, C.M.V., 2020. Animal welfare: impacts of pre-slaughter operations on the current poultry industry. Journal of Animal Behaviour and Biometeorology 8(2): 104–110. http://dx.doi.org/10.31893/jabb.20014.

Bello, A.U., Idrus, Z., Yong Meng, G., Awad, E.A. and Soleimani Farjam, A., 2018. Gut microbiota and transportation stress response affected by tryptophan supplementation in broiler chickens. Italian Journal of Animal Science 17(1): 107–113. https://doi.org/10.1080/1828051X.2017.1340814.

Berghaus, R.D., Thayer, S.G., Maurer, J.J., Hofacre, C.L., 2011. Effect of vaccinating breeder chickens with a killed *Salmonella* vaccine on *Salmonella* prevalences and loads in breeder and broiler chicken flocks. *Journal of Food Protection* 74 (5): 727–734. https://doi.org/10.4315/0362-028X.JFP-10-542.

Bergoug, H., Burel, C., Guinebretiere, M., Tong, Q., Roulston, N., Romanini, C.E.B., Exadaktylos, V., McGonnell, I.M., Demmers, T.G.M., Verhelst, R. and Bahr, C., 2013. Effect of pre-incubation and incubation conditions on hatchability, hatch time and hatch window, and effect of post-hatch handling on chick quality at placement. World's Poultry Science Journal, 69(2): 313–334. https://doi.org/10.1017/S0043933913000329.

Bindari, Y.R. and Gerber, P.F., 2022. Centennial Review: Factors affecting the chicken gastrointestinal microbial composition and their association with gut health and productive performance. Poultry Science, 101(1): 101612. https://doi.org/10.1016/j.psj.2021.101612.

Bokkers, E.A.M. and Koene, P., 2003. Eating behaviour, and preprandial and postprandial correlations in male broiler and layer chickens. British Poultry Science 44(4): 538–544. https://doi.org/10.1080/0007166031000161616 5.

Bowling, M., Forder, R., Hughes, R.J., Weaver, S. and Hynd, P.I., 2018. Effect of restricted feed intake in broiler breeder hens on their stress levels and the growth and immunology of their offspring. Translational Animal Science 2(3): 263–271. https://doi.org/10.1093/tas/txy064.

Broom, L.J. and Kogut, M.H., 2018. The role of the gut microbiome in shaping the immune system of chickens. Veterinary Immunology and Immunopathology 204: 44–51. https://doi.org/10.1016/j.vetimm.2018.10.002.

Brugaletta, G., Teyssier, J.R., Rochell, S.J., Dridi, S. and Sirri, F., 2022. A review of heat stress in chickens. Part I: Insights into physiology and gut health. Frontiers in Physiology 13: 934381. https://doi.org/10.3389/fphys.2022.934381.

Bucher, M.G., Zwirzitz, B., Oladeinde, A., Cook, K., Plymel, C., Zock, G., Lakin, S., Aggrey, S.E., Ritz, C., Looft, T. and Lipp, E., 2020. Reused poultry litter microbiome with competitive exclusion potential against *Salmonella Heidelberg*. Journal of Environmental Quality 49(4): 869–881. https://doi.org/10.1002/jeq2.20081.

Butterworth, A., 2019. Effects of high stocking density on broiler chicken and turkeys. CABI Reviews 1–16. https://doi.org/10.1079/PAVSNNR20191403.

Caekebeke, N., Ringenier, M., De Meyer, F., Ducatelle, R., Ongena, N., Van Immerseel, F. and Dewulf, J., 2020. A study on risk factors for macroscopic gut abnormalities in intensively reared broiler chickens. Avian Pathology 49(2): 193–201. https://doi.org/10.1080/03079457.2019.1711019.

Calefi, A.S., da Silva Fonseca, J.G., Cohn, D.W.H., Honda, B.T.B., Costola-de-Souza, C., Tsugiyama, L.E., Quinteiro-Filho, W.M., Ferreira, A.J.P. and Palermo-Neto, J., 2016. The gut-brain axis interactions during heat stress and avian necrotic enteritis. Poultry Science 95(5): 1005–1014. https://doi.org/10.3382/ps/pew021.

Cardeal, P.C., Araújo, I.C.S., Sousa, L.S., Melo, E.F., Carvalho, T.S.M., Triginelli, M.V., Pompeu, M.A. and Lara, L.J.C., 2021. Breeder age and post-hatch feed access influence physiology of residual yolk sac, intestinal development and performance of broiler chicks. Livestock Science 251: 104617. https://doi.org/10.1016/j.livsci.2021.104617.

Chi, Q., Wang, D., Hu, X., Li, S. and Li, S., 2019. Hydrogen sulfide gas exposure induces necroptosis and promotes inflammation through the MAPK/NF-kappaB pathway in broiler spleen. Oxidative Medicine and Cellular Longevity 2019: 8061823. https://doi.org/10.1155/2019/8061823.

Chinivasagam, H.N., Estella, W., Rodrigues, H., Mayer, D.G., Weyand, C., Tran, T., Onysk, A. and Diallo, I., 2016. On-farm *Campylobacter* and *Escherichia coli* in commercial broiler chickens: Re-used bedding does not influence *Campylobacter* emergence and levels across sequential farming cycles. Poultry Science 95(5): 1105–1115. https://doi.org/10.3382/ps/pew003.

Classen, H., Apajalahti, J., Svihus, B. and Choct, M., 2016. The role of the crop in poultry production. World's Poultry Science Journal 72(3): 459–472. https://doi:10.1017/S004393391600026X.

Cox, J.M. and Pavic, A., 2010. Advances in enteropathogen control in poultry production. Journal of Applied Microbiology 108(3): 745–755. https://doi.org/10.1111/j.1365-2672.2009.04456.x.

da Silva, C.S., Molenaar, R., Giersberg, M.F., Rodenburg, T.B., van Riel, J.W., De Baere, K., Van Dosselaer, I., Kemp, B., van den Brand, H. and de Jong, I.C., 2021. Day-old chicken quality and performance of broiler chickens from 3 different hatching systems. Poultry Science 100(3):100953. https://doi.org/10.1016/j.psj.2020.12.050.

Dai, D., Qi, G.H., Wang, J., Zhang, H.J., Qiu, K. and Wu, S.G., 2022. Intestinal microbiota of layer hens and its association with egg quality and safety. Poultry Science, 101(9): 102008.

Darras, V.M., 2019. The role of maternal thyroid hormones in avian embryonic development. Frontiers in Endocrinology 10: 66. https://doi.org/10.3389/fendo.2019.000.

Das, R., Mishra, P. and Jha, R., 2021. In ovo feeding as a tool for improving performance and gut health of poultry: a review. Frontiers in Veterinary Science 8: 754246. https://doi.org/10.3389/fvets.2021.754246.

Dayan, J., Reicher, N., Melkman-Zehavi, T. and Uni, Z., 2020. Incubation temperature affects yolk utilization through changes in expression of yolk sac tissue functional genes. Poultry Science 99(11): 6128–6138. https://doi.org/10.1016/j.psj.2020.07.037.

De Beer, M., McMurtry, J.P., Brocht, D.M. and Coon, C.N., 2008. An examination of the role of feeding regimens in regulating metabolism during the broiler breeder grower period. 2. Plasma hormones and metabolites. Poultry Science 87(2): 264–275. https://doi.org/10.3382/ps.2007-00196.

De Beer, M., Rosebrough, R.W., Russell, B.A., Poch, S.M., Richards, M.P. and Coon, C.N., 2007. An examination of the role of feeding regimens in regulating metabolism during the broiler breeder grower period. 1. Hepatic lipid metabolism. Poultry Science 86(8): 1726–1738. https://doi.org/10.1093/ps/86.8.1726.

De Jong, I., Gunnink, H., Van Hattum, T., Van Riel, J., Raaijmakers, M., Zoet, E. and Van den Brand, H., 2019. Comparison of performance, health and welfare aspects between commercially housed hatchery-hatched and on-farm hatched broiler flocks. Animal 13(6): 1269–1277. https://doi:10.1017/S1751731118002872.

de Jong, I., Schokker, D., Gunnink, H. and Rebel, J.M.J., 2022. Early life environment affects behavior, welfare, gut microbiome composition, and diversity in broiler chickens. Frontiers in Veterinary Science 9: 977359–977359. https://doi.org/10.3389/fvets.2022.977359.

de Jong, I.C., Fillerup, M., and Blokhuis, H.J., 2005. Effect of scattered feeding and feeding twice a day during rearing on parameters of hunger and frustration in broiler breeders. Applied Animal Behavior Science 92: 61–76. https://doi.org/10.1016/j.applanim.2004.10.022.

de Jong, I.C., van Riel, J., Bracke, M.B. and van den Brand, H., 2017. A meta-analysis of effects of post-hatch food and water deprivation on development, performance and welfare of chickens. PloS one 12(12): p.e0189350. https://doi.org/10.1371/journal.pone.0189350.

de Toledo, T.D.S., Roll, A.A.P., Rutz, F., Dallmann, H.M., Dai Prá, M.A., Leite, F.P.L. and Roll, V.F.B., 2020. An assessment of the impacts of litter treatments on the litter quality and broiler performance: A systematic review and meta-analysis. PloS one 15(5): p.e0232853. https://doi.org/10.1371/journal.pone.0232853.

Decuypere, E., Van As, P., Van der Geyten, S. and Darras, V.M., 2005. Thyroid hormone availability and activity in avian species: a review. Domestic Animal Endocrinology 29(1): 63–77. https://doi.org/10.1016/j.domaniend.2005.02.028.

Díaz Carrasco, J.M., Redondo, L.M., Casanova, N.A., and Fernández Miyakawa, M.E. 2022. The Role of Farm Environment and Management in Shaping the Gut Microbiota of Poultry. In: Kogut, M.H., Zhang, G. (eds) Gut Microbiota, Immunity, and Health in Production Animals. The Microbiomes of Humans, Animals, Plants, and the Environment, vol. 4. Springer, Cham. https://doi.org/10.1007/978-3-030-90303-9_10.

Ding, J., Dai, R., Yang, L., He, C., Xu, K., Liu, S., Zhao, W., Xiao, L., Luo, L., Zhang, Y. and Meng, H., 2017. Inheritance and establishment of gut microbiota in chickens. Frontiers in Microbiology 8: 1967. https://doi.10.3389/fmicb.2017.01967.

Ding, P., Liu, H., Tong, Y., He, X., Yin, X., Yin, Y., Zhang, H., and Song, Z., 2022. Developmental change of yolk microbiota and its role on early colonization of intestinal microbiota in chicken embryo. Animals 12(1): 16. https://doi.org/10.3390/ani12010016.

Dórea, F.C., Cole, D.J., Hofacre, C., Zamperini, K., Mathis, D., Doyle, M.P., Lee, M.D. and Maurer, J.J., 2010. Effect of *Salmonella* vaccination of breeder chickens on contamination of broiler chicken carcasses in integrated poultry operations. Applied and environmental microbiology 76(23): 7820–7825. https://doi.org/10.1128/AEM.01320-10.

Du, X., Qin, P., Liu, Y., Amevor, F.K., Shu, G., Li, D. and Zhao, X., 2022. Effects of key farm management practices on pullets welfare – A Review. Animals 12(6): 729. https://doi.org/10.3390/ani12060729.

Ebeid, T.A., Tůmová, E., Al-Homidan, I.H., Ketta, M. and Chodová, D., 2022. Recent advances in the role of feed restriction in poultry productivity: part I-performance, gut development, microbiota and immune response. World's Poultry Science Journal 1–18. https://doi.org/10.1080/00439339.2022.2097149.

Ekmay, R.D., De Beer, M., Rosebrough, R.W., Richards, M.P., McMurtry, J.P. and Coon, C.N., 2010. The role of feeding regimens in regulating metabolism of sexually mature broiler breeders. Poultry Science 89(6): 1171–1181. https://doi.org/10.3382/ps.2009-00465.

El-Hack, A., Mohamed, E., El-Saadony, M.T., Shehata, A.M., Arif, M., Paswan, V.K., Batiha, G.E.S., Khafaga, A.F. and Elbestawy, A.R., 2021. Approaches to prevent and control *Campylobacter* spp. colonization in broiler chickens: a review. Environmental Science and Pollution Research 28(5): 4989–5004. https://doi.org/10.1007/s11356-020-11747-3.

Farghly, M.F.A., Alagawany, M. and Abd El-Hack, M.E., 2018. Feeding time can alleviate negative effects of heat stress on performance, meat quality and health status of turkey. British Poultry Science 59(2): 205–210. https://doi.org/10.1080/00071668.2017.1413233.

Farhadi, D., Hosseini, S.M. and Dezfuli, B.T., 2016. Effect of house type on growth performance, litter quality and incidence of foot lesions in broiler chickens reared in varying stocking density. Journal of BioScience and Biotechnology: 5(1).

Ganjali, H., Raji, A.R. and Zarghi, H., 2015. Effect of post hatch delayed access to feed on performance, GIT physical and histological development and yolk absorption in young broiler chicks. Biomedical and Pharmacology Journal 8(2): 945–955.

Giersberg, M.F., Molenaar, R., de Jong, I.C., da Silva, C.S., van den Brand, H., Kemp, B. and Rodenburg, T.B., 2021. Effects of hatching system on the welfare of broiler chickens in early and later life. Poultry Science 100(3): 100946. https://doi.org/10.1016/j.psj.2020.12.043.

Goel, A., 2021. Heat stress management in poultry. Journal of Animal Physiology and Animal Nutrition 105(6): 1136–1145. https://doi.org/10.1111/jpn.13496.

Goliomytis, M., Tsipouzian, T. and Hager-Theodorides, A.L., 2015. Effects of egg storage on hatchability, chick quality, performance and immunocompetence parameters of broiler chickens. Poultry Science 94(9): 2257–2265. https://doi.org/10.3382/ps/pev200.

Goo, D., Kim, J.H., Choi, H.S., Park, G.H., Han, G.P. and Kil, D.Y. 2019. Effect of stocking density and sex on growth performance, meat quality, and intestinal barrier function in broiler chickens. Poultry Science 98: 1153–1160. https://doi.org/10.3382/ps/pey491.

Goossens, E., Van Erum, J., De Gussem, M., Van Limbergen, T., Callens, C., De Zutter, L., Ducatelle, R. and Van Immerseel, F., 2020. Incidence and associated risk factors of necrotic enteritis in Belgian layer pullet flocks. Avian Pathology 49(5): 476–485. https://doi.org/10.1080/03079457.2020.1772460.

Green, A.R. and Xin, H., 2009. Effects of stocking density and group size on heat and moisture production of laying hens under thermoneutral and heat-challenging conditions. Transactions of the ASABE 52(6): pp. 2027–2032. https://doi.10.13031/2013.29204 @2009.

Han, H., Zhou, Y., Liu, Q., Wang, G., Feng, J. and Zhang, M., 2021. Effects of ammonia on gut microbiota and growth performance of broiler chickens. Animals, 11(6): 1716. https://doi.org/10.3390/ani11061716.

Haussmann, M.F., Longenecker, A.S., Marchetto, N.M., Juliano, S.A. and Bowden, R.M., 2012. Embryonic exposure to corticosterone modifies the juvenile stress response, oxidative stress and telomere length. Proceedings of the Royal Society B: Biological Sciences 279(1732): 1447–1456. https://doi.org/10.1098/rspb.2011.1913.

He, Y., Maltecca, C. and Tiezzi, F., 2021. Potential use of gut microbiota composition as a biomarker of heat stress in monogastric species: a review. Animals 11(6): 1833. https://doi.org/10.3390/ani11061833.

Heidari, S. and Toghyani, M. 2018. Effect of stocking density and methionine levels on growth performance and immunity of broiler chicks. Iranian Journal of Applied Animal Science 8: 483–489.

Hieke, A.-S.C., Hubert, S.M., and Athrey, G., 2019. Circadian disruption and divergent microbiota acquisition under extended photoperiod regimens in chicken. *PeerJ* 7: e6592. https://doi.org/10.7717/peerj.6592.

Hofmann, T., Schmucker, S., Grashorn, M. and Stefanski, V., 2021. Short-and long-term consequences of stocking density during rearing on the immune system and welfare of laying hens. Poultry Science 100(8): E101243. https://doi.org/10.1016/j.psj.2021.101243.

Hofmann, T., Schmucker, S.S., Bessei, W., Grashorn, M. and Stefanski, V., 2020. Impact of housing environment on the immune system in chickens: a review. Animals 10(7): 1138. https://doi.org/10.3390/ani10071138.

Hosseindoust, A., Kang, H.K. and Kim, J., 2022. Quantifying heat stress; the roles on metabolic status and intestinal integrity in poultry, a review. Domestic Animal Endocrinology: 106745. https://doi.org/10.1016/j.domaniend.2022.106745.

Howie, J.A., Tolkamp, B.J., Bley, T. and Kyriazakis, I., 2010. Short-term feeding behaviour has a similar structure in broilers, turkeys and ducks. British Poultry Science, 51(6): 714–724. https://doi.org/10.1080/00071668.2010.528749.

Hu, X., Chi, Q., Wang, D., Chi, X., Teng, X. and Li, S. 2018. Hydrogen sulfide inhalation-induced immune damage is involved in oxidative stress, inflammation, apoptosis and the Th1/Th2 imbalance in broiler bursa of Fabricius. Ecotoxicology and Environmental Safety 164: 201–209. https://doi.org/10.1016/j.ecoenv.2018.08.029.

Iannetti, L., Neri, D., Santarelli, G.A., Cotturone, G., Vulpiani, M.P., Salini, R., Antoci, S., Di Serafino, G., Di Giannatale, E., Pomilio, F. and Messori, S., 2020. Animal welfare and microbiological safety of poultry meat: Impact of different at-farm animal welfare levels on at-slaughterhouse Campylobacter and Salmonella contamination. Food Control 109: 106921. https://doi.org/10.1016/j.foodcont.2019.106921.

Jobe, M.C., Ncobela, C.N., Kunene, N.W. and Opoku, A.R. 2019. Effects of *Cassia abbreviata* extract and stocking density on growth performance, oxidative stress and liver function of indigenous chickens. Tropical Animal Health and Production 51: 2567–2574. https://doi.org/10.1007/s11250-019-01979-y.

Kers, J.G., Velkers, F.C., Fischer, E.A., Hermes, G.D., Stegeman, J.A. and Smidt, H., 2018. Host and environmental factors affecting the intestinal microbiota in chickens. Frontiers in microbiology, 9, p. 235. doi.org/10.3389/fmicb.2018.00235.

Kers, J.G., Velkers, F.C., Fischer, E.A.J., Hermes, G.D.A., Lamot, D.M., Stegeman, J.A. and Smidt, H., 2019. Take care of the environment: housing conditions affect the interplay of nutritional interventions and intestinal microbiota in broiler chickens. Animal Microbiome 1: 10. https://doi.org/10.1186/s42523-019-0009-z.

Kierończyk, B., Rawski, M., Długosz, J., Świątkiewicz, S. and Józefiak, D., 2016. Avian crop function – a review. Annals of Animal Science 16(3): 653–678. https://doi.org/10.1515/aoas-2016-0032.

Kumari, K.N.R. and Nath, D.N., 2018. Ameliorative measures to counter heat stress in poultry. World's Poultry Science Journal 74(1): 117–130. doi:10.1017/S0043933917001003.

Lamot, D.M., Van De Linde, I.B., Molenaar, R., Van Der Pol, C.W., Wijtten, P.J.A., Kemp, B. and Van Den Brand, H., 2014. Effects of moment of hatch and feed access on chicken development. Poultry Science 93(10): 2604–2614. doi.org/10.3382/ps.2014-04123.

Lee, S., La, T.M., Lee, H.J., Choi, I.S., Song, C.S., Park, S.Y., Lee, J.B. and Lee, S.W., 2019. Characterization of microbial communities in the chicken oviduct and the origin of chicken embryo gut microbiota. Scientific Reports 9(1): 6838. https://doi.org/10.1038/s41598-019-43280-w.

Li, G.S., Zhu, F., Yang, F.X., Hao, J.P. and Hou, Z.C., 2020. Selection response and genetic parameter estimation of feeding behavior traits in Pekin ducks. Poultry Science 99(5): 2375–2384. https://doi.org/10.1016/j.psj.2020.01.013.

Li, J., Wang, Z., Cao, J., Dong, Y.L. and Chen, Y.X., 2014. Role of monochromatic light on development of cecal tonsil in young broilers. The Anatomical Record 297(7): 1331–1337. https://doi.org/10.1002/ar.22909.

Li, J.P., Wu, Q.F., Ma, S.C., Wang, J.M., Wei, B., Han, C.C., Li, L., He, H. and Liu, H.H., 2022. Effect of feed restriction on the intestinal microbial community structure of growing ducks. Archives of Microbiology 204: 85. https://doi.org/10.1007/s00203-021-02636-5.

Liu, C., Chaudhry, M.T., Zhao, D., Lin, T., Tian, Y. and Fu, J., 2019. Heat shock protein 70 protects the quail cecum against oxidant stress, inflammatory injury, and microbiota imbalance induced by cold stress. Poultry Science 98(11): 5432–5445. https://doi.org/10.3382/ps/pez327.

López-López, P., Sarmiento-Franco, L.A. and Santos-Ricalde, R., 2022. Effect of stocking density on performance, infection by *Eimeria* spp., intestinal lesions and foot pad injuries in broilers with outdoor access under tropical conditions. British Poultry Science 63(2): 108–114. https://doi.org/10.1080/00071668.2021.1966749.

Maki, J.J., Bobeck, E.A., Sylte, M.J. and Looft, T., 2020. Eggshell and environmental bacteria contribute to the intestinal microbiota of growing chickens. Journal of Animal Science and Biotechnology 11: 60. https://doi.org/10.1186/s40104-020-00459-w.

Mikec, M., Biđin, Z., Valentić, A., Savić, V., Zelenika, T.A., Raguž-Đurić, R., Novak, I.L. and Baleńovic, M., 2006. Influence of environmental and nutritional stressors on yolk sac utilization, development of chicken gastrointestinal system and its immune status. World's Poultry Science Journal 62(1): 31–40. https://doi.org/10.1079/WPS200582.

Moon, J., DuBien, J., Ramachandran, R., Liang, Y., Dridi, S. and Tabler, T., 2023. Effects of a sprinkler and cool cell combined system on cooling water usage, litter moisture, and indoor environment of broiler houses. Animals, 13(18): 2939. https://doi.org/10.3390/ani13182939.

Nangsuay, A., Meijerhof, R., Van den Anker, I., Heetkamp, M.J.W., Kemp, B. and Van den Brand, H., 2017. Effects of breeder age, strain, and eggshell temperature on nutrient metabolism of broiler embryos. Poultry Science 96(6): 1891–1900. https://doi.org/10.3382/ps/pew417.

Nasr, M.A., Alkhedaide, A.Q., Ramadan, A.A., Abd-El Salam, E.H. and Hussein, M.A., 2021. Potential impact of stocking density on growth, carcass traits, indicators of biochemical and oxidative stress and meat quality of different broiler breeds. Poultry Science 100(11): 101442. https://doi.org/10.1016/j.psj.2021.101442.

NCC. National Chicken Council. 2022. Broiler welfare guidelines and audit checklist. Accessed Dec. 2023. https://www.nationalchickencouncil.org/policy/animal-welfare/.

Nawab, A., Ibtisham, F., Li, G., Kieser, B., Wu, J., Liu, W., Zhao, Y., Nawab, Y., Li, K., Xiao, M. and An, L., 2018. Heat stress in poultry production: Mitigation strategies to overcome the future challenges facing the global poultry industry. Journal of Thermal Biology 1 78: 131–139. https://doi.org/10.1016/j.jtherbio.2018.08.010.

Nielsen, B.L., 2004. Behavioural aspects of feeding constraints: do broilers follow their gut feelings?. Applied Animal Behaviour Science 86(3–4): 251–260. https://doi.org/10.1016/j.applanim.2004.02.005.

Nijdam, E., Delezie, E., Lambooij, E., Nabuurs, M.J., Decuypere, E. and Stegeman, J.A., 2005. Feed withdrawal of broilers before transport changes plasma hormone and metabolite concentrations. Poultry Science 84(7): 1146–1152. https://doi.org/10.1093/ps/84.7.1146.

Oden, L.A., Lee, J.T., Pohl, S.K., Klein, A.E., Anderson, S.A., Dougherty, S.D., Broussard, C.T., Fitz-Coy, S.H., Newman, L.J. and Caldwell, D.J., 2012. Influence of diet on oocyst output and intestinal lesion development in replacement broiler breeders following live oocyst coccidiosis vaccination. Journal of Applied Poultry Research 21(3): 445–459. https://doi.org/10.3382/japr.2010-00264.

Oliveira, G.D.S., McManus, C., Salgado, C.B. and dos Santos, V.M., 2022. Effects of sanitizers on microbiological control of hatching eggshells and poultry health during embryogenesis and early stages after hatching in the last decade. Animals 12(20): 2826. https://doi.org/10.3390/ani12202826.

Olkowski, A.A., Nain, S., Laarveld, B. and Wojnarowicz, C., 2015. Changes in eggshell structure and predisposition of broilers to health problems: is there a common pathophysiology?. British Poultry Science 56(2): 267–274. https://doi.org/10.1080/00071668.2015.1008995.

Olsen, R., Kudirkiene, E., Thøfner, I., Pors, S., Karlskov-Mortensen, P., Li, L., Papasolomontos, S., Angastiniotou, C. and Christensen, J., 2017. Impact of egg disinfection of hatching eggs on the eggshell microbiome and bacterial load. Poultry Science 96(11): 3901–3911. https://doi.org/10.3382/ps/pex182.

Oviedo-Rondón, E.O., 2019. Holistic view of intestinal health in poultry. Animal Feed Science and Technology 250: 1–8. https://doi.org/10.1016/j.anifeedsci.2019.01.009.Rod.

Oznurlu, Y., Celik, I., Telatar, T. and Sur, E., 2010. Histochemical and histological evaluations of the effects of high incubation temperature on embryonic development of thymus and bursa of Fabricius in broiler chickens. British Poultry Science 51(1): 43–51. https://doi.org/10.1080/00071660903575558.

Panda, A.K., Bhanja, S.K. and Sunder, G.S., 2015. Early post hatch nutrition on immune system development and function in broiler chickens. World's Poultry Science Journal, 71(2), pp. 285–296. doi.org/10.1017/S004393391500029X.

Patra, A.K. and Kar, I., 2021. Heat stress affects microbiota composition, barrier integrity, nutrient transport in gut, production performance, and its amelioration in farm animals. Journal of Animal Science and Technology 63(2): 211. https://doi: 10.5187/jast.2021.e48.

Proszkowiec-Weglarz, M., Miska, K.B., Ellestad, L.E., Schreier, L.L., Kahl, S., Darwish, N., Campos, P. and Shao, J., 2022. Delayed access to feed early post-hatch affects the development and maturation of gastrointestinal tract microbiota in broiler chickens. BMC Microbiology 22: 206. https://doi.org/10.1186/s12866-022-02619-6.

Proszkowiec-Weglarz, M., Schreier, L.L., Kahl, S., Miska, K.B., Russell, B. and Elsasser, T.H., 2020. Effect of delayed feeding post-hatch on expression of tight junction- and gut barrier-related genes in the small intestine of broiler chickens during neonatal development. Poultry Science 99(10): 4714–4729. https://doi.org/10.1016/j.psj.2020.06.023.

Quinteiro-Filho, W.M., Ribeiro, A., Ferraz-de-Paula, V., Pinheiro, M.L., Sakai, M., Sá, L.R.M.D., Ferreira, A.J.P. and Palermo-Neto, J., 2010. Heat stress impairs performance parameters, induces intestinal injury, and decreases macrophage activity in broiler chickens. Poultry Science 89(9): 1905–1914. https://doi.org/10.1080/03079457.2012.709315.

Racicot, M., Venne, D., Durivage, A. and Vaillancourt, J.P., 2012a. Evaluation of strategies to enhance biosecurity compliance on poultry farms in Québec: Effect of audits and cameras. Preventative Veterinary Medicine 103(2–3): 208–218. https://doi.org/10.1016/j.prevetmed.2011.08.004.

Racicot, M., Venne, D., Durivage, A. and Vaillancourt, J., 2011b. Description of 44 biosecurity errors while entering and exiting poultry barns based on video surveillance in Quebec, Canada. Preventative Veterinary Medicine 100(3–4): 193–199.

Racicot, M., Venne, D., Durivage, A. and Vaillancourt, J.P., 2012c. Evaluation of the relationship between personality traits, experience, education and biosecurity compliance on poultry farms in Québec, Canada. Preventive Veterinary Medicine 103(2–3): 201–207. https://doi.org/10.1016/j.prevetmed.2011.08.011.

Rodrigues, I. and Choct, M., 2018. The foregut and its manipulation via feeding practices in the chicken. Poultry Science 97(9): 3188–3206. https://doi.org/10.3382/ps/pey191.

Rodrigues, I. and Choct, M., 2019. Feed intake pattern of broiler chickens under intermittent lighting: Do birds eat in the dark?. Animal Nutrition 5(2): 174–178. https://doi.org/10.1016/j.aninu.2018.12.002.

Rodrigues, I., Svihus, B., Bedford, M.R., Gous, R. and Choct, M., 2018. Intermittent lighting improves resilience of broilers during the peak phase of sub-clinical necrotic enteritis infection. Poultry Science 97(2): 438–446. https://doi.org/10.3382/ps/pex315.

Roll, V.F.B., Dai Prá, M.A. and Roll, A.P., 2011. Research on *Salmonella* in broiler litter reused for up to 14 consecutive flocks. Poultry Science 90(10): 2257–2262. https://doi.org/10.3382/ps.2011-01583.

Roque, K., Shin, K.M., Jo, J.H., Kim, H.A. and Heo, Y., 2015. Relationship between chicken cellular immunity and endotoxin levels in dust from chicken housing environments. Journal of Veterinary Science 16(2): 173–177. https://doi.org/10.4142/jvs.2015.16.2.173.

Rostagno, M.H., 2020. Effects of heat stress on the gut health of poultry. Journal of Animal Science 98(4): p.skaa090. https://doi.org/10.1093/jas/skaa090.

Rubio, L.A., 2019. Possibilities of early life programming in broiler chickens via intestinal microbiota modulation. Poultry Science 98(2): 695–706. https://doi.org/10.3382/ps/pey416.

Şahan, Ü., İpek, A. and Sözcü, A., 2014. Yolk sac fatty acid composition, yolk absorption, embryo development, and chick quality during incubation in eggs from young and old broiler breeders. Poultry Science 93(8): 2069–2077. https://doi.org/10.3382/ps.2013-03850.

Saino, N., Romano, M., Ferrari, R.P., Martinelli, R. and Møller, A.P., 2005. Stressed mothers lay eggs with high corticosterone levels which produce low-quality offspring. Journal of Experimental Zoology Part A: Comparative Experimental Biology 303(11): 998–1006. https://doi.org/10.1002/jez.a.224.

Santos, F.D.O., Santos Júnior, A.A., Oviedo-Rondon, E.O. and Ferket, P.R., 2012. Influence of housing system on growth performance and intestinal health of *Salmonella*-challenged broiler chickens. Current Research in Poultry Science. https://doi.10.3923/crps.2012.

Schokker, D., de Klerk, B., Borg, R., Bossers, A. and Rebel, J.M., 2021. Factors influencing the succession of the fecal microbiome in broilers. Livestock Science 247: 104486. https://doi.org/10.1016/j.livsci.2021.104486.

Schokker, D., Jansman, A.J.M., Veninga, G., Bruin, N., Vastenhouw, S.A., de Bree, F.M., Bossers, A., Rebel, J.M.J. and Smits, M.A., 2017. Perturbation of microbiota in one-day old broiler chickens with antibiotic for 24 hours negatively affects intestinal immune development. BMC Genomics 18: 241. https://doi.org/10.1186/s12864-017-3625-6.

Shi, D., Bai, L., Qu, Q., Zhou, S., Yang, M., Guo, S., Li, Q. and Liu, C., 2019. Impact of gut microbiota structure in heat-stressed broilers. Poultry Science 98(6): 2405–2413. https://doi.org/10.3382/ps/pez026.

Shinde, A.S., Goel, A., Mehra, M., Rokade, J., Bhadauria, P., Mandal, A.B. and Bhanja, S.K., 2015. Delayed post hatch feeding affects performance, intestinal morphology and expression pattern of nutrient transporter genes in egg type chickens. Journal of Nutrition & Food Sciences, 5(3): 372. https://doi.10.4172/2155-9600.1000372.

Shira, E.B., Sklan, D. and Friedman, A., 2005. Impaired immune responses in broiler hatchling hindgut following delayed access to feed. Veterinary Immunology and Immunopathology 105(1–2): 33–45. https://doi.org/10.1016/j.vetimm.2004.12.011.

Shterzer, N., Rothschild, N., Sbehat, Y., Stern, E., Nazarov, A. and Mills, E., 2020. Large overlap between the intestinal and reproductive tract microbiomes of chickens. Frontiers in Microbiology 11: 1508. https://doi.org/10.3389/fmicb.2020.015.

Shynkaruk, T., Buchynski, K. and Schwean-Lardner, K., 2022. Lighting program as a management tool for broilers raised without antibiotics – impact on productivity and welfare. British Poultry Science. https://doi.org/10.1080/00071668.2022.2083943.

Sibanda, N., McKenna, A., Richmond, A., Ricke, S.C., Callaway, T., Stratakos, A.C., Gundogdu, O. and Corcionivoschi, N., 2018. A review of the effect of management practices on *Campylobacter* prevalence in poultry farms. Frontiers in Microbiology 9: 2002. https://doi.org/10.3389/fmicb.2018.02002.

Singh, A.K. and Kim, W.K., 2021. Effects of dietary fiber on nutrients utilization and gut health of poultry: a review of challenges and opportunities. Animals 11(1): 181. https://doi.org/10.3390/ani11010181.

Simon, K., Verwoolde, M.B., Zhang, J., Smidt, H., de Vries Reilingh, G., Kemp, B. and Lammers, A., 2016. Long-term effects of early life microbiota disturbance on adaptive immunity in laying hens. Poultry Science 95(7): 1543–1554. https://doi.org/10.3382/ps/pew088.

Smith, J.A., 2011. Experiences with drug-free broiler production. Poultry Science 90(11): 2670–2678. https://doi.org/10.3382/ps.2010-01032.

Smith, J.A., 2019. Broiler production without antibiotics: United States field perspectives. Animal Feed Science and Technology 250: 93–98. https://doi.org/10.1016/j.anifeedsci.2018.04.027.

Su, Y., Ge, Y., Xu, Z., Zhang, D. and Li, D., 2021. The digestive and reproductive tract microbiotas and their association with body weight in laying hens. Poultry Science 100(11): 101422. https://doi.org/10.1016/j.psj.2021.101422.

Taha-Abdelaziz, K., Hodgins, D.C., Lammers, A., Alkie, T.N. and Sharif, S., 2018. Effects of early feeding and dietary interventions on development of lymphoid organs and immune competence in neonatal chickens: A review. Veterinary Immunology and Immunopathology 201: 1–11. https://doi.org/10.1016/j.vetimm.2018.05.001.

Tebrün, W., Motola, G., Hafez, M.H., Bachmeier, J., Schmidt, V., Renfert, K., Reichelt, C., Brüggemann-Schwarze, S. and Pees, M., 2020. Preliminary study: Health and performance assessment in broiler chicks following application of six different hatching egg disinfection protocols. PLoS One 15(5): p.e0232825. https://doi.org/10.1371/journal.pone.0232825.

Thanabalan, A. and Kiarie, E.G., 2021. Influence of feeding omega-3 polyunsaturated fatty acids to broiler breeders on indices of immunocompetence, gastrointestinal, and skeletal development in broiler chickens. Frontiers in Veterinary Science 8: 653152. https://doi.org/10.3389/fvets.2021.653152.

Thompson, K., Burkholder, K., Patterson, J. and Applegate, T.J., 2008. Microbial ecology shifts in the ileum of broilers during feed withdrawal and dietary manipulations. Poultry Science 87(8): 1624–1632. https://doi.org/10.3382/ps.2007-00324.

Tian, Y., Li, G., Chen, L., Bu, X., Shen, J., Tao, Z., Zeng, T., Du, X. and Lu, L., 2020. High-temperature exposure alters the community structure and functional features of the intestinal microbiota in Shaoxing ducks (*Anas platyrhynchos*). Poultry Science 99(5): 2662–2674. https://doi.org/10.1016/j.psj.2019.12.046.

Trudeau, S., Thibodeau, A., Côté, J-C., Gaucher, M.-L. and Fravalo, P., 2020. Contribution of the broiler breeders' fecal microbiota to the establishment of the eggshell microbiota. Frontiers in Microbiology 11: 666. https://doi: 10.3389/fmicb.2020.00666.

Tsiouris, V., 2016. Poultry management: a useful tool for the control of necrotic enteritis in poultry. Avian Pathology 45(3): 323–325. https://doi.org/10.1080/03079457.2016.1154502.

Tsiouris, V., Georgopoulou, I., Batzios, C., Pappaioannou, N., Ducatelle, R. and Fortomaris, P., 2015. High stocking density as a predisposing factor for necrotic enteritis in broiler chicks. Avian Pathology 44(2): 59–66. https://doi.org/10.1080/03079457.2014.1000820.

Tsiouris, V., Georgopoulou, I., Batzios, C., Pappaioannou, N., Ducatelle, R. and Fortomaris, P., 2018. Heat stress as a predisposing factor for necrotic enteritis in broiler chicks. Avian Pathology 47(6): 616–624. https://doi.org/10.1080/03079457.2018.1524574.

van der Wagt, I., de Jong, I.C., Mitchell, M.A., Molenaar, R. and van den Brand, H., 2020. A review on yolk sac utilization in poultry. Poultry Science 99(4): 2162–2175. https://doi.org/10.1016/j.psj.2019.11.041.

Van Meirhaeghe, H., Schwarz, A., Dewulf, J., Immerseel, F., Vanbeselaere, B. and De Gussem, M., 2019. Transmission of poultry diseases and biosecurity in poultry production. Biosecurity in animal production and veterinary medicine: from principles to practice. J. Dewulf and F. Van Inmmerseel. pp. 329–356.

Verwoolde, M.B., Arts, J., Jansen, C.A., Parmentier, H.K., and Lammers, A., 2022. Transgenerational effects of maternal immune activation on specific antibody responses in layer chickens. Front Veterinary Science. 9: 832130. https://doi.10.3389/fvets.2022.832130.

Wang, L., Lilburn, M., and Yu, Z. 2016. Intestinal microbiota of broiler chickens as affected by litter management regimens. Frontiers in Microbiology 7: 593. https://doi.org/10.3389/fmicb.2016.00593.

Wang, M., Lin, X., Jiao, H., Uyanga, V., Zhao, J., Wang, X., Li, H., Zhou, Y., Sun, S. and Lin, H., 2020. Mild heat stress changes the microbiota diversity in the respiratory tract and the cecum of layer-type pullets. Poultry Science 99(12): 7015–7026. https://doi.org/10.1016/j.psj.2020.09.024.

Wei, S., Gutek, A., Lilburn, M., and Yu, Z. 2013. Abundance of pathogens in the gut and litter of broiler chickens as affected by bacitracin and litter management. Veterinary Microbiology 166: 595–601. https://doi.org/10.1016/j.vetmic.2013.06.006.

Wen, C., Li, Q., Lan, F., Li, X., Li, G., Yan, Y., Wu, G., Yang, N. and Sun, C., 2021. Microbiota continuum along the chicken oviduct and its association with host genetics and egg formation. Poultry Science 100(7): 101104. https://doi.org/10.1016/j.psj.2021.101104.

Wickramasuriya, S.S., Park, I., Lee, K., Lee, Y., Kim, W.H., Nam, H. and Lillehoj, H.S., 2022. Role of physiology, immunity, microbiota, and infectious diseases in the gut health of poultry. Vaccines 10(2): 172. https://doi.org/10.3390/vaccines10020172.

Wiersema, M.L., Koester, L.R., Schmitz-Esser, S. and Koltes, D.A., 2021. Comparison of intestinal permeability, morphology, and ileal microbial communities of commercial hens housed in conventional cages and cage-free housing systems. Poultry Science 100(2): 1178–1191. https://doi.org/10.1016/j.psj.2020.10.052.

Wijnen, H.J., Molenaar, R., Kemp, B., van Roovert-Reijrink, I.A.M., van den Brand, H. and van der Pol, C.W., 2022. Effects of late incubation temperature and moment of first post-hatch feed access on neonatal broiler development, temperature preference, and stress response. Poultry Science 101(10): 102088. https://doi.org/10.1016/j.psj.2022.102088.

Wilkinson, N., Hughes, R.J., Bajagai, Y.S., Aspden, W.J., Van, T.T.H., Moore, R.J. and Stanley, D., 2020. Reduced environmental bacterial load during early development and gut colonisation has detrimental health consequences in Japanese quail. Heliyon 6(1): p.e03213. https://doi.org/10.1016/j.heliyon.2020.e03213.

Wilson, K.M., Bourassa, D.V., McLendon, B.L., Wilson, J.L. and Buhr, R.J., 2018. Impact of skip-a-day and every-day feeding programs for broiler breeder pullets on the recovery of *Salmonella* and *Campylobacter* following challenge. Poultry Science 97(8): 2775–2784. https://doi.org/10.3382/ps/pey150.

Xie, D., Li, J., Wang, Z.X., Cao, J., Li, T.T., Chen, J.L., and Chen, Y.X. 2011. Effects of monochromatic light on mucosal mechanical and immunological barriers in the small intestine of broilers. Poultry Science 90: 2697–2704. doi:10.3382/ps.2011-01416.

Xing, S., Wang, X., Diao, H., Zhang, M., Zhou, Y. and Feng, J., 2019. Changes in the cecal microbiota of laying hens during heat stress is mainly associated with reduced feed intake. Poultry Science 98(11): 5257–5264. https://doi.org/10.3382/ps/pez440.

Xiong, X., Yang, Y., Jiang, X., Yu, C., Peng, H., Chen, J., Xia, B., Du, H., Li, Q., Zhang, Z., Yang, L., Qiu, M., Hu, C., Song, X., Yan, H. and Yang, C., 2020. Effects of stocking density on performance, egg quality, reproductive hormones, and antioxidant capacity in egg-laying ducks, Journal of Applied Animal Research. 48: 454–459. https://doi.org/10.1080/09712119.2020.1824919.

Yalçın, S., Bağdatlıoğlu, N., Yenisey, C., Siegel, P.B., Özkan, S. and Akşit, M., 2012. Effect of manipulation of incubation temperature on fatty acid profiles and antioxidant enzyme activities in meat-type chicken embryos. Poultry Science 91(12): 3260–3270. https://doi.org/10.3382/ps.2012-02145.

Yalçin, S., Gursel, I., Bilgen, G., Horuluoglu, B.H., Gucluer, G. and Izzetoglu, G.T., 2017. Effect of egg storage duration and brooding temperatures on chick growth, intestine morphology and nutrient transporters. Animal 11(10): 1791–1797. https://doi.org/10.1017/S1751731117000404.

Yalçin, S., Gursel, I., Bilgen, G., Izzetoglu, G.T., Horuluoglu, B.H. and Gucluer, G., 2016. Egg storage duration and hatch window affect gene expression of nutrient transporters and intestine morphological parameters of early hatched broiler chicks. Animal 10(5): 805–811. https://doi.org/10.1017/S175173111500261X.

Yan, C., Xiao, J., Chen, D., Turner, S.P., Li, Z., Liu, H., Liu, W., Liu, J., Chen, S. and Zhao, X., 2021. Feed restriction induced changes in behavior, corticosterone, and microbial programming in slow-and fast-growing chicken breeds. Animals 11(1): 141. https://doi.org/10.3390/ani11010141.

Yilmaz-Dikmen, B.İ.L.G.E.H.A.N. and Sahan, U., 2009. The relationship among age, yolk fatty acids content, and incubation results of broiler breeders. Poultry Science 88(1): 185–190. https://doi.org/10.3382/ps.2008-00068.

Zhou, C., Gao, X., Cao, X., Tian, G., Huang, C., Guo, L., Zhao, Y., Hu, G., Liu, P. and Guo, X., 2022. Gut microbiota and serum metabolite potential interactions in growing layer hens exposed to high-ambient temperature. Frontiers in Nutrition 9: 877975. https://doi: 10.3389/fnut.2022.877975.

Zhou, Y., Liu, Q.X., Li, X.M., Ma, D.D., Xing, S., Feng, J.H. and Zhang, M.H. 2020. Effects of ammonia exposure on growth performance and cytokines in the serum, trachea, and ileum of broilers. Poultry Science 99: 2485–2493. https://doi.org/10.1016/j.psj.2019.12.063.

Zhu, L., Liao, R., Wu, N., Zhu, G. and Yang, C., 2019. Heat stress mediates changes in fecal microbiome and functional pathways of laying hens. Applied Microbiology Biotechnology. 103: 461–472. https://doi.org/10.1007/s00253-018-9465-8.

CHAPTER 3

Intersecting husbandry practices and piglet gut health and function

C.M. Nyachoti[1,*], *B. Jayaraman*[2] *and J. Lee*[1]

[1]Department of Animal Science, University of Manitoba, Winnipeg, MB, Canada R3T 2N2; [2]Evonik (SEA) Pte., Ltd., Singapore 609927; *Correspondence: Martin.Nyachoti@umanitoba.ca

Abstract

The immediate post-weaning period is one of the most stressful phases in a pig's life, and during this period, piglets are usually exposed to environmental, social and psychological stressors which have direct or indirect effects on gut health and overall growth performance. In this review, the impact of husbandry practices on gut health outcomes and performance of piglets is discussed. Husbandry practices in the swine barn generally include nutrition and management practices, maintenance of hygienic standards and disease prevention protocols, and animal welfare considerations. Poor husbandry practices could result in reduced feed intake, stress and disease conditions, and consequently affect gut health and performance in weaned piglets. Reduced feed intake is a major risk factor for impaired gut structure and function and therefore a key goal is to maximize feed intake in newly weaned piglets. In weaned piglets, crowding stress could reduce pig performance, favor the proliferation of pathogenic bacteria resulting in diarrhea, stimulate immune responses and interfere with beneficial microbial activities in the gut. Heat stress also negatively affects gut health and growth performance of young pigs. Sanitation conditions in the swine barn play an important role for optimal piglet performance because unclean conditions reduced growth performance, shifted nutrient requirements to support the immune system and negatively affected the gut morphology in weaned piglets. Appropriate biosecurity measures need to be designed to prevent disease entry and spread within a swine operation, which in turn helps keep all pigs and piglets in particular healthy. Collectively, husbandry practices relating to feeding and nutrition, animal welfare, biosecurity and disease prevention are important determinants of gut health and piglet performance. Thus, it is suggested that adopting high husbandry practices is a critical piece in strategies aimed at raising pigs without the use of in-feed antibiotics.

Keywords

gut health – husbandry practices – piglets

© C.M. NYACHOTI ET AL., 2024 | DOI:10.3920/9789004695467_004

1 Introduction

In commercial pork production piglets are weaned at a fairly young age than will otherwise occur naturally. At weaning piglets are subjected to a number of stressful situations, which includes being separated from the sow and mixed with piglets from other litters, moved to new locations and fed a new diet (Guevara *et al.*, 2022). These changes constitute significant stress in the piglet's life that often culminates to reduced growth performance and gut health challenges (Pluske *et al.*, 1997; Lallès *et al.*, 2007; Li *et al.*, 2017a). Newly weaned piglets do not readily consume solid feed leading to reduced nursery feed intake. Low feed intake, along with immature digestive (*i.e.*, low activity of digestive enzymes) and immune systems, are risk factors for enteric problems in weaned piglets (Hampson, 1986; Pluske *et al.*, 1997; Boudry *et al.*, 2004; Heo *et al.*, 2013). Switching from milk to solid feed deprives piglets the specific and non-specific immune factors present in the sow milk for immune protection, development and survival (Stokes, 2001) and causes imbalances in the gut microbiome, which often favours pathogen proliferation. Furthermore, the gastrointestinal tract immune system in the young pig takes 7 to 9 weeks to develop fully (King *et al.*, 2003; Bailey *et al.*, 2005). Thus, to effectively manage weaned pigs, all these aspects and the factors influencing them must be considered.

Basic goals in weaned pig management are centred around activities such as boosting feed intake and controlling incidences of disease so as to minimize post-weaning mortality and maximize post-weaning performance (Kirkden *et al.*, 2013). Recent legislative changes and changing consumer demands have dictated the removal of in-feed antimicrobials from nursery feeds in many jurisdictions, thus making it much more difficult to realize these goals. However, many nutritional technologies have been developed to help support gut health in young pigs thus maintaining acceptable mortality and performance targets. Another area that has received significant attention in nursery pig management is optimizing husbandry practices as these have been linked to gut health and function and therefore growth performance (Boyle *et al.*, 2022). Therefore, to effectively address growth performance and enteric problems associated with piglet weaning, effective husbandry management practices must be considered. Husbandry practices in the swine barn with direct implication on piglet gut health and function generally include nutrition and management practices for the sow and piglets, maintenance of high hygienic standards, disease prevention protocols, and paying attention to animal welfare needs such as providing comfortable thermal and physical settings for the piglets (Kirkden *et al.*, 2013; Canadian Pork Council (CPC), 2014; Boyle *et al.*, 2022). Specific factors that have been shown to impact gut health and growth performance in piglet and that can be influenced by husbandry practices includes feeding strategies (Dong and Pluske, 2007), exposure to stress (Khafipour *et al.*, 2014; Li *et al.*, 2017b; Ortega and Szabo, 2021; Vasquez *et al.*, 2022), sanitation (Jayaraman *et al.*, 2017; Boyle *et al.*, 2021; Lee *et al.*, 2022) and disease conditions (Opapeju *et al.*, 2009). These factors have been shown to influence gut structure and function, gut microbial population, incidences and severity of diarrhea, all of which have been used as indices of gut health outcomes (Li *et al.*, 2017b; Xia *et al.*, 2022; Vasquez *et al.*, 2022).

This chapter aims to provide an extensive review of the role of husbandry practices on weaned pig management in general and gut health and function in particular to optimize growth performance outcomes. As far as possible the mechanisms underlying the effects of husbandry practices on gut health and function in pigs are discussed. To achieve this goal, data from studies with other classes of pigs are included as appropriate. It is evident that to effectively manage gut health of piglets raised under commercial production systems with the prevailing legislative and consumer constraints, husbandry practices must be considered.

2 General farrow room management practices impact on gut health

Husbandry practices in the farrow room have significant implications on piglet survival and wellbeing. Sow feeding programs, although not discussed in this Chapter, influence not only the amount but also the quality of colostrum and therefore the quality of passive immunity passed on to piglets (*e.g.*, Reyes-Camacho *et al.*, 2020). Furthermore, as the main source of microorganisms that first colonize the piglet's gastrointestinal tract, manipulating the sow's diet has been shown to modulate the piglet's gut microbiome (Nowland *et al.*, 2021), which as discussed below has significant implications on gut health outcomes.

Another important husbandry practice in the farrow room with significant implications on gut health of piglets is the consumption of colostrum, which confers them with passive immunity from the sow. It has been amply demonstrated that colostrum intake is associated with colonization of the intestinal microbiota and development of the immune system in piglets (Le Dividich *et al.*, 2005; Nowland *et al.*, 2022).

Other husbandry practices in the farrow room that can affect piglet performance and subsequent gut health outcomes includes how the piglets are managed with regards to sufficient milk consumption and the age at which piglets are weaned. Strategies such as split-suckling, cross-fostering, and use of nurse sows are used to ensure that all piglets receive adequate nutrition, which is important for improved performance (Farmer and Edwards, 2022). Age of weaning is an important factor determining how well a piglet handles the stress associated with weaning, with older weaning leading to better gut health outcomes (Ming *et al.*, 2021; Wensley *et al.*, 2021).

3 Feed intake and its impact on gut health in weaned piglets

The effects of feed intake on gut morphology and growth performance in weaned piglets has been reviewed by Dong and Pluske (2007). For a detailed discussion on determinants of voluntary feed intake in swine, please refer to the review by Nyachoti *et al.* (2004). For the newly-weaned piglets various factors affecting feed intake have been identified and these include, creep feeding (Bruininx *et al.*, 2002a), weaning age (Davis *et al.*, 2006), mixing of different litters after weaning (McGlone and Cartis, 1985; Björk, 1989), dietary nutrient level and balance (D'Mello, 2003), diseases or the immune activation status (Edmonds *et al.*, 1997; Williams *et al.*, 1997; Jayaraman *et al.*, 2015), environmental factors (Bruininx *et al.*, 2002b), palatability of feed ingredients (Bell, 1984; van Heugten, 2001), physical form of diets (Patridge, 1989; Hancock and Behnke, 2001), feeding practices (Han *et al.*, 2006), and water supply and quality (Thacker, 2001; Dybkjaer *et al.*, 2006).

How each of these factors impact feed intake in weaned piglets and the implications of such impacts on gut health are discussed below.

3.1 *Creep feeding*

In commercial swine production, creep feeding during the suckling period prior to weaning has been a common husbandry practice because it has the potential to increase weaning weight of piglets and to lead to a smooth transition period as piglets are switched from the sow's milk to a dry nursery feed (Dong and Pluske, 2007; Cabrera *et al.*, 2013). Previous studies demonstrated that creep feed intake has a positive effect on post-weaning feed intake (Muns and Magowan, 2018; Bruininx *et al.*, 2002a; Kuller *et al.*, 2007; Cabrera *et al.*, 2013) and it is assumed that nursery piglets offered creep feed prompts them to get adapted to solid feed (Don and Pluske, 2007). In addition, Bruininx *et al.* (2002a) investigated whether consumption of creep feed prior to weaning stimulated the increase in post-weaning feed intake and performance. The authors showed that feed intake and daily gain during the first 8 days after weaning was higher for eaters than for non-eaters of creep feed in lactation or for those piglets not provided with creep feed during this period. Similar observations have been reported in a recent study by Sands *et al.* (2022). These studies indicate that piglets consuming more creep feed during lactation get adapted to solid diets, which promote gut development and therefore helps them more easily manage with a dietary change after weaning.

As recently reviewed by Huting *et al.* (2021), the benefits of creep feeding on gut health and function in piglets are not always evident. However, where observed, creep feeding has been shown to positively impact gut morphology, cell proliferation, and nutrient absorption. For example, Kuller *et al.* (2007) reported that creep feed intake during lactation improved net absorption in the small intestine after weaning and suggested that this could decrease the risk of post-weaning diarrhea. Also, Muns and Magowan (2018) observed higher villi height:crept dept in piglets due to creep feed consumption. By maintaining nutrient supply after weaning, creep feeding helps prevent villous atrophy, thereby reducing the chances of post-weaning diarrhea in piglets (Pluske *et al.*, 1996; Huting *et al.*, 2021). According to Metzler-Zebeli *et al.* (2021), creep feed intake may enhance fermentation in the hind gut with potential to positively impact gut homeostasis. Hence, adequate amount of creep feeding during lactation would enhance gut health of piglets after weaning (Kuller *et al.*, 2010; Muns and Magowan, 2018).

The management strategies for creep feeding include, offering fresh and palatable creep feed at less than two weeks of age of piglets (Appleby *et al.*, 1991), frequent feeding, feed should be accessible (Devillers *et al.*, 2004; Wattanakul *et al.*, 2005), increasing feeder space (Appleby *et al.*, 1992), and supplementation of feed additives in creep feed (Shim *et al.*, 2005; Cabrera *et al.*, 2013). For example, supplementation of the feed additive oligo-fructose in antibiotic-free creep feed favored the growth of beneficial bacteria (*Bifidobacterium* species) and reduced the harmful bacteria (Coliforms) in the colon (Shim *et al.*, 2005). Similarly, supplementation of creep feed with L-glutamine has been shown to improve feed conversion possibly due to improved intestinal health (Cabrera *et al.*, 2013). Recent work also shows that the physical characteristics of creep feed are important determinants of its acceptability by the piglet. For example, Chen *et al.* (2021) reported that a soft pellet creep feed promoted dry matter intake and mitigated the negative effects of weaning in piglets intestines. Collectively, creep feeding is beneficial in initiating and promoting gut health and function in weaned piglets, which in turn promotes growth performance in piglets through effective utilization of solid feed.

3.2 *Feeder space*

Pigs are highly social animals and their feeding behaviour is to eat in a group (Figueroa *et al.*, 2013), and therefore enough feeder space needs to be provided to allow group feeding. Hence, providing adequate feeder space is critical to weaned pig performance (CPC, 2014) because limited feeder space can increase competition at the feeder, which is likely to compromise feed intake thus leading to reduced growth rates (Georgsson and Svendsen, 2001; Averós *et al.*, 2012). In a growth study, Lindemann *et al.* (1987) demonstrated that weaned piglets offered 9 cm/pig feeder space allowance had significantly higher growth performance compared to those provided 3 cm/pig feeder space allowance. A study by He *et al.* (2016) indicated that limited feeder space is associated with increased risk of mortality or slow growth in weaned piglets. Overall, restricted feeder space can compromise growth performance and welfare of pigs, which may contribute to depressed performance in piglets. Because feed intake is an important factor in gut development (see Section 2.3 below), any factor such as space allowance, which influences feed intake is therefore potentially capable of impacting gut health outcomes in weaned piglets.

3.3 *Feed intake effects on gut health and function*

Reduced feed intake immediately after weaning could lead to adverse morphological and functional changes in the intestine (Dong and Pluske, 2007). The major adverse changes in intestinal morphology include shortening of the villi, hyperplasia of crypt cells, and increased epithelial cell mitosis (Nabuurs, 1993). Due to these changes in intestinal morphology, intestinal functions can be weakened, resulting in decreased brush-border enzyme activity and absorptive capacity (Hampson and Kidder, 1986; Nabuurs, 1993; Vente-Spreeuwenberg *et al.*, 2004). Furthermore, low feed intake and stress in piglets could lead to reduced gut mucosal integrity confirmed by an increase in paracellular transport, and a decrease in villous height (Spreeuwenberg *et al.*, 2001). Because of increased paracellular permeability, luminal antigens rather than bacteria may enter the lamina propria, resulting in inflammation (Spreeuwenberg *et al.*, 2001). Available evidence indicates that reduced feed intake is a major contributing factor to the abruptly-reduced villous height (Cera *et al.*, 1998). The brush border enzyme activities have been used as indicators of maturation and digestive capacity of the small intestine (Hampson and Kidder, 1986). Low feed intake associated with weaning negatively affects brush-border enzyme activity and absorption ability (Fig 3.1) (Dong and Pluske, 2007; Vente-Spreeuwenberg and Beyen, 2003). For optimal digestive and absorptive function of the small intestine, longer villus height is desirable (Pluske *et al.*, 1997). In this regard, Verdonk *et al.* (2007) demonstrated that restricted-fed piglets had lower villi height and deeper crypt depth compared to high-fed piglets. Similarly, Pluske *et al.* (1997) indicated that feed intake linearly and positively correlates with villi height in piglets. Moreover, low feed intake negatively affects intestinal barrier function in piglets (Verdonk, 2007; Groschwitz and Hogan, 2009). Furthermore, adequate feed intake prevents the loss of the barrier function of the tight junctions in piglets after weaning which indicates the significance of a sufficient luminal nutrient supply to maintain the barrier function. Hence, sufficient feed intake could be particularly essential for the proximal small intestine because this part

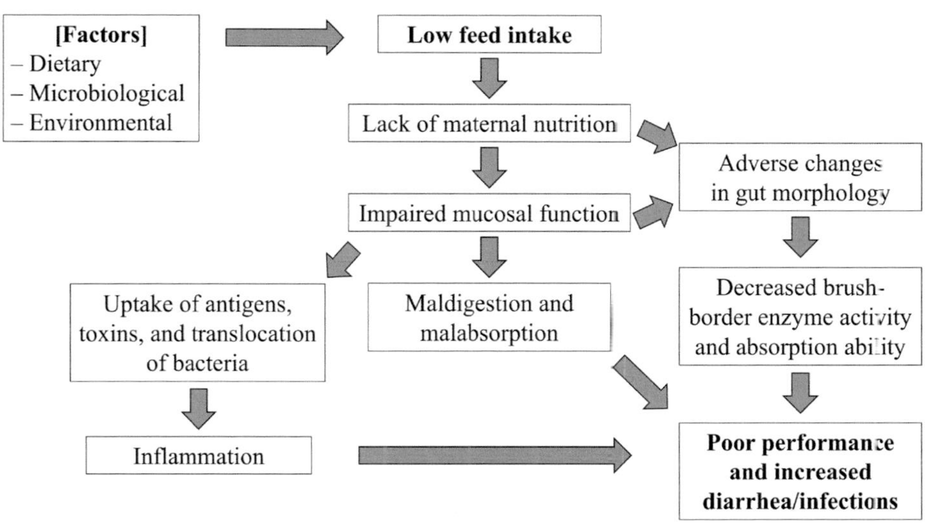

FIGURE 3.1 Effects of low feed intake on brush-border enzyme activity and absorption ability after weaning in piglets
ADAPTED FROM DONG AND PLUSKE (2007), VENTE-SPREEUWENBERG AND BEYEN (2003), AND SOMMER AND BÄCKHED (2013)

depends more on luminal nutrient supply than the distal small intestine (Stoll *et al.*, 2000). Verdonk *et al.* (2007) studied the effect of feed intake level on gut structure and permeability as determined using the Ussing chambers. The authors indicated that restricted feed intake negatively affected gut morphology compared to high levels of feed intake in weaned piglets but did not affect trans-epithelial transport.

Collectively, low feed intake is a major risk factor for impaired gut structure and function, which negatively affects gut morphology and barrier function. Hence, the key goal is to maximize feed intake in newly weaned piglets. Efforts to achieve this goal must start in the farrowing room with managing feed intake in the lactating sow. Sows nursing newborns require essentially full feeding during lactation, therefore, supplementing the sow's diet during late gestation and lactation periods may improve performance of the sow and her piglets (Kirkden *et al.*, 2013).

4 Stress factors affecting gut health and function in pigs

Pigs are subjected to various conditions with potential to cause stress. According to Mawdsley and Rampton (2005), a stressor is a factor or situation that can negatively impact the animal's homeostasis. In response to stress, the animal adopts behavioral changes and activates physiological mechanisms to maintain homeostasis. In modern commercial pork production settings, the main stressors for weaned pig can be classified into nutritional, thermal, socio-psychological, and immune stressors (Campbell *et al.*, 2013; Upadhaya and Kim, 2021; Guevara *et al.*, 2022). Many of these stressors have a drastic negative impact on feed intake as shown in Figure 3.2. Also, stress decreases immune cell numbers and increases immunosuppressive mechanisms which in turn suppresses the pig's immune competence ultimately leading to reduced growth rate (Khafipour *et al.*, 2014). With appropriate husbandry practices, however, many of these stressors can be mitigated so as to optimize piglet growth performance (Hyun *et al.* 1998). The following is a brief discussion on each of the stressor categories.

At weaning, the piglet is exposed to various stressors and pathogenic infections due to restricted space allowance, mingling with unfamiliar animals, unfavorable ambient temperature, contaminated air, and low biosecurity. The combined and additive effects of these stressors and pathogenic threats accelerate the detrimental effects on the growth performance of pigs. The health of the pig will, therefore, be improved if the number of stressors in the pigs' environment is reduced.

4.1 *Nutrition related stress and gut health in piglets*

During the immediate post-weaning period, piglets often reduce feed intake as they are presented with a dry mainly plant-based diet. This reduction in dry matter intake negatively impacts intestinal structure and function and favors the proliferation of pathogenic bacteria (Pluske *et al.*, 1997; Lallès *et al.*, 2004). It is for this reason

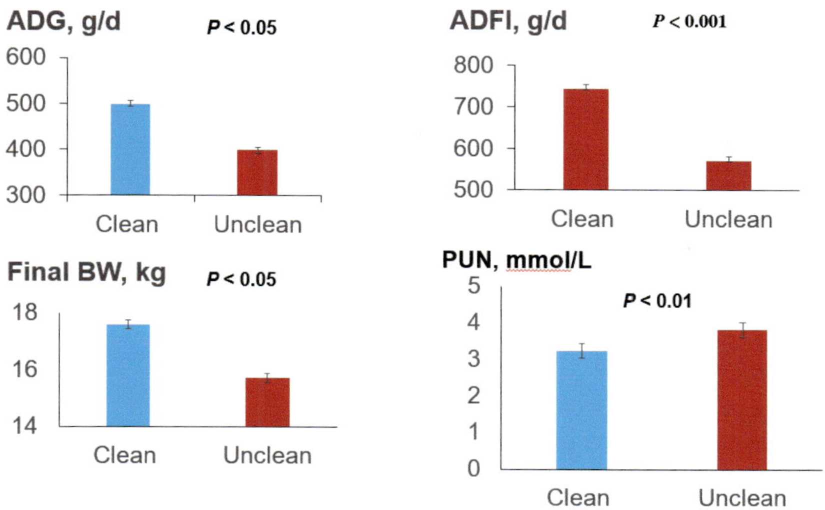

FIGURE 3.2 Effect of sanitation on growth performance and plasma urea nitrogen of weaned piglets[a] (d 0-21)
Adapted from Jayaraman *et al.* (2015)
[a] values are least square means ± Standard error of mean; n = 90 per treatment
Clean – clean sanitary conditions: piglets were housed in cleaned and disinfected rooms and fed antibiotic-free diets
Unclean – Unclean sanitary conditions: piglets were housed in a room and disinfection was not done, moreover, manure from swine herd was added to the pens on d 0 and 7 of the experiment

that a lot of emphasis is placed on encouraging newly weaned pigs to consume feed as soon as possible. Another nutritional stress that weaned piglets face is related to the consumption of diets containing allergens or antigens (*e.g.*, mycotoxins and peptidoglycans) that may damage the intestinal mucosa leading to the leaky gut syndrome. For example, diets containing high levels of regular soybean meal may cause mucosal damage due to the presence of trypsin inhibitors, which may impair digestion and absorption of capacity (D'Mello, 2000). Therefore, from a husbandry perspective, selection of hypoallergenic and antigen-free ingredients for nursery pig diet formulation is critical to addressing this type of nutritional stress.

4.2 *Socio-psychological stress effects on gut health and function*

In commercial farms, piglets are physically separated from the sow at weaning and mixed with piglets from other litters, and depending on the production system, they may be transported in large groups to another location. The separation from the sow and co-mingling with unfamiliar piglets causes socio-psychological stress (Gimsa *et al.*, 2018). Restricted space allowance is often one of the stressors that piglets experience at the time of weaning (Wolter *et al.*, 2003; Smith *et al.*, 2004). This results in crowding stress and limits access to feed which in turn negatively impacts growth performance (Hyun *et al.*, 1998; Khafipour *et al.*, 2014). Crowding

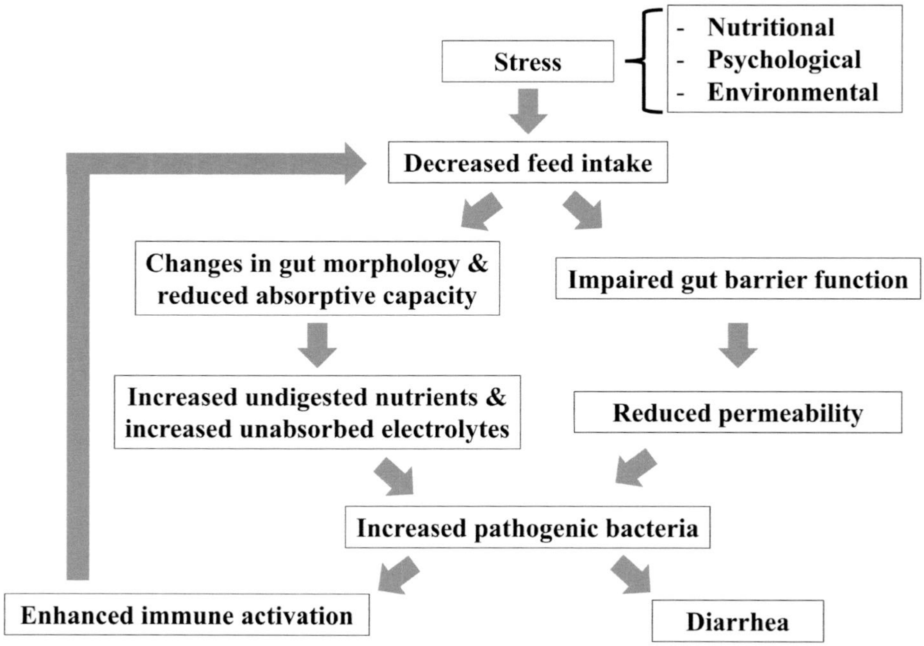

FIGURE 3.3 Schematic diagram illustrating the effects of stress in weaned piglets

stress also results in an increase in abnormal behaviour and levels of aggression, which increases non-growth energy expenditure (Turner *et al.*, 2003; Nyachoti *et al.*, 2004; Khafipour *et al.*, 2014). The current recommended space allowance for weaned pigs is at least 0.34 m^2 per pig in slatted pens (Holden *et al.*, 2002), however, in antibiotic-free feeding regimen, more space allowance might be required (Varley, 2004).

In relation to gut health, crowding stress reduced resistance to bacterial infection as well as increasing fecal shedding of *Escherichia coli* in weaned piglets (Jones *et al.*, 2001). Similarly, Khafipour *et al.* (2014) demonstrated that weaned piglets exposed to crowding stress and challenged with enterotoxigenic *Escherichia coli* (ETEC) had increased proliferation of ETEC in the ileal and colon digesta compared those provided with adequate space. Also, a chronic social stress model consisting of chronic mixing and crowding was shown to lead to significant alterations in intestinal barrier and nutrient transport functions in piglets (Li *et al.*, 2017b). Crowding stress leads to an increase in intestinal pH, which may create favorable conditions for ETEC colonization resulting in diarrhea. In general, low intestinal pH favors establishment of beneficial bacteria and inhibits harmful bacteria (Ohland and MacNaughton, 2010).

Stress has been shown to have direct effects on immunological parameters in pigs, including reduced lymphocyte and monocyte levels (Jones *et al.*, 2001). Crowding stress also resulted in an increase in stress indicators (*e.g.*, plasma cortisol) and indicators of systemic inflammation (higher fibrinogen and lower TTP:fibrinogen) in weaned piglets (Khafipour *et al.*, 2014). Moreover, crowding stress had a direct impact on immune response in pigs as indicated by an elevated serum cytokines and plasma cortisol (Khafipour *et al.*, 2014; Oh *et al.*, 2010).

In a review, Kil and Stein (2010) covered some of the measures that can be taken to reduce crowding stress, which include a) avoiding co-mingling of weaned pigs originating from different farms, as some piglets might bring pathogens which would result in disease outbreak, b) restricting mixing with non-littermates within the same farm in order to reduce stress of weaned pigs, c) avoiding mixing litters as this depresses growth rate and increases agonistic behaviour, d) grouping pigs by sex as this might minimize fighting and aggressive behaviour among piglets after mixing. Collectively, crowding stress can directly or indirectly deteriorate gut health and growth performance of weaned piglets.

4.3 *Thermal stress effects on gut health and function*

For optimal performance, pigs should be kept in a thermal environment that is within their comfort zone. Deviations from this requirement can not only negatively impact performance, but it can also harm gut health and function as has been demonstrated in several studies (*e.g.*, Ortego and Szabó, 2021; Wang *et al.*, 2022). Pigs experience cold stress or heat stress when environmental temperatures significantly fall below or rise above their thermal comfort zone, respectively.

Because of their large body surface area, piglets are very sensitive to cold temperatures, which predisposes them to various health challenges, including disease infections. For example, low temperature has been reported to be a risk factor for swine dysentery, whereas *Escherichia coli* infection is known to be severe in piglets exposed to cold temperatures (Wathes *et al.*, 1989; Jones *et al.*, 2001). Wathes *et al.* (1989) reported that weight gain and feed efficiency were decreased at 15 °C compared to at 20 or 30 °C and the incidence and severity of diarrhoea were greatest at 15 °C and least at 30 °C. Moreover, chronically exposing piglets to moderately cold temperatures was shown to increase incidences of diarrhea (Kelley *et al.*, 1982).

In mammals, the primary means for heat dissipation are through sweating, increased respiration and panting (Mortola and Frappell, 2000). However, because pigs lack functional sweat glands, they are highly susceptible to the negative impact of heat stress. Moreover, modern pig genetics are highly susceptible to heat stress because of the additional heat generation associated with fast growth (He *et al.*, 2021; Xia *et al.*, 2022).

Heat stress is one of the most detrimental factors affecting gut health and production performance in pigs, leading to significant economic losses to the swine industry (Mayorga *et al.*, 2019). The mechanism of how heat stress affects gut health and function in animals is presented in Fig 4. During heat stress conditions,

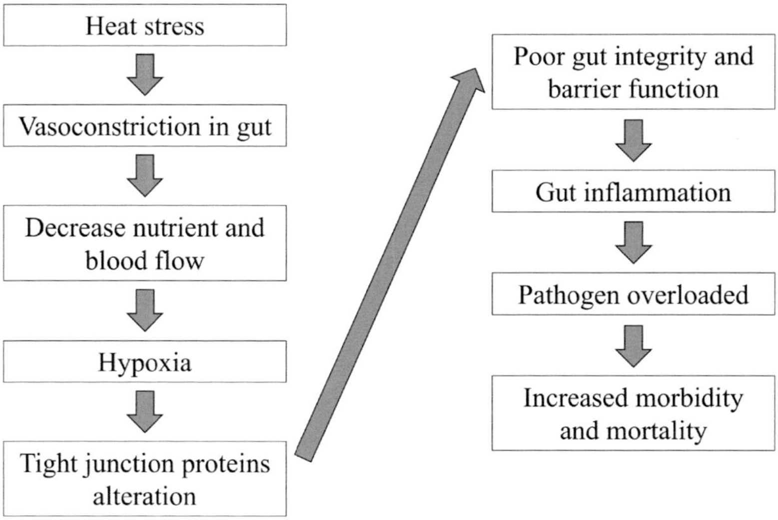

FIGURE 3.4 Prolonged heat stress negatively affects gut health in pigs

blood vessels in the gastrointestinal tract undergoes constriction to facilitate heat dissipation, which in turn reduces nutrient and blood flow leading to hypoxia. Consequently, it results in alterations of tight junction proteins which results in leaky gut (Xia *et al.*, 2022).

In pigs, maintenance of gut integrity and balanced microbiota composition is an indicator for the optimal gut health and functions, and any disturbance could result in increased permeability and infiltration of endotoxins (Liao, 2021). Pigs exposed to heat stress could be susceptible for disrupted gut integrity, dysbiosis and highly affected for pathogenic infections (*e.g.*, *Escherichia coli* K88 and Salmonella) (Xia *et al.*, 2022). In a recent review, Ortega and Szabó (2021) indicated that exposing pigs to chronic heat stress (temperature: 30 to 40 °C; relative humidity: 35 to 60%) deteriorated gut integrity by reducing villous height, crypt depth and VH:CD by 11, 3, and 9%, respectively.

There is evidence that heat stress impacts the gut microbiome, suggesting that its effects on gut health and function can partly be explained by this impact. Under thermoneutral conditionals, the order of abundance of bacteria phylum in the pig gut is as follows: Firmicutes > bacteriodes > Proteobacteria. Exposing pigs to heat stress even for a short duration could result in dysbiosis in the gut (Table 3.5; He *et al.*, 2021; Xiong *et al.*, 2020), thus potentially compromising gut health and function. For example, growing pigs subjected to acute heat stress (< 4 weeks) had increased Firmicutes-to-Bacteroidetes ratio in the gut microbiota (He *et al.*, 2021; Xiong *et al.*, 2020).

Another way in which heat stress impacts performance and gut health outcomes in swine is through increased oxidative stress (Li *et al.*, 2022). Adopting husbandry practices that minimize heat stress (*e.g.*, use of coolers and sprayers in the pig housing system) and selecting pig genotypes that are better equipped to handle oxidative stress, can help minimize incidences of oxidative stress in swine (Li *et al.*, 2022). Providing dietary antioxidants (vitamins, minerals, polyphenols) is considered an effective nutritional strategy to alleviate heat stress in pigs, thus consequently improving intestinal integrity and tight junction function (Ortaga and Szabó, 2021).

4.4 *Immunological stress effects on gut health and function*

In swine production systems, several factors cause immune-related stress to pigs, which has significant implications for gut health and function. The common husbandry-related stressors that cause an immune response in weaned pigs include dietary changes and exposure to pathogens. These stressors often challenge homeostasis of animals by inducing systemic or local inflammatory responses (Lee *et al.*, 2016). In this chapter, immunological stress as it relates to gut health and function in young pigs will be discussed.

During weaning, a major change in the degree and diversity of exposure to antigens derived from feed (see Section 3.1) and pathogenic organisms occur (Lallès *et al.*, 2007). In addition, the immuno-regulatory and immuno-protective portions of sow's milk are removed. At the time of weaning, the gut immune system stimulates immune responses, however, it is relatively immature (Stokes *et al.*, 2004). Mucus secretion by intestinal goblet cells is considered as a primary host defense mechanism and secretion may be either base-line or accelerated. Baseline secretions maintain the mucus coat by replacing mucus lost by erosion, digestion and digesta flow, whereas accelerated mucosal secretions are increased in response to pathogenic, physical, and toxic load (Blok, 2002).

As discussed in Section 3.5 below, the sanitary conditions piglets are subjected to can have a profound impact on performance and health, partly by eliciting inflammatory and immunological responses (Adewole *et al.*, 2016; Table 3.2). For example, subjecting piglets to unsanitary conditions negatively impacts growth performance and elicits an elevated immune response (Williams *et al.*, 1997). This phenomenon is known to provoke an inflammatory response (Le Floc'h *et al.*, 2006; Lee *et al.*, 2022) which leads to a repartitioning of energy and nutrient utilization towards supporting the immune system at the expense of growth performance (Humphrey and Klasing, 2004; Le Floc'h *et al.*, 2004, 2009). To this end, many studies have demonstrated that subjecting piglets to unsanitary conditions impacts dietary amino acid utilization, especially those amino acids that play a role in the immune defense system (Kahindi, 2014; Jayaraman *et al.*, 2015). This is because the immune system is highly dependent upon the adequate availability of amino acids for the synthesis of immunoglobulins and cytokines, and other molecules (*e.g.*, nitric oxide, superoxide, hydrogen peroxide, histamine, glutathione and anthranilic acid) with great biological importance (Kim *et al.*, 2007). For instance, the optimal standardized ileal digestible threonine:lysine for weaned pigs raised under unclean sanitary conditions was higher (66.5 vs. 65%) than for those raised under clean sanitary conditions (Jayaraman *et al.*, 2015). Ammonia exposure also negatively affects the immune system of pigs. Qin *et al.* (2022) and Li *et al.* (2023) found disrupted immune homeostasis and increased oxidative stress and lung injuries when pigs were exposed to NH_3 for 30 days.

4.5 *Effects of hygiene on performance and gut health and function*

An important husbandry practice in swine production is the maintenance of high hygienic conditions by cleaning and disinfecting the barn. Maintaining high hygiene standards in pig barns is essential to the production of healthy pigs as it can prevent and interrupt the spread of disease (CPC, 2014). High hygiene standards can also help minimize the emission and accumulation of noxious gases that can negatively impact pig health as discussed below.

Under poor hygienic conditions, pathogens have a greater opportunity to proliferate thus making infections that much more possible. For instance, pen hygiene was shown to be very important in preventing enteric infections in piglets (Madec *et al.*, 1998). Furthermore, in a meta-analysis, Pastorelli *et al.* (2012) demonstrated that weaned piglets raised under unclean sanitary conditions had higher incidences of soft feces and diarrhea compared to those raised under clean sanitary conditions. Also, Koo *et al.* (2022) reported increased intestinal abundance of ETEC and C. perfringens in weaned piglets subjected to poor sanitary conditions, which may have implications on gut health as these pathogens are known to cause enteric diseases.

In swine barns, ammonia (NH_3) and hydrogen sulphide (H_2S) are the most common noxious gaseous emissions (Kim *et al.*, 2008) and their concentrations depend on the manure management practices. Typical concentrations reported for NH_3 and H_2S in swine barns ranges from 1 to 40 ppm and 0.04 to 1.2 ppm, respectively (Crook *et al.*, 1991; Chang *et al.*, 2001; Kim *et al.*, 2008; Ni *et al.*, 2018). The maximum permissible levels of NH_3 and H_2S in the pig barn vary depending on the age of pigs (Table 3.1). As a strong irritant, H_2S at high levels can cause inflammation in the eyes and the respiratory system (Beauchamp *et al.*, 1984). Toxic levels of NH_3 and H_2S could drastically reduce feed intake, leading to depressed growth performance (Ni *et al.*, 2000) and possible damage to the gastrointestinal tract in the case of newly weaned pigs. For example, Jayaraman *et al.* (2017) and Kahindi *et al.* (2017) demonstrated that a room with unclean sanitary conditions had higher NH_3 (26.7 vs. 18.2 ppm and 35.0 vs. 28.5 ppm, respectively) and H_2S (0.10 vs. 0.01 ppm and 0.17 vs. 0.07 ppm, respectively) compared to a room with clean sanitary conditions. The authors speculated that the deterioration of air quality could have caused distress and a drop in feed intake, thus leading to the observed reduced growth performance of piglets reared under unclean sanitary conditions (Figure 3.3).

TABLE 3.1 Summary of the effects of ammonia (NH_3) and hydrogen sulfide (H_2S) exposure on pig health and productivity

Pigs	NH_3	Findings	Reference
3-week-old weaned pigs	20 ppm for 2 weeks	20 ppm of NH_3 inhalation has a potential as an exacerbating factor in respiratory disease	Hamilton *et al.* (1998)
3-week-old weaned pigs (8.4 kg)	0, 10, 19, or 37 ppm for 37 days	NH_3 was not associated with the incidence of respiratory or other disease Pigs had a delayed aversion to NH_3 at 10 ppm and above	Done *et al.* (2005)
4-week-old pigs (8.4 kg)	0, 50, 100, or 150 ppm for 4 weeks	Acute inflammatory reaction at 100 or 150 ppm, but not in 0 to 50 ppm Mild-to-heavy acute exudate in pigs exposed to 100 or 150 ppm NH_3	Drummond *et al.* (1980)
4-week-old pigs	5 or 20 ppm for 15 weeks	20 ppm of atmospheric NH_3 did not cause significant material risks to the health or metabolism of pigs	Cheng *et al.* (2014)
7-week-old pigs (8.5 kg)	0, 20, or 50 ppm for 30 days	NH_3 exposure induced oxidative stress and lung injuries Inhibited TCA cycle and activated glycolysis and lipolysis in lung tissues by NH_3 exposure	Qin *et al.* (2022)
7-week-old pigs (8.5 kg)	0, 20, or 50 ppm for 30 days	NH_3 exposure induces oxidative stress Increased activated T cells and induced pro-inflammatory response in 20 ppm NH_3 Oxidative damage to DNA and suppression of DNA repair process in 50 ppm NH_3	Li *et al.* (2023)
7-week-old pigs	5, 25, 50, or 100 ppm for 6 days	Maximum NH_3 concentration could be tolerated in pig housing is 9 ppm Nasal irritation and functional disturbances of the tracheal smooth-muscle concentrations at NH_3 concentrations as low as 25 ppm	Urbain *et al.* (1996)
10-week-old pigs (20 kg)	0, 10, 25, or 35 mg/m^3 for 25 days	Impaired the branched-chain amino acid catabolism above 25 mg/m^3 NH_3 exposure	Tang *et al.* (2021)

TABLE 3.1 Summary of the effects of ammonia (NH_3) (*cont.*)

Pigs	NH_3	Findings	Reference
Growing pigs (30 kg)	0, 5, 10, 15, 20, or 25 ppm for 4 weeks	Threshold concentration is 20 ppm. 25 ppm NH_3 caused respiratory tract injury and decreased respiratory immunity and growth performance, consistent with the increased presence of harmful bacteria identified by nasal microbiota analysis	Wang *et al.* (2019)
18-week-old pigs	0 or 90 mg/m^3 for 14 days	Imbalance of enteric microbial flora and the increased inflammatory damage in jejunum due to NH_3 toxicity	Li *et al.* (2021)
18-week-old pigs	0 or 90 mg/m^3 for 30 days	Pathological changes in liver, kidneys, lungs, hypothalamus, and jejunum of fattening pigs due to NH_3 exposure Changes in the solute vector family gene network which plays an important role in rheumatoid arthritis, bowel cancer, and drug metabolism	Xia *et al.* (2021)
18-week-old pigs	5 or 76 mg/m^3 for 30 days	NH_3 exposure induces oxidative stress and inflammatory response in spleen tissues	Chen *et al.* (2022)

Pigs	H_2S	Findings	Reference
3-week-old pigs	0, 0.5, 5, or 50 ppm for 12 days	Ambient H_2S increased the severity of respiratory distress and lung lesions	Santana *et al.* (2021)
Weaning pigs (11.7 kg)	0 or 20 ppm for 28 days	H_2S exposure induces inflammation, immune suppression, mucus accumulation and cell death in the porcine lung	Liu *et al.* (2020)
Weaning pigs (12.1 kg)	0 or 250 ppm for 5 h	Acute H_2S exposure induced liver dysfunction via affecting protein synthesis and lipid metabolism	Liu *et al.* (2022)

TABLE 3.2 Effect of different hygiene conditions (high vs. low sanitary) on growth performance and gut health in pigs

Pigs	High sanitary	Low sanitary	Findings	Reference
21-day-old weaned pigs (6.4 kg)	Disinfected and washed weekly	No disinfection or cleaning for 3 weeks, spread sow manure on pen floor, nylon manure bag mounted	– Pigs under low sanitation had reduced ADG and G:F in 2nd week, but contrary results that greater ADG and G:F were found in pigs under unclean conditions in week 3. – Pigs under low sanitation had higher VH and VH:CD ratio than pigs under clean conditions.	Lee *et al.* (2022)
21-day-old weaned pigs (7.1 kg)	Washed once a week	Spread sow manure on pen floor, not washed	Feeding nucleotide-rich yeast extract to piglets under unsanitary conditions improved ileal immune response and positively modulated proliferation of beneficial gut bacteria and suppression of harmful bacteria	Waititu *et al.* (2017)
21-day-old weaned pigs (7.2 kg)	Disinfected and cleaned once a week	No disinfection or cleaning for 3 weeks, spread sow manure on pen floor	Unclean conditions reduced ADFI Pigs under unclean conditions had a mild diarrhea and grew 50% less than those raised under clean conditions	Kahindi *et al.* (2014)

TABLE 3.2 Effect of different hygiene conditions (high vs. low sanitary) (*cont.*)

Pigs	High sanitary	Low sanitary	Findings	Reference
21-day-old weaned pigs (7.3 kg)	Disinfected and cleaned once a week	No disinfection or cleaning for 1 week, spread sow manure on pen floor	Depression of ADFI and ADG occurred under unclean condition. SAA requirement for gut tissue development could be higher in piglets when sanitary condition is poor	Kahindi *et al.* (2017)
21-day-old weaned pigs (7.5 kg)	Removed manure and washed daily	Daily sprayed of manure slurry mixtures	Exposure to a dirty environment reduced growth performance and increased manure gas emissions compared to clean environment	Li *et al.* (2017a)
27-day-old weaned pigs (7.4 kg)	Cleaned and disinfected before starting trial, cleaned once a week	Not cleaned and disinfected for 3 weeks, spread sow manure on pen floor	decreased growth performance when piglets subjected to poor sanitary conditions. Thr requirements increased for weaned pigs under unclean sanitary conditions.	Jayaraman *et al.* (2015)
27-day-old weaned pigs (7.4 kg)	Cleaned and disinfected before starting trial, cleaned once a week	Not cleaned and disinfected for 3 weeks, spread sow manure on pen floor	Unclean conditions increased aerial NH_3 and H_2S concentrations in the room Reduced growth performance of pigs in unclean conditions Pigs under unclean conditions had lower VH and shorter CD than clean conditions	Jayaraman *et al.* (2017)

TABLE 3.2 Effect of different hygiene conditions (high vs. low sanitary) (*cont.*)

Pigs	High sanitary	Low sanitary	Findings	Reference
28-day-old weaned pigs (8.3 kg)	Cleaned, disinfected, received a feed antibiotic everyday	Not disinfected or cleaned, no feed antibiotic	Decreased ADG, ADFI, and G:F, and higher faecal scores in pigs in poor sanitary conditions	Pastorelli *et al.* (2012)

Furthermore, poor sanitary conditions could affect gut structure which ultimately results in poor gut health. For example, Zhao *et al.* (2007) indicated that piglets raised under unclean conditions had shorter villous height and less crypt depth compared to those raised under clean conditions, which could be the influence of sanitary conditions. The underlying mechanism could be that, under unclean conditions, piglets had reduced feed intake which could have caused alterations in gut morphology as reported by Cera *et al.* (1988) and Pluske *et al.* (1996). Also, poor sanitation conditions could change the gut microbial composition of pigs, with significant impact on gut health and function given the important role of the gut microbiome in this regard. For example, Li *et al.* (2021) reported that exposure to excessive NH_3 levels increased *Pseudomonas* in the jejunum of fattening pigs, which inhibits the growth of *Lactobacillus* and *Streptococcus*. Results of these studies collectively demonstrate that maintaining high standards of cleanliness in the nursery is critical for optimal performance of piglets partly because of its direct or indirect effects on gut health and function.

5 Biosecurity measures to prevent disease

Biosecurity refers to the implementation of a set of measures that help reduce the risk of introducing and spreading pathogens. This is achieved through the installation of physical barriers, which may include doors, boot baths, hand washing stations, and changing boots between specific areas in the barn (Alarcón *et al.*, 2021). Development and implementation of procedural protocols is also instrumental in a robust biosecurity program. Biosecurity procedures are closely linked to the improvement of the health status and productivity of pigs (Levis and Baker, 2011) and it is one of the key components of reducing disease risk and maintaining the health status of a swine herd (Kyriazakis and Whittemore, 2006; Bottoms *et al.*, 2013). The components of a pig farm operation which need to be bio-secured are shown in Table 3.3.

TABLE 3.3 Components of a pig farm required for an adequate biosecurity

1.	Buildings and structures
2.	Consumable supplies and equipment
3.	Owners, workers, Veterinarians, Consultants, Visitors
4.	Transportation of pigs
5.	Management procedures
6.	Land area used for pig production

TABLE 3.4 Points to be considered for the implementation of biosecurity plans

1.	Prevalence of diseases that could affect the herd
2.	Mechanism of disease transmission
3.	Method of disease control
4.	Prevention of entry of pathogens into the farm
5.	Potential cost of a disease outbreak

Adapted from Levis and Baker (2011)

The prevention of pathogens from entering a farm or spreading from one section of the farm to the other is critical to enhancing animal health outcomes in any given herd. Pathogens are kept away from entering the farm using specific measures that constitutes what is known as bio-exclusion (also known as external biosecurity), whereas bio-containment (internal biosecurity) measures help stop the spread of pathogens within a facility (Levis and Baker, 2011). Although it may be difficult to link the impact of biosecurity measures to gut health outcomes, its role in creating an environment that makes it easier to achieve better gut health in young pigs cannot be denied. This is because several internal biosecurity measures help to minimize pathogen build-up and/or immunity system stimulation in pigs and therefore can have a direct impact on the prevention of gut health problems in piglets. For instance, pig management procedures that keep pigs of different age groups separated and pig movements going from young to old, in an all-in all-out manner, can reduce incidences of diarrhea in nursery pigs. Similarly, hygienic measures that involve effective cleaning and disinfection protocols are critical to the prevention of diarrheal diseases in piglets.

Another important component of internal biosecurity with significant implication on gut health outcomes in piglets, is the set of operational procedures that governs the activities and movements of personnel working in pig barns. These includes their movement among the different sections of the barn (*i.e.*, farrowing, nursery, and growing-finishing areas) and the steps taken to prevent disease transmission (*e.g.*, washing and disinfecting booths, changing clothes, washing and

TABLE 3.5 Heat stress alters gut microbiota in pigs (Adapted from He *et al.*, 2021)

Age of pig	Samples taken	Changes in phylum due to heat stress	Changes in genus due to heat stress	Reference
11 weeks	Rectum	Firmicutes (+) Proteobacteria (+) Spirochaetes (+) Actinobacteria (−) Bacteroidetes (−)		Le Sciellor *et al.*, 2019
29–31 kg BW	Rectum	Proteobacteria (+) Bacteroidetes (−)	Actinobacter (+)	Xiong *et al.*, 2020
1 st parity 85 d in gestation	Faeces	No difference	Coprococcus 3 (+) Coprostanoli-genes group (+) Halomonas (+) Ruminococcaceae UCG-005 (+) Ruminococcaceae UCG-013 (+) Bacteroidales RF16 group (−) Streptococcus (−) Treponema 2 (−)	He *et al.*, 2019

\+ indicates increase in abundance of microbe in pigs subjected to heat stress
− indicates decrease in abundance of microbe in pigs subjected to heat stress

disinfecting hands, handling sick animals, etc.) (Backhans *et al.*, 2015; Alarcón *et al.*, 2021).

5.1 *The cost of disease in pig industry*

Carr (2006) indicated that for unhealthy pigs, the price paid by a pig industry is the sum of the following: the costs of Veterinarian visits, veterinary medicine, vaccines, the cost of in-feed medicines, cost of lost efficiency pre-farm gate due to decrease in sow productivity, reduced growth rate and fee efficiency, loss of product quality, and loss of saleable product through meat processing condemnation. The points that need to be known to implement biosecurity plans are presented in Table 3.3.

5.2 *Pig flow – managing an all-in/all-out system*

The practice of an all-in/all-out management system breaks the disease cycle by preventing the sharing of air-space of pigs carrying clinical disease with pigs susceptible to the infection by that disease. The principle behind the all-in/all-out system is that between batches of pigs, the location is completely cleared, disinfected and rested to ensure the cycle of infection is broken and premises do not themselves serve as a reservoir for infective material (Kyriazakis and Whittemore, 1993).

The key reasons for using all-in/all-out systems are to minimize exposure levels to pathogenic organisms in the pig farm, to avoid the spread of diseases from adult pigs to younger pigs, and to increase feed efficiency and rate of gain by maintaining a high health status (Levis and Baker, 2011). Furthermore, it can be argued that by minimizing exposure to pathogenic organisms, and especially those that cause enteric diseases, this management strategy will be invaluable to production systems raising pigs without the use of in-feed antimicrobial agents.

6 Conclusions

Husbandry practices are the most important determinants of piglet performance. Poor husbandry practices could result in low feed intake, stress and disease conditions, and consequently affect gut health and performance of weaned piglets. Therefore, appropriate feeding and nutrition strategies, management practices and animal welfare activities, biosecurity and disease prevention measures are critical determinants of gut health and piglet performance. Thus, it is suggested that husbandry practices must be considered as a critical piece in the overall strategy of raising weaned piglets without in-feed antibiotics.

References

Adewole, D.I., Kim, I.H., and Nyachoti, C.M., 2016. Gut health of pigs: challenge models and response criteria with a critical analysis of the effectiveness of selected feed additives – a review. Asian-Australasian Journal of Animal Sciences 29: 909–924. https://doi.org/10.5713/ajas.15.0795.

Alarcón, L.V., Allepuz, A., and Mateu, E., 2021. Biosecurity in pig farms: a review. Porcine Health Management 7: 5. https://doi.org/10.1186/s40813-020-00181-z.

Appleby, M.C., Pajor, E.A., and Fraser, D., 1991. Effects of management options on creep feeding by piglets. Animal Production 53: 361–366. https://doi.org/10.1017/S0003356100020377.

Appleby, M.C., Pajor, E.A., and Fraser, D., 1992. Individual variation in feeding and growth of piglets: effects of increased access to creep food. Animal Production 55: 147–152. https://doi.org/10.1017/S0003356100037375.

Averós, X., Brossard, L., Dourmad, J., De Greef, K., Edwards, S., and Meunier-Salaün, M., 2012. Meta-analysis on the effects of the physical environment, animal traits, feeder and feed characteristics on the feeding behaviour and performance of growing-finishing pigs. Animal 6: 1275–1289. https://doi.org/10.1017/S1751731112000328.

Backhans, A., Sjölund, M., Lindberg, A., and Emanuelson, U., 2015. Biosecurity level and health management practices in 60 Swedish farrow-to-finish herds. Acta Veterinaria Scandinavica 57: 14. https://doi.org/10.1186/s13028-015-0103-5.

Bailey, M., Haverson, K., Inman, C., Harris, C., Jones, P., Corfield, G., Miller, B., and Stokes, C., 2005. The development of the mucosal immune system pre- and post-weaning: Balancing regulatory and effector function. Proceedings of the Nutrition Society 64: 451–457. https://doi.org/10.1079/PNS2005452.

Beauchamp, R.O., Bus, J.S., Popp, J.A., Boreiko, C.J., Andjelkovich, D.A., and Leber, P., 1984. A critical review of the literature on hydrogen sulfide toxicity. CRC Critical Reviews in Toxicology 13: 25–97. https://doi.org/10.3109/10408448409029321.

Bell, J.M., 1984. Nutrients and toxicants in rapeseed meal: a review. Journal of Animal Science 58: 996–1010. https://doi.org/10.2527/jas1984.584996.

Björk, A.K.K., 1989. Is social stress in pigs a detrimental factor to health and growth that can be avoided by amperozide treatment? Applied Animal Behaviour Science 23: 39–47. https://doi.org/10.1016/0168-1591(89)90005-1.

Bottoms, K., Poljak, Z., Dewey, C., Deardon, R., Holtkamp, D., and Friendship, R., 2013. Evaluation of external biosecurity practices on southern Ontario sow farms. Preventive veterinary medicine 109: 58–68. https://doi.org/10.1016/j.prevetmed.2012.08.013.

Boyle, L.A., Edwards, S.A., Bolhuis, J.E., Pol, F., Šemrov, M.Z., Schütze, S., Nordgreen, J., Bozakova, N., Sossidou, E.N., and Valros, A. 2022. The evidence for a causal link between disease and damaging behavior in pigs. Front. Vet. Sci. 8: 771682, doi: 10.3389/fvets.2021.771682.

Boudry, G., Péron, V., Le Huërou-Luron, I., Lallès, J.P., and Sève, B., 2004. Weaning induces both transient and long-lasting modifications of absorptive, secretory, and barrier properties of piglet intestine. The Journal of Nutrition 134: 2256–2262. https://doi.org/10.1093/jn/134.9.2256.

Bruininx, E.M.A.M., Binnendijk, G.P., van der Peet-Schwering, C.M.C., Schrama, J.W., den Hartog, L.A., Everts, H., and Beynen, A.C., 2002a. Effect of creep feed consumption on individual feed intake characteristics and performance of group-housed weanling pigs. Journal of Animal Science 80: 1413–1418. https://doi.org/10.2527/2002.8061413x.

Bruininx, E.M.A.M., Heetkamp, M.J.W., van den Bogaart, D., van der Peet-Schwering, C.M.C., Beynen, A.C., Everts, H., den Hartog, L.A., and Schrama, J.W., 2002b. A prolonged photoperiod improves feed intake and energy metabolism of weanling pigs. Journal of Animal Science 80: 1736–1745. https://doi.org/10.2527/2002.8071736x.

Cabrera, R.A., Usry, J.L., Arrellano, C., Nogueira, E.T., Kutschenko, M., Moeser, A.J., and Odle, J., 2013. Effects of creep feeding and supplemental glutamine or glutamine plus glutamate (Aminogut) on pre-and post-weaning growth performance and intestinal health of piglets. Journal of Animal Science and Biotechnology 4: 29. https://doi.org/10.1186/2049-1891-4-29.

Campbell, J.M., Crenshaw, J.D., and Polo, J., 2013. The biological stress of early weaned piglets. Journal of Animal Science and Biotechnology 4: 19.

Canadian Pork Council (CPC), 2014. Code of practice: for the care and handling of pigs. Canada Pork Council and National Farm Animal Care Council. Available at: https://www.nfacc.ca/pdfs/codes/pig_code_of_practice.pdf.

Carr, J., 2006. Chapter 7 The maintenance of health. In: Kyriazakis, I., and Whittemore, C.T. (eds). Whittemore's science and practice of pig production. Wiley-Blackwell: Hoboken, NJ, USA, pp. 263–315.

Cera, K.R., Mahan, D.C., Cross, R.F., Reinhart, G.A., and Whitmoyer, R.E., 1988. Effect of age, weaning and postweaning diet on small intestinal growth and jejunal morphology in young swine. Journal of Animal Science 66: 574–584. https://doi.org/10.2527/jas1988.662574x.

Chang, C.W., Chung, H., Huang, C.F., and Su, H.J.J., 2001. Exposure assessment to airborne endotoxin, dust, ammonia, hydrogen sulfide and carbon dioxide in open style swine houses. The Annals of Occupational Hygiene 45: 457–465. https://doi.org/10.1093/annhyg/45.6.457.

Chen, H., Wang, C., Wang, Y., Chen, Y., Wan, M., Zhu, J., and Zhu, A., 2021. Effects of soft pellet creep feed on pre-weaning and post-weaning performance and intestinal development in piglets. Animal Bioscience 34: 714–723. https://doi.org/10.5713/ajas.20.0034.

Chen, Y., Zhang, R., Ding, S., Nian, H., Zeng, X., Liu, H., Xing, H., Li, J., Bao, J., and Li, X., 2022. Transcriptome revealed exposure to the environmental ammonia induced oxidative stress and inflammatory injury in spleen of fattening pigs. Animals 12: 1204. https://doi.org/10.3390/ani12091204.

Cheng, Z., O'Connor, E.A., Jia, Q., Demmers, T.G.M., Wathes, C.M., and Wathes, D.C., 2014. Chronic ammonia exposure does not influence hepatic gene expression in growing pigs. Animal 8: 331–337. https://doi.org/10.1017/S1751731113002127.

Crook, B., Robertson, J.F., Glass, S.A.T., Botheroyd, E.M., Lacey, J., and Topping, M.D., 1991. Airborne dust, ammonia, microorganisms, and antigens in pig confinement houses and the respiratory health of exposed farm workers. American Industrial Hygiene Association Journal 52: 271–279. https://doi.org/10.1080/15298669191364721.

Davis, M.E., Sears, S.C., Apple, J.K., Maxwell, C.V., and Johnson, Z.B., 2006. Effect of weaning age and commingling after the nursery phase of pigs in a wean-to-finish facility on growth, humoral and behavioral indicators of well-being. Journal of Animal Science 84: 743–756. https://doi.org/10.2527/2006.843743x.

D'Mello, J.P.F., 2000. Anti-nutritional factors and mycotoxins. In: D'Mello, J.P.F., (eds). Farm animal metabolism and nutrition, CAB International, Wallingford, United Kingdom, pp. 383–403.

D'Mello, J.P.F., 2003. Adverse effects of amino acids. In: D'Mello, J.P.F., (eds). Amino acids in animal nutrition. CAB International, Wallingford, United Kingdom, pp. 125–142.

Dong, G.Z., and Pluske, J.R., 2007. The low feed intake in newly-weaned pigs: problems and possible solutions. Asian-Australasian Journal of Animal Sciences 20: 440–452. https://doi.org/10.5713/ajas.2007.440.

Done, S.H., Chennells, D.J., Gresham, A.C.J., Williamson, S., Hunt, B., Taylor, L.L., Bland, V., Jones, P., Armstrong, D., White, R.P., Demmers, T.G.M., Teer, N., and Wathes, C.M., 2005. Clinical and pathological responses of weaned pigs to atmospheric ammonia and dust. Veterinary Record 157: 71–80. https://doi.org/10.1136/vr.157.3.71.

Dosio, A., Mentaschi, L., Fischer, E.M., and Wyser, K., 2018. Extreme heat waves under 1.5 °C and 2 °C global warming. Environmental Research Letters 13: 054006. https://doi.org/10.1088/1748-9326/aab827.

Drummond, J.G., Curtis, S.E., Simon, J., and Norton, H.W., 1980. Effects of aerial ammonia on growth and health of young pigs. Journal of Animal Science 50: 1085–1091. https://doi.org/10.2527/jas1980.5061085x.

Dybkjær, L., Jacobsen, A.P., Tøgersen, F.A., and Poulsen, H.D., 2006. Eating and drinking activity of newly weaned piglets: effects of individual characteristics, social mixing, and addition of extra zinc to the feed. Journal of Animal Science 84: 702–711. https://doi.org/10.2527/2006.843702x.

Edmonds, M.S., Arentson, B.E., Nipper, W.A., and Froe, D.L., 1997. Segregated early weaning: effects of raising pigs on-site and off-site with and without vaccination on performance and economics. Journal of Animal Science 75(Suppl. 1): 195.

Farmer, C., and Edwards, S.A. Review: Improving the performance of neonatal pigs. Animal. 16 (2): 100350.

Figueroa, J., Solà-Oriol, D., Manteca, X., and Pérez, J.F., 2013. Social learning of feeding behaviour in pigs: Effects of neophobia and familiarity with the demonstrator conspecific. Applied Animal Behaviour Science 148: 120–127. https://doi.org/10.1016/j.applanim.2013.06.002.

Gan, F., Ren, F., Chen, X., Lv, C., Pan, C., Ye, G., Shi, J., Shi, X., Zhou, H., Shituleni, S.A., and Huang, K., 2013. Effects of selenium-enriched probiotics on heat shock protein mRNA levels in piglet under heat stress conditions. Journal of Agricultural and Food Chemistry 61: 2385–2391. https://doi.org/10.1021/jf300249j.

Georgsson, L., and Svendsen, J., 2001. One or two feeders for groups of 16 growing-finishing pigs: effects on health and production. Acta Agriculturae Scandinavica, Section A – Animal Science 51: 257–264. https://doi.org/10.1080/09064.

Gimsa, U., Tuchscherer, M. and Kanitz, E., 2018. Psychosocial Stress and Immunity – What Can We Learn From Pig Studies? Front. Behav. Neurosci. 12: 64. doi: 10.3389/fnbeh.2018.00064.

Groschwitz, K.R., and Hogan, S.P., 2009. Intestinal barrier function: molecular regulation and disease pathogenesis. Journal of Allergy and Clinical Immunology 124: 3–20. https://doi.org/10.1016/j.jaci.2009.05.038.

Guevara, R.D., Pastor, J.J., Manteca, X., Tedo, G., and Llonch, P., 2022. Systematic review of animal-based indicators to measure thermal, social, and immune-related stress in pigs. PLOS ONE 17: e0266524. https://doi.org/10.1371/journal.pone.0266524.

Hamilton, T.D.C., Roe, J.M., Hayes, C.M., and Webster, A.J.F., 1998. Effects of ammonia inhalation and acetic acid pretreatment on colonization kinetics of toxigenic *Pasteurella multocida* within upper respiratory tracts of swine. Journal of Clinical Microbiology 36: 1260–1265. https://doi.org/10.1128/JCM.36.5.1260-1265.1998.

Hampson, D.J., 1986. Alterations in piglet small intestinal structure at weaning. Research in Veterinary Science 40: 32–40. https://doi.org/10.1016/S0034-5288(18)30482-X.

Hampson, D.J., and Kidder, D.E., 1986. Influence of creep feeding and weaning on brush border enzyme activities in the piglet small intestine. Research in Veterinary Science 40: 24–31. https://doi.org/10.1016/S0034-5288(18)30481-8.

Han, Y.K., Thacker, P.A., and Yang, J.S., 2006. Effects of the duration of liquid feeding on performance and nutrient digestibility in weaned pigs. Asian-Australasian Journal of Animal Sciences 19: 396–401. https://doi.org/10.5713/ajas.2006.396.

Hancock, J.D., and Behnke, K.C., 2001. Use of ingredient and diet processing technologies (grinding, mixing, pelleting, and extruding) to produce quality feeds for pigs. In: Lewis, A.J., and Southern, L.L. (eds) Swine nutrition. CRC Press LLC, Boca Raton, FL, USA, pp. 469–498.

He, J., Guo, H., Zheng, W., Xue, Y., Zhao, R., and Yao, W., 2019. Heat stress affects fecal microbial and metabolic alterations of primiparous sows during late gestation. Journal of Animal Science and Biotechnology 10: 84. https://doi.org/10.1186/s40104-019-0391-0.

He, Y., Deen, J., Shurson, G.C., Wang, L., Chen, C., Keisler, D.H., and Li, Y.Z., 2016. Identifying factors contributing to slow growth in pigs. Journal of Animal Science 94: 2103–2116. https://doi.org/10.2527/jas.2015-0005.

He, Y., Maltecca, C., and Tiezzi, F., 2021. Potential use of gut microbiota composition as a biomarker of heat stress in monogastric species: a review. Animals 11: 1833. https://doi.org/10.3390/ani11061833.

Heo, J.M., Opapeju, F.O., Pluske, J.R., Kim, J.C., Hampson, D.J., and Nyachoti, C.M., 2013. Gastrointestinal health and function in weaned pigs: a review of feeding strategies to control post-weaning diarrhoea without using in-feed antimicrobial compounds. Journal of Animal Physiology and Animal Nutrition 97: 207–237. https://doi.org/10.1111/j.1439-0396.2012.01284.x.

Holden, P., Carr, J., Honeyman, M.S., Kliebestein, J.B., McKean, J.D., Harmon, J.D., Mabry, J.W., and Hoyer, S., 2002. Minimizing the use of antibiotics in pork production. Iowa Pork Industry Center. 8: 1–16.

Humphrey, B.D., and Klasing, K.C., 2004. Modification of nutrient metabolism and homeostasis by the immune system. World's Poultry Science Journal 60: 90–100. https://doi.org/10.1079/WPS20037.

Huting, A.M.S., Middelkoop, A., Guan, X., and Molist, F., 2021. Using nutritional strategies to shape the gastro-intestinal tracts of suckling and weaned piglets. Animals 11: 402. https://doi.org/10.3390/ani11020402.

Hyun, Y., Ellis, M., Riskowski, G., and Johnson, R.W., 1998. Growth performance of pigs subjected to multiple concurrent environmental stressors. Journal of Animal Science 76: 721–727. https://doi.org/10.2527/1998.763721x.

Jayaraman, B., Htoo, J., and Nyachoti, C.M., 2015. Effects of dietary threonine:lysine ratios and sanitary conditions on performance, plasma urea nitrogen, plasma-free threonine and lysine of weaned pigs. Animal Nutrition 1: 283–288. https://doi.org/10.1016/j.aninu.2015.09.003.

Jayaraman, B., Htoo, J.K., and Nyachoti, C.M., 2017. Effects of different dietary tryptophan: lysine ratios and sanitary conditions on growth performance, plasma urea nitrogen, serum haptoglobin and ileal histomorphology of weaned pigs. Animal Science Journal 88: 763–771. https://doi.org/10.1111/asj.12695.

Jones, P.H., Roe, J.M., and Miller, B.G., 2001. Effects of stressors on immune parameters and on the faecal shedding of enterotoxigenic *Escherichia coli* in piglets following experimental inoculation. Research in Veterinary Science 70: 9–17. https://doi.org/10.1053/rvsc.2000.0436.

Kahindi, R.K., Htoo, J.K., and Nyachoti, C.M., 2014. Effect of dietary lysine content and sanitation conditions on performance of weaned pigs fed antibiotic-free diets. Canadian Journal of Animal Science 94: 115–118. https://doi.org/10.4141/cjas2013-016.

Kahindi, R., Regassa, A., Htoo, J., and Nyachoti, M., 2017. Optimal sulfur amino acid to lysine ratio for post weaning piglets reared under clean or unclean sanitary conditions. Animal Nutrition 3: 380–385.

Kelley, K.W., Blecha, F., and Regnier, J.A. 1982. Cold exposure and absorption of colostral immunoglobulins by neonatal pigs. J. Anim. Sci. 55(2): 363–8. doi: 10.2527/jas1982.552363x.

Khafipour, E., Munyaka, P.M., Nyachoti, C.M., Krause, D.O., and Rodriguez-Lecompte, J.C., 2014. Effect of crowding stress and *Escherichia coli* K88+ challenge in nursery pigs supplemented with anti-*Escherichia coli* K88+ probiotics. Journal of Animal Science 92: 2017–2029. https://doi.org/10.2527/jas.2013-7043.

Kil, D.Y., and Stein, H.H., 2010. Board invited review: management and feeding strategies to ameliorate the impact of removing antibiotic growth promoters from diets fed to weanling pigs. Canadian Journal of Animal Science 90: 447–460. https://doi.org/10.4141/cjas10028.

Kim, K.Y., Ko, H.J., Choi, H.L., Kim, H.T., Kim, Y.S., Roh, Y.M., Lee, C.M., and Kim, C.N., 2007. Farmer and pig exposure to aerial contaminants in a pig confinement building. Transactions of the ASABE 50: 993–998. https://doi.org/10.13031/2013.23139.

Kim, K.Y., Ko, H.J., Kim, H.T., Kim, Y.S., Roh, Y.M., Lee, C.M., and Kim, C.N., 2008. Quantification of ammonia and hydrogen sulfide emitted from pig buildings in Korea. Journal of Environmental Management 88: 195–202. https://doi.org/10.1016/j.jenvman.2007.02.003.

King, M.R., Kelly, D., Morel, P.C.H., and Pluske, J.R., 2003. Aspects of intestinal immunity in the pig around weaning. In: Pluske, J.R., Le Dividich, J., and Verstegen, M.W.A. (eds). Weaning the pig: concepts and consequences. Wageningen Academic Publishers, Wageningen, the Netherlands, pp. 219–258. https://doi.org/10.3920/978-90-8686-513-0.

Kirkden, R.D., Broom, D.M., and Andersen, I.L., 2013. Invited review: piglet mortality: management solutions. Journal of Animal Science 91: 3361–3389. https://doi.org/10.2527/jas.2012-5637.

Koo, B., Choi, J. Holanda, D.M. Yang, C. and Nyachoti, C.M., 2022. Comparative effects of dietary methionine and cysteine supplementation on redox status and intestinal integrity in immunologically challenged-weaned pigs. *Amino Acids* https://doi.org/10.1007/s00726-022-03213-w.

Kuller, W.I., Soede, N.M., van Beers-Schreurs, H.M.G., Langendijk, P., Taverne, M.A.M., Kemp, B., Verheijden, J.H.M., 2007. Effects of intermittent suckling and creep feed intake on pig performance from birth to slaughter. Journal of Animal Science 85: 1295–1301. https://doi.org/10.2527/jas.2006-177.

Kuller, W.I., Tobias, T.J., and van Nes, A., 2010. Creep feeding intake in unweaned piglets is increased by exploration stimulation feeder. Livestock Science 129: 228–231. https://doi.org/10.1016/j.livsci.2010.01.003.

Kyriazakis I., Whittemore C.T. Chapter 7 the maintenance of health. Whittenmore's Science and Practice of Pig Production. 3rd ed. 2006. p. 263e315.

Labussière, E., Achard, C., Dubois, S., Combes, S., Castex, M., and Renaudeau, D., 2022. *Saccharomyces cerevisiae boulardii* CNCM I-1079 supplementation in finishing male pigs helps to cope with heat stress through feeding behaviour and gut microbiota modulation. British Journal of Nutrition 127: 353–368. https://doi.org/10.1017/S0007114521001756.

Lallès, J.P., Boudry, G., Favier, C., Le Floc'h, N., Lurona, I., Montagne, L., Oswald, I.P., Pie, S., Piel, C., and Sève, B., 2004. Gut function and dysfunction in young pigs: physiology. Animal Research 53: 301–316. https://doi.org/10.1051/animres:2004018.

Lallès J.P., Bosi, P., Smidt, H., and Stokes, C.R., 2007. Weaning – A challenge to gut physiologists. Livestock Science 108: 82–93. https://doi.org/10.1016/j.livsci.2007.01.091.

Le Dividich, J., Rooke, J.A., and Herpin, P. Nutritional and immunological importance of colostrum for the new-born pig. Journal of Agricultural Science, 143 (6), pp. 469–485. doi: 10.1017/S0021859605005642.

Lee, I.K., Kye, Y.C., Kim, G., Kim, H.W., Gu, M.J., Umboh, J., Maaruf, K., Kim, S.W., and Yun, C.H., 2016. Stress, nutrition, and intestinal immune responses in pigs – a review. Asian-Australasian Journal of Animal Sciences 29: 1075–1082. https://doi.org/10.5713/ajas.16.0118.

Lee, J., González-Vega, J.C., Htoo, J.K., Yang, C., and Nyachoti, C.M., 2022. Effects of dietary protein content and crystalline amino acid supplementation patterns on growth performance, intestinal histomorphology, and immune response in weaned pigs raised under different sanitary conditions. Journal of Animal Science 100:skac285. https://doi.org/10.1093/jas/skac285.

Le Floc'h, N., Jondreville, C., Matte, J.J., and Seve, B., 2006. Importance of sanitary environment for growth performance and plasma nutrient homeostasis during the post-weaning period in piglets. Archives of Animal Nutrition 60: 23–34. https://doi.org/10.1080/17450390500467810.

Le Floc'h, N., Lebellago, L., Matte, J.J., Melchior, D., and Sève, B., 2009. The effect of sanitary status degradation and dietary tryptophan content on growth rate and tryptophan metabolism in weaning pigs. Journal of Animal Science 87: 1686–1694. https://doi.org/10.2527/jas.2008-1348.

Le Floc'h, N., Melchior, D., and Obled, C., 2004. Modifications of protein and amino acid metabolism during inflammation and immune system activation. Livestock Production Science. 87: 37–45. https://doi.org/10.1016/j.livprodsci.2003.09.005.

Levis, D.G., and Baker, R.B., 2011. Biosecurity of pigs and farm security. University of Nebraska-Lincoln Extension. Available at: https://extensionpublications.unl.edu/assets/pdf/ec289.pdf.

Li, D., Shen, L., Zhang, D., Wang, X., Wang, Q., Qin, W., Gao, Y., and Li, X., 2023. Ammonia-induced oxidative stress triggered proinflammatory response and apoptosis in pig lungs. Journal of Environmental Sciences 126: 683–696. https://doi.org/10.1016/j.jes.2022.05.005.

Li, M.M., Seelenbinder, K.M., Ponder, M.A., Deng, L., Rhoads, R.P., Pelzer, K.D., Radcliffe, J.S., Maxwell, C.V., Ogejo, J.A., White, R.R., and Hanigan, M.D., 2017a. Effects of dirty housing and a *Salmonella* Typhimurium DT104 challenge on pig growth performance, diet utilization efficiency, and gas emissions from stored manure. Journal of Animal Science 95: 1264–1276. https://doi.org/10.2527/jas.2016.0863.

Li, Q., Yang, S., Chen, F., Guan, W., and Zhang, S., 2022. Nutritional strategies to alleviate oxidative stress in sows. Animal Nutrition 9: 60–73. https://doi.org/10.1016/j.aninu.2021.10.006.

Li, Y., Zhang, R., Li, X., Li, J., Ji, W., Zeng, X., and Bao, J., 2021. Exposure to the environmental pollutant ammonia causes changes in gut microbiota and inflammatory markers in fattening pigs. Ecotoxicology and Environmental Safety 208: 111564. https://doi.org/10.1016/j.ecoenv.2020.111564.

Li, Y.H., Song, Z., Kerr, K.A., and Moeser, A.J., 2017b. Chronic social stress in pigs impairs intestinal barrier and nutrient transporter function, and alters neuro-immune mediator and receptor expression. PLOS ONE 12: e0171617. https://doi.org/10.1371/journal.pone.0171617.

Liao, S.F., 2021. Invited review: maintain or improve piglet gut health around weanling: the fundamental effects of dietary amino acids. Animals 11: 1110. https://doi.org/10.3390/ani11041110.

Liao, S.F., and Nyachoti, M., 2017. Using probiotics to improve swine gut health and nutrient utilization. Animal Nutrition 3: 331–343. https://doi.org/10.1016/j.aninu.2017.06.007.

Lindemann, M.D., Kornegay, E.T., Meldrum, J.B., Shurig, G., and Gwazdauskas, F.C., 1987. The effect of feeder space allowance on weaned pig performance. Journal of Animal Science 64: 8–14. https://doi.org/10.2527/jas1987.6418.

Liu, F., Zhao, W., Le, H.H., Cottrell, J.J., Green, M.P., Leury, B.J., Dunshea, F.R., and Bell, A.W., 2022a. Review: what have we learned about the effects of heat stress on the pig industry? Animal 16: 100349. https://doi.org/10.1016/j.animal.2021.100349.

Liu, Z., Chen, L., Gao, X., Zou, R., Meng, Q., Fu, Q., Xie, Y., Miao, Q., Chen, L., Tang, X., Zhang, S., Zhang, H., and Schroyen, M., 2022. Quantitative proteomics reveals tissue-specific toxic mechanisms for acute hydrogen sulfide-induced injury of diverse organs in pig. Science of The Total Environment 806: 150365. https://doi.org/10.1016/j.scitotenv.2021.150365.

Liu, Z., Fu, Q., Tang, S., Xie, Y., Meng, Q., Tang, X., Zhang, S., Zhang, H., and Schroyen, M., 2020. Proteomics analysis of lung reveals inflammation and cell death induced by atmospheric H2S exposure in pig. Environmental Research 191: 110204. https://doi.org/10.1016/j.envres.2020.110204.

Madec, F., Bridoux, N., Bounaix, S., and Jestin, A., 1998. Measurement of digestive disorders in the piglet at weaning and related risk factors. Preventive Veterinary Medicine 35: 53–72. https://doi.org/10.1016/S0167-5877(97)00057-3.

Mawdsley, J.E., and Rampton, D.S., 2005. Psychological stress in IBD: new insights into pathogenic and therapeutic implications. Gut 54: 1481–1491. http://dx.doi.org/10.1136/gut.2005.064261.

Mayorga, E.J., Renaudeau, D., Ramirez, B.C., Ross, J.W., and Baumgard, L.H., 2019. Heat stress adaptations in pigs. Animal Frontiers 9: 54–61. https://doi.org/10.1093/af/vfy035.

McGlone, J.J., and Curtis, S.E., 1985. Behaviour and performance of weanling pigs in pens equipped with hide area. Journal of Animal Science 60: 20–24. https://doi.org/10.2527/jas1985.60120x.

Metzler-Zebeli, B.U., Sener-Aydemir, A., Sharma, S., and Lerch, F., 2021. Postnatal development of gut microbial activity and their importance for jejunal motility in piglets. Journal of Animal Science 99: skab171. https://doi.org/10.1093/jas/skab171.

Ming, D., Wang, W., Huang, C., Wang, Z., Shi, C., Ding, J., Liu, H., and Wang, F. Effects of Weaning Age at 21 and 28 Days on Growth Performance, Intestinal Morphology and Redox Status in Piglets. Animals 2021, 11, 2169. https://doi.org/ 10.3390/ani11082169.

Morales, A., Chávez, M., Vásquez, N., Htoo, J.K., Buenabad, L., Espinoza, S., and Cervantes, M., 2018. Increased dietary protein or free amino acids supply for heat stress pigs: effect on performance and carcass traits. Journal of Animal Science 96: 1419–1429. https://doi.org/10.1093/jas/sky044.

Mortola, J.P., and Frappell, P.B., 2000. Ventilatory responses to changes in temperature in mammals and other vertebrates. Annual Review of Physiology 62: 847–874. https://doi.org/10.1146/annurev.physiol.62.1.847.

Muns, R., and Magowan, E., 2018. The effect of creep feed intake and starter diet allowance on piglets' gut structure and growth performance after weaning. Journal of Animal Science 96: 3815–3823. https://doi.org/10.1093/jas/sky239.

Nabuurs, M.J.A., Hoogendoorn, A., Van Der Molden, E.J., and Van Osta, A.L.M., 1993. Villous height and crypt depth in weaned and unweaned pigs, reared under various circumstances in the Netherlands. Research in Veterinary Science 55: 78–84. https://doi.org/10.1016/0034-5288(93)90038-H.

Ni, J.Q., Heber, A.J., Diehl, C.A., and Lim, T.T., 2000. Ammonia, hydrogen sulphide and carbon dioxide release from pig manure in under-floor deep pits. Journal of Agricultural Engineering Research 77: 53–66. https://doi.org/10.1006/jaer.2000.0561.

Ni, J.Q., Heber, A.J., and Lim, T.T., 2018. Ammonia and hydrogen sulphide in swine production. In: Banhazi, T., Aland, A., and Hartung, J. (eds). Air Quality and Livestock Farming. CRC Press/Balkema, Leiden, The Netherlands, pp. 29–48. https://doi.org/10.1201/9781315738338.

Nowland, T.L., Kirkwood, R.N., and Pluske, J.R. 2022. Review: Can early-life establishment of the piglet intestinal microbiota influence production outcomes? Animal 16: 100368. https://doi.org/10.1016/j.animal.2021.100368.

Nyachoti, C.M., Zijlstra, R.T., de Lange, C.F.M., and Patience, J.F., 2004. Voluntary feed intake in growing-finishing pigs: a review of the main determining factors and potential approaches for accurate predictions. Canadian Journal of Animal Science 84: 549–566. https://doi.org/10.4141/A04-001.

Oh, H.K., Choi, H.B., Ju, W.S., Chung, C.S., and Kim, Y.Y., 2010. Effects of space allocation on growth performance and immune system in weaning pigs. Livestock Science 132: 113–118. https://doi.org/10.1016/j.livsci.2010.05.009.

Ohland, C.L., and MacNaughton, W.K., 2010. Probiotic bacteria and intestinal epithelial barrier function. American Journal of Physiology Gastrointestinal and Liver Physiology 298: G807–G819. https://doi.org/10.1152/ajpgi.00243.2009.

Opapeju, F.O., Krause, D.O., Payne, R.L., Rademacher, M., and Nyachoti, C.M., 2009. Effect of dietary protein level on growth performance, indicators of enteric health and gastrointestinal microbial ecology of weaned pigs induced with post-weaning colibacillosis. Journal of Animal Science 87: 2635–2643. https://doi.org/10.2527/jas.2008-1310.

Ortega, A.D.S.V., and Szabó, C., 2021. Adverse effects of heat stress on the intestinal integrity and function of pigs and the mitigation capacity of dietary antioxidants: a review. Animals 11: 1135. https://doi.org/10.3390/ani11041135.

Pastorelli, H., Le Floc'h, N., Merlot, E., Meunier-Salaün, M.C., van Milgen, J., and Montagne, L., 2012. Sanitary housing conditions modify the performance and behavioural response of weaned pigs to feed- and housing-related stressors. Animal 6: 1811–1820. https://doi.org/10.1017/S1751731112001231.

Patridge, I.G., 1989. Alternative feeding strategies for weaner pigs. In: Barnett, J.L., and Hennessy, D.P. (eds). Manipulation of pig production. Australasian Pig Science Association, Victoria, Australia, pp. 160–169.

Pluske, J.R., Williams, I.H., and Aherne, F.X., 1996. Maintenance of villous height and crypt depth in piglets by providing continuous nutrition after weaning. Animal Science 62: 131–144. https://doi.org/10.1017/S1357729800014417.

Pluske, J.R., Hampson, D.J., and Williams, I.H., 1997. Factors influencing the structure and function of the small intestine in the weaned pig: a review. Livestock Production Science 51: 215–236. https://doi.org/10.1016/S0301-6226(97)00057-2.

Qin, W., Shen, L., Wang, Q., Gao, Y., She, M., Li, X., and Tan, Z., 2022. Chronic exposure to ammonia induces oxidative stress and enhanced glycolysis in lung of piglets. Environmental Toxicology. 37: 179–191. https://doi.org/10.1002/tox.23382.

Reyes-Camacho, D., Vinyeta, E., Perez, J.F., Aumiller, T., Criado, L., Palade, L.M., Taranu, I., Folch, J.M., Calvo, M.A., Klis, J.D. van der, Sola-Oriol, D. 2020. Phytogenic actives supplemented in hyperprolific sows: effects on maternal transfer of phytogenic compounds, colostrum and milk features, performance and antioxidant status of sows and their offspring, and piglet intestinal gene expression. Journal of Animal Science 2020. 98(1).

Sands, J.M., Rodrigues, L.A., Wellington, M.O., Panisson, J.C., Columbus, D.A. Pre- and post-weaning performance of piglets offered different types of creep feed. Canadian Journal of Animal Science 2022. 102: 189–193.

Santana, C.M., Gauger, P., Vetger, A., Magstadt, D., Kim, D.S., Shrestha, D., Charavaryamath, C., and Rumbeiha, W.K., 2021. Ambient hydrogen sulfide exposure increases the severity of influenza A virus infection in swine. Archives of Environmental & Occupational Health 76: 526–538. https://doi.org/10.1080/19338244.2021.1896986.

Shim, S.B., Verstegen, M.W.A., Kim, I.H., Kwon, O.S., and Verdonk, J.M.A.J., 2005. Effects of feeding antibiotic-free creep feed supplemented with oligofructose, probiotics or synbiotics to suckling piglets increases the preweaning weight gain and composition of intestinal microbiota. Archives of Animal Nutrition 59: 419–427. https://doi.org/10.1080/17450390500353234.

Smith, L.F., Beaulieu, A.D., Patience, J.F., Gonyou, H.W., and Boyd, R.D., 2004. The impact of feeder adjustment and group size-floor space allowance on the performance of nursery pigs. Journal of Swine Health and Production 12: 111–118.

Sommer, F., and Bäckhed, F., 2013. The gut microbiota – masters of host development and physiology. Nature Reviews Microbiology 11: 227–238. https://doi.org/10.1038/nrmicro2974.

Spreeuwenberg, M.A.M., Verdonk, J.M.A.J., Gaskins, H.R., and Verstegen, M.W.A., 2001. Small intestine epithelial barrier function is compromised in pigs with low feed intake at weaning. The Journal of Nutrition 131: 1520–1527. https://doi.org/10.1093/jn/131.5.1520.

Stokes, C.R., Bailey, M., and Haverson, K., 2001. Development and function of the pig gastrointestinal immune system. In: Lindberg, J.E., and Ogle, B. (eds). Digestive physiology of pigs. CAB International, Wallingford, United Kingdom, pp. 59–65. https://doi.org/10.1079/9780851995175.0059.

Stokes, C.R., Bailey, M., Haverson, K., Harris, C., Jones, P., Inman, C., Pié, S., Oswald, I.P., Williams, B.A., Akkermans, A.D.L., Sowa, E., Rothkötter, H., and Miller, B.G., 2004. Postnatal development of intestinal immune system in piglets: implications for the process of weaning. Animal Research 53: 325–334. https://doi.org/10.1051/animres:2004020.

Stoll, B., Chang, X., Fan, M.Z., Reeds, P.J., and Burrin, D.G., 2000. Enteral nutrient intake level determines intestinal protein synthesis and accretion rates in neonatal pigs. American Journal of Physiology Gastrointestinal and Liver Physiology 279: G288–G294. https://doi.org/10.1152/ajpgi.2000.279.2.G288.

Tang, S., Yin, C., Xie, J., Jiao, J., Chen, L., Liu, L., Zhang, S., and Zhang, H., 2021. Aerial ammonia exposure induces the perturbation of the interorgan ammonia disposal and branched-chain amino acid catabolism in growing pigs. Animal Nutrition 7: 947–958. https://doi.org/10.1016/j.aninu.2021.04.004.

Thacker, P.A., 2001. Water in swine nutrition. In: Lewis, A.J., and Southern, L.L. (eds). Swine Nutrition. second ed. CRC Press, Boca Raton, FL, USA, pp. 381–398.

Turner, S.P., Allcroft, D.J., and Edwards, S.A., 2003. Housing pigs in large social groups: a review of implications for performance and other economic traits. Livestock Production Science 82: 39–51. https://doi.org/10.1016/S0301-6226(03)00008-3.

Upadhaya, S.D., Kim, I.H. 2021. The impact of weaning stress on gut health and the mechanistic aspects of several feed additives contributing to improved gut health function in weanling piglets – a review. Animals 2021. 11(8).

Urbain, B., Gustin, P., Charlier, G., Coignoul, F., Lambotte, J.L., Grignon, G., Foliguet, B., Vidic, B., Beerens, D., Prouvost, J.F., and Ansay, M., 1996. A morphometric and functional study of the toxicity of atmospheric ammonia in the extrathoracic airways in pigs. Veterinary Research Communications 20: 381–399. https://doi.org/10.1007/BF00366545.

van Heugten, E., 2001. Mycotoxins and other anti-nutritional factors in swine feeds. In: Lewis, A.J., and Southern, L.L. (eds). Swine Nutrition. second ed. CRC Press, Boca Raton, FL, USA, pp. 563–585.

Varley, M.A., 2004. Alternatives to antibiotics growth promoters for post-weaned piglets. The Pig Journal 54: 161–170.

Vasquez, N., Cervantes, M., Bernal-Barragan, H., Rodriguez-Tovar, L.E., and Morales, A., 2022. Short- and long-term exposure to heat stress differently affect performance, blood parameters, and integrity of intestinal epithelia of growing pigs. Animals 12: 2529. https://doi.org/10.3390/ani12192529.

Vente-Spreeuwenberg, M.A.M., and Beynen, A.C., 2003. Diet mediated modulation of small intestine integrity in weaned pigs. In: Pluske, J.R., Le Dividich, J., and Verstegen, M.W.A. (eds). Weaning the pig: concepts and consequences. Wageningen Academic Publishers, Wageningen, the Netherlands, pp. 145–198. https://doi.org/10.3920/978-90-8686-513-0.

Vente-Spreeuwenberg, M.A.M., Verdonk, J.M.A.J., Bakker, G.C.M., Beynen, A.C., and Verstegen, M.W.A., 2004. Effect of dietary protein source on feed intake and small intestinal morphology in newly weaned piglets. Livestock Production Science 86: 169–177. https://doi.org/10.1016/S0301-6226(03)00166-0.

Verdonk, J.M.A.J., Bruininx, E.M.A.M., van der Meulen, J., and Verstegen, M.W.A., 2007. Post-weaning feed intake level modulates gut morphology but not gut permeability in weaned piglets. Livestock Science 108: 146–149. https://doi.org/10.1016/j.livsci.2007.01.093.

Waititu, S.M., Yin, F., Patterson, R., Yitbarek, A., Rodriguez-Lecompte, J.C., and Nyachoti, C.M., 2017. Dietary supplementation with a nucleotide-rich yeast extract modulates gut immune response and microflora in weaned pigs in response to a sanitary challenge. Animal 11: 2156–2164. https://doi.org/10.1017/S1751731117001276.

Wang, T., He, Q., Yao, W., Shao, Y., Li, J., and Huang, F., 2019. The variation of nasal microbiota caused by low levels of gaseous ammonia exposure in growing pigs. Frontiers in Microbiology 10: 1083. https://doi.org/10.3389/fmicb.2019.01083.

Wathes, C.M., Miller, B.G., and Bourne, F.J., 1989. Cold stress and post-weaning diarrhoea in piglets inoculated orally or by aerosol. Animal Production 49: 483–496. https://doi.org/10.1017/S0003356100032694.

Wattanakul, W., Bulman, C.A., Edge, H.L., and Edwards, S.A., 2005. The effect of creep feed presentation method on feeding behaviour, intake and performance of suckling piglets. Applied Animal Behaviour Science 92: 27–36. https://doi.org/10.1016/j.applanim.2004.10.019.

Wensley, M.R., Tokach, M.D., Woodworth, J.C., Goodband, R.D., Gebhardt, J.T., Derouchey, J.M., and McKilligan, D. 2021. Maintaining continuity of nutrient intake after weaning. I. review of pre-weaning strategies. Translational Animal Science 2021. 5(1).

Williams, N.H., Stahly, T.S., and Zimmerman, D.R., 1997. Effect of chronic immune system activation on body nitrogen retention, partial efficiency of lysine utilization, and lysine needs of pigs. Journal of Animal Science 75: 2472–2480. https://doi.org/10.2527/1997.7592472x.

Wolter, B.F., Ellis, M., Corrigan, B.P., DeDecker, J.M., Curtis, S.E., Parr, E.N., and Webel, D.M., 2003. Impact of early postweaning growth rate as affected by diet complexity and space allocation on subsequent growth performance of pigsin a wean-to-finish production system. Journal of Animal Science 81: 353–359. https://doi.org/10.2527/2003.812353x.

Xia, B., Wu, W., Fang, W., Wen, X., Xie, J., and Zhang, H., 2022. Heat stress-induced mucosal barrier dysfunction is potentially associated with gut microbiota dysbiosis in pigs. Animal Nutrition 8: 289–299. https://doi.org/10.1016/j.aninu.2021.05.012.

Xiong, Y.X., Cao, S.T., Xiao, H., Wu, Q.W., Yi H.B., Jiang, Z.Y., Wang, L. 2022. Alterations in intestinal microbiota composition coincide with impaired intestinal morphology and dysfunctional ileal immune response in growing-finishing pigs under constant chronic heat stress. Journal of Animal Science and Biotechnology 2022. 13(1):(05 January 2022).

Xiong, Y.X., Yi, H., Wu, Q., Jiang, Z., and Wang, L., 2020. Effects of acute heat stress on intestinal microbiota in grow-finishing pigs, and associations with feed intake and serum profile. Journal of Applied Microbiology. 128: 840–852. https://doi.org/10.1111/jam.14504.

Zhao, J., Harper, A.F., Estienne, M.J., Webb, K.E. Jr., McElroy, A.P., and Denbow, D.M., 2007. Growth performance and intestinal morphology responses in early weaned pigs to supplementation of antibiotic-free diets with an organic copper complex and spray-dried plasma protein in sanitary and nonsanitary environments. Journal of Animal Science 85: 1302–1310. https://doi.org/10.2527/jas.2006-434.

CHAPTER 4

The impact of environmental conditions on the gut microbiome of broiler chickens

Jannigje G. Kers[1,2,*], *Hauke Smidt*[1,*] *and Francisca C. Velkers*[3,*]

[1]Laboratory of Microbiology, Wageningen University & Research, Wageningen, The Netherlands; [2]Department of Population Health Sciences, Institute for Risk Assessment Sciences (IRAS), Utrecht University, Utrecht, The Netherlands; [3]Department of Population Health Sciences, Division of Farm Animal Health, Utrecht University, Utrecht, The Netherlands; *Correspondence: annelies.kers@wur.nl; hauke.smidt@wur.nl; F.C.Velkers@uu.nl

Abstract

Knowledge of factors that influence the functioning of gut microbes is essential to improve health and reduce the use of antibiotics in poultry production. The environment, and more specifically housing conditions, can affect the gut microbiome of broiler chickens. The gut microbiome is defined as the collection of all the microorganisms and their "theatre of activity" in the gut environment. Previous research showed that the development of the gut microbiome is hampered in animals raised in high-hygiene environments, such as isolators. It has also been reported that human individuals and animals who live together show less variation in the gut microbiome compared to a group of random individuals. In a Dutch human population, it has been shown that the gut microbiome was primarily shaped by environmental factors and cohabitation. Therefore, a proper understanding of how the composition and functioning of the gut microbiome in broiler chickens is affected by interactions between hosts, and hosts and their environment, is needed. To this end, the effect of environmental factors should be considered in the design, analyses, and interpretation of study outcomes. In addition, knowledge of the development of intestinal microbiome composition and function would be of great value for optimizing the resilience of broiler flocks that can help in reducing the need for therapeutic antibiotics. Although the mechanisms driving gut microbiome development are not fully understood, there is data suggesting that this development is major in the first week post-hatch, and that around three weeks post-hatch, cecal microbiota composition can be considered mature in a commercial well-performing broiler flock. It is important to further study mechanisms underlying gut microbiome development and function to develop better diagnostics and management tools to improve broiler health.

© JANNIGJE G. KERS ET AL., 2024 | DOI:10.3920/9789004695467_005

Keywords

microbiome – gut – intestinal health – environment

1 Introduction

Knowledge of factors that influence the functioning of gut microbes is essential to improve health and reduce the use of antibiotics in poultry production. In this chapter, we will focus on how the environment, and more specifically housing conditions, can affect the gut microbiome of broiler chickens. The gut microbiome is defined as the collection of all the microorganisms and their "theatre of activity" in the gut environment. The development of the gut microbiome is hampered in animals raised in high-hygiene environments, such as isolators (Inman *et al.*, 2010; Mulder *et al.*, 2011). It has also been reported that human individuals and animals who live together show less variation in the gut microbiome compared to a group of random individuals (Laukens *et al.*, 2016; McCafferty *et al.*, 2013; Schloss *et al.*, 2014; Yatsunenko *et al.*, 2012). A large study in a Dutch human population showed that the gut microbiome was primarily shaped by environmental factors and cohabitation (Gacesa *et al.*, 2022). Therefore, for a proper understanding of how the composition and functioning of the gut microbiome in broiler chickens is affected by interactions between hosts, and hosts and their environment, the effect of environmental factors should be considered in the design, analyses and interpretation of study outcomes.

In addition, knowledge of the development of intestinal microbiome composition would be of great value for optimizing the resilience of broiler flocks and can reduce the need for therapeutic antibiotics. Across repeated experiments, a highly variable intestinal microbiota development has been observed in broilers (Stanley *et al.*, 2013). Early life colonization, feed additives, and antimicrobial drugs play a role in microbiome composition and might affect development (Ballou *et al.*, 2016; Gao *et al.*, 2017). Although the mechanisms driving gut microbiome development is not fully understood, there is data that suggests that the development is major in the first week of life and that after day 21 cecal microbiota composition can be considered as mature in a commercial well-performing broiler flock (Awad *et al.*, 2016; Gong *et al.*, 2008; Kers *et al.*, 2022; Wielen *et al.*, 2002). It is important to further study mechanisms underlying gut microbiome development and function to develop better diagnostic tools to improve broiler health.

2 The environment and an intervention

The concept of the relevance of the environment is not new, as the Dutch botanist and microbiologist Baas-Becking proposed in 1934 already: "*Everything is everywhere, but the environment selects*". However, we still have limited understanding of pathways and extent of transmission of microorganisms present in the living environment to the hosts (environment-to-host transmission) and exchange of microbiota between hosts sharing the same environment (host-to-host transmission), and how this shapes the intestinal microbial communities within hosts (Browne *et al.*, 2017; Valles-Colomer *et al.*, 2023). In general, meta-analyses provide insight into the available level of evidence from research outcomes by locating, appraising, and summarizing similar studies. Similar studies, however, are difficult to find within the research field of poultry gut microbiome. This is partly due to the technical variation in analytical methods used, and partly because of the broad range of host- and environmental factors that can affect the research outcome. To address this limitation, a study was performed where three different experimental housing conditions were compared with respect to their influence on cecal microbiota composition of broiler chickens. This study included a standard grow-out feed trial facility (housing condition 1, H1); a facility with floor pens for small-scale experiments (H2); and isolators (H3) (Kers, Velkers, *et al.*, 2019). Bird management conditions were kept as equal as possible across H1, H2 and H3. The wood shavings were transported from H1 to H2 and H3 three weeks before the start of the experiment and were stored under comparable conditions. Although the sizes of the pens were slightly different, the chick densities were the same. The temperature was monitored across housing conditions twice a day and corrected when needed. All chicks were derived from the same 42-week-old broiler breeder flock. At the hatchery, the chicks were randomly allocated to two different experimental facilities (H1 and H2 + H3). The chicks were transported to these two facilities in the same truck. As medium-chain fatty acids (MCFA) supplementation to feed can significantly change intestinal microbiota composition (Van der Hoeven-Hangoor *et al.*, 2013; Zentek *et al.*, 2012), a feed with and without MCFA was used as a tool to generate a biological (feed-induced) contrast in cecal microbiota composition between broilers in different pens, on which the additional effect of the housing condition could be tested. At 35 days of age, the cecal microbiota composition was determined based on 16S ribosomal RNA (rRNA) gene amplicon sequence analysis. The metabolic output of the microbiome was determined by measuring the production of acetate, butyrate, isobutyrate, lactate, and propionate, and effects on production performance were determined based on body weight on day 35. In H1 and in H2 different distances between pens were created, to assess if the distance between pens could influence the gut microbiome (Kers *et al.*, 2019), however, no correlation between the physical distance of pens and the percentage of shared gut microbiota was observed. In turn, the environment (*i.e.* the different housing conditions) affected microbiota composition and functionality to a larger extent than the MCFA feed intervention. When comparing the composition of the cecal microbiota of the broilers within the same housing condition, different effects on the alpha diversity (the number of different types of microorganisms in a sample) were observed in response to the MCFA intervention. In H1 no effect was observed, whereas in H2 and in H3 MCFA

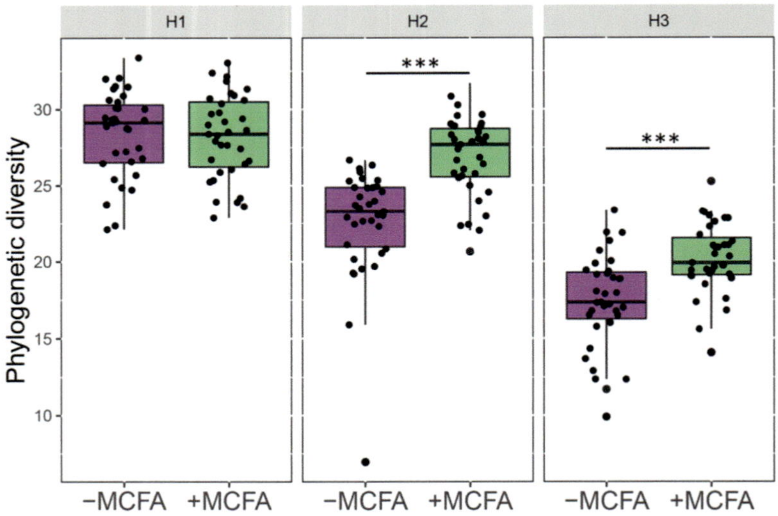

FIGURE 4.1 The alpha (phylogenetic) diversity (ASV level) was higher in the ceca of broilers fed +MCFA in H2 and H3 but not in H1. The diversity was highest in H1 and lowest in H3. n = 35 broilers per group, Kruskal-Wallis, * = $p < 0.05–0.01$, ** = $p < 0.01–0.001$, *** = $p < 0.001$. (J.G. Kers *et al.*, 2019)

supplementation resulted in higher microbial alpha diversity (Figure 4.1). Overall, around 10% of the variation between broilers could be explained by the feed intervention, whereas almost 30% of the variation could be explained by the housing condition. Broilers raised in the isolators (H3) had a higher body weight compared to the broilers raised in the feed research facility (H1), and the metabolic output was also significantly different between housing conditions. A study in layers, for example, showed a unique cloacal microbiota composition of layers per poultry house, whereas access to an outdoor range had no detectable effect on microbial community composition (Schreuder *et al.*, 2020). Another study on layers showed that cage-free systems resulted in a higher alpha diversity of the cecal microbiota compared to a cage system, and that the feed intervention (standard soy-based feed versus a soy-free feed) had a relatively lower effect on gut microbiota composition (Hubert *et al.*, 2019). Taken together, these studies reinforce the need to be aware of the potentially large impact of housing conditions on the development and effects on the gut microbiome.

3 Interpretation of experiments

Due to the large impact of the study design on the outcome of microbiome research (Knight *et al.*, 2018; Laukens *et al.*, 2016), it is important for poultry researchers to perform field experiments as soon as possible in order to facilitate the translation from experimental research to commercial poultry production. Nevertheless, it has also been shown that microbiome field studies have to deal with many hidden host and environmental variables, that are not all known (Kers *et al.*, 2018) and that limit the reproducibility of research outcomes. It has been proposed for rodent studies that the use of natural or "wild" gut microbiota can help to improve the reproducibility of research outcomes (Rosshart *et al.*, 2019; Rosshart *et al.*, 2017). Microbial communities encountered under field circumstances are not the same as those in a lab setting, and this can substantially affect how *e.g.* the immune system develops and functions. This in turn can lead to incorrect assumptions of how the immune system works under 'natural' circumstances (Rosshart *et al.*, 2019; Rosshart *et al.*, 2017). This reasoning applies also to poultry gut microbiome research. In commercial settings, no adult hens are present while this might positively influence the natural development of the cecal microbiome and immune system of chicken (Kubasova *et al.*, 2019; Meijerink *et al.*, 2020). Not all types of research can be done under field circumstances. One way to combat this problem is to study microbe-to-microbe interactions within the microbiome outside the host (Foster *et al.*, 2017). Tools, for example, intestinal organoids or a 'microbiome-on-a-chip', minimize or compensate for these differences when studying *e.g.* the gut microbiome (Oost *et al.*, 2022; Oost *et al.*, 2021; Rubert *et al.*, 2020; Stanley & van der Heijden, 2017). Those tools can improve knowledge on interactions between feed, microbiome and the intestinal epithelium to further decipher the difference between a healthy and unhealthy gut microbiome (Rubert *et al.*, 2020). Another benefit is that many compounds can be tested under the same condition. It should be kept in mind, however, that the use of such *in vitro* models can only complement, but not completely replace, *in vivo* studies that address the "normal" gut microbiome, the composition and function of which can strongly affect research outcomes. Therefore, experiments to test interventions should also be repeatedly performed in commercial poultry farms. Although this type of research is sometimes referred to as a 'microbial black box', as it might not provide much insight into what is taking place at the cellular and molecular level and how different microbes interact, it contains the complex and variable rearing environment of broilers which is lacking under lab settings.

4 Stochastic and deterministic processes in the establishment of the broiler cecal microbiome

In ecology, different processes are described that can shape a habitat. This includes stochastic processes such as dispersal, diversification, and ecological drift, and deterministic processes such as interactions between species, individuals, and environmental conditions (Burke *et al.*, 2011; Pereira & Berry, 2017; Stegen *et al.*, 2016; Stegen *et al.*, 2012; Zhou & Ning, 2017). The gut is also described as a habitat, or defined as a dynamic ecosystem, where those processes occur (Pereira & Berry,

2017). After 35 days the cecal microbiome composition and functionality was different in broilers raised in different housing environments (Kers *et al.*, 2019). However, when exactly the microbiome composition and functionality changed, at the start of the experiment or during the experiment, is unknown, and might have been a result of stochastic processes. Such stochastic changes over time can drive environmental effects, which was demonstrated in a mice study which explored the causes and consequences of the so-called cage effect (*i.e.* animals in the same cage developing similar microbial communities) (McCafferty *et al.*, 2013). Successional changes structuring the microbial community were studied longitudinally, and showed that a difference in initial composition, induced by an inoculation with a specific microbial community, was insufficient to eliminate the cage effect (McCafferty *et al.*, 2013). Stochastic differences in community assembly within each individual cage environment were shown to be the primary drivers of cage effects, rather that effects of the initial microbial composition (McCafferty *et al.*, 2013). This stresses the importance to move from cross-sectional studies to longitudinal studies at sufficient spatio-temporal resolution to further understand the mechanisms underlying the colonization and development of the gut microbiome and broiler health. Daily sampling of humans revealed that similar foods can have a different effects on the gut microbiome in different individuals (Johnson *et al.*, 2019). Therefore, non-invasive sampling methods are needed to measure the gut microbiome within the same broiler flock and preferably even within the same broilers across time. Examples of potential non-invasive samples are fecal and cecal droppings, cloacal swabs and boot socks samples (Kers, Fischer, *et al.*, 2019).

In a field study cecal microbiota samples were collected from different healthy broiler flocks (Kers *et al.*, 2022). The data was used to answer two questions. First, whether stratification of the cecal content samples into compositional clusters would help to provide insight into ecological processes related to microbiota succession and maturation, and second, which factors are important to explain the variation in microbial composition observed between broilers. A total of 10 well-performing poultry flocks on four farms were followed. Cluster 1 contained a total number of 34 'day-old' broilers (at day of chick placement, also referred to as day 0) (Figure 4.2). Cluster 2 contained two of the day-old broilers, all broilers of 2 days of age, and four broilers of 4 days of age (n = 97, Figure 4.2). Cluster 3 contained broilers of 4 days until 11 days of age, and around half of the broilers of 12 and 14 days of age (n = 183, Figure 4.2). The last cluster, cluster 4, contained the other half of the broilers that were 12 and 14 days of age and most of the broilers of 21 days and older (n = 243, Figure 4.2). The predominant family in cluster 1 was *Clostridiaceae*, in cluster 2 *Enterobacteriaceae*, in cluster 3 *Lachnospiraceae* and in cluster 4 *Ruminococcaceae* and *Lachnospiraceae*. It should be noted that this analysis was constrained by the fact that not the same individual broilers were followed over time considering that cecal sampling at different ages required sacrifice of animals housed together. However, the results indicate a conserved developmental

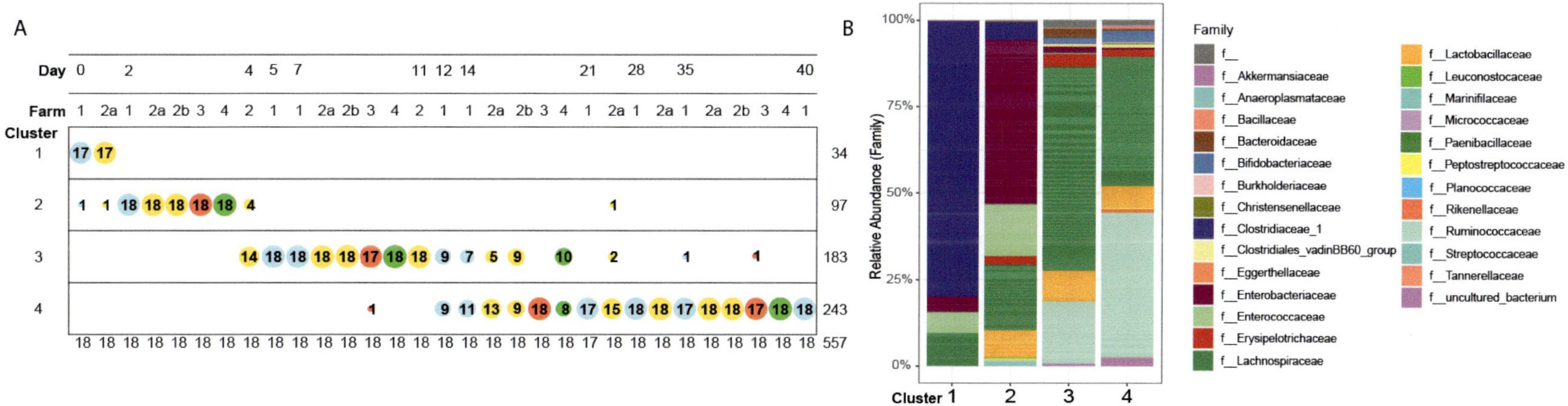

FIGURE 4.2 **A)** Transition model showing the progression of samples through each cluster (Dirichlet-multinomial modelling (DMM) method), performed on genus level, across age, stratified per farm (10 flocks). Farm 1 is blue, Farm 2 is yellow (a and b are different production cycles), Farm 3 is red, Farm 4 is green. **B)** The relative abundance at family level of the four clusters, only the labels of the 10 most abundant families are shown, the other families are not labeled for clarity (Kers, 2020).

trajectory of cecal microbiota composition that followed the same general trend across farms and that the variation in maturity across individuals was especially large around the second week of life (Kers *et al.*, 2022). This is in line with earlier studies that suggested that the assembly of the gut microbiota is largely deterministic (Pereira and Berry, 2016). Other researchers, however, suggested that stochastic factors might also play a significant role in shaping the structure of the microbial communities within the first days of life (Burke *et al.*, 2011; Schloss *et al.*, 2012). This is based on the 'lottery' theory that whoever gets there first in the gut 'wins' the space (Burke *et al.*, 2011; Munday, 2004).

Although there was a conserved developmental trajectory of cecal microbiota observed (Kers, 2020), further investigation is needed to understand, which factors are important to drive this maturation of the gut microbiome, especially also with respect to the observed variation along this process. In addition to the observed developmental clusters, the poultry house, body weight, and feed ingredients explained around 37% of the cecal microbiota variation between broilers (Kers *et al.*, 2022). This means that approximately 63% of the variation was not explained by the measured factors, indicating that either unmeasured (unknown) factors contribute, but also that stochastic effects can have a profound implication on the formation of the microbiome (Faust & Raes, 2012; Hildebrand *et al.*, 2013). Most likely, both processes deterministic and stochastic play an important role (McCafferty *et al.*, 2013), which is difficult to disentangle. Broilers within the same poultry house eat the same feed and peck in the same litter that contains feces of other broilers. Therefore, it remains difficult to separate biological interactions within the microbiome of an individual from those that occur via the interaction with other birds or the flock environment. Individual housing of broilers for experimental studies has substantial welfare implications, and is far from the commercial situation and therefore limits the real dynamics of the gut microbiota, as well as translatability of outcomes. One way to combat this challenge is to enlarge the sample size of studies/datasets, with broilers raised in different areas of the world, with different feed compositions, management conditions and different health status. This will increase the evidence and understanding of which factors are most important to explain the variation between broiler flocks. Nevertheless, even in large datasets it will be a challenge to disentangle the biological interactions due to the rearing conditions of commercial broiler flocks.

5 Conclusions

It is essential to be aware of the large impact of the study design on the results and thus on the interpretation of the outcomes with respect to the gut microbiome and its correlation with host-associated measurements such as health, growth performance, and immune development. In particular, environmental factors associated with housing conditions can affect the microbiota composition and functionality more extensively than a nutritional intervention. However, conserved development trajectories of cecal microbiota have been observed, despite strong influence of environmental factors. Non-invasive longitudinal sampling to monitor the development of the gut microbiome is needed to improve our understanding of the development of the gut microbiome and its interaction with broiler health.

References

Awad, W.A., Mann, E., Dzieciol, M., Hess, C., Schmitz-Esser, S., Wagner, M., & Hess, M. (2016). Age-Related Differences in the Luminal and Mucosa-Associated Gut Microbiome of Broiler Chickens and Shifts Associated with Campylobacter jejuni Infection. *Front Cell Infect Microbiol, 6*, 154. https://doi.org/10.3389/fcimb.2016.00154.

Ballou, A.L., Ali, R.A., Mendoza, M.A., Ellis, J.C., Hassan, H.M., Croom, W.J., & Koci, M.D. (2016). Development of the Chick Microbiome: How Early Exposure Influences Future Microbial Diversity. *Front Vet Sci, 3*, 2. https://doi.org/10.3389/fvets.2016.00002.

Berg, G., Rybakova, D., Fischer, D., Cernava, T., Vergès, M.-C.C., Charles, T., Chen, X., Cocolin, L., Eversole, K., & Corral, G.H. (2020). Microbiome definition re-visited: old concepts and new challenges. *Microbiome, 8*, 1–22.

Browne, H.P., Neville, B.A., Forster, S.C., & Lawley, T.D. (2017). Transmission of the gut microbiota: spreading of health. *Nature Reviews Microbiology, 15*(9), 531–543.

Burke, C., Steinberg, P., Rusch, D., Kjelleberg, S., & Thomas, T. (2011). Bacterial community assembly based on functional genes rather than species. *Proc Natl Acad Sci U S A, 108*(34), 14288–14293. https://doi.org/10.1073/pnas.1101591108.

Faust, K., & Raes, J. (2012). Microbial interactions: from networks to models. *Nature Reviews Microbiology, 10*(8), 538–550.

Foster, K.R., Schluter, J., Coyte, K.Z., & Rakoff-Nahoum, S. (2017). The evolution of the host microbiome as an ecosystem on a leash. *Nature, 548*(7665), 43–51.

Gacesa, R., Kurilshikov, A., Vich Vila, A., Sinha, T., Klaassen, M., Bolte, L., Andreu-Sánchez, S., Chen, L., Collij, V., & Hu, S. (2022). Environmental factors shaping the gut microbiome in a Dutch population. *Nature, 604*(7907), 732–739.

Gao, P., Ma, C., Sun, Z., Wang, L., Huang, S., Su, X., Xu, J., & Zhang, H. (2017). Feed-additive probiotics accelerate yet antibiotics delay intestinal microbiota maturation in broiler chicken. *Microbiome, 5*(1), 91.

Gong, J., Yu, H., Liu, T., Gill, J.J., Chambers, J.R., Wheatcroft, R., & Sabour, P.M. (2008). Effects of zinc bacitracin, bird age and access to range on bacterial microbiota in the ileum and caeca of broiler chickens. *J Appl Microbiol, 104*(5), 1372–1382. https://doi.org/10.1111/j.1365-2672.2007.03699.x.

Hildebrand, F., Nguyen, T.L., Brinkman, B., Yunta, R.G., Cauwe, B., Vandenabeele, P., Liston, A., & Raes, J. (2013). Inflammation-associated enterotypes, host genotype, cage and inter-individual effects drive gut microbiota variation in common laboratory mice. *Genome Biol, 14*(1), R4. https://doi.org/10.1186/gb-2013-14-1-r4.

Hubert, S.M., Al-Ajeeli, M., Bailey, C.A., & Athrey, G. (2019). The role of housing environment and dietary protein source on the gut microbiota of chicken. *Animals*, *9*(12), 1085.

Inman, C.F., Haverson, K., Konstantinov, S.R., Jones, P.H., Harris, C., Smidt, H., Miller, B., Bailey, M., & Stokes, C. (2010). Rearing environment affects development of the immune system in neonates. *Clin Exp Immunol*, *160*(3), 431–439. https://doi.org/10.1111/j.1365-2249.2010.04090.x.

Johnson, A.J., Vangay, P., Al-Ghalith, G.A., Hillmann, B.M., Ward, T.L., Shields-Cutler, R.R., Kim, A.D., Shmagel, A.K., Syed, A.N., Personalized Microbiome Class, S., Walter, J., Menon, R., Koecher, K., & Knights, D. (2019). Daily Sampling Reveals Personalized Diet-Microbiome Associations in Humans. *Cell Host Microbe*, *25*(6), 789–802 e785. https://doi.org/10.1016/j.chom.2019.05.005.

Kers, Fischer, Stegeman, Smidt, & Velkers. (2019). Comparison of Different Invasive and Non-Invasive Methods to Characterize Intestinal Microbiota throughout a Production Cycle of Broiler Chickens. *Microorganisms*, *7*(10). https://doi.org/10.3390/microorganisms7100431.

Kers, Velkers, Fischer, Hermes, Lamot, Stegeman, & Smidt. (2019). Take care of the environment: housing conditions affect the interplay of nutritional interventions and intestinal microbiota in broiler chickens [journal article]. *Animal Microbiome*, *1*(1), 10. https://doi.org/10.1186/s42523-019-0009-z.

Kers, J. (2020). *Intestinal microbiota and broiler health: hype or hope?* Utrecht University.

Kers, J.G., Velkers, F.C., Fischer, E.A., Hermes, G.D., Lamot, D.M., Stegeman, J.A., & Smidt, H. (2019). Take care of the environment: housing conditions affect the interplay of nutritional interventions and intestinal microbiota in broiler chickens. *Animal Microbiome*, *1*, 1–14.

Kers, J.G., Velkers, F.C., Fischer, E.A., Stegeman, J.A., Smidt, H., & Hermes, G.D. (2022). Conserved developmental trajectories of the cecal microbiota of broiler chickens in a field study. *FEMS Microbiology Ecology*, *98*(9), fiac090.

Kers, J.G., Velkers, F.C., Fischer, E.A.J., Hermes, G.D.A., Stegeman, J.A., & Smidt, H. (2018). Host and Environmental Factors Affecting the Intestinal Microbiota in Chickens. *Front Microbiol*, *9*, 235. https://doi.org/10.3389/fmicb.2018.00235.

Knight, R., Vrbanac, A., Taylor, B.C., Aksenov, A., Callewaert, C., Debelius, J., Gonzalez, A., Kosciolek, T., McCall, L.I., McDonald, D., Melnik, A.V., Morton, J.T., Navas, J., Quinn, R.A., Sanders, J.G., Swafford, A.D., Thompson, L.R., Tripathi, A., Xu, Z.Z., … Dorrestein, P.C. (2018). Best practices for analysing microbiomes. *Nat Rev Microbiol*, *16*(7), 410–422. https://doi.org/10.1038/s41579-018-0029-9.

Kubasova, T., Kollarcikova, M., Crhanova, M., Karasova, D., Cejkova, D., Sebkova, A., Matiasovicova, J., Faldynova, M., Pokorna, A., & Cizek, A. (2019). Contact with adult hen affects development of caecal microbiota in newly hatched chicks. *PLoS One*, *14*(3), e0212446.

Laukens, D., Brinkman, B.M., Raes, J., De Vos, M., & Vandenabeele, P. (2016). Heterogeneity of the gut microbiome in mice: guidelines for optimizing experimental design. *FEMS Microbiol Rev*, *40*(1), 117–132. https://doi.org/10.1093/femsre/fuv036.

McCafferty, J., Mühlbauer, M., Gharaibeh, R.Z., Arthur, J.C., Perez-Chanona, E., Sha, W., Jobin, C., & Fodor, A.A. (2013). Stochastic changes over time and not founder effects drive cage effects in microbial community assembly in a mouse model. *The ISME journal*, *7*(11), 2116–2125.

Meijerink, N., Kers, J.G., Velkers, F.C., van Haarlem, D.A., Lamot, D.M., de Oliveira, J.E., Smidt, H., Stegeman, J.A., Rutten, V.P., & Jansen, C.A. (2020). Early life inoculation with adult-derived microbiota accelerates maturation of intestinal microbiota and enhances NK cell activation in broiler chickens. *Frontiers in veterinary science*, *7*, 965.

Mulder, I.E., Schmidt, B., Lewis, M., Delday, M., Stokes, C.R., Bailey, M., Aminov, R.I., Gill, B.P., Pluske, J.R., & Mayer, C.-D. (2011). Restricting microbial exposure in early life negates the immune benefits associated with gut colonization in environments of high microbial diversity. *PLoS One*, *6*(12), e28279.

Munday, P.L. (2004). Competitive coexistence of coral-dwelling fishes: the lottery hypothesis revisited. *Ecology*, *85*(3), 623–628.

Oost, M.J., Ijaz, A., van Haarlem, D.A., van Summeren, K., Velkers, F.C., Kraneveld, A.D., Venema, K., Jansen, C.A., Pieters, R.H., & Ten Klooster, J.P. (2022). Chicken-derived RSPO1 and WNT3 contribute to maintaining longevity of chicken intestinal organoid cultures. *Scientific reports*, *12*(1), 10563.

Oost, M.J., Velkers, F.C., Kraneveld, A.D., & Venema, K. (2021). Development of the in vitro cecal chicken ALIMEntary tRact mOdel-2 to study microbiota composition and function. *Frontiers in Microbiology*, *12*, 726447.

Pereira, F.C., & Berry, D. (2017). Microbial nutrient niches in the gut. *Environmental microbiology*, *19*(4), 1366–1378.

Rosshart, S.P., Herz, J., Vassallo, B.G., Hunter, A., Wall, M.K., Badger, J.H., McCulloch, J.A., Anastasakis, D.G., Sarshad, A.A., Leonardi, I., Collins, N., Blatter, J.A., Han, S.J., Tamoutounour, S., Potapova, S., Foster St Claire, M.B., Yuan, W., Sen, S.K., Dreier, M.S., … Rehermann, B. (2019). Laboratory mice born to wild mice have natural microbiota and model human immune responses. *Science*, *365*(6452). https://doi.org/10.1126/science.aaw4361.

Rosshart, S.P., Vassallo, B.G., Angeletti, D., Hutchinson, D.S., Morgan, A.P., Takeda, K., Hickman, H.D., McCulloch, J.A., Badger, J.H., & Ajami, N.J. (2017). Wild mouse gut microbiota promotes host fitness and improves disease resistance. *Cell*, *171*(5), 1015–1028. e1013.

Rubert, J., Schweiger, P.J., Mattivi, F., Tuohy, K., Jensen, K.B., & Lunardi, A. (2020). Intestinal organoids: a tool for modelling diet-microbiome-host interactions. *Trends in Endocrinology & Metabolism*, *31*(11), 848–858.

Schloss, P.D., Iverson, K.D., Petrosino, J.F., & Schloss, S.J. (2014). The dynamics of a family's gut microbiota reveal variations on a theme. *Microbiome*, *2*, 1–13.

Schloss, P.D., Schubert, A.M., Zackular, J.P., Iverson, K.D., Young, V.B., & Petrosino, J.F. (2012). Stabilization of the murine gut microbiome following weaning. *Gut Microbes*, *3*(4), 383–393.

Schreuder, J., Velkers, F.C., Bouwstra, R.J., Beerens, N., Stegeman, J.A., de Boer, W.F., van Hooft, P., Elbers, A.R., Bossers, A., & Jurburg, S.D. (2020). An observational field study of the cloacal microbiota in adult laying hens with and without access to an outdoor range. *Animal Microbiome*, *2*(1), 1–11.

Stanley, C.E., & van der Heijden, M.G. (2017). Microbiome-on-a-chip: new frontiers in plant-microbiota research. *Trends in microbiology*, *25*(8), 610–613.

Stanley, D., Geier, M.S., Hughes, R.J., Denman, S.E., & Moore, R.J. (2013). Highly variable microbiota development in the chicken gastrointestinal tract. *PLoS One*, *8*(12), e84290. https://doi.org/10.1371/journal.pone.0084290.

Stegen, J.C., Fredrickson, J.K., Wilkins, M.J., Konopka, A.E., Nelson, W.C., Arntzen, E.V., Chrisler, W.B., Chu, R.K., Danczak, R.E., & Fansler, S.J. (2016). Groundwater-surface water mixing shifts ecological assembly processes and stimulates organic carbon turnover. *Nature communications*, *7*(1), 11237.

Stegen, J.C., Lin, X., Konopka, A.E., & Fredrickson, J.K. (2012). Stochastic and deterministic assembly processes in subsurface microbial communities. *ISME J*, *6*(9), 1653–1664. https://doi.org/10.1038/ismej.2012.22.

Valles-Colomer, M., Blanco-Míguez, A., Manghi, P., Asnicar, F., Dubois, L., Golzato, D., Armanini, F., Cumbo, F., Huang, K.D., & Manara, S. (2023). The person-to-person transmission landscape of the gut and oral microbiomes. *Nature*, 1–11.

Van der Hoeven-Hangoor, E., Van Der Vossen, J., Schuren, F., Verstegen, M., De Oliveira, J., Montijn, R., & Hendriks, W. (2013). Ileal microbiota composition of broilers fed various commercial diet compositions. *Poultry science*, *92*(10), 2713–2723.

Wielen, P., Keuzenkamp, D., Lipman, L.v., Knapen, F., & Biesterveld, S. (2002). Spatial and temporal variation of the intestinal bacterial community in commercially raised broiler chickens during growth. *Microbial ecology*, *44*, 286–293.

Yatsunenko, T., Rey, F.E., Manary, M.J., Trehan, I., Dominguez-Bello, M.G., Contreras, M., Magris, M., Hidalgo, G., Baldassano, R.N., Anokhin, A.P., Heath, A.C., Warner, B., Reeder, J., Kuczynski, J., Caporaso, J.G., Lozupone, C.A., Lauber, C., Clemente, J.C., Knights, D., … Gordon, J.I. (2012). Human gut microbiome viewed across age and geography. *Nature*, *486*(7402), 222–227. https://doi.org/10.1038/nature11053.

Zentek, J., Buchheit-Renko, S., Manner, K., Pieper, R., & Vahjen, W. (2012). Intestinal concentrations of free and encapsulated dietary medium-chain fatty acids and effects on gastric microbial ecology and bacterial metabolic products in the digestive tract of piglets. *Arch Anim Nutr*, *66*(1), 14–26. https://doi.org/10.1080/1745039X.2011.644916.

Zhou, J., & Ning, D. (2017). Stochastic community assembly: does it matter in microbial ecology? *Microbiology and Molecular Biology Reviews*, *81*(4), e00002–00017.

CHAPTER 5

The importance of early microbial exposures in pigs: the microbiome nexus

Kayla Law[1] *and Andres Gomez*[1,*]

[1]Department of Animal Science, University of Minnesota, St. Paul MN, 55108, USA; *Correspondence: gomeza@umn.edu

Abstract

Interest in the microbiome – the community of diverse bacteria, fungi, microeukaryotes and viruses that colonize animal and environmental surfaces – has become mainstream in the swine production field. The premise, given the critical influence of the microbiome in the animal physiological landscape, is that manipulating or targeting these microbial communities along the pig anatomy could favorably alter productive and sanitary goals in swine production systems. To that end, a myriad of dietary manipulation strategies (largely based on feed additives) have been proposed, which acquire significant value as the ban of growth promotion antibiotics becomes a reality worldwide. However, the success of such strategies in promoting pig growth and robust health has proven to be inconsistent, at least in comparison with the standard use of antibiotics for growth promotion. In this chapter, we take a step back to discuss the importance of the events that lead to the very first microbiome seeding (or establishment) in the pig, specifically, placing critical importance on the maternal environment. We discuss how different maternal (*e.g.* milk, vaginal canal, skin) and environmental (farrowing surfaces) microbial sources contribute to the primordial microbiome of newborn pigs and highlight the interplay between early microbial seeding events and an immature immune system as a key modulator of pig physiological performance along the life course. We also propose the need to consider manipulating early microbial seeding mechanisms, besides nutritional or feed additive strategies, to boost pig physiological performance along its productive cycle.

Keywords

Microbiome – swine production – early life – weaning – birth

© KAYLA LAW AND ANDRES GOMEZ, 2024 | DOI:10.3920/9789004695467_006

1 Introduction

The first microbes that colonize the pig anatomy are acquired from the farrowing and/or maternal environment. In this process, the newborn is mainly seeded with microbes present in the sow's vaginal canal, feces, skin, oral cavity, colostrum, milk and any environmental microbes present in the surfaces in which the farrowing takes place. The specific characteristics of the maternal and/or farrowing environment, determine which microbes first colonize the pig at diverse anatomical sites; including those with the most critical physiological relevance for the animal such as the gastrointestinal or respiratory tracts.

Although many of those firstly acquired microbes may have a transient presence in the pig, they can exert a fine molecular crosstalk with an immature immune repertoire in the first days of life, possibly shaping physiological performance along the life course. Moreover, the microbial colonization mechanisms characterizing birth and the first days of life may also determine long-term microbial succession dynamics at major developmental and productive stages of the animal. As such, early microbial seeding, and the circumstances in which this seeding takes place, may be important focal points for pig producers and stakeholders, as far as increasing the productive and sanitary efficiency of swine production systems.

In this chapter, we discuss the environmental and physiological events that characterize early microbiome seeding in the pig and the consequences of different colonization dynamics for the physiological landscape of the animal, at early and late developmental stages. We focus these discussion points mainly in the context of the swine gastrointestinal microbiome, but also considering other mammalian models, including humans.

2 Early microbial seeding

At birth, successful microbial colonization in the gastrointestinal tract of mammals requires a significant repression of an immature, innate immune system (Belkaid and Hand, 2014). In this process, tolerance to maternal (primarily vaginal canal rectum and skin) and environmental microbial sources is hypothesized to depend on blunted immune responses, such as lower production of proinflammatory cytokines, repression of cells involved in inflammatory responses (*e.g.* invariant natural killer T cells) and a shift to increased regulatory responses by T and B cells (Figure 5.1) (Schaub *et al.*, 2008; Basha *et al.*, 2014; Zhou *et al.*, 2022). Although the exact mechanisms involved are not completely understood yet, the main purpose of early innate immune repression is to welcome and tolerate new commensals without mounting an overt inflammatory response to those microbes.

However, these blunted immune responses could also render the newborn more susceptible to infections. As such, the role of those first microbes as agents applying colonization resistance, as well as the functions of maternal colostrum and milk emerge as critical factors to support a healthy microbiome colonization process that also exerts protection from pathogens. For example, breast milk can stimulate proliferation of gut bacteria characterized by production of class II antimicrobial peptides such as Lactobacillus and Bifidobacterium, which may come from milk or from the vaginal tract (Pang *et al.*, 2021; Qi *et al.*, 2021; Amiri *et al.*, 2022; Javvadi *et al.*, 2022) (Figure 5.1). Such stimulation may be mediated by the prebiotic effect of milk oligosaccharides. In this respect, it has been hypothesized that differences in reproductive performance between gilts and sows may be mediated by the diversity, structure and kind of milk oligosaccharides that the piglets are exposed to from farrowing to weaning (Wei *et al.*, 2018). These studies also show that the composition of milk oligosaccharides in the sow decreases from colostrum to transitional milk and that they are hardly detected in 1–2 day old pig feces (Difilippo *et al.*, 2016).

Altogether, these observations show that the piglet gut microbiome rapidly and actively degrades milk oligosaccharides, underscoring the importance of early lactation periods and potentially, of colostrum and milk microbes, to modulate the piglet's intestinal integrity. The lactation process can also trigger an acidic intestinal environment through production of lactic acid and short chain fatty acids by fermentative microbes, a process known to inhibit the proliferation of intestinal pathogens (Insoft *et al.*, 1996; Peng and Biswas, 2017). Vertically transmitted innate and passive immune factors from milk such soluble TLR-2 and 4, CD14 cells, IgA, lactoferrin, fatty acids and defensins also increase neonatal immunity against early life infections (Gardner *et al.*, 2017; Stelzer *et al.*, 2021).

Breast milk may further contribute to a tolerogenic environment conducive to early colonization; this effect may be exerted by directly seeding the gut with microbes (*e.g.* bifidobacteria and lactobacilli) and milk organic acids (*e.g.* lactate, citrate, acetate and formate) (Zhang *et al.*, 2018) that activate anti-inflammatory T-reg cells receptors (Mosconi *et al.*, 2010; Wood *et al.*, 2021). In turn, T-reg cells can stimulate the expression of tight junctions and mucus by mucin-producing goblet cells, which are essential to avoid inflammation through strengthening the intestinal epithelial barrier (Shen and Turner, 2006; McAuley *et al.*, 2007) (Figure 5.1). The importance of milk in feeding the immature gut has been demonstrated by source tracking of piglet microbiomes at birth, pointing to maternal colostrum as the most significant source responsible for early colonization (Liu *et al.*, 2019). Specifically, microbiome profiling in mother/piglet dyads show that maternal milk and piglet gut microbiomes are more closely associated in genetically related dyads than unrelated ones.

These observations may explain, in part, why all piglets in a litter, which would be exposed to the same pioneer microbes at birth, would have similar gut microbiome composition, particularly at pre-weaning stages, compared with non-related piglets (Gaio *et al.*, 2022). Pioneer microbiome composition in the human gut has also been primarily associated with maternal milk, with evidence of vertical transmission of specific strains between mother/infant dyads (Murphy *et al.*, 2017). Furthermore, when comparing specific strains of bacteria in piglet gut microbiomes, littermates harbored distinct strains not found in non-siblings (Berry *et al.*, 2021), further supporting the hypothesis that piglets born in the same litter are exposed to, and colonized by, the same pioneer microorganisms.

Hence, the factors that define the composition of sow milk and colostrum, including their microbial milieu, may offer clues to understand patterns of early microbiome colonization in piglets, future growth performance and the risk of infectious diseases later in life (Liu *et al.*, 2019). It has been well established that piglet growth performance is strongly associated with milk/colostrum, yield, intake, fat, protein and mineral content (Lewis *et al.*, 1978; Chen *et al.*, 2019; Miao *et al.*, 2019; Hojgaard *et al.*, 2020). It is also well-known that lactation and milk/colostrum constituents, particularly lactose and galactooligosaccharides, modulate the piglet gut microbiome during this period, influencing the high abundances of lactic acid bacteria (*e.g. Lactobacillus*) observed preweaning (Salcedo *et al.*, 2016; De Rodas *et al.*, 2018; Guevarra *et al.*, 2018; Mu *et al.*, 2019; Law *et al.*, 2021).

The realization that sow milk/colostrum microbes influence early colonization of the piglet gut (Liu *et al.*, 2019) motivates research on the maternal factors that influence this microbiome. Recent reports point to a rich and dynamic milk/colostrum microbiome that changes in composition between colostrum, transitional and mature milk (Chen *et al.*, 2018). Furthermore, recently, the possibility of an enteromammary route that populates milk microbes from the sow's gastrointestinal microbiome has been proposed (Greiner *et al.*, 2022). The microbial milieu of milk/colostrum has been shown to include known gut commensals such as *Clostridium Ruminococcus*, *Lactobacillus*, *Akkermansia*, *Prevotella* and *Campylobacter*, among others, as well as *Corynebacterium* and *Streptococcus*, which are commonly found in the skin. All of those taxa have also been found to populate the piglet gut in the early days of life (De Rodas *et al.*, 2018; Berry *et al.*, 2021; Law *et al.*, 2021; Gaio *et al.*, 2022).

It has also been shown that there are strong associations between the abundance of microbes possibly transmitted from sow's milk, the composition of mucosal associated-microbes in ileum and jejunum, and the expression of genes involved in barrier and immune function (Liu *et al.*, 2019). These observations indicate that modulating the sow milk microbiomes, for instance, by targeting their gut microbes through nutrition may also influence piglet microbial seeding and performance. However, the mechanisms involving the role of milk-associated microbes on intestinal integrity early in life remain unknown. Additionally, the way specific milk components affect the quality of the milk microbiome, as influenced by management and nutrition, is unclear.

The subsequent load of microbes from environmental (farrowing crates) and other maternal sources (vaginal, feces and skin), in addition to those in milk, could play an critical role in shaping immune responses from the first days and weeks of life, enhancing the size of immune-associated tissues (*e.g.* Peyer's patches) and strengthening innate and acquired immune responses along the life course (Round and Mazmanian, 2009; Round *et al.*, 2011). Thus, the first encounter between the microbiome and the immune system marks the beginning of effective responses to infectious insults and the establishment of a homeostatic relationship between host and its microbiome. Indeed, It has been shown that the kind and nature of early microbial exposures is associated with intestinal barrier function and subsequent systemic inflammatory status in humans and mice models (Ege *et al.*, 2011; Martinez *et al.*, 2017; Combellick *et al.*, 2018).

This phenom is partially recapitulated in the so-called "hygiene hypothesis", which states that lack of exposure to pathogens early in life was the culprit of immune-associated and inflammatory disorders seen in the western world (Strachan, 1989). However, the "hygiene hypothesis" has been reevaluated to state that proper immune regulation is independent from hygiene, and that early exposure to a wide range of 'friendly' microbes – not *just* potentially infectious microorganisms, is what determines proper immune responses along the life course (Scudellari, 2017). These observations have different implications for swine production systems, mainly because creating a gateway to the entrance of diverse microbes in farrowing crates, including non-pathogenic microorganisms, can pose biosecurity risks. The implications of early microbial seeding for swine production systems and ways to increase early microbial exposure is still a point of debate and contention.

3 Maternal and environmental seeding sources of piglet gut microbiomes

Transmission of microorganisms to offspring from the maternal environment from specific body site sources has been well established in humans (Ferretti *et al.*, 2018). Microbial colonization of the newborns' gastrointestinal tract is hypothesized to occur only after birth, being significantly influenced by delivery mode (Dominguez-Bello *et al.*, 2010; Dominguez-Bello *et al.*, 2016), dependance on breast milk or formula (Al-Shehri *et al.*, 2016), and delivery environment (Combellick *et al.*, 2018). For example, it has been established that natural birth and Cesarean-section babies exhibit significant distinctions in gut microbiome composition (Dominguez-Bello *et al.*, 2010; Kim *et al.*, 2020), with gut microbial communities in newborns resembling those of either the mother's vaginal tract, feces or skin surface depending on delivery method. However, the distinct microbial seeding signals as triggered by mother-specific microbial profiles and delivery mode (Dominguez-Bello *et al.*, 2010), may be lost after 6 weeks of life, and are further shaped by niche selection (Chu *et al.*, 2017; Ferretti *et al.*, 2018).

Thus, although delivery mode, and hence delivery environment, initially influence the primordial microbiome of the newborn at specific body sites (*e.g.* skin microbes initially colonizing the gut microbiome of C-section delivered babies), the microbiome later develops characteristics associated with the selective environment the host lives in (Chu *et al.*, 2017). However, an important question relevant for swine production systems focuses on elucidating the effect that those first microbial colonizers have along the productive life curse of the animal, independent of subsequent taxonomic succession of the microbiome. Specifically, more work is needed on determining whether the first molecular cross-talk between early microbes and the immune system has implications for immunological memory and resistance to pathogen colonization during subsequent and vulnerable life stages (*e.g.* weaning). In humans, large epidemiological studies have shown that the long-term risk of chronic inflammatory diseases (*e.g.* Inflammatory Bowel Disease, Rheumatoid Arthritis, Coeliac Disease, and Diabetes Mellitus), some of which may have an infectious etiology and are associated with altered immune function, is greater in C-section delivered people, even 40 years after delivery (Kristensen and Henriksen, 2016; Andersen *et al.*, 2020). Hence, in theory, by manipulating early microbial exposures in pigs, we may be able to modulate immune function later in life.

In pigs, a handful of studies have shown that the housing method selected during farrowing and lactation, and hence the degree of early microbial exposures, strongly influences acquisition of pioneer piglet microbes, as well as physiological performance. Some of the first experiments demonstrating the importance of increased exposure to microbes early in life for pigs have focused on evaluating the effects of raising pigs outdoors or indoors. In one particular experiment, microbial exposures were restricted at birth, by raising pigs indoors or in experimentally isolated environments with antibiotics and deprived from sow contact; this experiment showed that restriction of early of microbial exposures was associated with lower abundance of different *Lactobacillus* strains (*L. reuteri*, *L. delbrueckii*, *L. amylovorus*, *L. johnsonii* and *L. mucosae*) in both fecal samples and ileal mucosa at day 56 after farrowing, in comparison with pigs raised outdoors (Mulder *et al.*, 2009). Reduction in *Lactobacillus* also corresponded with increased expression of IFNα/β transcription/signaling factors and MHC class I genes, which was accompanied by upregulation of several chemokine-coding genes such as *CCL2*, *CCL8*, *CCL28*, *CCR1*, *CXCR4* and *CXCL12* in ileum tissue. The results indicate, therefore, that restriction of microbial exposures at birth is associated with an immune system prone to increased inflammatory responses later in life, with effects observed beyond weaning.

Along these lines, a similar experiment showed that when pigs are farrowed either outdoors or indoors, and then moved to individual isolator units at 24 h after birth, pigs of indoor origin showed greater immune activation in the ileum compared with pigs farrowed outdoors (Mulder *et al.*, 2011). Increased immune activation in pigs farrowed indoors was demonstrated by upregulation of Type I IFN, IL-22, IL-23 and Th17 pathways particularly in the first days of life, with less pronounced effects at days 28 and 56. These observations demonstrate that the microbial exposures at birth are necessary to shape the immune environment, particularly in the very early stages of pig life when susceptibility to disease is greater. However, these experiments did not assess specific phenotypic responses to early microbial exposures, in the context of productive performance or specific health outcomes. That is, whether more microbial exposures early in life lead to more resistance to infectious diseases, for instance, at weaning.

Other experiments have controlled specific microbial sources at birth or at the early days of pig life, assessing for subsequent microbial succession and effects on performance parameters. For instance, pigs exposed to agricultural soil 4 days after farrowing until weaning developed a distinct gut microbiome composition through the middle of nursery, compared to pigs raised in conventional conditions (Vo *et al.*, 2017). Although at the end of nursery these distinction became less clear, one of the main characteristics of soil-raised pigs was an apparent accelerated maturation of the gut microbiome, characterized by increased microbial richness and diversity at weaning and end of nursery, and greater abundance of bacteria typically associated with fermentation roles, such as *Prevotella*, *Ruminococcaceae* and *Faecalibacterium* through the end of nursery. A more recent study indicates that soil rearing is also associated with upregulation of genes in peripheral blood mononuclear cells involved in increased T-cell activation by antigen presenting cells before weaning (De Souza *et al.*, 2022). Soil-reared pigs were also reported to show increased average daily gain throughout the nursery period and greater weight gain at the end of nursery. However, soil-rearing was associated with lower weight at weaning, right after increased exposure to environmental microbial sources.

Along these lines, and considering that soil rearing is unattainable in production systems in the US, another recent report indicates that degree of disinfection in the farrowing environment (crate surfaces) is associated with microbiome development from farrowing to weaning (Law *et al.*, 2021). Omitting disinfection in farrowing crates and just cleaning with hot water power-washing was associated with greater nasal and gut microbiome diversity at birth, and distinct microbiome composition from farrowing to weaning, compared with pigs farrowed in crates sanitized with a widely used microbicidal along with hot water power-washing. Lack of disinfection was also associated with increased abundance of environmental microbes through weaning, particularly higher levels of *Enhydrobacter* in the piglet nasal cavity, and with significantly (more than 4 fold) greater abundance of fecal *Lactobacillus* from mid lactation to weaning. However, lack of disinfection was also correlated with lower birth and weaning weights, just as soil-reared pigs, but lower average daily gain the last 4 weeks of the nursery period.

Differences in growth performance in association with microbial exposures at birth, generate questions about the role of early microbial colonization in the growth process, and relative energetic contributions to maintain both growth and a healthy immune system. For instance, in the experiments aforementioned, both soil rearing and farrowing without disinfection correlated with greater microbial diversity and abundance of microbes potentially associated with fermentative and immunomodulating roles (*e.g. Prevotella, Lactobacillus*), which has also been observed in the context of outdoor rearing (Mulder *et al.*, 2009; Mulder *et al.*, 2011). These microbiome characteristics may potentially denote a gut environment conducive to more efficient energy harvest, favorable intestinal integrity and, potentially, disease resistance (Valeriano *et al.*, 2017; Yang *et al.*, 2018; Shin *et al.*, 2019), key traits at the point of weaning. However, both systems were also associated with lower growth performance at weaning; thus, are there any advantages of increasing microbial exposures early in life?

4 Importance of early microbial exposures in swine production: prospectus

Mammalian gut tissues, especially those associated with immune function (*e.g.* gut-associated lymphoid tissue, Peyer patches), are energetically expensive to develop and maintain (Reeds *et al.*, 1993; Choct, 2009; Kirschman and Milligan-Myhre, 2019). Thus, evolving tolerance to an increased source of colonizing microbes at an early stage by suppressing pro-inflammatory immune responses and mounting regulatory ones (Belkaid and Hand, 2014), may incur in increased energy expenditure, at the cost of allocating that same energy for growth in preweaning stages (Figure 5.1). This may be the principle behind the mode of action of growth promoting antibiotics, which suppress microbial presence and activity to allow animals to spend more energy on somatic growth rather than spending energy on a mucosal immunity capable of tolerating and/or recognizing early colonizing microbes, including those with pathogenic roles (Brüssow, 2015; Broom, 2017). For example, in line with evidence from disinfected farrowing environments, it has been shown that creating a less sanitized environment preweaning, for instance, by increasing access to maternal feces, can increase blood leukocyte counts in piglets along the lactation stage, compared to piglets in crates where maternal feces were regularly cleaned (Aviles-Rosa *et al.*, 2019); however, it has also been shown that access to feces preweaning can also result in lower weaning weight (Nowland *et al.*, 2021). These observations are in line with an early immune system that is costly to maintain, energetically speaking, especially in conditions in which more microbial exposures are allowed; conversely, restriction of microbial exposures should facilitate growth performance, at the cost of a less mature and developed immune system (Figure 5.1), potentially more susceptible to infection. Nonetheless, these hypotheses have not been systematically tested.

Considering the line of evidence aforementioned, it is worth thinking about possible productive advantages of manipulating (increasing) early microbial exposures at birth. These potential advantages may not immediately translate to growth

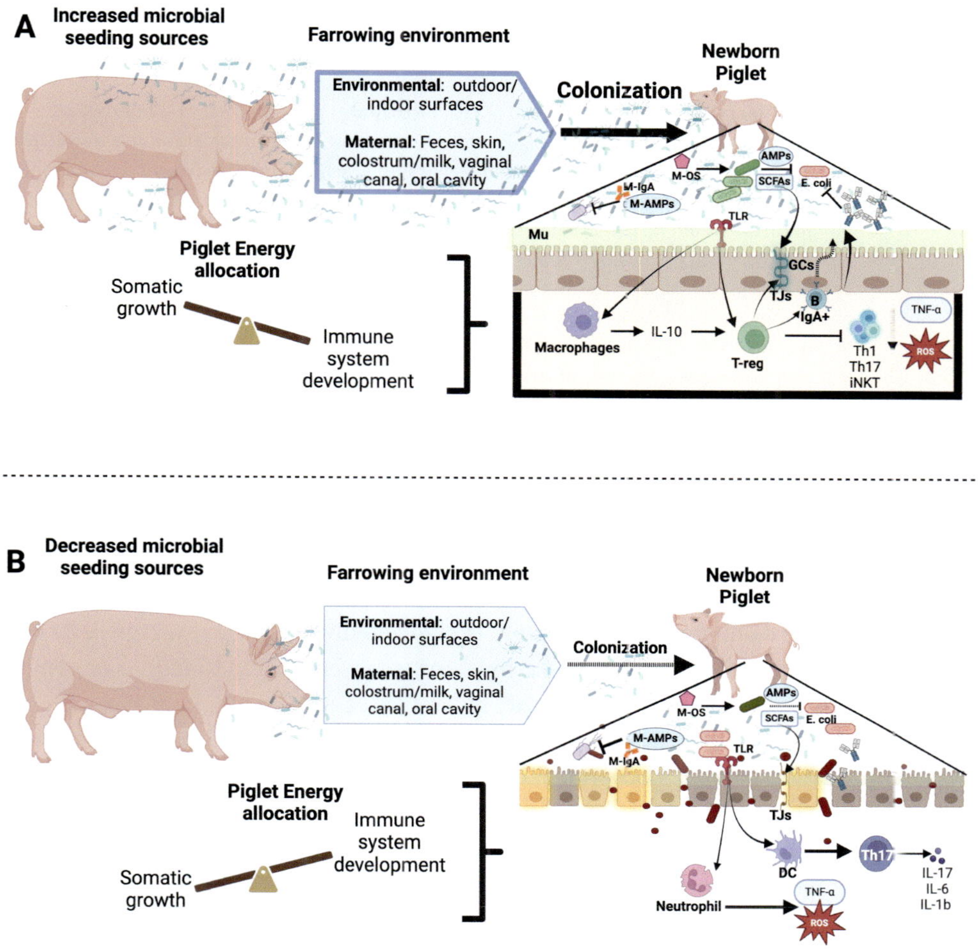

FIGURE 5.1 Early microbial conization and immune function in newborn piglets. **(A)** Increased microbial exposures at birth necessitate tolerance to new, incoming microbes, both potentially pathogenic and harmless/beneficial, but also containment to stop them from translocating and causing inflammation. Tolerance is mediated by blunting pro-inflammatory responses and promoting regulatory/anti-inflammatory ones. In this process, the role of interleukin 10 (IL-10) in stimulating T- regulatory (T_{reg}) cells and repressing cells that mediate inflammation, such as T helper cells 1 and 17 (Th1, Th17), and invariant natural killer T cells (iNKT). As a result, inflammatory mediators such as reactive oxygen species (ROS) and tumor necrosis factor alpha (TNF-α) are also repressed. T_{reg} cells also stimulate Immuno-globulin A producing B cells (IgA+B) to keep containing incoming colonizers, including potential pathogens such as E. coli. Milk constituents such as IgA (M-IgA) and antimicrobial peptides (M-AMPs) can further inhibit potential pathogens. Milk oligosaccharides (M-Os) can stimulate the growth of incoming beneficial commensals (in green), including those in milk such as lactobacilli and bifidobacteria. Incoming harmless commensals can also induce toll-like receptors (TLR) to shift immune responses to regulatory ones. These incoming microbes can produce AMPs that further inhibit pathogenic threats and can also produce short chain fatty acids (SCFAs) that result from metabolizing M-Os. SCFAs can induce intestinal integrity by stimulating the expression of barrier proteins such as tight junctions

(*cont.*)
(TJs) and mucin (Mu) by goblet cells (GCs). Overall increased microbial exposure and subsequent seeding ca alos exert colonization resistance against potential pathogens. This process of tolerance to increased microbial seeding and containment is energetically expensive; in theory, and given the high energetic costs of maintaining immune associated intestinal tissues, energy allocation would favor immune function at the expense of energy allocated for somatic growth. **(B)** In contrast, decreased microbial exposures in the farrowing environment would not necessitate the increased energetic effort of microbial tolerance and containment, and hence, more energy would be allocated to somatic growth by the piglet. However, less microbial exposures and seeding, and hence less microbial diversity in the gut, would result in less colonization resistance, fewer colonizing microbes that can produce SCFAs and AMPs, and potentially, selective proliferation of pathogenic microorganisms. Without protection of a more diverse microbial environment, sensing of these potentially pathogenic antigens can result in more translocation and also inflammation, mediated by factors such as TNF-α, ROS and shifts to Th1/Th17 to induce interleukin 17, 6 and 1 beta (1b), which further sustain a pro- inflammatory gut environment. Figure created with BioRender.com

promotion benefits at weaning; it is possible that those benefits are seen later along the productive cycle. For example, some studies have indicated that outdoor access pre-weaning or during nursery is associated with higher average daily gain, feed intake and body weight in the finishing stage, compared to indoor or conventional rearing systems (Gentry *et al.*, 2002; Lebret *et al.*, 2006; Lebret *et al.*, 2011). The advantages of increasing early microbial exposures may be analogous to those seen with nutritional restrictions in grower phases, which results in compensatory growth at finishing (Menegat *et al.*, 2020). In that sense, energetic tradeoffs between developing a robust immune system and emphasizing somatic growth, in favor of the former, may result in growth advantages later in life.

Early microbial seeding can also be viewed in the context of sanitary advantages. If animals spend more energy in developing a robust immune function at the expense of growth preweaning, in theory, they may be more resistant to infectious insults post weaning, than animals subjected to restriction of microbial exposures, as commonly done in farm standard operating procedures (*e.g.* through disinfection or conventional rearing). The basis for this potential advantage could lie in evidence that feeding dietary supplements with the potential of strengthening the immune system (*e.g.* natural organic acids and prebiotics) may help pigs overcome toxin challenges (*e.g.* lipopolysaccharide) more efficiently by dampening overt inflammation (Li *et al.*, 2006; Weber and Kerr, 2008; Wang *et al.*, 2011; Li and Kim, 2013). These protective effects have been shown to be greater in comparison with feeding growth promoting antibiotics (Wang *et al.*, 2011). However, these challenges are limited in scope as far as the immune responses involved and do not mimic natural disease spread conditions (*e.g.* lipopolysaccharide challenges) (Chen *et al.*, 2010; Pastorelli *et al.*, 2012).

In line with this evidence, it would be worthwhile to discuss the extent to which increasing microbial exposures in early life could render pigs more resistant to widespread disease outbreaks of importance today. It is clear that both increased sanitation and disinfection, which are rightly supported by biosecurity concerns, have helped control many diseases of importance today; but at the same time, it is also evident that threats from many viral, bacterial and protozoan infections, including some of importance to humans (*E.g.* Hepatitis E, Influenza, S. suis, Lawsonia, porcine reproductive and respiratory syndrome) have increased faster than expected in the last 15 years, regardless of increased vaccination efforts, or more strict biosecurity measures (VanderWaal and Deen, 2018). Also, although the ban on growth promoting antibiotics has triggered the commercialization a myriad of feed additives with potentially positive effects on health and performance, their reported effects is inconsistent across herds and studies, not to mention that the diverse mode of action among the multiple options available may target different productive or sanitary goals (Kiarie *et al.*, 2016; López-Gálvez *et al.*, 2021).

These issues open a window of opportunity to consider additional, out-of-the box strategies, such as manipulating early microbial exposures, to improve the efficiency of swine production systems and produce pigs that are more robust to disease outbreaks. However, systematic studies that evaluate the effect of alternative rearing systems, specifically on resistance to infection later in life are lacking. This prospective route of research does not advocate for eliminating disinfection nor lowering biosecurity measures in farrowing or rearing facilities. In addition, it is clear that outdoor or soil rearing is unattainable in most US production systems. Perhaps, the best of both worlds would be to allow animals to be exposed to sufficient microbial signals at birth, in conventional rearing systems, without neglecting biosecurity measures. In previous studies, elimination of disinfection from farrowing crates, still accompanied by hot power washing, was associated with potentially positive microbiome traits, such as higher bacterial diversity and and significantly increased abundance of potentially probiotic bacteria at weaning (*e.g.* *Lactobacillus*) (Law *et al.*, 2021). These results, obtained in university/experimental facilities and despite the lower weight performance outcomes as discussed above, indicate that it is actually possible to adjust current operating procedures in farms to stimulate piglet microbiomes at early life, without completely lowering the guard on biosecurity. However, we can learn more with additional studies that focus on alternative disinfection and cleaning procedures from commercial farms that are constantly exposed to pathogenic threats.

Fecal material transplant (FMT), could also be thought of as a way to introduce whole microbial communities in pigs, with the aim of improving specific performance parameters. FMT in recently weaned pigs is reported to alleviate post-weaning diarrhea, which was also associated with changes in the microbiome (*e.g.* higher diversity and prevalence of obligate anaerobes) and transcriptional changes in genes associated with immune function in diverse tissues (Su *et al.*, 2021). Other reports show that FMT, in suckling pigs (including addition of probiotics) are associated with greater concentrations of growth related hormones (IGF-1), greater concentrations of fecal short chain fatty acids, and increased numbers of secretory immunoglobulins (Cheng *et al.*, 2019; Xiang *et al.*, 2020). In weaned pigs, FMT is reported to have sanitary advantages as far as protective effects against African swine fever (Zhang *et al.*, 2020) and porcine circovirus (Niederwerder *et al.*, 2018). Positive effects of FMT on growth performance have also been reported (Wang *et al.*, 2019; Wang *et al.*, 2022). Interestingly, however, other reports indicate that when FMT is performed in preweaning stages, there could be negative impacts in intestinal morphology, nutrient absorption and even higher mortality, in addition to lowering microbiome diversity (McCormack *et al.*, 2018; Brunse *et al.*, 2019).

These observations may indicate that invasive microbial seeding or intentional introduction of complex microbial communities, when there is still an immature gastrointestinal tract and immune system, could be detrimental to suckling piglets. Implementing standard operating procedures for FMT also poses additional logistic and risk challenges, in the context of potential spread of pathogens and antibiotic resistance (Hu *et al.*, 2018; Canibe *et al.*, 2019). In contrast, natural exposure to environmental microbes or even allowing piglets to be more exposed to maternal microbial seeding sources may be less demanding from a logistics and even economic point of view, and pose less risks. Of course, biosecurity and other concerns related to the characteristics of the production system should prime.

5 Conclusion

The first microbes that colonize the piglet anatomy are of critical importance for the physiological development of the pig, primarily from an immune function perspective. In this process, the mother plays a fundamental role, as distinct microbial niches along her anatomy, and in milk and colostrum in particular, are the main source of this first seeding. However, the environment seems to also play a significant function. The importance of increasing sources of early microbial exposures, including environmental and intrinsically maternal factors, has been well documented; pointing to growth and immune benefits for the newborn pig and significant modulation of the microbiome. Although the physiological markers of these benefits can be detected from the first days after farrowing, growth performance gains appear to be delayed (compensatory). The evidence collected on this realm relies upon alternative rearing systems, including soil-exposure, lack of disinfection, access to feces and outdoor rearing. However, as many of these alternatives do not comply with the biosecurity and standards of most production systems, their application is still limited. Despite the existence of alternatives such as FMT or probiotics, which represent direct/intentional exposure to complex communities or simple microbial consortia, their effects remain inconsistent, including detrimental sanitary outcomes, or their application can pose economic, and logistic hurdles. Therefore, more research that focuses on ways to increase early microbial exposures in current and standard production facilities is needed. This work must illustrate the benefits of early exposures, not just in the context of growth performance, but perhaps, equally important, in terms of sanitary advantages when facing specific disease challenges.

Acknowledgements

We thank Dr. Lee Johnston at the University of Minnesota West Central Research and Outreach Center for useful discussions that lead to the development of this chapter. Authors were supported by funds provided by the Agricultural Research, Education and Extension Tech Transfer program (AGREETT) and by the MnDRIVE Global Food Ventures Graduate Student Professional Development Program.

References

Al-Shehri, S.S., E.L. Sweeney, D.M. Cowley, H.G. Liley, P.D. Ranasinghe, B.G. Charles, P.N. Shaw, D. Vagenas, J.A. Duley, and C.L. Knox. 2016. Deep sequencing of the 16S ribosomal RNA of the neonatal oral microbiome: a comparison of breast-fed and formula-fed infants. Sci. Rep. 6: 38309.

Amiri, S., R. Rezaei Mokarram, M. Sowti Khiabani, M. Rezazadeh Bari, and M. Alizadeh Khaledabad. 2022. Characterization of antimicrobial peptides produced by Lactobacillus acidophilus LA-5 and Bifidobacterium lactis BB-12 and their inhibitory effect against foodborne pathogens. LWT. 153: 112449.

Andersen, V., S. Möller, P.B. Jensen, F.T. Møller, and A. Green. 2020. Caesarean Delivery and Risk of Chronic Inflammatory Diseases (Inflammatory Bowel Disease, Rheumatoid Arthritis, Coeliac Disease, and Diabetes Mellitus): A Population Based Registry Study of 2,699,479 Births in Denmark During 1973–2016. Clin. Epidemiol. 12: 287–293.

Aviles-Rosa, E.O. 2019. The effect of swine's semiochemicals on pigs' behavior, physiology and production. Available from: https://ttu-ir.tdl.org/handle/2346/84994.

Aviles-Rosa, E.O., A. Rakhshandeh, and J.J. McGlone. 2019. Preliminary Study: Depriving Piglets of Maternal Feces for the First Seven Days Post-Partum Changes Piglet Physiology and Performance before and after Weaning. Animals (Basel). 9. doi:10.3390/ani9050268. Available from: http://dx.doi.org/10.3390/ani9050268.

Basha, S., N. Surendran, and M. Pichichero. 2014. Immune responses in neonates. Expert Rev. Clin. Immunol. 10: 1171–1184.

Belkaid, Y., and T.W. Hand. 2014. Role of the microbiota in immunity and inflammation. Cell. 157: 121–141.

Berry, A.S.F., M.K. Pierdon, A.M. Misic, M.C. Sullivan, K. O'Brien, Y. Chen, S.J. Murray, L.A. Ramharack, R.N. Baldassano, T.D. Parsons, and D.P. Beiting. 2021. Remodeling of the maternal gut microbiome during pregnancy is shaped by parity. Microbiome. 9: 146.

Broom, L.J. 2017. The sub-inhibitory theory for antibiotic growth promoters. Poult. Sci. 96: 3104–3108.

Brunse, A., L. Martin, T.S. Rasmussen, L. Christensen, M. Skovsted Cilieborg, M. Wiese, B. Khakimov, R. Pieper, D.S. Nielsen, P.T. Sangild, and T. Thymann. 2019. Effect of fecal microbiota transplantation route of administration on gut colonization and host response in preterm pigs. ISME J. 13: 720–733.

Brüssow, H. 2015. Growth promotion and gut microbiota: insights from antibiotic use. Environ. Microbiol. 17: 2216–2227.

Canibe, N., M. O'Dea, and S. Abraham. 2019. Potential relevance of pig gut content transplantation for production and research. J. Anim. Sci. Biotechnol. 10: 55.

Cheng, C.S., H.K. Wei, P. Wang, H.C. Yu, X.M. Zhang, S.W. Jiang, and J. Peng. 2019. Early intervention with faecal microbiota transplantation: an effective means to improve growth performance and the intestinal development of suckling piglets. Animal. 13: 533–541.

Chen, H., C. Bai, and X. Wang. 2010. The value of the lipopolysaccharide-induced acute lung injury model in respiratory medicine. Expert Rev. Respir. Med. 4: 773–783.

Chen, J., F. Zhang, W. Guan, H. Song, M. Tian, L. Cheng, K. Shi, J. Song, F. Chen, S. Zhang, F. Yang, C. Ren, and Y. Zhang. 2019. Increasing selenium supply for heat-stressed or actively cooled sows improves piglet preweaning survival, colostrum and milk composition, as well as maternal selenium, antioxidant status and immunoglobulin transfer. J. Trace Elem. Med. Biol. 52: 89–99.

Chen, W., J. Mi, N. Lv, J. Gao, J. Cheng, R. Wu, J. Ma, T. Lan, and X. Liao. 2018. Lactation Stage-Dependency of the Sow Milk Microbiota. Front. Microbiol. 9: 945.

Choct, M. 2009. Managing gut health through nutrition. Br. Poult. Sci. 50: 9–15.

Chu, D.M., J. Ma, A.L. Prince, K.M. Antony, M.D. Seferovic, and K.M. Aagaard. 2017. Maturation of the infant microbiome community structure and function across multiple body sites and in relation to mode of delivery. Nat. Med. 23: 314–326.

Combellick, J.L., H. Shin, D. Shin, Y. Cai, H. Hagan, C. Lacher, D.L. Lin, K. McCauley, S.V. Lynch, and M.G. Dominguez-Bello. 2018. Differences in the fecal microbiota of neonates born at home or in the hospital. Sci. Rep. 8: 15660.

De Rodas, B., B.P. Youmans, J.L. Danzeisen, H. Tran, and T.J. Johnson. 2018. Microbiome profiling of commercial pigs from farrow to finish. J. Anim. Sci. 96: 1778–1794.

De Souza, M.M., D.A. Koltes, H. Beiki, M.A. Sales, T. Tsai, C.V. Maxwell, J. Zhao, and J.E. Koltes. 2022. Early-Life Exposure of Pigs to Topsoil Alters miRNA and mRNA Expression in Peripheral Blood Mononuclear Cells. Front. Genet. 13: 886875.

Difilippo, E., F. Pan, M. Logtenberg, R.H.A.M. Willems, S. Braber, J. Fink-Gremmels, H.A. Schols, and H. Gruppen. 2016. Milk Oligosaccharide Variation in Sow Milk and Milk Oligosaccharide Fermentation in Piglet Intestine. J. Agric. Food Chem. 64: 2087–2093.

Dominguez-Bello, M.G., E.K. Costello, M. Contreras, M. Magris, G. Hidalgo, N. Fierer, and R. Knight. 2010. Delivery mode shapes the acquisition and structure of the initial microbiota across multiple body habitats in newborns. Proc. Natl. Acad. Sci. U.S.A. 107: 11971–11975.

Dominguez-Bello, M.G., K.M. De Jesus-Laboy, N. Shen, L.M. Cox, A. Amir, A. Gonzalez, N.A. Bokulich, S.J. Song, M. Hoashi, J.I. Rivera-Vinas, K. Mendez, R. Knight, and J.C. Clemente. 2016. Partial restoration of the microbiota of cesarean-born infants via vaginal microbial transfer. Nat. Med. 22: 250–253.

Ege, M.J., M. Mayer, A.-C. Normand, J. Genuneit, W.O.C.M. Cookson, C. Braun-Fahrländer, D. Heederik, R. Piarroux, E. von Mutius, and GABRIELA Transregio 22 Study Group. 2011. Exposure to environmental microorganisms and childhood asthma. N. Engl. J. Med. 364: 701–709.

Ferretti, P., E. Pasolli, A. Tett, F. Asnicar, V. Gorfer, S. Fedi, F. Armanini, D.T. Truong, S. Manara, M. Zolfo, F. Beghini, R. Bertorelli, V. De Sanctis, I. Bariletti, R. Canto, R. Clementi, M. Cologna, T. Crifò, G. Cusumano, S. Gottardi, C. Innamorati, C. Masè, D. Postai, D. Savoi, S. Duranti, G.A. Lugli, L. Mancabelli, F. Turroni, C. Ferrario, C. Milani, M. Mangifesta, R. Anzalone, A. Viappiani, M. Yassour, H. Vlamakis, R. Xavier, C.M. Collado, O. Koren, S. Tateo, M. Soffiati, A. Pedrotti, M. Ventura, C. Huttenhower, P. Bork, and N. Segata. 2018. Mother-to-Infant Microbial Transmission from Different Body Sites Shapes the Developing Infant Gut Microbiome. Cell Host Microbe. 24: 133–145. e5.

Gaio, D., M.Z. DeMaere, K. Anantanawat, G.J. Eamens, L. Falconer, T.A. Chapman, S. Djordjevic, and A.E. Darling. 2022. Phylogenetic diversity analysis of shotgun metagenomic reads describes gut microbiome development and treatment effects in the post-weaned pig. PLoS One. 17: e0270372.

Gardner, A.S., I.A. Rahman, C.T. Lai, A. Hepworth, N. Trengove, P.E. Hartmann, and D.T. Geddes. 2017. Changes in Fatty Acid Composition of Human Milk in Response to Cold-Like Symptoms in the Lactating Mother and Infant. Nutrients. 9. doi:10.3390/nu9091034. Available from: http://dx.doi.org/10.3390/nu9091034.

Gentry, J.G., J.J. McGlone, J.R. Blanton Jr, and M.F. Miller. 2002. Alternative housing systems for pigs: influences on growth, composition, and pork quality. J. Anim. Sci. 80: 1781–1790.

Greiner, L.L., D.C. Humphrey, S.N. Holland, C.J. Anderson, and S. Schmitz-Esser. 2022. The validation of the existence of the entero-mammary pathway and the assessment of the differences of the pathway between first and third parity sows. Transl Anim Sci. 6: txac047.

Guevarra, R.B., S.H. Hong, J.H. Cho, B.-R. Kim, J. Shin, J.H. Lee, B.N. Kang, Y.H. Kim, S. Wattanaphansak, R.E. Isaacson, M. Song, and H.B. Kim. 2018. The dynamics of the piglet gut microbiome during the weaning transition in association with health and nutrition. J. Anim. Sci. Biotechnol. 9: 54.

Hojgaard, C.K., T.S. Bruun, and P.K. Theil. 2020. Impact of milk and nutrient intake of piglets and sow milk composition on piglet growth and body composition at weaning. J. Anim. Sci. 98. doi:10.1093/jas/skaa060. Available from: http://dx.doi.org/10.1093/jas/skaa060.

Hu, J., L. Chen, Y. Tang, C. Xie, B. Xu, M. Shi, W. Zheng, S. Zhou, X. Wang, L. Liu, Y. Yan, T. Yang, Y. Niu, Q. Hou, X. Xu, and X. Yan. 2018. Standardized Preparation for Fecal Microbiota Transplantation in Pigs. Front. Microbiol. 9: 1328.

Insoft, R.M., I.R. Sanderson, and W.A. Walker. 1996. Development of immune function in the intestine and its role in neonatal diseases. Pediatr. Clin. North Am. 43: 551–571.

Javvadi, S.G., M. Kujawska, D. Papp, A.M. Gontarczyk, A. Jordan, M.A.E. Lawson, I.J. O'Neill, C. Alcon-Giner, R. Kiu, P. Clarke, N. Beraza, and L.J. Hall. 2022. A novel bacteriocin produced by Bifidobacterium longum subsp. infantis has dual antimicrobial and immunomodulatory activity. bioRxiv. 2022.01.27.477972. doi:10.1101/2022.01.27.477972. Available from: https://www.biorxiv.org/content/biorxiv/early/2022/02/25/2022.01.27.477972.

Kiarie, E., M.C. Walsh, and C.M. Nyachoti. 2016. Performance, digestive function, and mucosal responses to selected feed additives for pigs. J. Anim. Sci. 94: 169–180.

Kim, G., J. Bae, M.J. Kim, H. Kwon, G. Park, S.-J. Kim, Y.H. Choe, J. Kim, S.-H. Park, B.-H. Choe, H. Shin, and B. Kang. 2020. Delayed Establishment of Gut Microbiota in Infants Delivered by Cesarean Section. Front. Microbiol. 11: 2099.

Kirschman, L.J., and K.C. Milligan-Myhre. 2019. The Costs of Living Together: Immune Responses to the Microbiota and Chronic Gut Inflammation. Appl. Environ. Microbiol. 85. doi:10.1128/AEM.02147-18. Available from: http://dx.doi.org/10.1128/AEM.02147-18.

Kristensen, K., and L. Henriksen. 2016. Cesarean section and disease associated with immune function. J. Allergy Clin. Immunol. 137: 587–590.

Law, K., B. Lozinski, I. Torres, S. Davison, A. Hilbrands, E. Nelson, J. Parra-Suescun, L. Johnston, and A. Gomez. 2021. Disinfection of Maternal Environments Is Associated with Piglet Microbiome Composition from Birth to Weaning. mSphere. e0066321.

Lebret, B., M.C. Meunier-Salaün, A. Foury, P. Mormède, E. Dransfield, and J.Y. Dourmad. 2006. Influence of rearing conditions on performance, behavioral, and physiological responses of pigs to preslaughter handling, carcass traits, and meat quality. J. Anim. Sci. 84: 2436–2447.

Lebret, B., A. Prunier, N. Bonhomme, A. Foury, P. Mormède, and J.Y. Dourmad. 2011. Physiological traits and meat quality of pigs as affected by genotype and housing system. Meat Sci. 88: 14–22.

Lewis, A.J., V.C. Speer, and D.G. Haught. 1978. Relationship between Yield and Composition of Sows' Milk and Weight Gains of Nursing Pigs. J. Anim. Sci. 47: 634–638.

Li, J., and I.H. Kim. 2013. Effects of levan-type fructan supplementation on growth performance, digestibility, blood profile, fecal microbiota, and immune responses after lipopolysaccharide challenge in growing pigs. J. Anim. Sci. 91: 5336–5343.

Li, J., D.F. Li, J.J. Xing, Z.B. Cheng, and C.H. Lai. 2006. Effects of β-glucan extracted from Saccharomyces cerevisiae on growth performance, and immunological and somatotropic responses of pigs challenged with Escherichia coli lipopolysaccharide1. Journal of Animal Science. 84: 2374–2381. doi:10.2527/jas.2004-541. Available from: http://dx.doi.org/10.2527/jas.2004-541.

Liu, H., X. Zeng, G. Zhang, C. Hou, N. Li, H. Yu, L. Shang, X. Zhang, P. Trevisi, F. Yang, Z. Liu, and S. Qiao. 2019. Maternal milk and fecal microbes guide the spatiotemporal development of mucosa-associated microbiota and barrier function in the porcine neonatal gut. BMC Biol. 17: 106.

López-Gálvez, G., M. López-Alonso, A. Pechova, B. Mayo, N. Dierick, and J. Gropp. 2021. Alternatives to antibiotics and trace elements (copper and zinc) to improve gut health and zootechnical parameters in piglets: A review. Anim. Feed Sci. Technol. 271: 114727.

Martinez, K.A., 2nd, J.C. Devlin, C.R. Lacher, Y. Yin, Y. Cai, J. Wang, and M.G. Dominguez-Bello. 2017. Increased weight gain by C-section: Functional significance of the primordial microbiome. Sci Adv. 3: eaao1874.

McAuley, J.L., S.K. Linden, C.W. Png, R.M. King, H.L. Pennington, S.J. Gendler, T.H. Florin, G.R. Hill, V. Korolik, and M.A. McGuckin. 2007. MUC1 cell surface mucin is a critical element of the mucosal barrier to infection. J. Clin. Invest. 117: 2313–2324.

McCormack, U.M., T. Curião, T. Wilkinson, B.U. Metzler-Zebeli, H. Reyer, T. Ryan, J.A. Calderon-Diaz, F. Crispie, P.D. Cotter, C.J. Creevey, G.E. Gardiner, and P.G. Lawlor. 2018. Fecal Microbiota Transplantation in Gestating Sows and Neonatal Offspring Alters Lifetime Intestinal Microbiota and Growth in Offspring. mSystems. 3. doi:10.1128/mSystems.00134-17. Available from: http://dx.doi.org/10.1128/mSystems.00134-17.

Menegat, M.B., S.S. Dritz, M.D. Tokach, J.C. Woodworth, J.M. DeRouchey, and R.D. Goodband. 2020. A review of compensatory growth following lysine restriction in grow-finish pigs. Transl Anim Sci. 4: txaa014.

Miao, J., D. Adewole, S. Liu, P. Xi, C. Yang, and Y. Yin. 2019. Tryptophan Supplementation Increases Reproduction Performance, Milk Yield, and Milk Composition in Lactating Sows and Growth Performance of Their Piglets. J. Agric. Food Chem. 67: 5096–5104.

Mosconi, E., A. Rekima, B. Seitz-Polski, A. Kanda, S. Fleury, E. Tissandie, R. Monteiro, D.D. Dombrowicz, V. Julia, N. Glaichenhaus, and V. Verhasselt. 2010. Breast milk immune complexes are potent inducers of oral tolerance in neonates and prevent asthma development. Mucosal Immunol. 3: 461–474.

Mu, C., Z. Cai, G. Bian, Y. Du, S. Ma, Y. Su, L. Liu, J. Voglmeir, R. Huang, and W. Zhu. 2019. New Insights into Porcine Milk N-Glycome and the Potential Relation with Offspring Gut Microbiome. J. Proteome Res. 18: 1114–1124.

Mulder, I.E., B. Schmidt, M. Lewis, M. Delday, C.R. Stokes, M. Bailey, R.I. Aminov, B.P. Gill, J.R. Pluske, C.-D. Mayer, and D. Kelly. 2011. Restricting microbial exposure in early life negates the immune benefits associated with gut colonization in environments of high microbial diversity. PLoS One. 6: e28279.

Mulder, I.E., B. Schmidt, C.R. Stokes, M. Lewis, M. Bailey, R.I. Aminov, J.I. Prosser, B.P. Gill, J.R. Pluske, C.-D. Mayer, C.C. Musk, and D. Kelly. 2009. Environmentally-acquired bacteria influence microbial diversity and natural innate immune responses at gut surfaces. BMC Biol. 7: 79.

Murphy, K., D. Curley, T.F. O'Callaghan, C.-A. O'Shea, E.M. Dempsey, P.W. O'Toole, R.P. Ross, C.A. Ryan, and C. Stanton. 2017. The Composition of Human Milk and Infant Faecal Microbiota Over the First Three Months of Life: A Pilot Study. Sci. Rep. 7: 40597.

Niederwerder, M.C., L.A. Constance, R.R.R. Rowland, W. Abbas, S.C. Fernando, M.L. Potter, M.A. Sheahan, T.E. Burkey, R.A. Hesse, and A.G. Cino-Ozuna. 2018. Fecal Microbiota Transplantation Is Associated With Reduced Morbidity and Mortality in Porcine Circovirus Associated Disease. Front. Microbiol. 9: 1631.

Nowland, T.L., R.N. Kirkwood, K.J. Plush, M.D. Barton, and V.A. Torok. 2021. Exposure to maternal feces in lactation influences piglet enteric microbiota, growth, and survival preweaning. J. Anim. Sci. 99. doi:10.1093/jas/skab170. Available from: http://dx.doi.org/10.1093/jas/skab170.

Pang, B., H. Jin, N. Liao, J. Li, C. Jiang, D. Shao, and J. Shi. 2021. Lactobacillus rhamnosus from human breast milk ameliorates ulcerative colitis in mice via gut microbiota modulation. Food Funct. 12: 5171–5186.

Pastorelli, H., J. van Milgen, P. Lovatto, and L. Montagne. 2012. Meta-analysis of feed intake and growth responses of growing pigs after a sanitary challenge. Animal. 6: 952–961.

Peng, M., and D. Biswas. 2017. Short chain and polyunsaturated fatty acids in host gut health and foodborne bacterial pathogen inhibition. Crit. Rev. Food Sci. Nutr. 57: 3987–4002.

Qi, C., M. Ding, S. Li, Q. Zhou, D. Li, R. Yu, and J. Sun. 2021. Sex-dependent modulation of immune development in mice by secretory IgA-coated Lactobacillus reuteri isolated from breast milk. J. Dairy Sci. 104: 3863–3875.

Reeds, P.J., D.G. Burrin, T.A. Davis, and M.L. Fiorotto. 1993. Postnatal growth of gut and muscle: competitors or collaborators. Proc. Nutr. Soc. 52: 57–67.

Round, J.L., S.M. Lee, J. Li, G. Tran, B. Jabri, T.A. Chatila, and S.K. Mazmanian. 2011. The Toll-like receptor 2 pathway establishes colonization by a commensal of the human microbiota. Science. 332: 974–977.

Round, J.L., and S.K. Mazmanian. 2009. The gut microbiota shapes intestinal immune responses during health and disease. Nat. Rev. Immunol. 9: 313–323.

Salcedo, J., S.A. Frese, D.A. Mills, and D. Barile. 2016. Characterization of porcine milk oligosaccharides during early lactation and their relation to the fecal microbiome. J. Dairy Sci. 99: 7733–7743.

Schaub, B., J. Liu, I. Schleich, S. Höppler, C. Sattler, and E. von Mutius. 2008. Impairment of T helper and T regulatory cell responses at birth. Allergy. 63: 1438–1447.

Scudellari, M. 2017. News Feature: Cleaning up the hygiene hypothesis. Proc. Natl. Acad. Sci. U.S.A. 114: 1433–1436.

Shen, L., and J.R. Turner. 2006. Role of epithelial cells in initiation and propagation of intestinal inflammation. Eliminating the static: tight junction dynamics exposed. Am. J. Physiol. Gastrointest. Liver Physiol. 290:G577–82.

Shin, D., S.Y. Chang, P. Bogere, K. Won, J.-Y. Choi, Y.-J. Choi, H.K. Lee, J. Hur, B.-Y. Park, Y. Kim, and J. Heo. 2019. Beneficial roles of probiotics on the modulation of gut microbiota and immune response in pigs. PLoS One. 14: e0220843.

Stelzer, I.A., C. Urbschat, S. Schepanski, K. Thiele, I. Triviai, A. Wieczorek, M. Alawi, D. Ohnezeit, J. Kottlau, J. Huang, N. Fischer, H.-W. Mittrücker, M.E. Solano, B. Fehse, A. Diemert, F.R. Stahl, and P.C. Arck. 2021. Vertically transferred maternal immune cells promote neonatal immunity against early life infections. Nat. Commun. 12: 4706.

Strachan, D.P. 1989. Hay fever, hygiene, and household size. BMJ. 299: 1259–1260.

Su, Y., X. Li, D. Li, and J. Sun. 2021. Fecal Microbiota Transplantation Shows Marked Shifts in the Multi-Omic Profiles of Porcine Post-weaning Diarrhea. Front. Microbiol. 12: 619460.

Valeriano, V.D.V., M.P. Balolong, and D.-K. Kang. 2017. Probiotic roles of Lactobacillus sp. in swine: insights from gut microbiota. J. Appl. Microbiol. 122: 554–567.

VanderWaal, K., and J. Deen. 2018. Global trends in infectious diseases of swine. Proc. Natl. Acad. Sci. U.S.A. 115: 11495–11500.

Vo, N., T.C. Tsai, C. Maxwell, and F. Carbonero. 2017. Early exposure to agricultural soil accelerates the maturation of the early-life pig gut microbiota. Anaerobe. 45: 31–39.

Wang, J.P., J.S. Yoo, H.D. Jang, J.H. Lee, J.H. Cho, and I.H. Kim. 2011. Effect of dietary fermented garlic by Weissella koreensis powder on growth performance, blood characteristics, and immune response of growing pigs challenged with Escherichia coli lipopolysaccharide. J. Anim. Sci. 89: 2123–2131.

Wang, X., T. Tsai, F. Deng, X. Wei, J. Chai, J. Knapp, J. Apple, C.V. Maxwell, J.A. Lee, Y. Li, and J. Zhao. 2019. Longitudinal investigation of the swine gut microbiome from birth to market reveals stage and growth performance associated bacteria. Microbiome. 7: 109.

Wang, X., T. Tsai, B. Zuo, X. Wei, F. Deng, Y. Li, C.V. Maxwell, H. Yang, Y. Xiao, and J. Zhao. 2022. Donor age and body weight determine the effects of fecal microbiota transplantation on growth performance, and fecal microbiota development in recipient pigs. J. Anim. Sci. Biotechnol. 13: 49.

Weber, T.E., and B.J. Kerr. 2008. Effect of sodium butyrate on growth performance and response to lipopolysaccharide in weanling pigs. J. Anim. Sci. 86: 442–450.

Wei, J., Z.A. Wang, B. Wang, M. Jahan, Z. Wang, P.C. Wynn, and Y. Du. 2018. Characterization of porcine milk oligosaccharides over lactation between primiparous and multiparous female pigs. Sci. Rep. 8: 4688.

Wood, H., A. Acharjee, H. Pearce, M.N. Quraishi, R. Powell, A. Rossiter, A. Beggs, A. Ewer, P. Moss, and G. Toldi. 2021. Breastfeeding promotes early neonatal regulatory T-cell expansion and immune tolerance of non-inherited maternal antigens. Allergy. 76: 2447–2460.

Xiang, Q., X. Wu, Y. Pan, L. Wang, Y. Guo, C. Cui, L. Hu, L. Zhu, J. Peng, and H. Wei. 2020. Early Intervention Using Fecal Microbiota Transplantation Combined with Probiotics Influence the Growth Performance, Diarrhea, and Intestinal Barrier Function of Piglets. NATO Adv. Sci. Inst. Ser. E Appl. Sci. 10: 568.

Yang, J., K. Qian, C. Wang, and Y. Wu. 2018. Roles of Probiotic Lactobacilli Inclusion in Helping Piglets Establish Healthy Intestinal Inter-environment for Pathogen Defense. Probiotics Antimicrob. Proteins. 10: 243–250.

Zhang, J., F. Rodríguez, M.J. Navas, M. Costa-Hurtado, V. Almagro, L. Bosch-Camós, E. López, R. Cuadrado, F. Accensi, S. Pina-Pedrero, J. Martínez, and F. Correa-Fiz. 2020. Fecal microbiota transplantation from warthog to pig confirms the influence of the gut microbiota on African swine fever susceptibility. Sci. Rep. 10: 17605.

Zhang, S., F. Chen, Y. Zhang, Y. Lv, J. Heng, T. Min, L. Li, and W. Guan. 2018. Recent progress of porcine milk components and mammary gland function. J. Anim. Sci. Biotechnol. 9: 77.

Zhou, H., X. Lu, J. Huang, P. Jordan, S. Ma, L. Xu, F. Hu, H. Gui, H. Zhao, Z. Bai, H.P. Redmond, J.H. Wang, and J. Wang. 2022. Induction of Trained Immunity Protects Neonatal Mice Against Microbial Sepsis by Boosting Both the Inflammatory Response and Antimicrobial Activity. J. Inflamm. Res. 15: 3829–3845.

CHAPTER 6

Pesticides and preservatives: potentially harmful residues in feed for gut health in pigs

Barbara U. Metzler-Zebeli

Unit Nutritional Physiology, Institute of Physiology, Pathophysiology and Biosphysics, Department of Biomedical Sciences, University of Veterinary Medicine, Vienna, Austria; barbara.metzler@vetmeduni.ac.at

Abstract

Despite establishment of maximum tolerance levels, adverse effects of xenobiotic substances in pig feed rise questions about their safety for gut health. The gut represents the primary site of exposure to toxic agents and the gut microbiota may act as a crucial factor regulating xenobiotics' toxicity. Pesticides protect crop yields, lower the cost of feed production and reduce the carbon footprint. Their widespread use during the growing and harvesting season leads to residues in pig feed. Synthetic preservatives are added during diet mixing to prevent the formation of radical oxygen species and consequent lipid peroxidation and can be regularly found in pig diets. The aim of the present review is to address, based on the published literature, whether dietary pesticide residues and preservatives affect the gut homeostasis and function in pigs. However, literature on the potential effects of pesticide residues and synthetic preservatives on pig health and gut function are extremely scarce, demanding for experiments. Available evidence from pigs but also from other animals indicates that the selected insecticides (*i.e.* deltamethrin, cybermethrin and chlorpyrofos), the herbicide glyphosate as well as the most commonly used preservatives may alter the gut microbial community and compromise the gut barrier and antioxidative capacities.

Keywords

insecticides – herbicides – preservatives – gut microbiota – barrier function – inflammation – oxidative stress – pig

1 Introduction

Application of pesticides increases crop yields and economic benefits. As a consequence, different kinds of pesticides are extensively used worldwide in crop

© BARBARA U. METZLER-ZEBELI, 2024 | DOI:10.3920/9789004695467_007

production. Due to their ubiquitous application, humans, livestock and wildlife are exposed to different dosages throughout life (Donkor *et al.*, 2016; Sørensen *et al.*, 2021). Although xenobiotics enter the body also via lungs and skin, most exposure occurs after consumption of contaminated food and water. In modern conventional farming, the major ingredients of pig diets are cereals, high-protein legumes and/or legume or oil seed meals; all of which are frequently treated with pesticides during the growing and harvesting season. As a result, farm animals are chronically exposed to low levels during their lifetime. Pesticide residues have received considerable attention with regard to their entry into the food chain lately (Grados *et al.*, 2022), whereas their effects on animal-protein producing livestock animals, especially in pigs, have been insufficiently explored so far. In the feed sector, the presence of pesticides on cereals, legumes and oil seeds is mostly ignored during diet manufacturing, although animal protein (*e.g.* pork, beef and eggs) is regularly monitored for pesticide residues (EFSA, 2021). In addition to pesticides, other chemicals are added to the feed during diet mixing for preservation purposes to protect lipids from autoxidation (Xu *et al.*, 2021). In humans, evidence is growing that xenobiotic-induced changes in the gut microbiota can affect the development of the immune system and have been associated with an increased risk for the development of chronic immune mediated conditions, such as inflammatory bowel disease (Altajar and Moss, 2020; Grados *et al.*, 2022). In fact, pesticide residues constitute one of the main groups of chemicals by which the intestinal microbiota is regularly confronted (Yuan *et al.*, 2019). Accordingly, residues from pesticides and preservatives in complete feed for livestock animals may compromise gut health and trigger inflammation (Tejada, 2009; Wang *et al.*, 2022). This stands in contrast to dietary and management efforts to avert gut dysbiosis in young pigs to reduce the therapeutic use of antibiotics (Radcliffe, 2011; Metzler-Zebeli, 2022). The existing knowledge of the impact of pesticides and preservatives on the gut mostly originate from *in vitro* studies or studies with other nontarget animals, whereas evidence for pigs is still scarce. Therefore, the present review aims at raising the awareness about xenobiotics in pig diets. This chapter will summarize *in vivo* effects on the gut for a few selected pesticides and synthetic preservatives. Due to the lack of data for the target species, findings from other monogastric animals will be briefly presented instead.

2 Gut homeostasis

The gut epithelium forms the physical and biological barrier to prevent the entry of xenobiotics into the body and hence is exposed to potential xenobiotics in feed and water first. The impact that a lifelong exposure has on the gut of fattening pigs and reproductive sows still needs to be shown. The question that need to be addressed is whether pesticides and synthetic preservatives render the pigs more susceptible to gut dysbiosis and leaky gut. The homeostatic condition in the gut is primarily based on three components: 1) stable immune system, 2) well-functioning mucosal barrier and 3) balanced gut microbial ecosystem, which are vulnerable to external and internal stressors especially in the early life phases, such as the suckling and early postweaning period. The immediate time after weaning is generally associated with higher incidence of diarrhea due to loss of immune tolerance towards the newly developing gut microbiota and/or dietary compounds, which may be accompanied by an increased proliferation of opportunistic pathogens, such as enterotoxigenic *Escherichia coli* (Pié *et al.*, 2004; Heo *et al.*, 2013). Therefore, efforts are made to support gut barrier function and to supply nutrients rapidly for efficient absorption along the gastrointestinal tract (Radcliffe, 2011; Metzler-Zebeli, 2022). It can be assumed that xenobiotics, such as pesticides, raise the oxidative stress at the gut mucosa; as a consequence, susceptible individuals may be more prone to gut damage. Gut health-disrupting effects of pesticides are suspected to cause chronic low level inflammation in animals and humans (Giambo *et al.*, 2021). In general, infants are more vulnerable to pesticides than adults due to the possibility of disruption of developmental processes (Giambo *et al.*, 2021). The gut microbiota, barrier and immune functions co-develop after birth, establishing a homeostatic state (Arnaud *et al.*, 2020), which is a critical priming period where the interplay is most sensitive to environmental influence. These concerns are very relevant in pig production as fattening pigs barely reach adulthood. The mechanisms of interaction between pesticides and the gut microbiota depend on the nature of the pesticide. Theoretical modes of action how pesticides can interact with the host at gut level are illustrated in Figure 6.1.

Pesticides can affect gut functioning in multiple ways. They can alter the gut microbial composition and metabolic activity with consequences for the microbe-host interactions (Figure 6.1A). However, too little is still known about the relevance of the microbial perturbations for pig health and gut functioning. Pesticides may disrupt gut barrier function by opening up tight-junction proteins and desmosomal complexes between enterocytes (Figure 6.1B); thereby allowing the luminal antigens to enter the body which ultimately trigger inflammatory pathways. On the other hand, microbial activity in the gut lumen, and hepatic metabolism can help in the detoxification of pesticides via oxidation, reduction, hydrolysis and deconjugation. After hepatic metabolism, metabolites can be secreted into the gut lumen again as part of the bile, and by this influencing the microbial community and activity (Figure 6.1C).

2.1 *Gut microbiota*

In pigs, bacteria, archaea, fungi and viruses reside in the gut and have mutualistic relationships with their host. The porcine gut microbiome is complex and diverse; about 10^6 to 10^{12} symbiotic microbial cells live in the porcine gut, whereby species

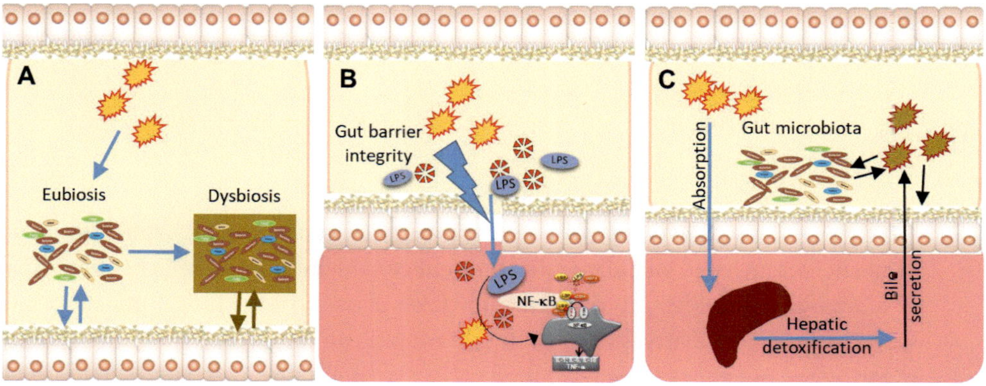

FIGURE 6.1 Interactions between pesticides, gut microbiota and the host. Pesticides can impair A) gut microbiota composition and activity; and B) barrier function with adverse effects on host. C) Absorption of pesticides, hepatic detoxification and secretion of metabolites with bile

abundances and diversity increase along the gastrointestinal tract (Metzler-Zebeli, 2022). Bacteria are the major component of the porcine microbiome and target of studies. A meta-analysis in 2017 divided the porcine gut bacteriome into 35 phyla and 887 genera across all gut segments from previously published sequences using high-throughput sequencing technologies (Holman *et al.*, 2017). The gut microbiota acts at the interface between food and the host, and is therefore vital for host health. In addition to their roles in nutrient assimilation, production of short-chain fatty acids (SCFA) and vitamins, inhibition and prevention of colonization by pathogens and translocation of toxins, the gut microbiota is indispensable for the gut development and maturation including the host immune system (Arnaud *et al.*, 2020; Metzler-Zebel., 2022). The loss of stability of a healthy gut microbiota leads to a dysbiotic state which renders the pig more susceptible to gut inflammation and loss of barrier function.

Strong evidence is emerging that the gut microbiota is an important factor regulating pesticides' toxicity in non-target organisms including humans (Giambo *et al.*, 2021). The toxicological relevance of the gut microbiota-xenobiotic interaction is of great concern as chemicals may disrupt gut microbiota functions, with a potential impairment of the gut homeostasis of the host (Defois *et al.*, 2018). Because exposure to foodborne chemicals occurs via their oral intake, the gut microbiota residing in the stomach and small intestines are likely the first to be exposed to the xenobiotic compounds before the communities in the large intestine are affected by the toxin either directly or indirectly after host or microbial metabolism. Nevertheless, literature is mainly available for large intestinal and fecal bacterial communities, neglecting effects that occur in the proximal segments of the gut and their interference in digestive processes. Dietary components, such as soluble fiber, fat and protein, can modify stomach emptying (Metzler-Zebeli *et al.*, 2019), thereby modulating the exposure time to the toxin.

3 Pesticides

To increase crop yields and improve the quality of plant products, different kinds of pesticides are used globally which brought considerable economic benefit (Yuan *et al.*, 2019). Pesticides are divided into three classes: herbicides, fungicides and insecticides. Recent surveys for residues have indicated that amounts of pesticides in food/feed are at or below legal limits [*i.e.* maximum residue levels (MRLs)] and actual intakes for humans or livestock are much lower than these conservative exposure limits (Vicini *et al.*, 2019). Nevertheless and despite not surpassing legal limits, plasma and urine levels have lately demonstrated the accumulation of pesticides in the human body (Taira *et al.*, 2013; Velmurugan *et al.*, 2017). The establishment of upper limits, derived after a comprehensive assessment of the active substances and intended uses of the pesticide, should prevent harmful exposure to pesticides. However, evidence is accumulating about the negative impact of lifelong exposure to low levels of residual pesticides in food and feed. The maximal allowances of pesticides in food and feed are, for instance, regulated in the European Union by the Regulation European Commission No 396/2005; (EFSA, 2021). In the US, it is the FDA's Center for Veterinary Medicine program that monitors animal food consumed by livestock and poultry animals which become or produce food for human consumption (FDA, 2022). The pyrethroid insecticides cypermethrin and deltamethrin, the organochloride dieldrin, the phosphorothioate pirimiphos-methyl and the organophosphate chlorpyrifos were detected in pork samples (fat and tissues) in 2019 (EFSA, 2021), probably due to carry over of their use in feed. The EFSA recommended in 2021 to continue the monitoring of animal fat products partly based on the fact that the rate of residual pesticides in swine fat remained steady in the surveillance periods 2016 and 2019 in the European Union (EFSA, 2021). The FDA reported that 72% of cereals aimed for animal food contained pesticide residues, though in legal limits (FDA, 2022). Although the detected levels were nonviolative, glyphosate was found in the greatest number of samples (14.2%), followed by piperonyl butoxide (13.3% of samples; a synergist used in combination with pyrethrins for control of insects), ethoxyquin (10.8% of samples; specific uses as preservative in animal foods), and malathion (11.8% of samples; organophosphate insecticide; FDA, 2022). Notably, due to their toxicity to the environment, dieldrin has been banned in 2001 by the Stockholm convention, whereas chlorpyrifos was banned by the European Union in 2020 and in the US in 2022. Due to their long half-life, they are still present in the environment or feedstuffs are imported from other countries. In the following, effects of the insecticides cypermethrin, deltamethrin, chlorpyrofos and the herbicide glyphosate on the gut homeostasis will be discussed in more details.

3.1 *Insecticides*

3.1.1 Cypermethrin and deltamethrin

Pyrethroid insecticides have been extensively used in agriculture and urban areas because of their high efficiency and low mammalian toxicity (McCoy *et al.*, 2012). Despite their use in large-scale commercial agricultural applications, physiological effects of cypermethrin and deltamethrin have been mainly studied in freshwater ecosystems due to their high toxicity for nontarget aquatic organisms (Lidova *et al.*, 2016; Zhao *et al.*, 2019; Wu *et al.*, 2022). Both compounds are type II pyrethroid insecticides, which have received substantial attention due to their neurotoxicity and their property to cross the blood-brain barrier (Ali *et al.*, 2020). Deltamethrin also finds application against ecto-parasites in veterinary medicine, including livestock, with the skin as the entry route. Nevertheless, knowledge on the impact of these insecticides in higher vertebrates, such as livestock animals, is still rare, and available literature from other nontarget animals will be reported instead. The toxic effect of pyrethroids in insects is due to their prolonged activation of voltage-gated sodium channels in target and nontarget organisms, which are critical for the electrical transmission in the nervous system (Narahashi *et al.*, 1996; Spencer *et al.*, 2001; Davis *et al.*, 2007; Zhorov and Dong, 2022). Mechanistically, cypermethrin seems to act via nuclear factor erythrocyte derived-2 like 2 (Nrf2) activation in carps, thereby upregulating cellular protection pathways and apoptosis (Zhao *et al.*, 2019); a receptor also expressed at the gut mucosa of pigs (Meng *et al.*, 2022). After consumption, deltamethrin is in part rapidly absorbed, whereas the other part is intestinally hydrolyzed to metabolites 3-(2,2-dibromvinyl)-2,2-dimethylcyclopropic acid und 3-phenoxybenzoic acid (Bundesgesundheitsamt, 2022). Both metabolites have a role as xenobiotic on their own. It can be assumed that the gut microbiota is involved in the intestinal metabolism of the pyrethroids. Evidence for microbial detoxification of the pyrethroids exists for bacterial strains in soil, such as *Bacillus*, *Pseudomonas* and *Stenotrophomonas* (Halden *et al.*, 1999; Chen *et al.*, 2011, 2012). Alterations in gut mucosal functioning and the taxonomic composition of the microbiota after pyrethroid exposure have been reported for freshwater fish (*e.g.* Wu *et al.*, 2022) and bats (*e.g.* Oliveira *et al.*, 2022). In crucian carp, chronic exposure to deltamethrin affected all major bacterial phyla (Table 6.1; Wu *et al.*, 2022). Deltamethrin exposure increased Proteobacteria and Verrucomicrobiota, whereas Fusobacteria decreased, disturbing the gut microbial-host communication and contributing to the pesticide-induced mucosal inflammation (Wu *et al.*, 2022). These phyla are dominant in the porcine gut (Holman *et al.*, 2017); therefore, similar effects on the gut microbiome can be supposed. Moreover, deltamethrin raised the oxidative stress and apoptosis at the gut wall that triggered proinflammatory signaling cascades and compromised barrier function in the carps (Wu *et al.*, 2022). Observations made in fruit-eating bats (*Artibeus lituratus*; Oliveira *et al.*, 2022) support the findings and the gut toxicity of deltamethrin in mammals. The gut of bats exposed to 0.02 and 0.04 mg deltamethrin/kg of papaya responded with inflammatory infiltration, degeneration and necrosis at the mucosal lining and goblet cell hyperplasia as the most frequent pathologies. The latter was accompanied with increased secretion of acid mucins to protect the gut mucosa from exposure to the pesticide or assumingly to pesticide-related metabolites (Oliveira *et al.*, 2022).

3.1.2 Chlorpyrifos

The highly-lipophilic organophosphorus compound chlorpyrifos is one of the most frequently used non-persistent insecticides worldwide (Gilani *et al.*, 2016). Aside from its neurotoxic effects, chorpyrifos can cross the placental barrier and appears in amniotic fluid. Its neonatal exposure has been associated with metabolic syndrome in humans due to permanent changes in physiology and metabolism (Reygner *et al.*, 2016b). In rat pups, perinatal exposure to chlorpyrifos induced perturbation in the development of the gut microbiota, structure and function, which led to impaired nutrient assimilation and growth, microbial dysbiosis and increased bacterial translocation into spleen and liver (Joly Condette *et al.*, 2015, 2022). Clorpyriphos reduced villus height and villosity, compromising the secretion of digestive enzymes and nutrient absorption in the rat pups. Using targeted approaches (*i.e.* culturing and quantitative PCR), almost all investigated genera were altered by chlorpyrifos either in the ileum (Joly Condette *et al.*, 2015) or colon of pups of chlorpyrifos-exposed dams at weaning (Reygner *et al.*, 2018) compared to pups from non-exposed litters. Specifically, chlorpyrifos interfered in the postnatal bacterial colonization as indicated by decreased abundances of lactobacilli and bifidobacteria in ileal digesta and butyrate-producing clostridial clusters in colonic digesta in rat pups at weaning (Joly Condette *et al.*, 2015; Reygner *et al.*, 2018); taxa that exhibit similar niches in the porcine gut and are often associated with high gut health in pigs (Metzler-Zebeli *et al.*, 2013). The abnormal microbiota development in ileum, cecum and colon was associated with an impaired gut epithelial protection via mucin-2 production and raised innate immune response due to increased expression of Toll-like receptors-2 and -4; receptors which bind bacterial lipopolysaccharide (Joly Condette *et al.*, 2015). Combined with its detrimental effects on the integrity of the gut barrier, more lipopolysaccharides could enter the body, triggering systemic endotoxemia and finally low-grade inflammation (Reygner *et al.*, 2016b; Requile *et al.*, 2018; Liang *et al.*, 2019). Notably, there was no clear dose-response relationship in the rat pups when exposed to 1 and 3.5 or 5 mg/kg body weight and day. Moreover, the impact of chlorpyrifos on the gut differed to that before weaning, when the exposure continued until adulthood (day 60 of life) (Table 6.1; Joly Condette *et al.*, 2015). Generally, effects were greater in the ileum than colon, demonstrating the gradient of chlorpyrifos and its metabolites in gastrointestinal digesta. Bioactivation of chlorpyriphos and first clearance occur from the duodenum onwards (Leoni *et al.* 2012). Lower dosages may be fully absorbed or microbially metabolized cranially to the colon (Joly Condette *et al.*, 2015).

TABLE 6.1 Effect of insecticides on the gut homeostasis in pigs

Name	Chemical categorization	Test organism or system	Concentration	Effect on gut function	Effect on gut microbiota	Reference
Deltamethrin	Pyrethroid	Crucian carp	Non-toxic (0.3 µg/L water)	Intestinal gene expression: Antioxidant enzymes: *SOD*, *GPX1*, *CAT* ↑; Proinflammatory factors: *IFNG* ↑; Barrier function: *CLDN12* ↑; Apoptosis and autophagy: *CASP9*. ↑ Intestinal histology: Goblet cells ↑.	Proteobacteria ↑ Verrucomicrobiota ↑ Proteobacteria ↓ Actinobacteria ↓ Firmicutes ↓	Wu *et al.*, 2022
Deltamethrin	Pyrethroid	Crucian carp	Non-toxic (0.6 µg/L water)	Intestinal gene expression: Antioxidant enzymes: *GPX1*, *CAT* ↑; Proinflammatory factors: *TNFA*, *IFNG*, *IL1B* ↑; Barrier function: *ZO1* ↑; Apoptosis and autophagy: *CASP9*, *BCL2*, *BAX*, *ATG5*, *ATG12*, *BECN1*. ↑ Intestinal histology: Goblet cells ↑.	Proteobacteria ↑ Verrucomicrobiota ↑ Proteobacteria ↓ Actinobacteria ↓ Firmicutes ↓ Bacteroidetes ↓	Wu *et al.*, 2022

TABLE 6.1 Effect of insecticides on the gut homeostasis in pigs (*cont.*)

Name	Chemical categorization	Test organism or system	Concentration	Effect on gut function	Effect on gut microbiota	Reference
Chlorpyrifos	Pyrethroid	In vitro (SHIME)	1 mg/day	–	After 30 days: Ascending colon: Total bacteria ↑ *Escherichia coli* ↑ *Bifidobacterium* spp. ↓ Transverse colon: Total bacteria ↑ *Clostridium leptum* ↓ *Bacteroides* ↑ Descending colon: *Clostridium leptum* ↓ *Lactobacillus* spp. ↑	Joly Condette *et al.*, 2022
Chlorpyrifos	Pyrethroid	In vitro (SHIME; CaCo2/TC7 cells)	0.35 mg	CaCo2/TC7 cells: Gene expression: *CAV1, VIL1, CDH1, CLDN4* ↓	SHIME colon: *Bifidobacterium* spp. ↓; acetate ↑	Réquilé *et al.*, 2018
Chlorpyrifos	Organo-phosphorus	Wistar rat	1 and 3.5 mg/kg/day (perinatal exposure)	[metabolic disorder: hyperinsulinemia]	Colon: Firmicutes ↓ *Clostridium coccoides* group ↓ *Clostridium leptum* group (↓) No effect on SCFA	Reygner *et al.*, 2016

TABLE 6.1 Effect of insecticides on the gut homeostasis in pigs (*cont.*)

Name	Chemical categorization	Test organism or system	Concentration	Effect on gut function	Effect on gut microbiota	Reference
Chlorpyrifos	Organo-phosphorus	Wistar rat pups	1 or 5 mg/kg/day (peri- and post-natal exposure)	Day 21 of life, duodenum, jejunum, ileum: Villus height ↓ Villosity width ↓ Epithelial thickness ↓ Circular/longitudinal muscle thickness ↓ Colon: crypt depth ↓ Gene expression: *MUC2* (ileum, colon) ↑ *BD2* (only 1 mg, ileum) ↑ *TLR4* (ileum, colon) ↓ Day 60 of life: Villus height ↓ Villosity width ↓ Circular muscle thickness ↓ Colon: crypt depth ↓ Gene expression: *MUC2* (ileum, colon) ↑ *BD2* (only 1 mg, ileum) ↑ *TLR4* (ileum, colon) ↓	Day 21 of life (Culture), ileum: Aerobic bacteria ↑ *Enterococcus* spp. ↑ *Clostridium* spp. ↑ *Staphylococcus* spp. (ileum, cecum, colon) ↑ *Bacteroides* spp. ↑ *Lactobacillus* spp. (ileum, cecum, colon) ↓ *Bifidobacterium* spp. ↓ Day 60 of life (Culture): *Bacteroides* spp. (colon) ↑ *Lactobacillus* spp. (ileum, cecum, colon) ↓ *Bifidobacterium* spp. (colon)↓	Joly Condette *et al.*, 2015

Abbreviations: *ATG5*, autophagy related 5; *ATG12*, autophagy related 12; *BAX*, BLC2 associated X; *BCL2*, B-cell lymphoma 2; *BD2*, beta-defensin-2; *BECN1*, beglin-1; *CASP9*, caspase-9; *CAT*, catalase; *CAV1*, *caveolin-1*; *CDH1*, cadherin-1; *CLDN4*, claudin-4; *CLDN12*, claudin-12; *GPX1*, glutathione peroxidase 1; IL, interleukin; *IL1B*, interleukin-1β; *INFG*, interferon-γ; *MUC2*, mucin-2; SCFA, short-chain fatty acids; *SOD*, superoxide dismutase; *TLR4*, toll-like receptor-4; *TNFA*, tumor-necrosis factor-α; *VIL1*, villin-1; *ZO1*, zona occludens-1.

3.2 *Herbicides*

3.2.1 Glyphosate

From the utilized herbicides, glyphosate is the world's most widely used 'weed killer' to eradicate plants and grasses during cultivation season (Benbrook, 2016; Sørensen *et al.*, 2021). It is a nonselective herbicide which primarily acts via inhibition of the 5-enolpyruvylshikimate-3-phosphate (EPSP) synthase, an enzyme in the shikimate pathway found in the chloroplasts of plants and some microorganisms but not in animal cells (Vicini *et al.*, 2019; Sørensen *et al.*, 2021). In addition to its herbicidal property, glyphosate has antimicrobial and mineral-chelating properties and acts as an oxidative stressor in the gut (Figure 6.2; Tejada, 2009; Qiu *et al.*, 2020; Sørensen *et al.*, 2021; Wang *et al.*, 2022); features that can contribute to loss of barrier function in the gut (Wang *et al.*, 2022). The copper-chelating activity of glyphosate may affect a range of metabolic processes in prokaryotic and eukaryotes. Copper atoms are a central component in numerous metabolic enzymes in the body, such as copper/zinc superoxide dismutase, cytochrome c oxidase, mitogen-activated protein kinase, and cAMP-degrading phosphodiesterase (Broom *et al.*, 2021). Therefore, chelation of copper may interfere in cellular respiration, antioxidant activity, iron transport, pigmentation, and connective tissue development.

Since herbicides are applied multiple times during the growing and harvesting season, residues, even when below the upper legal limits, frequently remain on the crop. In addition, glyphosate-tolerant crops have been developed using a transgene from a resistant bacterial EPSP synthase that renders the crops insensitive to glyphosate (Vicini *et al.*, 2019). Much of the grain from these crops are consumed by farm animals, like pigs (Vicini *et al.*, 2019). Consequently, pigs are exposed to residues of glyphosate in higher amounts than humans. In accordance, glyphosate was the most detected pesticide in animal feed in the US in 2021 (FDA, 2022). Effects of glyphosate residues in Roundup Ready resistant crops, like soybean meal, on

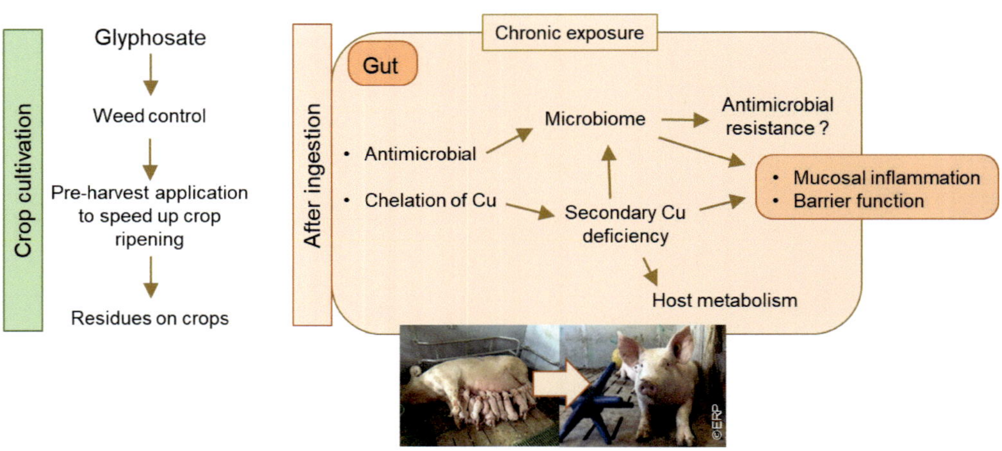

FIGURE 6.2 Modes of action of glyphosate from field to gut

growth performance in fattening pigs were small, leading to the conclusion that Roundup Ready soybean meal was essentially equivalent in composition and nutritional value to conventional soybean meal for growing-finishing pigs (Cromwell *et al.*, 2002). Glyphosate is rapidly absorbed in the upper part of the gut and eliminated from the body via the kidneys without being metabolized. In lack of scientific data, Sørensen *et al.* (2021) estimated the average level of glyphosate in stomach, small intestinal and colonic digesta of pigs of being 0.005, 0.005 and 0.02 mg/ml wet digesta, respectively, when pigs were fed a barley-soybean meal based diet with residual glyphosate of 20 mg/kg (the upper MRL in the European Union).

Glyphosate affects the gut microbiota by the inhibition of the EPSP synthase and hence by blocking the synthesis of the aromatic amino acids tyrosine, phenylalanine and tryptophan (Steinrücken and Amrhein *et al.*, 1980). While data for the porcine gut are still inconclusive, mostly due to a lack of data, scientific evidence exists that glyphosate residues alter the abundance of the predominant phyla in the gut of rodents, honey bees and quail as reviewed by Giambo *et al.* (2022). The studies consistently demonstrated a depression in Firmicutes (*e.g. Lactobacillaceae, Eubacteriaceae, Lachnospiraceae, Ruminococcaceae* and *Clostridiaceae*) and Actinobacteria, whereas Bacteroidetes increased in feces and intestinal digesta (Table 6.2; Tang *et al.*, 2020; Giambo *et al.*, 2022). Moreover, the authors associated the glyphosate-related dybiosis with an increase in the prevalence of neurobehavioral abnormalities in rodents (Tang *et al.*, 2020; Giambo *et al.*, 2022). In the study of Lozano *et al.* (2018) in rats, glyphosate did not only cause dysbiosis with increased Bacteroidetes and less *Lactobacillaceae*. It also raised the abundance of a highly tolerant or resistant strain of *Escherichia coli* which did not carry EPSP synthase gene, explaining the resistance. However, the dosages of glyphosate applied in the referenced studies were often above the MRL for glyphosate residues in the EU and much higher than the estimated chronic exposure in growing pigs (Aitbali *et al.*, 2018; Tang *et al.*, 2020); thus, results may not be representative for the low level of exposure that pigs experience. In the conducted short-term *in vitro* study on the porcine gut microbiota, colonic digesta from eight to nine week-old pig was incubated with 228 mg/d glyphosate, which related to the maximum residue level of glyphosate on animal feed, for four days in a continuous culture system (Table 6.2; Krause *et al.*, 2020). Results indicated that the short-term exposure of the colonic microbiota to glyphosate did not lead to alterations in taxonomy and functional enzyme abundances, including the expression of the glyphosate target enzyme EPSP synthase. Nevertheless, untargeted metabolomics revealed functional changes after this short period of time in that microbial pyridoxamine, a form of vitamin B6, and imidazolepropionic acid, a histidine intermediate, as well as cholic acid metabolism seemed to be altered by the pesticide exposure (Krause *et al.*, 2020). Unfortunately, the authors missed to report whether the donor pigs were exposed to glyphosate residues during their lifetime including perinatal and early postnatal exposure, which may have led to adaptations in the colonic microbial abundances and metabolism.

Also, it can be assumed that the effect of residual glyphosate on the microbial communities is greater in the stomach or small intestine of pigs, where *Lactobacillus* is a dominant genus (Newman *et al.*, 2018) which has been shown to be susceptible to glyphosate (Aitbali *et al.*, 2018). Albeit glyphosate exposure in weaned pigs did not alter the morphological structure in the duodenum and jejunum (Qiu *et al.*, 2020), it compromised the mucosal antioxidant capacities and barrier function while upregulating pro-inflammatory pathways in the small intestines of piglets exposed to 0, 10, 20 and 40 mg/kg feed as indicated by corresponding mRNA expression levels (Table 6.2). Results for IPEC-J2 cells support increased oxidative stress due to compromised activities of antioxidant enzymes (*i.e.* superoxide dismutase and plasma glutathione peroxidase) and increased inflammation after glyphosate exposure (Bai *et al.*, 2022). Mechanistically, inflammation seemed to be related to an increased activation of nuclear factor kappa-light-chain-enhancer of activated B cells (NFκB) which then led to the increased release of pro-inflammatory factors [interleukin (IL)-1β, IL-6, tumor necrosing factor-α] and the mRNA expression of inducible nitric oxide synthase and cyclooxygenase-2 (Bai *et al.*, 2022). Simultaneously, glyphosate induced the activation of the Nrf2/heme oxygenase 1 signaling pathway to defend against this stress. Due to its copper-chelating properties, glyphosate can directly interfere in the synthesis of superoxide dismutase – a copper- and zinc-containing metallo-protein (Broom *et al.*, 2021), which can explain the compromised antioxidative enzyme capacities after glyphosate exposure (Wang *et al.*, 2022). Larsen *et al.* (2022) investigated whether glyphosate residues in feed led to epigenetic modifications in the small intestine. The authors expected changes; however, glyphosate had only little effects on DNA methylation in the small intestine. The authors, however, stated that they could not rule out that glyphosate contamination of their basal diet may have compromised their baseline measurements. This statements highlights the importance to control the exposure to pesticides prior to experiments due to potential programming effects and epigenetic modifications on the gut (Grados *et al.*, 2022).

4 Feed preservatives

Preservatives are primarily added to feed to prevent the formation of radical oxygen species (ROS) that form during storage and heating and cause lipid peroxidation. In order to protect the natural antioxidants, such as the supplemented tocopherols and vitamin C from getting oxidized, synthetic antioxidants are widely used, among which butylated hydroxytoluene (BHT), butylated hydroxyanisole (BHA), and ethoxyquin (EQ) are the most frequently used (Blaszczyk *et al.*, 2013). Propyl gallate and tert-butylhydroquinone are less frequently added (Carocho and Feirera, 2013). Risk assessments of feed preservatives are mainly conducted from the consumer point of view; missing that the chemicals may have an impact on the animal

TABLE 6.2 Effect of glyphosate on the gut homeostasis in pigs

Pig/gut site/in vitro system (number of observation)	Glyphosate level	Exposure time	Effects on gut function	Effects on gut microbiota	Reference
Continuous culture system using colonic digesta of 8 to 9 week-old pigs (n = 4)	228 mg/day	4 days; bioreactor fluid collected on day 2, 3 and 4 and analyzed	–	No effects on metagenome, metaproteome, SCFA concentration	Krause *et al.*, 2020
IPEC-J2 cells (number of observations not provided)	0, 30, 60 mg/L culture media	4, 12, 24 hours	Cell viability ↓ Activities of oxidative enzymes (SOD-1, GSH-PX) ↓ Release of IL-1β, IL-6, IL-8, NF-κB, COX-2, NOS ↑	–	Bai *et al.*, 2022
Weaned pigs (n = 26/group) Small intestine	20 and 200 ppm	9 days	No effect on global methylation; IL-18 promoter methylation: CpG1, CpG2, CpG5 ↓ No effect on DNA methylation gene expression: *DNMT1*, *DNMT3A*, *DNMT3B*	–	Larsen *et al.*, 2022

TABLE 6.2 Effect of glyphosate on the gut homeostasis in pigs (*cont.*)

Pig/gut site/in vitro system (number of observation)	Glyphosate level	Exposure time	Effects on gut function	Effects on gut microbiota	Reference
Weaned pigs (n = 7/group) Duodenum Jejunum	0, 10, 20, 40 ppm	35 days	No effect on histo-morphology in duodenum and jejunum (villus height, crypt depth, mucosal thickness); Duodenum: expression of *SOD1*, *GPX1*, *HO1* ↓ Duodenum and jejunum: expression of *GCLM* ↓ No effect on secretion of IL-1β, IL-6, TNF-α, sIgA in duodenum and jejunum	–	Qiu *et al.*, 2020

Abbreviations: COX-2, cyclooxygenase-2; *DNMT*, DNA methyltransferase; *GCLM*, glutamate-cysteine ligase modifier subunit; GSH-PX, glutathione peroxidase; *HO1*, heme oxygenase-1; IL, interleukin; NF-κB, nuclear factor kappa-light-chain-enhancer of activated B cells; NOS; nitric oxide synthase; sIgA, secretory immunoglobulin A; SCFA, short-chain fatty acids; SOD-1, superoxide dismutase-1; TNF-α, tumor-necrosis-factor-α.

directly. Theoretically, they can interfere in interventions to promote gut homeostasis. In this regard, the toxicity of BHA, BHT and EQ has been investigated extensively using a variety of experimental conditions (EFSA, 2018, 2022; Xu *et al.*, 2021). In general, a scarcity of information exists about effects of synthetic preservatives on the gut and the main absorption sites, not to mention the gut microbiota. The question that needs to be addressed in the future is how stable the preservatives remain in the gastrointestinal tract, especially after the passage of the stomach, being exposed to very low pH, which may cause oxidation and/or hydrolysis of the compounds on its own. Another thinkable scenario is that the preservatives fulfill a similar function in the gut, contributing to the antioxidative capacity at the gut mucosa and hence supporting the gut barrier. Although the activity of antioxidants is the topic of current research, little information has been published on their transformation and metabolic products (Xu *et al.*, 2021), not to mention their impact on the gut microbiota and metabolism in enterocytes. The transformation of preservatives occurs in the feed and in the body, whereby the role of the gut microbiota in the transformation of the preservatives has not been studied in detail so far.

4.1 *Ethoxyquin*

Ethoxyquin (6-ethoxy-1,2-dihydro-2,2,4- trimethylquinoline) is used in livestock, aquaculture and pet feed to slow the development of rancidity of fats and stabilize fat soluble vitamins (*i.e.* vitamin A and E; EFSA, 2022). It is synthesized from p-phenetidine, a recognized potential mutagen, which remains in the additive as an impurity, at concentrations of < 2.5 mg/kg additive. Due to concerns about its carcinogenicity and toxicity, the safety of EQ use in pet foods has been questioned (Blaszczyk *et al.*, 2013). Concerns have been also raised in relation to accumulation of EQ transformation products in meat, eggs and dairy. In Europe, the authorization for use of EQ as a feed additive has been suspended, due to gaps in knowledge with regard to the presence and toxicity of the EQ transformation products, whereas it is still widely used in other parts of the World. The FDA listed EQ as the third most abundant pesticide in animal feed in the US in 2020 (FDA, 2022). In 2022, the EFSA reassessed the safety of the utilization of EQ in feed for livestock animals and declared that a maximum total supplementation level of 50 mg/kg in complete feed for all animal species, except dairy animals, would be safe (EFSA, 2022). Below this level, residues of EQ or its metabolites/transformation products would not cause a risk for the consumer. However, this included mainly growing animals, as the EFSA did not positively assess EQ as safe at any level for long-living and reproductive animals due to the residual p-phenetidine. Ethoxiquin is rapidly absorbed from the gastrointestinal tract and quickly metabolized and eliminated via the kidneys (Blaszczyk *et al.*, 2013). Safety assessment studies in (weaned) pigs, like the most recently done by the EFSA (EFSA, 2022), mainly focused on organs, *i.e.* spleen, liver and kidneys, and blood hematological and biochemical parameters, which showed no aberrations compared to the control. Evidence for microbiota-modulating capacities of EQ, however, exist from soil studies where the application of quinone imine, like EQ, is of interest as nitrification inhibitor to improve nitrogen use efficiency and decelerate nitrogen cycling (Papadopoulou *et al.*, 2022). Accordingly, the transformation product 2,6-dihydro-2,2,4-trimethyl-6-quinone imine of EQ has a strong inhibitory effect on ammonia-oxidizing bacteria, like *Sphingomonas*, and nitrite-oxidizing bacteria with *Nitrobacter* being more sensitive than *Nitrospira* (Papadopoulou *et al.*, 2022); taxa that also inhabit the porcine gut (Holman *et al.*, 2017; Newman *et al.*, 2018). Moreover, the fact that 'off-target' bacterial and fungal communities at the two pH values tested (pH 5.4 versus 7.4) were affected in the study of Papadopoulou *et al.* (2022) is notable, considering that these pH ranges can be found in the small and large intestines of pigs. Therefore, it is valid to speculate about perturbations in the porcine gut microbiota after EQ ingestion. In this line of reasoning, Papadopoulou *et al.* (2022) stressed the need for monitoring the impact on soil microorganisms to ensure soil ecosystem functioning.

4.2 *Phenolic compounds*

The other two mostly used preservatives in feed, BHT and BHA, are lipophilic phenolic compounds which applications ranges from preservation of the nutritive value, flavor, freshness and color of food, animal feed and cosmetics (EFSA, 2018, 2022). Butylated hydroxyanisole (BHA) is a mixture of 3-tert-butyl-4-hydroxyanisole and 2-tert-butyl-4-hydroxyanisole. When intended for feed for livestock, BHA can be added alone or in combination with BHT and/or EQ and has been considered as safe for livestock animal species at a dosage of up to 150 mg/kg complete feed (EFSA, 2018). Likewise, the additive BHT has been considered safe for weaned pigs and fattening pigs up to a maximum concentration of 150 mg/kg complete feed (EFSA, 2022). Similar to EQ, BHA and BHT are rapidly absorbed from the gastrointestinal tract. Evidence exists for hepatic metabolism of BHA and BHT, whereby the metabolites are excreted through the gut via mixing with bile to be excreted in feces and/or in urine via the kidneys. The microbiota residing the gut caudal to the Ductus choledochus may be affected and interact with these metabolites.

For both preservatives, carcinogenicity, cytotoxicity, oxidative stress induction and endocrine disrupting effects have been reported at levels above the recommended dietary levels (EFSA, 2018, 2022) under in vitro and in vivo experimental conditions (*e.g.* Jayalakshmi and Sharma, 1986; Xu *et al.*, 2021). The recommendation of the EFSA (2018), for instance, to prove the non-toxicity of BHA in pig feed have been performed in the 1980 (Hansen *et al.*, 1982; Wurtzen and Olsen, 1986), which included an examination of the gut lining but no microbial culturing. The EFSA reported for BHA proliferative and parakeratotic changes of the stomach lining and oesophageal epithelium in a few pigs at a dietary level of 200 and 400 mg BHA/kg complete feed (EFSA, 2018), but results were controversial. In terms of their modes of action, BHA induced apoptosis being mediated via calcium and zinc ions and caspase-3 activity which increased vulnerability of the cells to oxidative stress in rat thymocytes, whereas BHT induced non-apoptotic cell death (Mizobuchi *et al.*, 2022). Controversially, BHT has been shown to protect lab rodents against hepatotoxins (Muntane *et al.*, 2000), whereas similar studies for enterocytes are missing. Similar observations have been reported for BHA using a dextran sodium sulfate-induced inflammatory bowel disease mouse model (Wang *et al.*, 2021). In this model, BHA counteracted the dextran sulfate sodium-induced gut permeability and increased the antioxidative capacity via regulating the Nrf2- Kelch like erythroid cell-derived protein with CNC homology associated protein 1 pathway (Wang *et al.*, 2021). Moreover, phenolic compounds are applied in the food and feed industry to the control of pathogenic microorganisms in foods. In this regard, BHA has been shown to possess potent antimicrobial activity against *Enterococcus faecalis* and *Staphylococcus aureus* (Gutierrez-Fernandez *et al.*, 2013; Fernandez-Alvarez *et al.* 2014). *Enterococcus faecalis* is a commensal in the porcine gut (Holman *et al.*, 2017); therefore, similar antimicrobial actions of BHA are thinkable in the porcine gut.

5 Concluding remarks and outlook

Pesticides are widely used for protecting plants from pests; hence, crops used for pig feed from conventional but also organic farming, are contaminated with pesticide residues (EFSA, 2021). Maximum levels have been established for pesticides and synthetic preservatives, below which these chemicals have been declared as 'harmless'. Nevertheless, scientific evidence increasingly demonstrates that pesticides and certain preservatives need to be re-assessed for their effects on the body, especially in relation to gut function and the gut microbiota-host interactions. Notably, it is alarming how little information in the literature exists for the effects of preservatives and pesticides in pigs despite their ubiquitous utilization and relevance for animal product safety. In fact, pork (*i.e.* meat, organs and fat) is regularly monitored for pesticide residues (EFSA, 2021). The fate of pesticides and preservatives in the porcine gut is mostly unknown; including the microbial capabilities to metabolize them. Literature addressing their impact on the microbial community and metabolism is the stomach and small intestine largely does not exist, not only for pigs but also in other animal species, albeit these communities are first affected by the chemicals and their important role in the microbiota-host interplay. This is even more alarming considering the massive attention that porcine gut health and gut microbiota-host interactions momentarily receive based on the necessity to develop feeding strategies to abrogate gut disorders and the progressing ban of in-feed antibiotics worldwide. The toxicity of pesticides for the host may be modified through the activity on the gut microbiota; therefore, the pesticide-induced dysbiosis may increase its toxicity for the host as indicated from rat studies (Tang *et al.*, 2020). Therefore, a better understanding of the bidirectional relationship between pesticides, but also preservatives, and the microbiota especially in the upper gut is needed for an improved understanding of the toxicokinetics of pesticides in pigs upon ingestion. Moreover, animal feed often contains a 'cocktail' of pesticides. Therefore, it is critical to assess potential interactions between the various types of pesticides as well as between pesticides and preservatives in the porcine gut cannot be ruled out. For studies, it will be crucial that actual *in vivo* exposure conditions that correspond to practical levels of chemical exposure and length of exposure time are taken into consideration (Sørensen *et al.*, 2021). As many xenobiotics (*e.g.* glyphosate) are readily absorbed from the small intestine, measurements of the compound or their metabolites in urine may be used to get a realistic estimate of the actual amount ingested (Vicini *et al.*, 2019). In humans, transmission of pesticides into breast milk have been reported (Mekonen *et al.*, 2021; Dong *et al.*, 2022), whereas the carryover of pesticides and preservatives or their metabolites into porcine milk and a potential epigenetic influence *in utero* have not been studied in pigs. The importance of the perinatal time period for gut priming, microbial colonization and immune competence is more and more recognized. If transmission occurs, xenobiotics may permanently influence the gut development and function in pigs. Lastly, there is evidence from other animal species that the negative impact of pesticides can be moderated by modifying the animal's diet, which would be valid to follow up in relation to porcine gut health. For instance, adding prebiotics in the form of inulin to the rat chow alleviated the effect of chlorpyrifos on the gut microbiota and proinflammatory signaling *in vitro* (Reygner *et al.*, 2018; Joly Condette *et al.*, 2022). Moreover, the provision of adequate amounts of aromatic amino acids via the diet may alleviate the need for the microbes to synthesize these amino acids, which may prevent the antimicrobial effect of glyphosate (Nielsen *et al.*, 2017).

Abbreviations

BHA	(butylated hydroxyanisole)
BHT	butylated hydroxytoluene
EPSP	5-enolpyruvylshikimate-3-phosphate
EQ	ethoxyquin
MRL	maximum residue level
Nrf2	nuclear factor erythroid 2-related factor 2
ROS	radical oxygen species
SCFA	short-chain fatty acids

References

Aitbali, Y., Ba-Mhamed, S., Elhidar, N., Nafis, A., Soraa, N., Bennis, M. (2018) Glyphosate based- herbicide exposure affects gut microbiota, anxiety and depression-like behaviors in mice. Neurotoxicol. Teratol. 67: 44–49. doi: 10.1016/j.ntt.2018.04.002.

Ali, H., El-Sayed, N.M., Khodeer, D.M., Ahmed, A., Hanna, P.A., Moustafa, Y. (2020) Nano selenium ameliorates oxidative stress and inflammatory response associated with cypermethrin-induced neurotoxicity in rats. Ecotoxicol. Environ. Saf. 195: 110479.

Altajar, S., and Moss, A. (2020) Inflammatory Bowel Disease Environmental Risk Factors: diet and Gut Microbiota. Curr. Gastroenterol. Rep. 22: 57. doi: 10.1007/s11894-020-00794-y.

Arnaud, A.P., Rome, V., Richard, M., Formal, M., David-Le Gall, S., Boudry, G. (2020) Post-natal co-development of the microbiota and gut barrier function follows different paths in the small and large intestine in piglets. FASEB J. 34(1): 1430–1446. doi: 10.1096/fj.201902514R.

Bai, G., Zhou, R., Jiang, X., Zou, Y., Shi, B. (2022) Glyphosate-based herbicides induces autophagy in IPEC-J2 cells and the intervention of N-acetylcysteine. Environ. Toxicol. 37(8): 1878–1890. doi: 10.1002/tox.23534.

Benbrook, C.M. (2016) Trends in glyphosate herbicide use in the United States and globally. Environ. Sci. Eur. 28(1): 3. doi: 10.1186/s12302-016-0070-0.

Błaszczyk, A., Augustyniak, A., Skolimowski, J. (2013) Ethoxyquin: an antioxidant used in animal feed. Int. J. Food Sci. 2013: 585931. doi: 10.1155/2013/585931.

Broom, L.J., Monteiro, A., Piñon, A. (2021) Recent advances in understanding the influence of zinc, copper, and manganese on the gastrointestinal environment of pigs and poultry. Animals (Basel) 11(5): 1276. doi: 10.3390/ani11051276.

Bundesgesundheitsamt (2022) Ableitung von HBM-I-Werten für Deltamethrin und Cyfluthrin – Stellungnahme der Kommission Human-Biomonitoring des Umweltbundesamtes. Bundesgesundheitsblatt Gesundheitsforschung Gesundheitsschutz. 65(9): 942–945. German. doi: 10.1007/s00103-022-03568-1.

Carocho, M., Ferreiram I.C. (2013) A review on antioxidants, prooxidants and related controversy: natural and synthetic compounds, screening and analysis methodologies and future perspectives. Food Chem. Toxicol. 51: 15–25. doi: 10.1016/j.fct.2012.09.021.

Chen, S.H., Yang, L., Hu, M.Y., Liu, J.J. (2011) Biodegradation of fenvalerate and 3-phenoxybenzoic acid by a novel *Stenotrophomonas* sp. strain ZS-S-01 and its use in bioremediation of contaminated soils. Appl. Microbiol. Biotechnol 90: 755–767. doi: 10.1007/s00253-010-3035-z.

Chen, S., Hu, W., Xiao, Y., Deng, Y., Jia, J., Hu, M. (2012) Degradation of 3-Phenoxybenzoic Acid by a *Bacillus* sp. PLoS ONE 7(11): e50456. doi: 10.1371/journal.pone.0050456.

Cromwell, G.L., Lindemann, M.D., Randolph, J.H., Parker, G.R., Coffey, R.D., Laurent, K.M., Armstrong, C.L., Mikel, W.B., Stanisiewski, E.P., Hartnell, G.F. (2002) Soybean meal from roundup ready or conventional soybeans in diets for growing-finishing swine. J. Anim. Sci. 80(3): 708–715. doi: 10.2527/2002.803708x.

Davies, T.G., Field, L.M., Usherwood, P.N., Williamson, M.S. (2007) DDT, pyrethrins, pyrethroids and insect sodium channels. IUBMB Life 59: 151–162. doi: 10.1080/15216540701352042.

Defois, C., Ratel, J., Garrait, G., Denis, S., Le Goff, O., Talvas, J., Mosoni, P., Engel, E., Peyret, P. (2018) Food chemicals disrupt human gut microbiota activity and impact intestinal homeostasis as revealed by in vitro systems. Sci. Rep. 8: 11006. doi: 10.1038/s41598-018-29376-9.

Dong, Y., Yin, S., Zhang, J., Guo, F., Aamir, M., Liu, S., Liu, K., Liu, W. (2022) Exposure patterns, chemical structural signatures, and health risks of pesticides in breast milk: A multicenter study in China. Sci. Total Environ. 830: 154617. doi: 10.1016/j.scitotenv.2022.154617.

Donkor, A., Osei-Fosu, P., Dubey, B., Kingsford-Adaboh, R., Ziwu, C., Asante, I. (2016) Pesticide residues in fruits and vegetables in Ghana: a review. Environ. Sci. Pollut. Res. Int. 23(19): 18966–18987. doi: 10.1007/s11356-016-7317-6.

EFSA (2018a) Evaluation of the impact of glyphosate and its residues in feed on animal health. EFSA J. 16(5): 5283. doi:10.2903/j.efsa.2018.5283.

EFSA (2018b) Safety and efficacy of butylated hydroxyanisole (BHA) as a feed additive for all animal species. EFSA J. 16(3): 5215 doi: 10.2903/j.efsa.2018.5215.

EFSA (2021) The 2019 European Union report on pesticide residues in food. EFSA J. 19(4): 6491 doi: 10.2903/j.efsa.2021.6491.

EFSA (2022) Safety and efficacy of a feed additive consisting of butylated hydroxytoluene (BHT) for all animal species (Lanxess Deutschland GmbH). EFSA J. 20(5): 7286 doi: 10.2903/j.efsa.2022.7286.

FDA (2022) Pesticide residue monitoring program fiscal year 2020 pesticide report. https://www.fda.gov/food/chemicals-metals-pesticides-food/pesticides.

Fernandez-Alvarez, L., del Valle, P., de Arriaga, D., Garcia-Armesto, M.R., Rua, J. (2014) Binary combinations of BHA and other natural and synthetic phenolics: Antimicrobial activity against *Staphylococcus aureus* and antioxidant capacity. Food Contr. 422: 303–309. doi: 10.1016/j.foodcont.2014.02.020.

Giambò, F., Teodoro, M., Costa, C., Fenga, C. (2021) Toxicology and microbiota: how do pesticides influence gut microbiota? A review. Int. J. Environ. Res. Public Health 18(11): 5510. doi: 10.3390/ijerph18115510.

Gilani, R.A., Rafique, M., Rehman, A., Munis, M.F.H., Rehman, S.U., Chaudhary, H.J. (2016) Biodegradation of chlorpyrifos by bacterial genus Pseudomonas. J. Basic Microbiol. 56: 105–119. doi:10.1002/jobm.201500336.

Grados, L., Pérot, M., Barbezier, N., Delayre-Orthez, C., Bach, V., Fumery, M., Anton, P.M., Gay-Quéheillard, J. (2022) How advanced are we on the consequences of oral exposure to food contaminants on the occurrence of chronic non communicable diseases? Chemosphere 303(Pt 3): 135260. doi: 10.1016/j.chemosphere.2022.135260.

Gutiérrez-Fernández, J., García-Armesto, M.R., Alvarez-Alonso, R., del Valle, P., de Arriaga, D., Rúa, J. (2013) Antimicrobial activity of binary combinations of natural and synthetic phenolic antioxidants against *Enterococcus faecalis*. J. Dairy Sci. 96(8): 4912–4920. doi: 10.3168/jds.2013-6643.

Halden, R.U., Tepp, S.M., Halden, B.G., Dwyer, D.F. (1999) Degradation of 3-phenoxybenzoic acid in soil by *Pseudomonas pseudoalcaligenes* POB310 (pPOB) and two modified *Pseudomonas* strains. Appl. Environ. Microbiol 65: 3354–3359. doi: 10.1128/AEM.65.8.3354-3359.1999.

Hansen, E.V., Meyer, O., Olsen, P. (1982) Study of toxicity of butylated hydroxyanisole (BHA) in pregnant gilts and their foetuses. Toxicology 23(1): 79–83. doi: 10.1016/0300-483x(82)90043-9.

Heo, J.M., Opapeju, F.O., Pluske, J.R., Kim, J.C., Hampson, D.J., Nyachoti, C.M. (2013) Gastrointestinal health and function in weaned pigs: a review of feeding strategies to control post-weaning diarrhoea without using in-feed antimicrobial compounds. J. Anim. Physiol. Anim. Nutr. (Berl.) 97(2): 207–237. doi: 10.1111/j.1439-0396.2012.01284.x.

Holman, D.B., Brunelle, B.W., Trachsel, J., Allen, H.K. (2017) Meta-analysis to define a core microbiota in the swine gut. mSystems 2(3): e00004-17. doi: 10.1128/mSystems.00004-17.

Jayalakshmi, C.P., Sharma, J.D. (1986) Effect of butylated hydroxyanisole (BHA) and butylated hydroxytoluene (BHT) on rat erythrocytes. Environ. Res. 41(1): 235–238. doi: 10.1016/s0013-9351(86)80185-2.

Joly Condette, C., Bach, V., Mayeur, C., Gay-Quéheillard, J., Khorsi-Cauet, H. (2015) Chlorpyrifos exposure during perinatal period affects intestinal microbiota associated with delay of maturation of figestive tract in rats. J. Pediatr. Gastroenterol. Nutr. 61(1): 30–40. doi: 10.1097/MPG.0000000000000734.

Joly Condette, C., Djekkoun, N., Reygner, J., Depeint, F., Delanaud, S., Rhazi, L., Bach, V., Khorsi-Cauet, H. (2022) Effect of daily co-exposure to inulin and chlorpyrifos on selected microbiota endpoints in the SHIME® model. Environ. Pollut. 302: 118961. doi: 10.1016/j.envpol.2022.118961.

Krause, J.L., Haange, S.B., Schäpe, S.S., Engelmann, B., Rolle-Kampczyk, U., Fritz-Wallace, K., Wang, Z., Jehmlich, N., Türkowsky, D., Schubert, K., Pöppe, J., Bote, K., Rösler, U., Herberth, G., von Bergen, M. (2020) The glyphosate formulation Roundup® LB plus influences the global metabolome of pig gut microbiota in vitro. Sci. Total Environ. 745: 140932. doi: 10.1016/j.scitotenv.2020.140932.

Larsen, K., Christensen, T.B., Højberg, O., Sørensen, M.T. (2022) Exposure of pigs to glyphosate affects gene-specific DNA methylation and gene expression. Toxicol. Rep. 9: 298–310. doi: 10.1016/j.toxrep.2022.02.007.

Leoni, C., Balduzzi, M., Buratti, F.M., Testai, E. (2012) The contribution of human small intestine to chlorpyrifos biotransformation. Toxicol. Lett. 215(1): 42–48. doi: 10.1016/j.toxlet.2012.09.016.

Liang, Y., Zhan, J., Liu, D., Luo, M., Han, J., Liu, X., Liu, C., Cheng, Z., Zhou, Z., Wang, P. (2019) Organophosphorus pesticide chlorpyrifos intake promotes obesity and insulin resistance through impacting gut and gut microbiota. Microbiome 7(1): 19. doi: 10.1186/s40168-019-0635-4.

Lidova, J., Stara, A., Kouba, A., Velisek, J. (2016) The effects of cypermethrin on oxidative stress and antioxidant biomarkers in marbled crayfish (Procambarus fallax f. virginalis). Neuro. Endocrinol. Lett. 37(Suppl1): 53–59.

Lozano, V.L., Defarge, N., Rocque, L.-M., Mesnage, R., Hennequin, D., Cassier, R., de Vendomois, J.S., Panoff, J.-M., Séralini, G.-E., Amiel, C. (2018) Sex-dependent impact of Roundup on the rat gut microbiome. Toxicol. Rep. 5: 96–107. doi: 10.1016/j.toxrep.2017.12.005.

McCoy, M.R., Yang, Z., Fu, X., Ahn, K.C., Gee, S.J., Bom, D.C., Zhong, P., Chang, D., Hammock, B.D. (2012) Monitoring of total type ii pyrethroid pesticides in citrus oils and water by converting to a common product 3-phenoxybenzoic acid. J. Agric. Food Chem. 60(20): 5065–5070. doi: 10.1021/jf2051653.

Mekonen, S., Ambelu, A., Wondafrash, M., Kolsteren, P., and Spanoghe, P. (2021) Exposure of infants to organochlorine pesticides from breast milk consumption in southwestern Ethiopia. Sci. Rep. 11: 22053. doi: 10.1038/s41598-021-01656-x.

Meng, Q., Zhang, Y., Li, J., Shi, B., Ma, Q., Shan, A. (2022) Lycopene affects intestinal barrier function and the gut microbiota in weaned piglets via antioxidant signaling regulation. J. Nutr. 9: nxac208. doi: 10.1093/jn/nxac208.

Metzler-Zebeli, B.U., Mann, E., Schmitz-Esser, S., Wagner, M., Ritzmann, M., Zebeli, Q. Changing dietary calcium-phosphorus level and cereal source selectively alters abundance of bacteria and metabolites in the upper gastrointestinal tracts of weaned pigs. Appl. Environ. Microbiol. 79(23): 7264–7272. doi: 10.1128/AEM.02691-13.

Metzler-Zebeli, B.U., Newman, M.A., Ladinig, A., Kandler, W., Grüll, D., Zebeli, Q. Transglycosylated starch accelerated intestinal transit and enhanced bacterial fermentation in the large intestine using a pig model. Br. J. Nutr. 122(1): 1–13. doi: 10.1017/S0007114519000849.

Metzler-Zebeli, B.U. (2022) Chapter 8: Porcine Gut Microbiota and Host Interactions during the Transition from the Suckling to Post-Weaning Phase. In: Gut Microbiota, Immunity, and Health in Production Animals (Kogut, M.H., Zhang, G., eds.). Springer International Publishing.

Mizobuchi, M., Ishidoh, K., Kamemura, N. (2022) A comparison of cell death mechanisms of antioxidants, butylated hydroxyanisole and butylated hydroxytoluene. Drug Chem. Toxicol. 45(4): 1899–1906. doi: 10.1080/01480545.2021.1894701.

Muntané, J., Rodríguez, F.J., Segado, O., Quintero, A., Lozano, J.M., Siendones, E., Pedraza, C.A., Delgado, M., O'Valle, F., García, R., Montero, J.L., De La Mata, M., Miño, G. (2000) TNF-alpha dependent production of inducible nitric oxide is involved in PGE(1) protection against acute liver injury. Gut 47(4): 553–562. doi: 10.1136/gut.47.4.553.

Narahashi, T. (1996) Neuronal ion channels as the target sites of insecticides. Pharmacol. Toxicol. 79(1): 1–14. doi: 10.1111/j.1600-0773.1996.tb00234.x.

Newman, M.A., Petri, R.M., Grüll, D., Zebeli, Q., Metzler-Zebeli, B.U. (2018) Transglycosylated starch modulates the gut microbiome and expression of genes related to lipid synthesis in liver and adipose tissue of pigs. Front. Microbiol. 9: 224. doi: 10.3389/fmicb.2018.00224.

Nielsen, L.N., Roager, H.M., Casas, M.E., Frandsen, H.L., Gosewinkel, U., Bester, K., Licht, T.R., Hendriksen, N.B., Bahl, M.I. (2018) Glyphosate has limited short-term effects on commensal bacterial community composition in the gut environment due to sufficient aromatic amino acid levels. Environ. Pollut. 233: 364–376. doi: 10.1016/j.envpol.2017.10.016.

Oliveira, J.M., Condessa, S.S., Destro, A.L.F., Lima, G.D.A., do Carmo Cupertino, M., Cardoso, S.A., Freitas, M.B., de Oliveira, L.L. (2022) Morphophysiological alterations in fruit-eating bats after oral exposure to deltamethrin. Int. J. Exp. Pathol. 103(5): 219–230. doi: 10.1111/iep.12455.

Papadopoulou, E.S., Bachtsevani, E., Papazlatani, C.V., Rousidou, C., Brouziotis, A., Lampronikou, E., Tsiknia, M., Vasileiadis, S., Ipsilantis, I., Menkissoglu-Spiroudi, U.,

Ehaliotis, C., Philippot, L., Nicol, G.W., Karpouzas, D.G. (2022) The effects of quinone imine, a new potent nitrification inhibitor, dicyandiamide, and nitrapyrin on target and off-target soil microbiota. Microbiol. Spectr. 10(4): e0240321. doi: 10.1128/spectrum.02403-21.

Pié, S., Lallès, J.P., Blazy, F., Laffitte, J., Sève, B., Oswald, I.P. (2004) Weaning is associated with an upregulation of expression of inflammatory cytokines in the intestine of piglets. J. Nutr. 134(3): 641–647. doi: 10.1093/jn/134.3.641.

Qiu, S., Fu, H., Zhou, R., Yang, Z., Bai, G., Shi, B. (2019) Toxic effects of glyphosate on intestinal morphology, antioxidant capacity and barrier function in weaned piglets. Ecotoxicol. Environ. Saf. 187: 109846. doi: 10.1016/j.ecoenv.2019.109846.

Radcliffe, J.S. (2011) Nonruminant Nutrition Symposium: nutrient and non-nutrient sensing and signaling in the gastrointestinal tract. J. Anim. Sci. 89(6): 1853. doi:10.2527/jas.2011-4117.

Réquilé, M., Gonzàlez Alvarez, D.O., Delanaud, S., Rhazi, L., Bach, V., Depeint, F., Khorsi-Cauet, H. Use of a combination of in vitro models to investigate the impact of chlorpyrifos and inulin on the intestinal microbiota and the permeability of the intestinal mucosa. Environ. Sci. Pollut. Res. Int. 25(23): 22529–22540. doi: 10.1007/s11356-018-2332-4.

Reygner, J., Joly Condette, C., Bruneau, A., Delanaud, S., Rhazi, L., Depeint, F., Abdennebi-Najar, L., Bach, V., Mayeur, C., Khorsi-Cauet, H. (2016) Changes in composition and function of human intestinal microbiota exposed to chlorpyrifos in oil as assessed by the SHIME® model. Int. J. Environ. Res. Public. Health 13(11): 1088. doi: 10.3390/ijerph13111088.

Reygner, J., Lichtenberger, L., Elmhiri, G., Dou, S., Bahi-Jaber, N., Rhazi, L., Depeint, F., Bach, V., Khorsi-Cauet, H., Abdennebi-Najar, L. (2016) Inulin supplementation lowered the metabolic defects of prolonged exposure to chlorpyrifos from gestation to young adult stage in offspring rats. PLoS One 11(10): e0164614. doi: 10.1371/journal.pone.0164614.

Spencer, C.I., Yuill, K.H., Borg, J.J., Hancox, J.C., Kozlowski, R.Z. Actions of pyrethroid insecticides on sodium currents, action potentials, and contractile rhythm in isolated mammalian ventricular myocytes and perfused hearts. J. Pharmacol. Exp. Ther. 298(3): 1067–1082. Erratum in: J. Pharmacol. Exp. Ther. (2001) 299(1): 399.

Steinrücken HC, Amrhein N. The herbicide glyphosate is a potent inhibitor of 5-enolpyruvylshikimic acid-3-phosphate synthase. Biochem Biophys Res Commun. 1980 Jun 30; 94(4): 1207–12. doi: 10.1016/0006-291x(80)90547-1. PMID: 7396959.

Taira, K., Fujioka, K., Aoyama, Y. (2013) Qualitative profiling and quantification of neonicotinoid metabolites in human urine by liquid chromatography coupled with mass spectrometry. PLoS One 8(11): e80332. doi: 10.1371/journal.pone.0080332.

Tang, Q., Tang, J., Ren, X., Li, C. (2020) Glyphosate exposure induces inflammatory responses in the small intestine and alters gut microbial composition in rats. Environ. Pollut. 261: 114129. doi: 10.1016/j.envpol.2020.114129.

Tejada, M. 2009. Evolution of soil biological properties after addition of glyphosate, diflufenican and glyphosate+diflufenican herbicides. Chemosphere 76: 365–373. doi: 10.1016/j.chemosphere.2009.03.040.

Velmurugan, G., Ramprasath, T., Swaminathan, K., Mithieux, G., Rajendhran, J., Dhivakar, M., Parthasarathy, A., Babu, D.D., Thumburaj, L.J., Freddy, A.J., Dinakaran, V.,

Puhari, S.S., Rekha, B., Christy, Y.J., Anusha, S., Divya, G., Suganya, K., Meganathan, B., Kalyanaraman, N., Vasudevan, V., Kamaraj, R., Karthik, M., Jeyakumar, B., Abhishek, A., Paul, E., Pushpanathan, M., Rajmohan, R.K., Velayutham, K., Lyon, A.R., Ramasamy, S. (2017) Gut microbial degradation of organophosphate insecticides-induces glucose intolerance via gluconeogenesis. Genome Biol. 18(1): 8. doi: 10.1186/s13059-016-1134-6.

Vicini, J.L., Reeves, W.R., Swarthout, J.T., Karberg, K.A. (2019) Glyphosate in livestock: feed residues and animal health1. J. Anim. Sci. 97(11): 4509–4518. doi: 10.1093/jas/skz295.

Wang, W., Zhai, D., Bai, Y., Xue, K., Deng, L., Ma, L., Du, T., Ye, Z., Qu, D., Xiang, A., Chen, G., Zhao, Y., Wang, L., Lu, Z. (2021) Loss of QKI in macrophage aggravates inflammatory bowel disease through amplified ROS signaling and microbiota disproportion. Cell Death Discov. 7(1): 58. doi: 10.1038/s41420-021-00444-w.

Wang, X., Lu, Q., Guo, J., Ares, I., Martínez, M., Martínez-Larrañaga, M.R., Wang, X., Anadón, A., Martínez, M.A. (2022) Oxidative stress and metabolism: A mechanistic insight for glyphosate toxicology. Ann. Rev. Pharmacol. Toxicol. 62: 617–639. doi: 10.1146/annurev-pharmtox-020821-111552.

Wu, H., Gao, J., Xie, M., Wu, J., Song, R., Yuan, X., Wu, Y., Ou, D. (2022) Chronic exposure to deltamethrin disrupts intestinal health and intestinal microbiota in juvenile crucian carp. Ecotoxicol. Environ. Saf. 241: 113732. doi: 10.1016/j.ecoenv.2022.113732.

Würtzen, G., Olsen, P. 1986. BHA study in pigs. Food Chem. Toxicol. 24(10–11): 1229–1233. doi: 10.1016/0278-6915(86)90311-x.

Xu, X., Liu, A., Hu, S., Ares, I., Martínez-Larrañaga, M.R., Wang, X., Martínez, M., Anadón, A., Martínez, M.A. (2021) Synthetic phenolic antioxidants: Metabolism, hazards and mechanism of action. Food Chem. 353: 129488. doi: 10.1016/j.foodchem.2021.129488.

Zhao, H., Wang, Y., Guo, M., Liu, Y., Yu, H., Xing, M. (2021) Environmentally relevant concentration of cypermethrin or/and sulfamethoxazole induce neurotoxicity of grass carp: Involvement of blood-brain barrier, oxidative stress and apoptosis. Sci. Total Environ. 762: 143054. doi: 10.1016/j.scitotenv.2020.143054.

Zhorov, B.S., Dong, K. (2022) Pyrethroids in an alpha fold2 model of the insect sodium channel. Insects. 2022 Aug 18;13(8): 745. doi: 10.3390/insects13080745.

CHAPTER 7

Gut health: environment, climate change and agriculture sustainability

Elizabeth Santin

I See Inside Institute, Curitiba, Parana, Brazil – 80540-300;
santin.elizabeth21@gmail.com

Abstract

Animal production have a significant impact on sustainability, and gut health plays a crucial role in this context. Issues related to digestion and absorption in the gut can lead to decreased feed efficiency, resulting in increased feed and water consumption as well as higher land use and production costs. Additionally, the accumulation of undigested feed can contaminate soil and contribute to greenhouse gas emissions, leading to environmental problems. There is a close relationship between the animal and its microbial community, known as the holobiont concept. Climate change-induced shifts in temperature and precipitation patterns can alter soil microbiota, which, in turn, affects the nutrient and phytochemical composition of plants. These changes can have profound implications for the gut microbiota of both animals and humans. By taking a comprehensive holistic approach, we can strive to improve the quality of animal products, protect animal welfare, and ultimately, enhance the well-being of our planet.

Keywords

holobiont – microbial community – animal feed efficiency – phytochemical – animal well-being

1 Introduction

> The things of nature do not really belong to us, we should leave them to our children as we have received them. (Oscar Wilde 1882)

This old citation of Wilde, describes the concept of sustainability, a word that has been present in all scientific public media recently. In fact, all society sectors are very interested in reducing the human impact on the Earth planet, after

© ELIZABETH SANTIN, 2024 | DOI:10.3920/9789004695467_008

the constant climate changes tragedies observed in different geographic areas (Glanemann, 2020). With this there is a need for a collective action to became sustainable in every human activity, especially in food production. The stakeholders face a growing need to conceptualize the environmental limits of human activities and determine the earth's carrying capacity (Stepphen *et al*, 2015). To be sustainable means that we should attend today human needs without compromise the resource for future generation (Report of WCED, 1987), and it becomes more evident that there are limited resources that we should preserve to maintain the human life in this planet.

The livestock sector is fundamental to produce the food, textile and other products for human live, but is also an important user of natural resources such as land, water and nutrients as contributes to both greenhouse gases and nutrient losses, the latter often resulting in water and air pollution (FAO, 2018). In the agricultural sector, the livestock industry contributes 53% of the total emissions (Reisinger and Clark, 2018) and a recent consumer report by the International Food Information Council 2021 Food & Health Survey suggests that almost 60% of consumers recognize the need for food products they purchase/consume to be produced in a sustainable way (Sollid *et al*, 2022).

Applying science-based targets for planetary boundaries or safe operating space in human activity is of paramount importance to make changes to improve life on Earth. And even being a very developed area, there is still room to increase resource efficiency in livestock production to reduce the competition for resources with other sectors, limit the emissions and preventing pollution (FAO, 2018). This should be the mission for professionals that work in of agriculture and livestock.

Normally we described the sustainability in three pillars: economic, social, and environmental equity. However, if we open the social pillar of animal production chain, we should expand it to animal welfare, as well as food safety and worker safety as important part of sustainability project. In animal production, according to FAO (2018), any change in feed efficiency can have an impact on the environmental footprint of animal-sourced products, considering that more than 50 percent of the animal production footprint is related to feed ingredients. As an example, the N intake reduction or N digestibility increase will have a proportional effect of reducing excreted N. In this way, feed efficiency can be improved through better feed digestibility and absorption which is related to a healthy gut in animals. In addition, the better gut health is also associated to better animal welfare and more control of foodborne pathogens. Normally the sustainability of any production or service is evaluate through the life cycle assessment (LCA) methodology, which imply in considering all the input (resources) and output (waste and final product) of the production chain up to final disposal to calculate it impacts on environmental, society and economy. From cradle to grave, life cycle assessment (LCA) represents an internationally standardized method (ISO 14040:2006) that quantifies the environmental pressure related to the production of goods and commodities, and its use can result in an improved production process as a strategic management and decision

(Palmieri *et al*, 2017).Considering the animal feed efficiency, it is possible calculate some impact of the animal production on environment and in economic aspect, but, the abroad context of sustainability that included the social aspect, need still to be considered in that calculation. In fact, most of the information of safe operating space at the animal production chain are limited to the farm gate, considering the live animal as a final product. At this point will not be possible to include the lost due to condemnation at slaughterhouse, food safety and most of time, the animal welfare and safety of workers are not even mentioned. Including a more precise parameters to address the economic and social impact is fundamental to stablish decision-making processes that drives in more sustainable products.

There is also scientific evidence regarding the impact of climate changes in environmental microbiota of the soil (Hayden *et al* 2012), and it can affect the crop production and nutritional value, that contribute to the animal gut health, behavior and feed efficiency.

In this scenario, the objective of this chapter will be to instigate the reader to think holistic about the concept of gut health on animal efficiency, health, welfare and food safety as important player in the sustainability aspect. At same point, we will consider the impact of the climate changes in the ecology of soil and crop, showing that all is interconnected and that all should be evaluate as one world and one health.

2 Impact of gut health on sustainability (environmental, social and economic)

2.1 *The concept of gut health*

The gut is a vital organ required for nutrient digestion and absorption and is directly correlate with feed efficiency. In addition, has an extensive area of immune tissue that is fundamental barrier against foreign antigens (microorganisms, toxins, injuries, etc.) and that control the animal metabolism.

Currently, there are many different definitions of gut health, but it can be broadly defined as a balance in the symbiosis between microbiota and the host. It is very appropriate to mention the concept of holobiont defined as a biomolecular network composed of the host plus its associated microbes (Margulis, 1990). The holobiont theory consider the host and microbiota as a unique entity where one can interact with each other in cooperation or mutualism to stablish a balance in order to have the better fitness of both systems (Simon *et al*, 2019). This balance can be affected by the host immune reaction to the microbiome (not only their microorganism but as well their genes and metabolites) and vice-versa. In a balanced relationship, the microbiota will be compartmentalized by the mucus layer and other barriers in the intestinal lumen and a host tolerance will be developed that will be defined as eubiosis (healthy). When there are changes in microbiota and this tolerance is broken, an inflammatory immune reaction (acute or chronic) occurs (dysbiosis).

The acute phase inflammatory response (innate) is a part of the initial defense mechanism which involves systemic and metabolic changes. This inflammatory response has the function of activating additional effector cells and molecules at the site of infection, providing a physical barrier for the spread of infection and promoting the repair of tissues that have undergone changes. Inflammation is characterized by pain, redness, heat, swelling, and loss of function (even temporary) of the affected tissue, which in this case, means less digestion and absorption. The release of chemical mediators (cytokines) by the intestinal cells will coordinate all these responses as well as some changes in the normal metabolism of carbohydrates, proteins and lipids, with a decrease in anabolic pathways and an increase in catabolic pathways. The expected results are alteration in intestinal digestion and absorption rate, associated to a change in metabolic pathways that drift the resources from the growing to immune response.

The activation of immune responses during dysbiosis to promote inflammation in the intestinal mucosa can be measured by histological alterations in intestinal morphology as increased immune cell infiltration, congestion and increase of lamina propria and epithelia thickness through the "I See Inside" methodology (https://isiinstitute.com/) that translates all this intestinal mucosa alteration in numbers, allowing to quantify the impact in animal performance (Belote *et al*, 20).

In the regular way, the primary acute systemic inflammation switches from the proinflammatory phase to the adaptation phase with catabolic state of low-energy requirements and eventually progresses to resolution (Liu *et al*, 2012). In a recent review Soares *et al* (2022) present the concept of low-grade chronical inflammation induced by many environmental, nutritional and infection factors. The exposure to any of these triggers for an extended period (days to weeks) can lead to the above-mentioned chronic intestinal inflammatory state, resulting in a disrupted digestive function, increased oxidative stress, microbiota dysbiosis, loss of barrier functionality, and immune dysfunctions. In practical way, the dysbiosis can be translated as poor feed intake, weight gain and or feed conversion rate or feed efficiency. Also, the metabolic changes in immune response can affect other tissues as muscle and affect the quality of meat.

2.2 *The dysbiosis and sustainability*

As mentioned before, the relationship between microbiota and animals is very sensitive and there are many factors that can affect it including diet, environmental conditions (temperature, humidity, etc.) and management (vaccination, transport, any stress); all of which can affect gut health and poultry sustainability. Considering all these factors, a specific case of dietary proteins affecting gut health issue and sustainability can be cited as example.

2.2.1 Excess of protein and social impact

When there is an excess of dietary crude protein/AA or undigested proteins that reach the low intestine, it will be fermented by some microorganisms producing highly pro-inflammatory metabolites (brain chain fatty acids, biogenic amines, ammonia, etc.) to the intestinal mucosa inducing a dysbiosis. The increase of ammonia and other gases in the barn will affect the well-being of the workers for the odor, toxicity, dirty litter and poor environmental conditions, aspect that should be considered in social sustainability pillar. In addition, there are studies showing that inflammation on intestinal can favor Salmonella (Winter *et al*, 2012), what can also be true for other potent protein fermenters as E. coli, Clostridia and Campylobacter. These microorganisms are related to foodborne disease, and will impact animal health, animal welfare and food safety that are also related to social sustainability impact.

2.2.2 Excess of dietary protein and environmental impact

A study of Professor Kebreab *et al* (2016), from UC Davis, showed that a negative impact of excess of dietary crude protein on the environment for monogastric production occurs mainly from 1) excretion of excess nitrogen and phosphorus, leading to the deterioration of aquatic systems 2) direct green-house gas (**GHG**) emissions from manure storage and application to the field and 3) ammonia emissions responsible for acidification and eutrophication of N-limited ecosystems. In addition, this study also presented that reducing intake of protein and phosphor is the most effective way to reduce environmental impacts; however, this must be achieved without impairing animal performance or having a negative environmental impact. Utilization of supplemental AA to meet protein requirements can reduce N excretion by 8% for every 1%-unit reduction in dietary protein intake.

2.2.3 Excess of dietary protein and Economic impact

One important consequence of the dysbiosis is the poor feed efficiency. This means, we will need more feed to grow the same amount of animal protein, resulting in poor FCR increasing the cost of animal production (economic impact).

2.3 *Rethinking the concept of Feed Conversion Ration (FCR) for sustainability*

As consequence of the dysbiosis, the poor FCR means high impact on LCA of animal production because will need more feed ingredients, more land to grow crops, more energy in the feed mill. There is also an increase excretion of nitrogen, direct **GHG** emissions and ammonia emissions. The improved FCR (2.30 instead of 2.75) reduced Global Warming Potential (GWP) by up to 15% in pig production systems (Kebreab *et al*, 2016).

Normally, the FCR is the equation of amount of feed intake by final weight gain of live animal. But the final product is meat, egg, milk, plus the subproduct. To consider the FCR as valuable parameter for sustainability in animal chain, the concept of it should be reviewed and be a equation of feed intake and final product (kg of feed: kg of saleable product, including all the subproduct of it as feather and other similar subproduct as a example. This will be a huge change in the way we calculate the efficiency of livestock sector.

TABLE 7.1 Gut health issues impact sustainability

Pillar of sustainability	Parameters that gut health issue impact
1 Environmental equality	Increased excretion of nitrogen Direct GHG and ammonia emissions Increased land need for more feed ingredients Increased energy consumption at feed mills (more feed to be produced) acidification and eutrophication of N-limited ecosystems.
2 Animal welfare	Inflammation, intestinal gas, foot lesion (pain? discomfort?).
3 Food safety	Inflammation and surplus of protein are advantageous for food pathogens as *Salmonella, Campylobacter* colonization of the gut
4 Well-being of workers	Odor, gas, dirty litter, poor environmental conditions inside the barn zoonosis by direct transmission from animal to human
5 Financial sustainability	Cost of feed (ingredients and energy) Cost of medication of birds (if needed) More slaughter condemnation (food lesion, food safety, etc.)

Apply precise parameters to evaluate sustainability is a key to really make difference in the decision-making process towards more sustainable products throughout their life cycle. The impact of lost by foodborne pathogens, for example, will probably add another number that could described better the real FCR and could bring more opportunity to improve the livestock production.

In the table 7.1, it is a summary of the most important points that could be considered as sustainability impact of gut health issues.

3 The effect of environmental and climate changes in soil, crops and gut health

Considering the Earth as interconnection of many holobionts, it is possible to understand the cascade impact of the climate change in the environmental, microorganism, animal and human life. There is, of course, the impact of the great weather alteration that results in tragedies as inundation, hurricanes and other that may have direct effect on life, but the mainly topic here will be the mild consequences that have long term effect.

Climate changes will affect the biodiversity in different system and however this could be imperceptive for animal and human health at the begin, it would increase the proportion in the time.

A very interesting study by Greenspan *et al* (2020) presenting the impact of warming in the ecological community, can illustrate what is the climate change effect on microbiota and gut health. These authors, showed in a bromelia ecosystem that climate warming increase the reproduction of mosquitos, and their larvae consume and reduce the population of the bacterial genus *Zoogloea*, which is known for its role in water purification from organic pollutants. When this happened, they observed dispersion in gut microbiota of tadpole associated to poor grow of this last organism. The complex interaction between the different form of life in that community, and its impact in the overall organism in that ecosystem clarify the dynamic that we can expect with the climate changes.

This also point out the holistic view need for sustainability. When it was mentioned that a product is sustainable, it means that each sector involved in that production cycle needs to look closely how to reduce the use of resources and waste delivered at the end. With this, each part is able to reduce their footprint in the ecosystem and this could stop the consequent changes in the ecosystem overall.

Soil is fundamental for life in earth and considered a non-renewable natural resource (Stockman *et al*, 2014; Dazzi and Papa, 2022). The degradation practices in agriculture and climate changes affects the soil and results in cascade events that will change the soil ecosystem that may affect the crop quality (Drissner *et al*, 2007; Austin *et al*, 2009) as well all life in the planet.

Analyze of Australian native grassland soil demonstrates that warming and elevate CO_2 resulted in increased mineral N and NH_4^+, while affecting the soil microbiota (Hayden *et al* 2012).

The reduction in crop nutrients have also been reported by Myers *et al* (2014) showed that crops like wheat, corn, rice and soy growth in high levels of atmospheric CO_2 reduce 10% of zinc, 5% of iron, and 8% of protein (Medek *et al*, 2017) in their content. Zhu *et al* (2018) also find that significant reductions in vitamins B1 (thiamine), B2 (riboflavin), B5 (pantothenic acid), and B9 (folate) in response to projected CO_2 levels in rice.

It brings the concern related to nutrition deficiency of these minerals in humans but also animals. As these nutrients are very important for immune response, it is also expected some impact in the interaction of the animal/human host and different infections challenges and relationship with microbiota. The climate changes are also related to increase of mycotoxins (Warnatzsch *et al*, 2020).

Catania *et al* (2021) described a very interesting hypothesis that the climate changes affect the crops and consequently the diet, that will affect the gut microbiota and cause dysbiosis. The dysbiotic holobiont environmental make changes in microbiota-host immune axis and the abnormal immune response can induce different metabolic disease in the host as low-grade chronical inflammation for example.

However, this hypothesis needs still be proved by trial, all this information brings important aspects that can be affected by climate changes in a cascade of the events that will affect the animal and human life in the planet.

Final remarks

The climate changes will affect all the holobiots system;

The cascate events that will ocurre in soil, plants, animals and human health are still under investigation;

Intestinal Low-grade chronical inflammation without clinic signs is the host response for the changes in holobiont symbiosis;

The metabolic cost of chronic inflammation caused by dysbiosis will affect all five pillar of sustainability including animal and human welfare and food safety;

Calculate the safe operating space for planetary boundaries considering other parameters as FCR adjusted to final saleable product and subproduct, adding the foodborne pathogens and slaughterhouse lost could be more appropriate to decision-making process for sustainable products.

References

Austin, E.E., Castro, H.F., Sides, K.E., Schadt, C.W., and Classen, A.T. (2009) Assessment of 10 years of CO2 fumigation on soil microbial communities and function in a sweetgum plantation. *Soil Biol Biochem* 41: 514–520.

Belote, B.L., Tujimoto-Silva, A., Hümmelgen, P.H., Sanches, A.W.D., Wammes, J.C.S., Hayashi, R.M., Santin, E. Histological Parameters to Evaluate Intestinal Health on Broilers Challenged with Eimeria and C Lostridium Perfringens with or without Enramycin as Growth Promoter. *Poult Sci* 2018, *97*, 2287–2294, doi:10.3382/ps/pey064.

Catania, F., Baedke, J., Fabregas-Tejeda, A., Delgado, A.N. Vitali, V., Long, L.A.N. Global climate change, diet, and the complex relationship between human host and microbiome: Towards and integrated picture. *BioEssays*. 2021;43: 2100049. wileyonlinelibrary.com /journal/bies **1 of 12** https://doi.org/10.1002/bies.202100049.

Dazzi, C., Papa, G. (2022) A new definition of soil to promote soil awareness, sustainability, security and governance International Soil and Water Conservation Research 10 (2022) 99e108.

Drissner, D., Blum, H., Tscherko, D., and Kandeler, E. (2007) Nine years of enriched CO2 changes the function and structural diversity of soil microorganisms in a grassland. *Eur J Soil Sci* 58: 260–269.

Belote, B.L., Tujimoto-Silva, A., Hümmelgen, P.H., Sanches, A.W.D., Wammes, J.C.S., Hayashi, R.M., Santin, E. Histological Parameters to Evaluate Intestinal Health on Broilers Challenged with Eimeria and C Lostridium Perfringens with or without Enramycin as Growth Promoter. *Poult Sci* 2018, *97*, 2287–2294, doi:10.3382/ps/pey064.

FAO, FAOSTAT, Food and agriculture organization of the United Nations. Rome, URL: http://faostat.fao.org, 2018.

Glanemann, N., Willner, S.N. & Levermann, A. Paris Climate Agreement passes the cost-benefit test. *Nat Commun* **11**, 110 (2020). https://doi.org/10.1038/s41467-019-13961-1.

Greenspan, S.E., Migliorini, G.H., Lyra, M.L., Pontes, M.R., Carvalho, T., Ribeiro, L.P., Moura-Campos, D., Haddad, C.F.B., Toledo, L.F., Romero, G.Q., Becker, G. (2020) Warming drives ecological community changes linked to host-associated microbiome dysbiosis. Nature Climate Changes VOL. 10 | NOVEMBER 2020. https://doi.org/10.1038/s41558-020-0899-5.

Hayden, H.L., Mele, P.M., Bougoure, D.S., Allan, C.Y., Norng, S., Piceno, Y.M., Brodie, E.L., DeSantis, T.Z., Andersen, L, Williams, A.L., Hovenden, M.J. Changes in the microbialcommunity structure of bacteria, archae and fungi in response to elevated CO2 and warming in an Australian native grassland soil. 2012. Environmental Microbiology (2012) **14**(12), 3081–3096.

Kebreab, E. , Liedke, A ., Caro, D., Deimling, S. Binder, M., Finkbeiner, M. Environmental impact of using specialty feed ingredients in swine and poultry production: A life cycle assessment. J. Anim. Sci. 2016.94:2664–2681 doi:10.2527/jas2015-9036.

Liu, T.F., Brown, C.M., Gazzar, M. McPhail, L. Millet, P., Rao, A., Vachharajani, V.T., Yoza, B.K., McCall, C.E. Fueling the flame: bioenergy couples metabolism and inflammation. Journal of Leukocyte Biology. Volume 92, September 2012.

Margulis, L. (1990). Words as battle cries: symbiogenesis and the new field of endocytobiology. *Bioscience* 40, 673–677. doi: 10.2307/1311435.

Medek D.E., Schwartz, J., Myers S.S. 2017. Estimated effects of future atmospheric CO2 concentrations on protein intake and the risk of protein deficiency by country and region. Environ Health Perspect 125(8): 087002, PMID: 28885977, https://doi.org/ 10.1289/EHP41.

Myers, S., Zanobetti, A., Kloog, I. *et al.* Increasing CO2 threatens human nutrition. *Nature* 510, 139–142 (2014). https://doi.org/10.1038/nature13179.

Palmieri N., Forleo, M.B., Salimei, E. Environmental impacts of a dairy cheese chain including whey feeding: An italian case study. *J. Clean. Prod.* 2017, *140*, 881–889.

Reisinger A, Clark H. How much do direct livestock emissions actually contribute to global warming? Glob Chang Biol. 2018 Apr;24(4):1749-1761. doi: 10.1111/gcb.13975. Epub 2017 Nov 27. PMID: 29105912.

Simon, JC., Marchesi, J.R., Mougel, C. *et al.* Host-microbiota interactions: from holobiont theory to analysis. *Microbiome* 7, 5 (2019). https://doi.org/10.1186/s40168-019-0619-4

Soares, I., Belote, B.L., Santin, E., Dal Pont, G.C., Kogut, M.H. Morphological Assessment and Biomarkers of Low-Grade, Chronic Intestinal Inflammation in Production Animals. *Animals* 2022, *12*,3036. https://doi.org/10.3390/ani12213036.

Sollid, K.; Webster, A.D.; Paipongna, M.; Smith, K. Food perceptions, beliefs, and behaviors amid a global pandemic: Results of the International Food Information Council 2021 Food & Health Survey. *Nutr. Today* 2022, *57*, 26–33.

Steffen W, Richardson K, Rockström J, Cornell SE, Fetzer I, Bennett EM, Biggs R, Carpenter SR, de Vries W, de Wit CA, Folke C, Gerten D, Heinke J, Mace GM, Persson LM, Ramanathan V, Reyers B, Sörlin S. Sustainability. Planetary boundaries: guiding human development on a changing planet. Science. 2015 Feb 13;347(6223):1259855. doi: 10.1126/science.1259855. Epub 2015 Jan 15. PMID: 25592418.

Stockman, U., Minasny, B., & McBratney, A.B. (2014). How fast does soil grow? Geoderma, 216, 48e61. https://doi.org/10.1016/j.geoderma.2013.10.007.

Warnatzsch. E.A., Reay, D.S, Leggieri, M.C, Battilani, P. Climate Change Impact on Aflatoxin Contamination Risk in Malawi's Maize Crops. Front. Sustain. Food Syst., 2020 Sec. Agro-Food Safety https://doi.org/10.3389/fsufs.2020.591792.

WCED – Report of the World Commission on Environment and Development: Our Common Future, 1987 (https://sustainabledevelopment.un.org/content/documents/5987our-common-future.pdf).

Wilde, O. – 1882 May 17, Ottawa Citizen, Oscar Wilde: Lecture in the Grand Opera House, Quote Page 1, Column 4, Ottawa, Ontario, Canada.

Winter, S.E., Thiennimitr, P., Winter, M.G., Butler, B.P., Huseby, D.L., Crawford, R.W., Russell, J.M., Bevins, C.L., Adams, L.G., Tsolis, R.M., Roth, J.R., Bäumler, A.J. Gut inflammation provides a respiratory electron acceptor for Salmonella. Nature. 2010 Sep 23; 467(7314): 426–9. doi: 10.1038/nature09415. PMID: 20864996; PMCID: PMC2946174.

Zhu, C., Kobayashi, K., Loladze, I., Zhu, J, Jiang, Q. Xu, X., Liu, G., Seneweera, S., Ebi, K., L., Drewnowski, A., Fukagawa, N.K., Ziska, L.H. (2018) Carbon Dioxide (CO2) levels this century will alter the protein, micronutrientes, and vitamin contente of rice grains with potential health consequences for poorest rice-dependent countries. Science Advances Vol. 4., 5 DOI: 10.1126/sciadv.aaq1012.

CHAPTER 8

Mycotoxin interactions, the poultry intestine and the potential impact of climate change

L.J. Broom

Gut Health Consultancy, Exeter, UK; guthealthconsultancy@gmail.com

Abstract

Fungi are ubiquitous and some are known to infect plants and animals. Fungi produce both primary and secondary metabolites, with the latter not considered essential for their normal functioning, but can be toxic to animal cells, and thus are referred to as mycotoxins. Studies report that mycotoxin contamination of plant-based feed raw materials, or finished animal feeds, is virtually inevitable, often occurring as a combination of toxins. Deoxynivalenol, fumonisins, and/or zearalenone are most frequently detected as co-contaminants in finished feed surveys. Although poultry are relatively tolerant of these mycotoxins due to their specific toxicokinetics, these toxins are known to have toxic consequences for poultry, although species differences exist. In various regions of the world, there are regulatory or guidance levels for these contaminants in feeds. However, whilst the individual effects of these mycotoxins on the poultry host, particularly chickens, are relatively well known, less is understood about the impact of their interactions for the host, which is very pertinent due to being common co-contaminants. In this chapter, we explore these mycotoxin combinations, at realistic feed concentrations, and their interactions, in poultry, particularly chickens, focusing on the intestine. Although effects are clearly reported to be occurring at the intestinal (and systemic) level, generally, the bird seems to be able to adapt to, or compensate for, these in most studies reviewed, with minimal or no significant impact on bird health or growth performance. However, these alterations could be of significance in other contexts, for example commercial production, and this needs to be further considered or evaluated. Moreover, climate change is expected to alter the frequency and distribution of mycotoxin contamination of feed (raw materials) and thus the poultry industry needs to remain vigilant of the expanding range of known mycotoxins, knowledge of their interactions, and the potential consequences for commercial poultry production.

Keywords

deoxynivalenol – fumonisins – zearalenone – avian – health

© L.J. Broom, 2024 | DOI:10.3920/9789004695467_009

1 Introduction

The fungal kingdom comprises millions of eukaryotic species that are ubiquitous. A proportion of these fungal species are known to infect plants and animals (Fisher *et al.* 2020). In the case of plant-fungal interactions, these can be detrimental to the host plant, but are often beneficial (Rai and Agarkar, 2016). Fungi produce both primary (required for normal growth, development and reproduction) and secondary metabolites, with the latter not considered necessary for essential host functions. Fungal secondary metabolites do, however, show biological activities, which can include toxicity to animal cells (*i.e.*, mycotoxins). Some mystery remains around the relevance of these secondary metabolites to the fungus, but the common belief is that they provide protection and/or give them a competitive advantage within their ecological setting (Venkatesh and Keller, 2019). In support of this, growth in mixed communities (*i.e.*, ≥ 2 microbes) is a strong activator of gene clusters responsible for the production of these secondary metabolites. Mycotoxins themselves can have direct antimicrobial activity and/or interfere with quorum sensing microbial communication, thus influencing the growth of microbes, expression of virulence proteins, the formation of biofilms, etc., (Venkatesh and Keller, 2019).

Numerous factors influence fungal growth (and mycotoxin production), with temperature, water activity and humidity particularly well recognised, and pH, substrate(s) and fungal strain also important (Awuchi *et al.* 2022). Stress factors (*e.g.*, competition, drought, plant damage) can also promote fungal activity and mycotoxin production. Most fungi are mesophilic, with growth between 5 and 35 °C, and growth optima between 25 and 30 °C (Daou *et al.* 2021). The optimal temperature for mycotoxin production is considered to be between 20 to 30 °C, indicating that fungal growth and mycotoxin production are unlikely to be perfectly, positively, correlated. Moreover, there can be interactions between fungal growth- and/or mycotoxin production-dependent factors (Medina *et al.* 2013). If conditions permit, fungi can grow and/or produce their mycotoxins in the field or during the harvest, transport or storage of feed raw materials. Generally, aflatoxins have tended to be a particular concern in the tropics and subtropics, whereas fusariotoxins (produced by *Fusarium* spp.) have been more significant in temperate climates (Gruber-Dorninger *et al.* 2019). Climate change is expected to potentially influence the geographical distribution of crops and the regional occurrence of fungi and mycotoxins (Zingales *et al.* 2022). Elevated global temperatures associated with climate change, and the associated increased frequency of 'extreme' weather events (*e.g.*, drought or prolonged rainfall), is likely to increase the global distribution of fungi and the conditions that induce their production of mycotoxins. A 10-year mycotoxin survey of feed and feed raw material samples was able to correlate extreme weather conditions with much higher mycotoxin contamination levels than those typically observed in that particular region (Gruber-Dorninger *et al.* 2019). Moreover, global trade flows of feed raw materials also mean that environmental impacts on mycotoxin contamination in one region of the world can impact poultry production in another. Due to climate change, it can be predicted that mycotoxin contamination of animal feedstuffs is highly likely to be more prevalent moving forward and thus an even more significant challenge to the health, productivity and welfare of livestock.

This association of fungi with plants, invariably as part of mixed microbial communities (Berg *et al.* 2021), means that mycotoxin contamination of plant-based feed raw materials, or finished animal feeds, is virtually inevitable. Analytical studies confirm this. Using LC-MS/MS analysis, Streit *et al.* (2013) found that all 83 tested feed and feed ingredient samples were contaminated. Similarly, but utilising various analytical methods (*i.e* ELISA, HPLC, and LC-MS/MS), 88% of 74,821 feed and feed raw material samples were contaminated in a recent survey spanning a ten-year period (Gruber-Dorninger *et al.* 2019). Deoxynivalenol (DON) (64% of samples), fumonisins (FUM) (60%), and zearalenone (ZEN) (45%) were most frequently detected, followed by aflatoxin B1 (23%), T-2 (19%), and ochratoxin A (15%). These mycotoxins, produced by *Aspergillus, Fusarium* and *Penicillium* species, have, traditionally, been a main focus due to their established toxicity, recognised occurrence, and, therefore, the implementation of guidance or regulated limits in food and feedstuffs. There are, however, over 500 currently known mycotoxins, the majority of which we are only just beginning to understand (Khoshal *et al.* 2019). For reference, current EU regulatory limits for DON, FUM and ZEN are 5, 20, and 0.5 mg/kg feed, respectively.

Various fungal taxa, each potentially producing multiple mycotoxins, can be present on a particular feed raw material, which are likely to be mixed with other similarly contaminated ingredients prior to feeding, almost certainly exposing animals to a cocktail of mycotoxins (Gruber-Dorninger *et al.* 2019, Khoshal *et al.* 2019, Streit *et al.* 2013). The study by Gruber-Dorninger *et al.* (2019) probably represents one of, if not, the largest mycotoxin surveys of feed and feed raw material samples to date. From this work, they reported that the most frequent co-contaminants of finished feeds were DON and FUM (48% of samples), DON and ZEN (48%), or FUM and ZEN (43%), identifying these combinations as deserving of particular focus under practical conditions (Table 8.1). For the reasons outlined (*e.g.*, predicted increase in mycotoxin contamination due to climate change and the co-occurrence of mycotoxins), this review will explore these DON, FUM and/or ZEN combinations, at real-world concentrations, and their interactions, in poultry, particularly chickens.

2 Toxicokinetics

DON-3-sulphate (major DON metabolite in chickens), FUM, and ZEN are detected in chicken plasma within 30 minutes of oral administration, suggesting rapid absorption primarily from the proximal GIT (Antonissen *et al.* 2015a, Devreese *et al.* 2015, Lauwers *et al.* 2019). It is noteworthy that Lauwers *et al.* (2019)

TABLE 8.1 Percentage of global feed samples positive for deoxynivalenol, fumonisins, zearalenone, and their co-occurrences

Mycotoxin	Positive finished feed samples (%)	Co-occurrence with deoxynivalenol in feed (%)	Co-occurrence with fumonisins in feed (%)	Co-occurrence with zearalenone in feed (%)
Deoxynivalenol	70	–	48	48
Fumonisins	73	48	–	43
Zearalenone	56	48	43	–

Sample positive data from Gruber-Dorninger *et al.* 2019

detected only DON-3-sulphate in chicken plasma, and not DON itself, which may help guide appropriate metabolite analysis in relevant matrices. There is limited information on how the co-occurrence of fusariotoxins in samples, and thus the poultry gastrointestinal tract, may alter their absorption, but work by Antonissen *et al.* (2015a) indicated that chronic exposure to DON (3.12 mg DON/kg feed) for 3 weeks did not affect the rate at which FUM B1 was detected in chicken plasma, or other toxicokinetic parameters, following administration of a single oral bolus of 2.5 mg FUM/kg bodyweight. Ingested mycotoxins may be metabolised to more or less toxic compounds by gut microbes, enteric tissues, and/or, upon leaving intestinal tissues, the liver (principally) and/or kidneys (Guerre, 2020). Biotransformation of mycotoxins typically occurs in two stages (phase I and phase II), mediated in large part by cellular cytochrome P450 enzymes (Tran *et al.* 2020), which have been identified in chickens (Antonissen *et al.* 2017a). Based on mycotoxin structure, phase I can include hydroxylation, oxidation, or reduction, followed by phase II conjugation reactions (*e.g.*, covalent binding of glucuronic acid or sulphate). In chickens, DON is notably converted to DON-3-sulphate, FUM to hydrolysed forms, and ZEN to α-zearalenol. Generally, the oral bioavailability of most mycotoxins is low in poultry when compared with other animals due to rapid and efficient elimination into excreta (via bile and/or urine), typically limiting tissue accumulation and persistence, but this may obscure 'true' exposure (den Hollander *et al.* 2021). Such aspects, including enterohepatic recirculation, feed intake, affinity for plasma proteins, etc., determine the toxicokinetics (*i.e* absorption, distribution, metabolism, or excretion (ADME)) of a particular mycotoxin within the host and help explain relative sensitivities between species and individuals (Devreese *et al.* 2018). In general, due to their specific toxicokinetics, poultry are considered relatively tolerant of DON, FUM, and ZEN, than say swine, although differences in sensitivity exist between avian species, and assessments may be revised as new data emerge (Escrivá *et al.* 2015, EFSA, 2022). This is a concise overview of the toxicokinetics of these mycotoxins in poultry, having been thoroughly reviewed relatively recently (Guerre *et al.* 2015).

3 Basic toxicology

Studies have confirmed the toxic effects of mycotoxins on poultry cells and organs. Intestinal epithelial cells, hepatocytes, and nephron cells are an obvious focus for ingested mycotoxin toxicity, being principal sites in their absorption, metabolism, and/or excretion. Trichothecene mycotoxins (*e.g.* deoxynivalenol) have been found to bind to, and inhibit, the A-site of the peptidyl transferase centre on the large ribosomal subunit (60S) of eukaryotes, impairing peptide bond formation during the elongation cycle of protein synthesis (Garreau de Loubresse *et al.* 2014). This inhibition induces a ribotoxic stress response, which can result in the activation of numerous protein kinases, changes in gene expression, inhibited protein synthesis, and impaired cell functioning (Awad and Zentek, 2015, Maresca, 2013). FUM B1 is structurally similar to sphingolipids (So) and sphinganine (Sa) and thus can inhibit ceramide synthase and sphingolipid biosynthesis, leading to a reduction in ceramide and the accumulation of free So and Sa in tissues, serum and urine, which are considered cytotoxic and pro-apoptotic (Escrivá *et al.* 2015). Importantly, sphingolipids are membrane lipids involved in cell integrity and regulation, as well as participating in signal transduction pathways as secondary messengers. Similarly, ZEN has a similar structure to 17β-oestradiol, a principal female sex hormone, allowing it to bind to host cells expressing oestrogen receptors, activate related cellular pathways, and thus replicate the physiological effects of the hormone (Devreese *et al.* 2015). Although not chicken-specific, a recent, comprehensive, metabolic profiling study generally confirmed these, as well as related and/or broader, effects of the major mycotoxins using a human hepatic cancer cell line (Gerdemann *et al.* 2022). Indeed, these mycotoxins have all been shown to not only affect the focal cells and related organs mentioned previously, but also to induce changes in the structure and function of other organs and cells (notably immune-related), serum biochemistry, productive parameters, and/or pathogen susceptibility of poultry (Awad *et al.* 2013, Ochieng *et al.* 2021).

4 Deoxynivalenol and fumonisins combinations

As highlighted previously, DON and FUM are relatively common co-contaminants of finished feeds. Feeding one-day-old broilers with diets contaminated with 25.4 mg FUM B1 + B2, 4.6 mg DON, or their combination (4.3 mg DON and 22.9 mg FUM B1 + B2, per kg of feed), significantly down-regulated mucin 2 (*MUC2*) expression in the duodenum after 15 days of feeding, with no affect in the jejunum or ileum, and no differences between the mycotoxin treatments (Antonissen *et al.* 2015b). *MUC2* is the principal mucin secreted by goblet cells into the gastrointestinal tract, which forms the structural component of mucus (Bar Shira and Friedman, 2018). Mucus performs multiple functions within the gut, notably providing a protective layer for the epithelium (Duangnumsawang *et al.* 2021). These results could support the notion that these mycotoxins are absorbed from the proximal gut and may, therefore, have more impact in these regions. In addition, although DON and FUM are reported to have differing primary toxicological effects, it is noteworthy that both had similar effects on *MUC2* expression, which was not exacerbated or diminished by their combination. Similarly, in the study, mucin monosaccharide composition was also only altered in the duodenum, and influenced more by FUM, with no DON interactions. N-acetylgalactosamine (increased), galactose (decreased), and N-acetylneuraminic acid (increased) were most affected. The mucin monosaccharide composition determines its properties (*e.g.*, charge, microbial degradation resistance) and thus mucus functionality (Duangnumsawang *et al.* 2021). If the down-regulated MUC2 expression actually translates to a relative decrease in protein secretion, then the quantity of mucus will be diminished and contribute to altered functional characteristics. Further work would be required to understand the significance of these modifications to the bird (*e.g.*, enteric pathogen susceptibility). Antonissen *et al.* (2015b) also investigated the expression of a panel of genes related to oxidative stress, digestive enzymes or nutrient transporters in the intestine. Interestingly, the expression of some of these genes were altered in the jejunum or ileum, which indicates the impact of these mycotoxins, at these levels, extended beyond the duodenum. Similarly, there was little evidence that the mycotoxin combination was more influential than either alone, and, for a number of transporters in the jejunum, their co-administration seemed to actually negate the effect of one of the mycotoxins (*i.e.* FUM) alone, which indicates the potential of one mycotoxin (*e.g.* DON) to interfere with, or block, the action of the other. Curiously though, the changes in transporter expression seen in the jejunum struggle to explain the increased apparent ileal methionine digestibility in the DON plus FUM group compared to control, which was not observed for either DON or FUM alone, indicating other factors were at play. Another study reported that similar concentrations of DON and FUM (5.0 mg/kg and 20.0 mg/kg feed, respectively), in combination but not separately, reduced apparent ileal energy digestibility from 3,126 to 2,895 kcal/kg feed and reduced bodyweight gain in broiler chickens to day 21 (Liu *et al.* 2020), further suggesting the potential for this mycotoxin combination to affect nutrient digestibility. Other work has reported that there were no effects of feeding close to EU guidance levels of DON and/or FUM for 15 days on the intestinal expression of the cytochrome P450 enzymes that the author's assessed, other than up-regulation of CYP1A4 in the jejunum with the FUM only diet (Antonissen *et al.* 2017a). This suggests that the jejunum could be a relevant site for FUM metabolism and that combined feeding with DON shifts this dynamic. However, in another study that was mentioned previously, chronic exposure to DON below EU maximum guidance levels (3.12 mg/kg feed) did not appear to influence the toxicokinetics of a single administration of FUM (2.5 mg FUM/kg bodyweight), indicating that intestinal barrier function as it relates to FUM, and (thus) FUM absorption, were not altered by DON (Antonissen *et al.* 2015a). It is also worth noting that chronic exposure to DON did not affect bird growth performance or macroscopic pathology. Similarly, these parameters (including (zero) mortality) were not significantly different between mycotoxin treatment groups in Antonissen *et al.* (2015b), suggesting that the intestinal changes observed were not detrimental under the conditions of the study.

Grenier *et al.* (2016) investigated the effects of DON, FUM or their combination, at realistic concentrations (1.5 mg DON/kg and/or 20 mg FUM/kg, respectively), in broiler chickens challenged with a high dose (×25 recommended) of a coccidiosis vaccine, or not. In unchallenged birds, there were little, if any, effects of the mycotoxins on the assessed parameters. However, in challenged birds there were notable differences. Both DON or FUM alone increased lesion scores (LS) in the jejunum and caecum, but their combination attenuated LS in the jejunum, but not the caecum. Given lesions at these two sites invariably result from different components of the overdosed vaccine (*Eimeria maxima* vs. *E. tenella*), these results suggest differing interactions between the mycotoxin combination, infecting *Eimeria* spp., and intestinal region, being potentially protective in the jejunum, but this was not supported by oocyst counts in the jejunal mucosa. In faeces, FUM increased oocyst counts, but DON did not, and DON + FUM was intermediate. Importantly, oocyst counts clearly help indicate infection status, but do not fully capture infection dynamics (Soutter *et al.* 2020). In challenged birds, jejunal expression of key genes were elevated by DON, such as IL-1β, IL-6, IL-8, whereas the DON + FUM combination moderated these increases and increased IL-10 and SOCS-1, which may have contributed to the reduced jejunal pathology seen in these birds. However, there were no significant differences in the performance or mortality (none) of challenged birds between the diets, which proposes that the mycotoxin-related changes that were detected were not detrimental or beneficial to them in the context of the study challenge, but could otherwise be notable (*e.g.*, secondary infections). Overall, based on 20 endpoints, the authors identified 3, 9, 2 or 6 endpoints with antagonistic, additive, synergistic or no interactions, respectively, between DON and FUM. In a similar manner, with plasma corticosterone concentration as an endpoint after feeding broiler chickens a DON and/or FUM contaminated diet (close to EU guidance levels) for 15 days, all diets increased this variable (trend for FUM), but there was no evidence of additive or synergistic effects on this marker of stress response, highlighting the nature of the interaction between DON and FUM depends on the parameter(s) under assessment (Antonissen *et al.* 2017b).

The combined effect of DON (4 mg/kg) and FUM (3 mg/kg) on necrotic enteritis challenged broiler chickens has also been assessed. Their combination increased NE lesion scores, serum FITC-D concentration, and caecal tonsil lipopolysaccharide-induced tumor necrosis factor-alpha expression (day 21), but reduced jejunal claudin-1 (day 28 and 35), claudin-2 (day 21, 28 and 35), and zona occludens-1 (day 35) expression, ileum crypt depth (d35), and caecal tonsil (day 35) and spleen (day 21) CD8+: CD4+ ratios compared to the NE challenged control group (Shanmugasundaram *et al.* 2022). The results indicate that the combination of DON and FUM exacerbated the compromised gut barrier integrity induced by NE, possibly through heightening the inflammatory response to NE and impairing cellular functions, such as epithelial tight junction formation, and altered T cell profiles in the caecal tonsil and spleen, leading to notable, numerical (but not significant) differences in bird performance. Unfortunately, there were no individual DON or FUM dietary treatments in the study, which prevents understanding the nature of their interactions when combined, on the basis of their individual effects.

A recent study by Wang *et al.* (2021) is also worthy of attention. FUM B1 and DON were found to be key contaminants of the egg yolk and albumen of breeder hen eggs in China. Inoculating embryonated, specific pathogen free, breeder eggs with combinations of FUM and DON showed some potential associations with reduced hatching rate and pathologies of the gizzard and lungs in their progeny, with residual FUM B1 and, less so, DON detected in these organs. Transfer of mycotoxins from hens to their eggs and progeny, and the consequences of this, deserve further investigation. Future studies should assess whether the presence of mycotoxin combinations in other, particularly intestinal, tissues of the progeny cause any pathologies or affect gut function. Strategies that seek to reduce the use of antimicrobials in poultry production advocate for robust chicks and so any factors that potentially compromise the young bird's (gut) health need to be determined and mitigated.

5 Deoxynivalenol and zearalenone combinations

DON and ZEN are also common co-contaminants of finished feeds. However, there are, generally, a shortage of studies investigating their interactions in poultry, which is likely to be due to ZEN's effects being considered primarily related to reproductive function, and thus less relevant to growing birds, as well as the fact that poultry are considered relatively insensitive to ZEN. However, increasing concentrations of DON (0.265–12.209 mg/kg) and ZEN (0.013–1.094 mg/kg) in Ross broiler chicken diets were reported to linearly decrease the relative density of the duodenum, but linearly increase its length, during week 2 of exposure, while the highest DON and ZEN diet reduced duodenal villus height (Yunus *et al.* 2012). During the 4th week of exposure, the relative density and villus height of both the duodenum and jejunum were linearly reduced, whereas jejunal length was linearly increased. The short circuit current of the jejunum was reduced by increased mycotoxin concentrations during the 5th week of exposure, indicative of reduced ion transport, and thus decreased active absorption of nutrients across the jejunal epithelium, but potentially compensated for by increased length. However, bird growth performance was only affected by the mycotoxin contaminated diets during/until the 3rd week of feeding, but not thereafter, suggesting that the chickens adapted to, or even compensated for, exposure (even though the highest doses exceeded EU regulatory limits), possibly through increased microbial detoxification in the intestine and/or reduced absorption of the mycotoxins (Yunus *et al.* 2012). There have also been reports of DON and ZEN in combination affecting immune cell markers in broiler chickens. Feeding a diet contaminated with DON (2.95 mg/kg) and ZEN (1.59 mg/kg) for two weeks reduced CD4+ cells in the duodenum, but increased IgM-bearing cells in peripheral blood, highlighting the potential for the combination to affect both localised and systemic immune cell populations (Revajová *et al.* 2013).

Other studies have investigated the effects of DON and ZEN in the context of an intestinal challenge, such as coccidiosis. 3.8 mg DON and 0.2 mg ZEN/kg of feed reduced LS in the duodenum, elevated IFN-Y expression by caecal tonsils (post-challenge), increased (post-challenge), then decreased (post-recovery) serum IgA, while IgG remained elevated, decreased (post-recovery) blood CD8+ cells, increased blood B cells (post-recovery), and attenuated the increase in blood monocyte numbers (post-challenge) in Ross 308 broiler breeder female chickens challenged at 8 weeks of age with *E. acervulina*, *E. maxima*, and *E. tenella* at doses to induce lesions without mortality (Girgis *et al.* 2008). These results suggest immunomodulatory effects of the mycotoxin combination in *Eimeria* challenged birds. Decreased numbers of immune cells in the blood could be indicative of their reduced generation and/or increased recruitment to, for example, the intestine, while changes in serum immunoglobulins could reflect alterations in the differentiation of Ig-secreting cells and/or their ability to actually secrete these proteins. The up-regulation of IFN-Y could be potentially beneficial as it has been associated with resistance to coccidia, but there were no differences between diets in terms of post-challenge bodyweight gain or oocyst output, suggesting none of these immune-related changes were either beneficial or detrimental in the context of the coccidial challenge, and despite likely negative impacts on duodenal morphology (*e.g.*, lower villus height and surface area) (Girgis *et al.* 2010a). Girgis *et al.* (2010b) subsequently reported that this combination of mycotoxins, but at higher concentrations (6.5 mg/kg DON and 0.73 mg/kg ZEN), altered the recruitment pattern of CD4+ and CD8+ cells, and macrophages, possibly reflecting delayed or impaired recruitment, to the jejunum in birds challenged with only *E. maxima* at 2 and 4 weeks of age. However, again, it was not apparent that the detected changes (adversely or otherwise) affected the response of the birds to primary, or immunity, to secondary coccidial challenge. It is apparent that the combination of DON and ZEN can alter intestinal characteristics and immune parameters, both in the gut and systemically, including in response to coccidial challenge, but these changes may not have significance and/or the birds can employ compensatory mechanisms. However, it remains to be confirmed whether the changes observed could be relevant for other concurrent, or subsequent, challenges.

6 Fumonisins and zearalenone combinations

Although FUM and ZEN were found together in 43% of the finished feeds analysed (Gruber-Dorninger *et al.* 2019), there is a serious lack of data relating to the effect of this combination in poultry, probably due to the perception that FUM and ZEN (outlined above) have not been considered a significant concern for relevant species. Feeding 7.5 and 0.6 mg/kg feed of FUM and ZEN, respectively, for 14 days to male Ross broiler chickens or Grade Maker turkeys did not cause mortality, signs of mycotoxicosis, impair their growth performance, or impact relative organ weights, including the gizzard, intestine or caecum (Tardieu *et al.* 2021). FUM B1 and α-ZOL were the most abundant metabolites found in the liver. These results suggest that feeding FUM and ZEN at close to, or within, EU regulatory limits do not appear to be detrimental to either chickens or turkeys. However, there were no diets with either FUM or ZEN alone, preventing better appreciation and understanding of their interactions, and there was some evidence of mycotoxin accumulation or persistence in edible tissues, which could represent a food safety issue. Clearly, there is a need for this mycotoxin combination to be further investigated to better appreciate its relevance for poultry species and the food chain.

7 Deoxynivalenol, fumonisins, and zearalenone combinations

Although estimates of the percentage of finished feed samples contaminated with all of DON, FUM and ZEN are hard to come by, co-occurrence figures suggest this combination is likely to be relatively common. Thus, a number of published studies have sought to understand their interactions in poultry species. Feeding male Ross broiler chickens with diets contaminated with either DON (5 mg/kg), FUM (20 mg FUM B1 + B2/kg), ZEN (0.5 mg/kg), or their combination (5, 20, and 0.5 mg/kg of DON, FUM B1 + B2, and ZEN, respectively), did not result in mortality or any signs of mycotoxicosis in any of the birds during a 35-day study (Metayer *et al.* 2019). Relative weights of the gizzard, intestine, and caecum were not different between groups at day 35. Examination of the small intestine segments revealed no effects on measured parameters in the duodenum or ileum, but increased crypt depth and area, and reduced villus:crypt ratio in the jejunum for birds fed the mycotoxins individually or in combination. In terms of performance, however, the only significant differences were for both the ZEN and combined diets to increase feed intake between 11 and 35 days, although these groups were notably heavier (~150 g) than control birds at day 35. This study indicated that the jejunum could be most sensitive to ZEN or the three mycotoxins fed in combination at concentrations in accordance with EU regulatory limits, but there was a general lack of toxic effects and no indications of additive, synergistic or antagonist interactions. Conversely, under commercial conditions, a relatively large, longitudinal study identified a three-way interaction for combinations of DON, FUM and ZEN (below EU regulatory limits) to increase broiler chicken FCR (Kolawole *et al.* 2020), showing their combined potential to impair the efficiency of feed utilisation, which may be more apparent when birds are subjected to infectious and non-infectious stressors in practical production. Moreover, feeding a combined DON, FUM and ZEN diet (5, 20, and 0.5 mg/kg of DON, FUM B1 + B2, and ZEN, respectively) had modest effects on the concentration of FUM in the liver and breast muscle of male, Ross, broiler chickens compared to a FUM only diet (20 mg FUM B1 + B2/kg) (Tardieu *et al.* 2019), while a similar combination diet increased the amount of ZEN metabolites in the liver of chickens versus ZEN alone (Tardieu *et al.* 2020). ZEN and its metabolites, however, were not detected in breast muscle. These results do suggest that the combination of DON, FUM and ZEN at EU relevant levels can alter the function of the gut, oral bioavailability and/or metabolism of the individual mycotoxins in chickens and that the apparent consequences of their combination could be dependent on, for example, experimental facility- vs. commercial-type rearing conditions.

Some studies have also looked at the effect of the DON, FUM and ZEN combination in other poultry species. Similar to chickens, DON, FUM, or ZEN diets (5, 20, or 0.5 mg/kg of DON, FUM B1 + B2, and ZEN, respectively), or their combination, fed to male Grade Maker turkeys between 55 and 70 days of age, had no appreciably negative effects on the birds, with no apparent interactions (Travel *et al.* 2019). Likewise, the mycotoxin combination may have the potential to slightly increase the concentration of FUM in the breast muscle and liver of male Grade Maker turkeys, but with no real effect on ZEN or its metabolites in their livers, compared to diets with either FUM or ZEN alone, respectively (Tardieu *et al.* 2019, Tardieu *et al.* 2020). The dietary combination of DON, FUM, and ZEN diets (5, 20, and 0.5 mg/kg of DON, FUM B1 + B2, and ZEN, respectively) did, however, reduce the growth performance of male mule ducks compared to diets with either mycotoxin alone (Peillod *et al.* 2021). None of the other measured variables indicated interactions between the three mycotoxins that could help explain the impaired performance. These results, perhaps not unexpectedly, reveal some similarities and differences between the avian species in their response to diets with the combination of DON, FUM, and ZEN at EU regulatory limits. For example, their combination appears more detrimental to duck growth performance, while, in chickens, they appear to adversely affect the oral bioavailability and/or metabolism of ZEN. More studies are required to better understand the implications of this fusariotoxin combination on gut function, physiology and growth performance of poultry species, perhaps considering any interactions with their microbial environment.

8 Mycotoxin countermeasures

As fungi can grow and/or produce their mycotoxins in the field or during the harvest, transport or storage of feed raw materials, effective countermeasures during all these stages prior to feed raw materials being fed to animals are important. Prudent agronomic practices, such as preventing insect damage and drought conditions in crops, as well as transporting and storing raw materials appropriately, can all contribute to reduced contamination with mycotoxins (Bryden, 2011). Feed mills usually have minimal control over the management of these raw materials prior to arrival at their facilities, and so they also seek to reduce risk to animals through strategic screening policies with, for example, rapid test strips or NIR to provide qualitative or quantitative assessment (Fumagalli *et al.* 2021). However, surveys of finished feeds still clearly show routine contamination with multiple mycotoxins at variable concentrations (Gruber-Dorninger *et al.* 2019). Therefore, it is common to incorporate 'additives' into feeds to seek further reassurance that the impact of any ingested mycotoxins on the host will be minimised, typically by seeking to either absorb or bind the mycotoxins, or enzymatically convert them to less toxic metabolites, in the intestine (Ochieng *et al.* 2021). However, the various products available will all have differing characteristics and variable efficacy/specificities against the range of possible mycotoxins (Kolawole *et al.* 2019, Ochieng *et al.* 2021). For this reason, monitoring temporal, as well as specific yearly, trends in mycotoxin occurrence, and seeking to utilise additives with relevant, proven efficacy, is currently an appropriate strategy against feed mycotoxins. Such a strategy does, however, need to be practical (*e.g.*, cost-effective, potentially handling multiple products) and should be reviewed on the basis of actual flock health and performance.

9 Conclusions

DON, FUM and ZEN are common contaminants of feed raw materials and finished feeds, and frequently occur in combinations. Recent studies have begun to investigate the impact of these mycotoxin combinations and their interactions at concentrations more relevant to regulated or guidance levels and/or that are reflective of those more normally detected in global raw material and feed samples. Although fusariotoxins are considered to be rapidly absorbed from the proximal GIT, these studies have reported their direct or indirect effects throughout the gut, including on morphology, gene expression, immune cell recruitment, integrity or

TABLE 8.2 Summary of key findings regarding the effects of deoxynivalenol, fumonisins and/or zearalenone combinations in poultry

	Species	Gender, age, mycotoxin concentrations and administration duration	Key findings relating to mycotoxin treatments from relevant studies	References
Deoxynivalenol and fumonisins	Chicken (broiler)	Mixed gender 1-day-old 3.12 mg DON/kg feed for 3 weeks followed by single oral bolus of 2.5 mg FBs/kg bodyweight.	Chronic exposure to DON did not affect the toxicokinetics of a single dose of FUM. No difference in bird growth performance.	Antonissen *et al.* 2015a
	Chicken (broiler)	Mixed gender 1-day-old ~5 mg or ~25 mg DON and/or FUM/kg feed for 15 days	Reduced *MUC2* expression and altered monosaccharide composition in duodenum. Changes in the expression of some genes related to oxidative stress, digestive enzymes or nutrient transporters in the jejunum and ileum. Increased apparent ileal methionine digestibility with DON plus FUM. No differences in growth performance.	Antonissen *et al.* 2015b

TABLE 8.2 Summary of key findings regarding the effects of deoxynivalenol (*cont.*)

	Species	Gender, age, mycotoxin concentrations and administration duration	Key findings relating to mycotoxin treatments from relevant studies	References
	Chicken (broiler)	Male Day of hatch DON plus FUM at 5.0 mg/kg and 20.0 mg/kg feed, respectively, for 21 days.	DON plus FUM reduced apparent ileal energy digestibility and bodyweight gain.	Liu *et al.* 2020
	Chicken (broiler)	Male 1-day-old 1.5 mg DON/kg and/or 20 mg FB/kg, respectively ×25 recommended dose of a coccidiosis vaccine (day 14)	DON or FUM alone increased lesion scores in the jejunum, but their combination reduced. DON + FUM potentially moderated inflammatory response. No significant differences in performance or mortality. Of 20 endpoints, 3, 9, 2 or 6 were identified as antagonistic, additive, synergistic or no interactions, respectively.	Grenier *et al.* 2016

TABLE 8.2 Summary of key findings regarding the effects of deoxynivalenol (*cont.*)

Species	Gender, age, mycotoxin concentrations and administration duration	Key findings relating to mycotoxin treatments from relevant studies	References
Chicken (broiler)	Mixed gender 1-day-old ~4.5 mg or 17.5 mg DON and/or FUM/kg feed for 14 days	No significant indication of interaction between DON and FUM for plasma corticosterone concentration.	Antonissen *et al.* 2017b
Chicken (broiler)	Male 1-day-old DON (4 mg/kg) and FUM (3 mg/kg) to 35 days Necrotic enteritis challenge (days 14–21)	Increased NE lesion scores and gut permeability, and reduced tight junction protein expression, ileal morphology, and caecal tonsil and spleen CD8+: CD4+ ratios. Numerical (but not significant) differences in bird performance.	Shanmugasundaram *et al.* 2022
Breeder eggs	Specific pathogen feed, embryonated eggs (day 11). 24 µg of FB1 and 0.1 µg of DON.	Egg DON and FUM associated with reduced hatching rate and pathologies of the gizzard and lungs in progeny	Wang *et al.* 2021

TABLE 8.2 Summary of key findings regarding the effects of deoxynivalenol (*cont.*)

	Species	**Gender, age, mycotoxin concentrations and administration duration**	**Key findings relating to mycotoxin treatments from relevant studies**	**References**
Deoxynivalenol and zearalenone	Chicken (broiler)	Male 1-day-old 0.265–12.209 mg DON/kg DON and 0.013–1.094 mg ZEN/kg feed for 35 days.	Reduced duodenal or jejunal relative densities or increased length, or decreased jejunal short circuit current, with increasing mycotoxin concentrations. Growth performance similar after 21 days.	Yunus *et al.* 2012
	Chicken (broiler)	Mixed gender 1-day-old 2.95 mg DON/kg and 1.59 mg ZEN/kg for 28 days.	Reduced CD4+ cells in the duodenum, but increased IgM-bearing cells in peripheral blood	Revajová *et al.* 2013
	Chicken (broiler breeder)	Female 1-day-old 3.8 mg DON and 0.2 mg ZEN/kg of feed for 70 days. *E. acervulina, E. maxima,* and *E. tenella* dosed to induce lesions without mortality at 8 weeks of age	Reduced lesion scores in duodenum, changes in blood immune cell populations and antibodies. Reduced duodenal villus height and surface area. No differences in post-challenge bodyweight gain or oocyst output.	Girgis *et al.* 2008 Girgis *et al.* 2010a

TABLE 8.2 Summary of key findings regarding the effects of deoxynivalenol (*cont.*)

	Species	**Gender, age, mycotoxin concentrations and administration duration**	**Key findings relating to mycotoxin treatments from relevant studies**	**References**
	Chicken (broiler breeder)	Female 1-day-old 6.5 mg/kg DON and 0.73 mg/kg ZEN for 42 days. Challenged with *E. maxima* at 2 and 4 weeks of age	Altered recruitment pattern of CD4+, CD8+ cells, and macrophages to the jejunum. Bird responses to primary or secondary challenge apparently not different.	Girgis *et al.* 2010b
Fumonisins and zearalenone	Chicken (broiler) Turkey	Male 20- or 55-day-old 7.5 and 0.6 mg/kg feed of FUM and ZEN, respectively, for 14 days	Potential for FUM or ZEN residues in chicken and turkey livers	Tardieu *et al.* 2021
Deoxynivalenol, fumonisins and zearalenone	Chicken (broiler)	Male 1-day-old 5 mg DON, 20 mg FB1 + FB2, 0.5 mg ZEN, or combination in feed to 35 days	Combination or individual mycotoxins changed jejunal crypt morphology Notable, but non-significant, increases in bodyweight with FUM, ZEN or combo.	Metayer *et al.* 2019
	Chicken (broiler)	Likely mixed sex 1-day-old Variable mycotoxin concentrations (but below EU regulatory limits) ~30 days	Three-way interaction for combinations of DON, FUM and ZEN to increase broiler chicken FCR.	Kolawole *et al.* 2020

TABLE 8.2 Summary of key findings regarding the effects of deoxynivalenol (*cont.*)

Species	Gender, age, mycotoxin concentrations and administration duration	Key findings relating to mycotoxin treatments from relevant studies	References
Ducks	Male 84-day-old 5 mg DON, 20 mg FB1 + FB2, 0.5 mg ZEN, or combination in feed for 12 days	Combination had poorer weight gain and FCR	Peillod *et al.* 2021

pathogen-induced pathology (Table 8.2). Some of these effects could be regarded as favourable and it is clear that these mycotoxins may have no, additive, synergistic, or even, antagonistic interactions, depending on the parameter(s) under investigation. Interestingly though, overall, the effects observed were, generally, not associated with significant differences in bird health or growth performance. These studies do broadly support EU regulatory limits for DON, FUM and ZEN, either alone or in combination. One does wonder though whether the effects of these mycotoxin combinations and concentrations would have more notable impact under field conditions, where the environment is invariably less well controlled and birds invariably encounter more numerous stressors (*e.g.*, pathogens, vaccines). It is noteworthy that a longitudinal field study did find a three-way interaction for the combination of DON, FUM and ZEN, at real-world levels, to impair conversion of feed to chicken liveweight (Kolawole *et al.* 2020). Moreover, the potential for these mycotoxins to accumulate in edible poultry tissues (*e.g.*, liver), or be transferred to the egg and potentially compromise progeny health warrants further consideration. Additional studies are clearly needed, particularly in understanding the impact of interactions between moderate concentrations of prominent mycotoxins on additional gut health parameters, including the microbiome, and better understanding their impact in practical production.

A challenge with these studies is the near impossibility of producing a basal or control diet free of mycotoxins. In addition, the method used to contaminate the 'mycotoxin diet' is important (*i.e.* naturally contaminated feed raw materials, culture material or pure toxin(s)). Naturally contaminated material is often more toxic in research studies, perhaps because it's a complex cocktail of compounds (Metayer *et al.* 2019). The mycotoxins reported in studies may only reflect the focal ones, or only the ones (or a subset of those) analysed for. It is possible that these other mycotoxins are of negligible relevance, but we should be mindful that the 'mycotoxin contaminated' diet is probably a broader cocktail than stated, and often compared to a 'control' diet with some mycotoxin contamination.

Some key determinants of fungal growth and/or their production of mycotoxins are temperature, moisture and stressors. Climate change is causing rising global temperatures, as well as more extreme weather events, such as droughts or extended rainfall, which are predicted to increase mycotoxin (co-) occurrence in globally traded feed raw materials. In the tropics and subtropics, where drought stress during plant growth or high temperatures and humidity around harvest promote aflatoxin B1 contamination, its co-occurrence with FUM or ochratoxin A is noteworthy (Gruber-Dorninger *et al.* 2019). Altered weather patterns associated with climate change are likely to increase aflatoxin B1 contamination of feed raw materials in regions currently considered to have temperate climates, and so more research should be focused on its interactions with fusarotoxins and poultry (gut) health and performance. Similarly, 'emerging' mycotoxins, such as beauvericin, enniatins and moniliformin, are frequently detected in finished feeds, and their interactions with the better studied mycotoxins also needs to be understood (Khoshal *et al.* 2019, Kolawole *et al.* 2020).

Climate change can also be expected to increase the degree and frequency of heat stress experienced by global poultry, which can also impact their gut health (Rostagno, 2020). Therefore, there is a pressing need to better investigate interactions between heat stress and feed mycotoxins. Although mycotoxins can also be considered a stressor and, in conjunction with heat stress, could overwhelm poultry stress response pathways, the relationship between heat stress and realistic dietary mycotoxin concentrations may not be so straightforward and remains to be clearly established. As reviewed here, typical dietary DON, FUM, and/or ZEN concentrations may not exacerbate intestinal disease severity. Similarly, heat stress can actually mitigate the pathogenesis of *Eimeria* spp. infection in chickens, potentially by retarding parasite development and, therefore, altering their interaction with the host (Schneiders *et al.* 2020). This will be an interesting and fruitful avenue for further investigation.

Typical dietary concentrations of DON, FUM, ZEN, and their interactions, as reported here, are not (generally) clearly toxological or performance repressive, although some species differences exist. However, there are clearly effects occurring at the intestinal (and systemic) level that the bird appears largely able to adapt to, or compensate for, in most studies reviewed. It is less clear, however, whether these alterations would have significance in other contexts, for example commercial production. Moreover, the type of interactions that occur between these mycotoxins can range from antagonistic to synergistic, which makes predicting possible outcomes very challenging. The range of identified mycotoxins is expanding, as is our understanding of gut health, and, as the impact of climate change continues to emerge, poultry professionals will need to remain engaged about how these factors are changing and interacting to implement strategies that promote the health, productivity and welfare of poultry.

References

Antonissen, G., Devreese, M., Van Immerseel, F., De Baere, S., Hessenberger, S., Martel, A. and Croubels, S., 2015a. Chronic exposure to deoxynivalenol has no influence on the oral bioavailability of fumonisin B1 in broiler chickens. Toxins 7: 560–571. https://doi.org/10.3390/toxins7020560.

Antonissen, G., Van Immerseel, F., Pasmans, F., Ducatelle, R., Janssens, G.P., De Baere, S., Mountzouris, K.C., Su, S., Wong, E.A., De Meulenaer, B., Verlinden, M., Devreese, M., Haesebrouck, F., Novak, B., Dohnal, I., Martel, A. and Croubels, S., 2015b. Mycotoxins deoxynivalenol and fumonisins alter the extrinsic component of intestinal barrier in broiler chickens. Journal of Agricultural Food Chemistry 63: 10846–55. https://doi.org/10.1021/acs.jafc.5b04119.

Antonissen, G., Devreese, M., De Baere, S., Martel, A., Van Immerseel, F. and Croubels, S., 2017a. Impact of *Fusarium* mycotoxins on hepatic and intestinal mRNA expression of cytochrome P450 enzymes and drug transporters, and on the pharmacokinetics of oral enrofloxacin in broiler chickens. Food and Chemical Toxicology 101: 75–83. https://doi.org/10.1016/j.fct.2017.01.006.

Antonissen, G., De Baere, S., Devreese, M., Van Immerseel, F., Martel, A. and Croubels, S., 2017b. Feed contamination with *Fusarium* mycotoxins induces a corticosterone stress response in broiler chickens. Poultry Science 96: 14–17. https://doi.org/10.3382/ps/pew280.

Awad, W., Ghareeb, K., Böhm J. and Zentek, J., 2013. The toxicological impacts of the *Fusarium* mycotoxin, deoxynivalenol, in poultry flocks with special reference to immunotoxicity. Toxins 5: 912–925. https://doi.org/10.3390/toxins5050912.

Awad, W.A. and Zentek, J., 2015. The feed contaminant deoxynivalenol affects the intestinal barrier permeability through inhibition of protein synthesis. Archives of Toxicology 89: 961–965. https://doi.org/10.1007/s00204-014-1284-9.

Awuchi, C.G., Ondari, E.N., Eseoghene, I.J., Twinomuhwezi, H., Amagwula, I.O. and Morya, S., 2021. Fungal growth and mycotoxins production: types, toxicities, control strategies, and detoxification. In: Sultan, S. and Singh, G.K.S. (eds.) Fungal reproduction and growth. IntechOpen, London, UK, pp. 19–38. https://doi.org/10.5772/intechopen.100207.

Bar Shira, E. and Friedman, A., 2018. Innate immune functions of avian intestinal epithelial cells: response to bacterial stimuli and localization of responding cells in the developing avian digestive tract. PLoS ONE 13: e0200393. https://doi.org/10.1371/journal.pone.0200393.

Berg, G., Kusstatscher, P., Abdelfattah, A., Cernava, T. and Smalla, K., 2021. Microbiome modulation – toward a better understanding of plant microbiome response to microbial inoculants. Frontiers in Microbiology 12: 650610. https://doi.org/10.3389/fmicb.2021.650610.

Bryden, W.L., 2009. Mycotoxins and mycotoxicoses: significance, occurrence and mitigation in the food chain. In: Ballantyne, B., Marrs, T. and Syversen, T. (eds.) General and Applied Toxicology. John Wiley and Sons Ltd, Chichester, UK, pp. 3529–3553. https://doi.org/10.1002/9780470744307.gat157.

Daou, R., Joubrane, K., Maroun, R.G., Khabbaz, L.R., Ismail, A. and El Khoury, A., 2021. Mycotoxins: factors influencing production and control strategies. AIMS Agriculture and Food 6: 416–447. https://doi.org/10.3934/agrfood.2021025.

den Hollander, D., Croubels, S., Lauwers, M., Caekebeke, N., Ringenier, M., De Meyer, F., Reisinger, N., Van Immerseel, F., Dewulf, J. and Antonissen, G., 2021. Applied research note: biomonitoring of mycotoxins in blood serum and feed to assess exposure of broiler chickens. Journal of Applied Poultry Research 30: 100111. https://doi.org/10.1016/j.japr.2020.10.010.

Devreese, M., Antonissen, G., Broekaert, N., De Baere, S., Vanhaecke, L., De Backer, P. and Croubels, S., 2015. Comparative toxicokinetics, absolute oral bioavailability, and biotransformation of zearalenone in different poultry species. Journal of Agricultural Food Chemistry 63: 5092–8. https://doi.org/10.1021/acs.jafc.5b01608.

Devreese, M., Croubels, S., De Baere, S., Gehring, R. and Antonissen, G., 2018. Comparative toxicokinetics and plasma protein binding of ochratoxin A in four avian species. Journal of Agricultural Food Chemistry 66: 2129–2135. https://doi.org/10.1021/acs.jafc.7b06048.

Duangnumsawang, Y., Zentek, J. and Goodarzi Boroojeni, F., 2021. Development and functional properties of intestinal mucus layer in poultry. Frontiers in immunology 12: 745849. https://doi.org/10.3389/fimmu.2021.745849.

EFSA Panel on Contaminants in the Food Chain (CONTAM), 2022. Assessment of information as regards the toxicity of fumonisins for pigs, poultry and horses. EFSA Journal 20: e07534. https://doi.org/10.2903/j.efsa.2022.7534.

Escrivá, L., Font, G. and Manyes, L., 2015. *In vivo* toxicity studies of *Fusarium* mycotoxins in the last decade: A review. Food and Chemical Toxicology 78: 185–206. https://doi.org/10.1016/j.fct.2015.02.005.

Fisher, M.C., Gurr, S.J., Cuomo, C.A., Blehert, D.S., Jin, H., Stukenbrock, E.H., Stajich, J.E., Kahmann, R., Boone, C., Denning, D.W., Gow, N.A.R., Klein, B.S., Kronstad, J.W., Sheppard, D.C., Taylor, J.W., Wright, G.D., Heitman, J., Casadevall, A. and Cowen, L.E., 2020. Threats posed by the fungal kingdom to humans, wildlife, and agriculture. mBio: 11:e00449-20. https://doi.org/10.1128/mBio.00449-20.

Fumagalli, F., Ottoboni, M., Pinotti, L. and Cheli, F., 2021. Integrated mycotoxin management system in the feed supply chain: innovative approaches. Toxins 13: 572. https://doi.org/10.3390/toxins13080572.

Garreau de Loubresse, N., Prokhorova, I., Holtkamp, W., Rodnina, M.V., Yusupova, G. and Yusupov, M., 2014. Structural basis for the inhibition of the eukaryotic ribosome. Nature 513: 517–522. https://doi.org/10.1038/nature13737.

Gerdemann, A., Behrens, M., Esselen, M. and Humpf, H.U., 2022. Metabolic profiling as a powerful tool for the analysis of cellular alterations caused by 20 mycotoxins in HepG2 cells. Archives of Toxicology 96: 2983–2998. https://doi.org/10.1007/s00204-022-03348-5.

Girgis, G.N., Sharif, S., Barta, J.R., Boermans, H.J. and Smith, T.K., 2008. Immunomodulatory effects of feed-borne *Fusarium* mycotoxins in chickens infected with coccidia. Experimental Biology and Medicine 233: 1411–1420. https://doi.org/10.3181/0805-RM-173.

Girgis, G.N., Barta, J.R., Brash, M. and Smith, T.K., 2010a. Morphologic changes in the intestine of broiler breeder pullets fed diets naturally contaminated with *Fusarium* mycotoxins with or without coccidial challenge. Avian Diseases 54: 67–73. https://doi.org/10.1637/8945-052809-Reg.1.

Girgis, G.N., Barta, J.R., Girish, C.K., Karrow, N.A., Boermans, H.J. and Smith, T.K., 2010b. Effects of feed-borne Fusarium mycotoxins and an organic mycotoxin adsorbent on immune cell dynamics in the jejunum of chickens infected with *Eimeria maxima*. Veterinary Immunology and Immunopathology 138: 218–23. https://doi.org/10.1016/j.vetimm.2010.07.018.

Grenier, B., Dohnal, I., Shanmugasundaram, R., Eicher, S.D., Selvaraj, R.K., Schatzmayr, G. and Applegate, T.J., 2016. Susceptibility of broiler chickens to coccidiosis when fed subclinical doses of deoxynivalenol and fumonisins – special emphasis on the immunological response and the mycotoxin interaction. Toxins 8: 231. https://doi.org/10.3390/toxins8080231.

Gruber-Dorninger, C., Jenkins, T. and Schatzmayr, G., 2019. Global mycotoxin occurrence in feed: a ten-year survey. Toxins: 11: 375. https://doi.org/10.3390/toxins11070375.

Guerre, P., 2015. Fusariotoxins in avian species: toxicokinetics, metabolism and persistence in tissues. Toxins 7: 2289–2305. https://doi.org/10.3390/toxins7062289.

Guerre, P., 2020. Mycotoxin and gut microbiota interactions. Toxins 12: 769. https://doi.org/10.3390/toxins12120769.

Khoshal, A.K., Novak, B., Martin, P.G.P., Jenkins, T., Neves, M., Schatzmayr, G., Oswald, I.P. and Pinton, P., 2019. Co-occurrence of DON and emerging mycotoxins in worldwide finished pig feed and their combined toxicity in intestinal cells. Toxins 11: 727. https://doi.org/10.3390/toxins11120727.

Kolawole, O., Meneely, J., Greer, B., Chevallier, O., Jones, D.S., Connolly, L. and Elliott, C., 2019. Comparative *in vitro* assessment of a range of commercial feed additives with multiple mycotoxin binding claims. Toxins 11: 659. https://doi.org/10.3390/toxins11110659.

Kolawole, O., Graham, A., Donaldson, C., Owens, B., Abia, W.A., Meneely, J., Alcorn, M.J., Connolly, L. and Elliott, C.T., 2020. Low doses of mycotoxin mixtures below EU regulatory limits can negatively affect the performance of broiler chickens: a longitudinal study. Toxins 12: 433. https://doi.org/10.3390/toxins12070433.

Lauwers, M., Croubels, S., Letor, B., Gougoulias, C. and Devreese, M., 2019. Biomarkers for exposure as a tool for efficacy testing of a mycotoxin detoxifier in broiler chickens and pigs. Toxins 11: 187. https://doi.org/10.3390/toxins11040187.

Liu, J.D., Doupovec, B., Schatzmayr, D., Murugesan, G.R., Bortoluzzi, C., Villegas, A.M. and Applegate, T.J., 2020. The impact of deoxynivalenol, fumonisins, and their combination on performance, nutrient, and energy digestibility in broiler chickens. Poultry Science 99: 272–279. https://doi.org/10.3382/ps/pez484.

Maresca, M., 2013. From the gut to the brain: journey and pathophysiological effects of the food-associated trichothecene mycotoxin deoxynivalenol. Toxins 5: 784–820. https://doi.org/10.3390/toxins5040784.

Medina, A., Schmidt-Heydt, M., Cárdenas-Chávez, D.L., Parra, R., Geisen, R. and Magan, N., 2013. Integrating toxin gene expression, growth and fumonisin B1 and B2 production by a strain of *Fusarium verticillioides* under different environmental factors. Journal of the Royal Society Interface. http://doi.org/10.1098/rsif.2013.0320.

Metayer, J-P., Travel, A., Mika, A., Bailly, J-D., Cleva, D., Boissieu, C., Guennec, J.L., Froment, P., Albaric, O., Labrut, S., Lepivert, G., Marengue, E., Tardieu, D. and Guerre, P., 2019. Lack of toxic interaction between fusariotoxins in broiler chickens fed throughout their life at the highest level tolerated in the European Union. Toxins 11: 455. https://doi.org/10.3390/toxins11080455.

Ochieng, P.E., Scippo, M.L., Kemboi, D.C., Croubels, S., Okoth, S., Kang'ethe, E.K., Doupovec, B., Gathumbi, J.K., Lindahl, J.F. and Antonissen, G., 2021. Mycotoxins in poultry feed and feed ingredients from sub-Saharan Africa and their impact on the production of broiler and layer chickens: a review. Toxins 13: 633. https://doi.org/10.3390/toxins13090633.

Peillod, C., Laborde, M., Travel, A., Mika, A., Bailly, J.D., Cleva, D., Boissieu, C., Le Guennec, J., Albaric, O., Labrut, S., Froment, P., Tardieu, D. and Guerre, P., 2021. Toxic effects of fumonisins, deoxynivalenol and zearalenone alone and in combination in ducks fed the maximum EU tolerated level. Toxins 13: 152. https://doi.org/10.3390/toxins13020152.

Rai, M. and Agarkar, G., 2016. Plant-fungal interactions: what triggers the fungi to switch among lifestyles? Critical Reviews in Microbiology 42: 428–38. https://doi.org/10.3109/1040841X.2014.958052.

Revajová, V., Levkut, M., Levkutová, M., Bořutová, R., Grešaková, L., Košiková, B. and Leng, L., 2013. Effect of lignin supplementation of a diet contaminated with *Fusarium* mycotoxins on blood and intestinal lymphocyte subpopulations in chickens. Acta Veterinaria Hungarica 61: 354–65. https://doi.org/10.1556/AVet.2013.023.

Rostagno, M.H., 2020. Effects of heat stress on the gut health of poultry. Journal of Animal Science 98: skaa090, https://doi.org/10.1093/jas/skaa090.

Schneiders, G.H., Foutz, J.C., Milfort, M.C., Ghareeb, A.F.A., Fuller, A.L., Rekaya, R., Williams, S.M. and Aggrey, S.E., 2020. Heat stress reduces sexual development and affects pathogenesis of *Eimeria* maxima in meat-type chickens. Scientific Reports 10: 10736. https://doi.org/10.1038/s41598-020-67330-w.

Shanmugasundaram, R., Adams, D., Ramirez, S., Murugesan, G.R., Applegate, T.J., Cunningham, S., Pokoo-Aikins, A. and Glenn, A.E., 2022. Subclinical doses of combined fumonisins and deoxynivalenol predispose *Clostridium perfringens*-inoculated broilers to necrotic enteritis. Frontiers in Physiology 13: 934660. https://doi.org/10.3389/fphys.2022.934660.

Soutter, F., Werling, D., Tomley, F.M. and Blake, D.P., 2020. Poultry coccidiosis: design and interpretation of vaccine studies. Frontiers in Veterinary Science 7: 101. https://doi.org/10.3389/fvets.2020.00101.

Streit, E., Schwab, C., Sulyok, M., Naehrer, K., Krska, R. and Schatzmayr, G. 2013. Multi-mycotoxin screening reveals the occurrence of 139 different secondary metabolites in feed and feed ingredients. Toxins 5: 504–523. https://doi.org/10.3390/toxins5030504.

Tardieu, D., Travel, A., Metayer, J-P., Le Bourhis, C. and Guerre, P., 2019. Fumonisin B1, B2 and B3 in muscle and liver of broiler chickens and turkey poults fed with diets containing fusariotoxins at the EU maximum tolerable level. Toxins 11: 590. https://doi.org/10.3390/toxins11100590.

Tardieu, D., Travel, A., Metayer, J-P., Le Bourhis, C. and Guerre, P., 2020. Zearalenone and metabolites in livers of turkey poults and broiler chickens fed with diets containing fusariotoxins. Toxins 12: 525. https://doi.org/10.3390/toxins12080525.

Tardieu, D., Travel, A., Le Bourhis, C., Metayer, J.P., Mika, A., Cleva, D., Boissieu, C. and Guerre, P., 2021. Fumonisins and zearalenone fed at low levels can persist several days in the liver of turkeys and broiler chickens after exposure to the contaminated diet was stopped. Food and Chemical Toxicology 148: 111968. https://doi.org/10.1016/j.fct.2021.111968.

Tran, V.N., Viktorová, J. and Ruml, T., 2020. Mycotoxins: biotransformation and bioavailability assessment using caco-2 cell monolayer. Toxins: 628. https://doi.org/10.3390/toxins12100628.

Travel, A., Metayer, J.P., Mika, A., Bailly, J.D., Cleva, D., Boissieu, C., Le Guennec, J., Albaric, O., Labrut, S., Lepivert, G., Marengue, E., Froment, P., Tardieu, D. and Guerre, P. 2019. Toxicity of fumonisins, deoxynivalenol, and zearalenone alone and in combination in turkeys fed with the maximum European Union-tolerated level. Avian Diseases 63: 703–712. https://doi.org/10.1637/aviandiseases-D-19-00073.

Venkatesh, N. and Keller, N.P., 2019. Mycotoxins in conversation with bacteria and fungi. Frontiers in Microbiology 10: 403. https://doi.org/10.3389/fmicb.2019.00403.

Wang, Y., Quan, H., Li, X., Li, Q., Haque, M.A., Shi, Q., Fu, Q. and He, C., 2021. Contamination with fumonisin B and deoxynivalenol is a threat to egg safety and contributes to gizzard ulcerations of newborn chickens. Frontiers in Microbiology 12: 676671. https://doi.org/10.3389/fmicb.2021.676671.

Yunus, A.W., Blajet-Kosicka, A., Kosicki, R., Khan, M.Z., Rehman, H. and Böhm, J., 2012. Deoxynivalenol as a contaminant of broiler feed: intestinal development, absorptive functionality, and metabolism of the mycotoxin. Poultry Science 91: 852–61. https://doi.org/10.3382/ps.2011-01903.

Zingales, V., Taroncher, M., Martino, P.A., Ruiz, M.J. and Caloni, F., 2022. Climate change and effects on molds and mycotoxins. Toxins: 14: 445. https://doi.org/10.3390/toxins14070445.

CHAPTER 9

Mycotoxins and poultry gut health

Haitham M. Yakout

Global technical coordinator, Mono-gastric, Adisseo, USA;
haitham.yakout@adisseo.com

Abstract

Mycotoxins are a crucial problem facing poultry producers worldwide, recognized and documented but overlooked for decades. These natural feed contaminants constitute a significant risk factor for human and animal health, as up to 25% of the world's crop production may be exposed and contaminated. These secondary metabolites of fungi found in human foods and animal feeds can cause serious health problems in animals, eventually leading to severe economic losses. A census in the scientific community agrees on the harmful impacts of mycotoxins on farm animals, specifically poultry. Mycotoxin exposure has many adverse effects ranging from impaired growth, feed refusal, lower pathogens resistance, and hepato- and nephrotoxicity to higher mortality rates. Although the physiological effects of mycotoxins have been studied and researched, there is still a need to translate this wealth of knowledge into quantifiable performance effects. Studies on the impact of intestinal functionality and mode of action on impairment at the intestinal cell level are still limited. As the Gastrointestinal tract (GIT) serves as a massive physical interface between the outside environment and different barriers down in the proximal part, it inevitably hosts many colonies of a multitude of micro-organisms that must be controlled and prevented from entering the sterile body along with required nutrients. With its physical, chemical, immunological, and microbial barriers, the gut eliminates the uptake, absorption, and distribution of foreign materials, including toxins, from targeting highly active tissues and organs. In this chapter, we discuss the global effects of mycotoxins and their mode of action. We also dive deeper into published data for compromised intestinal barriers due to mycotoxins exposure or insult. Lastly, we tackle issues related to impaired and leaky gut, thin mucus layer, overloaded inflammatory factors, and debilitated bacterial homeostasis.

Keywords

Natural feed contaminants – mycotoxicosis – secondary metabolites – gut barrier – bacterial homeostasis

© HAITHAM M. YAKOUT, 2024 | DOI:10.3920/9789004695467_010

1 Introduction

Globally, poultry's extensive production and consumption position it with significant importance. Poultry product demand will rise further to ensure enough supply of the animal protein requirement of the ever-increasing human population. As a result, the challenge for poultry scientists, nutritionists, and the whole industry is to produce enough healthy chicken meat and eggs in the most efficient method possible. As a result, food safety and quality are becoming increasingly detrimental when positioning poultry products' market value. One of the most common contaminants of food and feed is mycotoxins, considered a top risk factor for human and animal health. Fungi, yeast or bacteria, produce these toxic secondary metabolites. Those critical to poultry are produced by fungi of the genera *Aspergillus*, *Fusarium*, or *Penicillium* and can contaminate grains in the field "pre-harvest", during harvest, or in storage "post harvest" as favorable conditions are available. The effects of mycotoxins in animal husbandry and production are well recognized. Many adverse health effects in humans and animals can be induced by mycotoxins acute and chronic dietary exposures, making these chemically diverse substances highly relevant to agricultural contaminants. Many published papers and reports have referenced the food and Agriculture Organization's (FAO) estimation of the overall global prevalence of mycotoxins in crops at 25% (Eskola *et al.*, 2020; Kermanshahi *et al.*, 2007; Verma *et al.*, 2007). Most of those reports were conducted worldwide in every region, regardless of climate and environmental conditions.

Additionally, many of these mycotoxins can cause serious health problems in humans and livestock, resulting in the condemnation of agricultural products. Thus, mycotoxins contamination is responsible for significant economic losses associated with human and animal welfare, production, and domestic and international trade, counting to several hundred million USD annually worldwide. The clinical response is dose-dependent and time-dependent and can vary from acute to chronic. The reaction to known mycotoxins of clinical importance tends to be sub-acute or chronic, and symptoms often are subtle and vague. Severe egg production reduction in layers and breeders and growth depression in broilers, turkeys, and ducks represent mycotoxin contamination's negative impact on performance. Exposure to mycotoxins by humans and animals is unavoidable, so it is essential to find practical mitigation approaches to limit or reduce their harmful impacts.

2 Mycotoxins

The med of the last century, the early 1960s, marked the discovery of aflatoxins (AFLA). An unknown disease, "turkey X," killed 100,000 turkey poults in England. Only in the last 50 years has it evident that common fungi and molds found in foods and feeds can develop mycotoxins. 1960 to 1975 was known as the mycotoxin gold rush, as a good body of well-funded research attracted many scientists to search and discover these toxic agents. Throughout history, these toxins have caused catastrophic epidemics in humans and animals. For example, ergotism killed hundreds of thousands of people in Europe, and stachbotryotoxicosis in Russia killed thousands of horses in the 1930s. In the med 40s, alimentary aleukia killed 100,000 Russians, and in England in the 1960s aflatoxicosis killed 100,000 turkey poults (Pitt, 2000).

Mycotoxin is a term driven from "*mykes*" which means fungi, and "*Toxicon*," which means poison, venomous or toxic. Fungal metabolic pathways can produce a variety of substances that are not required for their proliferation and are hence known as secondary metabolites (Andretta *et al.*, 2011; Broom, 2015). These metabolites are generated as an adaptive reaction to severe environmental conditions and have no significance in steady fungal growth (Pitt, 2000; R. Xu *et al.*, 2022). Molds that mainly belong to a broad range of fungi such as *Aspergillus*, *Penicillium*, and *Fusarium* genera secret this structurally diverse low-molecular-weight (MW~700 Da), non-enzymatic poisonous secondary metabolites (Grenier & Applegate, 2013; Murugesan *et al.*, 2015; Gao *et al.*, 2020). Mycotoxins can vary structurally from simple four-C molecules in moniliformins to very complex substances in phomopsins (Zain, 2011; Liew & Mohd-Redzwan, 2018;). The environment necessary for fungal growth and the subsequent mycotoxin production varies greatly depending on the substrate on which the fungal species grows on. Some molds are known to be toxigenic, so that they can produce one or more toxic metabolites. However, it is essential to note that not all molds are toxigenic, and not all secondary metabolites from molds are toxic (Zain, 2011). Climate conditions also have a significant impact and differences in the presence of mycotoxins, thus posing a significant health risk for humans and animals consuming contaminated crop materials. This risk depends on location and harvest year (Gruber-Dorninger *et al.*, 2019).

Due to its origin diversity, mycotoxin's different structure leads to significantly different physical, chemical, and biological properties (Guerre, 2020). These molecules' structure can vary from a single heterocyclic ring (molecular weights of > 50 Da) up to groups of irregularly organized six or 8-membered rings with (< 500 Da total molecular weights). A census in the research community found more than 200 mold species can produce mycotoxins (Murugesan *et al.*, 2015). In addition, fungal species and others can produce approximately 500+ mycotoxins (Broom, 2015; Gao *et al.*, 2020). Thanks to the latest analytical techniques, using liquid chromatography-tandem mass spectrometry (LC-MS/MS) led to the detecting of many mycotoxin metabolites. A decade ago, Streit *et al.* (2013) examined animal feed and raw materials and reported that all samples contained a wide range of 139 mycotoxins secondary metabolites. Some of these mycotoxins that affect human and animal health and economic growth have received significant attention from researchers, nutritionists, and veterinarians. These include but are not limited to AFLA, ochratoxin A (OTA), deoxynivalenol (DON), fumonisins (FUM), zearalenone (ZEN), patulin (PAT), nivalenol (NIV), and citrinin (CTN) (Murugesan *et al.*, 2015; Adegbeye *et al.*, 2020; Yang *et al.*, 2020).

Farm animals ingesting mycotoxin-contaminated feed would show either acute or chronic toxicity. However, chronic exposure to mycotoxin is the leading cause of almost all significant problems in livestock production, seen in growth retardation, impaired production, and sub-optimal immune functions. The ingestion of low levels of single or mycotoxins combination is always linked to a wide range of metabolic, physiological, and immunological distributions and leads to economic losses (Xu *et al.*, 2022). Pitt (2000) describes four basic types of mycotoxin toxicity: acute, chronic, mutagenic, and teratogenic. Acute mycotoxin toxicity is considered the most known and described by deteriorating liver and kidney functions, which may lead to death in some extreme cases. Another type is mycotoxins interference with protein synthesis, leading to necrosis and severe immunodeficiency. Some other toxins affect the replication of DNA, resulting in mutagenic or teratogenic effects (Pitt, 2000). Finally, long-term exposure to low levels of mycotoxins "chronic toxicity" showed cancer induction, especially in the liver, as a prime effect.

3 Mycotoxicosis in poultry production

Due to harsh environmental conditions and as an adaptive response, fungal metabolic pathways can produce a variety of toxic compounds that are not essential for their growth and development, known as secondary metabolites. These toxins can induce diverse and harmful effects at the cellular level with catastrophic interference with pathways and processes. For example, many published studies documented economic losses in all poultry production sectors due to Aflatoxicosis. To put it in perspective, two decades ago, Desjardins *et al.* (2003) reported that AFLA, FUM, and DON combined were responsible for a USD 900 million economic loss annually. The magnitude of response to AFLA contamination would greatly vary depending on age, sex, weight, diet, duration of exposure, and presence of other mycotoxins (Zain, 2011). Down-regulation of protein expression and synthesis is considered one of the most understood effects of mycotoxins on targeted tissues (Broom, 2015). Therefore, rapidly proliferating and differentiating cells and tissues with high protein synthesis, secretions, and/or turnover (*e.g.*, intestine, liver, immune system, etc.) are constantly targeted by different individuals or combined mycotoxins. Although the physiological effects of mycotoxins have been extensively studied and researched through the years, there is a need to translate this wealth of knowledge into quantifiable performance effects. To narrow this knowledge gap, extensive published meta-analysis studies in pigs (Andretta, Kipper, Lehnen, Hauschild, *et al.*, 2012a) and poultry (Andretta *et al.*, 2011; Andretta 2012b) employing different techniques to compose and analyze data sets. On the broiler side, dietary mycotoxins presence, specifically AFLA, and OTA, are responsible for a significant reduction in feed intake (12%) and body weight gain (14%) compared to the control diet. These two mycotoxins significantly reduced feed ingestion by 20 and 11%, respectively. Residual data analysis showed that 65% of the variability in body weight gain was explained by feed intake, as challenged broilers ingested fewer nutrients, such as protein and methionine, than non-challenged broilers. Furthermore, younger birds were more sensitive to different mycotoxins effects and showed more significant growth impairment than older birds.

Aflatoxicosis can be acute; for a short period of contamination consuming very high levels of the toxin that leads to death, and chronic, which is more prolonged exposure to lower concentrations of the toxin leading to cancer, immune suppression, and other slow pathological conditions (Murugesan *et al.*, 2015). AFLA is one of the well-absorbed and quickest-mobilized mycotoxins, regardless of the species. As a result, reactive epoxides are produced and responsible for producing carcinogenic metabolites. Aflatoxicosis compromises production, as seen in reduced feed intake, body weight gain, and even higher mortality in broilers (Andretta *et al.*, 2011). In laying hens, performance losses such as reduced egg production parameters, changes in organ weight, and liver damage interfere with biochemical and hematological profiles (Applegate *et al.*, 2009; Raj *et al.*, 2023). Reducing serum protein levels while compromising essential metabolic functions, such as lowering several enzymatic activities controlling starch, proteins, and lipids, the breakdown and digestion, and increased liver fats were also reported (Murugesan *et al.*, 2015). The overall health of AFLA-infected birds would be compromised due to immunosuppression, resulting in disease outbreaks, vaccination failures, and poor antibody titers (Devegowda, 2005; Murugesan *et al.*, 2015). When poultry consume contaminated or mold feed, AFLA appears in poultry products (*i.e.*, meat and eggs). In a study analyzing 115 samples of chicken meat and 80 egg samples, Iqbal *et al.* (2014) found that 35% of the meat samples and 28% of the egg samples were positive for AFLA contamination. Furthermore, laying hens' livers showed the highest AFLA-B_1 (2.98 μg/kg) and total AFLA (3.23 μg/kg). Lastly, some mycotoxins may indicate a level of competitiveness as reported between AFLA and OTA in feed ingredients or even in absorption rate in the GIT. When AFLA is present at shallow levels or even absent, OTA tends to be at higher concentrations and vice versa (Zain, 2011).

The FUM group has six toxins identified as A_1, A_2, B_1, B_2, B_3, and B_4 (Murugesan *et al.*, 2015). Fumonisins (B_1 and B_2) are metabolites of *Fusarium proliferatum* and *verticillioides*, considered caner-promoting metabolites (Zain, 2011). The fungus is known to be highly toxic in many animal studies. Along with AFLA, and OTA, FUM is one of those few mycotoxins known to be carcinogenic and possibly patulin (Desjardins A. *et al.*, 2003). Avian species are known to express relative resistance to FUM toxicity. Exposing broilers, turkeys, or ducks to high levels of FUM, poor performance, reduction in body weight gains, liver pathology, and alteration in enzymatic activities and serum constituents were reported (Tardieu *et al.*, 2007;

Tran *et al.*, 2005). Multifocal hepatic necrosis, biliary hyperplasia, and increased extramedullary hematopoiesis were some hepatic changes in liver pathology in chicks resulting from consuming FUM-contaminated diets (Murugesan *et al.*, 2015; Tardieu *et al.*, 2007; Tran *et al.*, 2005; Voss *et al.*, 2001). Ducks from day 7 to 63 of age (Tran *et al.* 2005) reported poor performance measured in significantly reduced body weights due to FUM B_1 (128 mg/kg feed). These data sets strongly correlate with previously published results on turkeys and broilers fed FUM B_1 more than 50 and 75 mg/kg feed for 21 days, respectively (Ledoux *et al.*, 1996; Broomhead *et al.*, 2002). In contrast, studies published by (Kubena *et al.*, 1999; Broomhead *et al.*, 2002) concluded that the adverse FUM effects seemed to diminish as birds grew older, as body weights of ducks fed FUM B_1 after 70 days of age were similar to the control group. Similar results were reported in laying hens when fed 100 and 200 mg/kg feed FUM B_1 for 420 days; broilers fed 25 and 50 mg/kg feed, FUM B_1 for 49 days, and male turkeys when fed 50 and 75 mg/kg feed, FUM B_1 for 98 and 126 days, respectively (Ledoux *et al.*, 1996; Kubena *et al.*, 1999; Broomhead *et al.*, 2002). All these results collectively confirm the higher resistance of avian species to FUM B_1 compared to other mycotoxins (Tran *et al.*, 2005; Tardieu *et al.*, 2007).

Trichothecenes mycotoxins (TCT) are a group of fungal metabolites with similar backbone structures that belong to several fungal genera (> 100). However, most of them are isolated from *Fusarium pp*. The TCT group toxins may include but are not limited to; T-2, HT-2, DAS, nivalenol, DON (*vomitoxin*), and 3-acetyl deoxynivalenol (Zain, 2011; Murugesan *et al.*, 2015). Feed refusal is the primary and most observed physiological response to trichothecenes ingestion; thus, this group earned its name, feed refusal toxins. In addition, these toxins are tissue irritants, and as animals ingest them, oral lesions, dermatitis, and intestinal irritation are some of the significant observations of their toxicities.

Poultry is more susceptible to T-2 and DAS toxins within the monogastric group, whereas swine tend to be more sensitive to DON (Devegowda, 2005). The main effect of TCT at the cellular level is primarily protein synthesis inhibition followed by a secondary disturbance of DNA and RNA synthesis. TCT affects cells actively dividing, proliferating, and differentiating, such as the epithelial lining of the GIT, intestine, skin, immune system, etc. Most TCT toxic effects would be extensive necrosis and inflammation in the oral cavity, specifically the mucosa (Devegowda, 2005; Zain, 2011). Consequently, poultry performance is negatively affected by TCT and noted in impaired growth, poor feathering; Helicopter disease; reduced egg production in layers, and poor shell quality (Devegowda, 2005; Murugesan *et al.*, 2015). After AFLA, TCT toxins are the 2nd most immunosuppressive toxins. Further, regression of the bursa of Fabricius, leucopenia, proteinemia, abnormal blood coagulation, and immunosuppression are some of the TCT toxicity documented clinical signs (Dänicke, 2002). There is a variation between poultry species' sensitivity to TCT toxins; broilers compromised their performance as dietary T-2 toxins were 3–4 mg/kg, while ducks tend to be more sensitive as lower T-2 levels as

0.4 mg/kg negatively affected their performance (Dänicke, 2002; Murugesan *et al.*, 2015). Acute DON contamination, on the other hand, significantly induces negative performance in broiler chickens and alters critical functions of the intestinal tract, reducing absorption surface area and increasing the permeability of the GIT (Awad *et al.*, 2011; Huff *et al.*, 1981; Murugesan *et al.*, 2015).

As one of the most common contaminating mycotoxins, Zearalenone is produced by Fusarium molds, specifically by *Fusarium graminearum*. It is very stable thermally, highly toxic, and contaminates corn, barley, wheat, and many other grains. It is a non-steroidal compound that exerts estrogen like activity in some farm animals causing reproductive toxicity. As ZEA and its metabolites possess these estrogenic activities, they could impose implications like hypoestrogenic syndrome in humans (Devegowda, 2005; Zinedine *et al.*, 2007). Moreover, liver, kidney, and immunotoxicity are common clinical signs of ZEA toxicity (Zain, 2011; Wu *et al.*, 2021). Following oral administration, the absorption of ZEA is quick, and it is difficult to measure to some degree due to the extensive biliary excretion. In pigs and humans, ZEA adsorption and metabolization in the intestinal cells is rapid, and it degrades to its metabolites; α-ZEA, β-ZEA, α-ZAL, and β-ZAL, which ends up conjugated with glucuronic acid (Zinedine *et al.*, 2007). In sheep, higher levels of these metabolites are excreted in the urine, indicating that both biliary excretion and enterohepatic cycling are crucial in determining the fate of ZEA. These reports suggest almost total excretion of ZEA glucuronide in the bile. It endures further re-absorption and is metabolized by the intestinal mucosal cells, entering the liver and systemic circulation via the portal blood (Biehl *et al.*, 1993; Zinedine *et al.*, 2007).

It has been reported by (Zinedine *et al.*, 2007) that average ZEA contamination in agricultural commodities ranges from concentrations as low as 5 to 50 mg/kg to a maximum of 120 to 180 mg/kg. Although ZEA often co-occurs with DON, it does cause a lesser impact on poultry than DON. Adverse effects on performance and impaired immunity usually correlate with higher DON contamination levels above 5 mg/kg feed. However, as DON contamination levels are below 5 mg/kg feed, the "combined carry-over" effects of both toxins on poultry are minimal to none (Dänicke, 2002). In comparison to pigs and cattle, poultry is relatively resistant to ZEA. However, exposing turkeys and chickens to higher dietary ZEA concentrations shows vent enlargement and enhanced secondary sex characteristics (Devegowda, 2005). In a report by (Allen, *et al.*, 1981a, 1981b), authors indicated that increasing ZEA dietary concentrations up to 800 mg/kg was not a determinant of laying hens' performance or even their progeny. In the same line, broilers (6 to 9 weeks of age) and turkey poults (3.5 to 6.5 weeks of age) sustained production, maintained organ weights, and had standard immune organs functionality as they were fed diets with purified ZEA up to 800 mg/kg diet (Allen, Mirocha, *et al.*, 1981a, 1981b). On the other hand, feeding ZEA-contaminated feed (10 mg/kg) to a laying hen caused an accumulation of ZEA metabolites in tissues (liver), egg yolk, and excreta. Older turkeys showed similar negative results on egg production as

(Allen *et al.*, 1983) reported a 20% egg production reduction due to feeding adult female turkeys diets contaminated with 10 mg/kg ZEA.

Ochratoxin is one of the most prevalent mycotoxins in a group of structurally related metabolites produced by one of two genres: *Aspergillus* and *Penicillium*. OTA can induce many toxin effects in poultry, including nephrotoxic, hepatotoxic, teratogenic, and immunotoxic (Iqbal *et al.*, 2014; Zhai *et al.*, 2021). Moreover, depending on the geographical location and climate, significant producers of OTA would be *Aspergillus ochraceous* in tropical and subtropical countries and *Penicillium verrucosum* in temperate countries (Devegowda, 2005). Poultry OTA toxicity would be noted in general weakness, anemia, and reduced feed intake, resulting in poor performance and impaired egg production. At higher concentrations, excessive mortality had been reported (Huff & Hamilton, 1975; Hamilton *et al.*, 1982; Burditt *et al.*, 1984). Poultry is more sensitive to OTA when compared to mammals; however, they possess a lower absorption rate (*e.g.*, pigs 66%, rabbits 56%, and 40% in chickens; Wang *et al.*, 2019), with a greater affinity for OTA to bind to proteins. Absorption of OTA begins at the stomach, usually referred to as the 1st OTA peak in the intestine, with a prominent presence in the proximal part of the GIT. Due to enterohepatic circulation, which closely connects the gut with other tissues, and low OTA absorption, a significant portion of the unabsorbed toxin remains circulating in the GIT lumen. About 99% of OTA binds to serum protein due to passive absorption; once it enters the bloodstream, the unabsorbed toxin can reach and interact with the microbiota, explicitly residing pathogens in the hindgut (Grenier & Applegate, 2013; Zhai *et al.*, 2021). At this region, the intestinal microbiota releases the conjugated OTA by triggering a 2nd peak of higher OTA levels in the intestinal contents, which is re-absorbed another time by the intestine, allowing a prolonged retention time in the GIT. This entero-toxicity is responsible for higher OTA residuals in the intestine and other tissues, specifically in the cecum. Wang *et al.* (2019) reported that OTA exerts toxic effects on its primary target organs, such as the kidney and liver, where it is metabolized and accumulated, thus inducing inflammation and even cancer in the liver. In the liver, the expected sequence would be a disruption of both tight junctions (TJ) and gut microbiota diversity due to OTA contamination. This leads to an impairment of the local immune response and dysregulation of intestinal functions, which results in systematic toxicity and chronic mycotoxicosis, as in liver inflammation (Zhai *et al.*, 2021). As the toll-like (TLR)-4 expression in the cell membrane increases, it generates inflammation, a vital component of the TLR4/myeloid (MyD) 88 dependent signaling pathway (Xu *et al.*, 2017); TLR4 plays a significant role in the initiation and acceleration of inflammation. Based on the higher mRNA expression and protein abundance of TLR4 Myd88, the NFαB signaling pathway as a downstream pathway of the TLR4/myeloid (MyD) 88 dependent signaling pathways, crucially amplify inflammatory responses by up-regulating the expression of various pro-inflammatory cytokines genes, and a higher ratio of p-IKBα/

IKBα (Xu *et al.*, 2017; Wang *et al.*, 2019). Pathophysiological changes in the kidney include reduced urine concentration and glomerular filtration rates, impairment of the proximal tubular functions, and degeneration and alterations in renal integrity (Huff & Hamilton, 1975; Murugesan *et al.*, 2015). Thus, higher residue concentrations in the kidney, cecum, liver, pectoral muscles, duodenum, jejunum, ileum, and leg muscles 1.59, 0.94, 0.92, 0.54, 0.42, 0.41, 0.32, and 0.20 ng/kg, respectively were reported after two weeks administration of OTA to ducklings (Wang *et al.*, 2019).

As in other mycotoxins, OTA would appear in poultry products (*i.e.*, meat and eggs) of poultry fed contaminated or mold feed. Out of 115 chicken meat and 80 egg samples, Iqbal *et al.* (2014) reported that 41% of the meat samples (averaged 1.41 mg/kg OTA) and 35% of the egg samples (averaged 1.17 mg/kg OTA) were positive for OTA. Scientists noted a level of competitiveness between OTA and AFLA, with the detection of higher concentrations of OTA. At the same time, as AFLA is present at deficient levels or even absent, OTA tends to be detected at higher concentrations and vice versa either in feed ingredients or in the absorption rate in the lower GIT (Zain, 2011).

3.1 *Relevant mycotoxin in poultry lifecycle*

The complete poultry lifecycle would be from day 1 (day of hatch; DOH), laying period, and post-peak of production up to 65 weeks of egg production, if not more. This lifecycle can be differentiated into 3 or 4 critical periods.

- Critical period 1: young birds from DOH up to 6 to 8 weeks old
- Critical period 2: growing birds (pullets) up to lay onset at 16 to 18 weeks of age
- Critical period 3: early laying period from 19 to 20 weeks old up to 40 to 45 weeks old

During these critical periods, birds face multiple stressors and stress-induced factors that will consequently have negative impacts on birds' health, performance, and production. Critical period 1, up to 8 weeks of age, would be considered the whole lifecycle of broilers and younger birds-phase for older birds such as laying hens or breeders. During this early age, the most significant effect of mycotoxins would be on the adverse alteration of the GIT growth detected in decreased mucosa, reduced intestinal secretions, impaired nutrient utilization, reduced diversity of the gastrointestinal microflora, and reduced gut motility. GIT is the primary site governing mycotoxin uptake, absorption, and subsequent distribution in different tissues. Some mycotoxin passage across the mucosal barrier can be rapid, as with aflatoxins (AF), or very limited, as with fumonisins FUM; (Grenier & Applegate, 2013). Other factors should also be considered regarding mycotoxins severity, including diet composition, dietary feed contamination, genetics, and mycotoxin absorption rate. Regardless, the intestinal epithelium constantly interacts with the contaminated feed and is considered the first target of these contaminants. The fast absorption in the GIT and the quick appearance of most mycotoxins in circulation indicate that most of the ingested toxin is absorbed in the proximal part of the GIT. Therefore, the intestinal epithelium could be compromised by Mycotoxins either before absorption in the upper part GIT or throughout the entire intestine by non-absorbed toxins resulting in deleterious effects causing gut barrier impairments and dysfunctions (Riahi *et al.*, 2021).

The main target of TCT mycotoxins is rapidly proliferating and differentiating cells and tissues with a high protein turnover, simply the GIT, including the small intestine, liver, and immune system (Yegani *et al.*, 2006). Inhibition of nucleic acid protein synthesis directly results from contamination by *Fusarium* mycotoxins, and their metabolites act negatively, affecting such high protein turnover cells (Lucke *et al.*, 2018; Paraskeuas *et al.*, 2021). The toxic effects of *Fusarium* mycotoxins in animals and poultry include reduced growth and feed refusal (Fitzpatrick *et al.*, 1988). One proposed mechanism would be alterations in brain neurotransmitter concentrations responsible for reduced feed intake induced by *Fusarium* mycotoxins (Boyd *et al.*, 1988). Vomiting in pigs (Prelusky and Trenholm, 1993), immunosuppression (Riahi *et al.*, 2021), gastrointestinal lesions, and neurological and reproductive disorders (Prelusky & Trenholm, 1993; Swamy *et al.*, 2002, 2004) are among other chronic effects of Fusarium toxins containing poultry.

In poultry, the most frequently found mycotoxins in cereal grains are DON and FUM, produced from *Fusarium* fungi (Antonissen *et al.*, 2015). The European Union has set "*safe*" limits for DON and FUM in poultry feed, up to 5 and 20 mg/kg, respectively. However, previous studies indicated that DON and FUM could cause mycotoxicosis and impaired broiler gut health function and, as a result, impair broiler growth performance, even at lower concentrations than the EU recommendations (Grenier *et al.*, 2016; Paraskeuas *et al.*, 2021; Riahi *et al.*, 2021). It is also established that feeding DON or FUM to broilers at levels below those that cause adverse effects on health and performance may affect small intestinal morphology (Antonissen *et al.*, 2015; Grenier *et al.*, 2016). Fitzpatrick *et al.* (1988) reported that nervous system disturbances are considered a criterion for acute DON toxicity in poultry. As broilers fed DON at 10 mg/kg of feed for six weeks resulted in shorter and thinner villi in the duodenum and jejunum. Medium and moderate levels of DON reduced the absorption area due to decreased villi height in the duodenum and jejunum. In turkey poults, small intestinal morphology was adversely influenced when Fed pure T-2 toxin or diacetoxyscirpenol (DAS) at levels up to 1 mg/kg of feed for 32 days.

Overall gut health is another negative mycotoxins effect, as DON and FUM may act as predisposing factors for the proliferation of certain bacteria, *i.e.*, *C. perfringens*, to induce necrotic enteritis in broilers. Both reduce villus height and transepithelial electrical resistance, which leads to a decreased absorption of dietary nutrients and an increased intestinal barrier permeability, higher nutrient uptake, and leakage of plasma amino acids into the intestinal lumen, providing the necessary growth substrate for *C. perfringens* growth and proliferation.

It is estimated that up to 70–80% of the immune defenses reside in the gastrointestinal tract (GIT), which possesses its immune system. However, it is well-accepted that mycotoxins can modulate immune responses, and each can act differently from other mycotoxins. The entrance of most enteric pathogens to the animal's body is through the GIT, which is also a standard route for vaccination in poultry. Activation of the intestinal immune system is induced by vaccination or antigen delivery, resulting in the division and proliferation of immune cells. As mentioned above, high turnover cells are mycotoxins' main targets; therefore, higher susceptibility to enteric infections directly impacts feeding animals mycotoxin-contaminated diets. For example, Aflatoxin B1 suppresses leukocyte migration and lymphoblastogenesis, reduces lymphocyte proliferation, alters cytokine synthesis by macrophages and/or T cells due to inhibiting DNA, RNA, and protein synthesis, and affects the developing immune system during embryonic development. DON elevates serum immunoglobulin A (IgA) and causes concurrent depression in IgM and IgG (Cheng *et al.*, 2018). A combination of ochratoxin A (OTA) and AFB1 inhibits macrophage activity.

Due to the harmful effects of mycotoxins on the gastrointestinal tract described above, immunity acquired through vaccination can be impaired by ingesting contaminated feed. It is well-accepted that following mycotoxin contamination of the feed, vaccine efficacy is reduced, leading to lower titer levels. Lohmann male broilers showed significantly lower NDV titers at week 5 of age which was linearly correlated with increasing dietary DON levels of 3.5 to 14 ppm, and Ross broiler breeders fed 6.4 to 12.6 ppm DON for 12 weeks dropped IBV titers (Yegani *et al.*, 2006). Mycotoxins reduce the level of antibodies following infection or vaccination and reduce the activity of phagocytic cells. A broiler study showed that mycotoxicosis might cause vaccination failure against the Avian Influenza virus (Ravikumar *et al.*, 2022). Further, mycotoxin-induced immunosuppression may be manifested as depressed T- or B-lymphocyte activity and suppressed production and impaired macrophage/neutrophil-effector functions.

During the 2nd critical period, (Yegani *et al.*, 2006) different mycotoxins show a wide array of effects on growing pullets and males. Affects range from feed refusal, deteriorated sperm quality, impaired testis and ovaries development to impaired growth of reproductive organs. Zearalenone (ZEN) and its metabolites exhibit estrogenic properties, which have a biological structure that can bind to estrogen receptors in the reproductive organs. Therefore, Zen can cause reproductive disorders in poultry, such as reductions in testis growth in males and cysts in ovaries in females. FUM also exhibits feed refusal due to altered brain neurotransmitter concentrations, mainly an increase in 5-hydroxytryptamine (5-HT; serotonin). Fusarium mycotoxins alter serotonergic activity in the brain and cause subsequent reductions in feed intake. Aflatoxin exposure in males lowered semen volume, serum proteins concentration, cholesterol, and triglycerides and exhibited delays in physiological and behavioral sexual maturation in poults.

Older birds during critical period 3, from the early laying period from 19–20 weeks old up to 40–45 weeks old, are also affected by different mycotoxins. Symptoms range from hemorrhagic ovaries, reduced egg production, reduced egg quality, cessation of egg production, and lower embryonic development and hatchability. Poor eggshell integrity is one of the significant production parameters affected during this critical period. As GIT is negatively altered by mycotoxin contamination, nutrient absorption is significantly impaired, especially for Ca. This is reflected in lower nutrient levels in the liver and poor vit. D3 conversion, which leads to lower Ca-binding protein synthesis. As a result, low Ca levels are detected in plasma, and more significant mobilization of Ca from bone is correlated with inadequate bone Ca deposition, which leads to poor shell deposition. Laying hens exhibited a 22% drop in egg production 12% increase in cracked eggs, an incidence of blood spots, and a further 18% egg breakage during transit as they were fed T-2 contaminated diets. Moreover, DON caused increased hatching malformation and early embryonic mortality associated with reduced eggshell thickness. Aflatoxin B1 showed similar results, increasing embryonic mortality by up to 9% and reducing hatchability by up to 11%. Zearalenone reduced egg-specific gravity, shell thickness, and interior egg quality. Concurrent serum calcium reduction and increased serum phosphorus have also been reported.

4 Nutritional implications of mycotoxicosis

A large body of literature recognizes that major mycotoxins would adversely affect feed intake, efficiency, and growth rate and negatively impact production. Although there is an enduring relationship investigating various nutrients and mycotoxicosis, especially aflatoxin, the interaction is compounded and needs to be fully acknowledged (Girardet *et al.*, 2011; Wang *et al.*, 2019; Xu *et al.*, 2022; Zhang *et al.*, 2020). Proteins, lipids, and carbohydrates are the building blocks for animals' growth and development. Once absorbed due to normal functional digestion, cells, and tissue can efficiently utilize their ingested end-products and function properly. However, many factors may affect the proper utilization of feed ingredients and subsequent nutrients, including but not limited to; mycotoxin type, species, toxin concentration, and ingestion of single or multi-toxin-contaminated feed. Non-ruminants are generally more susceptible to mycotoxins' nutritional effects than ruminants (Grenier & Applegate, 2013; Xu *et al.*, 2022). Usually, feed consumption correlates with nutrient availability, which is negatively impacted by different mycotoxins, resulting in feed refusal. Studies report ingesting *Fusarium* mycotoxins-contaminated feed led to a higher incidence of feed refusal (Boyd *et al.*, 1988; Wang *et al.*, 2019; Wang & Hogan, 2019). Additionally, this has been well documented for AFLA, DON, and ZEA while considerably reported for FUM and OTA (Xu *et al.*, 2022). Regardless of the species, single mycotoxin contamination has been shown to reduce feed intake, as in aflatoxicosis in broilers (Andretta *et al.*, 2011), DON (Wang *et al.*, 2019;

Wang & Hogan, 2019), OTA (Huff & Hamilton, 1975; Hamilton *et al.*, 1982; Burditt *et al.*, 1984), or FUM in broilers, turkeys or ducks (Tran *et al.*, 2005; Tardieu *et al.*, 2007). In contrast, Grenier & Applegate (2013) suggested that DON lowers feed intake in broilers, but feed refusal was not consistently noted. Authors noted that broiler diets must be contaminated with 15 mg/kg diet DON to elicit feed refusal under experimental conditions. Furthermore, extensive research focused on the co-occurrence of multiple mycotoxin impacts on reducing feed intake. In poultry on the broiler side, (Andretta *et al.*, 2011; Andretta *et al*, 2012b) reported that dietary mycotoxins' presence, specifically AFLA, and OTA, was responsible for a significant reduction in feed intake. A combination of DON + FUM + ZEA (5 + 20 + 5 mg/kg diet) was accountable for decreased feed intake that was more pronounced than that caused by DON 5 mg/kg diet alone (Metayer *et al.*, 2019). Wang *et al.* (2019) noted that broilers' feed refusal might be attributed to changes in feeding preference or feeding behavior. The latter has never been thoroughly investigated in poultry fed-DON contaminated diets. Reduction in feed intake may result from changing feeding behavior, such as slower feeding rate, reduced meal size, and frequency (Girardet *et al.*, 2011). When broilers are fed DON-contaminated diets below or equal to recommended dietary levels, birds exhibit immediate and prolonged feed aversion (Wang *et al.*, 2019). These toxins are known to be tissue irritants, and as animals ingest them, oral lesions and intestinal irritation are significant clinical observations of their toxicities. Trichothecenes, especially DON, are responsible for inducing anorexia (Xu *et al.*, 2022) due to activating neurochemical pathways responsible for suppressing feed intake and terminating feeding behavior in broilers (Richards & Proszkowiec-Weglarz, 2007). Additionally, DON would affect cells that are actively dividing, proliferating, and differentiating, such as the epithelial lining of the GIT, intestine, skin, immune system, etc. Most TCT toxic effects would be extensive necrosis and inflammation in the oral cavity, specifically the mucosa (Devegowda, 2005; Zain, 2011). Another proposed mechanism would be alterations in brain neurotransmitter concentrations responsible for reduced feed intake induced by *Fusarium* mycotoxins (Boyd *et al.*, 1988). Deoxynivalenol alters brain neurotransmitter concentrations and causes subsequent reductions in feed intake (Boyd *et al.*, 1988; Fitzpatrick *et al.*, 1988; Girardet *et al.*, 2011; Mansour & Abeer, 2013).

4.1 *Nutrient digestibility and energy utilization*

Farm animals' inability to utilize feed efficiently, thereby reducing nutrients and energy digestibility, has been well-researched and documented. The generally accepted knowledge is that impaired absorption, distribution, and utilization processes are recurrent during mycotoxicosis across almost all nutrient classes. Inhibition of protein expression and synthesis is known to occur during mycotoxicosis (Broom, 2015); thus, limiting amino acid nutrition is extremely important during this critical time. It is acceptable to encounter higher methionine requirements for chickens experiencing aflatoxicosis to attain maximal growth than birds fed normal non-contaminated diets (Devegowda, 2005). It has been reported that 65% of the variability in broilers' body weight gain was explained by feed intake, as challenged broilers ingested fewer nutrients, such as protein and methionine, than non-challenged (Andretta *et al.*, 2011; Andretta *et al.*, 2012b). Exposing monogastric to either AFLA, DON, or FUM alone as well as a combination have been reported to reduce dry matter, gross energy and/or metabolizable energy, crude protein/amino acids, and crude fat digestibilities in many published reports (Grenier & Applegate, 2013; Xu *et al.*, 2022). Earlier studies (Zou *et al.*, 2012) showed no adverse effects of low FUM (5.5 mg/kg diet) on broilers' dry matter, crude protein, and gross energy. Similarly, DON did not suppress macronutrient digestibility and retention in the broiler's GIT. The broiler's intestine may have expressed morphological and functional modifications to adapt to the DON challenge (Yunus *et al.*, 2012). Similarly, Dänicke *et al.* (2002) reported minor reductions in nutrient digestibility and metabolizable energy depression as laying hens were fed ZEA-contaminated corn. Han *et al.* (2008) also reported that feeding ducks a very low AFLA-B1 at 20 and 40 μg/kg resulted in lower protein digestibility, 8 and 13%, respectively. It was proposed by (Gbore *et al.*, 2010) that these adverse effects might be due to alterations in albumin synthesis and serum protein concentrations consistently observed with intoxicated animals. However, (Xu *et al.*, 2022) reported that the lowest observed effects of toxins were at concentrations above recommended levels, significantly reducing nutrient digestibility coefficients. This is true for natural AFLA-B_1 (0.1 mg/kg) in ducks, artificial AFLA-B_1 (0.04 and 1.5 mg/kg) in broilers, artificial DON (5 mg/kg) in broilers, artificial FUM (20 mg/kg) in broilers, artificial OTA (2 mg/kg) in broilers, artificial DON + FUM (1.5 + 20 mg/kg) in broilers, and purified ZEA+ natural FUM (1 + 3.15 mg/kg) in broilers, respectively (Zou *et al.*, 2012; Chen *et al.*, 2016; Feng *et al.*, 2017; Liu *et al.*, 2020; Xu *et al.*, 2022). Similar results were also reported by (Verma *et al.*, 2007; Applegate *et al.*, 2009; Kermanshahi *et al.*, 2007; Murugesan *et al.*, 2015;) for broilers and laying hens when fed diets contaminated with higher AFLA, OTA, or their combinations. In addition to reducing energy utilization for laying hens with AFLA, energy requirements for maintenance increased significantly (Verma *et al.*, 2007; Grenier & Applegate, 2013). On the other hand, other articles reported surprisingly promising results with mycotoxicosis. Feeding broilers DON-contaminated wheat positively improved performance, and nutrient digestibility was seen in a reduction of intestinal viscosity (Dänicke *et al.*, 2007). A reasonable explanation would be the increase in degrading grain cell walls penetrated by degrading enzymes produced by the growing fungi. Thus, cell wall degradation, and structural modulation of nutrients, proteins, and other fractions, indicate an abundance of nutrients in animals (Grenier & Applegate, 2013). Another suggestion would be morphological adaptation and epithelial conditioning, noted in reducing the absorptive area while conserving the total GIT digestibility, which can control and minimize mycotoxin transfer to the systemic circulation (Yunus *et al.*, 2012). Changing feeding behavior, such as slower feeding rate, reduced meal size, and frequency, are direct results of reduced feed intake (Girardet *et al.*, 2011). Feed refusal and reduced feed intake lower mycotoxin exposure, minimizing the bulk passing through the GIT, and increasing retention time, nutrient digestibility, and absorption (Grenier & Applegate, 2013). Inconsistent results reported when investigating mycotoxins' effects on nutrient digestibility might be attributed to many factors including mycotoxin type, source, concentration, and individual or combined exposure. It might also be animal related based on age, species, breed, gender, production type, and stage. Finally, the length of mycotoxin exposure facilitates higher or lower mycotoxins exposure (Grenier & Applegate, 2013; Xu *et al.*, 2022).

4.2 *Digestive capacity*

A highly functional GIT could digest and absorb nutrients, electrolytes, and water in tissues along the digestive tract. Any interference with digestive enzymes' functionality and efficacy of digesting protein, lipids, and carbohydrates ultimately results in irritated and debilitated GIT. Impaired absorption, distribution, and utilization processes across all nutrient classes are frequently noted with mycotoxicosis, concurrent with low enzymatic activities at the lower GIT (Devegowda, 2005). So, attention must be directed to fungal metabolites and how they may disrupt digestion and cause nutrient and energy losses (Zou *et al.*, 2012; Chen *et al.*, 2016; Feng *et al.*, 2017; Liu *et al.*, 2020; Xu *et al.*, 2022). Complimentary to nutrient digestibility studies, many articles agree that mycotoxin assault on the GIT modulates enzymatic activities. Enzymatic modulation results from enzyme synthesis failure rather than inhibition (Devegowda, 2005; Grenier & Applegate, 2013; Chen *et al.*, 2016). Digestive enzymes modulation of pancreatic α-amylase, lipase, trypsin, and chymotrypsin, was reported among poultry species, broilers, breeders, and ducks due to AFLA-B_1 exposure (Devegowda, 2005; Xu *et al.*, 2022). Aflatoxicosis would cause an increase in α-amylase and lipase activities attributed to higher proenzyme levels from the damaged pancreas that could be dose and time-dependent (Grenier & Applegate, 2013; Chen *et al.*, 2016; Xu *et al.*, 2022). Increased proenzyme levels in the duodenum might be how birds compensate for reduced feed intake to increase nutrient absorption (Grenier & Applegate, 2013). Further, mycotoxicosis results in poor digestion due to low concentrations of available amino acids, fatty acids, sugars, etc. Without enough nutrients during the digestive process, impaired growth, production, and reproduction is evident (Devegowda, 2005). In laying hens, jejunum disaccharides and maltase activities were higher during aflatoxicosis, while alkaline phosphatase and aminopeptidase levels were lowered in mice. This enzymatic modulation is a direct result of changes in intestinal villi morphology. Sucrase and maltase are expressed 30 to 40% more per enterocyte at the upper 40% of well-developed villi than the lower 60% of the crypt-villus axis. So, as mycotoxin metabolites reside in the lumen forcing villi deformation, immature or damaged villi cannot express their total digestive capacity (Applegate *et al.*, 2009; Grenier & Applegate, 2013; Chen *et al.*, 2016; Feng *et al.*, 2017).

5 The gastrointestinal tract

The GIT is a muscular tube that forms a physical route responsible for feed ingestion, passing down the entire GIT, digestion, nutrients, electrolytes and water absorption, immunity, and excretion of wasted undigested materials (feces). It also serves as a massive physical interface between the outside environment and different barriers in the proximal part. This includes acids in the stomach and antimicrobial factors in saliva and other excreta, and in the hindgut or lower GIT; it is inevitably colonized by a multitude of micro-organisms which must be prevented from entering the sterile body-proper along with the nutrients. Inside, it is lined by a mucus membrane composed of four layers; the most inner is the mucosa resting on the submucosa, and muscularis propria lays underneath the serosa would be the outermost layer. Notably, all segments of the GIT have all four layers (Zhai *et al.*, 2021). The higher susceptibility of those high protein synthesis and turnover cells was the focus of many researchers and publications. A particular emphasis focuses on exposure to different mycotoxins at occasional and realistic mycotoxin levels, compromising GIT functionality (Grenier & Applegate, 2013; Antonissen *et al.*, 2015; Murugesan *et al.*, 2015; Grenier *et al.*, 2016). There is a considerable variation of mycotoxins (or their metabolites) absorption affected by many factors, but it plays a crucial role in determining systematic exposure and tissue distribution. A meta-analysis by (Grenier & Applegate 2013) illustrates that a significant portion of ingested toxins is absorbed at the proximal part of the GIT, as in AFLA (80%). Still, the remaining unabsorbed percentage of AFLA remains in the lumen, implying that the gut epithelium is exposed to a high proportion of the ingested, non-absorbed toxin. The effect is similar to an extent with those minimally absorbed mycotoxins; OTA (40%), DON (5 to 20%), and FUM (1%), respectively. The increased exposure time and concentrations can occur due to enterohepatic cycling (Murugesan *et al.*, 2015). This adversely affects birds' overall health, growth, and production. Cells in the GIT will be continuously exposed to higher ingested mycotoxin concentrations, whether absorbed in the systemic circulation or not. Furthermore, those unabsorbed toxins will eventually move down onto the GIT and interact more with intestinal cells, which would be more exposed to toxins (Broom, 2015). As mycotoxins are discussed, almost always lower exposure levels are usually neglected for their detrimental effects on animals' health and production. However, it has been confirmed repeatedly that intestinal cells are more vulnerable to lower "*chronic*" than higher "*acute*" mycotoxin concentrations (Awad *et al.*, 2011; Huff *et al.*, 1981; Swamy *et al.*, 2002, 2004; Zhai *et al.*, 2021). Indeed, toxicosis caused by more than one mycotoxin was found to have a synergistic or at least additive effect on exposed cells and tissues in virtually all cases, reinforcing the conceptualization that cell viability is critical in sustaining good avian health and performance.

6 Intestinal barrier

6.1 *Components of the intestinal barrier*

A highly functional GIT can uptake, absorb and distribute nutrients, electrolytes, and water in tissues. At the same time, it eliminates the uptake, absorption, and distribution of foreign materials, including toxins, to target tissues and organs. As the first interface with the outer environment and feed particles, its mucosa layer is constantly under stress performing its selective function of allowing the uptake of needed nutrients, electrolytes, and water (Broom, 2015), while preventing harmful substances from passing from the lumen to the systematic circulation (Antonissen *et al.*, 2015). This particular barrier can be divided into two components: intrinsic and extrinsic. The intrinsic barrier is the single-cell epithelial layer that covers the gut wall. It is pivotal to maintaining this physical barrier and the TJ that seals any voids among the epithelial cells and ties them together. Other secretions not a physical component of the epithelium are considered extrinsic barriers. However, these secretions, antimicrobial peptides, cytokines, and reactive oxygen species (ROS) significantly impact the epithelium's functionality and the barrier's maintenance (Antonissen *et al.*, 2015; Bouhet & Oswald, 2005; Gao *et al.*, 2020).

Gao *et al.* (2020) defined the intestinal barrier as a multi-interconnected barrier composed of physical, chemical, immunological, and microbial barriers. The chemical barrier would be those protein contacts, known as tight junctions, that interconnect epithelial cells to each other from the basolateral to the apical side of the epithelial cells (Odenwald & Turner, 2017). Tight junctions can selectively allow molecules and nutrients through the intestinal epithelium via the paracellular route, thus considered the primary determinant of barrier functionality in intact epithelia.

Mucins and anti-microbial peptides secreted by the intestinal epithelial cells are anti-microbial related proteins that form the chemical barrier. Goblet cells produce mucins, while Paneth cells, mainly located in the small intestine, primarily produce anti-microbial proteins (Gao *et al.*, 2020; Kim & Ho, 2010). Furthermore, intestinal epithelial cells mediate interactions between the mucosal immune system and luminal materials. The immunological barrier has immune cells residing in the lamina propria; such as macrophages, dendritic cells, B and T cells, and other immune regulators; such as cytokines. These immune mediators are secreted by immune and epithelial cells and play a significant role in regulating bird's both inflammatory and immune responses. The commensal bacteria, known as the endogenous or intestinal microbiota, colonize and reside in the intestinal tract, representing many different species that shape the microbial barrier. Depending on other environmental factors such as nutrition and stress, these microbes can have a crucial role in both immune, nutritional, and physiological balance (Kogut, 2013; Gao *et al.*, 2020).

Managing different nutrients or toxins and triggering the immune and inflammatory pathways responses that affect physiological and pathological states can be adopted by recognized and sensed by different networks. Kogut (2013) explained that in the innate immune system, Toll-like receptors are found in immune cells, intestinal cells, and adipocytes, recognize specific microbial compounds, and sense nutritional signals. As a result, they can modulate certain activities to ensure the host's ability to respond under different stress conditions.

7 Mycotoxicosis and impaired intestinal function

7.1 *Effects of mycotoxins on the epithelial barrier, cellular junctions, renewal, and repairs*

Subsequently, after ingesting mycotoxin-contaminated feed, the entire GIT is exposed to potentially higher concentrations of at least one toxin. The intestinal barrier is the first line of defense and functions as an innate or adaptive component of immunity. Within this barrier, the epithelial cells perform different mechanisms as mitigation methods reducing the infection risks by these fungal toxic metabolites. Defense mechanisms may be directed to the intrinsic barrier; epithelial cells, TJ, or the extrinsic barrier; mucus layer, anti-microbial peptides, and cytokines (Antonissen *et al.*, 2015; Bouhet & Oswald, 2005; Gao *et al.*, 2020).

Along with epithelial cells, TJ including occludin, claudins, junctional adhesion molecules (JAMs), and tricellulin, are of pivotal importance in maintaining the integrity of the intestinal physical barrier and are constantly targeted by fungal toxic metabolites (Zhang *et al.*, 2020). Failure to maintain intestinal integrity results in increasing permeability, consequently growing bacteria and toxins infiltration to the basolateral part of the enterocytes inducing inflammation (Taranu *et al.*, 2015). This would increase susceptibility to infectious diseases like *Clostridium perfringens*-induced necrotic enteritis in broilers and colibacillosis in pigs (Jia *et al.*, 2020). Exposure to DON, FB_1, and AFLA-B_1 reduces the trans-epithelial electrical resistance (TEER) and changes the integrity of the epithelium, inducing inflammation and increasing paracellular permeability, leading to nutrient leakage into the lumen (Antonissen *et al.*, 2015; Taranu *et al.*, 2015; Zhang *et al.*, 2020). Moreover, OTA induced gut permeability, which was noted as a direct result of the reduction of two isoforms of the claudin protein (McLaughlin *et al.*, 2004). The same toxin affects plasma membrane microdomain protein content, which regulates TJ assembly and intestinal transport activity (Bouhet & Oswald, 2005; Mansour & Abeer, 2013; Odenwald & Turner, 2017). Additionally, FUM toxic effects inhibit sphingolipid biosynthesis and glycosphingolipids, which may play a crucial role in regulating the epithelial cell electrical properties (Bouhet *et al.*, 2004; Bouhet & Oswald, 2005). Protein synthesis inhibition and increased oxidative stress are mainly responsible for altering the TJ integrity and increased permeability. This directly results from reducing protein expression and activation of MAPKs; p44/42 ERK, p38, and SAPK/JNK (Taranu *et al.*, 2015). Earlier publications indicated that the MAPK signaling pathways in the intestine are activated by DON exposure. This results in intestinal dysfunction associated with pro-inflammatory cytokines production (Jia *et al.*, 2020). As mycotoxins are absorbed in the intestine, pro-inflammatory cytokines induce TJ modulation. It is reported that both IFN-γ and TNF-α synergistically instigate the epithelial barrier reduction through reducing TJ proteins expression, including occluding. Increasing the exposure time to FB1 prevented porcine intestinal cells from establishing a proper TEER while altering the already established monolayer (Bouhet *et al.*, 2004). In piglets, DON and ZEA contamination upregulated the expression of pro-inflammatory cytokines (TNF-α, IL-1β, and IL-6) in the jejunum. Meanwhile, TJ proteins, including claudin-4 and E-cadherin expression, were decreased (Jia *et al.*, 2020). It is suggested (Suzuki, 2013) that mycotoxicosis by DON and/or ZEA overproduces pro-inflammatory cytokines around the epithelial cells, creating a pro-inflammatory environment. These pro-inflammatory cytokines alter junctions localization and reorganize the cytoskeleton, ending in modulating TJ proteins.

A highly functional GIT would have an effective barrier with epithelial cells that are constantly regenerating. Generally, intestinal epithelial cells, the main constituents of the intestinal physical barrier, are re-generated and renewed every 3 to 5 days (Bouhet & Oswald, 2005; Gao *et al.*, 2020). A fully functional and developed villi express more enzymatic activities at the upper 40% compared to the lower 60% of the crypt-villus axis. Fungal toxic metabolites induce necrosis, suppress villi proliferation, and prevent enterocytes from reaching total digestive capacity (Applegate *et al.*, 2009; Grenier & Applegate, 2013; Chen *et al.*, 2016; Feng *et al.*, 2017). When weaned piglets were fed DON-contaminated diets, lower saccharide activities were expressed, suggesting detrimental effects on gut health. Trechocences, T-2, and treatments in turkeys and mice disrupted intestinal integrity in all segments of the lower GIT (Bouhet & Oswald, 2005; Gao *et al.*, 2020). The intestine epithelial can sustain short-term mycotoxin exposure and self-repair to maintain the integrity of the physical barrier. However, in the case of prolonged mycotoxicosis exposure, the damage is more pronounced to the point that it exceeds self-regulating abilities, then the intestinal wall would be compromised as a result of epithelial cells and TJ disruption (Turner, 2009; Suzuki, 2013; Gao *et al.*, 2020; Zhang *et al.*, 2020).

7.2 *Effects of mycotoxins on the chemical barrier and altered functionality*

Aside from the physical epithelial barrier, the intestinal epithelial cells secret a mucus layer of anti-microbial proteins, mainly of mucins, antimicrobial peptides (AMPs), chemokines and cytokines and reactive oxygen species (ROS) that collectively from the chemical barrier (Bouhet & Oswald, 2005; Antonissen *et al.*, 2015; Gao *et al.*, 2020). Although these mediums are not an actual physical constituent of the epithelium, they are essential for epithelium, functionality and maintaining a healthy barrier. Mucins are polymeric glycoproteins synthesized through a baseline and active exocytosis by specialized differentiated cells, goblet cells (Ferez-Vilar & Hill, 1999) and function mainly to keep a neutral pH at the apical surface of the epithelium with their secretion of a continuous gel which is a bicarbonate-rich fluid (Bouhet & Oswald, 2005). Further, they help maintain the epithelium layer lubricated to protect it against the acidic environment and proteases formed endogenously or from bacterial origin (Montagne *et al.*, 2004). In a healthy gut, erosion of the luminal side is protected by the synthesis and these secretions of the goblet cells that usually form a very tight adherence of mucins and the apical side of the epithelium. As a result, they form a specific complex between mucin oligosaccharides and mucin-binding proteins on the apical side of the epithelial cells. This is how mucins contribute to gut physiology by acting as a selective diffusion barrier permeable to nutrients, electrolytes, and water, not for macromolecules and harmful substances, including microorganisms and their toxins, as well as protecting the gastrointestinal tract associated with harmful microflora (Antonissen *et al.*, 2015; Montagne *et al.*, 2004). Lastly, mucins perform as media or substrates for endogenous bacteria fermentation, producing different products, mainly SCFAs, and ammonium, that significantly maintain a healthy gut (Montagne *et al.*, 2004). Antimicrobial peptides are another group of substances, mainly defensins (type α and β) and cathelicidins secreted by the mucosal epithelial cells and are considered another mucosal defense agent. They exert their effects by disrupting the microbial membrane layer utilizing their net-positive charge and folding into the microbial amphipathic structure (Bouhet & Oswald, 2005).

Mycotoxin exposure would cascade negatively in different layers at a cellular level. The two most common *Fusarium* mycotoxin for poultry, DON and FUM, have been extensively studied and shown to biochemically increase lipid peroxidation (Awad *et al.*, 2012; Cheng *et al.*, 2018; Jiang *et al.*, 2014), reduce antioxidant enzyme activity (Cheng *et al.*, 2018; Jiang *et al.*, 2014), and eventually compromise broilers antioxidant functionality and alter performance (Paraskeuas *et al.*, 2021). In a study feeding broilers feed with corn naturally contaminated with Fusarium mycotoxins, authors reported higher jejunal malonaldehyde, ileal superoxide dismutase (**SOD**) concentrations at 21 days of age as well as elevated levels of SOD, jejunal malonaldehyde, and serum diamine oxidase at 42 days of age (Cheng *et al.*, 2018). One aspect would be through inducing and modulating various antioxidant mechanisms, including altering mRNA expression of specific proteins crucial to balance the antioxidant/pro-oxidant equilibrium. This involves modulating the expression of proteins such as heme-oxygenase (HMOX), xanthine oxidoreductase (XOR), xanthine dehydrogenase (XDH), and xanthine oxidase (XO); (Antonissen *et al.*, 2015). Hypoxanthine degradation to uric acid which is an antioxidant is mediated by XOR, but this reaction is known to produce ROS as well, including peroxide ad superoxide. The latter group play a crucial role as antimicrobial agents in the GIT. So, it is suggested that if less hypoxanthine is converted to xanthine by the epithelial cells, this would result in less uric acid formation and consequently lead to less ROS generation (Carro *et al.*, 2009; Harrison, 2002). Additionally, reduction on cellular content of certain pro-oxidants (*i.e.*, heme and iron) and the increase in antioxidant biliverdin is a direct outcome of HMOX upregulation (Jansen & Daiber, 2012). Maintaining minimum intracellular level of certain functional minerals such as zinc by slowing down its efflux would counteract oxidative damage produced by mycotoxins, thus preserving critical antioxidant activity. Lastly, mycotoxins-induced oxidative stress can be overturned by increasing the intestine's intracellular absorption of neutral amino acids like methionine. Different mechanisms include stabilizing sulfhydryls from oxidant-induced disulfide bond formation and membrane phospholipids stabilization (Faa *et al.*, 2008; Su *et al.*, 2014).

Although mycotoxin-induced mucins *in vitro* studies delineated variable responses, the effect of mycotoxin feed contamination on the secretion of AMPs is poorly researched and studied (Bouhet & Oswald, 2005; Gao *et al.*, 2020). Toxic metabolites for AFLA, OTA, and ZEA reportedly down-regulated protein expression of intestinal mucins MUC2, MUC4AC, and MUC5B (Wan *et al.*, 2014; Huang *et al.*, 2019). Additionally, exposure to T-2 toxin reduced MUC2 protein expression in Caco-2 cells and a thin MUC2 layer in HT-29 cells. In contrast, the same authors reported that exposure to DON, ZEA, and FUM B1 led to the up-regulation of porcine β-definsins 1 and the β-definsins 2 mRNA expression. At the same time, no changes were noted in β-defensins 1 and β-defensins 2 secretions in IPEC-J2 cells (Lam-Yim Wan *et al.*, 2013). Variabilities of *in vitro* studies may be due to discrepancies in the quantification techniques utilized because protein quantification sensitivity is very low compared to transcript measurements (Gao *et al.*, 2020).

Very limited published research has investigated how mycotoxins and metabolites influence intestinal chemical barriers. Mycotoxin-induced mucins and goblet cells show different *in vivo* responses, from no change to reduction to an increase. DON exposure led to inhibiting AMPs and MUC2 mRNA expression in broiler chickens (Antonissen *et al.*, 2015). The same authors reported that DON-challenged broilers, in conjunction with reducing MUC2 mRNA expression in the duodenum, mucin monosaccharides composition varied significantly. In another study, trichothecenes increased intestinal permeability due to reductions in tight junction proteins and reduced mucin due to a reduction in goblet cell numbers (Pinton & Oswald, 2014). Moreover, a reduction in goblet cell numbers was also reported during OTA contamination (Gao *et al.*, 2020). Mucins are polymeric glycoproteins with a high proportion of carbohydrates (*i.e.*, GlcNac, GalNac, Gal, Fuc, Man, and NeuAc) built on a protein domain rich in amino acids proline, threonine, and serine (Johansson & Hansson, 2013; Tailford *et al.*, 2015). In mucins, the O-glycan structures are different and complex and predominantly consist of α-and β-linked GlcNac, GalNac, and Gal (Tailford *et al.*, 2015). In the mucus layer of the duodenal region, the proportion of GalNac and Gal, but not GlcNac, are significantly altered by mycotoxins exposure. As a result, elongation of the O-glycan core structures is noted and modified frequently by fucose and sialic acid sugar residues (Tailford *et al.*, 2015). In all mycotoxin-exposed boilers, a significantly increased proportion of NeuAc was found with a trend of fucose fraction reduction by DON exposure. These changes in mucin O-linked oligosaccharide composition can directly influence the integrity mucus layer to a significant level and modulate the intestinal microbiota diversity (Antonissen *et al.*, 2015). The same group reported a down regulation of XDH in the jejunum as broilers were fed DON and DON + FUM-B1.

On the other hand, in laying hens, no effects were observed on goblet cells or mucin secretion due to exposure to AFLA (Applegate *et al.*, 2009). Goblet cells hyperplasia (increasing goblet cells numbers) is generally responsible for higher mucin secretions and was reported in broilers challenged with higher FUM B1 (Antonissen *et al.*, 2015) and AFLA (Jahanian *et al.*, 2016). AFLA B1 is reported to cause genetic damage due to cell growth inhibition and increased lactate dehydrogenase activity. This AFLA B1 cytotoxicity is connected to ROS generation, eventually damaging the cell membrane and DNA (Zhang *et al.*, 2015). Similarly, cellular homeostasis is negatively affected by exposure to DON and FUM-B1 due to the induction of oxidative stress on a cellular level (Antonissen *et al.*, 2015). Accordingly, Fusarium toxins, DON and FUM, acting as stressors on the intestinal cells, reportedly upregulated heat shock proteins HSP70 and HSP90 which play a critical role in protecting and repairing broilers intestinal cells (Paraskeuas *et al.*, 2021). Together, these results suggest that a depletion of the goblet cell number would result in the devastation of the mucus barrier due to the hypersecretion of mucins (Johansson & Hansson, 2016).

7.3 *Effects of mycotoxins on the immunological barrier and altered functionality*

Up to 70–80% of the immune defenses reside in the GIT representing the intestinal immune system, which is a collective result of interrelated actions of both immune (lymphocytes, macrophages, etc.), and non-immune cells as in intestinal epithelial cells (Pinton & Oswald, 2014). As foreign substances invade the GIT, the host intestinal inflammation is triggered as way of defending itself. When broilers were fed diets containing corn that was naturally contaminated with Fusarium mycotoxins, the relative weight of bursa Fabricius was decreased indicating suppressive immune function. This comes in agreement with previous research studies indicating immune organs development disruption shown in swell or atrophy due to Fusarium mycotoxins contamination (Li *et al.*, 2012; Cheng *et al.*, 2018). One of the most critical mediators regulating the immune system and inflammatory response is a group of small peptides known as cytokines. Of these cytokines, TGF-α, IL-1, IL-10, IL-15, and IL-18 play a key function controlling the immune cells influx into the mucosa, healthy epithelial cells growth and homeostasis as they are continuously expressed by the epithelial cells. While on the other hand, other cytokines are also expressed by intestinal epithelial cells, but they are vigilantly upregulated as a reaction to bacterial insult. This group includes, IL-1α or β, IL-6, IL-8, TNF-α, MCP-1 and CCL20 (Bouhet & Oswald, 2005). It is well-accepted that mycotoxins can modulate immune responses, and each can act differently from other mycotoxins. Indeed, mycotoxin exposure, even at low or moderate concentrations, can induce an inflammatory response by up-regulating cytokines, chemokines, and gene expressions (Pinton & Oswald, 2014). Both immunosuppression or even immunestimulation are noted with tricoherences, especially DON depending on dosage, and exposure time (Riahi *et al.*, 2021). In chickens, measuring inflammatory responses is through the expression levels of nuclear factor kappa B ($NF\text{-}_kB$) pathway-related genes. Translocation of $NF\text{-}_kB$ that controls the expression of certain pro- and anti-inflammatory genes due to the activation of toll-like receptors (TLRs; TLR2 and TLR4) would trigger an inflammatory response (Keestra *et al.*, 2013). Further, nuclear factor kappa B kinase subunit alpha (IKKa) plays a crucial role in modulating other inflammatory genes by being a primary regulator of the ($NF\text{-}_kB$) pathway and tumor necrosis alpha (TNFa) (Wu *et al.*, 2018; Paraskeuas *et al.*, 2021).

Several *in vivio* studies reported that *Fusarium* mycotoxins are responsible for upregulating TLRs and pro- and anti-inflammatory cytokines. the toxicity of feed-born Fusarium mycotoxins was reported in broilers at 42 days of age with higher liver and bursa Fabricius and thymes relative weights. Also, increased spleen mRNA expression of IL-1β, IL-6, and IFN-γ, respectively (Li *et al.*, 2012). The exposure to OTA, up-regulation of IL-1β, and TNF-α mRNA expression was reported in jejunal mucosa in ducks and broiler chickens (Ruan *et al.*, 2019; Tong *et al.*, 2020). DON increased mRNA expression of $NF\text{-}_kB$, while FUM elevated $NF\text{-}_kB$ and TLR4 expression in the duodenum of the small intestine of broilers and Pekin ducklings (Paraskeuas *et al.*, 2021). Contrary, feeding broilers DON at 10 mg/kg for 35 days resulted in downregulation of the expression of IL-1β and transformed growth factor beta receptor I (TGFBRI), while no response was noted on TNF-α, $NF\text{-}_kB$, and IL-8 mRNA expression in the jejunum (Ghareeb *et al.*, 2013). Further, exposure to mycotoxin contamination also reduces CD4+ cell numbers, which are involved in cytokine production, and $CD8^+$ cell numbers, which play a crucial role in host target cell protection. Feeding broilers AFLA-B1 contaminated diets reduced $CD4^+$ cell numbers in the jejunum (Kraieski *et al.*, 2017). Generally, there is a lack of research investigating mycotoxin's interference with the intestinal immunological barrier. Although some mycotoxins are considerably studied, as in AFLA, OTA, or DON, others need more exploration, as in FUM, indicating a need for further future focus on their effects on immune system disruption.

7.4 *Effects of mycotoxins on the microbial barrier*

The gut serves as the main interface between the host immune system and other microorganisms, whereas a highly functional GIT is the primary site to interact efficiently with the outer environment and feed particles. To maintain poultry health and productivity, it is extremely crucial to maintain gut microbiota at high evenness and richness levels (Carrasco *et al.*, 2019), and to maintain GIT homeostasis, it is essential for the host-microbiota to crosstalk, this include thousands of bacterial species residing in the gut lumen forming gut microbiota that are distributed heterogeneously across different GIT regions (Marchix *et al.*, 2018). These residing GIT bacterial communities are dominated at the phylum level by *Firmicutes, Bacteroidetes, proteobacteria* and *Actinobacteria* and they differ through the chicken GIT with a specific microbial profile identifying each region as in crop, gizzard, ileum, cecum and colon (Oakley *et al.*, 2014; Stanley *et al.*, 2014). These diverse communities play an integral role in maintaining the host GIT homeostasis by maintaining morphological integrity and by modulating many physiological functions. As the digestive efficiency is significantly linked to gut microbiota, nutrient absorption, and energy balance, the bird productivity is influenced by gut microbiota and diversity. Several host variables can impact the development and evolution of the gastrointestinal microbiota, although interlinkage between the gut microbiota and the immune defense mechanisms are obviously dominating at specific stages. Infections of the gastrointestinal tract are immediately identified by the host immune system, which reacts through a complex interconnected network of pathways including the innate and adaptive immune systems (Carrasco *et al.*, 2019). Antimicrobial peptides presented on the intestine epithelial surface are a group of substances, mainly defensins (type α and β) and cathelicidins secreted by the mucosal epithelial cells and are considered a mucosal defense agent which comprise an essential part of innate immunity. Microbial exposure affects defensin expression in the gut, and reducing microbial exposure resulted in reduced expression of duodenal and ceca beta-defensins in young birds (Carrasco *et al.*, 2019). Mycotoxin exposure would also influence the innate immunity indicated by different expressions of antimicrobial peptides. *In vitro* exposure to DON, ZEA, and FUM B1 led to the up-regulation of porcine β-definsins 1 and the β-definsins 2 mRNA expression (Wan *et al.*, 2014; Huang *et al.*, 2019). At the same time, no changes were noted in β-defensins 1 and β-defensins 2 secretions in IPEC-J2 cells (Lam-Yim Wan *et al.*, 2013). Dietary exposure of AFLA to broiler chicks resulted in a significant increase in ileal bacterial counts of *E. coli, Klebsiella, Salmonella,* and total negative bacteria (Galarza-Seeber *et al.*, 2016; Jahanian *et al.*, 2017). Further, as broilers were exposed to DON, cecal species' richness and evenness were reduced showing increased *Firmicutes*, and *Bacteroidetes* while reducing *proteobacteria* indicating an overall weakened GIT microbiota community that collectively has a negative effect on bird's immunity and overall health (Gao *et al.*, 2020; Lucke *et al.*, 2018). On the other hand, other research groups did not detect any effects on gut microbiota diversity and relative abundance as broilers were exposed to AFLA- B1 (Liu *et al.*, 2018). With the census in the scientific communities that microbial communities in the GIT are representative of the intestinal bacterial flora, there is still unclear understanding of the microbiota differences at various GIT regions. Studies focusing on modifications to the intestinal microbial barrier generated by mycotoxins exposure (*i.e.*, FUM-B1 and ZEA) are considerably available in other species (Wang *et al.*, 2016; Yang *et al.*, 2017), but limited in poultry (Gao *et al.*, 2020), therefore, more research is required to close this knowledge gap.

8 Conclusion

Although the physiological effects of mycotoxins have been extensively studied and researched, there is still a need to translate this wealth of knowledge into quantifiable performance effects. As mycotoxin-contaminated feed is ingested, acute or chronic toxicity symptoms would affect the bird's production, immune response, and overall health. Symptoms vary based on many factors including the severity of toxin contamination and exposure period. Still, they are always associated with a wide range of metabolic, physiological, and immunological disturbances that result in economic losses because of ingesting single or sometimes multiple mycotoxins. GIT is the primary site governing mycotoxin uptake and absorption, among other nutrients and water, and subsequent distribution in different tissues. The fast absorption in the GIT and most mycotoxins' quick appearance in the circulation indicate that most of the ingested toxin is absorbed in the proximal part of the GIT. Therefore, mycotoxins compromise the intestinal epithelium either before absorption in the upper part GIT or throughout the entire intestine by non-absorbed toxins resulting in harmful effects causing gut barrier impairments and dysfunctions. The intestinal barrier, a significant component of the GIT is composed of a multi-interconnected layer of physical, chemical, immunological, and microbial barriers. Detrimental effects of mycotoxicosis on any of these interconnected barriers would have a crucial negative impact on their integrity and subsequently would have deleterious impact on bird's overall health, performance, and productively. Research results indicate that mycotoxin exposure even at a lower level will result in impaired performance, and debilitated health. This bring more need to focus research on effects of neglected low contamination levels of mycotoxins that can be encountered occasionally during the bird's whole life cycle. In this respect, dietary strategies to mitigate mycotoxicosis and preserve intestinal integrity, control oxidative stress and inflammation need to consider the mycotoxin-relevant and intestinal site-specific effects seen on a wide range of molecular markers. As a result, overall gut health if compromised by mycotoxin contaminated feed, may provide the necessary environment for growth and proliferation of certain pathogens, *i.e.*, *C. perfringens* inducing necrotic enteritis. It is clearly obvious that mycotoxins are frequently predominant in feed ingredients. This is a living problem facing poultry producers daily, so continued efforts to get feed ingredients tested for toxin contamination levels, effective implementation of prevention, control strategies, and good biosecurity practices will be the major concerns and will shape the reality of current and future modern poultry production.

References

Adegbeye, M.J., Reddy, P.R.K., Chilaka, C.A., Balogun, O.B., Elghandour, M.M.M.Y., Rivas-Caceres, R.R., & Salem, A.Z.M. (2020). Mycotoxin toxicity and residue in animal products: Prevalence, consumer exposure and reduction strategies – A review. *Toxicon, 177*, 96–108. https://doi.org/10.1016/J.TOXICON.2020.01.007.

Allen, N.K., Mirocha, C.J., Aakhus-Allen, S., Bitgood, J.J., Weaver, G., & Bates, F. (1981a). Effect of dietary zearalenone on reproduction of chickens. *Poultry Science, 60*(6), 1165–1174. https://doi.org/10.3382/ps.0601165.

Allen, N.K., Mirocha, C.J., Weaver, G., Aakhus-Allen, S., & Bates, F. (1981b). Effects of dietary zearalenone on finishing broiler chickens and young turkey poults. *Poultry Science, 60*(1), 124–131. https://doi.org/10.3382/ps.0600124.

Allen, N.K., Peguri, A., Mirocha, C.J., & Newman, J.A. (1983). Effects of fusarium cultures, T-2 toxin, and zearalenone on reproduction of turkey females. *Poultry Science, 62*(2), 282–289. https://doi.org/10.3382/ps.0620282.

Andretta, I., Kipper, M., Lehnen, C.R., Hauschild, L., Vale, M.M., & Lovatto, P.A. (2011). Meta-analytical study of productive and nutritional interactions of mycotoxins in broilers. *Poultry Science, 90*(9), 1934–1940. https://doi.org/10.3382/ps.2011-01470.

Andretta, I., Kipper, M., Lehnen, C.R., Hauschild, L., Vale, M.M., & Lovatto, P.A. (2012a). Meta-analytical study of productive and nutritional interactions of mycotoxins in growing pigs. *Animal, 6*(9), 1476–1482. https://doi.org/10.1017/S1751731111002278.

Andretta, I., Kipper, M., Lehnen, C.R., & Lovatto, P.A. (2012b). Meta-analysis of the relationship of mycotoxins with biochemical and hematological parameters in broilers. *Poultry Science, 91*(2), 376–382. https://doi.org/10.3382/ps.2011-01813.

Antonissen, G., van Immerseel, F., Pasmans, F., Ducatelle, R., Janssens, G.P.J., de Baere, S., Mountzouris, K.C., Su, S., Wong, E.A., de Meulenaer, B., Verlinden, M., Devreese, M., Haesebrouck, F., Novak, B., Dohnal, I., Martel, A., & Croubels, S. (2015). Mycotoxins deoxynivalenol and fumonisins alter the extrinsic component of intestinal barrier in broiler chickens. *Journal of Agricultural and Food Chemistry, 63*(50), 10846–10855. https://doi.org/10.1021/acs.jafc.5b04119.

Applegate, T.J., Schatzmayr, G., Pricket, K., Troche, C., & Jiang, Z. (2009). Effect of aflatoxin culture on intestinal function and nutrient loss in laying hens. *Poultry Science, 88*(6), 1235–1241. https://doi.org/10.3382/ps.2008-00494.

Awad, W.A., Ghareeb, K., Dadak, A., Gille, L., Staniek, K., Hess, M., & Böhm, J. (2012). Genotoxic effects of deoxynivalenol in broiler chickens fed low-protein feeds. *Poultry Science, 91*(3), 550–555. https://doi.org/10.3382/ps.2011-01742.

Awad, W.A., Hess, M., Twaruzek, M., Grajewski, J., Kosicki, R., Böhm, J., & Zentek, J. (2011). The impact of the Fusarium mycotoxin deoxynivalenol on the health and performance of broiler chickens. *International Journal of Molecular Sciences, 12*(11), 7996–8012. https://doi.org/10.3390/ijms12117996.

Biehl, M.L., Prelusky, D.B., Koritz, G.D., Hartin, K.E., Buck, W.B., & Trenholm, H.L. (1993). 1993 – Toxci and Appl Pharmacology – Biliary excretion and enterohepatic cycling of ZEA in pigs. *Toxicology and Applied Pharmacology, 121*, 152–159.

Bouhet, S., Hourcade, E., Loiseau, N., Fikry, A., Martinez, S., Roselli, M., Galtier, P., Mengheri, E., & Oswald, I.P. (2004). The mycotoxin fumonisin B1 alters the proliferation and the barrier function of porcine intestinal epithelial cells. *Toxicological Sciences*, *77*(1), 165–171. https://doi.org/10.1093/toxsci/kfh006.

Bouhet, S., & Oswald, I.P. (2005). The effects of mycotoxins, fungal food contaminants, on the intestinal epithelial cell-derived innate immune response. *Veterinary Immunology and Immunopathology*, *108*(1–2 SPEC. ISS.), 199–209. https://doi.org/10.1016/j.vetimm.2005.08.010.

Boyd, K.E., Fitzpatrick, D.W., Wilson, J.R., & Wilson, L.M. (1988). Effect of T-2 Toxin on Brain Biogenic Monoamines in Rats and Chickens. *Can. J. Vet. Res.*, *52*, 181–185.

Broom, L. (2015). Mycotoxins and the intestine. *Animal Nutrition*, *1*(4), 262–265. https://doi.org/10.1016/j.aninu.2015.11.001.

Broomhead, J.N., Ledoux, D.R., Bermudez, A.J., & Rottinghaust, G.E. (2002). Chronic effects of fumonisin B1 in broilers and turkeys fed dietary treatments to market age. *Poultry Science*, *81*(1), 56–61. https://doi.org/10.1093/ps/81.1.56.

Burditt, S.J., Hagler, W.M., & Hamilton, P.B. (1984). Feed refusal during ochratoxicosis in turkeys. *Poultry Science*, *63*(11), 2172–2174. https://doi.org/10.3382/ps.0632172.

Carrasco, J.M.D., Casanova, N.A., & Miyakawa, M.E.F. (2019). Microbiota, gut health and chicken productivity: What is the connection? In *Microorganisms* (Vol. 7, Issue 10). MDPI AG. https://doi.org/10.3390/microorganisms7100374.

Carro, M.D., Falkenstein, E., Blemings, K.P., & Klandorf, H. (2009). Determination of xanthine oxidoreductase activity in broilers: Effect of pH and temperature of the assay and distribution in tissues. *Poultry Science*, *88*(11), 2406–2414. https://doi.org/10.3382/ps.2009-00278.

Chen, X., Murdoch, R., Zhang, Q., Shafer, D.J., & Applegate, T.J. (2016). Effects of dietary protein concentration on performance and nutrient digestibility in Pekin ducks during aflatoxicosis. *Poultry Science*, *95*(4), 834–841. https://doi.org/10.3382/ps/pev378.

Chen, X., Naehrer, K., & Applegate, T.J. (2016). Interactive effects of dietary protein concentration and aflatoxin B1 on performance, nutrient digestibility, and gut health in broiler chicks. *Poultry Science*, *95*(6), 1312–1325. https://doi.org/10.3382/ps/pew022.

Cheng, Y., Xu, Q., Chen, Y., Su, Y., Wen, C., & Zhou, Y. (2018). Modified palygorskite improves immunity, antioxidant ability, intestinal morphology, and barrier function in broiler chickens fed naturally contaminated diet with permitted feed concentrations of fusarium mycotoxins. *Toxins*, *10*(11). https://doi.org/10.3390/toxins10110482.

Dänicke, S. (2002). Prevention and control of mycotoxins in the poultry production chain: A European view. In *World's Poultry Science Journal* (Vol. 58, Issue 4, pp. 451–474). https://doi.org/10.1079/WPS20020033.

Dänicke, S., Ueberschär, K.H., Halle, I., Matthes, S., Valenta, H., & Flachowsky, G. (2002). Effect of addition of a detoxifying agent to laying hen diets containing uncontaminated or Fusarium toxin-contaminated maize on performance of hens and on carryover of zearalenone. *Poultry Science*, *81*(11), 1671–1680. https://doi.org/10.1093/ps/81.11.1671.

Dänicke, S., Valenta, H., & Matthes, S. (2007). On the interactions between Fusarium toxin-contaminated wheat and nonstarch polysaccharide hydrolyzing enzymes in diets of broilers on performance, intestinal viscosity, and carryover of deoxynivalenol. *Poultry Science, 86*(2), 291–298. https://doi.org/10.1093/ps/86.2.291.

Desjardins, A., Maragos, C., Norred, W., Pestka, J., Philips, T., Vardon, P., Whitaker, T., Wood, G., & van Egmond, H. (2003). *Mycotoxins: risks in plant, animal, and human systems*. Council for Agricultural Science and Technology.

Devegowda, G., and M.T.N.K. (2005). *The Mycotoxin Blue Book* (Diaz, D.E., Ed.). British Library.

Eskola, M., Kos, G., Elliott, C.T., Hajšlová, J., Mayar, S., & Krska, R. (2020). Worldwide contamination of food-crops with mycotoxins: Validity of the widely cited 'FAO estimate' of 25%. In *Critical Reviews in Food Science and Nutrition* (Vol. 60, Issue 16, pp. 2773–2789). Taylor and Francis Inc. https://doi.org/10.1080/10408398.2019.1658570.

Faa, G., Nurchi, V.M., Ravarino, A., Fanni, D., Nemolato, S., Gerosa, C., van Eyken, P., & Geboes, K. (2008). Zinc in gastrointestinal and liver disease. *Coordination Chemistry Reviews, 252*(10–11), 1257–1269. https://doi.org/10.1016/J.CCR.2007.09.024.

Feng, G.D., He, J., Ao, X., & Chen, D.W. (2017). Effects of maize naturally contaminated with aflatoxin B1 on growth performance, intestinal morphology, and digestive physiology in ducks. *Poultry Science, 96*(6), 1948–1955. https://doi.org/10.3382/ps/pew420.

Ferez-Vilar, J., & Hill, R.L. (1999). The structure and assembly of secreted mucins. In *Journal of Biological Chemistry* (Vol. 274, Issue 45, pp. 31751–31754). https://doi.org/10.1074/jbc.274.45.31751.

Fitzpatrick, D.W., Boyd, K.E., Wilson, L.M., & Wilson, J.R. (1988). Effect Of The Trichothecene Deokynivalenol On Brain Biogenic Monoamines Concentrations In Rats And Chickens. *Journal of Environmental Science and Health, Part B, 23*(2), 159–170. https://doi.org/10.1080/03601238809372594.

Galarza-Seeber, R., Latorre, J.D., Bielke, L.R., Kuttappan, V.A., Wolfenden, A.D., Hernandez-Velasco, X., Merino-Guzman, R., Vicente, J.L., Donoghue, A., Cross, D., Hargis, B.M., & Tellez, G. (2016). Leaky Gut and Mycotoxins: Aflatoxin B1 does not increase gut permeability in broiler chickens. *Frontiers in Veterinary Science, 3*(FEB). https://doi.org/10.3389/fvets.2016.00010.

Gao, Y., Meng, L., Liu, H., Wang, J., & Zheng, N. (2020). The compromised intestinal barrier induced by mycotoxins. In *Toxins* (Vol. 12, Issue 10). MDPI AG. https://doi.org/10.3390/toxins12100619.

Gbore, F.A., Yinusa, R.I., & Salleh. (2010). Evaluation of subchronic dietary fumonisin B 1 on nutrient digestibility and growth performance of rats. *African Journal of Biotechnology, 9*(38), 6442–6447. http://www.academicjournals.org/AJB.

Ghareeb, K., Awad, W.A., Soodoi, C., Sasgary, S., Strasser, A., & Böhm, J. (2013). Effects of Feed Contaminant Deoxynivalenol on Plasma Cytokines and mRNA Expression of Immune Genes in the Intestine of Broiler Chickens. *PLoS ONE, 8*(8). https://doi.org/10.1371/journal.pone.0071492.

Girardet, C., Bonnet, M.S., Jdir, R., Sadoud, M., Thirion, S., Tardivel, C., Roux, J., Lebrun, B., Wanaverbecq, N., Mounien, L., Trouslard, J., Jean, A., Dallaporta, M., & Troadec, J.D. (2011).

The food-contaminant deoxynivalenol modifies eating by targeting anorexigenic neurocircuitry. *PLoS ONE*, *6*(10). https://doi.org/10.1371/journal.pone.0026134.

Grenier, B., & Applegate, T.J. (2013a). Modulation of intestinal functions following mycotoxin ingestion: Meta-analysis of published experiments in animals. In *Toxins* (Vol 5, Issue 2, pp. 396–430). https://doi.org/10.3390/toxins5020396.

Grenier, B., & Applegate, T.J. (2013b). Modulation of intestinal functions following mycotoxin ingestion: Meta-analysis of published experiments in animals. In *Toxins* (Vol 5, Issue 2, pp. 396–430). https://doi.org/10.3390/toxins5020396.

Grenier, B., Dohnal, I., Shanmugasundaram, R., Eicher, S.D., Selvaraj, R.K., Schatzmayr, G., & Applegate, T.J. (2016). Susceptibility of broiler chickens to coccidiosis when fed subclinical doses of deoxynivalenol and fumonisins – special emphasis on the immunological response and themycotoxin interaction. *Toxins*, *8*(8). https://doi.org/10.3390/toxins8080231.

Gruber-Dorninger, C., Jenkins, T., & Schatzmayr, G. (2019). Global mycotoxin occurrence in feed: A ten-year survey. *Toxins*, *11*(7). https://doi.org/10.3390/toxins11070375.

Guerre, P. (2020). Mycotoxin and Gut Microbiota Interactions. *Toxins*, *12*(12). https://doi.org/10.3390/toxins12120769.

Hamilton, P.B., Huff, W.E., Harris, J.R., & Wyatt, R.D. (1982). Natural occurrences of ochratoxicosis in poultry. *Poultry Science*, *61*(9), 1832–1841. https://doi.org/10.3382/ps.0611832.

Han, X.Y., Huang, Q.C., Li, W.F., Jiang, J.F., & Xu, Z.R. (2008). Changes in growth performance, digestive enzyme activities and nutrient digestibility of cherry valley ducks in response to aflatoxin B1 levels. *Livestock Science*, *119*(1–3), 216–220. https://doi.org/10.1016/j.livsci.2008.04.006.

Harrison, R. (2002). Structure and function of xanthine oxidoreductase: where are we now? *Free Radical Biology and Medicine*, *33*(6), 774–797. https://doi.org/10.1016/S0891-5849(02)00956-5.

Huang, X., Gao, Y., Li, S., Wu, C., Wang, J., & Zheng, N. (2019). Modulation of Mucin (MUC2, MUC5AC AND MUC5B) mRNA expression and protein production and secretion in Caco-2/HT29-MTX co-cultures following exposure to individual and combined aflatoxin M1 and ochratoxin A. *Toxins*, *11*(2). https://doi.org/10.3390/toxins11020132.

Huff, W.E., Doerr, J.A., Hamilton, P.B., & Vesonder, R.F. (1981). Acute toxicity of vomitoxin (Deoxynivalenol) in broiler chickens. *Poultry Science*, *60*(7), 1412–1414. https://doi.org/10.3382/ps.0601412.

Huff, W.E., & Hamilton, P.B. (1975). Decreased plasma carotenoids during ochratoxicosis. *Poultry Science*, *54*(4), 1308–1310. https://doi.org/10.3382/ps.0541308.

Iqbal, S.Z., Nisar, S., Asi, M.R., & Jinap, S. (2014). Natural incidence of aflatoxins, ochratoxin A and zearalenone in chicken meat and eggs. *Food Control*, *43*, 98–103. https://doi.org/10.1016/j.foodcont.2014.02.046.

Jahanian, E., Mahdavi, A.H., Asgary, S., & Jahanian, R. (2016). Effect of dietary supplementation of mannanoligosaccharides on growth performance, ileal microbial counts, and jejunal morphology in broiler chicks exposed to aflatoxins. *Livestock Science*, *190*, 123–130. https://doi.org/10.1016/J.LIVSCI.2016.05.008.

Jahanian, E., Mahdavi, A.H., Asgary, S., & Jahanian, R. (2017). Effects of dietary inclusion of silymarin on performance, intestinal morphology and ileal bacterial count in aflatoxin-challenged broiler chicks. *Journal of Animal Physiology and Animal Nutrition, 101*(5), e43–e54. https://doi.org/10.1111/JPN.12556.

Jansen, T., & Daiber, A. (2012). Direct antioxidant properties of bilirubin and biliverdin. Is there a role for biliverdin reductase? *Frontiers in Pharmacology, 3 MAR*. https://doi.org/10.3389/fphar.2012.00030.

Jia, R., Liu, W., Zhao, L., Cao, L., & Shen, Z. (2020). Low doses of individual and combined deoxynivalenol and zearalenone in naturally moldy diets impair intestinal functions via inducing inflammation and disrupting epithelial barrier in the intestine of piglets. *Toxicology Letters, 333*, 159–169. https://doi.org/10.1016/j.toxlet.2020.07.032.

Jiang, S.Z., Li, Z., Wang, G.Y., Yang, Z.B., Yang, W.R., Zhang, G.G., & Wu, Y.B. (2014). Effects of Fusarium mycotoxins with yeast cell wall absorbent on hematology, serum biochemistry, and oxidative stress in broiler chickens. *Journal of Applied Poultry Research, 23*(2), 165–173. https://doi.org/10.3382/japr.2013-00830.

Johansson, M.E.V., & Hansson, G.C. (2013). Mucus and the goblet cell. *Digestive Diseases, 31*(3–4), 305–309. https://doi.org/10.1159/000354683.

Johansson, M.E.V., & Hansson, G.C. (2016). Immunological aspects of intestinal mucus and mucins. In *Nature Reviews Immunology* (Vol. 16, Issue 10, pp. 639–649). Nature Publishing Group. https://doi.org/10.1038/nri.2016.88.

Keestra, A.M., de Zoete, M.R., Bouwman, L.I., Vaezirad, M.M., & van Putten, J.P.M. (2013). Unique features of chicken Toll-like receptors. *Developmental and Comparative Immunology, 41*(3), 316–323. https://doi.org/10.1016/j.dci.2013.04.009.

Kermanshahi, H., Akbari, M.R., Maleki, M., & Behgar, M. (2007). Effect of Prolonged Low Level Inclusion of Aflatoxin B_1 into Diet on Performance, Nutrient Digestibility, Histopathology and Blood Enzymes of Broiler Chickens Effect of Organic Acid and Probiotic on Performance and GIT Health of Broiler Chickens View project. *Journal of Animal and Veterinary Advances, 6*(5), 686–692. https://www.researchgate.net/publication/26590532.

Kim, Y.S., & Ho, S.B. (2010). Intestinal goblet cells and mucins in health and disease: Recent insights and progress. In *Current Gastroenterology Reports* (Vol. 12, Issue 5, pp. 319–330). https://doi.org/10.1007/s11894-010-0131-2.

Kogut, M.H. (2013). The gut microbiota and host innate immunity: Regulators of host metabolism and metabolic diseases in poultry? In *Journal of Applied Poultry Research* (Vol. 22, Issue 3, pp. 637–646). Oxford University Press. https://doi.org/10.3382/japr.2013-00741.

Kraieski, A.L., Hayashi, R.M., Sanches, A., Almeida, G.C., & Santin, E. (2017). Effect of aflatoxin experimental ingestion and Eimeira vaccine challenges on intestinal histopathology and immune cellular dynamic of broilers: Applying an Intestinal Health Index. *Poultry Science, 96*(5), 1078–1087. https://doi.org/10.3382/ps/pew397.

Kubena, L.F., Harvey, R.B., Buckley, S.A., Bailey, R.H., & Rottinghaus, G.E. (1999). Effects of long-term feeding of diets containing moniliformin, supplied by Fusarium fujikuroi culture material, and fumonisin, supplied by Fusarium moniliforme culture material to laying hens. *Poultry Science*, *78*(11), 1499–1505. https://doi.org/10.1093/ps/78.11.1499.

Lam-Yim Wan, M., Jackson Woo, C.S., Allen, K.J., Turner, P.C., & El-Nezamia, H. (2013). Modulation of porcine β-defensins 1 and 2 upon individual and combined Fusarium toxin exposure in a swine jejunal epithelial cell line. *Applied and Environmental Microbiology*, *79*(7), 2225–2232. https://doi.org/10.1128/AEM.03277-12.

Ledoux, D.R., Bermudez, A.J., & Rottinghaus, G.E. (1996). Effects of Feeding Fusarium moniliforme Culture Material, Containing Known Levels of Fumonisin B1, in the Young Turkey Poult. *Poultry Science*, *75*(12), 1472–1478. https://doi.org/10.3382/ps.0751472.

Li, Z., Yang, Z.B., Yang, W.R., Wang, S.J., Jiang, S.Z., & Wu, Y.B. (2012). Effects of feed-borne Fusarium mycotoxins with or without yeast cell wall adsorbent on organ weight, serum biochemistry, and immunological parameters of broiler chickens. *Poultry Science*, *91*(10), 2487–2495. https://doi.org/10.3382/ps.2012-02437.

Liew, W.P.P., & Mohd-Redzwan, S. (2018). Mycotoxin: Its impact on gut health and microbiota. *Frontiers in Cellular and Infection Microbiology*, *8*(FEB). https://doi.org/10.3389/fcimb.2018.00060.

Liu, J.D., Doupovec, B., Schatzmayr, D., Murugesan, G.R., Bortoluzzi, C., Villegas, A.M., & Applegate, T.J. (2020). The impact of deoxynivalenol, fumonisins, and their combination on performance, nutrient, and energy digestibility in broiler chickens. *Poultry Science*, *99*(1), 272–279. https://doi.org/10.3382/ps/pez484.

Liu, N., Wang, J.Q., Jia, S.C., Chen, Y.K., & Wang, J.P. (2018). Effect of yeast cell wall on the growth performance and gut health of broilers challenged with aflatoxin B 1 and necrotic enteritis. *Poultry Science*, *97*(2), 477–484. https://doi.org/10.3382/ps/pex342.

Lucke, A., Böhm, J., Zebeli, Q., & Metzler-Zebeli, B.U. (2018). Dietary deoxynivalenol contamination and oral lipopolysaccharide challenge alters the cecal microbiota of broiler chickens. *Frontiers in Microbiology*, *9*(APR). https://doi.org/10.3389/fmicb.2018.00804.

Marchix, J., Goddard, G., & Helmrath, M.A. (2018). Host-Gut Microbiota Crosstalk in Intestinal Adaptation. In *CMGH* (Vol. 6, Issue 2, pp. 149–162). Elsevier Inc. https://doi.org/10.1016/j.jcmgh.2018.01.024.

McLaughlin, J., Padfield, P.J., Burt, J.P.H., & O'Neill, C.A. (2004). Ochratoxin A increases permeability through tight junctions by removal of specific claudin isoforms. *American Journal of Physiology – Cell Physiology*, *287*(5 56-5). https://doi.org/10.1152/ajpcell.00007.2004.

Metayer, J.P., Travel, A., Mika, A., Bailly, J.D., Cleva, D., Boissieu, C., Guennec, J.le, Froment, P., Albaric, O., Labrut, S., Lepivert, G., Marengue, E., Tardieu, D., & Guerre, P. (2019). Lack of toxic interaction between fusariotoxins in broiler chickens fed throughout their life at the highest level tolerated in the european union. *Toxins*, *11*(8). https://doi.org/10.3390/toxins11080455.

Montagne, L., Piel, C., & Lallès, J.P. (2004). Effect of Diet on Mucin Kinetics and Composition: Nutrition and Health Implications. In *Nutrition Reviews* (Vol. 62, Issue 3, pp. 105–114). https://doi.org/10.1301/nr.2004.mar.105-114.

Murugesan, G.R., Ledoux, D.R., Naehrer, K., Berthiller, F., Applegate, T.J., Grenier, B., Phillips, T.D., & Schatzmayr, G. (2015). Prevalence and effects of mycotoxins on poultry health and performance, and recent development in mycotoxin counteracting strategies. *Poultry Science*, *94*(6), 1298–1315. https://doi.org/10.3382/ps/pev075.

Oakley, B.B., Lillehoj, H.S., Kogut, M.H., Kim, W.K., Maurer, J.J., Pedroso, A., Lee, M.D., Collett, S.R., Johnson, T.J., & Cox, N.A. (2014). The chicken gastrointestinal microbiome. *FEMS Microbiology Letters*, *360*(2), 100–112. https://doi.org/10.1111/1574-6968.12608.

Odenwald, M.A., & Turner, J.R. (2017). The intestinal epithelial barrier: A therapeutic target? *Nature Reviews Gastroenterology and Hepatology*, *14*(1), 9–21. https://doi.org/10.1038/nrgastro.2016.169.

Paraskeuas, V., Griela, E., Bouziotis, D., Fegeros, K., Antonissen, G., & Mountzouris, K.C. (2021). Effects of deoxynivalenol and fumonisins on broiler gut cytoprotective capacity. *Toxins*, *13*(10). https://doi.org/10.3390/toxins13100729.

Pinton, P., & Oswald, I.P. (2014). Effect of deoxynivalenol and other type B trichothecenes on the intestine: A review. In *Toxins* (Vol. 6, Issue 5, pp. 1615–1643). MDPI AG. https://doi.org/10.3390/toxins6051615.

Pitt, J.I. (2000). Toxigenic fungi and mycotoxins. *British Medical Bulletin*, *56*(1), 184–192. https://academic.oup.com/bmb/article/56/1/184/388023.

Prelusky, D.B., & Trenholm, H.L. (1993). The efficacy of various classes of anti-emetics in preventing deoxynivalenol-induced vomiting in swine. *Natural Toxins*, *1*(5), 296–302. https://doi.org/10.1002/nt.2620010508.

Raj, J., Farkaš, H., Jakovčević, Z., Vasiljević, M., Kumar, R., & Asrani, R. K. (2023). Effects of supplemented multicomponent mycotoxin detoxifying agent in laying hens fed aflatoxin B1 and T2-toxin contaminated feeds. *Poultry Science*, *102*(8). https://doi.org/10.1016/j.psj.2023.102795.

Ravikumar, R., Chan, J., & Prabakaran, M. (2022). Vaccines against Major Poultry Viral Diseases: Strategies to Improve the Breadth and Protective Efficacy. In *Viruses* (Vol. 14, Issue 6). MDPI. https://doi.org/10.3390/v14061195.

Riahi, I., Marquis, V., Pérez-Vendrell, A.M., Brufau, J., Esteve-Garcia, E., & Ramos, A.J. (2021). Effects of deoxynivalenol-contaminated diets on metabolic and immunological parameters in broiler chickens. *Animals*, *11*(1), 1–15. https://doi.org/10.3390/ani11010147.

Richards, M.P., & Proszkowiec-Weglarz, M. (2007). Mechanisms regulating feed intake, energy expenditure, and body weight in poultry. *Poultry Science*, *86*(7), 1478–1490. https://doi.org/10.1093/ps/86.7.1478.

Ruan, D., Wang, W.C., Lin, C.X., Fouad, A.M., Chen, W., Xia, W.G., Wang, S., Luo, X., Zhang, W.H., Yan, S.J., Zheng, C.T., & Yang, L. (2019). Effects of curcumin on performance, antioxidation, intestinal barrier and mitochondrial function in ducks fed corn contaminated with ochratoxin A. *Animal*, *13*(1), 42–52. https://doi.org/10.1017/S1751731118000678.

Stanley, D., Hughes, R.J., & Moore, R.J. (2014). Microbiota of the chicken gastrointestinal tract: Influence on health, productivity and disease. In *Applied Microbiology and Biotechnology* (Vol. 98, Issue 10, pp. 4301–4310). Springer Verlag. https://doi.org/10.1007/s00253-014-5646-2.

Streit, E., Schwab, C., Sulyok, M., Naehrer, K., Krska, R., & Schatzmayr, G. (2013). Multi-mycotoxin screening reveals the occurrence of 139 different secondary metabolites in feed and feed ingredients. *Toxins*, *5*(3), 504–523. https://doi.org/10.3390/toxins5030504.

Su, S., Miska, K.B., Fetterer, R.H., Jenkins, M.C., & Wong, E.A. (2014). Expression of digestive enzymes and nutrient transporters in Eimeria acervulina-challenged layers and broilers. *Poultry Science, 93*(5), 1217–1226. https://doi.org/10.3382/ps.2013-03807.

Suzuki, T. (2013). Regulation of Intestinal Barrier Function by Dietary Polyphenols. *Current Nutrition & Food Science, 9*(2), 85–92. https://doi.org/10.2174/1573401311309020002.

Swamy, H.V.L.N., Smith, T.K., Cotter, P.F., Boermans, H.J., & Sefton, A.E. (2002). Effects of feeding blends of grains naturally contaminated with Fusarium mycotoxins on production and metabolism in broilers. *Poultry Science, 81*(7), 966–975. https://doi.org/10.1093/ps/81.7.966.

Swamy, H.V.L.N., Smith, T.K., Karrow, N.A., & Boermans, H.J. (2004). Effects of feeding blends of grains naturally contaminated with Fusarium mycotoxins on growth and immunological parameters of broiler chickens. *Poultry Science, 83*(4), 533–543. https://doi.org/10.1093/ps/83.4.533.

Tailford, L.E., Crost, E.H., Kavanaugh, D., & Juge, N. (2015). Mucin glycan foraging in the human gut microbiome. *Frontiers in Genetics, 5*(FEB). https://doi.org/10.3389/fgene.2015.00081.

Taranu, I., Braicu, C., Marin, D.E., Pistol, G.C., Motiu, M., Balacescu, L., Beridan Neagoe, I., & Burlacu, R. (2015). Exposure to zearalenone mycotoxin alters in vitro porcine intestinal epithelial cells by differential gene expression. *Toxicology Letters, 232*(1), 310–325. https://doi.org/10.1016/j.toxlet.2014.10.022.

Tardieu, D., Bailly, J.D., Skiba, F., Métayer, J.P., Grosjean, F., & Guerre, P. (2007). Chronic toxicity of fumonisins in turkeys. *Poultry Science, 86*(9), 1887–1893. https://doi.org/10.1093/ps/86.9.1887.

Tong, C., Li, P., Yu, L.H., Li, L., Li, K., Chen, Y., Yang, S.H., & Long, M. (2020). Selenium-rich yeast attenuates ochratoxin A-induced small intestinal injury in broiler chickens by activating the Nrf2 pathway and inhibiting NF-KB activation. *Journal of Functional Foods, 66*. https://doi.org/10.1016/j.jff.2020.103784.

Tran, S.T., Auvergne, A., Benard, G., Bailly, J.D., Tardieu, D., Babilé, R., & Guerre, P. (2005). Chronic effects of fumonisin B1 on ducks. *Poultry Science, 84*(1), 22–28. https://doi.org/10.1093/ps/84.1.22.

Turner, J.R. (2009). Intestinal mucosal barrier function in health and disease. *Nature Reviews Immunology, 9*(11), 799–809. https://doi.org/10.1038/nri2653.

Verma, J., Johri, T.S., & Swain, B.K. (2007). Effect of aflatoxin, ochratoxin and their combination on protein and energy utilisation in white leghorn laying hens. *Journal of the Science of Food and Agriculture, 87*(5), 760–764. https://doi.org/10.1002/jsfa.2655.

Voss, K.A., Riley, R.T., Norred, W.P., Bacon, C.W., Meredith, F.I., Howard, P.C., Plattner, R.D., Collins, T.F.X., Hansen, D.K., & Porter, J.K. (2001). An overview of rodent toxicities: Liver and kidney effects of Fumonisins and Fusarium moniliforme. *Environmental Health Perspect, 109*, 259–266.

Wan, L.Y.M., Allen, K.J., Turner, P.C., & El-nezami, H. (2014). Modulation of mucin mRNA (MUC5AC AND MUC5B) expression and protein production and secretion in Caco-2/HT29-MTX Co-cultures following exposure to individual and combined Fusarium mycotoxins. *Toxicological Sciences, 139*(1), 83–98. https://doi.org/10.1093/toxsci/kfu019.

Wang, A., & Hogan, N.S. (2019). Performance effects of feed-borne Fusarium mycotoxins on broiler chickens: Influences of timing and duration of exposure. *Animal Nutrition, 5*(1), 32–40. https://doi.org/10.1016/j.aninu.2018.08.001.

Wang, A., Schwean-Lardner, K., & Hogan, N.S. (2019). Feed preference and feeding behaviours in grower broilers fed diets containing wheat naturally contaminated with fusarium mycotoxins. *British Poultry Science, 60*(3), 309–316. https://doi.org/10.1080/00071668.2019.1596224.

Wang, J., Tang, L., Glenn, T.C., & Wang, J.S. (2016). Aflatoxin B1 induced compositional changes in gut microbial communities of male F344 rats. *Toxicological Sciences, 150*(1), 54–63. https://doi.org/10.1093/toxsci/kfv259.

Wang, W., Zhai, S., Xia, Y., Wang, H., Ruan, D., Zhou, T., Zhu, Y., Zhang, H., Zhang, M., Ye, H., Ren, W., & Yang, L. (2019). Ochratoxin A induces liver inflammation: Involvement of intestinal microbiota. *Microbiome, 7*(1). https://doi.org/10.1186/s40168-019-0761-z.

Wu, K., Ren, C., Gong, Y., Gao, X., Rajput, S.A., Qi, D., & Wang, S. (2021). The insensitive mechanism of poultry to zearalenone: A review. In *Animal Nutrition* (Vol. 7, Issue 3, pp. 587–594). KeAi Communications Co. https://doi.org/10.1016/j.aninu.2021.01.002.

Wu, S., Liu, Y., Duan, Y., Wang, F., Guo, F., Yan, F., Yang, X., & Yang, X. (2018). Intestinal toxicity of deoxynivalenol is limited by supplementation with Lactobacillus plantarum JM113 and consequentially altered gut microbiota in broiler chickens. *Journal of Animal Science and Biotechnology, 9*(1). https://doi.org/10.1186/s40104-018-0286-5.

Xu, H., Hao, S., Gan, F., Wang, H., Xu, J., Liu, D., & Huang, K. (2017). In vitro immune toxicity of ochratoxin A in porcine alveolar macrophages: A role for the ROS-relative TLR4/MyD88 signaling pathway. *Chemico-Biological Interactions, 272*, 107–116. https://doi.org/10.1016/j.cbi.2017.05.016.

Xu, R., Kiarie, E.G., Yiannikouris, A., Sun, L., & Karrow, N.A. (2022). Nutritional impact of mycotoxins in food animal production and strategies for mitigation. *Journal of Animal Science and Biotechnology, 13*(1). https://doi.org/10.1186/s40104-022-00714-2.

Yang, X., Liu, L., Chen, J., & Xiao, A. (2017). Response of intestinal bacterial flora to the long-term feeding of aflatoxin B1 (AFB1) in mice. *Toxins, 9*(10). https://doi.org/10.3390/toxins9100317.

Yang, Y., Li, G., Wu, D., Liu, J., Li, X., Luo, P., Hu, N., Wang, H., & Wu, Y. (2020). Recent advances on toxicity and determination methods of mycotoxins in foodstuffs. *Trends in Food Science & Technology, 96*, 233–252. https://doi.org/10.1016/J.TIFS.2019.12.021.

Yegani, M., Smith, T.K., Leeson, S., & Boermans, H.J. (2006). Effects of feeding grains naturally contaminated with Fusarium mycotoxins on performance and metabolism of broiler breeders. *Poultry Science, 85*(9), 1541–1549. https://doi.org/10.1093/ps/85.9.1541.

Yunus, A.W., Blajet-Kosicka, A., Kosicki, R., Khan, M.Z., Rehman, H., & Böhm, J. (2012). Deoxynivalenol as a contaminant of broiler feed: Intestinal development, absorptive functionality, and metabolism of the mycotoxin. *Poultry Science, 91*(4), 852–861. https://doi.org/10.3382/ps.2011-01903.

Zain, M.E. (2011). Impact of mycotoxins on humans and animals. *Journal of Saudi Chemical Society, 15*(2), 129–144. https://doi.org/10.1016/j.jscs.2010.06.006.

Zhai, S., Zhu, Y., Feng, P., Li, M., Wang, W., Yang, L., & Yang, Y. (2021). Ochratoxin A: its impact on poultry gut health and microbiota, an overview. In *Poultry Science* (Vol. 100, Issue 5). Elsevier Inc. https://doi.org/10.1016/j.psj.2021.101037.

Zhang, J., Zheng, N., Liu, J., Li, F.D., Li, S.L., & Wang, J.Q. (2015). Aflatoxin B1 and aflatoxin M1 induced cytotoxicity and DNA damage in differentiated and undifferentiated Caco-2 cells. *Food and Chemical Toxicology, 83*, 54–60. https://doi.org/10.1016/J.FCT.2015.05.020.

Zhang, W., Zhang, S., Wang, J., Shan, A., & Xu, L. (2020). Changes in intestinal barrier functions and gut microbiota in rats exposed to zearalenone. *Ecotoxicology and Environmental Safety, 204*. https://doi.org/10.1016/j.ecoenv.2020.111072.

Zinedine, A., Soriano, J.M., Moltó, J.C., & Mañes, J. (2007). Review on the toxicity, occurrence, metabolism, detoxification, regulations and intake of zearalenone: An oestrogenic mycotoxin. In *Food and Chemical Toxicology* (Vol. 45, Issue 1, pp. 1–18). https://doi.org/10.1016/j.fct.2006.07.030.

Zou, Y., Yang, Z.B., Yang, W.R., Jiang, S.Z., Zhang, G.G., & Chi, F. (2012). Effect of purified zearalenone on nutrient digestibility in broilers fed 2 levels of fumonisin from naturally contaminated corn (zea mays). *Journal of Applied Poultry Research, 21*(2), 251–258. https://doi.org/10.3382/japr.2011-00359.

CHAPTER 10

Mycotoxins and swine gut health

K.L. Summers

Animal Biosciences and Biotechnology Laboratory, Beltsville Agricultural Research Center, Agricultural Research Center, United States Department of Agriculture, Beltsville, Maryland 20705, USA; katie.mason.summers@gmail.com

Abstract

Mycotoxins are secondary metabolites made by diverse fungi that can have toxic effects when ingested by animals or humans. Often mycotoxins are found as feed contamination and when animals ingest them, the economic impact can be great due to animal illness, stunted growth, reduced reproductive ability, and even death. The fungi that produce mycotoxins are varied, but here we examine three significant fungal genera that predominantly effect swine health through their mycotoxin production: *Aspergillus*, *Fusarium*, and *Penicillium*. In this chapter, we will review the major swine-effecting mycotoxins, aflatoxins, deoxynivalenol, zearalenone, fumonisin, and ochratoxin A. We will investigate their general symptoms and focus on the effect of each mycotoxin on gut health and the indigenous microbiome.

Keywords

mycotoxin – swine – pig – fungi – porcine

1 Mycotoxins and swine

Mycotoxins have a long history of causing disease when ingested by agricultural animals. This is a significant concern for farmers and producers as mycotoxins can have estrogenic, carcinogenic, anabolic, and teratogenic effects (reviewed in [1]). Monogastric animals, including pigs and poultry, are especially susceptible to mycotoxins due to increased mycotoxin prevalence in dietary components such as cereals [2, 3]. Each mycotoxin is produced by different fungal species, and thus, have different environmental conditions necessary for optimal mycotoxin production [4, 5]; however, high moisture and elevated temperatures during feed growth and/or storage are often associated with the fungal growth resulting in elevated mycotoxin production [1, 6–9]. Due to the ubiquitous nature of fungi, it is impossible

© K.L. SUMMERS, 2024 | DOI:10.3920/9789004695467_011

TABLE 10.1 Common pig mycotoxins and their effects

Mycotoxin	Main fungal producer(s)	Common effects in pigs
Aflatoxins (B1, B2, G1, G2)	*Aspergillus flavus* *Aspergillus parasiticus*	Feed reduction, reduced weight gain, tremors, hepatopathies, immunosuppression, jaundice
Fumonisins (B1, B2, B3)	*Fusarium verticillioides*	Pulmonary edema, hepatopathies, Feed reduction, tumors, immunosuppression, reduced growth rate, pulmonary lesions, cyanosis
Zearalenone	*Fusarium graminearum*	Hyperestrogenism, edematose vulva, prolapsed rectum, anestrus, reduced litter size, pseudogestation, infertility, spontaneous abortions
Ochratoxins (Ochratoxin A)	*Aspergillus ochraceus* *Penicillium viridicatum* *Penicillium verrucosum*	Nephropathy, hepatotoxicity, immunosuppression
Deoxynivalenol	*Fusarium graminearum*	Reduced feed intake, reduced growth rate, vomiting, diarrhea, digestive lesions, sudden death, ulcers

to eliminate all mycotoxins from feed sources, but regulatory standards have been established to assist in reducing animal illness [10]. The most common reported mycotoxins in swine feed are aflatoxins, deoxynivalenol, zearalenone, fumonisin, and ochratoxin A [11–13] (Table 10.1).

1.1 *Aflatoxins*

Aflatoxins are a family of fungal-produced toxins found on agricultural crops including peanuts, tree nuts, corn, wheat, and cottonseed [14]. This family includes over 20 different aflatoxins, but the most common pig-effecting aflatoxins are Aflatoxin B1, B2, G1, and G2 and are natural contaminants of agricultural feed and grains [1, 15]. Aflatoxins are difurocoumarolactones that are considered carcinogenic, hepatotoxic and immunosuppressive (reviewed in [11]) and are produced by *Aspergillus* genera. *Aspergillus flavus* and *Aspergillus parasiticus* produce the highest levels of aflatoxins and produce aflatoxin B1 which is considered the most toxic to pigs [16–18].

In swine, aflatoxins are considered the most toxic and can result in multiple pathologies. Aflatoxins are liposoluble and can be absorbed by mucus membranes such as the lungs and gastrointestinal tract but are also absorbed by the skin. Overall the effect of aflatoxin is most severe in young pigs and is a concern in nursing piglets as aflatoxins have been found in sow milk, resulting in reduced piglet growth and high death rates [6]. Piglets *in utero* are very susceptible to the effects of aflatoxins and demonstrate reduced immune response after mitogenic stimulation [19]. High levels of ingestion can lead to animal death at any age, but other pathologies result in more overall loss to producers. Aflatoxins lead to immunosuppression, feed intake reduction, tremors, jaundice, and reduced liver function [20, 21].

Aflatoxicosis can be acute but is most often chronic (~90% of cases). After pig ingestion, the toxin travels through the circulatory and lymphatic systems to multiple organs and tissues. The highest concentrations are often found in the liver and kidneys. Symptoms of acute aflatoxicosis include liver lesions, jaundice, and sudden death [20]. Chronic aflatoxicosis is more common with less severe symptoms including reduced feed intake and growth rates [22]. Often diagnosis comes after a combination of vague health effects are noted, including slow growth in combination with secondary infections that are not responsive to treatment. Immunosuppression due to aflatoxicosis has been found to alter porcine response to vaccinations and increased susceptibility to infectious disease outbreaks [23]. Aflatoxin has been found to inhibit both the innate and adaptive immune responses [23–26]. While toxin effects are not completely understood, the effects appear to be broad including the dysregulation of dendritic cells [27], reduction in pro-inflammatory cytokines and an increase in anti-inflammatory [18], and altered macrophage and neutrophil function [19].

1.2 *Zearalenone*

Zearalenone, also known as F2 toxin, is a phenolic resorcyclic acid lactone and acts as a non-steroid estrogenic toxin [28]. Fusarium species including *Fusarium graminearum*, *Fusarium culmorum*, and *Fusarium crookwellense* produce zearalenone. Zearalenone is produced by fungi during high humidity conditions and is a common contaminant of cereal grains including corn, wheat, barley, sorghum, and rye [29]. As an estrogen mimic, the effects of zearalenone are often seen in the reproductive tract of pigs. While fertility and reproduction are often reduced after exposure to zearalenone, reduction in nutrient digestion, reduced growth, and hepatotoxicity have also been reported [30, 31].

After ingestion, zearalenone is reduced to α- and β-zearalenol in the intestine and liver cells [32, 33]. These metabolites have different levels of efficacy on estrogen receptors in comparison to zearalenone. α-zearalenol has higher biological activity on estrogen receptors while β-zearalenol has lower [34]. The resulting increase in α-hydroxylation results in increased toxicity in pigs.

In gilts this presents as vulvar swelling and redness, irregular estrus, and rectal and/or vaginal prolapses [20]. In sows, false heat, false pregnancy, reduced piglet birth size, and abortions are seen [35]. Zearalenone is found in sow milk and can result in vulvar swelling and redness in nursing gilts [36]. Young boars also exhibit symptoms such as reduced testosterone and sperm production [37].

1.3 *Ochratoxin A*

Aspergillus ochraceus, Penicillium verrucosum, and *Penicillium viridicatum* are the main producers of ochratoxin A (OTA), a pentaketide derived from the dihydrocoumarin family couple to beta-phenylalanine [38]. OTA has many effects and is carcinogenic, hepatotoxic, nephrotoxic, teratogenic, and immunosuppressive [20, 39]. In agricultural species, pigs are the most susceptible to OTA ingestion and the resulting toxicity [40, 41]. One common effect is nephrotoxicity with kidney lesions (mycotoxic porcine nephropathy) and damage to the proximal tubules of the kidney [42, 43]. Chronic, low-dose symptoms are often subtle in swine and can include reduced feed efficiency and growth rate [44]. One study found that after a year of elevated OTA exposure, histological examination demonstrated only proliferative changes in the interstitium and mild nephropathy overall [45]. OTA ingestion also results in immunosuppression and increased susceptibility to secondary bacterial infections, especially in young pigs [40].

Interestingly, ruminants have microbial members of the microbiome that can degrade OTA to a less toxic ochratoxin α [46]. These microbes are lacking in the porcine microbiome and likely play a role in their elevated susceptibility to OTA compared to other ruminant agricultural animals. Further, OTA is the most common mycotoxin found in processed pork meat products, making it of significant concern to human health as well [39, 47, 48].

1.4 *Deoxynivalenol*

Deoxynivalenol (DON) is also known as vomitoxin and is a type B trichothecene. Trichothecenes are a large family of chemically related mycotoxins that are made by numerous fungal species including *Fusarium, Cephalosporium, Trichothecium,* and *Stachybotrys*. DON is a sesquiterpene produced by *Fusarium graminearum* and is a significant contaminant of corn and wheat [49, 50]. While sometimes referred to as vomitoxin, vomiting is not the most common symptom. Ingestion of DON is usually at low, chronic levels and leads to immunosuppression, altered protein synthesis, altered neurotransmitter activity, reduced intestinal nutrient absorption, intestinal barrier damage, and altered intestinal epithelium function [20, 23, 51]. These effects can lead to reduced feed intake and growth rate. Pigs that ingest DON often refuse food but also have a range of symptoms due to the neuroendocrine changes and immunotoxicity [52]. One such effect is the altered tryptophan metabolism after DON consumption that alters pig behavior including time spent standing and drinking water [53].

Acute DON exposure results in digestive lesions and ulcers, anorexia, vomiting, and diarrhea. If extremely high levels are ingested, pig death can result [54]. The digestive injuries can lead to altered gut immune responses as microbial antigens pass through cellular tight junctions and activate the innate immune response in the lamina propria [55]. Further, the immune effects of DON ingestion depend upon dose, frequency, and duration of exposure, but can result in immunostimulatory or immunosuppressive effects [56, 57]. Low levels of DON seem to induce pro-inflammatory genes while high levels of DON result in immune dysregulation and leukocyte apoptosis [57, 58] While less common, DON can also have teratologic and reproductive effects [23, 59]. Although pigs can adapt to low levels of mycotoxin ingestion, even small levels of DON can result in reduced final body weights, making this an important consideration for producers [60].

1.5 *Fumonisins*

Fumonisins are a group of structurally related mycotoxins produced predominantly by *Fusarium verticillioides* and *Fusarium proliferatum*. These mycotoxins are known for causing species-specific diseases and symptoms, such as kidney toxicities in rodent, leukoencephalomalacia in horses, and pulmonary edema in pigs [61]. Porcine pulmonary edema (PPE) is specific to pigs and has not been seen as a disease in other animals fed fumonisin. *F. verticillioides* produces Fumonisin B1 most often and is the most abundant of the fumonisins in nature. Fumonisin B1 ingestion can lead to PPE, but also causes hepatic lesions, elevated serum cholesterol, altered heart and liver health, and while not specifically immunosuppressive, causes an inhibition in phagocytosis by pulmonary macrophages [23, 61–64].

As seen in other mycotoxins, chronic exposure is more common in pigs and results in reduced feed intake and growth rate. Acute toxicity causes PPE and pigs display severe respiratory signs, cyanosis and death [20, 61]; however, non-lethal pulmonary edema in pigs has been seen with lower levels of exposure [65]. Fumonisins are poorly absorbed by the gastrointestinal tract but mycotoxin that is absorbed remains in the liver and kidneys for an extended time [66]. Fumonisins have been found to alter immunity through decreased vaccine efficacy following ingestion of fumonisin B1, resulting in increased susceptibility to secondary infections [23, 67, 68].

2 Co-mycotoxin effects

Fungi are ubiquitous in the environment, and as such, multiple fungi can contaminate feed sources simultaneously, resulting in co-mycotoxin exposure. While studies conflict on how often multiple mycotoxins infect feed, it is generally accepted that this is a common occurrence. One study assessing agricultural feed found that 48% of feed and feedstuff samples are contaminated by 2 or more mycotoxins [50]. Other studies found that co-contamination rates ranged from 75–100% [69, 70]. Shockingly, a recent study assessed the prevalence of 19 mycotoxins in over 200 pig feed samples from different regions in Spain and found 40% of the samples were contaminated with more than five mycotoxins [71]. Based on these studies, the possibility for co-occurrence of mycotoxins must always be considered when assessing pig feed.

A single fungus can produce multiple mycotoxins which can result in co-contamination of feed products. One example is *Fusarium* species such as *F. graminearum* that can produce DON and zearalenone [71]. Further, different mycotoxins can be produced by multiple fungi all found in the same feed material. These numerous mycotoxins can have antagonistic, synergistic, or additive effects. These effects were recently defined by Grenier *et al.* [72] and examples include synergistic nephrotoxicity effects such as citrinin modifying ochratoxin distribution, resulting in a more toxic effect in the kidney [73]. Co-exposure to mycotoxins has also been demonstrated to alter immune responses in an additive way. One example is a study demonstrating DON and fumonisin co-ingestion for 35 days results in increased inflammatory cytokine expression and mitogenic stimulation [74]. Taken together, the potential for multiple mycotoxin exposure in pigs is a complex problem that necessitates better mycotoxin biomarkers and research to understand mechanisms behind these effects.

3 Mycotoxin and gut effects

3.1 *Effect on the pig intestine*

While parameters such as dose are obvious in altering animal toxicity, other variables include pig age, sex, and duration of exposure [4, 75, 76]. Overall, regardless of mycotoxin type, the effects are greatest in young pigs and male pigs [68, 75]. Following ingestion, the gut is the first target of the mycotoxins. Intestinal epithelial cells act as the first barrier, protecting the pig from toxins, but over time intestinal damage and altered function can occur [74, 77, 78]. As damage occurs, barrier function is reduced and microbial antigens can pass through tight junctions and alter immune responses [74, 77–80]. These changes can explain the increase in susceptibility to enteric infections, especially in combination with the overall immunosuppressive actions of mycotoxins. Studies in piglets fed fumonisin and then challenged with *Escherichia coli* were found to have bacterial translocation to the mesenteric lymph nodes and dissemination to multiple organs [55, 81].

One study in poultry suggests that DON alters the function of the transporter for glucose uptake in the small intestine, sodium-glucose dependent transporter (SGLT-1) [82]. While this study was in poultry, similar effects in pigs are suspected, as diarrhea is seen in both species and is suspected to be affected by this change in glucose transport. Other reduced absorption pathways are the palmitate and monocarboxylate in the small intestine [83]. However, fumonisin exposure in pigs results in up-regulated sodium-dependent glucose absorption, underscoring the numerous effects mycotoxins can have on intestinal health [84–86]. Fumonisin can also reduce folate uptake due to sphingolipid depletion in the pig intestine [87]. Further studies in swine are needed to understand the species specific effect these mycotoxins have on the intestinal epithelium.

3.2 *Mycotoxins and the microbiome*

As pigs ingest mycotoxins, the gastrointestinal tract becomes the site of multiple toxin effects. One important and often overlooked effect is the antimicrobial properties of some mycotoxins. Due to this inherent antimicrobial effect, mycotoxins can alter the gut microbiota in pigs. Due to the intricate link between the gut microbiome and many facets of animal health, mycotoxin changes to the microbiome can result in secondary changes in the animal that alter health. Mycotoxins have been shown to have antimicrobial properties, and thus, are able to alter the gut microbiota in pigs [88–92]. However, little research has been done to date to understand the potential impact on the microbiota and its overall effects on pig health and growth, with most studies being performed in rodents and with culture-based techniques [93].

Some ruminants have been found to have indigenous members of the microbiome that can degrade toxins such as OTA, to reduce the effects of this mycotoxin to animal health [46]. However, these microbes have found to be missing in the pig small intestine, allowing the mycotoxins to be absorbed and to cause toxic effects [78, 90, 94]. One *in vitro* study found that the microbiota of the small intestine, cecum, and colon could hydrolyze masked mycotoxin DON, but could not further metabolize active DON [95]. Based on the microbes in ruminants, supplementing pigs with microbes that could break down or reduce the efficacy of mycotoxins would be a potential targeted approach towards mycotoxin control.

Waché *et al.* studied the effect of DON on the culturable bacteriome in the pig intestine and found that ingestion of DON-contaminated feed resulting in altered bacteriome composition, suggesting that the mycotoxin could alter animal health through destabilizing the intestinal microbiome [92]. One proposed mechanism is through microbiome dysbiosis, mycotoxins alter intestinal barrier integrity, and can therefore change the mucosal immune response [96].

Recently a study investigated the effect of DON and zearalenone on the gut microbiome in weaned piglets through metaproteomics [97]. Both toxins resulted in elevated ribosome proteins and pentose-phosphate pathways, but also decreased glycolysis and carbohydrate metabolism pathways. High levels of the mycotoxins increased thioredoxin-dependent peroxiredoxin, an antioxidant enzyme. Taken together, these results suggest that ingestion of these two mycotoxins results in altered microbial metabolism, oxidative stress responses, and changed the composition and structure of the gut microbiome. Interestingly, piglets fed a low level of these two toxins had microbiomes and metaproteomes similar to control animals that were fed no toxins, suggesting there is some resilience to toxin effects in pigs at low levels.

The mixture of DON and zearalenone was also assessed in its ability to alter the colonic microbiome during experimental mycotoxicosis. Piotrowska *et al.* found that mycotoxins altered the diversity of Enterobacteriaceae in the colon and functional diversity [98]. Zearalenone reduced mesophilic aerobic bacteria levels in the presence or absence of DON through unknown mechanism. Another group investigated the colonic microbiome in pigs fed DON- and zearalenone-contaminated feed and found that mycotoxin ingestion led to the increase in two operational taxonomic units (OTUs) classified as *Lactobacillus* to dominating the colon microbiome [99]. While not currently known, the authors hypothesized that these specific lactobacilli may play a role in mycotoxin detoxification in the colon of pigs.

Another study investigated the mucosal microbiome of the jejunum in pigs during mycotoxin challenge with aflatoxin and DON. While this study also took into account the role of weaning weight on growth performance, piglets fed mycotoxin contaminated feed led to reduced growth rates and gut health regardless of weight at weaning [100]. When looking at the microbiome composition by 16S-based high throughput sequencing, they found that Proteobacteria were altered by mycotoxin exposure, but other phyla were not significantly disrupted. The authors suggested that *Prevotella copri* may be a species of interest as an indicator of microbiota diversity during mycotoxin challenge.

While studies investigating the effects of mycotoxins and multiple mycotoxins on the microbiome of pigs have started to describe changes in the composition and diversity of the microbes, little remains known about the mechanism behind these changes and the effects these have on other members of the microbiome and pig health and growth. Future work is needed, utilizing high throughput sequencing technologies and more accurate biomarkers, to begin to understand these microbial changes to determine better targeted approaches to reducing the effects of mycotoxins.

4 Prevention

Mycotoxins appear in feed due to contamination of an ingredient. This contamination frequently originates in the field but can become worse during the processing and storage phases of feed production. Prevention of mycotoxin contamination can be approached through the minimizing of fungal infection of dietary components, such as corn. Corn that is stressed by drought or heat stress is less resistant to fungi such as *A. flavus* or *A. parasiticus* [101]. The use of irrigation, adequate fertilization, and the selection of appropriate varieties of corn for the geography are important steps producers can take to minimize crop stress. Storage conditions of dietary components are also very important as keeping grain cool and dry are critical in preventing toxin production. Thorough cleaning and drying between use of grains is also an important task in prevention. Antifungal treatments can be used to reduce fungal growth on stored grains but has no efficacy on mycotoxins already present in the grains at the time of treatment [6].

Geographic regions have varying levels of mycotoxin types based on their environmental conditions. In addition to temperature, humidity is a critical factor altering crop susceptibility to fungal infection, both in the field and during storage. Studies investigating humidity and corn contamination with aflatoxins demonstrated the importance of initial moisture content and different relative humidity on mycotoxin levels [12, 102]. High moisture content or high humidity resulted in moldy corn at the end of experimental storage (30 days or less) for multiple mycotoxins. Suggestions for real-world applicability suggest storing grains or other dietary components for less than two weeks when heat and humidity are elevated [6]. Interestingly, it has been suggested that zearalenone and DON may not require as much humidity to cause significant contamination compared to aflatoxins [102]. However, *Aspergillus* spp. produce more aflatoxins with warmer conditions compared to zearalenone. The environmental conditions of the region where a food crop is grown must be considered to reduce as many fungal toxins as possible, specifically for that geographic area.

Overall, producers cannot completely eliminate, but can take multiple steps in reducing, mycotoxin contamination of crops including storing commodities at low temperatures whenever possible, utilizing fungicides as necessary, reducing insect infestation in storage facilities, and through the development of fungal resistant varieties of crops. If crops do become infected, removing contaminated seeds or products should be done promptly, or if contamination is extensive, the contaminated product should be destroyed. While some products, such as acetic acid, ammonia gas, and organize solvents, have been used to decontaminate mycotoxins on commodities, these chemicals are somewhat limited due to the necessity of chemical safety for animals and humans exposed to the decontaminating chemical.

5 Conclusion

Mycotoxins are secondary metabolites produced by a variety of fungi. Pigs are highly susceptible to the effects of mycotoxins and their ingestion results in financial loss to producers and farmers, but also becomes a health risk to the human population that eats their meat products. Fungi such as *Aspergillus*, *Fusarium*, and *Penicillium* are important mycotoxin-producing genera that can contaminate pig feed components such as corn. Mycotoxins in pigs have diverse negative effects including immunosuppression, altered reproduction, and organ effects such as nephrotoxicity. While the mycotoxin mechanisms leading to symptoms are not completely understood, much remains to be understood about their local effect on the indigenous microbiome and its role in susceptibility to secondary diseases and immunosuppression. As mycotoxins are ubiquitous and cannot be eliminated, more research must be done to better understand their effect on pig biology.

Works cited

(CAST), C.f.A.S.T., *Mycotoxins: risks in plant, animal, and human systems*. 2003, CAST.

Abdel-Wahhab, M.A., *et al.*, *Zizyphus spina-christi extract protects against aflatoxin B1-initiated hepatic carcinogenicity*. Afr J Tradit Complement Altern Med. 2007. **4**(3): pp. 248–56.

Alassane-Kpembi, I., O. Puel, and I.P. Oswald, *Toxicological interactions between the mycotoxins deoxynivalenol, nivalenol and their acetylated derivatives in intestinal epithelial cells*. Arch Toxicol, 2015. **89**(8): pp. 1337–46.

Ali-Vehmas, T., *et al.*, *Measurement of antibacterial activities of T-2 toxin, deoxynivalenol, ochratoxin A, aflatoxin B1 and fumonisin B1 using microtitration tray-based turbidimetric techniques*. Zentralbl Veterinarmed A, 1998. **45**(8): pp. 453–8.

Andretta, I., *et al.*, *Meta-analytical study of productive and nutritional interactions of mycotoxins in growing pigs*. Animal, 2012. **6**(9): pp. 1476–82.

Arroyo-Manzanares, N., *et al.*, *Occurrence of Mycotoxins in Swine Feeding from Spain*. Toxins, 2019. **11**(342).

Atroshi, F., *et al.*, *Effects of tamoxifen, melatonin, coenzyme Q10, and L-carnitine supplementation on bacterial growth in the presence of mycotoxins*. Pharmacol Res, 1998. **38**(4): pp. 289–95.

Awad, W.A., *et al.*, *In vitro effects of deoxynivalenol on small intestinal D-glucose uptake and absorption of deoxynivalenol across the isolated jejunal epithelium of laying hens*. Poult Sci, 2007. **86**(1): pp. 15–20.

Benzoni, E., *et al.*, *Influence of in vitro exposure to mycotoxin zearalenone and its derivatives on swine sperm quality*. Reprod Toxicol, 2008. **25**(4): pp. 461–7.

Binder, E.M., *et al.*, *Worldwide occurrence of mycotoxins in commodities, feeds and feed ingredients*. Anim. Feed Sci. Technol., 2007. **137**: pp. 265–282.

Bouhet, S. and I.P. Oswald, *The intestine as a possible target for fumonisin toxicity*. Mol Nutr Food Res, 2007. **51**(8): pp. 925–31.

Bryden, W.L. *Mycotoxin contamination of the feed supply chain: Implications for animal productivity and feed security*. Animal Feed Science and Technology, 2012. **173**(1–2): pp. 134–158.

Bryden, W.L., *Mycotoxins in the food chain: human health implications*. Asia Pac J Clin Nutr, 2007. 16 Suppl 1: pp. 95–101.

Burel, C., *et al.*, *Effect of Low Dose of Fumonisins on Pig Health: Immune Status, Intestinal Microbiota and Sensitivity to Salmonella*. Toxins (Basel), 2013. 5(4): pp. 841–864.

Chaytor, A.C., *et al.*, *Effects of chronic exposure of diets with reduced concentrations of aflatoxin and deoxynivalenol on growth and immune status of pigs*. J Anim Sci, 2011. **89**(1): pp. 124–35.

Chaytor, A.C., *et al.*, *Occurrence and Decontamination of Mycotoxins in Swine Feed*. Asian-Aust. J. Anim. Sci., 2011. **24**(5): pp. 723–738.

Devreese, M., P. De Backer, and S. Croubels, *Overview of the most important mycotoxins for the pig and poultry husbandry*. Vlaams Diergeneeskundig Tijdschrift., 2013. **82**: pp. 171–180.

Dietrich, B., *et al.*, *Fusarium mycotoxin-contaminated wheat containing deoxynivalenol alters the gene expression in the liver and the jejunum of broilers*. Animal, 2012. **6**(2): pp. 278–91.

Duarte, S.C., C.M. Lino, and A. Pena, *Ochratoxin A in feed of food-producing animals: an undesirable mycotoxin with health and performance effects*. Vet Microbiol, 2011. **154**(1–2): pp. 1–13.

el Khoury, A. and A. Atoui, *Ochratoxin a: general overview and actual molecular status*. Toxins (Basel), 2010. 2(4): pp. 461–93.

Escriva, L., G. Font, and L. Manyes, *In vivo toxicity studies of fusarium mycotoxins in the last decade: a review*. Food Chem Toxicol, 2015. **78**: pp. 185–206.

Ferrara M, *et al.*, *Effect of Penicillium nodicum contamination rates on ochratoxin A accumulation in dry-cured salami*. Food Control, 2016. **67**: pp. 235–239.

Fitzpatrick, D.W., *et al.*, *Measurement of the relative binding affinity of zearalenone, alpha-zearalenol and beta-zearalenol for uterine and oviduct estrogen receptors in swine, rats and chickens: an indicator of estrogenic potencies*. Comp Biochem Physiol C Comp Pharmacol Toxicol, 1989. **94**(2): pp. 691–4.

Ghareeb, K., *et al.*, *Impacts of the feed contaminant deoxynivalenol on the intestine of monogastric animals: poultry and swine*. J Appl Toxicol, 2015. **35**(4): pp. 327–37.

Gratz, S.W., *et al.*, *Porcine Small and Large Intestinal Microbiota Rapidly Hydrolyze the Masked Mycotoxin Deoxynivalenol-3-Glucoside and Release Deoxynivalenol in Spiked Batch Cultures In Vitro*. Appl Environ Microbiol, 2018. **84**(2).

Grenier, B. and I. Oswald, *Mycotoxin co-contamination of food and feed: meta-analysis of publications describing toxicological interactions*. World Mycotoxin Journal, 2011. **4**: pp. 285–313.

Grenier, B. and T.J. Applegate, *Modulation of intestinal functions following mycotoxin ingestion: meta-analysis of published experiments in animals*. Toxins (Basel), 2013. **5**(2): pp. 396–430.

Guerre, P., *Worldwide Mycotoxins Exposure in Pig and Poultry Feed Formulations*. Toxins (Basel), 2016. **8**(12).

Haschek, W.M., *et al.*, *Fumonisin toxicosis in swine: an overview of porcine pulmonary edema and current perspectives*. Environ Health Perspect, 2001. **109 Suppl 2**: pp. 251–7.

Hendel, E.G., *et al.*, *Survey of mycotoxins in 2016 United States corn*. Journal of Animal Science, 2017. **95**: pp. 16–17.

Hennig-Pauka, I., *et al.*, *Current challenges in the diagnosis of zearalenone toxicosis as illustrated by a field case of hyperestrogenism in suckling piglets*. Porcine Health Manag, 2018. **4**: p. 18.

Holanda, D.M. and S.W. Kim, *Impacts of weaning weights and mycotoxin challenges on jejunal mucosa-associated microbiota, intestinal and systemic health, and growth performance of nursery pigs*. J Anim Sci Biotechnol, 2022. **13**(1): p. 43.

Houghton E., *Mycotoxins*. 2018 [cited 2022 07/18/2022]; Available from: https://www.thepigsite.com/disease-guide/mycotoxins.

Jelinek, C.F., A.E. Pohland, and G.E. Wood, *Worldwide occurrence of mycotoxins in foods and feeds – an update*. J Assoc Off Anal Chem, 1989. **72**(2): pp. 223–30.

Kebede, H., *et al.*, *Relationship between aflatoxin contamination and physiological responses of corn plants under drought and heat stress*. Toxins (Basel), 2012. **4**(11): pp. 1385–403.

Kordic, B., *et al.*, *Experimental study of the effects of known quantities of zearalenone on swine reproduction.* J Environ Pathol Toxicol Oncol, 1992. **11**(2): pp. 53–5.

Kumar, P., *et al.*, *Aflatoxins: A Global Concern for Food Safety, Human Health and Their Management.* Front Microbiol, 2016. 7: p. 2170.

Lahouar, A., *et al.*, *Effects of temperature, water activity and incubation time on fungal growth and aflatoxin B1 production by toxinogenic Aspergillus flavus isolates on sorghum seeds.* Rev Argent Microbiol, 2016. **48**(1): pp. 78–85.

Lalles, J.P., *et al.*, *Consumption of fumonisin B1 for 9 days induces stress proteins along the gastrointestinal tract of pigs.* Toxicon, 2010. **55**(2–3): pp. 244–9.

Lalles, J.P., M. Lessard, and G. Boudry, *Intestinal barrier function is modulated by short-term exposure to fumonisin B(1) in Ussing chambers.* Vet Res Commun, 2009. **33**(8): pp. 1039–43.

Lessard, M., *et al.*, *Intestinal physiology and peptidase activity in male pigs are modulated by consumption of corn culture extracts containing fumonisins.* J Nutr, 2009. **139**(7): pp. 1303–7.

Lewczuk, B., *et al.*, *Histological structure of duodenum in gilts receiving low doses of zearalenone and deoxynivalenol in feed.* Exp Toxicol Pathol, 2016. **68**(2–3): pp. 157–66.

Liew, W.P. and S. Mohd-Redzwan, *Mycotoxin: Its Impact on Gut Health and Microbiota.* Front Cell Infect Microbiol, 2018. **8**: p. 60.

Loh, Z.H., *et al.*, *Toxin Degradation by Rumen Microorganisms: A Review.* Toxins (Basel), 2020. **12**(10).

Loiseau, N., *et al.*, *Fumonisin B1 exposure and its selective effect on porcine jejunal segment: sphingolipids, glycolipids and trans-epithelial passage disturbance.* Biochem Pharmacol, 2007. **74**(1): pp. 144–52.

Lucioli, J., *et al.*, *The food contaminant deoxynivalenol activates the mitogen activated protein kinases in the intestine: interest of ex vivo models as an alternative to in vivo experiments.* Toxicon, 2013. **66**: pp. 31–6.

Ma, R., *et al.*, *Individual and Combined Occurrence of Mycotoxins in Feed Ingredients and Complete Feeds in China.* Toxins (Basel), 2018. **10**(3).

Malagutti, L., *et al.*, *Effects of ochratoxin A on heavy pig production.* Animal Research, 2005. **54**: pp. 179–184.

Malekinejad, H., R. Maas-Bakker, and J. Fink-Gremmels, *Species differences in the hepatic biotransformation of zearalenone.* Vet J, 2006. **172**(1): pp. 96–102.

Mallmann, C.A. and P. Dilkin, *Mycotoxins and Mycotoxicosis in Swine.* Special Nutrients ed. 2011: LAMIC.

Maresca, M., *et al.*, *Both direct and indirect effects account for the pro-inflammatory activity of enteropathogenic mycotoxins on the human intestinal epithelium: stimulation of interleukin-8 secretion, potentiation of interleukin-1beta effect and increase in the transepithelial passage of commensal bacteria.* Toxicol Appl Pharmacol, 2008. **228**(1): pp. 84–92.

Maresca, M., *From the gut to the brain: journey and pathophysiological effects of the food-associated trichothecene mycotoxin deoxynivalenol.* Toxins (Basel), 2013. **5**(4): pp. 784–820.

Marin, D.E., *et al.*, *Changes in performance, blood parameters, humoral and cellular immune responses in weanling piglets exposed to low doses of aflatoxin.* J Anim Sci, 2002. **80**(5): pp. 1250–7.

Marin, D.E., *et al.*, *Fumonisin B1 alters cell cycle progression and interleukin-2 synthesis in swine peripheral blood mononuclear cells*. Mol Nutr Food Res, 2007. **51**(11): pp. 1406–12.

Marin, D.E., *et al.*, *Sex-related differences in the immune response of weanling piglets exposed to low doses of fumonisin extract*. Br J Nutr, 2006. **95**(6): pp. 1185–92.

Mehrzad, J., *et al.*, *Aflatoxin B(1) interferes with the antigen-presenting capacity of porcine dendritic cells*. Toxicol In Vitro, 2014. **28**(4): pp. 531–7.

Molinie, A. and A. Leszkowicz, *Toxic effects of co-contamination of Ochratoxin A and citrinin*. Drug Metabolism Reviews, 2003. **35**: pp. 77–77.

Mousa, W., *et al.*, *Modelling the effect of water activity and temperature on growth rate and aflatoxin production by two isolates of Aspergillus flavus on paddy*. J Appl Microbiol, 2011. **111**(5): pp. 1262–74.

Ogundero, V.W., *Temperature and aflatoxin production by Aspergillus flavus and A. parasiticus strains from Nigerian groundnuts*. Journal of Basic Microbiology, 1987. **27**(9): pp. 511–514.

Olsen, M., *et al.*, *Metabolism of zearalenone by sow intestinal mucosa in vitro*. Food Chem Toxicol, 1987. **25**(9): pp. 681–3.

Ostry, V., *et al.*, *Mycotoxins as human carcinogens-the IARC Monographs classification*. Mycotoxin Res, 2017. **33**(1): pp. 65–73.

Oswald, I.P., *et al.*, *Mycotoxin fumonisin B1 increases intestinal colonization by pathogenic Escherichia coli in pigs*. Appl Environ Microbiol, 2003. **69**(10): pp. 5870–4.

Osweiler, G.D. and S.M. Ensley, *Mycotoxins in grains and feed*. Diseases of Swine, ed. J.J. Zimmerman, *et al.* Vol. 10. 2012, Chichester, West-Sussex, UK: Wiley-Blackwell.

Parussolo, G., *et al.*, *Ochratoxin A production by Aspergillus westerdijkiae in Italian-type salami*. Food Microbiol, 2019. **83**: pp. 134–140.

Patience, J.F., *et al.*, *Evaluation of two mycotoxin mitigation strategies in grow-finish swine diets containing corn dried distillers grains with solubles naturally contaminated with deoxynivalenol*. J Anim Sci, 2014. **92**(2): pp. 620–6.

Patience J.F., *Just how much mycotoxin can a pig handle?* 2019 [7/18/2022]; Available from: https://www.nationalhogfarmer.com/nutrition/just-how-much-mycotoxin-can-pig-handle.

Payne, G.A. and M.P. Brown, *Genetics and physiology of aflatoxin biosynthesis*. Annu Rev Phytopathol, 1998. **36**: pp. 329–62.

Pestka, J.J., *Deoxynivalenol: mechanisms of action, human exposure, and toxicological relevance*. Arch Toxicol, 2010. **84**(9): pp. 663–79.

Pestka, J.J., *Deoxynivalenol-induced proinflammatory gene expression: mechanisms and pathological sequelae*. Toxins (Basel), 2010. **2**(6): pp. 1300–17.

Pestka, J.J., *et al.*, *Cellular and molecular mechanisms for immune modulation by deoxynivalenol and other trichothecenes: unraveling a paradox*. Toxicol Lett, 2004. **153**(1): pp. 61–73.

Pierron, A., I. Alassane-Kpembi, and I.P. Oswald, *Impact of mycotoxin on immune response and consequences for pig health*. Anim Nutr, 2016. **2**(2): pp. 63–68.

Pierron, A., I. Alassane-Kpembi, and I.P. Oswald, *Impact of two mycotoxins deoxynivalenol and fumonisin on pig intestinal health*. Porcine Health Manag, 2016. **2**: p. 21.

Piotrowska, M., *et al.*, *The effect of experimental fusarium mycotoxicosis on microbiota diversity in porcine ascending colon contents.* Toxins (Basel), 2014. **6**(7): pp. 2064–81.

Pistol, G.C., *et al.*, *Natural feed contaminant zearalenone decreases the expressions of important pro- and anti-inflammatory mediators and mitogen-activated protein kinase/NF-kappaB signalling molecules in pigs.* Br J Nutr, 2014. **111**(3): pp. 452–64.

Prelusky, D.B., *et al.*, *Biological fate of fumonisin B1 in food-producing animals.* Adv Exp Med Biol, 1996. **392**: pp. 265–78.

Reddy, K.E., *et al.*, *Colon Microbiome of Pigs Fed Diet Contaminated with Commercial Purified Deoxynivalenol and Zearalenone.* Toxins (Basel), 2018. **10**(9).

Robert, H., *et al.*, *Impact of mycotoxins on the intestine: are mucus and microbiota new targets?* J Toxicol Environ Health B Crit Rev, 2017. **20**(5): pp. 249–275.

Rodrigues, I. and K. Naehrer, *A three-year survey on the worldwide occurrence of mycotoxins in feedstuffs and feed.* Toxins, 2012. **4**: pp. 663–675.

Romero, A., *et al.*, *Mycotoxins modify the barrier function of Caco-2 cells through differential gene expression of specific claudin isoforms: Protective effect of illite mineral clay.* Toxicology, 2016. **353–354**: pp. 21–33.

Saenz, J.S., *et al.*, *Metaproteomics Reveals Alteration of the Gut Microbiome in Weaned Piglets Due to the Ingestion of the Mycotoxins Deoxynivalenol and Zearalenone.* Toxins (Basel), 2021. **13**(8).

Shen, Y.B., A.C. Weaver, and S.W. Kim, *Physiological Effects of Deoxynivalenol from Naturally Contaminated Corn on Cerebral Tryptophan Metabolism, Behavioral Response, Gastrointestinal Immune Status and Health in Pigs Following a Pair-Feeding Model.* Toxins (Basel), 2021. **13**(6).

Silvotti, L., *et al.*, *Immunotoxicological effects on piglets of feeding sows diets containing aflatoxins.* Vet Rec, 1997. **141**(18): pp. 469–72.

Smith, G.W., *et al.*, *Sequence of cardiovascular changes leading to pulmonary edema in swine fed culture material containing fumonisin.* Am J Vet Res, 1999. **60**(10): pp. 1292–300.

Stoev, S.D., *et al.*, *Experimental mycotoxic nephropathy in pigs provoked by a mouldy diet containing ochratoxin A and fumonisin B1.* Exp Toxicol Pathol, 2012. **64**(7–8): pp. 733–41.

Stoev, S.D., *et al.*, *Experimental one year ochratoxin A toxicosis in pigs.* Exp Toxicol Pathol, 2002. **53**(6): pp. 481–7.

Stoev, S.D., *et al.*, *Mycotoxic nephropathy in Bulgarian pigs and chickens: complex aetiology and similarity to Balkan endemic nephropathy.* Food Addit Contam Part A Chem Anal Control Expo Risk Assess, 2010. **27**(1): pp. 72–88.

Stoev, S.D., *et al.*, *Susceptibility to secondary bacterial infections in growing pigs as an early response in ochratoxicosis.* Exp Toxicol Pathol, 2000. **52**(4): pp. 287–96.

Streit, E., *et al.*, *Current situation of mycotoxin contamination and co-occurrence in animal feed – focus on Europe.* Toxins (Basel), 2012. **4**(10): pp. 788–809.

Sun, X.M., *et al.*, *Effects of sterigmatocystin, deoxynivalenol and aflatoxin G1 on apoptosis of human peripheral blood lymphocytes in vitro.* Biomed Environ Sci, 2002. **15**(2): pp. 145–52.

Theumer, M.G., *et al.*, *Immunobiological effects of fumonisin B1 in experimental subchronic mycotoxicoses in rats.* Clin Diagn Lab Immunol, 2002. **9**(1): pp. 149–55.

Tkaczyk, A. and P. Jedziniak, *Mycotoxin Biomarkers in Pigs-Current State of Knowledge and Analytics*. Toxins (Basel), 2021. **13**(8).

Vlachou, M., *et al.*, *Ochratoxin A in Slaughtered Pigs and Pork Products*. Toxins (Basel), 2022. **14**(2).

Wache, Y.J., *et al.*, *Impact of deoxynivalenol on the intestinal microflora of pigs*. Int J Mol Sci, 2009. **10**(1): pp. 1–17.

Wache, Y.J., *et al.*, *The mycotoxin deoxynivalenol inhibits the cell surface expression of activation markers in human macrophages*. Toxicology, 2009. **262**(3): pp. 239–44.

Wang, D.F., *et al.*, *Interaction of zearalenone and soybean isoflavone in diets on the growth performance, organ development and serum parameters in prepubertal gilts*. J Anim Physiol Anim Nutr (Berl), 2012. **96**(5): pp. 939–46.

Weaver, A.C., *et al.*, *The use of feed additives to reduce the effects of aflatoxin and deoxynivalenol on pig growth, organ health and immune status during chronic exposure*. Toxins (Basel), 2013. **5**(7): pp. 1261–81.

Wellington, M.O., *et al.*, *Effect of long-term feeding of graded levels of deoxynivalenol (DON) on growth performance, nutrient utilization, and organ health in finishing pigs and DON content in biological samples*. J Anim Sci, 2020. **98**(12).

Wen, Y.Q., *et al.*, *Effect of Stored Humidity and Initial Moisture Content on the Qualities and Mycotoxin Levels of Maize Germ and Its Processing Products*. Toxins (Basel), 2020. **12**(9).

Wild, C.P., *Aflatoxin exposure in developing countries: the critical interface of agriculture and health*. Food Nutr Bull, 2007. **28**(2 Suppl): pp. S372–80.

Young, L.G., *et al.*, *Vomitoxin in corn fed to young pigs*. J Anim Sci, 1983. **57**(3): pp. 655–64.

Zinedine, A., *et al.*, *Review on the toxicity, occurrence, metabolism, detoxification, regulations and intake of zearalenone: an oestrogenic mycotoxin*. Food Chem Toxicol, 2007. **45**(1): pp. 1–18.

Zomborszky-Kovacs, M., *et al.*, *Preliminary communication: examination of the harmful effect to fetuses of fumonisin B(1) in pregnant sows*. Teratog Carcinog Mutagen, 2000. **20**(5): pp. 293–9.

CHAPTER 11

Environmental heat stress and avian gut integrity: more than a second brain feeling

Jalila S. Dridi[1], *Walter Bottje*[2] *and Sami Dridi*[2,*]

[1]École Universitaire de Kinésithérapie, Université d'Orléans, Rue de Chartres, 45100 Orléans, France; [2]Department of Poultry Science, University of Arkansas, Fayetteville, AR 72701, USA; *Center of Excellence for Poultry Science, University of Arkansas, 1260 W. Maple Street, Fayetteville AR 72701, USA; *Correspondence: dridi@uark.edu

Abstract

Poultry production sustainability is facing several challenges from a steep projected increase (almost the double) in global demand for meat and egg, and the need to adapt to unprecedented climate change and global warming along with increasing pressure on natural resource availability and security. Heat stress is one of the most challenging stressors to poultry in general, and to the avian gastrointestinal tract (GIT) in particular. GIT, which earned the nickname "second brain" because it contains the enteric nervous system, comprises millions of neurons, and produces many different neurotransmitters, is the foundation of the organism's overall health. This chapter aims to provide a current and critical overview on the effects of HS on avian gut integrity. After setting the stage by describing HS, thermoregulatory mechanisms and the thermogenic curve in chickens, the chapter focuses next on the effects of HS on intestinal hemodynamics, blood flow responses, and gut integrity, offering a review of current knowledges on its underlying molecular mechanisms. The chapter ends with brief revision of some nutritional and management approaches to mitigate HS adverse effects on leaky gut syndrome.

Keywords

heat stress – gut integrity – organelle dysfunction – autophagy – ER stress – signaling pathways

© JALILA S. DRIDI ET AL., 2024 | DOI:10.3920/9789004695467_012

1 Introduction

The poultry (meat and egg) productions support the livelihood and food security of billions of people worldwide. In fact, by means of intensive groundbreaking genetic selection programs (Dridi, 2020; Preisinger, 2018; Rubin *et al.*, 2010), the egg- and poultry meat-production was outstandingly increased during the last decades to reach an estimated total of about 84.36- (average of 50 g egg weight) and 97.39-million metric tonnes of egg and meat, respectively (Preisinger, 2018; Shahbandeh, 2022). However, and despite this spectacular progress, poultry production sustainability will face many serious challenges.

First, as human population is predicted to reach about 9.1 billion by 2050, the Food and Agriculture Organization (FAO) of the United Nation estimated an increased demand in animal proteins by 73% to feed the future and satisfy these nutritional needs ((FAO), 2019). In another word, annual meat production will need to rise by over 200 million tonnes, and at least 50 million hens will have to be added each year, assuming management conditions to support the genetic potentials (Preisinger, 2018). This will be arduous because approximately 70% of the world's population will be urban and crop (cereal) growth and yield, that is steadily declining, will need to increase by about 3 billion tonnes ((FAO), 2019) under growing pressure on already scarce agricultural and natural resources (water, land, and energy) (Crew, 2022; DeFries *et al.*, 2015; He *et al.*, 2021; Jury and Vaux, 2005; Kwan, 2022; Lambin and Meyfroidt, 2011; Liu, 2022; Mekonnen and Hoekstra, 2016). Furthermore, increased use of food crops for biofuel production, which currently represent about 10% (110 million tonnes) of global production, could add more complexity and have serious implications for future food security ((FAO), 2019).

Second, the shift to "no antibiotics ever-NAE" or "raised without antibiotics, RWA" era, due to increasing concerns for antibiotic resistance and consumer health and preference, could be challenging for chickens, growers, and poultry industry. For instance, the number of birds produced in NAE programs in the U.S. today is now at more than 50% (> 4.5 billion broiler chickens), and this paralleled an increased incidence of intestinal disease outbreaks such coccidiosis and necrotic enteritis (Gyorke *et al.*, 2013; Karaer *et al.*, 2012) as well as escalated occurrence of metabolic disorders (Emami *et al.*, 2021a; Greene *et al.*, 2020; Greene *et al.*, 2019; Julian, 2005). Although it is still a matter of debate that begs science-based data, it is generally acknowledged that production efficiency (weight gain, feed conversion ratio, mortality and yield) is adversely impacted in the antibiotic-free production system. Thus, antibiotic-free production seems to result in a greater carbon footprint and is less efficient, which raise the question to whether it is sustainable or not (Cervantes, 2015).

Third, climate change is one of the most challenging and key determinant of poultry production sustainability and food security. Unusually warm and longer season temperatures have increased markedly over the past decades (Alley *et al.*,

2005; Mora *et al.*, 2013). Predicted warming during the 21st century will be even larger than during the 20th century. Indeed, based on scenarios and models developed by climate researchers (Figure 11.1), the predicted 'high emissions' pathway for CO_2, coupled with rises in the other greenhouse gases, would be expected to result in a global average warming of around 4.5 °C by 2100 (Tollefson, 2020). This global warming is expected to cause further increases in atmospheric moisture content, more extreme heatwaves, fewer frosts, further decreases in the extent and thickness of sea ice, further melting of mountain glaciers and ice sheets, shifts in rainfall, and further ocean warming (Tierney *et al.*, 2020). Global warming and heat stress (HS) are already affecting animals, insects and crops (Chen *et al.*, 2011). Of particular

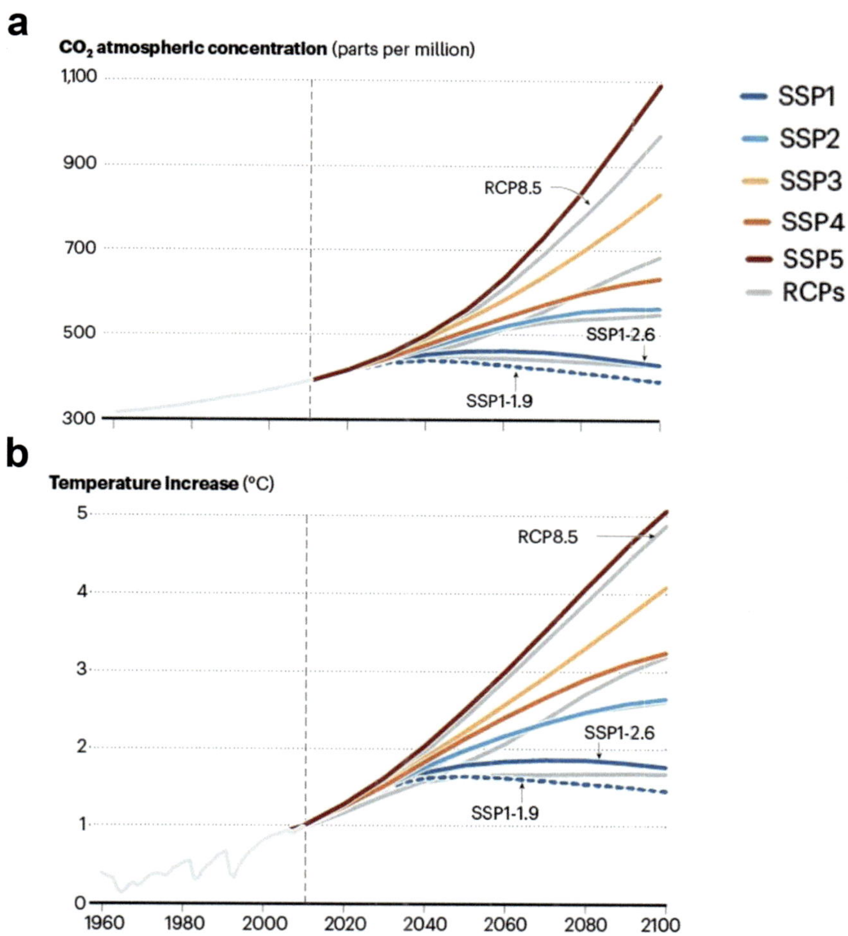

FIGURE 11.1 Prediction and scenario model for future global warming. Researchers have developed Socioeconomic pathways (SSPs) to explore different ranges of development and how they would alter the climate, including CO_2 concentrations (a) and temperatures (b). Figure was adapted from (Tollefson, 2020) and (O'Neill, 2016). RCP, representative concentration pathway; SSP, Socioeconomic pathway

interest, in modern broiler (meat-type)- and layer (egg-type)-chickens, the strong adverse effects of HS on growth, feed efficiency, meat yield, mortality, welfare, egg production, and meat and egg quality are well documented (Barrett *et al.*, 2019; Cahaner and Leenstra, 1992; Caulfield *et al.*, 2014; Dale and Fuller, 1980; Deeb and Cahaner, 2002; Leenstra and Cahaner, 1992; Mack *et al.*, 2013; Song, 2015; Wang *et al.*, 2017; Zaboli *et al.*, 2019), and such effects will take a heavy toll and large amplitude during the next decades as the heat anomalies will continue to be larger and more frequent (Tollefson, 2020).

Although they are homeothermic endotherms, chickens are susceptible to HS, and due to covering feathers, high core body temperature (~41 °C), high metabolic rate, and lack of sweat gland they cannot cope with high ambient temperature. In fact, beside stress and discomfort, HS can damage cells, induce organ dysfunction, and in extreme case it can lead to death by spiraling hyperthermia (Emami *et al.*, 2020). One of the most vulnerable avian-organ to HS is the gastrointestinal tract (GIT), which earned the nickname "second brain" because it contains the enteric nervous system (ENS), comprises millions of neurons, and produces many different neurotransmitters (Han *et al.*, 2018; Muller *et al.*, 2020; Zhang *et al.*, 2022). In this chapter, after overviewing heat stress, thermoregulatory mechanisms and the thermogenic curve in chickens, we will focus next on the effects of heat stress on intestinal hemodynamics, blood flow responses, and gut integrity, offering a review of current knowledges on its underlying molecular mechanisms. It then turns, without being exhaustive, to revise some nutritional and management approaches to mitigate HS adverse effects on leaky gut syndrome. The main goal is to facilitate and encourage further progress by asking new questions, which may help for developing further mechanism-based strategies to improve chicken gut health and integrity under global warming and HS conditions.

2 Overview of heat stress and thermogenic zones in avian species

Stress is an ambiguous and wide concept which is attributed to various definitions depending on every-day-life experiences and can be classified as physiological, psychological, emotional, positive (eustress), negative (distress), short, cyclic, or chronic (Shahsavarani, 2015). Generally and comprehensively, stress is defined as any magnitude and influence of internal and/or external (surrounding environment) forces to the body, which tends to displace its systems from their resting or ground state, and thereby disturbing its homeostasis (Silverman *et al.*, 2010). Heat stress is a response to high environmental temperature and humidity. It happens when birds are outside of their comfort temperature (thermoneutral) zone and have difficulty achieving a balance between body heat production and body heat loss. As depicted in Figure 11.2, there are three zones of thermogenic responses of chickens, comfort zone (thermoneutral zone) where body temperature is controlled

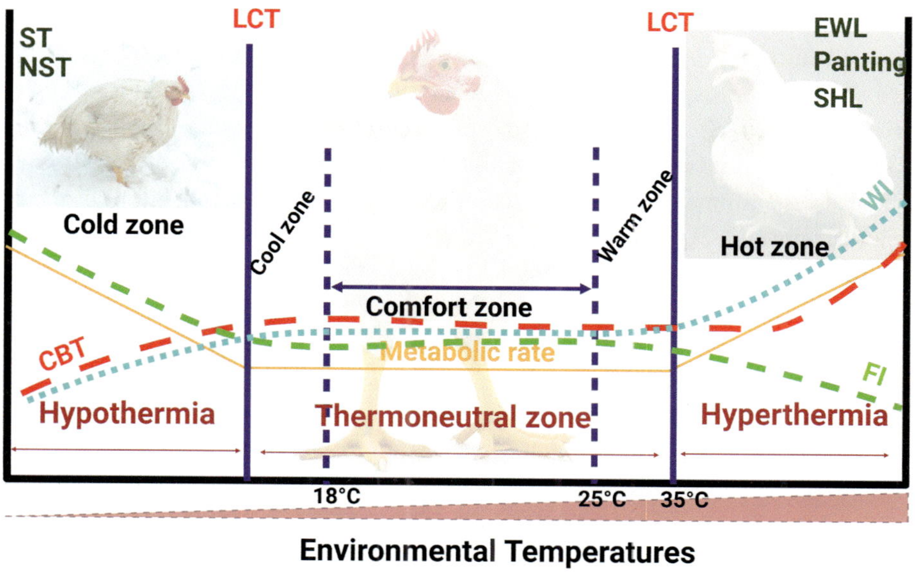

FIGURE 11.2 Thermogenic curve of chickens. The thermoneutral and comfort zone ranged between the lower and upper critical temperature. Below the LCT and under cold stress exposure heat is generated by ST and NST and food intake is stimulated. Under heat stress condition and above the UCT, thermogenesis increases to permit EWL mainly via panting and SHL. CBT, core body temperature; EWL, evaporative water loss; FI, feed intake; LCT, lower critical temperature; NST, non-shivering thermogenesis; SHL, sensible heat loss; ST, shivering thermogenesis; UCT, upper critical temperature; WI, water intake

mainly by conduction, zone of evaporative regulation where the body temperature is regulated by vasodilation, followed by evaporation and sensible heat loss, and zone of body cooling where the body temperature is maintained by vasoconstriction for heat retention, followed by shivering and non-shivering thermogenesis (Dridi, 2022a).

McKechnie reported that the body temperature in avians can differ from the skin temperature by several degrees (McKechnie, 2022). He found that the mean deep body temperature measured in 202 species spanning most avian orders varied between 38.5 °C and 41 °C, and can reach up to 45 °C under heat stress exposure (McKechnie, 2022). Transfers of heat between birds and their surrounding environments happens through conduction, radiation, convection, and evaporation. In response to heat stress, it is well known that animals will initiate several 'layers' of physiological events in order to maintain body temperature homeostasis. As body temperature begins to rise, the first layers of response are passive and/or behavioral adjustments that do not require energy input. Birds will spread their wings and lie on the floor of the barn so that body heat will be conducted to the floor. Birds will also decrease feed intake in order to reduce heat generation due to absorption and metabolism of nutrients. As body temperature continues to increase in response to heat stress, birds will initiate adjustments of the cardiovascular system that will shut blood from the body core (*e.g.* GIT, reproductive tract) to vasculature where heat dissipation can occur (*e.g.* comb, wattle, legs, and respiratory tract). This shunting of blood flow from body core regions to the periphery do not require appreciable amounts of energy input and therefore do not add additional amounts of heat to the animal, however it creates a hypoxia-like status in internal organs including the GIT.

If body temperature continues to rise, birds will enter into active thermoregulation that helps the animal respond to heat stress but increases the heat load on the animal. The major form of heat dissipation takes place in the upper respiratory tract. As blood passes through the capillaries of the respiratory tract, convective heat loss occurs by evaporation (latent heat loss), which helps the bird in resisting to increases in body temperature. In this stage of heat stress response, birds will initiate a rapid rate of respiration with a shallow breathing pattern termed gular flutter, which involves rapid pulsation of the hyoid apparatus and the gular membranes (Bartholomew, 1968). Gular flutter allows air to move over the respiratory surfaces to allow evaporative cooling without serious disruption of blood acid-base balance that could occur when air passes across gas exchanging surfaces in the parabronchi and air-capillaries. With further elevation in body temperature, birds will shift from gular flutter to very deep breathing (panting or thermal polypnea) that will result in severe disruption of acid-base balance due to loss of carbon dioxide, leading to respiratory alkalosis and hypocapnia (Bech and Johansen, 1980; Marder and Arad, 1989). It has been reported that some other organs could contribute to the evaporative cooling in birds. For instance, cornea has been found to contribute to the maintenance of brain temperature (Pinshow *et al.*, 1982). Pinshow *et al.* (1982) demonstrated that ocular circulation represents another source of cooled venous blood to the rete ophthalamicum, increasing cooler blood coming from capillaries adjacent to the moist membranes of the buccopharyngeal cavity. Hoffman and co-workers have shown that cloaca is also an important component of evaporative cooling in birds (Hoffman *et al.*, 2007). Janse van Vuuren and colleagues, on the other hand, have shown that beak and unfeathered skin are heat radiators (Janse van Vuuren, 2020).

3 Brain-peripheral axis and thermoregulation mechanisms in avian species

Bech and Midtgard reported that the avian thermoregulation is characterized by the maintenance of brain temperature below core body temperature (Bech, 1981; Pinshow *et al.*, 1982). The long-standing question is, then, whether birds have a static (set-point) or dynamic body temperature. First, the avian body temperature is marked by a circadian rhythm where it is higher during the active phase compared to the rest phase (Prinzinger, 1991). Second, Tzschentke and Nichelmann showed that birds have a thermoregulation set-point (Tzschentke, 1999), although several studies in precocial birds have shown that the core body temperature in hatchlings is lower than that in adult birds (Myhre *et al.*, 1975). It seems that the lower core body temperature in juvenile birds is associated with a lower thermoregulatory set-point, and not a thermoregulation failure. It has been suggested that this lower thermoregulatory set-point is a mechanism for energy saving (Pis, 2002).

As described above, heat load induces hyperthermia which is different from fever or fibrile. The etiology, structures, and mechanisms involved in each state are different [for review see (Mota-Rojas *et al.*, 2021)]. Briefly, fever can be caused by pathogen infections and inflammation, which in turn generate a defense strategy by the organism to stop and repair the damage (Kluger *et al.*, 1975). Hyperthermia, however, is caused mainly by the increase in heat production with the simultaneous decrease of the mechanisms involved in heat loss (Fischer *et al.*, 2020).

Although the underlying physiological and molecular pathways of thermoregulation set-point control in birds need to be fully defined, our current understanding is rooted to the 1965-Hammel model (Boulant, 2006), which is based on a synaptic network that encompasses four different types of hypothalamic neurons (warm-sensitive neurons, temperature-insensitive neurons, heat loss- and heat production-effector neurons) and the peripheral afferent pathways. Under heat stress exposure, thermal energy is sensed by diverse thermoreceptors mainly in the spinal cord, which has been shown to be a high thermosensitive region, particularly the thoracic area (Ostnes, 1992). In fact, heating of the spinal cord induces rapid panting (thermal polypnea) in domestic chickens (Avery, 1983). Thermoreceptors are also distributed in the brain, deep tissues, and peripheral regions such as the skin, beak, and tongue. As panting consists of an increase in respiratory rate and a decrease in respiratory volume, which is believed to restrict hyperventilation to surfaces of the respiratory system, osmo- and baro-receptors are also involved in detecting changes in the environmental- and body-temperatures (Ruuskanen *et al.*, 2021). These peripheral thermosensory signals are then transmitted in a feedforward manner to a second-order thermal sensory neurons in the spinal and trigeminal dorsal horn, which project to third-order sensory neurons in the dorsal subnucleus of the lateral parabrachial nucleus (LPB) area. The thermosensory signals for thermoregulation responses are, subsequently, dispatched from the LPB to the preoptic anterior hypothalamic (POAH) area (Simon, 2000; Simon *et al.*, 1981), which via (de)polarization and firing rate of the above-described neurons, provide command signals descending to peripheral effectors and evoke behavioral, autonomic, somatic, and hormonal responses counteracting changes in environmental temperatures before they impact body core temperature (Boulant and Gonzalez, 1977; Nakamura and Morrison, 2008) (Figure 11.3). It worth noting that when the temperatures approach near-extreme levels, temperature guardian neurons are activated and to start more robust thermoregulatory mechanisms.

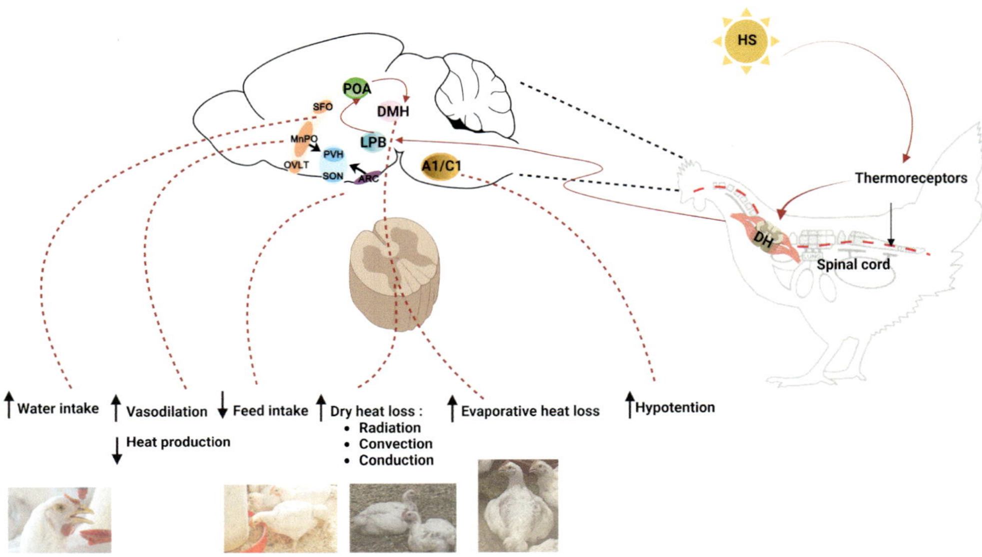

FIGURE 11.3 Thermoregulation mechanisms in avian species. The thermoregulatory system that enables maintenance of a relatively constant body temperature over a wide range of ambient temperatures in avian species consists of a sensory part containing thermo-, osmo-, and baro-receptors that detect thermal energy changes. These peripheral thermosensory signals are integrated and transmitted to the hypothalamic LPB and POA, in which temperature-sensitive neurons monitor local temperature changes and temperature information received from peripheral thermoreceptors. Via (de)polarization and firing rate of specific neurons, POA provides command neurological and endocrine signals descending to peripheral effectors and evoke behavioral, autonomic, somatic, and hormonal responses counteracting changes in environmental temperatures before they impact body core temperature. A1/C1, catecholaminergic neurons; DH, dorsal horn; DMH, dorsomedial nuclei; HS, heat stress; LPB, lateral parabrachial nucleus; MnPO, median preoptic nucleus; OVLT, organum vasculosum lamina terminalis; PVH, paraventricular hypothalamus; SFO, subfornical organ; SON, supraoptic nuclei. The image is not for scale, and it is adapted from (Kim *et al.*, 2021; Wedel, 2007).

At cellular levels, the molecular entity of thermoreceptors was discovered first by David Julius, 2021 Nobel Prize in Physiology or Medicine, who made a breakthrough by cloning the capsaicin receptor (Caterina *et al.*, 1997). This discovery stimulated considerable subsequent research in hunting for other temperature- and pain-sensitive ion channels leading to the identification of thermally sensitive transient receptor potential vanilloid (TRPV2, TRPV3, TRPV4), transient receptor potential ankyrin (TRPA1), and transient receptor potential cation channel subfamily M member 8 (TRPM3, TRPM8) (Caterina *et al.*, 1999; Liedtke *et al.*, 2000; McKemy *et al.*, 2002; Story *et al.*, 2003; Vriens *et al.*, 2011; Xu *et al.*, 2002). As depicted in the schematic diagram (Figure 11.4), the relative activation of these receptors depends on temperature threshold. In avian species, there is little information regarding molecular

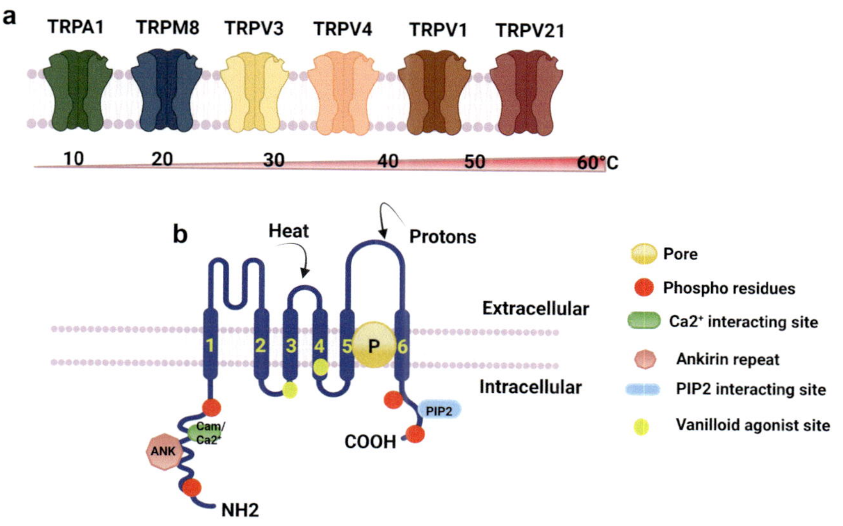

FIGURE 11.4 Temperature-sensitive ion channels and structure of TRPV1. A schematic diagram (not for scale) depicting the relative activation of TRP ion channels to temperature thresholds (a). Structure of TRPV1 that contains 6 transmembrane domains and pore formed between the 5th and the 6th transmembrane domain. The phosphorylated sites can be phosphorylated by both PKC and PKA.

thermosensation, although TRPV1 (NM_204572), TRPV4 (NM_205513), TRPA1 (NM_001318460), TRPM7 (NM_001398044), and TRPM8 (NM_001007082) were identified and several other TRPMs were predicted in chicken genome. TRPA1 is expressed in chicken skin and it has been shown to be responsive to temperature marginally below the average core body temperature (Saito *et al.*, 2014). TRPV1 is expressed in the majority of heat-sensitive neurons in the dorsal root ganglia, and activated by high temperature (> 45 °C) (Mahmoud *et al.*, 2007).

4 Heat stress and intestinal blood flow in avian species

As previously indicated, the cardiovascular system shunts blood away from organs in the body core to areas in the periphery to help facilitate heat dissipation. To a certain extent, blood flow in an organ can dictate or impact organ function. Thus, a reduction of blood flow in the intestines can diminish absorption of nutrients during digestion.

An investigation of blood flow in the celiac artery of broilers in response to heat stress was reported (Bottje and Harrison, 1986). In that study, broilers were implanted with a blood flow probe on the celiac artery, which is the first major branch from the abdominal aorta delivering blood to the duodenum, spleen and hepatic artery. Broilers that were chronically instrumented were placed inside a small environmental chamber in which air temperature could be adjusted and birds could be monitored without disturbance.

Birds were monitored initially to establish the celiac blood flow response to feeding alone in response to thermal neutral conditions (25 °C). Blood flow in the celiac artery prior to feeding was approximately 25 mL/min. In response to feeding in which broilers consumed approximately 20 g of feed in a 20 min feeding-period, blood flow roughly doubled and remained elevated for a 3 and a half hour post-prandial feeding-period.

To investigate the effects of HS on celiac blood flow (MBF), birds chronically instrumented with a blood flow probe, chest transducer (to measure respiratory rate) and a cannula in the carotid artery (to take blood samples for blood gas analysis) were placed in a cage in the environmental chambers. Changes in MBF, respiratory rate (RR), blood pH, blood partial pressure of carbon dioxide (PCO_2) and body temperature in response to HS (indicated by the increase in ambient temperature (Ta)) in the chamber, are shown in Figure 11.5a. This figure illustrates the hemodynamic changes that were described in the previous sections.

The elevation of intestinal blood flow during thermoneutral conditions (dashed line), was significantly decreased within 10 minutes of the onset of HS and a significant increase in body temperature. This adjustment occurred before a significant increase in respiratory rate (RR) and changes in blood pH and PCO_2. Values for each of the variables that were monitored (with the exception of the rectal temperature Tr) rapidly returned to values that were not different from those obtained prior to HS when ambient temperature was lowered back to thermoneutral. Although not measured in that study, there are numerous reports that would indicate that at the same time blood flow decreased in the celiac artery in these birds, blood flow would have increased to tissues where heat dissipation would occur. The vasoconstriction occurring in arterioles throughout the splanchnic vasculature must occur to maintain blood pressure homeostasis in the animal. The decrease in blood flow would diminish the uptake of nutrients, particularly lipids, in which absorption is blood-flow dependent.

The shift of visceral blood flow towards peripheral circulation under heat stress conditions has been reported to cause GIT hypoperfusion and hypoxia (Lambert *et al.*, 2002). It is well established that in a well-oxygenated cell, hypoxia-inducible factor alpha (HIF-α) subunits are hydroxylated on proline residues by the oxygen-dependent prolyl-4-hydroxylases (PHDs), which facilitates its binding to the E3 ubiquitin-ligase von Hippel-Lindau protein (pVHL), and targets HIF-α for proteasomal degradation (Huang *et al.*, 1998). In addition and under normoxic condition, HIF-α undergoes asparaginyl hydroxylation by factor inhibiting HIF (FIH), which inhibits its binding to p300/CBP co-activators (Semenza, 2002). Under hypoxia, however, hydroxylation is inhibited and PHDs and FIH are inactivated, resulting in HIF-α nuclear translocation, heterodimerization with HIF-1β, and binding to hypoxia-response element (HREs) in the regulatory regions of target genes (Figure 11.5b). This HIF system activation is a key regulator of a broad range of cellular responses from metabolic adaptation, cell growth, to survival and apoptosis (Chandel and Simon, 2008). Among the downstream signaling of hypoxia and HIF, are the activation of NF-ϰB and cytokine production. In fact, hypoxia activates NF-ϰB via the canonical pathway, which involves the activation of transforming growth factor-B activating kinase (TAK1) and the inhibition of ϰB kinase complex (IKK) and/or NF-ϰB essential modulator (Nemo) (Culver *et al.*, 2010; Wu *et al.*, 2010). Although, it is still under extensive investigations, hypoxia could activate NF-ϰB via non-canonical and atypical pathways (Bandarra *et al.*, 2014). Hypoxia has also been shown to induce the expression of proinflammatory cytokines such as TNFα (Taylor *et al.*, 1998; Yun *et al.*, 1997), leading to intestinal epithelial permeability (see section 5). In poultry, heat stress has been reported to aggravate intestinal inflammation via NF-ϰB signaling pathway in E-coli infected chickens (Tang *et al.*, 2021). Although investigating studies are underway in the gut, Dridi's group has demonstrated that heat stress alters the expression of inflammasome and (chemo)cytokine expression in isolated red and white blood cells (Greene *et al.*, 2022), and that of hypoxia- and oxygen homeostasis-related genes in the hypothalamus of broiler chickens (Brugaletta *et al.*, 2022).

5 HS alters intestinal barrier integrity, nutrient transporters, and induces leaky gut syndrome

Maintaining a healthy gut is of a primordial importance because gut function is inextricably associated with overall health and welfare of birds. Home of trillions of microorganisms, the gut's primary function is the digestion, absorption of nutrients, and the excretion of waste. it also has a major influence on both the development and function of the immune system, as well as on gut-brain communications (Flint *et al.*, 2012; Sommer and Backhed, 2013). Additionally, and importantly, GIT provides a protection against pathogens and toxins through a highly organized and tightly regulated intestinal barrier integrity, which allows selective permeability of essential ions, nutrients, and water but restricts the entry of bacterial toxins and pathogens (Vancamelbeke and Vermeire, 2017).

The intestinal epithelial barrier is a one-cell-thick internal lining of the gut that contains various specialized cell types including enterocytes, goblet cells, enteroendocrine cells, Paneth cells, stem cells, microfold, cup and tuft cells (Figure 11.6a). Summarily, enterocytes are the most abundant cells with nutrient absorption function and maintenance of the epithelial barrier integrity (Kagnoff, 2014). Goblet cells secrete mucous, trefoil peptides, and resistin-like molecule-β, which are central to both the defense and repair of the epithelial layer and have significant roles in epithelial homeostasis (Specian and Oliver, 1991). Enteroendocrine cells secrete

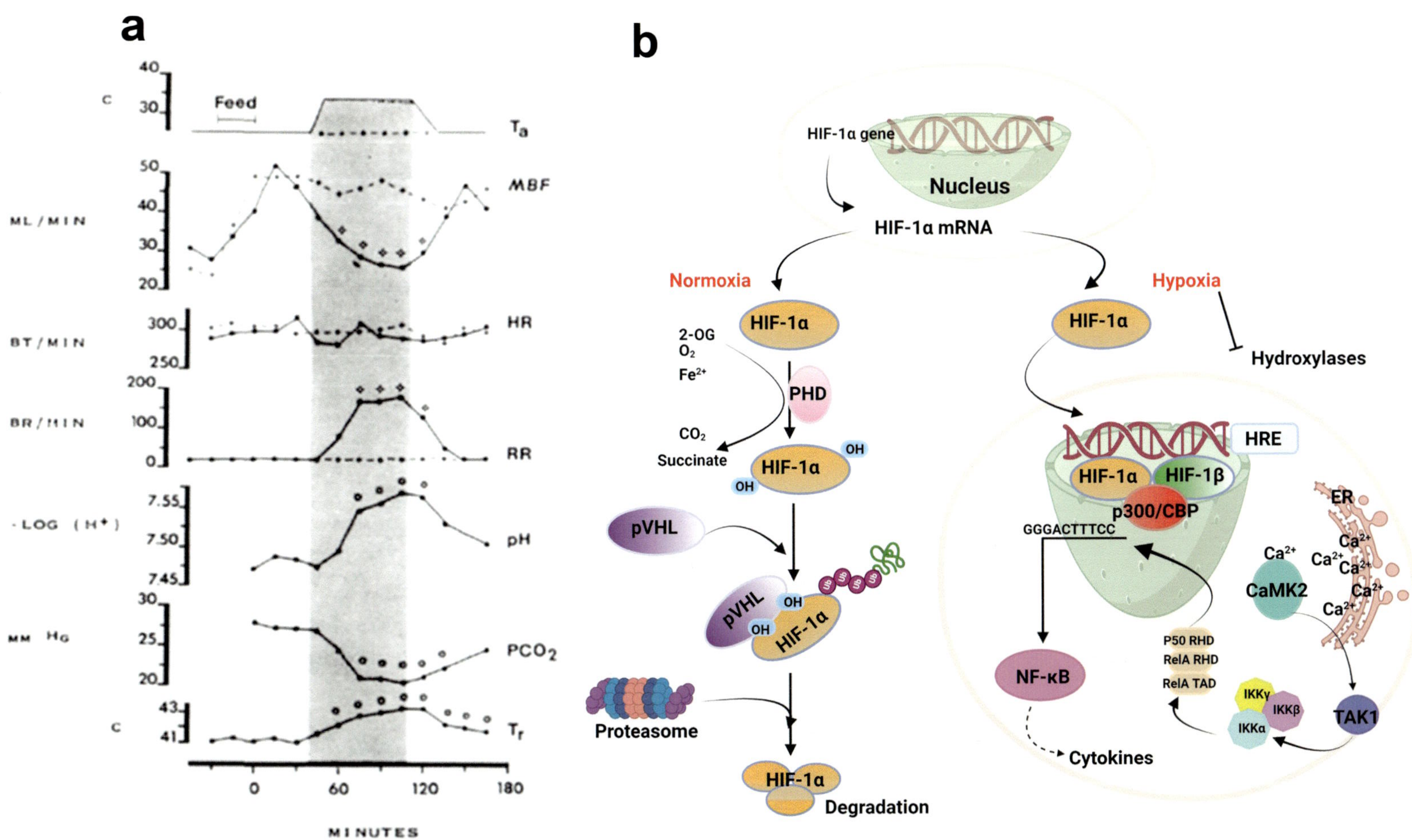

a
b
Feed
C
ML / MIN
BT / MIN
BR / MIN
-LOG (H+)
MM HG
T_a
MBF
HR
RR
pH
PCO_2
T_r
MINUTES
HIF-1α gene
Nucleus
HIF-1α mRNA
Normoxia
Hypoxia
HIF-1α
2-OG
O_2
Fe^{2+}
PHD
CO_2
Succinate
OH
pVHL
Ub
Proteasome
Degradation
Hydroxylases
HRE
HIF-1β
p300/CBP
GGGACTTTCC
NF-κB
Cytokines
P50 RHD
RelA RHD
RelA TAD
IKKγ
IKKβ
IKKα
TAK1
CaMK2
Ca^{2+}
ER

FIGURE 11.5 ← Effect of HS on celiac mean blood flow and hypoxia-induced NF-ϰB activation. (a) Changes in celiac mean blood flow (MBF), heart rate (HR), respiratory rate (RR), arterial blood pH (pH), and partial pressure of carbon dioxide (PCO2), and rectal temperatures (Tr) in response to feeding (~20 g feed consumed in 20 min) and heat stress (Ta) in chronically instrumented broilers. From (Bottje and Harrison, 1986). (b) Under normoxia, HIF-1α is hydroxylated by PHD protein and targeted for proteasome-mediated degradation after being polyubiquitinated by VHL protein. Under hypoxia, HIF-1α is stabilized and translocated to the nucleus to for heterodimer with HIF-1β and target NF-ϰB activation. Hypoxia induces also CaMK2, which in turn activates TAK1, a member of MAPK family, leading to activation of NF-ϰB via IKK complex. 2-OG, 2-oxoglutaric acid (also: alpha-ketoglutaric acid); CaMK2, calcium/calmodulin-dependent kinase 2; CBP, CREB-binding protein; HIF, hypoxia-inducible factor; HRE, HIF response element; IKK, Inhibitor of ϰB kinase; PHD, prolyl hydroxylase; TAK1, Transforming Growth Factor-B activating kinase; Rel, v-rel avian reticuloendotheliosis viral oncogene homolog; VHL, von Hippel Lindau

humoral mediators with immunoregulatory effects (Gribble and Reimann, 2019). Paneth cells reside in the base of crypts, secrete antimicrobial peptides, such as α-defensin, to impede microbial entry to the intestinal lumen, and maintain the stem cell niche (Bevins and Salzman, 2011). Microfold cells secrete immunoglobulin A, which, in addition to goblet cells, helps present bacterial antigens to dendritic cells and play an important role in transporting lumen antigens into the associated intestinal lymphoid structures (Dillon and Lo, 2019). Tuft cells play an important innate role against helminth and protozoan infections (Schneider *et al.*, 2019).

Underneath the above-described epithelial layer, there is a thin layer of connective tissue which is the lamina propria that has a pivotal role in nurturing healthy communication between the microbiome and the immune cells (Figure 11.6a).

As described above, the transport of molecules across the epithelial layer occurs through different pathways including a passive diffusion via trans-cellular, or carrier-mediated, and/or paracellular through the adjacent cells. The latter pathway is controlled by the tight junction proteins that seal the paracellular space between the cells and selectively restrict the passage of hydrophilic molecules (Suzuki, 2013). This tight junction is constituted of a protein complex containing occludin (OCLN), claudins (CLDNs), zonula occludens (ZOs), tricellulin (TRIC), cingulin (CGN), and junctional adhesion molecules (JAMs), which interact with each other as well as with the desmosomes (cadherins, armadillo, desmoplakin) and the cytoskeleton, and form a complex architectural gate [for reviews see (Tsukita *et al.*, 2001; Zihni *et al.*, 2016)] (Figure 11.6b).

Although further mechanistic and functional studies are warranted, a growing body of scientific evidence has documented the deleterious effects of HS on avian gut integrity. In fact, it has been shown that HS impaired the intestinal morphology by shortening the villus and deepening the crypt (He *et al.*, 2018), which indicate a damage to the intestinal epithelium as well as a decrease in epithelial turnover. In

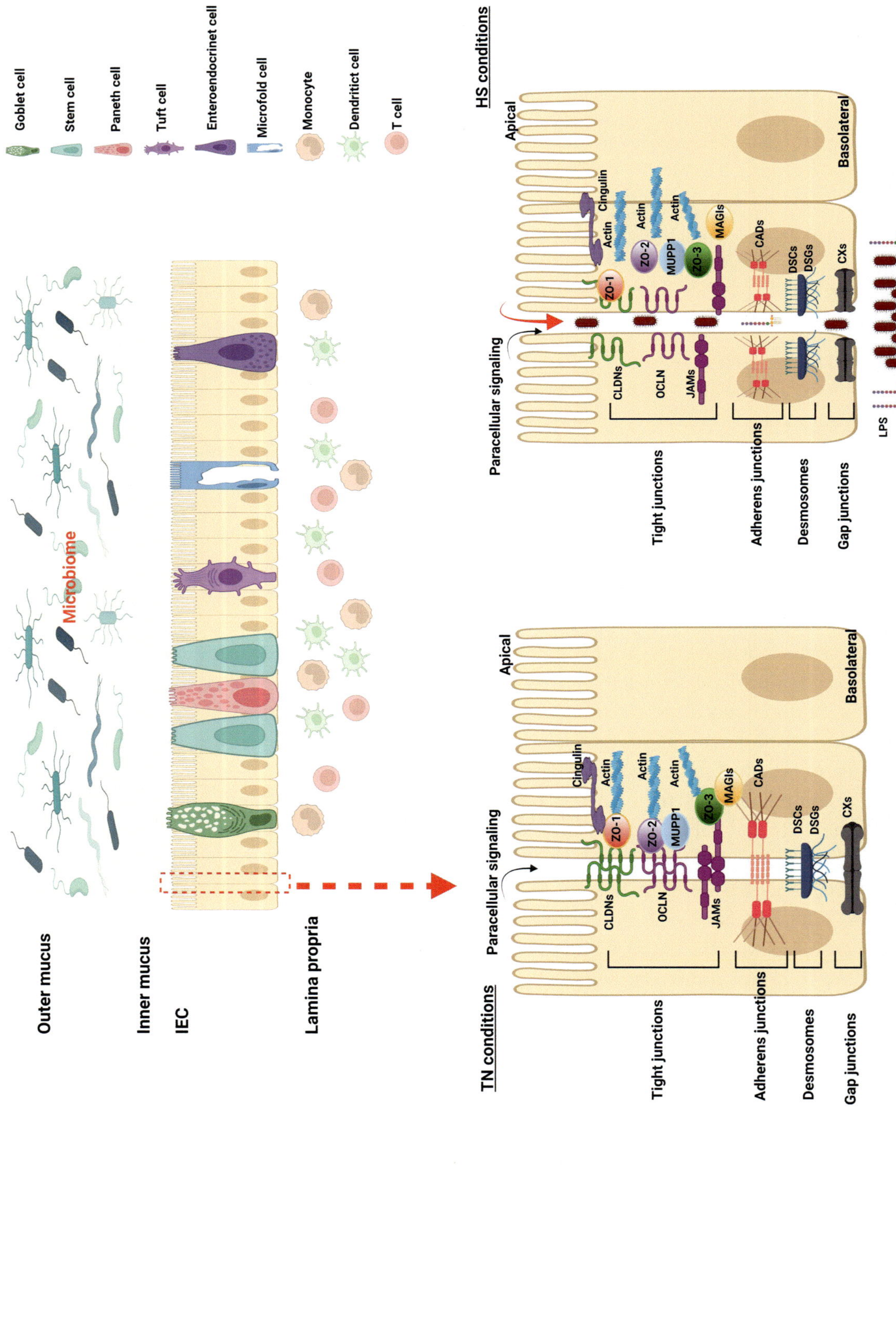
Goblet cell
Stem cell
Paneth cell
Tuft cell
Enteroendocrinet cell
Microfold cell
Monocyte
Dendritict cell
T cell
Outer mucus
Inner mucus
IEC
Lamina propria
Microbiome
TN conditions
HS conditions
Paracellular signaling
Apical
Basolateral
Tight junctions
Adherens junctions
Desmosomes
Gap junctions
CLDNs
OCLN
JAMs
ZO-1
ZO-2
ZO-3
MUPP1
MAGIs
Cingulin
Actin
CADs
DSCs
DSGs
CXs
LPS

FIGURE 11.6 ← Effect of HS on intestinal barrier integrity. The intestinal epithelial barrier contains various specialized cell types, including goblet cells, enteroendocrine cells, Paneth cells, stem cells, microfold cells, and cup and tuft cells (a). The intestinal mucus, secreted by goblet cells, forms an extra intestinal protective layer which separates epithelial cells from bacterial penetration (a). The epithelial cells function as a physical barrier or gate that is facilitated by tight connections between adjacent cells. Under TN conditions, this tight junction, constituted of a protein complex containing occludin, claudins, zonula occludens, cingulin, junctional adhesion molecules, adheren junctions, gap junctions, and desmosomes, which interact with each other as well as with the cytoskeleton, and form a complex architectural gate to seal the paracellular space between the cells and selectively restrict the passage of molecules (b). Under HS conditions, however, the tight junction is weakened, leading to leaky gut syndrome (c). CLDNs, claudins; CADs, cadherins; CXs, connexins; DSCs, desmocollins; DSGc, desmogleins; JAMs, junctional adhesion molecules; MAGIs, membrane-associated guanylate kinase (MAGUKS) with inverted domain structure; MUPP1, multi-PDZ domain protein 1, ZO, zonula occludens

rodents, it has been shown that HS induced intestinal epithelial necrosis and villi desquamation (Ji *et al.*, 2018). Several research, including our owns, showed that HS induced leaky gut in chickens as demonstrated by an increased levels of FITC-D in serum (Emami *et al.*, 2021b; Ruff *et al.*, 2020; Tabler *et al.*, 2020). This HS-induced intestinal permeability was accompanied by a dysregulation of tight junction and gup junction gene and protein expression (Song *et al.*, 2014; Tabler *et al.*, 2020; Uerlings *et al.*, 2018; Varasteh *et al.*, 2015; Zhang *et al.*, 2017) (Figure 11.6c). *In vitro* study using co-culture of two mammalian colonic epithelial cell lines (Caco-2 and HT-29) showed that HS alone or combined with hypoxia disrupted the cell monolayer and increased the paracellular permeability through dysregulation of ZO-1, CLDN3, and OCLN at both mRNA and protein levels (Lian *et al.*, 2021). Studies with transgenic mice showed that deletion of CLDN2 decreased the transepithelial conductance and paracellular permeability for Na^+, however specific overexpression of CLDN2 in the intestinal epithelia resulted in increased length of colonic crypts and an overall enlargement of the intestine that is associated with increased paracellular permeability to the Na^+ ion and solute FITC-dextran (Muto *et al.*, 2010). Intestinal epithelial-specific knockout of CLDN7 resulted in increased paracellular permeability to solutes of molecular weight between 457 and 4000 Da (Tanaka *et al.*, 2015). CLDN15-KO mice, on the other hand, exhibited megaintestine, which is associated with nutritional defects as a result of impaired intestinal Na^+ and K^+ paracellular permeability and glucose malabsorption (Tamura *et al.*, 2011; Tamura *et al.*, 2008). It is worth mentioning that the composition of epithelial tight junction proteins not only varies spatially along the length of the GIT but also in the crypt-luminal axis as epithelial cells transition from a proliferative to differentiated phenotype. In addition, the spatiotemporal distribution and probably the function of the tight junction proteins might vary between species.

In a recent study, Dridi's group have shown that cyclic chronic HS altered the expression of nutrient (carbohydrate, fatty acids, amino acids) transporters in chickens and this effect seemed to be affected by genetic selection and varied between jungle fowl, slow-, and fast-growing chickens (Abdelli *et al.*, 2021). Similar effects have been reported by other studies in chickens as well as in other large-farm animals (Al-Zghoul *et al.*, 2019; Habashy *et al.*, 2017; Sun *et al.*, 2015; Tang *et al.*, 2022). This is in line with compromised digestive functionality, nutrient digestibility, absorption and reduced intestinal enzymatic secretory activity (trypsin, lipase, amylase, etc.) during HS (Alhotan *et al.*, 2021).

Furthermore, in addition to intestinal epithelial cell disruption, studies in broilers have shown that HS reduced the number of goblet cells which in turn decreased mucin production via down regulation of the glycosylated protein MUC1 and MUC2 (Zhang *et al.*, 2017). HS has been reported to alter also the enteroendocrine cells and intestinal hormones in broilers (Mazzoni *et al.*, 2022).

6 HS alters the immune system and induces intestinal inflammation

The leaky gut syndrome and/or dysbiosis induced by heat stress can lead to bacterial translocation and endotoxemia, and thereby disturb the gut-liver axis (Alhenaky *et al.*, 2017; Nanto-Hara *et al.*, 2020). Dridi's group has recently shown that heat stress modulated the expression of cyto(chemo)kine genes in red and white blood cells as well as in chicken gut segments (Emami *et al.*, 2021b; Greene *et al.*, 2022), indicating a systemic and local inflammation. Looking into the underlying molecular mechanism(s) that need further in-depth investigations to be fully defined at least in avian species, indicate that HS perturbs commensal gut microbiota (see section 8), causes thinning of the protective mucin layer, and increases the contact of pathogens with the intestinal epithelial cells (IPC). This bacterial penetration of the inner mucin layer has been shown to evoke an inflammatory response, which may result in exaggeration of intestinal barrier dysfunction and increased IPC apoptosis and shedding. Although the upregulation of heat shock proteins, particularly HSP70, under HS conditions is known to be involved in the anti-inflammatory mechanism through inhibition of nuclear translocation of nuclear factor kappa B (NF-κB) (Shi *et al.*, 2006), recent studies reported that extracellular HSP70 induced inflammation and modulated LPS-stimulated inflammatory response in monocyte THP-1 cells via induction of pro-inflammatory cytokine secretion (Asea *et al.*, 2000; Hulina *et al.*, 2018). Anti-inflammatory properties and mechanisms of HSP are beyond the scope of this chapter and readers can find a complete description in different elegant reviews (Borges *et al.*, 2012).

It is believed that in case of elevated cellular stress, such as the case of chronic HS, HSP70 can be released from cells via endolysosomal pathway (plasma membrane translocation and release associated with lipid vesicles) or a passive release after cell death by necrosis (Johnson and Fleshner, 2006). The released-HSP70 has chaperone and cytokine regulating activity (chaperokine), acts as damage-associated molecular pattern (DAMP) and danger signal for the immune system, and can stimulate an inflammatory response via engaging toll-like receptor 2 and 4 (TLR2/4) and myeloid differentiation factor 88 (MyD88) (Asea *et al.*, 2002). This activates, in turn, mitogen-activated protein kinases (MAPK) and induces release of pro-inflammatory cytokines (Piccinini and Midwood, 2010). Furthermore, activation of TLR4/MyD88 has been shown to upregulate myosin light chain kinase (MLCK), which induces opening of tight junctions and aggravates intestinal permeability (Nighot *et al.*, 2017).

It has been previously shown that MyD88 targeted inflammasome activation (Tarallo *et al.*, 2012), and inflammasome activation has been recently linked to intestinal epithelial barrier function through multiple mechanisms, including antimicrobial peptide production and mucus secretion (Huber *et al.*, 2012; Wlodarska *et al.*, 2014). The inflammasome complex comprises of a sensor/receptor, the adaptor protein apoptosis-associated speck-like protein containing a CARD (ASC) and the enzymatic component, caspase-1. Inflammasome-forming receptors are cytosolic sensors that include nucleotide-binding oligomerization domain-like receptors (NLRs, NLRP3, NLRC4, NLRP6, NLRP1b), the absent in melanoma (AIM)2-like receptors (ALRs) and pyrin. Pyrin sequentially recruits ASC and caspase-1 to form an inflammasome complex in response to bacterial toxins (Xu *et al.*, 2014b). Sharma and co-workers showed that pyrin is required for inflammasome activation and IL18 maturation, which promote intestinal barrier permeability (Sharma *et al.*, 2018). The NLRP3 inflammasome is expressed in both gut immune and epithelial cells, and its activation is triggered by diverse insults ranging from pathogen-associated molecular pattern (PAMP), DAMP, bacterial and viral nucleic acids, organelle stress such as endoplasmic reticulum (ER) stress, mitochondria-derived components (mtDNA, reactive oxygen species), and exogenous stimuli or environmental irritants (Mitoma *et al.*, 2013; Murakami *et al.*, 2012; Zaki *et al.*, 2011). Although the role of the NLRP3 inflammasome in gut inflammation is still unsettled, increasing number of studies reported that NLRP3-deficient mice exhibited breached epithelial integrity and increased translocation of commensal bacteria from the gut to the systemic circulation (Zaki *et al.*, 2010). In addition to inducing pro-inflammatory cytokine (IL-1β and IL-18) secretion, the NLRP3 inflammasome regulates the phagosome acidification that controls the immune process (Sokolovska *et al.*, 2013).

The NLRC4 inflammasome, on the other hand, has been found to have an important role in maintaining the balance between tolerance towards intestinal commensals and inflammatory response to pathogens (Franchi *et al.*, 2012). NLRP6 has been shown to be expressed in mammalian gut and implicated, through the helicase DHX15, interferon, and mitochondrial adapter protein MAVS, in intestinal anti-viral innate immunity (Wang *et al.*, 2015). Furthermore, NLRP6 also contributes to the formation of a mucus layer that protects the epithelium against invasion of pathogens and physical damage (Wlodarska *et al.*, 2014). Moreover, NLRP6 participates in the secretion of antimicrobial peptides from epithelial cells (Levy *et al.*, 2015).

A recent study showed that NLRP3 inflammasome activation altered the protein expression of ZO-1, OCLN, CLDN1, and CLDN2 in Caco-2 cells (Feng *et al.*, 2018), and specific inhibition of NLRP3 resulted in increased expression of tight-junction-related proteins in JPEG-J2 cells (Kang *et al.*, 2022). In rodents, HS activated NLRP3 inflammsome, which in turn lead to splenic lymphocyte pyroptosis (Wang *et al.*, 2019a). In broilers, although there is no data today related to GIT to the best of our knowledge, chronic HS has been shown to promote hepatocyte inflammation via NF-κB and NLRP3 inflammasome activation (Liu *et al.*, 2022). Dridi's group reported that chronic HS modulated the expression of inflammasome (NLRP3, NLRC3, NLRC5) genes in isolated white and red blood cells (Greene *et al.*, 2022). Although the above-mentioned data suggest that inflammasome might mediate the deleterious effect of HS on gut integrity, further mechanistic and functional studies are warranted.

Pro-inflammatory cytokines, including ILs, TNFα, and INFγ are well-known for their undeniable role in the regulation of intestinal tight junction integrity (Capaldo and Nusrat, 2009). TNFα activated NLRP3 and dysregulated tight junction protein (ZO-1, CLDN1, OCLN) expression (Wu *et al.*, 2018). Protein kinase C (PKC) and NF-κB pathways are other major mechanisms mediating the effects of TNFα on intestinal tight junctions (Al-Sadi *et al.*, 2016; Aveleira *et al.*, 2010). INFγ induced an increase in gut barrier permeability via the alteration of ZO-1 and OCLN expression in an adenosine monophosphate-activated protein kinase (AMPK)-dependent manner, irrespective of the cellular energy status (Scharl *et al.*, 2009). IL-1β induced a rapid phosphorylation and activation of p38 kinase and downstream activation of a p38-dependent transcription factor, activating transcription factor-2 (ATF-2), which translocated to the nucleus and attached to the binding motif on the MLCK promoter region, leading to MLCK gene activation and protein synthesis, and thereby MLCK-dependent opening of the intestinal tight junction barrier (Al-Sadi *et al.*, 2013). In addition, the transcription factor ETS like-1 protein (ELK1) plays a key role in mediating the effect of IL-1β on intestinal barrier permeability (Al-Sadi *et al.*, 2013).

7 HS dysregulates cellular organelles

7.1 *Mitochondria*

HS increases energy metabolism and oxygen demand, particularly in internal organs such the GIT. Mitochondria are the cell powerhouses producing energy in the form of adenosine triphosphate (ATP) (McBride *et al.*, 2006) via the electron transport chain (ETC), a series of five multi-protein complexes, and oxidative phosphorylation (OXPHOS) (Zhao *et al.*, 2019). Briefly, in OXPHOS, NADH- and FADH-linked energy substrates provide electrons to the ETC at Complex I and Complex II, respectively. The electrons are passed along the ETC to the terminal electron acceptor, oxygen, which is fully reduced to water. The passage of electrons is associated with proton pumping across the inner mitochondrial membrane providing a proton motive force that is used to drive ATP synthesis as protons flow back into the mitochondrial matrix through ATP synthase (Complex V). Two inefficiencies in mitochondrial OXPHOS are leakage of the electrons from the ETC due to site specific defects in electron transport and proton leak in which protons enter the matrix across the inner mitochondrial membrane at sites other than through ATP synthase, which dissipates proton motive force without concomitant synthesis of ATP. Leakage of electrons from the ETC causes univalent reduction of oxygen to form superoxide and other reactive oxygen species (ROS). The production of ROS by mitochondria makes this organelle a major source of endogenous oxidative stress in the cell (Murphy, 2009). The passage of electrons along the ETC is not always 100% efficient. Instead, electrons may 'leak' from the ETC that causes a univalent reduction (incomplete reduction) of oxygen to form superoxide (O_2-). Superoxide is reduced to hydrogen peroxide (H_2O_2). Superoxide and hydrogen peroxide along with other oxygen containing radicals are collectively called ROS. Under normal conditions, a low level of ROS generation can be beneficial to cell function and important in signal transduction of gene expression. However, if ROS levels rise in the cells because overall antioxidant activity is insufficient in the cell, elevations of ROS can lead to oxidation of critical cellular components (*e.g.* lipids, proteins, DNA) that can be detrimental to cell function.

HS has been demonstrated to cause morphological and functional changes to mitochondria, leading to impaired performance, increased ROS production, induced oxidative stress, decreased mitochondrial efficiency, and induced cellular injury (Wang *et al.*, 2019b; Yu *et al.*, 2013). Although data related to the effects of HS on avian gut mitochondria are scarce, studies in other tissues showed that HS in isolated broiler skeletal muscle mitochondria resulted in an elevation in superoxide production which was accompanied by decreased avian uncoupling protein (av-UCP) gene and protein expression along with lipid and protein oxidation (Mujahid *et al.*, 2007). Studies in rodents have shown that HS reduced the number of cristae membranes, induced mitochondria swelling, increases the permeability of the mitochondrial inner membrane and impairs oxidative phosphorylation (Qian *et al.*, 2004). In addition, HS altered mitochondrial bioenergy metabolism as indicated by a decrease in the respiratory control rate (RCR) and dysfunction of H^+- and Ca^{2+}-ATPase activities. Moreover, HS induced mitochondrial Ca^{2+} overload and altered membrane permeability transition (MPT), indicating a disruption of mitochondrial homeostasis and membrane potential, as well as an abnormal opening of the mitochondrial membrane permeability transition pore (MPTP) (Qian *et al.*, 2004). Together, these mitochondrial damages due to HS induced the

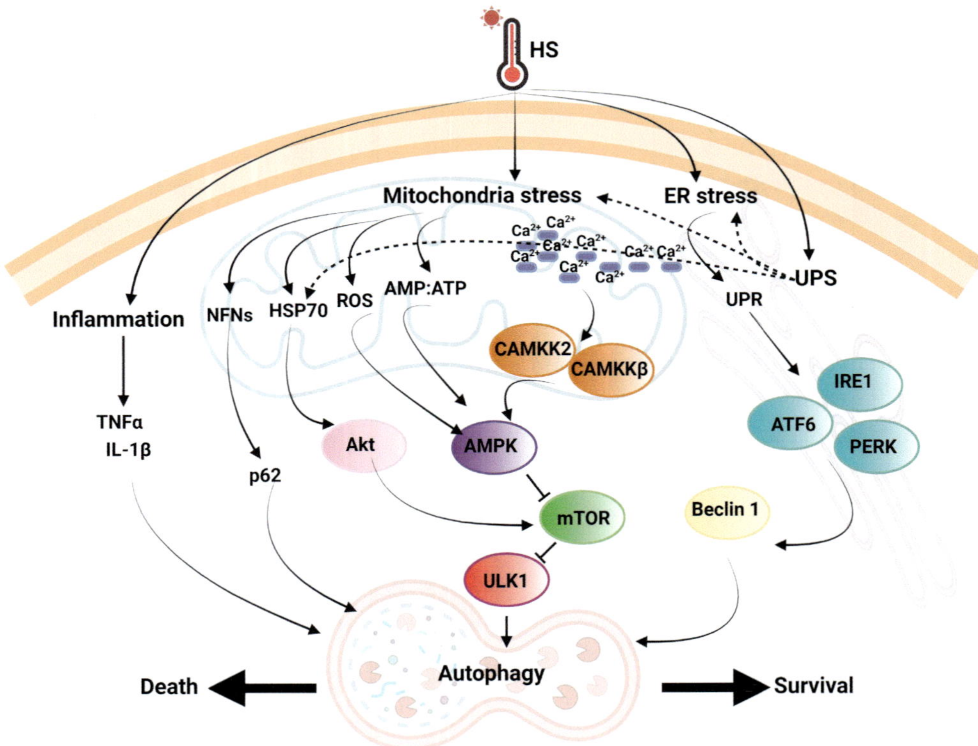

FIGURE 11.7 Schematic overview of the effects of HS on cellular organelles. HS induces organelles dysfunction via various pathways, including energy (ATP-AMPK-Akt) pathway, mitochondrial calcium, ROS, and mtUPR pathway, proinflammation pathway, HSP pathway, and ER stress pathway, all of which can activate autophagy for survival or death, depending of stress condition, severity, and duration. AMP, adenosine monophosphate; AMPK, AMP-activated protein kinase; ATP, adenosine triphosphate; ATF6, activating transcription factor; CAMMK, calcium/calmodulin-dependent protein kinase kinase; HSPs, heat shock proteins; IRE1, inositol-requiring protein 1; MFNs, mitofusins; PERK, double-stranded RNA-activated protein kinase-like ER kinase; mTOR, mechanistic target of rapamycin; ULK1, Unc-51 Like Autophagy Activating Kinase 1; UPR, unfolded protein response; UPS, ubiquitin-proteasome system

release of cytochrome c from mitochondria, which in turn activated caspase 3 leading to apoptosis (Figure 11.7). This indicates that the mitochondria is a critical checkpoint in regulating cell life, and it is a key target for HS. In *C. elegans*, HS disrupted mitochondrial dynamics (fission and fusion)-associated genes and caused, in turn, mitochondrial fragmentation (Machiela *et al.*, 2020). This sequentially triggered an upregulation of stress response pathways as evidenced by the activation of mitochondrial unfolded protein response (mtUPR) (Machiela *et al.*, 2020). MtUPR, one of the key mitochondrial feature and mechanism for quality control, is a molecular (nucleus-mitochondria) network and a transcriptional program associated with activation of mitochondrial chaperones, proteases, and antioxidant enzymes to repair defective mitochondria and stabilize mitochondrial homeostasis (Shpilka and Haynes, 2018).

7.2 *Endoplasmic reticulum*

It is now widely accepted that subcellular organelles communicate with each other to regulate function and coordinate responses to changing cellular conditions (Gottschling and Nystrom, 2017). One such communication is ER-mitochondrial contact and coupling to enhance Ca^{2+} transfer and promote mitochondrial respiration and bioenergetics (Bravo *et al.*, 2011). Although data are sparse, it is conceivable that HS would affect ER-mitochondria coupling.

Homma and Fujii showed that heat load induced ER stress, which elicited unfolded protein response (UPR) in cells (Homma and Fujii, 2016). Under normal conditions, the ER chaperone binding immunoglobulin protein (BiP), also known as glucose-regulated protein 78 (GRP78), binds the ER-resident sensor proteins [inositol-requiring protein 1α (IRE1α), double-stranded RNA-activated protein kinase-like ER kinase (PERK), and activating transcription factor 6 (ATF6)] and renders them inactive (Ron and Walter, 2007). However, when unfolded and misfolded proteins accumulate in the ER under stress conditions, GRP78 dissociates from the ER sensors to perform its chaperone function, and consequently UPR branches are activated [for detailed review see (Hetz *et al.*, 2020)]. In brief, first the activation of PERK induces the phosphorylation of eukaryotic translation initiation factor 2 (eIF2α), resulting in the global suppression of protein translation (Harding *et al.*, 1999), but it increased ATF4 expression (Harding *et al.*, 2003), which activates the proapoptotic transcription factor C/EBP-homologous protein (CHOP) (Zinszner *et al.*, 1998). Second, ATF6 is activated and translocated to the Golgi apparatus membrane to be proteolytically processed and cleaved before it relocates to the nucleus where it induces the transcription of ER chaperone genes (Schindler and Schekman, 2009; Ye *et al.*, 2000). Third, IRE1α dimerizes, transphosphorylates itself, and cleaves the X-box binding protein 1 (XBP1) to produce spliced sXBP1, which is then translocated to the nucleus where it induces the expression of genes involved in ER-associated degradation (ERAD) and chaperones (Calfon *et al.*, 2002; Lee *et al.*, 2003; Yoshida *et al.*, 2001). Furthermore, activated IRE1α induces the degradation of IRE1-dependent decay (RIDD) mRNA, which plays a key role in cell death (Han *et al.*, 2009; Hollien and Weissman, 2006).

HS has been reported to induce ER-mediated apoptosis in mammalian cells (Hou *et al.*, 2014; Kim *et al.*, 2013; Shellman *et al.*, 2008). HS altered the expression of IRE1α in various mammalian cells (Homma and Fujii, 2016). In rodents, HS has also been shown to induce muscle apoptosis through ER stress and IRE1-signaling pathways (Sharma *et al.*, 2021) (Figure 11.7). In intestinal porcine epithelial cells (IPEC-J2), Cui *et al.* (2021) have shown that HS exposure for 12h increased ROS, decreased superoxide dismutase 2 (SOD2) expression, increased cytochrome c release and caspase 3 activation, and induced apoptosis via the ATF4-CHOP-CHAC1 signal pathway (Cui *et al.*, 2021).

7.3 *Lysosomes*

Growing literature indicated that ROS are arguably one of the most important endogenous inducer for lysosomal membrane permeabilization, which plays a pivotal role in cell death (Cesen *et al.*, 2012). Lysosomes are membrane-bound organelles that contain an arsenal of different hydrolases, enabling them to act as the terminal degradative compartment of the endocytotic, phagocytic and autophagic pathways (Luzio *et al.*, 2007). The lysosomal-mitochondrial axis-apoptosis signal pathway is widely accepted and suggested an interplay between the two organelles in apoptosis (Repnik and Turk, 2010). It has been shown that lysosome released cathepsin B in the cytosol, which cleaved BH3 interacting domain death agonist (BID) and degraded B-cell lymphoma 2 (Bcl2), leading in turn to activation of caspases and subsequent mitochondrial depolarization which triggered mitochondrial apoptotic pathway (Hsu *et al.*, 2014). By using *in vivo* and *in vitro* intestinal SW480 cells, Yi and co-workers have shown that HS induced intestinal injury through lysosome-mitochondria-dependent pathway (Yi *et al.*, 2017).

7.4 *Ubquitine-proteasome system*

Organelle quality control and proteostasis are essential for proper cellular homeostasis under stressed conditions. As described above, the damaging effect of HS, high reactive free radicals, and hypoxia can alter the folding capacity of a cell leading to translational arrest and accumulation/aggregation of misfolded or unfolded proteins (Tyedmers *et al.*, 2010). Cells possess various mechanisms for stress response and intracellular removal of misfolded proteins and/or dysfunctional organelles, involving cellular chaperones, the ubiquitin-proteasome system (UPS), and autophagy, however if the system is overwhelmed and not controlled properly, the cell death pathway may be activated.

The UPSs mostly degrade single and unfolded polypeptides that are able to enter into the narrow channel of the proteasome. The degradation of proteins by the UPS is initiated by the sequential conjugation of ubiquitin, a conserved polypeptide of 76 amino acids, to the substrate protein via three enzymes (Dridi, 2022b; Finley, 2009). The ubiquitine activating enzyme E1 activates the carboxyl-terminal glycine residue of the ubiquitin in an ATP-dependent manner. The activated ubiquitin is, next transferred to a cysteine site of the ubiquitin-conjugating enzyme E2. Next, the ubiquitin ligase E3 enzyme links the ubiquitin from the E2 enzyme to a lysine residue of the target protein. When substrates are progressively modified with Ub, either at the N terminus or at a lysine side chain of Ub, various linear or branched Ub chains are built. Thus, Ub tags can be diverse in structure and will dictate the outcome of protein modification, because they can recruit accessory factors or receptors harboring Ub-binding domains (UBDs), including shuttle factors that deliver substrates to the proteasome (Braten *et al.*, 2016). Once ubiquitinated, the protein is recognized and degraded by the proteasome. The selective destruction of many short-lived regulatory proteins as well as damaged proteins is mediated by the 26S proteasome (Ciechanover, 1998). Bush and co-workers have shown that proteasome inhibition, using the peptide aldehyde MG132, increased the mRNA abundance of cytosolic HSP70 and ER chaperones, and conferred thermotolerance to MCDK cells (Bush *et al.*, 1997). In avian and mammalian cells, proteasome inhibition has been shown to activate heat-shock-inducible factors (HSFs), indicating specific roles of HSFs in the regulation of the ubiquitin-proteasome-mediated stress response (Kawazoe *et al.*, 1998; Kim *et al.*, 1999; Mathew *et al.*, 1998; Pirkkala *et al.*, 2000). In *C. elegans*, HS increased the UPS activity in the intestine, but not in the body wall muscle tissue, indicating a potential tissue-specific response of the UPS to high ambient temperature (Pispa *et al.*, 2020). In both mammals and avians, ubiquitins, certain Ub conjugating enzymes, and 26S proteasome have been reported to be major stress-induced transcripts (Bond and Schlesinger, 1985; Fornace *et al.*, 1989; Seufert and Jentsch, 1990). Rossi *et al.* (2010) found that the proteasome activating protein AIRAP was a heat shock protein (Rossi *et al.*, 2010). HS also causes a rapid build-up of K48-linked ubiquitinated proteins, which presumably indicates more rapid ubiquitination and proteasomal degradation of thermally damaged proteins (Medicherla and Goldberg, 2008; Parag *et al.*, 1987). Two additional Ub ligases Hul5/HECT and Rsp5/Nedd4 were identified as major players in the ubiquitylation and turnover of cytosolic misfolded proteins upon HS (Fang *et al.*, 2014; Fang *et al.*, 2011).

7.5 *Autophagy*

Autophagy, cellular self-eating/self-digestion, is a conserved intracellular catabolic and degradation process that is essential for recycling cytoplasmic material and eliminating damaged or dysfunctional cell constituents, such as proteins, aggregates or even entire organelles (Dikic and Elazar, 2018). The term autophagy was first coined by Christian de Duve (1974 Nobel Prize in Physiology or Medicine) for the discovery of the lysosome and autophagosome. Defining the mechanisms of autophagy using yeast as a model organism was developed by Yoshinori Ohsumi (2016 Nobel Prize in Physiology or Medicine). There are three types of autophagy, including chaperone-mediated (which involves the selective degradation of specific cargo proteins, that are recognized and delivered to the lysosome by a chaperone complex (Massey *et al.*, 2006)), microautophagy (which involves the nonselective sequestration of cytoplasmic components directly into the lysosome through invaginations in their limiting membrane (Wang *et al.*, 2022)), and macroautophagy (herein referred to as autophagy, which involves the sequestration within an autophagosome, a unique double-membrane cytosolic vesicle, that fuse with late endosomes and lysosomes, promoting the delivery of organelles, aggregated proteins and cytoplasm to the luminal acidic degradative milieu that enables their breakdown into constituent molecular building blocks that can be recycled by the cell (Hamasaki *et al.*, 2013)).

Although it has been acknowledged that HS activate autophagy, the underlying mechanisms need to be explored and fully defined. The intracellular energy sensing (ATP-AMPK-Akt) pathways is one of the major mechanism involved. As autophagy is a catabolic process that plays a key role in the generation of nutrient and energy to the starved cells, and as AMPK is the master energy sensor, it is obvious that AMPK regulates autophagy. As described above, HS reduced ATP levels, leading to an increase in the AMP:ATP ratio and subsequent activation of AMPK (Hafen *et al.*, 2018). Jiang and colleagues have shown that HS inhibited mechanistic target of rapamycin (mTOR), activated AMPK, and increased autophagy flux in human hepatocyte cell lines (Jiang *et al.*, 2019). Yoshihara and co-workers, on the other hand, have demonstrated that HS activated the Akt/mTOR signaling pathway in rodent skeletal muscle (Yoshihara *et al.*, 2013). This differential effect is probably due to tissue- or specie-specific effect of HS, and/or experimental conditions. Interestingly, Akt has been shown to act as a negative regulator of autophagy through activation of mTOR and phosphorylation of beclin 1 (Wang *et al.*, 2012), and HS has been reported to activate both Akt and AMPK, introducing a complex relationship and network between nutrient sensing, energy sensing, and autophagy regulation during HS (Figure 11.7).

The second potential mediator of HS effect on autophagy is mitochondrial-ER stress, mitochondrial ROS and intracellular calcium signaling. As described above, HS induces ROS production and ROS has been found to activate autophagy specifically through Atg4 (Scherz-Shouval *et al.*, 2007). Ashrafi *et al.* (2013) showed that ROS, via parkin PBR E3 Ub ligase (PRKN), induced mitophagy, which is a specialized form of autophagy that selectively target and degrade dysfunctional mitochondria (Ashrafi and Schwarz, 2013). Furthermore, Alexander *et al.* (2010) showed that ROS inhibited mTOR in an AMPK-dependent manner and consequently induced autophagy (Alexander *et al.*, 2010). It has been also shown that ubiquitylation of mitochondrial fusion proteins, mitofusin 1/2 (MFN1/2) and the voltage dependent anion channel 1 (VDAC1), by parkin, facilitated the recruitment of the autophagy adaptor p62 and eventually induced autophagosomal degradation of the damaged mitochondria (Geisler *et al.*, 2010). HS is known to induce calcium overload, which in turn increased ROS (Duchen *et al.*, 1998), activated calcium/calmodulin-dependent protein kinase kinase (CAMKK) 2 and CAMKKβ, leading to AMPK activation and autophagy induction (Cardenas and Foskett, 2012).

An additional potential mechanism by which HS affects autophagy machinery is cytokine production, where the interaction is still complex and continuously evolving. As described above HS induced pro-inflammatory cytokine production, and TNFα (as an example) has been shown to activate autophagy in melanoma cells

(Tyciakova *et al.*, 2021). Similarly, IL-1β has been reported to induce autophagy by affecting calcium homeostasis in pancreatic acinar cells (Xu *et al.*, 2014a).

Accumulating evidence suggests a connection between the HSP system and autophagy, in which HSP may represent an intracellular control mechanism that switches the cell from an initial catabolic degradation phase to an anabolic protein synthesis and repair phase (Dokladny *et al.*, 2015). In fact, it is well known that HS induced HSP expression and it has been shown that overexpression of HSP70 suppressed starvation-induced autophagy in A549 cell lines (Dokladny *et al.*, 2013). In C2C12 mouse myotubes, HS activated HSPs, HSF1, and altered autophagy machinery (Summers and Valentine, 2019). Mechanistic studies revealed that HSF1 regulated autophagy by directly binding to ATG7 promoter and transcriptionally up-regulating its expression (Desai *et al.*, 2013). It has been shown, also, that HSF1 facilitated casein kinase 1-mediated phosphorylation of p62, whereas inhibition of HSF1 impaired protein aggregate-induced autophagosome formation (Watanabe *et al.*, 2017). It is worth mentioning that in *C. elegans* and under mild stress, ATG genes were recently demonstrated to be required for longevity and increased thermotolerance caused by hermetic HS or HSF1 overexpression (Kumsta *et al.*, 2017), indicating that autophagy can be a survival or death pathways depending on stress duration, condition, and/or severity.

As mentioned above, HS is associated with an accumulation of unfolded and misfolded proteins in the lumen of the ER, which resulted in ER stress and activation of UPR. The UPR can be cytoprotective and prosurvival response; however, if the stress is chronic, the UPR can become cytotoxic and induce apoptotic cell death (Gorman *et al.*, 2012). Zheng and colleagues showed that UPR, via PERK signaling, coordinated with autophagy (ATG5, ATG7, beclin-1, and MAP1LC3B) to alleviate protein misfolding in human chondrocyte C2812 cells during ER stress (Zheng *et al.*, 2019). Gade *et al.* (2012), have shown that ATF6 activated autophagy via death-associated kinase 1 (DAPK1) (Gade *et al.*, 2012). Additionally, IRE1 has been reported to induce autophagy via MAPK pathways in mouse embryonic fibroblast cells (Ogata *et al.*, 2006). Of particular interest, Homma and Fujii (2016) showed that HS (*in vitro* incubation at 40 °C and 43 °C) activated all three UPR branches, leading to autophagic stimulation in several human cancer cell lines and in mouse embryonic fibroblasts.

8 HS modulates intestinal microbiota

Increasing data from diverse species (mammals, rodents, dairy cattle, swine, etc.) indicated that HS perturbs the gut microbiota, leading to dysbiosis (Contreras-Jodar *et al.*, 2019; Correia Sales *et al.*, 2021; Kim *et al.*, 2022; Le Sciellour *et al.*, 2019; Qu *et al.*, 2021; Wen *et al.*, 2021).

In avian species, it has been shown that HS reduced the viable counts of *Lactobacillus* and *Bifidobacterium*, and increased that of the opportunistic pathogens, including *Escherichia coli*, *Coliforms*, *Clostridium*, *Salmonella*, resulting in an imbalanced gut microflora in broiler (meat-type) chickens (Shi *et al.*, 2019; Song *et al.*, 2014; Zhang *et al.*, 2017). In layer (egg-type) chickens, Zhu *et al.* (2019) reported that HS increased *Bacteroidetes*, and decreased *Firmicutes* in fecal samples. Functional predictions indicated that cysteine- and methionine-metabolism as well as benzoate degradation were more abundant, however retinol metabolism and phenyl-propanoid biosynthesis were decreased by HS (Zhu *et al.*, 2019). Xing and co-workers, on the other hand, reported that the effect of HS on laying hen cecal microbiota was mainly associated with decreased feed intake (Xing *et al.*, 2019). Interestingly, these studies suggest differing effects along the intestinal tract, with the small intestine being more sensitive in comparison to the ceca. Dridi's group showed that microbial composition in ileal luminal and mucosal contents varied between chicken population selected for slow-, medium-, or high-growth rate as well as between HS and thermoneutral conditions (Emami *et al.*, 2022).

Although further mechanistic studies are needed, several potential biological and physiological pathways may explain the effect of HS on intestinal microbiota. Modification of feed and/or water consumption might affect the intestinal activity and motility as well as digesta viscosity during HS exposure, and thereby may modulate the substrate availability needed for the microbiota (Metzler-Zebeli *et al.*, 2019). Also, HS is well known to activate the hypothalamic-pituitary-adrenal (HPA) axis, which in turn alters the immune system and the hos/microbiota interaction (Galley and Bailey, 2014). Recent studies showed that CLDN7 plays a key role in the microbiota diversity in mice (Ding *et al.*, 2022). Although significant studies and publications showed effects of HS on avian gut microbiota, most of these studies are descriptive and associative, and mechanistic and functional studies are warranted.

9 Management and/or nutritional strategies to alleviate HS adverse effects

A multitude of strategies, from management to nutritional interventions, have been employed to ameliorate the HS productivity losses in poultry. Without being exhaustive, we decide to describe few strategies in this section, and the readers can find several elegant detailed reviews elsewhere (Lian *et al.*, 2020; Shakeri, 2022). In addition to diverse poultry house managements, the use of the sprinkler technology seemed to reduce water usage and improve growth performance in heat-stressed broilers via reducing core body temperature and systemic/local stress (Liang *et al.*, 2020). Although the effects of the sprinkler on gut integrity and microbial composition need to be explored, Dridi's group have shown that HS activated the hypothalamic AMPKα1/2 at Thr172 site, indicating a reduced levels of cellular ATP, and the sprinkler technology reversed this effects along with reduced stress as evidenced by a decrease in corticosterone and HSP levels (Rajaei-Sharifabadi *et al.*, 2017b). Additionally, the sprinkler increased feed intake and improved growth performance via modulation of feeding-related hypothalamic neuropeptides (Rajaei-Sharifabadi *et al.*, 2017b).

Supplementation of prebiotics, probiotics, postbiotics, and synbiotics has been shown to contribute to the maintenance of the intestinal barrier integrity, reduced the concentration of zonulin in fecal samples, and improved over all the gut health in humans (Shing *et al.*, 2014). *In vitro* study using Caco-2 cells showed that probiotic treatment enhanced TEER value, OCLN gene expression, and mucus production (Barnett *et al.*, 2018). In chickens, these additives have also been shown to alleviate the adverse effects of HS through enhancing the gut microstructure and morphology, increased expression of OCLN, and ZO-1 proteins, thereby improved gut integrity (Ashraf *et al.*, 2013; Song *et al.*, 2014). Several downstream signaling cascades have been suggested to mediate the beneficial effects of these supplements, including PKC (Zyrek *et al.*, 2007), MAPK (Segawa *et al.*, 2011), HSPs, chemo(cyto)kines, TLRs, and NF-κB [for review see (Plaza-Diaz *et al.*, 2019)]. These downstream pathways need to be explored in avian species under different types of stress.

Recently, our group showed that phytogenic feed/water additives improved growth performance via modulation of hypothalamic feeding- and drinking-related pathways, systemic and local inflammatory systems, HSPs, and metabolic-related genes in chronic cyclic heat-stressed broilers (Flees *et al.*, 2017; Flees *et al.*, 2021; Greene *et al.*, 2021a; Greene *et al.*, 2021b; Rajaei-Sharifabadi *et al.*, 2017a). More recently, Dridi's group found that in feed-herbal adaptogen supplementation, plant extracts known as stress response modifiers (Panossian and Wikman, 2010), improved growth performance, and reduced serum FITC (indicator of gut integrity) independently of thermoregulatory pathway in heat-stressed broilers (Greene *et al.*, 2021c). However, the exact underlying molecular mechanisms are still not known and beg further in-depth investigations.

References

(FAO), F.a.A.O., 2019. How to feed the world in 2050. FAO, Rome, Italy.

Abdelli, N., Ramser, A., Greene, E.S., Beer, L., Tabler, T.W., Orlowski, S.K., Perez, J.F., Sola-Oriol, D., Anthony, N.B. and Dridi, S., 2021. Effects of Cyclic Chronic Heat Stress on the Expression of Nutrient Transporters in the Jejunum of Modern Broilers and Their Ancestor Wild Jungle Fowl. Front Physiol 12: 733134. 10.3389/fphys.2021.733134.

Al-Sadi, R., Guo, S., Ye, D., Dokladny, K., Alhmoud, T., Ereifej, L., Said, H.M. and Ma, T.Y., 2013. Mechanism of IL-1beta modulation of intestinal epithelial barrier involves p38 kinase and activating transcription factor-2 activation. J Immunol 190: 6596–6606. 10.4049/jimmunol.1201876.

Al-Sadi, R., Guo, S., Ye, D., Rawat, M. and Ma, T.Y., 2016. TNF-alpha Modulation of Intestinal Tight Junction Permeability Is Mediated by NIK/IKK-alpha Axis Activation of the Canonical NF-kappaB Pathway. Am J Pathol 186: 1151–1165. 10.1016/j.ajpath.2015.12.016.

Al-Zghoul, M.B., Alliftawi, A.R.S., Saleh, K.M.M. and Jaradat, Z.W., 2019. Expression of digestive enzyme and intestinal transporter genes during chronic heat stress in the thermally manipulated broiler chicken. Poult Sci 98: 4113–4122. 10.3382/ps/pez249.

Alexander, A., Cai, S.L., Kim, J., Nanez, A., Sahin, M., MacLean, K.H., Inoki, K., Guan, K.L., Shen, J., Person, M.D., Kusewitt, D., Mills, G.B., Kastan, M.B. and Walker, C.L., 2010. ATM signals to TSC2 in the cytoplasm to regulate mTORC1 in response to ROS. Proc Natl Acad Sci USA 107: 4153–4158. 10.1073/pnas.0913860107.

Alhenaky, A., Abdelqader, A., Abuajamieh, M. and Al-Fataftah, A.R., 2017. The effect of heat stress on intestinal integrity and Salmonella invasion in broiler birds. J therm Biol 70: 9–14. 10.1016/j.jtherbio.2017.10.015.

Alhotan, R.A., Al-Sagan, A.A., Al-Abdullatif, A.A., Hussein, E.O.S., Saadeldin, I.M., Azzam, M.M. and Swelum, A.A., 2021. Interactive effects of dietary amino acid density and environmental temperature on growth performance and expression of selected amino acid transporters, water channels, and stress-related transcripts. Poult Sci 100: 101333. 10.1016/j.psj.2021.101333.

Alley, R.B., Clark, P.U., Huybrechts, P. and Joughin, I., 2005. Ice-sheet and sea-level changes. Science 310: 456–460. 10.1126/science.1114613.

Asea, A., Kraeft, S.K., Kurt-Jones, E.A., Stevenson, M.A., Chen, L.B., Finberg, R.W., Koo, G.C. and Calderwood, S.K., 2000. HSP70 stimulates cytokine production through a CD14-dependant pathway, demonstrating its dual role as a chaperone and cytokine. Nat Med 6: 435–442. 10.1038/74697.

Asea, A., Rehli, M., Kabingu, E., Boch, J.A., Bare, O., Auron, P.E., Stevenson, M.A. and Calderwood, S.K., 2002. Novel signal transduction pathway utilized by extracellular HSP70: role of toll-like receptor (TLR) 2 and TLR4. J Biol Chem 277: 15028–15034. 10.1074/jbc.M200497200.

Ashraf, S., Zaneb, H., Yousaf, M.S., Ijaz, A., Sohail, M.U., Muti, S., Usman, M.M., Ijaz, S. and Rehman, H., 2013. Effect of dietary supplementation of prebiotics and probiotics on intestinal microarchitecture in broilers reared under cyclic heat stress. J Anim Physiol Anim Nutr (Berl) 97 Suppl 1: 68–73. 10.1111/jpn.12041.

Ashrafi, G. and Schwarz, T.L., 2013. The pathways of mitophagy for quality control and clearance of mitochondria. Cell Death Differ 20: 31–42. 10.1038/cdd.2012.81.

Aveleira, C.A., Lin, C.M., Abcouwer, S.F., Ambrosio, A.F. and Antonetti, D.A., 2010. TNF-alpha signals through PKCzeta/NF-kappaB to alter the tight junction complex and increase retinal endothelial cell permeability. Diabetes 59: 2872–2882. 10.2337/db09-1606.

Avery, P., Richards, S., 1983. Thermosensitivity of the hypothalamus and spinal cord in the domestic fowl. J therm Biol 8: 237–239.

Bandarra, D., Biddlestone, J., Mudie, S., Muller, H.A. and Rocha, S., 2014. Hypoxia activates IKK-NF-kappaB and the immune response in Drosophila melanogaster. Biosci Rep 34. 10.1042/BSR20140095.

Barnett, A.M., Roy, N.C., Cookson, A.L. and McNabb, W.C., 2018. Metabolism of Caprine Milk Carbohydrates by Probiotic Bacteria and Caco-2:HT29(-)MTX Epithelial Co-Cultures and Their Impact on Intestinal Barrier Integrity. Nutrients 10. 10.3390/nu10070949.

Barrett, N.W., Rowland, K., Schmidt, C.J., Lamont, S.J., Rothschild, M.F., Ashwell, C.M. and Persia, M.E., 2019. Effects of acute and chronic heat stress on the performance, egg quality, body temperature, and blood gas parameters of laying hens. Poult Sci 98: 6684–6692. 10.3382/ps/pez541.

Bartholomew, G.A., Lasiewski, R.C., Crawford, E.C., 1968. Patterns of panting and gular flutter in cormorants, pelicans, owls, and doves. The Condor 70: 31–34.

Bech, C. and Johansen, K., 1980. Ventilatory and circulatory responses to hyperthermia in the mute swan (Cygnus olor). J Exp Biol 88: 195–204. 10.1242/jeb.88.1.195.

Bech, C., Midtgård, U., 1981. Brain temperature and the rete mirabile opthalmicum in the Zebra Finch (Poephila guttata). J Comp Physiol B 145: 89–93.

Bevins, C.L. and Salzman, N.H., 2011. Paneth cells, antimicrobial peptides and maintenance of intestinal homeostasis. Nat Rev Microbiol 9: 356–368. 10.1038/nrmicro2546.

Bond, U. and Schlesinger, M.J., 1985. Ubiquitin is a heat shock protein in chicken embryo fibroblasts. Mol Cell Biol 5: 949–956. 10.1128/mcb.5.5.949-956.1985.

Bottje, W.G. and Harrison, P.C., 1986. The effect of high ambient temperature and hypercapnia on postprandial intestinal hyperemia in domestic cockerels. Poult Sci 65: 1606–1612. 10.3382/ps.0651606.

Boulant, J.A., 2006. Neuronal basis of Hammel's model for set-point thermoregulation. J Appl Physiol (1985) 100: 1347–1354. 10.1152/japplphysiol.01064.2005.

Boulant, J.A. and Gonzalez, R.R., 1977. The effect of skin temperature on the hypothalamic control of heat loss and heat production. Brain Res 120: 367–372. 10.1016/0006-8993(77)90916-7.

Braten, O., Livneh, I., Ziv, T., Admon, A., Kehat, I., Caspi, L.H., Gonen, H., Bercovich, B., Godzik, A., Jahandideh, S., Jaroszewski, L., Sommer, T., Kwon, Y.T., Guharoy, M., Tompa, P. and Ciechanover, A., 2016. Numerous proteins with unique characteristics are degraded by the 26S proteasome following monoubiquitination. Proc Natl Acad Sci USA 113: E4639–4647. 10.1073/pnas.1608644113.

Bravo, R., Vicencio, J.M., Parra, V., Troncoso, R., Munoz, J.P., Bui, M., Quiroga, C., Rodriguez, A.E., Verdejo, H.E., Ferreira, J., Iglewski, M., Chiong, M., Simmen, T., Zorzano, A., Hill, J.A. Rothermel, B.A., Szabadkai, G. and Lavandero, S., 2011. Increased ER-mitochondrial coupling promotes mitochondrial respiration and bioenergetics during early phases of ER stress. J Cell Sci 124: 2143–2152. 10.1242/jcs.080762.

Brugaletta, G., Greene, E., Ramser, A., Maynard, C.W., Tabler, T.W., Sirri, F., Anthony, N.B., Orlowski, S. and Dridi, S., 2022. Effect of Cyclic Heat Stress on Hypothalamic Oxygen Homeostasis and Inflammatory State in the Jungle Fowl and Three Broiler-Based Research Lines. Front Vet Sci 9: 905225. 10.3389/fvets.2022.905225.

Bush, K.T., Goldberg, A.L. and Nigam, S.K., 1997. Proteasome inhibition leads to a heat-shock response, induction of endoplasmic reticulum chaperones, and thermotolerance. J Biol Chem 272: 9086–9092. 10.1074/jbc.272.14.9086.

Cahaner, A. and Leenstra, F., 1992. Effects of high temperature on growth and efficiency of male and female broilers from lines selected for high weight gain, favorable feed conversion, and high or low fat content. Poult Sci 71: 1237–1250. 10.3382/ps.0711237.

Calfon, M., Zeng, H., Urano, F., Till, J.H., Hubbard, S.R., Harding, H.P., Clark, S.G. and Ron, D., 2002. IRE1 couples endoplasmic reticulum load to secretory capacity by processing the XBP-1 mRNA. Nature 415: 92–96. 10.1038/415092a.

Capaldo, C.T. and Nusrat, A., 2009. Cytokine regulation of tight junctions. Biochim Biophys Acta 1788: 864–871. 10.1016/j.bbamem.2008.08.027.

Cardenas, C. and Foskett, J.K., 2012. Mitochondrial Ca(2+) signals in autophagy. Cell Calcium 52: 44–51. 10.1016/j.ceca.2012.03.001.

Caterina, M.J., Rosen, T.A., Tominaga, M., Brake, A.J. and Julius, D., 1999. A capsaicin-receptor homologue with a high threshold for noxious heat. Nature 398: 436–441. 10.1038/18906.

Caterina, M.J., Schumacher, M.A., Tominaga, M., Rosen, T.A., Levine, J.D. and Julius, D., 1997. The capsaicin receptor: a heat-activated ion channel in the pain pathway. Nature 389: 816–824. 10.1038/39807.

Caulfield, M.P., Cambridge, H., Foster, S.F. and McGreevy, P.D., 2014. Heat stress: a major contributor to poor animal welfare associated with long-haul live export voyages. Vet J 199: 223–228. 10.1016/j.tvjl.2013.09.018.

Cervantes, H.M., 2015. Antibiotic-free poultry production: Is it sustainable? J. Appl. Poult. Res 24: 91–97. doi: 10.3382/japr/pfv006.

Cesen, M.H., Pegan, K., Spes, A. and Turk, B., 2012. Lysosomal pathways to cell death and their therapeutic applications. Exp Cell Res 318: 1245–1251. 10.1016/j.yexcr.2012.03.005.

Chandel, N.S. and Simon, M.C., 2008. Hypoxia-inducible factor: roles in development, physiology, and disease. Cell Death Differ 15: 619–620. 10.1038/cdd.2008.11.

Chen, I.C., Hill, J.K., Ohlemuller, R., Roy, D.B. and Thomas, C.D., 2011. Rapid range shifts of species associated with high levels of climate warming. Science 333: 1024–1026. 10.1126/science.1206432.

Ciechanover, A., 1998. The ubiquitin-proteasome pathway: on protein death and cell life. EMBO J 17: 7151–7160. 10.1093/emboj/17.24.7151.

Contreras-Jodar, A., Nayan, N.H., Hamzaoui, S., Caja, G. and Salama, A.A.K., 2019. Heat stress modifies the lactational performances and the urinary metabolomic profile related to gastrointestinal microbiota of dairy goats. PLoS One 14: e0202457. 10.1371/journal.pone.0202457.

Correia Sales, G.F., Carvalho, B.F., Schwan, R.F., de Figueiredo Vilela, L., Moreno Meneses, J.A., Gionbelli, M.P. and Luiza da Silva Avila, C., 2021. Heat stress influence the microbiota and organic acids concentration in beef cattle rumen. J therm Biol 97: 102897. 10.1016/j.jtherbio.2021.102897.

Crew, B., 2022. Solving the energy crisis. Nature 609: S1. 10.1038/d41586-022-02830-5.

Cui, Y., Zhou, X., Chen, L., Tang, Z., Mo, F., Li, X.C., Mao, H., Wei, X., Wang, C. and Wang, H., 2021. Crosstalk between Endoplasmic Reticulum Stress and Oxidative Stress in Heat Exposure-Induced Apoptosis Is Dependent on the ATF4-CHOP-CHAC1 Signal Pathway in IPEC-J2 Cells. J Agric Food Chem 69: 15495–15511. 10.1021/acs.jafc.1c03361.

Culver, C., Sundqvist, A., Mudie, S., Melvin, A., Xirodimas, D. and Rocha, S., 2010. Mechanism of hypoxia-induced NF-kappaB. Mol Cell Biol 30: 4901–4921. 10.1128/MCB.00409-10.

Dale, N.M. and Fuller, H.L., 1980. Effect of diet composition on feed intake and growth of chicks under heat stress. II. Constant vs. cycling temperatures. Poult Sci 59: 1434–1441. 10.3382/ps.0591434.

Deeb, N. and Cahaner, A., 2002. Genotype-by-environment interaction with broiler genotypes differing in growth rate. 3. Growth rate and water consumption of broiler progeny from weight-selected versus nonselected parents under normal and high ambient temperatures. Poult Sci 81: 293–301. 10.1093/ps/81.3.293.

DeFries, R., Fanzo, J., Remans, R., Palm, C., Wood, S. and Anderman, T.L., 2015. Global nutrition. Metrics for land-scarce agriculture. Science 349: 238–240. 10.1126/science.aaa5766

Desai, S., Liu, Z., Yao, J., Patel, N., Chen, J., Wu, Y., Ahn, E.E., Fodstad, O. and Tan, M., 2013. Heat shock factor 1 (HSF1) controls chemoresistance and autophagy through transcriptional regulation of autophagy-related protein 7 (ATG7). J Biol Chem 288: 9165–9176. 10.1074/jbc.M112.422071.

Dikic, I. and Elazar, Z., 2018. Mechanism and medical implications of mammalian autophagy. Nat Rev Mol Cell Biol 19: 349–364. 10.1038/s41580-018-0003-4.

Dillon, A. and Lo, D.D., 2019. M Cells: Intelligent Engineering of Mucosal Immune Surveillance. Front Immunol 10: 1499. 10.3389/fimmu.2019.01499.

Ding, Y., Wang, K., Xu, C., Hao, M., Li, H. and Ding, L., 2022. Intestinal Claudin-7 deficiency impacts the intestinal microbiota in mice with colitis. BMC Gastroenterol 22: 24. 10.1186/s12876-022-02100-8.

Dokladny, K., Myers, O.B. and Moseley, P.L., 2015. Heat shock response and autophagy – cooperation and control. Autophagy 11: 200–213. 10.1080/15548627.2015.1009776.

Dokladny, K., Zuhl, M.N., Mandell, M., Bhattacharya, D., Schneider, S., Deretic, V. and Moseley, P.L., 2013. Regulatory coordination between two major intracellular homeostatic systems: heat shock response and autophagy. J Biol Chem 288: 14959–14972. 10.1074/jbc.M113.462408.

Dridi, S., 2020. The use of nutrigenomics in poultry breeding. In: H.Z. Sammy, E. Aggrey, Michèle Tixier-Boichard, Douglas D. Rhoads (Ed.), Advances in poultry genetics and genomics. Burleigh Dodds Science Publishing, UK, pp. 580.

Dridi, S., 2022a. Food Intake Regulation: Factors Influencing Food Intake- chapter 3 in Nutritional Biochemistry: From the Classroom to the Research Bench Bentham Science Publishers, Sharjah, U.A.E., 272 pp.

Dridi, S., 2022b. Proteins, Amino acids, and Nitrogen Metabolism. In: S. Dridi (Ed.), Nutritional Biochemistry: From the Classroom to the Research Bench. Bentham Science, U.A.E., pp. 272. 10.2174/97898150515751220101.

Duchen, M.R., Leyssens, A. and Crompton, M., 1998. Transient mitochondrial depolarizations reflect focal sarcoplasmic reticular calcium release in single rat cardiomyocytes. J Cell Biol 142: 975–988. 10.1083/jcb.142.4.975.

Emami, N.K., Cauble, R.N., Dhamad, A.E., Greene, E.S., Coy, C.S., Velleman, S.G., Orlowski, S., Anthony, N., Bedford, M. and Dridi, S., 2021a. Hypoxia further exacerbates woody breast myopathy in broilers via alteration of satellite cell fate. Poult Sci 100: 101167. 10.1016/j.psj.2021.101167.

Emami, N.K., Greene, E.S., Kogut, M.H. and Dridi, S., 2021b. Heat Stress and Feed Restriction Distinctly Affect Performance, Carcass and Meat Yield, Intestinal Integrity, and Inflammatory (Chemo)Cytokines in Broiler Chickens. Front Physiol 12: 707757. 10.3389/fphys.2021.707757.

Emami, N.K., Jung, U., Voy, B. and Dridi, S., 2020. Radical Response: Effects of Heat Stress-Induced Oxidative Stress on Lipid Metabolism in the Avian Liver. Antioxidants (Basel) 10. 10.3390/antiox10010035.

Emami, N.K., Schreier, L.L., Greene, E., Tabler, T., Orlowski, S.K., Anthony, N.B., Proszkowiec-Weglarz, M. and Dridi, S., 2022. Ileal microbial composition in genetically distinct chicken lines reared under normal or high ambient temperatures. Anim Microbiome 4: 28. 10.1186/s42523-022-00183-y.

Fang, N.N., Chan, G.T., Zhu, M., Comyn, S.A., Persaud, A., Deshaies, R.J., Rotin, D., Gsponer, J. and Mayor, T., 2014. Rsp5/Nedd4 is the main ubiquitin ligase that targets cytosolic misfolded proteins following heat stress. Nat Cell Biol 16: 1227–1237. 10.1038/ncb3054.

Fang, N.N., Ng, A.H., Measday, V. and Mayor, T., 2011. Hul5 HECT ubiquitin ligase plays a major role in the ubiquitylation and turnover of cytosolic misfolded proteins. Nat Cell Biol 13: 1344–1352. 10.1038/ncb2343.

Feng, Y., Wang, Y., Wang, P., Huang, Y. and Wang, F., 2018. Short-Chain Fatty Acids Manifest Stimulative and Protective Effects on Intestinal Barrier Function Through the Inhibition of NLRP3 Inflammasome and Autophagy. Cell Physiol Biochem 49: 190–205. 10.1159/000492853.

Finley, D., 2009. Recognition and processing of ubiquitin-protein conjugates by the proteasome. Annu Rev Biochem 78: 477–513. 10.1146/annurev.biochem.78.081507.101607.

Fischer, M., Tamariz-Ellemann, A. and Gliemann, L., 2020. Hyperthermia: The hotter the body, the hungrier the brain? J Physiol 598: 2053–2054. 10.1113/JP279745.

Flees, J., Rajaei-Sharifabadi, H., Greene, E., Beer, L., Hargis, B.M., Ellestad, L., Porter, T., Donoghue, A., Bottje, W.G. and Dridi, S., 2017. Effect of Morinda citrifolia (Noni)-Enriched Diet on Hepatic Heat Shock Protein and Lipid Metabolism-Related Genes in Heat Stressed Broiler Chickens. Front Physiol 8: 919. 10.3389/fphys.2017.00919.

Flees, J.J., Emami, N.K., Greene, E., Ganguly, B. and Dridi, S., 2021. Phytogenic Water Additives Improve Broiler Growth Performance via Modulation of Intermediary Metabolism-Related Signaling Pathways. Animals (Basel) 11. 10.3390/ani11030750.

Flint, H.J., Scott, K.P., Louis, P. and Duncan, S.H., 2012. The role of the gut microbiota in nutrition and health. Nat Rev Gastroenterol Hepatol 9: 577–589. 10.1038/nrgastro.2012.156.

Fornace, A.J., Jr., Alamo, I., Jr., Hollander, M.C. and Lamoreaux, E., 1989. Ubiquitin mRNA is a major stress-induced transcript in mammalian cells. Nucleic Acids Res 17: 1215–1230. 10.1093/nar/17.3.1215.

Franchi, L., Kamada, N., Nakamura, Y., Burberry, A., Kuffa, P., Suzuki, S., Shaw, M.H., Kim, Y.G. and Nunez, G., 2012. NLRC4-driven production of IL-1beta discriminates between pathogenic and commensal bacteria and promotes host intestinal defense. Nat Immunol 13: 449–456. 10.1038/ni.2263.

Gade, P., Ramachandran, G., Maachani, U.B., Rizzo, M.A., Okada, T., Prywes, R., Cross, A.S., Mori, K. and Kalvakolanu, D.V., 2012. An IFN-gamma-stimulated ATF6-C/EBP-beta-signaling pathway critical for the expression of Death Associated Protein Kinase 1 and induction of autophagy. Proc Natl Acad Sci USA 109: 10316–10321. 10.1073/pnas.1119273109.

Galley, J.D. and Bailey, M.T., 2014. Impact of stressor exposure on the interplay between commensal microbiota and host inflammation. Gut Microbes 5: 390–396. 10.4161/gmic.28683.

Geisler, S., Holmstrom, K.M., Skujat, D., Fiesel, F.C., Rothfuss, O.C., Kahle, P.J. and Springer, W., 2010. PINK1/Parkin-mediated mitophagy is dependent on VDAC1 and p62/SQSTM1. Nat Cell Biol 12: 119–131. 10.1038/ncb2012.

Gorman, A.M., Healy, S.J., Jager, R. and Samali, A., 2012. Stress management at the ER: regulators of ER stress-induced apoptosis. Pharmacol Ther 134: 306–316. 10.1016/j.pharmthera.2012.02.003.

Gottschling, D.E. and Nystrom, T., 2017. The Upsides and Downsides of Organelle Interconnectivity. Cell 169: 24–34. 10.1016/j.cell.2017.02.030.

Greene, E., Cauble, R., Dhamad, A.E., Kidd, M.T., Kong, B., Howard, S.M., Castro, H.F., Campagna, S.R., Bedford, M. and Dridi, S., 2020. Muscle Metabolome Profiles in Woody Breast-(un)Affected Broilers: Effects of Quantum Blue Phytase-Enriched Diet. Front Vet Sci 7: 458. 10.3389/fvets.2020.00458.

Greene, E., Flees, J., Dadgar, S., Mallmann, B., Orlowski, S., Dhamad, A., Rochell, S., Kidd, M., Laurendon, C., Whitfield, H., Brearley, C., Rajaram, N., Walk, C. and Dridi, S., 2019. Quantum Blue Reduces the Severity of Woody Breast Myopathy via Modulation of Oxygen Homeostasis-Related Genes in Broiler Chickens. Front Physiol 10: 1251. 10.3389/fphys.2019.01251.

Greene, E.S., Adeogun, E., Orlowski, S.K., Nayani, K. and Dridi, S., 2022. Effects of heat stress on cyto(chemo)kine and inflammasome gene expression and mechanical properties in isolated red and white blood cells from 4 commercial broiler lines and their ancestor jungle fowl. Poult Sci 101: 101827. 10.1016/j.psj.2022.101827.

Greene, E.S., Cauble, R., Kadhim, H., de Almeida Mallmann, B., Gu, I., Lee, S.O., Orlowski, S. and Dridi, S., 2021a. Protective effects of the phytogenic feed additive "comfort" on growth performance via modulation of hypothalamic feeding- and drinking-related neuropeptides in cyclic heat-stressed broilers. Domest Anim Endocrinol 74: 106487. 10.1016/j.domaniend.2020.106487.

Greene, E.S., Emami, N.K. and Dridi, S., 2021b. Research Note: Phytobiotics modulate the expression profile of circulating inflammasome and cyto(chemo)kine in whole blood of broilers exposed to cyclic heat stress. Poult Sci 100: 100801. 10.1016/j.psj.2020.10.055.

Greene, E.S., Maynard, C., Owens, C.M., Meullenet, J.F. and Dridi, S., 2021c. Effects of Herbal Adaptogen Feed-Additive on Growth Performance, Carcass Parameters, and Muscle Amino Acid Profile in Heat-Stressed Modern Broilers. Front Physiol 12: 784952. 10.3389/fphys.2021.784952.

Gribble, F.M. and Reimann, F., 2019. Function and mechanisms of enteroendocrine cells and gut hormones in metabolism. Nat Rev Endocrinol 15: 226–237. 10.1038/s41574-019-0168-8.

Gyorke, A., Pop, L. and Cozma, V., 2013. Prevalence and distribution of Eimeria species in broiler chicken farms of different capacities. Parasite 20: 50. 10.1051/parasite/2013052.

Habashy, W.S., Milfort, M.C., Fuller, A.L., Attia, Y.A., Rekaya, R. and Aggrey, S.E., 2017. Effect of heat stress on protein utilization and nutrient transporters in meat-type chickens. Int J Biometeorol 61: 2111–2118. 10.1007/s00484-017-1414-1.

Hafen, P.S., Preece, C.N., Sorensen, J.R., Hancock, C.R. and Hyldahl, R.D., 2018. Repeated exposure to heat stress induces mitochondrial adaptation in human skeletal muscle. J Appl Physiol (1985) 125: 1447–1455. 10.1152/japplphysiol.00383.2018.

Hamasaki, M., Furuta, N., Matsuda, A., Nezu, A., Yamamoto, A., Fujita, N., Oomori, H., Noda, T., Haraguchi, T., Hiraoka, Y., Amano, A. and Yoshimori, T., 2013. Autophagosomes form at ER-mitochondria contact sites. Nature 495: 389–393. 10.1038/nature11910.

Han, D., Lerner, A.G., Vande Walle, L., Upton, J.P., Xu, W., Hagen, A., Backes, B.J., Oakes, S.A. and Papa, F.R., 2009. IRE1alpha kinase activation modes control alternate endoribonuclease outputs to determine divergent cell fates. Cell 138: 562–575. 10.1016/j.cell.2009.07.017.

Han, W., Tellez, L.A., Perkins, M.H., Perez, I.O., Qu, T., Ferreira, J., Ferreira, T.L., Quinn, D., Liu, Z.W., Gao, X.B., Kaelberer, M.M., Bohorquez, D.V., Shammah-Lagnado, S.J., de Lartigue, G. and de Araujo, I.E., 2018. A Neural Circuit for Gut-Induced Reward. Cell 175: 665–678 e623. 10.1016/j.cell.2018.08.049.

Harding, H.P., Zhang, Y. and Ron, D., 1999. Protein translation and folding are coupled by an endoplasmic-reticulum-resident kinase. Nature 397: 271–274. 10.1038/16729.

Harding, H.P., Zhang, Y., Zeng, H., Novoa, I., Lu, P.D., Calfon, M., Sadri, N., Yun, C., Popko, B., Paules, R., Stojdl, D.F., Bell, J.C., Hettmann, T., Leiden, J.M. and Ron, D., 2003. An integrated stress response regulates amino acid metabolism and resistance to oxidative stress. Mol Cell 11: 619–633. 10.1016/s1097-2765(03)00105-9.

He, C., Liu, Z., Wu, J., Pan, X., Fang, Z., Li, J. and Bryan, B.A., 2021. Future global urban water scarcity and potential solutions. Nat Commun 12: 4667. 10.1038/s41467-021-25026-3.

He, X., Lu, Z., Ma, B., Zhang, L., Li, J., Jiang, Y., Zhou, G. and Gao, F., 2018. Effects of chronic heat exposure on growth performance, intestinal epithelial histology, appetite-related hormones and genes expression in broilers. J Sci Food Agric 98: 4471–4478. 10.1002/jsfa.8971.

Hetz, C., Zhang, K. and Kaufman, R.J., 2020. Mechanisms, regulation and functions of the unfolded protein response. Nat Rev Mol Cell Biol 21: 421–438. 10.1038/s41580-020-0250-z.

Hoffman, T.C., Walsberg, G.E. and DeNardo, D.F., 2007. Cloacal evaporation: an important and previously undescribed mechanism for avian thermoregulation. J Exp Biol 210: 741–749. 10.1242/jeb.02705.

Hollien, J. and Weissman, J.S., 2006. Decay of endoplasmic reticulum-localized mRNAs during the unfolded protein response. Science 313: 104–107. 10.1126/science.1129631.

Homma, T. and Fujii, J., 2016. Heat stress promotes the down-regulation of IRE1alpha in cells: An atypical modulation of the UPR pathway. Exp Cell Res 349: 128–138. 10.1016/j.yexcr.2016.10.006.

Hou, C.H., Lin, F.L., Hou, S.M. and Liu, J.F., 2014. Hyperthermia induces apoptosis through endoplasmic reticulum and reactive oxygen species in human osteosarcoma cells. Int J Mol Sci 15: 17380–17395. 10.3390/ijms151017380.

Hsu, S.F., Hsu, C.C., Cheng, B.C. and Lin, C.H., 2014. Cathepsin B is involved in the heat shock induced cardiomyocytes apoptosis as well as the anti-apoptosis effect of HSP-70. Apoptosis 19: 1571–1580. 10.1007/s10495-014-1033-9.

Huang, L.E., Gu, J., Schau, M. and Bunn, H.F., 1998. Regulation of hypoxia-inducible factor 1alpha is mediated by an O2-dependent degradation domain via the ubiquitin-proteasome pathway. Proc Natl Acad Sci USA 95: 7987–7992. 10.1073/pnas.95.14.7987.

Huber, S., Gagliani, N., Zenewicz, L.A., Huber, F.J., Bosurgi, L., Hu, B., Hedl, M., Zhang, W., O'Connor, W., Jr., Murphy, A.J., Valenzuela, D.M., Yancopoulos, G.D., Booth, C.J., Cho, J.H., Ouyang, W., Abraham, C. and Flavell, R.A., 2012. IL-22BP is regulated by the inflammasome and modulates tumorigenesis in the intestine. Nature 491: 259–263. 10.1038/nature11535.

Hulina, A., Grdic Rajkovic, M., Jaksic Despot, D., Jelic, D., Dojder, A., Cepelak, I. and Rumora, L., 2018. Extracellular Hsp70 induces inflammation and modulates LPS/LTA-stimulated inflammatory response in THP-1 cells. Cell Stress Chaperones 23: 373–384. 10.1007/s12192-017-0847-0.

Janse van Vuuren, A.K., Kemp, L.V., McKechnie, A.E., 2020. The beak and unfeathered skin as heat radiators in the Southern Ground-hornbill. J Avian Biol 51: e02457. doi:10.1111/jav.02457.

Ji, J., Gu, Z., Li, H., Su, L. and Liu, Z., 2018. Cryptdin-2 predicts intestinal injury during heatstroke in mice. Int J Mol Med 41: 137–146. 10.3892/ijmm.2017.3229.

Jiang, J., Chen, S., Li, K., Zhang, C., Tan, Y., Deng, Q., Chai, Y., Wang, X., Chen, G., Feng, K., Zhang, L., Xie, C.M. and Ma, K., 2019. Targeting autophagy enhances heat stress-induced apoptosis via the ATP-AMPK-mTOR axis for hepatocellular carcinoma. Int J Hyperthermia 36: 499–510. 10.1080/02656736.2019.1600052.

Johnson, J.D. and Fleshner, M., 2006. Releasing signals, secretory pathways, and immune function of endogenous extracellular heat shock protein 72. J Leukoc Biol 79: 425–434. 10.1189/jlb.0905523.

Julian, R.J., 2005. Production and growth related disorders and other metabolic diseases of poultry – a review. Vet J 169: 350–369. 10.1016/j.tvjl.2004.04.015.

Jury, W.A. and Vaux, H., Jr., 2005. The role of science in solving the world's emerging water problems. Proc Natl Acad Sci USA 102: 15715–15720. 10.1073/pnas.0506467102.

Kagnoff, M.F., 2014. The intestinal epithelium is an integral component of a communications network. J Clin Invest 124: 2841–2843. 10.1172/JCI75225.

Kang, J., Zhou, Y., Zhu, C., Ren, T., Zhang, Y., Xiao, L. and Fang, B., 2022. Ginsenoside Rg1 Mitigates Porcine Intestinal Tight Junction Disruptions Induced by LPS through the p38 MAPK/NLRP3 Inflammasome Pathway. Toxics 10. 10.3390/toxics10060285.

Karaer, Z., Guven, E., Akcay, A., Kar, S., Nalbantoglu, S. and Cakmak, A., 2012. Prevalence of subclinical coccidiosis in broiler farms in Turkey. Trop Anim Health Prod 44: 589–594. 10.1007/s11250-011-9940-z.

Kawazoe, Y., Nakai, A., Tanabe, M. and Nagata, K., 1998. Proteasome inhibition leads to the activation of all members of the heat-shock-factor family. Eur J Biochem 255: 356–362. 10.1046/j.1432-1327.1998.2550356.x.

Kim, A., Madara, J.C., Wu, C., Andermann, M.L. and Lowell, B.B., 2021. Neural basis for regulation of vasopressin secretion by anticipated disturbances in osmolality. Elife 10. 10.7554/eLife.66609.

Kim, D., Kim, S.H. and Li, G.C., 1999. Proteasome inhibitors MG132 and lactacystin hyperphosphorylate HSF1 and induce hsp70 and hsp27 expression. Biochem Biophys Res Commun 254: 264–268. 10.1006/bbrc.1998.9840.

Kim, J.H., Park, S.J., Kim, T.S., Park, H.J., Park, J., Kim, B.K., Kim, G.R., Kim, J.M., Huang, S.M., Chae, J.I., Park, C.K. and Lee, D.S., 2013. Testicular hyperthermia induces Unfolded Protein Response signaling activation in spermatocyte. Biochem Biophys Res Commun 434: 861–866. 10.1016/j.bbrc.2013.04.032.

Kim, S.H., Ramos, S.C., Valencia, R.A., Cho, Y.I. and Lee, S.S., 2022. Heat Stress: Effects on Rumen Microbes and Host Physiology, and Strategies to Alleviate the Negative Impacts on Lactating Dairy Cows. Front Microbiol 13: 804562. 10.3389/fmicb.2022.804562.

Kluger, M.J., Ringler, D.H. and Anver, M.R., 1975. Fever and survival. Science 188: 166–168.

Kumsta, C., Chang, J.T., Schmalz, J. and Hansen, M., 2017. Hormetic heat stress and HSF-1 induce autophagy to improve survival and proteostasis in C. elegans. Nat Commun 8: 14337. 10.1038/ncomms14337.

Kwan, J., 2022. Europe's energy crisis hits science hard. Science 377: 1244–1245. 10.1126/science.ade8708.

Lambert, G.P., Gisolfi, C.V., Berg, D.J., Moseley, P.L., Oberley, L.W. and Kregel, K.C., 2002. Selected contribution: Hyperthermia-induced intestinal permeability and the role of oxidative and nitrosative stress. J Appl Physiol (1985) 92: 1750–1761; discussion 1749. 10.1152/japplphysiol.00787.2001.

Lambin, E.F. and Meyfroidt, P., 2011. Global land use change, economic globalization, and the looming land scarcity. Proc Natl Acad Sci USA 108: 3465–3472. 10.1073/pnas.1100480108.

Le Sciellour, M., Zemb, O., Hochu, I., Riquet, J., Gilbert, H., Giorgi, M., Billon, Y., Gourdine, J.L. and Renaudeau, D., 2019. Effect of chronic and acute heat challenges on fecal microbiota composition, production, and thermoregulation traits in growing pigs1,2. J Anim Sci 97: 3845–3858. 10.1093/jas/skz222.

Lee, A.H., Iwakoshi, N.N. and Glimcher, L.H., 2003. XBP-1 regulates a subset of endoplasmic reticulum resident chaperone genes in the unfolded protein response. Mol Cell Biol 23: 7448–7459. 10.1128/MCB.23.21.7448-7459.2003.

Leenstra, F. and Cahaner, A., 1992. Effects of low, normal, and high temperatures on slaughter yield of broilers from lines selected for high weight gain, favorable feed conversion, and high or low fat content. Poult Sci 71: 1994–2006. 10.3382/ps.0711994.

Levy, M., Thaiss, C.A., Zeevi, D., Dohnalova, L., Zilberman-Schapira, G., Mahdi, J.A., David, E., Savidor, A., Korem, T., Herzig, Y., Pevsner-Fischer, M., Shapiro, H., Christ, A., Harmelin, A., Halpern, Z., Latz, E., Flavell, R.A., Amit, I., Segal, E. and Elinav, E., 2015. Microbiota-Modulated Metabolites Shape the Intestinal Microenvironment by Regulating NLRP6 Inflammasome Signaling. Cell 163: 1428–1443. 10.1016/j.cell.2015.10.048.

Lian, P., Braber, S., Garssen, J., Wichers, H.J., Folkerts, G., Fink-Gremmels, J. and Varasteh, S., 2020. Beyond Heat Stress: Intestinal Integrity Disruption and Mechanism-Based Intervention Strategies. Nutrients 12. 10.3390/nu12030734.

Lian, P., Braber, S., Varasteh, S., Wichers, H.J. and Folkerts, G., 2021. Hypoxia and heat stress affect epithelial integrity in a Caco-2/HT-29 co-culture. Sci Rep 11: 13186. 10.1038/s41598-021-92574-5.

Liang, Y., Tabler, G.T. and Dridi, S., 2020. Sprinkler Technology Improves Broiler Production Sustainability: From Stress Alleviation to Water Usage Conservation: A Mini Review. Front Vet Sci 7: 544814. 10.3389/fvets.2020.544814.

Liedtke, W., Choe, Y., Marti-Renom, M.A., Bell, A.M., Denis, C.S., Sali, A., Hudspeth, A.J., Friedman, J.M. and Heller, S., 2000. Vanilloid receptor-related osmotically activated channel (VR-OAC), a candidate vertebrate osmoreceptor. Cell 103: 525–535. 10.1016/s0092-8674(00)00143-4.

Liu, X., Liu, W., Tang, Q., Liu, B., Wada, Y., Yang, H., 2022. Global Agricultural Water Scarcity Assessment Incorporating Blue and Green Water Availability Under Future Climate Change. Earth's Future 10: e2021EF002567. https://doi.org/10.1029/2021EF002567.

Liu, Y.L., Ding, K.N., Shen, X.L., Liu, H.X., Zhang, Y.A., Liu, Y.Q., He, Y.M. and Tang, L.P., 2022. Chronic heat stress promotes liver inflammation in broilers via enhancing NF-kappaB and NLRP3 signaling pathway. BMC Vet Res 18: 289. 10.1186/s12917-022-03388-0.

Luzio, J.P., Pryor, P.R. and Bright, N.A., 2007. Lysosomes: fusion and function. Nat Rev Mol Cell Biol 8: 622–632. 10.1038/nrm2217.

Machiela, E., Liontis, T., Dues, D.J., Rudich, P.D., Traa, A., Wyman, L., Kaufman, C., Cooper, J.F., Lew, L., Nadarajan, S., Senchuk, M.M. and Van Raamsdonk, J.M., 2020. Disruption of mitochondrial dynamics increases stress resistance through activation of multiple stress response pathways. FASEB J 34: 8475–8492. 10.1096/fj.201903235R.

Mack, L.A., Felver-Gant, J.N., Dennis, R.L. and Cheng, H.W., 2013. Genetic variations alter production and behavioral responses following heat stress in 2 strains of laying hens. Poult Sci 92: 285–294. 10.3382/ps.2012-02589.

Mahmoud, M.E., Shimizu, Y., Shiina, T., Nikami, H., Dosoky, R.M., Ahmed, M.M. and Takewaki, T., 2007. Involvement of a capsaicin-sensitive TRPV1-independent mechanism in lipopolysaccharide-induced fever in chickens. Comp Biochem Physiol A Mol Integr Physiol 148: 578–583. 10.1016/j.cbpa.2007.07.012.

Marder, J. and Arad, Z., 1989. Panting and acid-base regulation in heat stressed birds. Comp Biochem Physiol A Comp Physiol 94: 395–400. 10.1016/0300-9629(89)90112-6.

Massey, A.C., Zhang, C. and Cuervo, A.M., 2006. Chaperone-mediated autophagy in aging and disease. Curr Top Dev Biol 73: 205–235. 10.1016/S0070-2153(05)73007-6.

Mathew, A., Mathur, S.K. and Morimoto, R.I., 1998. Heat shock response and protein degradation: regulation of HSF2 by the ubiquitin-proteasome pathway. Mol Cell Biol 18: 5091–5098. 10.1128/MCB.18.9.5091.

Mazzoni, M., Zampiga, M., Clavenzani, P., Lattanzio, G., Tagliavia, C. and Sirri, F., 2022. Effect of chronic heat stress on gastrointestinal histology and expression of feed intake-regulatory hormones in broiler chickens. Animal 16: 100600. 10.1016/j.animal.2022.100600.

McBride, H.M., Neuspiel, M. and Wasiak, S., 2006. Mitochondria: more than just a powerhouse. Curr Biol 16: R551–560. 10.1016/j.cub.2006.06.054.

McKechnie, A.E., 2022. Regulation of body temperature: patterns and processes Chapter 45 in Sturkie's Avian Physiology 7th Edition Elsevier Publishing Group, USA, 1462 pp.

McKemy, D.D., Neuhausser, W.M. and Julius, D., 2002. Identification of a cold receptor reveals a general role for TRP channels in thermosensation. Nature 416: 52–58. 10.1038/nature719.

Medicherla, B. and Goldberg, A.L., 2008. Heat shock and oxygen radicals stimulate ubiquitin-dependent degradation mainly of newly synthesized proteins. J Cell Biol 182: 663–673. 10.1083/jcb.200803022.

Mekonnen, M.M. and Hoekstra, A.Y., 2016. Four billion people facing severe water scarcity. Sci Adv 2: e1500323. 10.1126/sciadv.1500323.

Metzler-Zebeli, B.U., Siegerstetter, S.C., Magowan, E., Lawlor, P.G., Petri, R.M., N.E., O.C. and Zebeli, Q., 2019. Feed Restriction Modifies Intestinal Microbiota-Host Mucosal Networking in Chickens Divergent in Residual Feed Intake. mSystems 4. 10.1128/mSystems.00261-18.

Mitoma, H., Hanabuchi, S., Kim, T., Bao, M., Zhang, Z., Sugimoto, N. and Liu, Y.J., 2013. The DHX33 RNA helicase senses cytosolic RNA and activates the NLRP3 inflammasome. Immunity 39: 123–135. 10.1016/j.immuni.2013.07.001.

Mora, C., Frazier, A.G., Longman, R.J., Dacks, R.S., Walton, M.M., Tong, E.J., Sanchez, J.J., Kaiser, L.R., Stender, Y.O., Anderson, J.M., Ambrosino, C.M., Fernandez-Silva, I., Giuseffi, L.M. and Giambelluca, T.W., 2013. The projected timing of climate departure from recent variability. Nature 502: 183–187. 10.1038/nature12540.

Mota-Rojas, D., Wang, D., Titto, C.G., Gomez-Prado, J., Carvajal-de la Fuente, V., Ghezzi, M., Boscato-Funes, L., Barrios-Garcia, H., Torres-Bernal, F., Casas-Alvarado, A. and Martinez-Burnes, J., 2021. Pathophysiology of Fever and Application of Infrared Thermography (IRT) in the Detection of Sick Domestic Animals: Recent Advances. Animals (Basel) 11. 10.3390/ani11082316.

Mujahid, A., Akiba, Y., Warden, C.H. and Toyomizu, M., 2007. Sequential changes in superoxide production, anion carriers and substrate oxidation in skeletal muscle mitochondria of heat-stressed chickens. FEBS Lett 581: 3461–3467. 10.1016/j.febslet.2007.06.051.

Muller, P.A., Schneeberger, M., Matheis, F., Wang, P., Kerner, Z., Ilanges, A., Pellegrino, K., Del Marmol, J., Castro, T.B.R., Furuichi, M., Perkins, M., Han, W., Rao, A., Pickard, A.J., Cross, J.R., Honda, K., de Araujo, I. and Mucida, D., 2020. Microbiota modulate sympathetic neurons via a gut-brain circuit. Nature 583: 441–446. 10.1038/s41586-020-2474-7.

Murakami, T., Ockinger, J., Yu, J., Byles, V., McColl, A., Hofer, A.M. and Horng, T., 2012. Critical role for calcium mobilization in activation of the NLRP3 inflammasome. Proc Natl Acad Sci USA 109: 11282–11287. 10.1073/pnas.1117765109.

Murphy, M.P., 2009. How mitochondria produce reactive oxygen species. Biochem J 417: 1–13. 10.1042/BJ20081386.

Muto, S., Hata, M., Taniguchi, J., Tsuruoka, S., Moriwaki, K., Saitou, M., Furuse, K., Sasaki, H., Fujimura, A., Imai, M., Kusano, E., Tsukita, S. and Furuse, M., 2010. Claudin-2-deficient mice are defective in the leaky and cation-selective paracellular permeability properties of renal proximal tubules. Proc Natl Acad Sci USA 107: 8011–8016. 10.1073/pnas.0912901107.

Myhre, K., Cabanac, M. and Myhre, G., 1975. Thermoregulatory behavior and body temperature in chicks of willow grouse (Lagopus lagopus lagopus). Poult Sci 54: 1174–1179. 10.3382/ps.0541174.

Nakamura, K. and Morrison, S.F., 2008. A thermosensory pathway that controls body temperature. Nat Neurosci 11: 62–71. 10.1038/nn2027.

Nanto-Hara, F., Kikusato, M., Ohwada, S. and Toyomizu, M., 2020. Heat Stress Directly Affects Intestinal Integrity in Broiler Chickens. J Poult Sci 57: 284–290. 10.2141/jpsa.0190004.

Nighot, M., Al-Sadi, R., Guo, S., Rawat, M., Nighot, P., Watterson, M.D. and Ma, T.Y., 2017. Lipopolysaccharide-Induced Increase in Intestinal Epithelial Tight Permeability Is Mediated by Toll-Like Receptor 4/Myeloid Differentiation Primary Response 88 (MyD88) Activation of Myosin Light Chain Kinase Expression. Am J Pathol 187: 2698–2710. 10.1016/j.ajpath.2017.08.005.

O'Neill, B.C., Tebaldi, C., van Vuuren, D.P., Eyring, V., Friedlingstein, P., Hurtt, G., Knutti, R., Kriegler, E., Lamarque, J.-F., Lowe, J., Meehl, G.A., Moss, R., Riahi, K., and Sanderson, B.M., 2016. The Scenario Model Intercomparison Project (ScenarioMIP) for CMIP6. Geosci. Model Dev 9: 3461–3482.

Ogata, M., Hino, S., Saito, A., Morikawa, K., Kondo, S., Kanemoto, S., Murakami, T., Taniguchi, M., Tanii, I., Yoshinaga, K., Shiosaka, S., Hammarback, J.A., Urano, F. and Imaizumi, K., 2006. Autophagy is activated for cell survival after endoplasmic reticulum stress. Mol Cell Biol 26: 9220–9231. 10.1128/MCB.01453-06.

Ostnes, J.E., Bech, C., 1992. Thermosensitivity of different parts of the spinal cord of the pigeon (Columba livia). I Exp Biol 162: 185–196.

Panossian, A. and Wikman, G., 2010. Effects of Adaptogens on the Central Nervous System and the Molecular Mechanisms Associated with Their Stress-Protective Activity. Pharmaceuticals (Basel) 3: 188–224. 10.3390/ph3010188.

Parag, H.A., Raboy, B. and Kulka, R.G., 1987. Effect of heat shock on protein degradation in mammalian cells: involvement of the ubiquitin system. EMBO J 6: 55–61. 10.1002/j.1460-2075.1987.tb04718.x.

Piccinini, A.M. and Midwood, K.S., 2010. DAMPening inflammation by modulating TLR signalling. Mediators Inflamm 2010. 10.1155/2010/672395.

Pinshow, B., Bernstein, M.H., Lopez, G.E. and Kleinhaus, S., 1982. Regulation of brain temperature in pigeons: effects of corneal convection. Am J Physiol 242: R577–581. 10.1152/ajpregu.1982.242.5.R577.

Pirkkala, L., Alastalo, T.P., Zuo, X., Benjamin, I.J. and Sistonen, L., 2000. Disruption of heat shock factor 1 reveals an essential role in the ubiquitin proteolytic pathway. Mol Cell Biol 20: 2670–2675. 10.1128/MCB.20.8.2670-2675.2000.

Pis, T., 2002. The body temperature and energy metabolism in growing chicks of capercaillie (Tetrao urogallus). Journal of Thermal Biology 27: 191–198.

Pispa, J., Matilainen, O. and Holmberg, C.I., 2020. Tissue-specific effects of temperature on proteasome function. Cell Stress Chaperones 25: 563–572. 10.1007/s12192-020-01107-y.

Plaza-Diaz, J., Ruiz-Ojeda, F.J., Gil-Campos, M. and Gil, A., 2019. Mechanisms of Action of Probiotics. Adv Nutr 10: S49–S66. 10.1093/advances/nmy063.

Preisinger, R., 2018. Innovative layer genetics to handle global challenges in egg production. Br Poult Sci 59: 1–6. 10.1080/00071668.2018.1401828.

Prinzinger, R., Pressmar, A., Schleucher, E., 1991. Body temperature in birds. Comparative Biochemistry and Physiology Part A 99: 499–506.

Qian, L., Song, X., Ren, H., Gong, J. and Cheng, S., 2004. Mitochondrial mechanism of heat stress-induced injury in rat cardiomyocyte. Cell Stress Chaperones 9: 281–293. 10.1379/csc-20r.1.

Qu, Q., Li, H., Bai, L., Zhang, S., Sun, J., Lv, W., Ye, C., Liu, C. and Shi, D., 2021. Effects of Heat Stress on Gut Microbiome in Rats. Indian J Microbiol 61: 338–347. 10.1007/s12088-021-00948-0.

Rajaei-Sharifabadi, H., Ellestad, L., Porter, T., Donoghue, A., Bottje, W.G. and Dridi, S., 2017a. Noni (Morinda citrifolia) Modulates the Hypothalamic Expression of Stress- and Metabolic-Related Genes in Broilers Exposed to Acute Heat Stress. Front Genet 8: 192. 10.3389/fgene.2017.00192.

Rajaei-Sharifabadi, H., Greene, E., Piekarski, A., Falcon, D., Ellestad, L., Donoghue, A., Bottje, W., Porter, T., Liang, Y. and Dridi, S., 2017b. Surface wetting strategy prevents acute heat exposure-induced alterations of hypothalamic stress- and metabolic-related genes in broiler chickens. J Anim Sci 95: 1132–1143. 10.2527/jas.2016.1290.

Repnik, U. and Turk, B., 2010. Lysosomal-mitochondrial cross-talk during cell death. Mitochondrion 10: 662–669. 10.1016/j.mito.2010.07.008.

Ron, D. and Walter, P., 2007. Signal integration in the endoplasmic reticulum unfolded protein response. Nat Rev Mol Cell Biol 8: 519–529. 10.1038/nrm2199.

Rossi, A., Trotta, E., Brandi, R., Arisi, I., Coccia, M. and Santoro, M.G., 2010. AIRAP, a new human heat shock gene regulated by heat shock factor 1. J Biol Chem 285: 13607–13615. 10.1074/jbc.M109.082693.

Rubin, C.J., Zody, M.C., Eriksson, J., Meadows, J.R., Sherwood, E., Webster, M.T., Jiang, L., Ingman, M., Sharpe, T., Ka, S., Hallbook, F., Besnier, F., Carlborg, O., Bed'hom, B., Tixier-Boichard, M., Jensen, P., Siegel, P., Lindblad-Toh, K. and Andersson, L., 2010. Whole-genome resequencing reveals loci under selection during chicken domestication. Nature 464: 587–591. 10.1038/nature08832.

Ruff, J., Barros, T.L., Tellez, G., Jr., Blankenship, J., Lester, H., Graham, B.D., Selby, C.A.M., Vuong, C.N., Dridi, S., Greene, E.S., Hernandez-Velasco, X., Hargis, B.M. and Tellez-Isaias, G., 2020. Research Note: Evaluation of a heat stress model to induce gastrointestinal leakage in broiler chickens. Poult Sci 99: 1687–1692. 10.1016/j.psj.2019.10.075.

Ruuskanen, S., Hsu, B.Y. and Nord, A., 2021. Endocrinology of thermoregulation in birds in a changing climate. Mol Cell Endocrinol 519: 111088. 10.1016/j.mce.2020.111088.

Saito, S., Banzawa, N., Fukuta, N., Saito, C.T., Takahashi, K., Imagawa, T., Ohta, T. and Tominaga, M., 2014. Heat and noxious chemical sensor, chicken TRPA1, as a target of bird repellents and identification of its structural determinants by multispecies functional comparison. Mol Biol Evol 31: 708–722. 10.1093/molbev/msu001.

Scharl, M., Paul, G., Barrett, K.E. and McCole, D.F., 2009. AMP-activated protein kinase mediates the interferon-gamma-induced decrease in intestinal epithelial barrier function. J Biol Chem 284: 27952–27963. 10.1074/jbc.M109.046292.

Scherz-Shouval, R., Shvets, E., Fass, E., Shorer, H., Gil, L. and Elazar, Z., 2007. Reactive oxygen species are essential for autophagy and specifically regulate the activity of Atg4. EMBO J 26: 1749–1760. 10.1038/sj.emboj.7601623.

Schindler, A.J. and Schekman, R., 2009. In vitro reconstitution of ER-stress induced ATF6 transport in COPII vesicles. Proc Natl Acad Sci USA 106: 17775–17780. 10.1073/pnas.0910342106.

Schneider, C., O'Leary, C.E. and Locksley, R.M., 2019. Regulation of immune responses by tuft cells. Nat Rev Immunol 19: 584–593. 10.1038/s41577-019-0176-x.

Segawa, S., Fujiya, M., Konishi, H., Ueno, N., Kobayashi, N., Shigyo, T. and Kohgo, Y., 2011. Probiotic-derived polyphosphate enhances the epithelial barrier function and maintains intestinal homeostasis through integrin-p38 MAPK pathway. PLoS One 6: e23278. 10.1371/journal.pone.0023278.

Semenza, G.L., 2002. Physiology meets biophysics: visualizing the interaction of hypoxia-inducible factor 1 alpha with p300 and CBP. Proc Natl Acad Sci USA 99: 11570–11572. 10.1073/pnas.192442299.

Seufert, W. and Jentsch, S., 1990. Ubiquitin-conjugating enzymes UBC4 and UBC5 mediate selective degradation of short-lived and abnormal proteins. EMBO J 9: 543–550. 10.1002/j.1460-2075.1990.tb08141.x.

Shahbandeh, M., 2022. Global chicken meat production 2012–2022, Statista, New York, USA.

Shahsavarani, A.M., Abadi, E.A., Halkhoran, M.H., 2015. Stress: Facts and Theories through Literature Review International Journal of Medical Reviews 2: 230–241.

Shakeri, M., Le, H.H., 2022. Deleterious effects of heat stress on poultry production: unveiling the benefits of betaine and polyphenols. Poultry 1: 147–156.

Sharma, D., Malik, A., Guy, C.S., Karki, R., Vogel, P. and Kanneganti, T.D., 2018. Pyrin Inflammasome Regulates Tight Junction Integrity to Restrict Colitis and Tumorigenesis. Gastroenterology 154: 948–964 e948. 10.1053/j.gastro.2017.11.276.

Sharma, S., Chaudhary, P., Sandhir, R., Bharadwaj, A., Gupta, R.K., Khatri, R., Bajaj, A.C., Baburaj, T.P., Kumar, S., Pal, M.S., Reddy, P.K. and Kumar, B., 2021. Heat-induced endoplasmic reticulum stress in soleus and gastrocnemius muscles and differential response to UPR pathway in rats. Cell Stress Chaperones 26: 323–339. 10.1007/s12192-020-01178-x.

Shellman, Y.G., Howe, W.R., Miller, L.A., Goldstein, N.B., Pacheco, T.R., Mahajan, R.L., LaRue, S.M. and Norris, D.A., 2008. Hyperthermia induces endoplasmic reticulum-mediated apoptosis in melanoma and non-melanoma skin cancer cells. J Invest Dermatol 128: 949–956. 10.1038/sj.jid.5701114.

Shi, D., Bai, L., Qu, Q., Zhou, S., Yang, M., Guo, S., Li, Q. and Liu, C., 2019. Impact of gut microbiota structure in heat-stressed broilers. Poult Sci 98: 2405–2413. 10.3382/ps/pez026.

Shi, Y., Tu, Z., Tang, D., Zhang, H., Liu, M., Wang, K., Calderwood, S.K. and Xiao, X., 2006. The inhibition of LPS-induced production of inflammatory cytokines by HSP70 involves inactivation of the NF-kappaB pathway but not the MAPK pathways. Shock 26: 277–284. 10.1097/01.shk.0000223134.17877.ad.

Shing, C.M., Peake, J.M., Lim, C.L., Briskey, D., Walsh, N.P., Fortes, M.B., Ahuja, K.D. and Vitetta, L., 2014. Effects of probiotics supplementation on gastrointestinal permeability, inflammation and exercise performance in the heat. Eur J Appl Physiol 114: 93–103. 10.1007/s00421-013-2748-y.

Shpilka, T. and Haynes, C.M., 2018. The mitochondrial UPR: mechanisms, physiological functions and implications in ageing. Nat Rev Mol Cell Biol 19: 109–120. 10.1038/nrm.2017.110.

Silverman, M.N., Heim, C.M., Nater, U.M., Marques, A.H. and Sternberg, E.M., 2010. Neuroendocrine and immune contributors to fatigue. PM R 2: 338–346. 10.1016/j.pmrj.2010.04.008.

Simon, E., 2000. The enigma of deep-body thermosensory specificity. Int J Biometeorol 44: 105–120. 10.1007/s004840000060.

Simon, E., Martin, R. and Simon-Oppermann, C., 1981. Central nervous versus total body thermosensitivity of the duck. Int J Biometeorol 25: 249–256. 10.1007/BF02184526.

Sokolovska, A., Becker, C.E., Ip, W.K., Rathinam, V.A., Brudner, M., Paquette, N., Tanne, A., Vanaja, S.K., Moore, K.J., Fitzgerald, K.A., Lacy-Hulbert, A. and Stuart, L.M., 2013. Activation of caspase-1 by the NLRP3 inflammasome regulates the NADPH oxidase NOX2 to control phagosome function. Nat Immunol 14: 543–553. 10.1038/ni.2595.

Sommer, F. and Backhed, F., 2013. The gut microbiota – masters of host development and physiology. Nat Rev Microbiol 11: 227–238. 10.1038/nrmicro2974.

Song, D.J., and King, A.J., 2015. Effects of heat stress on broiler meat quality. World's. Poult. Sci. J 71: 701–709.

Song, J., Xiao, K., Ke, Y.L., Jiao, L.F., Hu, C.H., Diao, Q.Y., Shi, B. and Zou, X.T., 2014. Effect of a probiotic mixture on intestinal microflora, morphology, and barrier integrity of broilers subjected to heat stress. Poult Sci 93: 581–588. 10.3382/ps.2013-03455.

Specian, R.D. and Oliver, M.G., 1991. Functional biology of intestinal goblet cells. Am J Physiol 260: C183–193. 10.1152/ajpcell.1991.260.2.C183.

Story, G.M., Peier, A.M., Reeve, A.J., Eid, S.R., Mosbacher, J., Hricik, T.R., Earley, T.J., Hergarden, A.C., Andersson, D.A., Hwang, S.W., McIntyre, P., Jegla, T., Bevan, S. and Patapoutian, A., 2003. ANKTM1, a TRP-like channel expressed in nociceptive neurons, is activated by cold temperatures. Cell 112: 819–829. 10.1016/s0092-8674(03)00158-2.

Summers, C.M. and Valentine, R.J., 2019. Acute Heat Exposure Alters Autophagy Signaling in C2C12 Myotubes. Front Physiol 10: 1521. 10.3389/fphys.2019.01521.

Sun, X., Zhang, H., Sheikhahmadi, A., Wang, Y., Jiao, H., Lin, H. and Song, Z., 2015. Effects of heat stress on the gene expression of nutrient transporters in the jejunum of broiler chickens (Gallus gallus domesticus). Int J Biometeorol 59: 127–135. 10.1007/s00484-014-0829-1.

Suzuki, T., 2013. Regulation of intestinal epithelial permeability by tight junctions. Cell Mol Life Sci 70: 631–659. 10.1007/s00018-012-1070-x.

Tabler, T.W., Greene, E.S., Orlowski, S.K., Hiltz, J.Z., Anthony, N.B. and Dridi, S., 2020. Intestinal Barrier Integrity in Heat-Stressed Modern Broilers and Their Ancestor Wild Jungle Fowl. Front Vet Sci 7: 249. 10.3389/fvets.2020.00249.

Tamura, A., Hayashi, H., Imasato, M., Yamazaki, Y., Hagiwara, A., Wada, M., Noda, T., Watanabe, M., Suzuki, Y. and Tsukita, S., 2011. Loss of claudin-15, but not claudin-2, causes Na+ deficiency and glucose malabsorption in mouse small intestine. Gastroenterology 140: 913–923. 10.1053/j.gastro.2010.08.006.

Tamura, A., Kitano, Y., Hata, M., Katsuno, T., Moriwaki, K., Sasaki, H., Hayashi, H., Suzuki, Y., Noda, T., Furuse, M., Tsukita, S. and Tsukita, S., 2008. Megaintestine in claudin-15-deficient mice. Gastroenterology 134: 523–534. 10.1053/j.gastro.2007.11.040.

Tanaka, H., Takechi, M., Kiyonari, H., Shioi, G., Tamura, A. and Tsukita, S., 2015. Intestinal deletion of Claudin-7 enhances paracellular organic solute flux and initiates colonic inflammation in mice. Gut 64: 1529–1538. 10.1136/gutjnl-2014-308419.

Tang, L.P., Li, W.H., Liu, Y.L., Lun, J.C. and He, Y.M., 2021. Heat stress aggravates intestinal inflammation through TLR4-NF-kappaB signaling pathway in Ma chickens infected with Escherichia coli O157:H7. Poult Sci 100: 101030. 10.1016/j.psj.2021.101030.

Tang, S., Xie, J., Fang, W., Wen, X., Yin, C., Meng, Q., Zhong, R., Chen, L. and Zhang, H., 2022. Chronic heat stress induces the disorder of gut transport and immune function associated with endoplasmic reticulum stress in growing pigs. Anim Nutr 11: 228–241. 10.1016/j.aninu.2022.08.008.

Tarallo, V., Hirano, Y., Gelfand, B.D., Dridi, S., Kerur, N., Kim, Y., Cho, W.G., Kaneko, H., Fowler, B.J., Bogdanovich, S., Albuquerque, R.J., Hauswirth, W.W., Chiodo, V.A., Kugel, J.F., Goodrich, J.A., Ponicsan, S.L., Chaudhuri, G., Murphy, M.P., Dunaief, J.L., Ambati, B.K., Ogura, Y., Yoo, J.W., Lee, D.K., Provost, P., Hinton, D.R., Nunez, G., Baffi, J.Z., Kleinman, M.E. and Ambati, J., 2012. DICER1 loss and Alu RNA induce age-related macular degeneration via the NLRP3 inflammasome and MyD88. Cell 149: 847–859. 10.1016/j.cell.2012.03.036.

Taylor, C.T., Dzus, A.L. and Colgan, S.P., 1998. Autocrine regulation of epithelial permeability by hypoxia: role for polarized release of tumor necrosis factor alpha. Gastroenterology 114: 657–668. 10.1016/s0016-5085(98)70579-7.

Tierney, J.E., Poulsen, C.J., Montanez, I.P., Bhattacharya, T., Feng, R., Ford, H.L., Honisch, B., Inglis, G.N., Petersen, S.V., Sagoo, N., Tabor, C.R., Thirumalai, K., Zhu, J., Burls, N.J., Foster, G.L., Godderis, Y., Huber, B.T., Ivany, L.C., Kirtland Turner, S., Lunt, D.J., McElwain, J.C., Mills, B.J.W., Otto-Bliesner, B.L., Ridgwell, A. and Zhang, Y.G., 2020. Past climates inform our future. Science 370. 10.1126/science.aay3701.

Tollefson, J., 2020. How hot will Earth get by 2100? Nature 580: 443–445. 10.1038/d41586-020-01125-x.

Tsukita, S., Furuse, M. and Itoh, M., 2001. Multifunctional strands in tight junctions. Nat Rev Mol Cell Biol 2: 285–293. 10.1038/35067088.

Tyciakova, S., Valova, V., Svitkova, B. and Matuskova, M., 2021. Overexpression of TNFalpha induces senescence, autophagy and mitochondrial dysfunctions in melanoma cells. BMC Cancer 21: 507. 10.1186/s12885-021-08237-1.

Tyedmers, J., Mogk, A. and Bukau, B., 2010. Cellular strategies for controlling protein aggregation. Nat Rev Mol Cell Biol 11: 777–788. 10.1038/nrm2993.

Tzschentke, B., Nichelmann, M., 1999. Development of avian thermoregulatory system during the early postnatal period: development of the thermoregulatory set-point. Ornis Fennica 76: 189–198.

Uerlings, J., Song, Z.G., Hu, X.Y., Wang, S.K., Lin, H., Buyse, J. and Everaert, N., 2018. Heat exposure affects jejunal tight junction remodeling independently of adenosine monophosphate-activated protein kinase in 9-day-old broiler chicks. Poult Sci 97: 3681–3690. 10.3382/ps/pey229.

Vancamelbeke, M. and Vermeire, S., 2017. The intestinal barrier: a fundamental role in health and disease. Expert Rev Gastroenterol Hepatol 11: 821–834. 10.1080/17474124.2017.1343143.

Varasteh, S., Braber, S., Akbari, P., Garssen, J. and Fink-Gremmels, J., 2015. Differences in Susceptibility to Heat Stress along the Chicken Intestine and the Protective Effects of Galacto-Oligosaccharides. PLoS One 10: e0138975. 10.1371/journal.pone.0138975.

Vriens, J., Owsianik, G., Hofmann, T., Philipp, S.E., Stab, J., Chen, X., Benoit, M., Xue, F., Janssens, A., Kerselaers, S., Oberwinkler, J., Vennekens, R., Gudermann, T., Nilius, B. and Voets, T., 2011. TRPM3 is a nociceptor channel involved in the detection of noxious heat. Neuron 70: 482–494. 10.1016/j.neuron.2011.02.051.

Wang, G., Shen, T., Li, P., Luo, Z., Tan, Y., He, G., Zhang, X., Yang, J., Liu, J., Wang, Y., Tang, H., Luo, X. and Yang, X., 2019a. The Increase in IL-1beta in the Early Stage of Heatstroke Might Be Caused by Splenic Lymphocyte Pyroptosis Induced by mtROS-Mediated Activation of the NLRP3 Inflammasome. Front Immunol 10: 2862. 10.3389/fimmu.2019.02862.

Wang, J., Xue, X., Liu, Q., Zhang, S., Peng, M., Zhou, J., Chen, L. and Fang, F., 2019b. Effects of duration of thermal stress on growth performance, serum oxidative stress indices, the expression and localization of ABCG2 and mitochondria ROS production of skeletal muscle, small intestine and immune organs in broilers. J therm Biol 85: 102420. 10.1016/j.jtherbio.2019.102420.

Wang, L., Klionsky, D.J. and Shen, H.M., 2022. The emerging mechanisms and functions of microautophagy. Nat Rev Mol Cell Biol. 10.1038/s41580-022-00529-z.

Wang, P., Zhu, S., Yang, L., Cui, S., Pan, W., Jackson, R., Zheng, Y., Rongvaux, A., Sun, Q., Yang, G., Gao, S., Lin, R., You, F., Flavell, R. and Fikrig, E., 2015. Nlrp6 regulates intestinal antiviral innate immunity. Science 350: 826–830. 10.1126/science.aab3145.

Wang, R.C., Wei, Y., An, Z., Zou, Z., Xiao, G., Bhagat, G., White, M., Reichelt, J. and Levine, B., 2012. Akt-mediated regulation of autophagy and tumorigenesis through Beclin 1 phosphorylation. Science 338: 956–959. 10.1126/science.1225967.

Wang, R.H., Liang, R.R., Lin, H., Zhu, L.X., Zhang, Y.M., Mao, Y.W., Dong, P.C., Niu, L.B., Zhang, M.H. and Luo, X., 2017. Effect of acute heat stress and slaughter processing on poultry meat quality and postmortem carbohydrate metabolism. Poult Sci 96: 738–746. 10.3382/ps/pew329.

Watanabe, Y., Tsujimura, A., Taguchi, K. and Tanaka, M., 2017. HSF1 stress response pathway regulates autophagy receptor SQSTM1/p62-associated proteostasis. Autophagy 13: 133–148. 10.1080/15548627.2016.1248018.

Wedel, M., 2007. Evidence for bird-like air sacs in saurischian dinosaurs. Journal of Experimental Zoology 268: 1147–1147.

Wen, C., Li, S., Wang, J., Zhu, Y., Zong, X., Wang, Y. and Jin, M., 2021. Heat Stress Alters the Intestinal Microbiota and Metabolomic Profiles in Mice. Front Microbiol 12: 706772. 10.3389/fmicb.2021.706772.

Wlodarska, M., Thaiss, C.A., Nowarski, R., Henao-Mejia, J., Zhang, J.P., Brown, E.M., Frankel, G., Levy, M., Katz, M.N., Philbrick, W.M., Elinav, E., Finlay, B.B. and Flavell, R.A., 2014. NLRP6 inflammasome orchestrates the colonic host-microbial interface by regulating goblet cell mucus secretion. Cell 156: 1045–1059. 10.1016/j.cell.2014.01.026.

Wu, D., Wu, K., Zhu, Q., Xiao, W., Shan, Q., Yan, Z., Wu, J., Deng, B., Xue, Y., Gong, W., Lu, G. and Ding, Y., 2018. Formononetin Administration Ameliorates Dextran Sulfate Sodium-Induced Acute Colitis by Inhibiting NLRP3 Inflammasome Signaling Pathway. Mediators Inflamm 2018: 3048532. 10.1155/2018/3048532.

Wu, Z.H., Wong, E.T., Shi, Y., Niu, J., Chen, Z., Miyamoto, S. and Tergaonkar, V., 2010. ATM- and NEMO-dependent ELKS ubiquitination coordinates TAK1-mediated IKK activation in response to genotoxic stress. Mol Cell 40: 75–86. 10.1016/j.molcel.2010.09.010.

Xing, S., Wang, X., Diao, H., Zhang, M., Zhou, Y. and Feng, J., 2019. Changes in the cecal microbiota of laying hens during heat stress is mainly associated with reduced feed intake. Poult Sci 98: 5257–5264. 10.3382/ps/pez440.

Xu, B., Bai, B., Sha, S., Yu, P., An, Y., Wang, S., Kong, X., Liu, C., Wei, N., Feng, Q. and Zhao, Q., 2014a. Interleukin-1beta induces autophagy by affecting calcium homeostasis and trypsinogen activation in pancreatic acinar cells. Int J Clin Exp Pathol 7: 3620–3631.

Xu, H., Ramsey, I.S., Kotecha, S.A., Moran, M.M., Chong, J.A., Lawson, D., Ge, P., Lilly, J., Silos-Santiago, I., Xie, Y., DiStefano, P.S., Curtis, R. and Clapham, D.E., 2002. TRPV3 is a calcium-permeable temperature-sensitive cation channel. Nature 418: 181–186. 10.1038/nature00882.

Xu, H., Yang, J., Gao, W., Li, L., Li, P., Zhang, L., Gong, Y.N., Peng, X., Xi, J.J., Chen, S., Wang, F. and Shao, F., 2014b. Innate immune sensing of bacterial modifications of Rho GTPases by the Pyrin inflammasome. Nature 513: 237–241. 10.1038/nature13449.

Ye, J., Rawson, R.B., Komuro, R., Chen, X., Dave, U.P., Prywes, R., Brown, M.S. and Goldstein, J.L., 2000. ER stress induces cleavage of membrane-bound ATF6 by the same proteases that process SREBPs. Mol Cell 6: 1355–1364. 10.1016/s1097-2765(00)00133-7.

Yi, G., Li, L., Luo, M., He, X., Zou, Z., Gu, Z. and Su, L., 2017. Heat stress induces intestinal injury through lysosome- and mitochondria-dependent pathway in vivo and in vitro. Oncotarget 8: 40741–40755. 10.18632/oncotarget.16580.

Yoshida, H., Matsui, T., Yamamoto, A., Okada, T. and Mori, K., 2001. XBP1 mRNA is induced by ATF6 and spliced by IRE1 in response to ER stress to produce a highly active transcription factor. Cell 107: 881–891. 10.1016/s0092-8674(01)00611-0.

Yoshihara, T., Naito, H., Kakigi, R., Ichinoseki-Sekine, N., Ogura, Y., Sugiura, T. and Katamoto, S., 2013. Heat stress activates the Akt/mTOR signalling pathway in rat skeletal muscle. Acta Physiol (Oxf) 207: 416–426. 10.1111/apha.12040.

Yu, J., Liu, F., Yin, P., Zhao, H., Luan, W., Hou, X., Zhong, Y., Jia, D., Zan, J., Ma, W., Shu, B. and Xu, J., 2013. Involvement of oxidative stress and mitogen-activated protein kinase signaling pathways in heat stress-induced injury in the rat small intestine. Stress 16: 99–113. 10.3109/10253890.2012.680526.

Yun, J.K., McCormick, T.S., Villabona, C., Judware, R.R., Espinosa, M.B. and Lapetina, E.G., 1997. Inflammatory mediators are perpetuated in macrophages resistant to apoptosis induced by hypoxia. Proc Natl Acad Sci USA 94: 13903–13908. 10.1073/pnas.94.25.13903.

Zaboli, G., Huang, X., Feng, X. and Ahn, D.U., 2019. How can heat stress affect chicken meat quality? – a review. Poult Sci 98: 1551–1556. 10.3382/ps/pey399.

Zaki, M.H., Boyd, K.L., Vogel, P., Kastan, M.B., Lamkanfi, M. and Kanneganti, T.D., 2010. The NLRP3 inflammasome protects against loss of epithelial integrity and mortality during experimental colitis. Immunity 32: 379–391. 10.1016/j.immuni.2010.03.003.

Zaki, M.H., Lamkanfi, M. and Kanneganti, T.D., 2011. The Nlrp3 inflammasome: contributions to intestinal homeostasis. Trends Immunol 32: 171–179. 10.1016/j.it.2011.02.002.

Zhang, C., Zhao, X.H., Yang, L., Chen, X.Y., Jiang, R.S., Jin, S.H. and Geng, Z.Y., 2017. Resveratrol alleviates heat stress-induced impairment of intestinal morphology, microflora, and barrier integrity in broilers. Poult Sci 96: 4325–4332. 10.3382/ps/pex266.

Zhang, T., Perkins, M.H., Chang, H., Han, W. and de Araujo, I.E., 2022. An inter-organ neural circuit for appetite suppression. Cell 185: 2478–2494 e2428. 10.1016/j.cell.2022.05.007.

Zhao, R.Z., Jiang, S., Zhang, L. and Yu, Z.B., 2019. Mitochondrial electron transport chain, ROS generation and uncoupling (Review). Int J Mol Med 44: 3–15. 10.3892/ijmm.2019.4188.

Zheng, W., Xie, W., Yin, D., Luo, R., Liu, M. and Guo, F., 2019. ATG5 and ATG7 induced autophagy interplays with UPR via PERK signaling. Cell Commun Signal 17: 42. 10.1186/s12964-019-0353-3.

Zhu, L., Liao, R., Wu, N., Zhu, G. and Yang, C., 2019. Heat stress mediates changes in fecal microbiome and functional pathways of laying hens. Appl Microbiol Biotechnol 103: 461–472. 10.1007/s00253-018-9465-8.

Zihni, C., Mills, C., Matter, K. and Balda, M.S., 2016. Tight junctions: from simple barriers to multifunctional molecular gates. Nat Rev Mol Cell Biol 17: 564–580. 10.1038/nrm.2016.80.

Zinszner, H., Kuroda, M., Wang, X., Batchvarova, N., Lightfoot, R.T., Remotti, H., Stevens, J.L. and Ron, D., 1998. CHOP is implicated in programmed cell death in response to impaired function of the endoplasmic reticulum. Genes Dev 12: 982–995. 10.1101/gad.12.7.982.

Zyrek, A.A., Cichon, C., Helms, S., Enders, C., Sonnenborn, U. and Schmidt, M.A., 2007. Molecular mechanisms underlying the probiotic effects of Escherichia coli Nissle 1917 involve ZO-2 and PKCzeta redistribution resulting in tight junction and epithelial barrier repair. Cell Microbiol 9: 804–816. 10.1111/j.1462-5822.2006.00836.x.

CHAPTER 12

Environmental effects on gut health in production animals

The impact of thermal stress on intestinal health and productivity of dairy cows

D.E. Rico[1,*], *A. Razzaghi*[2] *and J.E. Rico*[3,*]

[1]Centre De Recherche En Sciences Animales de Deschambault (CRSAD), Deschambault, QC, Canada, G0A 1S0; [2]Department of Animal Science, Agricultural and Natural Resources Research and Education Center of Razavi Khorasan, Mashhad 91735-488, Iran; [3]Department of Clinical Studies, New Bolton Center, School of Veterinary Medicine, University of Pennsylvania, Kennett Square 19348, USA; *Correspondence: daniel.rico@crsad.qc.ca; ricoe@vet.upenn.edu

Abstract

A series of physiological changes take place in order to allow animals to adapt to temperatures outside their comfort zone. Allowing for optimal temperature conditions maximizes the availability of nutrients for productive processes, whereas moving out of this zone results in reduced performance. Although cold and heat stress result in strikingly different phenotypes, some commonalities are present, such as the activation of the heat shock protein response to protect cells, and also the possibility of death in severe cases. Animal performance losses under cold stress are observed mainly due to use of energy stores to produce heat, which can be observed concomitantly with increased feed intake, depending on the degree of stress. On the other hand, heat stress results in reduced feed intake and further performance losses in response to other factors, such as nutrient partitioning to support immune responses observed under heat stress-induced leaky gut. It is thought that nutrient partitioning changes are mediated by immune activation, as it can directly influence insulin secretion, increasing glucose uptake by immune cells and simultaneously limiting the mobilization of adipose tissue-origin lipids. Intestinal hypoxia under heat stress is thought to be a major factor in leaky gut development, although changes in microbial composition observed under these conditions may also pay a role. Understanding the role of thermal stress on intestinal health may thus be a key factor in the future development of management practices aiming to ameliorate intestinal health and animal performance.

Keywords

thermal stress – leaky gut – adaptation – gut permeability

© D.E. RICO ET AL., 2024 | DOI:10.3920/9789004695467_013

1 Introduction

External stimuli that challenge an animal's ability to sustain homeostasis can be thought of as system stressors. Thermal stress is an environmental challenge that induces a number of counter responses from the animal, and typically include changes in body temperature, metabolic rate, and an associated activation of mechanisms to dissipate or preserve heat. Such stress occurs when the environmental temperature is above or below the thresholds for upper or lower critical temperatures, and can be ultimately manifested as reduced productivity, and compromise health and comfort/welfare. Although much of the recent focus around thermal stress in dairy cows is placed on the effects of ever higher temperatures (*i.e.*, heat stress), it is worth noting that animals can be similarly challenged by low temperatures (*i.e.*, cold stress). Indeed, in many latitudes in the northern and southern hemispheres, cold stress continues to be an important consideration in the winter months. In principle, the closer that thermal conditions are to physiological optimums, the lower the "effort" undergone by the system for counter-regulation, and the lower the negative impacts of thermal stress on the animal. In this context, aiming to guarantee environmental conditions that do not stray far from the physiological optimum is paramount to maximize productivity and welfare. The present chapter reviews the available literature on the impact of both cold and heat stresses on intestinal health and productivity of dairy cows.

1.1 *Physiological limits of thermal tolerance in dairy cattle*

The thermoneutral zone (TNZ) is a narrow thermal range where the animal does not have to expend energy to maintain normal body temperature (Kadzere *et al.*, 2002). Dairy cattle show maximum genetic potential with minimal physiological costs and maximum productivity only within their TNZ. (Folk, 1974; Tao *et al.*, 2018). More generally, the TNZ would be situated under environmental conditions that do not trigger temperature-regulating adaptations, and consequently, that do not alter nutrient partitioning away from production-related processes. Cattle are able to regulate body temperature within an upper critical temperature and a lower critical temperature, commonly known as the homeothermic zone. Brody (1956) proposed the "comfort zone" of the TNZ was found between −1 and 15 °C for European cattle. For lactating dairy cows, the TNZ may be in the range of 0 to 25 °C (Bianca, 1976; Kadzere *et al.*, 2002), although others have suggested that the range of neutral temperatures for dairy cattle may be even broader, ranging from −5 °C to 25 °C (Knizkova *et al.*, 2002). For younger animals, the TNZ is known to be higher and narrower (~13–25 °C for calves under 3 weeks of age; Bianca, 1971), which is associated both with a lower capacity to generate body heat heat and a larger exposed surface area, relative to their body size.

Thermal stress will occur when environmental temperature reaches values above or below the critical thresholds of the TNZ, and can be predicted on the basis of interacting environmental factors such as temperature, radiation, humidity, and wind, as well as intrinsic animal factors, such as age, breed, production level, reproductive status, skin pigmentation, and coat cover (*i.e.*, degree of insulation; Yousef, 1985; Collier *et al.*, 2017). For instance, if cattle are wet and unsheltered from wind, their lower critical temperature can rise (*i.e.*, increased cold stress sensitivity; Gasparrini *et al.*, 2015), whereas having a dry, heavy winter coat, can reduce the lower critical temperature to be as low as −8 °C (Berman and Meltzer, 1973).

As for environmental predictors of thermal stress, fluctuations of air temperature within a range of −0.5 to 20 °C and relative humidity between 60 to 80% are considered as thermo neutral for most breeds. A practical example of these relationships is the temperature-humidity index (THI), which was proposed as an indirect index for assessing the severity of thermal stress (particularly heat stress) in cows. When exposed to heat above the TNZ, dairy cows will experience increased body temperature, and will commonly respond by increasing heat dissipation mechanisms, while reducing appetite (Collier *et al.*, 2017). When the upper critical temperature is attained, the ability to effectively regulate the internal body temperature is compromised (*i.e.*, hyperthermia), resulting in uncontrolled heat stress and subsequent alterations in health, productivity, and behavior (Figure 12.1; Berman *et al.*, 1985; West, 2003; Tao *et al.*, 2018). As reported by Vitali *et al.* (2009), dairy cow mortality will sharply increase at a "breakpoint" of 79.6 for maximum THI; however, negative impacts on productivity can already be observed at THI of 68.

On the other side of this thermal spectrum, when weather conditions put the animal below the lower critical temperature, cold stress ensues, and the ability of

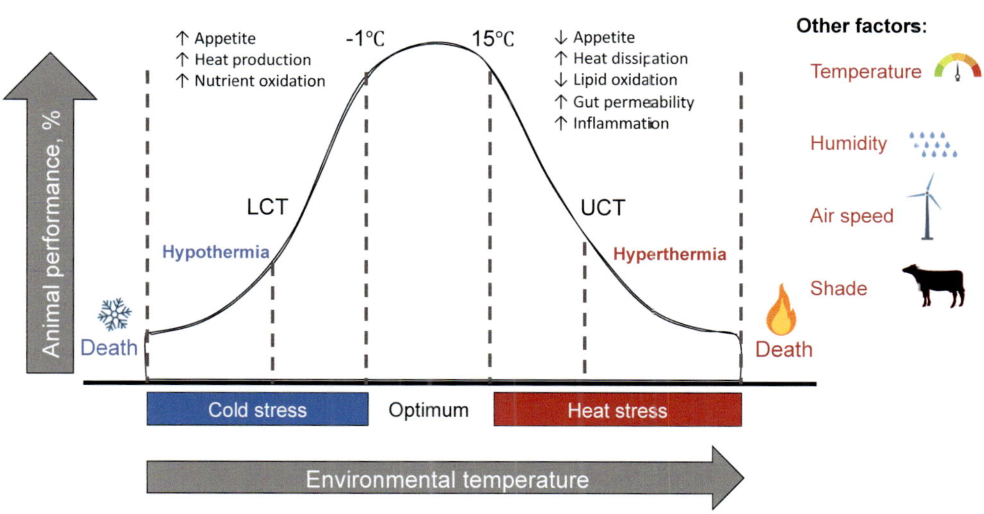

FIGURE 12.1 Main features of the adaptation to cold and heat stress in dairy cows. The impact of ambient temperature on the animal is modulated by other factors, such as humidity, air speed and access to shade. LCT: lower critical temperature; UCT: upper critical temperature

the cow to maintain normal body temperature is compromised (*i.e.*, hypothermia; Kleiber, 1961; Hamada, 1971). Similar to the concept of THI during heat stress, windchill is a relevant parameter for assessing cold stress in cattle. Improving on the windchill factor used in humans, Ames and Insley (1975), proposed a windchill factor for cattle, which considered the cubic association between heat loss and wind speed in hairy animals, such as cows, where wind speeds higher than 40 km/h significantly perturb hair insolation. Furthermore, cold sensation from air movement can be exacerbated when the coat is wet, as indicated by increased heat production of animals exposed to wind and rain. Overall, cold exposure results in reduced body temperature, and leads to enhancements in heat production mechanisms and reductions in heat dissipation mechanisms. These responses are associated with hormonal and behavioral changes, such as increases in thyroxine secretion and in feed intake (*i.e.*, hyperphagia). Early work from Kibler and Brody (1949) showed that body heat production increased up to 35% in dairy cows experiencing temperatures below the lower critical temperature (−15 °C). Interestingly, the increase in heat production varied between cows of the same breed, as well as between Jersey and Holstein cows, which highlights the influence of cow-intrinsic factors, such as body size and milk production level, the latter being associated to metabolic heat derived from milk production (Kibler and Brody, 1949). In addition, there is variation in the capacity of animals to generate body heat, and younger animals are known to exhibit lower cold tolerance (LCT; Bianca, 1971), which is associated to the larger surface area relative to their body size. Illustrating this point, the LCT for non-lactating cows and growing calves, is typically higher (−10 °C and −25 °C) relative to that of lactating cows. Furthermore, young calves from birth to 3 weeks of age have a LCT of +13 °C, whereas calves from 3 weeks old to weaning have a LCT of +1 °C.

1.2 *Physiological adaptations to thermal stress*

Herein, we provide a high-level view of the main changes taking place as animals cope with thermal stress. For an in-depth review of the biology of adaption to the thermal stress in cattle, readers are referred to the excellent work by Collier and Gebremedhin (2015).

In homeothermic animals, such as cows, body temperature is maintained relatively constant during rapid environmental changes, which is necessary for cellular and molecular homeostasis (Horowitz, 2002; Nakamura and Morrison, 2008). Indeed, maintenance of euthermia requires significant alterations in the short and long term, that is, a combination of homeostatic and homeorhetic responses.

Perception and regulation of core temperature in the short term are known to be mediated by the nervous system via thermosensory pathways such as the spinothalamocortical pathway (*i.e.*, temperature perception) and the parabrachial neurons-preoptic area pathway (*i.e.*, thermal homeostasis), which are triggered by detection of temperature changes at the level of the skin, before other organs are implicated (Nakamura and Morrison, 2008). Afferent responses, such as shivering and cardiac output changes in response to cold are regulated by the parabrachial neurons-preoptic area pathway (Nakamura and Morrison, 2008).

At the cellular level, thermal stress causes a series of adaptive changes in gene expression, which are observed on a wide range of organisms including prokaryotes and eukaryotes. Indeed, many cell-level alterations (*i.e.*, cellular stress) are common to cold and heat stress, which broadly involve the disruption of gene transcription, translation, protein misfolding, and cell energetics; however, some features such as altered lipid bilayer properties and fatty acid composition are more typical under cold stress condition (Sonna *et al.*, 2002).

Globally involved in the response to different types of stress (*e.g.*, heat, cold and oxidative stress), the heat shock protein (HSP) response refers to an increased expression of ubiquitous proteins (*e.g.*, HSP60 and HSP70) that act as molecular chaperones involved generally in the improvement of protein interactions and folding (Lindquist and Craig, 1988; Morimoto *et al.*, 1994; Feder and Hofmann, 1999).

Both cold and heat stress can have a significant effect on the expression and activity of heat shock proteins (Sonna *et al.*, 2002), a response known to occur within minutes of initiation of the thermal stress (Lindquist, 1986). The transcription of HSPs is regulated via heat shock factors (HSFs), which interact with specific DNA sequences (*i.e.*, heat shock elements; HSE) in their promoter regions (Pirkkala *et al.*, 2001). Importantly, HSE are also present in proteins other than HSPs, for instance C-reactive protein, super oxide dismutase and albumin (Sonna *et al.*, 2002).

As reviewed by Fujita (1999) even mild hypothermia can elicit direct HSP responses and alter the response to future cold stress events (*i.e.*, thermal tolerance). According to Horowitz, (2002), beyond the homeostatic response to heat challenges, two main mechanisms are known to specifically counter the damaging effects of heat overload, namely the fast-acting *heat shock response* and the long-acting *chronic exposure response, or heat acclimation*. Similar to cold, heat stress, can also induce thermal tolerance, preparing animals for future heat challenges.

Of note, the commonalities in the cellular response to stress can lead to cross-protection across different types of stress (Sonna *et al.*, 2002); however, to our knowledge, there is currently no evidence that animals exposed to alternating bouts of cold and heat stress can develop such cross-tolerance.

2 Thermal stress: effects on performance in dairy cows

Whenever an animal's homeostatic mechanisms are activated by thermal stress, energy is used in the process of thermoregulation that is no longer available for productive processes (Bianca, 1976). Thus, the more the organism combats environmental stress, the less is its capacity for allocating nutrients to productive processes (*e.g.*, growth, lactation). As stated earlier in this chapter, thermal stress can compromise several aspects of metabolism, productivity, and health of dairy animals. In this section, we will summarize the effects of such stress across the cow life cycle, that is, from birth to adulthood.

2.1 *Heat stress*

2.1.1 Effects in growing animals

As mentioned earlier, the TNZ of young calves is narrower and higher relative to that of adult cows (13 to 25 °C *vs.* −5 to 25 °C, respectively). Physiological and behavioral impairments are typically observed above during heat stress, including increased maintenance energy needs, reduced feed intake, and compromised immunity, all of which can lead to poor growth, higher susceptibility to infection and disease, and in extreme cases, death (Bianca, 1976; Beede and Collier, 1986). Under this condition, growing cattle divert energy that is typically partitioned for growth towards maintaining thermal homeostasis (Maurya *et al.*, 2015), resulting in a reduction in growth and growth efficiency (Baumgard and Rhoads, 2012). Generally, heat stress in calves will result in slow growth rates and a higher age at first calving in heifers. Of note, heifers are less impacted by heat stress than lactating cows, which is associated with the lower rates of heat production of growing animals relative to lactating ones (*i.e.*, lower rates of metabolic heat production).

In addition to the immediate effects of heat stress on growing calves and heifers, these young animals may also experience negative impacts when the stress takes place while *in utero* (Collier *et al.*, 1980). Reduced birth weights and altered endocrine dynamics in the offspring of cows subjected to heat stress during pregnancy have also been observed (Collier *et al.*, 1982). The work of Laporta *et al.* (2020) indicates that maternal hyperthermia during late gestation is associated with dramatic reductions in lactation performance of the offspring, and that these effects can be transgenerational, as indicated by reduced milk production and productive life of F1 offspring, and compromised lactation performance of the granddaughters (F2). Interestingly, heifers born to heat-stressed dams during late gestation are born smaller and produce less milk in their first lactation relative to heifers born to cooled dams, even though being of the same age and weight at calving (Monteiro *et al.*, 2016; Skibiel *et al.*, 2018). These observations illustrate the principle that heat stress can "program" animals *in utero* to induce negative trajectories of productivity and health, with effects that can span generations. However, whether these transgenerational effects are entirely due to heat stress or to the differences in nutrient intake, is not currently known. Pair-feeding experiments (*i.e.*, accounting for teh impact of heats stress on appetite) may shed some light on this question.

2.1.2 Effects in lactating animals

Heat stress results in body hyperthermia when cows fail to sufficiently dissipate body heat, including heat from enteric fermentation of feeds, digestion, and nutrient metabolism (West, 2003). Generally, for most cow breeds, fluctuations of air temperature between −0.5 to 20 °C and a relative humidity between 60 to 80% correspond to thermo neutral values. Heat stress typically becomes evident at THI values higher than 72 (Ravagnolo and Misztal, 2000), however, heat stress can be observed at THI of 68 in cows producing more than 35 kg of milk per day (Collier *et al.*, 2012). This observation suggests that sensitivity to heat stress may increase with the level of production (Berman, 2005), and implies that modern cows may be increasingly susceptible to hyperthermia (*i.e.*, lower THI stress threshold as production level increases) as more metabolic heat is produced in relation to milk synthesis. Illustrating this principle, Collier *et al.* (2019) summarized the relationship between feed intake and daily averaged THI in lactating cows. The data indicate that appetite of low-producing cows (*i.e.*, < 20 kg/day) is unaffected by increased until at least a 77 = THI, whereas in high-producing cows (*i.e.*, > 40 kg/day), appetite is linearly reduced in the THI range of 68 to 77, with a reduction > 50% in daily dry matter intake.

Reductions in feed intake will directly limit nutrient supply and will consequently lead to reduced milk yield in heat-stressed dairy cows (Bernabucci *et al.* 2010). Interestingly, only 35–50% of the reduction in milk production is attributable to changes in feed intake, while the remaining proportion is likely explained by metabolic adaptations to heat stress, which includes reduced rumination and nutrient absorption, increased nutrient requirements for maintenance, and changes in nutrient partitioning between tissues (Collier *et al.*, 2006; Rhoads *et al.*, 2009; Wheelock *et al.*, 2010), and increased glucose utilization by the immune system (Kvidera *et al.*, 2017; see section 1), all of which will ultimately reduce nutrient availability for milk synthesis in the mammary gland.

Other production effects associated with heat stress include alterations in milk composition and quality during the summer months. Although milk lactose concentration is typically unaffected, protein and fat contents are commonly reduced, and somatic cell counts are increased during the months with elevated heat load (Heck *et al.*, 2009). Because increased somatic cell count in milk occurs with a concomitant reduction in circulating immunoglobulin G (IgG) and inflammatory cytokines, it is believed that immune function is adversely affected during heat stress. Although milk fat concentration is often unchanged or slightly increased under experimentally induced heat stress, milk fat yield is consistently reduced (Rhoads *et al.*, 2009; Wheelock *et al.*, 2010; Ruiz-Gonzalez *et al.*, 2020), indicating a net negative effect on mammary milk fat synthesis. Because adipose tissue lipolysis is commonly reduced during heat tress (increased insulin concentrations despite reduced feed intake; Rhoads *et al.*, 2009; Wheelock *et al.*, 2010; Ruiz-Gonzalez *et al.*, 2023), reduced milk fat synthesis may be the result of limited availability of preformed fatty acids from dietary origin and those from adipose tissue lipolysis (*i.e.*, free fatty acids). Similar to changes in fat yield, reductions in the yields of protein and lactose are also observed during heat stress (Ruiz-Gonzalez *et al.*, 2023), and can only be partly explained by reductions in feed intake, underscoring the importance of changes in nutrient partitioning as a major factor driving reduced synthesis of milk components in the mammary gland (Rhoads *et al.*, 2009; Wheelock *et al.*, 2010).

2.2 *Cold stress*

In principle, exposure to cold environments increases the energy expenditure necessary for maintaining homeothermy (Kibler and Brody; 1949). Diverting energy for heat production means that the animal will shift priorities for nutrient allocation, partitioning less nutrients toward heating processes such as growth and milk synthesis (Young, 1975).

As detailed by Bianca (1976), cold normally represents a smaller problem than heat, a situation clearly observed when comparing the lethality of low and high temperatures at the extreme: "*The difference between normal and lethal body temperature is of the order of 15–25 °C in the cold, but of only about 3–6 °C in the heat*". The LCT represents the point of environmental temperature below which the animal needs to increase its heat production to prevent its body temperature from falling; however, this point will vary depending upon the acclimation, weight, condition, and energy intake of the animal. Indeed, high rates of growth, high yields of milk, and state of pregnancy, are all associated with a high metabolic rate and higher rates of heat production, which results in a higher tolerance to cold. While heat stress can be exacerbated by the consumption of fermentable of feeds and nutrient metabolism, these factors may play a positive role in thermal tolerance under cold conditions. For example, dairy cows fed high grain diets may have LCTs (as low as −30 °C to −45 °C).

As explained earlier in this chapter, cold environments commonly stimulate appetite, reflecting the increased metabolic demands of homeothermic regulation. In addition to the hyperphagic response, rumination activity, reticulorumen motility and rate of passage of digesta are all increased (Kennedy *et al.*, 1976; Westra and Christopherson, 1976; Gonyou *et al.*, 1979).

2.2.1 Cold stress effects in growing animals

Limited information is available on the effects of cold stress during prenatal life, and its consequences during subsequent development. However, rodent research suggests that prenatal cold stress can impact maternal metabolism and modulate the activity of brain centers responsible for the release of glucocorticoids such as cortisol (Wang *et al.*, 2019) as well as altering maternal metabolism (Lian *et al.*, 2020). Because maternal cold stress can influence the physiology of the placenta, the possibility exists that maternal cold stress can affect fetal development by reducing nutrient delivery (*i.e.*, undernutrition).

2.2.2 Cold stress effects in lactating animals

Cold stress usually stimulates appetite because the metabolic demands of the animal increase. Feeding rate increases by 22–31% for chopped hay and by up to 10%

for pellets at fixed intakes, thereby reducing the duration of eating (Broucek *et al.*, 1991). Rumination time also increases under the cold conditions, perhaps to compensate for the decreased eating time. Thus, it appears that increased feed intake in lactating animals is accommodated by a proportional increase in chewing effort rather than by an increase in chewing efficiency. Cold exposure decreases the proportion of cardiac output that flows to the gastrointestinal tract, lower limb hoof bone, and diaphragm, the latter of which is associated with a decreased respiratory rate. Cold stress also increases secretion of the catabolic hormones *i.e.*, cortisol and catecholamine and the more modest changes in secretion of the anabolic hormone insulin, resulting in a net catabolic response. The combined effect of these endocrine changes is to increase the mobilization of energy sources for use by muscles. Further, elevated thyroid hormone levels in a cold environment may enhance the effect of rapid lipolysis.

Cold stress causes a reduction in milk secretion in lactating cows. Local cooling of the mammary gland and reduced mammary blood flow may contribute to the decline in yield. It was demonstrated that 24-h cold exposure decreased glucose uptake by the udder, whereas glucose utilization tended to increase in the non-mammary tissues. The increased responsiveness of non-mammary tissues to insulin during prolonged cold exposure would therefore decrease the availability of glucose to the udder and result in a lower substrate for milk fat synthesis. In conjunction with changes in blood flow, this could be the cause of the decreased glucose uptake and lactose excretion by the mammary gland (Broucek *et al.*, 1991).

3 Thermal stress, leaky gut and immune system activation in dairy cattle

As previously mentioned, to dissipate body heat during the adaptation to heat stress, animals experience increased peripheral vasodilation and a compensatory decrease in intestinal blood flow, which results in hypoxia and increased intestinal permeability (Lambert *et al.*, 2002; Johnson *et al.*, 2016). Koch *et al.* (2019) reported that heat stress impairs gut barrier integrity and results in recruitment of immune cells into the intestine. Using a Cr-EDTA transfer test, Fontoura *et al.* (2022) reported increased intestinal permeability in dairy cows 3 days after initiation of a heat stress challenge. Similarly, Ruiz-Gonzalez *et al.*, (2023) reported increased concentrations of fecal calprotectin and inflammation markers in plasma (*i.e.* lipopolysaccharide-binding protein, C-reactive protein, and tumor necrosis factor-alpha) in lactating dairy cows under heat stress.

Activation of the immune response is thought to be the result of increased transfer of large molecules of bacterial origin, such as lipopolysaccharides (LPS), into the blood stream (*i.e.*, endotoxemia; Kvidera *et al.*, 2017). Such response to endotoxemia is characterized by increased interleukin (IL1-β) release from immune cells (Dalmas, 2019), triggering insulin secretion and reducing lipolysis. There is evidence that the composition of gut microbiota can be altered during heat stress (Uyeno *et al.*, 2010), which could have important implications on gut permeability, the immune system and thus on nutrient partitioning (Emmanuel *et al.*, 2007; Garcia-Hernandez *et al.*, 2017). Importantly, changes in the fecal metabolome, as a proxy for intestinal metabolites, have been reported in cows under heat stress, which suggest functional changes in microbiota which could also impact the host (Ruiz-Gonzlez *et al.*, 2022). Furthermore, gut microbiota alterations have been linked to activation of the endocannabinoid system, including some lipid mediators which could further increase leaky gut, circulating LPS and adipogenesis (Muccioli *et al.*, 2010), thus potentially reducing the flow of nutrients for milk synthesis. However, a causal role of intestinal microbiota composition in the fore-mentioned increase in gut permeability has not yet been established during stress.

Contrary to the impact of heat stress, cold stress was shown to reduce blood flow not only to the splanchnic tissues, but also to the skin in new-born lambs, whereas tissues capable of thermogenesis (*i.e.* muscle and brown adipose tissues) exhibited increased blood flow (Alexander *et al.*, 1973). Cold stress has been shown to alter digestive function by delaying gastric emptying and modulating gastric and intestinal secretions in human subjects (Thompson *et al.*, 1983), although its impact on intestinal barrier function has not been extensively studied. However, one study in young broilers subjected to a 72-h cold stress reported increased endotoxin and heat shock factor-1, suggesting a negative impact on barrier function.

Gut microbiota changes in response to cold stress have not been extensively investigated, although a role of intestinal microbiota in facilitating brown adipose tissue thermogenesis has been reported (Zhang *et al.*, 2022). Guo *et al.* (2021) reported moderate changes in rumen microbiota, composition in sheep undergoing cold stress, which was mainly characterized by an increase in the relative abundance of the Lachnospiraceae group. How cold environments may affect ruminal and intestinal microbiota composition of growing and lactating cattle needs further investigation, as it may impact the capacity of animals to thermoregulate under those conditions.

References

Alexander, G., Bell, A.W., and Hales, J.R.S. 1973. Effects of cold exposure on tissue blood flow in the new-born lamb. J. Physiol. 234:65–77.

Ames, D. R., and L. W. Insley. 1975. Wind-chill effect for cattle and sheep. J. Anim. Sci. 40: 161–165.

Baumgard, L.H.; Rhoads, R.P. 2012. RUMINANT NUTRITION SYMPOSIUM: Ruminant Production and Metabolic Responses to Heat Stress. J. Anim. Sci. 90, 1855–1865.

Beede, D.K.; Collier, R.J. 1986. Potential nutritional strategies for intensively managed cattle during thermal stress. J. Anim. Sci. 62: 543–554.

Berman, A. 2005. Estimates of heat stress relief needs for Holstein dairy cows. J. Anim. Sci. 83:1377–1384.

Berman, A., and Meltzer, A. 1973. Critical temperatures in lactating dairy cattle: A new approach to an old program. Int. J. Biometeorol. 17:167–176.

Berman, A., Folman, Y., Kaim, M., Mamen, M., Herz, Z., Wolfenson, D., Arieli, A., and Graber, Y. 1985. Upper critical temperatures and forced ventilation effects for high-yielding cows in a subtropical climate. J. Dairy Sci. 68:1488–1495.

Bernabucci U., Lacetera N., Baumgard L.H., Rhoads R.P., Ronchi B., Nardone A. 2010. Metabolic and hormonal acclimation to heat stress in domesticated ruminants. Animal 4: 1167–83.

Bianca, W. 1971. Die anpassung des haustieres an seine klimatische umgebung. Schweiz. Landw. Forsch. 10:155–205.

Bianca, W. 1976. The significance of meteorology in animal production. Int, J. Biometeor. 20:139–156.

Brody, S. 1956. Climatic physiology of cattle. J. Dairy Sci. 39:715–725.

Broucek, J., M. Letkovicovfi, and K. Kovalcuj. 1991. Estimation of cold stress effect on dairy cows. Int J Biometeorol 35: 29–32.

Collier R.J., Dahl G.E., Van Baale M.J. 2006. Major advances associated with environmental effects on dairy cattle. J Dairy Sci. 89: 1244–1253.

Collier R.J., and Gebremedhin K.G. 2015. Thermal biology of domestic animals. Annu Rev Anim Biosci. 2015. 3: 513–32.

Collier, R.J., Doelger, S.G., Head, H.H., Thatcher, W.W., & Wilcox, C.J. 1982. Effects of heat stress during pregnancy on maternal hormone concentrations, calf birth weight and postpartum milk yield of Holstein cows. J. Animal Sci. 54: 309–319.

Collier. R.J., Hall, L.H., Baumgard, L.H., Zimbelman, R.B., and Xiao, Y. 2019. Feed and water intake of heat stressed cattle. Western Dairy Management Conference.

Collier, R. J., Renquist, B.J., and Xiao, Y. 2017. A 100-Year Review: Stress physiology including heat stress. J. Dairy Sci. 100:10367–10380.

Dalmas, E., 2019. Innate immune priming of insulin secretion. Current Opinion Immunol. 56:44–49.

Emmanuel, D.G.V., Madsen, K.L., Churchill, T.A., Dunn, S.M., and Ametaj, B.M. 2007. Acidosis and lipopolysaccharide from Escherichia coli B:055 cause hyperpermeability of rumen and colon tissues. J. Dairy Sci. 90:5552–5557.

Feder, M. E., and G. E. Hoffmann. 1999. “Heat-shock proteins, molecular chaperones, and the stress response: evolutionary and ecological physiology.” Ann Rev Physiol. 61.1: 243–282.

Folk, G.E. 1974. Temperature regulation. Pages 87–132 in the Textbook of Environmental Physiology. G.E. Folk, ed. Lea and Febiger, Philadelphia, PA.

Fujita, Jun. 1999. “Cold shock response in mammalian cells.” J Mol Microbiol Biotechnol. 1.2: 243–255.

Gasparrini, A., Guo, Y., Hashizume, M., Lavigne, E., Zanobetti, A., Schwartz, J., Tobias, A., Tong, S., Rocklöv, J., Forsberg, B., Leone, M., De Sario, M., Bell, M.L., Guo, Y.L.L., Wu, C., Kan, H., Yi, S.M., de Sousa Zanotti Stagliorio Coelho, M., Saldiva, P.H.N., Honda, Y., Kim, H., and Armstrong, B. 2015. Mortality risk attributable to high and low ambient temperature: a multicountry observational study. Lancet 386, 369–375.

Garcia-Hernandez, V., Quiros, M., Nusrat, A. 2017. Intestinal epithelial claudins: expression and requlation in homeostasis and inflammation. Ann N Y Acad Sci. 1397:66–79.

Gonyou, H.W., Christopherson, R.J., and Young, B.A. 1979. Effects of cold temperature and winter conditions on some aspects of behavior of feedlot cattle. J. Appl. Anim. Ethol. 5:113–124.

Guo, H., Zhou, G., Tian, G., Liu, Y., Dong, N., Li, L., Zhang, S., Chai, H., Chen, Y., and Yang, Y. 2021. Changes in rumen microbiota affect metabolites, immune responses and antioxidant enzyme activities of sheep under cold stimulation. Animals. 11: 712.

Hamada, T. 1971. Estimation of lower critical temperatures for dry and lactating dairy cows. J. Dairy Sci. 54:1704–1705.

Heck, J.M.L., Van Valenberg, H.J.F., Dijkstra, J., & Van Hooijdonk, A.C.M. 2009. Seasonal variation in the Dutch bovine raw milk composition. Journal of dairy science, 92: 4745–4755.

Horowitz, Michal. 2002. "From molecular and cellular to integrative heat defense during exposure to chronic heat." *Comparative Biochemistry and Physiology Part A: Molecular & Integrative Physiology* 131.3: 475–483.

Johnson, J.S., Martin, K.L., Pohler, K.G., and Stewart, K.R. 2016. Effects of rapid temperature fluctuations prior to breeding on reproductive efficiency in replacement gilts. J. Therm. Biol. 61:29–37.

Kadzere, C.T., Murphy, M.R., Silanikove, N., Maltz, E., 2002. Heat stress in lactating dairy cows: a review. Livest. Prod. Sci. 77: 59–91.

Kennedy, P.M., Christopherson, R.J., and Milligan, L.P. 1976. The effect of cold exposure of sheep on digestion, rumen turnover time and efficiency of nmicrobial synthesis. Brit J. Nutr. 36:231–242.

Kibler, H. H., and S. Brody. 1949. Environmental physiology with special reference to domestic animals. 7. Influence of temperature, 50° to 5° F and 50° to 95° F, on heat production and cardiorespiratory activities of dairy cattle. Univ. Missouri Agric. Exp. Stn. Res. Bull. 450:3–28.

Kleiber, M. 1961. The Fire of Life. John Wiley and Sons, New York, NY.

Knizkova, I., Kunc, P., Koubkova, M., Flusser, J., and Dolezal, O. 2002. Evaluation of naturally ventilated dairy barn management by a thermographic method. Livest Prod Sci. 77:349–353.

Koch, F., Thom, U., Albrecht, E., Weikard, R., Nolte, W., Kuhla, B., & Kuehn, C. 2019. Heat stress directly impairs gut integrity and recruits distinct immune cell populations into the bovine intestine. Proceedings of the National Academy of Sciences, 116(21), 10333–10338.

Kvidera, S.K., Horst, E.A., Abuajamieh, M., Mayorga, E.J., Sanz Fernandez, M.V., and Baumgard, L.H. 2017. Glucose requirements of an activated immune system in lactating Holstein cows. J. Dairy Sci. 100:2360–2374.

Lambert, G.P., Gisolfi, C.V., Berg, D.J., Moseley, P.L., Oberley, L.W., Kregel, K.C. 2002. Selected contribution: hyperthermia induced intestinal permeability and the role of oxidative and nitrosative stress. J Appl Physiol. 92: 1750–1761.

Laporta, J., Ferreira, F.C., Quellet, V., Dado-Senn, B., Almeida, A.K., De Vries, A., Dahl, G.E. 2020. Late-gestation heat stress impairs daughter and granddaughter lifetime performance. J. Dairy Sci. 103: 7555–7568.

Lian S, Li, W., Wang, D., Xu, B., Guo, X., Yang, H., and Wang, J. 2020. Effects of prenatal cold stress on maternal serum metabolomics in rats. Life Sci. 246:117432.

Lian, S., Guo J., Wang, L., Li, W., Wang, J., Ji, H., Kong, F., Xu, B., Li, S., and Yang, H. 2017. Impact of prenatal cold stress on placental physiology, inflammatory response, and apoptosis in rats. Oncotarget. 8(70):115304–115314.

Lindquist, S. 1986. The heat-shock response. Annu. Rev. Biochem. 55:1151–1191.

Lindquist, S., and Craig, E.A. 1988. The heat-shock proteins. Ann. Rev. Genet. 22: 631–677.

Maurya, V.P., Sejian, V., Gupta, M., Dangi, S.S., Kushwaha, A., Singh, G., Sarkar, M. 2015. Adaptive Mechanisms of Livestock to Changing Climate. In Climate Change Impact on Livestock: Adaptation and Mitigation; Sejian, V., Gaughan, J., Baumgard, L., Prasad, C., Eds.; Springer: New Delhi, India, Volume 9, pp. 123–138.

Monteiro, A.P.A., Guo, J.R., Weng, X.S., Ahmed, B.M., Hayen, M.J., Dahl, G.E., Bernard, J.K., and Tao, S. 2016. Effect of maternal heat stress during the dry period on growth and metabolism of calves. J. Dairy Sci. 99:3896–3907.

Morimoto, R.I., Tissieres, A., and Georgopoulos, C. 1994. Progress and perspectives on the biology of heat shock proteins and molecular chaperones. In: The Biology of Heat Shock Proteins and Molecular Chaperones. R.I. Morimoto, A. Tissieres, A., and Georgopoulos, C., eds. Cold Spring Harbor Lab. Press, New York. pp. 1–30.

Muccioli, G.G., Naslain, D., Backhed, F., Reigstad, C.S., Lambert, D.M., Delzenne, N.M., Cani, P.D. 2010. The endocannabinoid system links gut microbiota to adipogenesis. Mole Sys Biol. 6:392.

Nakamura, K., and S.F. Morrison. 2008. A thermosensory pathway that controls body temperature. *Nature neuroscience* 11.1: 62–71.

Pirkkala, L., Nykänen, P., & Sistonen, L.E.A. 2001. Roles of the heat shock transcription factors in regulation of the heat shock response and beyond. The FASEB Journal, 15(7), 1118–1131.

Ravagnolo, O., and Misztal, I. 2000. Genetic component of heat stress in dairy cattle, parameter estimation. J. Dairy Sci. 83:2126–2130.

Rhoads M.L., Rhoads, R.P., VanBaale, M.J., Collier, R.J., Sanders, S.R., Weber, W.J., Crooker, B.A., Baumgard, L.H. 2009. Effects of heat stress and plane of nutrition on lactating Holstein cows: I. Production, metabolism, and aspects of circulating somatotropin. J Dairy Sci. 92: 1986–1997.

Ruiz-Gonzalez, A., Rico, D.E., and Rico, J.E. 2022. Modulation of fecal metabolites by heat stress and diet, and their association with inflammation and leaky gut markers in dairy cows. Metabolites. 12:142.

Ruiz-González, A., Suissi, W., Baumgard, L. H., Martel-Kennes, Y., Chouinard, P. Y., Gervais R., and Rico, D. E. (2023). Increased dietary vitamin D3 and calcium partially alleviate heat stress symptoms and inflammation in lactating Holstein cows independent of dietary concentrations of vitamin E and selenium. J Dairy Sci, 106:3984–4001.

Skibiel, A.L., Penagaricano, F., Amorin, R., Ahmed, B.M., Dahl, G.E., and Laporta, J. 2018. In utero heat stress alters the offspring epigenome. Sci Rep. 8:14609.

Sonna, L.A., Fujita, J., Gaffin, S.L., & Lilly, C.M. 2002. Invited review: effects of heat and cold stress on mammalian gene expression.J. App. Physiol., *92*(4), 1725–1742.

Tao, S., R. Orellana, X. Weng, T. Marins, G. Dahl, and J. Bernard. 2018. Symposium review: The influences of heat stress on bovine mammary gland function. J. Dairy Sci. 101:5642–5654.

Thompson, D.G., Richelson, E., and Malagelada, J-R. 1983. Perturbation of upper gastrointestinal function by cold stress. Gut. 24:277–283.

Uyeno, Y., Sekiguchi, Y., Tajima, K., Takenaka, A., Kurihara, M., and Kamagata, Y. 2010. An rRNA-based analysis for evaluating the effect of heat stress on the rumen microbial composition of Holstein heifers. Anaerobe. 16:27–33.

Vitali, A., M. Segnalini, L. Bertocchi, U. Bernabucci, A. Nardone, and N. Lacetera. 2009. Seasonal pattern of mortality and relationships between mortality and temperature-humidity index in dairy cows. J. Dairy Sci. 92:3781–3790.

Wang, D., Xu, B., Wang, J., Wang, H., Guo, J., Ji, H., *et al.* 2019. Response of the maternal hypothalamus to cold stress during late pregnancy in rats. Brain Res. 1722:146354.

West, J.W. 2003. Effects of heat-stress on production in dairy cattle. J Dairy Sci. 86(6): 2131–2144.

Westra, R., and Christopherson, R.J. 1976. Effects of cold on digestibility, retention time of digesta, reticulum motility and thyroid hormones in sheep. Can J. Anim Sci. 56:699–708.

Wheelock, J.B., R.P. Rhoads, M.J. Van Baale, S.R. Sanders, and L.H. Baumgard. 2010. Effects of heat stress on energetic metabolism in lactating Holstein cows. J Dairy Sci 93: 644–655.

Young, B.A., 1975. Temperature induced changes in metabolism and body weight of cattle (Bos taurus). Can J. Anim Sci. 55:619.

Yousef, M. K. 1985. Stress Physiology in Livestock. 1st ed. CRC press, Boca Raton, FL.

Zhang, Y., Sun, L., Zhu, R., Zhang, S., and Liu, S. 2022. Porcine gut microbiota in mediating host metabolic adaptation to cold stress. NPJ Biofilms Microbioms. 8:18.

CHAPTER 13

Epithelial water transport and environment moisture in poultry gut health and disease

Mariano E. Fernández-Miyakawa[1,*] *and Michael H. Kogut*[2]

[1]Institute of Pathobiology, Instituto Nacional de Tecnología Agropecuaria, Argentina; [2]USDA-ARS, Southern Plains Agricultural Research Center, College Station, Texas, USA; *Correspondence: Fernandezmiyakawa.m@inta.gob.ar

Abstract

Modern poultry production has reached very high levels of efficiency and competitiveness that largely depend on more demanding management. The digestive system, and particularly the intestine, is key to reaching optimal efficiency levels. Water, the main nutrient of animals, and its homeostasis, has direct and indirect influences on animal health since it affects the establishment of the intestinal and environmental microbiota, the development of diseases, immunity, and the digestive and nutrient absorption processes, and strongly influencing the productivity of the animals. The gastrointestinal tract absorbs a large volume of fluids daily and this transport of water through the intestine is closely related to the movement of solutes, primarily associated with the transport of ions. Any disturbance in the gastrointestinal tract creates the potential for a large loss of total body water and electrolytes. In birds, the available information suggests that intestinal ion transport adapts efficiently to alterations in nutrition and the environment, but for example there is little information regarding the pathophysiological processes in birds of many infections that produce intestinal imbalances, including common pathogens such as *Salmonella* or *C. perfringens*. Dysregulation of normal ion transport in the bird intestine manifests as diarrhea, malabsorption, and intestinal inflammation, resulting in poor production efficiency. The role of the tight junction is important in all these processes, although there are discrepancies regarding its role in the initial step of development of inflammatory processes that can lead to deregulation of intestinal homeostasis, as well as its importance in the processes of nutrient absorption in birds compared to mammals. A better understanding of the normal and pathological interaction of water and ion balance in the intestine, together with the regulation of the epithelial barrier during digestive, infectious, and inflammatory processes, will help us to optimize the management of poultry in intensive farming to maximize gut health, well-being, and production.

© MARIANO E. FERNÁNDEZ-MIYAKAWA AND MICHAEL H. KOGUT, 2024 | DOI:10.3920/9789004695467_014

Keywords

water – epithelial cells – ion transport – gut health – tight junctions – pore pathway – aquaporin

1 Introduction

Water is the most important nutrient for animals, accounts for 50 to 80 percent of an animal's weight and is involved in every physiological process. The gastrointestinal tract of chickens, as many other vertebrates, plays an important role in the maintenance of fluid homeostasis, absorbing a large amount of water and electrolytes daily. Epithelial ion transport is the main driver of the water absorption in the gut, and it is affected by nutrition, microbiota, infectious diseases, stress and environment, which are also the most important factors that affect poultry intestinal health. Any disturbance in the gastrointestinal physiology can produce a large loss of total body electrolytes, and simultaneously water, that will disturb the environment of animals. In intensively confined farm animals, gastrointestinal water imbalance would have a profound impact in the environmental moisture, including litter and soil, influencing environmental dynamics of microbial communities. In turn, the chicken gastrointestinal microbial community will be indirectly affected because of the influence produced by this environmental microbiota, and eventually would increase gut stressors and pathogens that can trigger physio-pathological processes that affect animal health, productivity, and food safety.

2 Ion transport and water balance in the chicken gut

Water transport by epithelia of the digestive tract is of considerable importance in animal physiology, health, and productivity. The reabsorption of water in the gastrointestinal tract basically depends on a simple epithelium that fulfils the selective uptake of nutrients and ions and prevent the pathogen invasion. The intestinal epithelium is the single cell layer that form the luminal surface of both the small and large intestine of the gastrointestinal tract. This unique layer of cell represents the largest area of the body that is in direct contact with environmental contaminants and microbes.

The gut plays a relatively more important role in water and salt regulation in birds than in mammals. In birds, the small intestine is important for net absorption of Ca^{2+} and K^{+}, but not for Na^{+} nor water but there is a tightly regulated control of salt and water balance in the avian lower intestine (Laverty *et al.*, 2006). The hindgut (coprodeum, colon, caeca) of birds is capable of active sodium reabsorption and solute-linked water absorption (Braun, 2003) (Lehmann *et al.*, 2016) and the epithelium of the lower intestine reabsorbs isotonic Na^+, Cl^-,and water by mechanisms apparently similar to those identified in mammals (D.L. Goldstein, 1989); (GOLDSTEIN & SKADHAUGE, 2000). Epithelial permeability decreases along the length of the intestine, the distal colon of mammals and coprodeum of birds are the least permeable regions. (GOLDSTEIN & SKADHAUGE, 2000). The cecum is a key organ for fluid reabsorption and caecal transport capacity for NaCl and water from chymus and refluxed urine is therefore probably large, perhaps larger than that of the colon plus coprodeum. Net water absorption in birds occurs in caeca > rectum > coprodeum, but net Na^+ absorption (an active process) occurs in rectum > caeca and coprodeum.

Anatomically, the most notable features of the gut are the segmental division, associated with clear functional differences, and the mucosa organized into the crypt-villus axis. The length of villi decreases from the proximal to the distal end of the intestine (Leeson *et al.*, 2005; Nasrin *et al.*, 2012) . Villi are capillary-rich protrusions into the intestinal lumen of 1.8 mm (proximal) to 0.5 mm (distal) length. The villi increase the epithelial surface by a factor of 30 in the small intestine but are completely absent in the colon. The villi enlarge surface area, and the surface enlargement factor for villi is greater in birds than in mammals (Karasov & Caviedes-Vidal, 2021). The presence of microvilli on each enterocyte increases the absorptive surface another 600 times (Kiela & Ghishan, 2016a). Crypts form pockets in the mucosa and connect to the intestinal lumen only via a small opening due to dense packing of cells (Bonis *et al.*, 2021). The microenvironment within the crypt is additionally isolated from the digestive process by a constant flow of mucus and anti-microbial products.

Despite its apparently simple architecture, the epithelium and the underlying mesenchyme is exquisitely structured. The intestinal lining contains a highly distinct cell populations that are continuously renewed from the crypts by the same stem cell population and moves from the crypt up the villus to be eventually shed into the lumen at the villus tip. In chickens, in a manner similar to mice and humans, stem cells are restricted to the crypts and enterocytes are found along the villi (Zhang & Wong, 2017) (Zhang & Wong, 2018). However, in contrast to mice and humans, proliferating and immune cells in chicken are observed along the villi as well as within the crypts (Shira & Friedman, 2018). The continuous flow of cells from villus crypt to tip allows phenotypically flexible responses but also creates high turnover and likely high maintenance costs for the intestine, which have been estimated to be 20–30% of basal metabolism and 28–46% of whole-body protein synthesis (Starck & Rahmaan, 2003). Luminal stimulus as feed influences the cell sub-type quantities and distribution in small intestinal epithelium, which determine and maintain intestinal morphology and functionality (Reicher *et al.*, 2020) Regardless of these effects on intestinal development and as a general concept, the overall spatial organization in the epithelium remains constant and along the crypt-villus axis, with certain cell types and functions appearing only in specific positions and regarding electrolyte balance, while the immature crypt epithelium is primarily involved in ion secretion, the mature villus epithelium derived from the crypts is involved in water and nutrient absorption. Therefore, cells of the intestinal epithelium must remain plastic enough and decidedly reactive to environmental signals that drives their destiny and function throughout their lifetime. Clearly, there is a need to maintain a balanced diversity of cell types in a constantly regenerating tissue with intricate mechanisms of spatial or temporal cell fate control. This control can be quickly affected by any of the challenges present in the modern poultry system, including components of diet, pathogens, or toxins.

In the intestine, fluid is moved either across the plasma membranes of the cells that comprise the epithelial layer (transcellular transport) or between these cells, through the tight junction complexes that form a barrier. The passage of molecules across the epithelial layer of gut involves two types of transport mechanisms called paracellular and transcellular transport. Paracellular transport is the movement of substances (fluids, electrolytes, small water-soluble compounds) across the epithelium through the intercellular space between the cells. This movement is a type of passive diffusion or convection that does not requires energy, and it is driven by a transepithelial concentration gradient. Paracellular transport does not appear to contain any transporters that have the ability to be saturated and the pathways involved tend to be relatively unselective, primarily selecting permeant molecules on the basis of size and charge. The rate-limiting barrier for paracellular transport is the tight junction, a complex containing a large number of proteins (see below). Paracellular transport is driven by transepithelial electrochemical gradients that are created by transcellular transport. The tight junction of epithelial cells allows differential permeation of ions but tends to exclude macromolecules and although paracellular diffusion has a minor role in the regular absorption of nutrients in the intestine, it would be involved in the potential absorption of determined nutrients in birds, as explained in the next "Diet, feed additives and epithelial water transport" section. Transcellular transport is the predominant route for most of the transepithelial transport, and it occurs through the epithelial cell, involving transmembrane proteins in the apical and basal plasma membranes, and the intervening cytoplasm. The transcellular pathway is considered as an important aspect in the context of absorption of nutrients (amino acids, sugars, fatty acids, minerals) in the small intestine and it can be highly specific. Transcellular transport is divided into simple passive transport, passive diffusion with an efflux pump, active transport, and endocytosis.

As described above, tight junction barrier of the epithelial cell monolayer does not form a perfect seal as permeability is required for water, nutrient, and ion transport. Water and solute flux across tight junctions follow two distinct pathways based on their size- and charge-selectivity and capacity (Shen *et al.*, 2011); (Zuo *et al.*, 2020). The "pore pathway" is a high-capacity, charge- and size-selective route with a maximum diameter of 6 Å; whereas, the "leak pathway" has low conductance and is relatively nonselective and accommodates molecules with diameters up to ~100 Å that involves a transient and reversible rearrangement of the tight junctions and actin cytoskeleton (Zuo *et al.*, 2020). Lastly, water, ions, macromolecules, viruses, and bacteria substances can cross the epithelial barrier through an "unrestricted pathway" that results from epithelial damage. The "unrestricted pathway" is tight junction independent process and is commonly referred to as 'leaky gut' (Wageha A. Awad *et al.*, 2017). The concept of "leaky gut" is associated with inflammatory states that increase paracellular permeability to endotoxins and bacteria. However, no direct evidence exists supporting bacterial or endotoxin transport through the paracellular pathway; rather, the translocation of bacteria appears to take place only when the intestinal structure is altered, and it is a distinct process that is different from the normal paracellular intestinal permeability of small water-soluble compounds. The currently used term "leaky gut" is a hypothetical condition and it has to be differentiated from the concept of "leaky" epithelia previously proposed (Frömter & Diamond, 1972) and historically used to define any epithelium with a substantially greater paracellular small ion permeability as compared to their transcellular small ion permeability. The opposite concept is "tight" epithelia, which exhibit an equivalent or even greater transcellular small ion permeability as compared to the paracellular permeability.

3 Tight junctions and intestinal barrier

At the cellular level, the epithelial lining is constituted by a single layer of columnar epithelial cells and interepithelial tight junctions present at the apical-most region of the paracellular space between adjacent enterocytes. The epithelial cells require special carriers on the cell membrane to facilitate the transport of ions from the intestinal lumen into the epithelial cells and vice versa since the membrane lipid bilayers are relatively impermeable to ions. These singular proteins are predominantly either transporters or ion channels which permits efficient ion homeostasis of the intestine. The tight regulation of ion transport by epithelial cells can occur exclusively because of the polarity of the epithelial cells conferred by the presence of the apical tight junctions which guarantees that the stable location in the cell membrane of apical and basolateral ion transporters. Tight junctions involve an array of membrane-spanning proteins (*e.g.*, barrier forming occludin and pore forming claudins) that are linked to a cytoplasmic plaque and the actin cytoskeleton. Tight junction are multi-protein complexes that are vital for the maintenance and functional integrity of the epithelial barrier of the intestine. Tight junction proteins include the transmembrane proteins claudins, occludins, junctional adhesion molecules and the cytoplasmic zonula occludens proteins (Paradis *et al.*, 2021). Tight junction proteins regulate paracellular permeability, and their disruption is considered to be followed by the permeation of luminal noxious molecules that may cause an excessive activation of the immune system and inflammation (Suzuki, 2013) (John *et al.*, 2011) (S.H. Lee, 2015) suggesting that tight junction impairment plays a major role in the progression and duration of intestinal inflammation. A defective intestinal epithelial tight junction barrier, resulting in an increased intestinal permeability, appears to be a crucial mechanistic feature in the pathogenesis of infectious and non-infectious stimuli induction of barrier dysfunction in the intestine (Camilleri, 2019).

Claudins are the major components of tight junctions and control the paracellular permeability to water, ions, and larger solute molecules via the "pore pathway" (Rosenthal *et al.*, 2017). Based on their composition and expression patterns within the tight junction complex, claudins can be classified as either barrier-forming (claudins 10 and 17) which prevent the unlimited passage of water, solutes, and luminal antigens and dominate the intestinal epithelial barrier (Barmeyer *et al.*, 2016) or channel-forming that form passages for water and monovalent cations (Rosenthal *et al.*, 2017) (Amasheh *et al.*, 2002). Claudins 2 and 15 have been specifically identified as 'water channels' (Rosenthal *et al.*, 2017) (Rosenthal *et al.*, 2020). Occludin and other proteins as ZO, tricellulin, and actin filaments have been implicated in mediating the "leak pathway" for large solutes but the proteins comprising this pathway remain unresolved. Moreover, many aspects of the "leak pathway" are still under debate including molecular mechanism, properties, and regulation (Monaco *et al.*, 2021).

4 Intestinal barrier function and the microbiota

The intestinal barrier integrity can be affected by multiple intrinsic and extrinsic factors, including diet, antibiotics, heat stress, enteric infections, mycotoxins, chronic inflammation and gut microbial population (Martel *et al.*, 2022) (Yegani & Korver, 2008) (Pan & Yu, 2014). The gut microbiota help maintain gut functional homeostasis in multiple ways including colonization resistance against pathogens, educating the development of the mucosal immune system, inducing the production of mucus and antimicrobial peptides by specialized epithelial cells, and converting diet products into metabolites which protect the barrier in multiple ways (Ghosh *et al.*, 2021). Therefore, disruption of the gut microbiota (dysbiosis) affects the gut barrier function and immune response resulting detrimental effects on the intestine including inflammation and oxidative stress in the epithelium that results in the breakdown of the integrity of the gut barrier and increases the exposure of colonocytes to enteric infections and toxins from the lumenal milieu further enhancing inflammatory processes and release of reactive oxygen species (ROS) that alter tight junctions causing diarrhea and water and electrolyte loss into the gut lumen (Guttman & Finlay, 2009) (O'Hara & Buret, 2008) (Paradis *et al.*, 2021).

The gut microbiota, due to their close proximity to the intestinal epithelium, produce a diverse range of metabolites from dietary components that affect gut permeability and impact both host health and disease via gut barrier function and mucosal immune responses. Short chain fatty acids (SCFAs), including acetate, proprionate, and butyrate, have all been shown to facilitate the assembly and expression of tight junction proteins, increase paracellular pathway transport of water and solutes, and increase water and anion release to maintain osmotic equilibrium in the lumen (Diao *et al.*, 2019; Feng *et al.*, 2018; Sivaprakasam *et al.*, 2017; Wang *et al.*, 2012). Further, butyrate has anti-inflammatory activities that protects the intestine against inflammation-mediated barrier disruption (). Likewise, tryptophan metabolites (indole derivatives) have been shown to enhance gut barrier integrity and induce immune responses through the induction of IL-22 (Shimada *et al.*, 2013) (Yu *et al.*, 2018). Bile acid metabolites have been shown to both enhance or decrease gut barrier tight junction functions (Pavlidis *et al.*, 2015; Sarathy *et al.*, 2017; Ticho *et al.*, 2019) depending upon the suitable ratio between primary (host-derived) and secondary (microbial-derived) bile acids to mediate a protective gut barrier.

Microbiota structural components (LPS, flagellin, peptidoglycan, muramyl dipeptide, lipotechoic acid, zymosan, bacteria CpG) have also been shown to play a role in gut barrier maintenance and integrity by activating the Toll-like and NOD-like receptor pathways in the epithelial barrier cells (Marques & Boneca, 2011; Wells *et al.*, 2011). These structural components directly or indirectly by regulating mucosal immunity regulate gut barrier functions.

5 Water balance in gastrointestinal infectious disease

Transport of water, electrolytes, and solutes across the intestinal epithelial barrier occurs through both the transcellular and paracellular pathways described previously. However, the barrier is constantly regulated in reaction to extracellular elements that activate different intracellular signaling (Ma *et al.*, 2018). The fluid composition of the luminal and serosal surfaces of the intestinal barrier endures persistent change between digestive and non-digestive phases. The fluid compositions along the barrier surfaces varies depending upon the feed ingredients, luminal microbiota composition, the ionic and solute contents of the luminal and serosal fluid, the type of proteolytic enzymes secreted, the inflammatory state of the mucosa, and the presence of inflammatory mediators and cytokines. All of which are associated with the intestinal health or predisposition to disease (Nusrat *et al.*, 2000).

The exposure to microorganisms significantly affects intestinal physiology and intestinal ion transport, and this dysregulation results in hypersecretion or malabsorption, and intestinal inflammation, which clearly affect production efficiency. In most obvious events of dysregulation, diarrhea is observed. Gastrointestinal infections are the result of the interaction between various pathogen virulence factors and host defense mechanisms that can result in altered gut transport or impaired barrier function, both of which can manifest clinical symptoms such as diarrhea.

Many microbial metabolites are known to directly interact with individual intestinal ion transporters, increasing or decreasing their activity and/or their presence on the brush border membrane of intestinal epithelial cells. Microbial pathogens infecting intestinal epithelial cells or stimulation of epithelial cells with proinflammatory mediators generated by cells in the underlying mucosa, resulted in the rapid upregulation of gene expression and ensuing production of proinflammatory cytokines and chemokines by intestinal epithelial cells (Eckmann *et al.*, 1993). Microorganisms can interact also with tight junctions and the surrounding actin cytoskeleton to decrease intestinal barrier function. Therefore, the epithelial cells layer is a fundamental component of a sophisticated communications organization that transmits signals between the external environment and the host's mucosal immune system including both innate and adaptive responses and regulates nutrients absorption and ion/water balance.

"Leaky gut" is a condition in which the small intestine lining becomes altered, allowing lumenal contents such as commensal bacteria, pathogens and their metabolite or toxins to pass the epithelial barrier, resulting in cell damage or inflammation of the intestine. Once these materials pass between cells, they are recognized by the immune system triggering the activation of an immune response. This primary defense against luminal microorganism in the intestine of vertebrates is the innate immunity which includes a combination of chemical, mechanical and environmental factors (Campbell *et al.*, 2011). Water transport functions are considered essential components of this protective barrier (Glenn *et al.*, 2009) (Hallstrom & McCormick, 2011). Therefore, any change in ion transport and water homeostasis would have an influence in this primary defense of the host. However, the direct influence of water transport on bacterial-epithelial interactions are usually not recognized or not considered when evaluating the protection of animals against gut infections. During a leaky gut episode, the paracellular pathway becomes the primary route for the passive flux of water and solutes in the intestine (Ma *et al.*, 2018). This water flux can be bidirectional and is controlled by differences in hydrostatic pressure or the osmotic gradient across the epithelial barrier.

Functional disturbances in intestinal permeability, affecting the "pore pathway", can be caused by different environmental factors such as a high mycotoxin levels in the diet (Gao *et al.*, 2020; von Buchholz *et al.*, 2022), heat stress (Brugaletta *et al.*, 2022; Emami *et al.*, 2021), feed restriction (Baxter *et al.*, 2017; Emami *et al.*, 2021), non-invasive, enteric pathogen colonization such as *Campylobacter jejuni* (Wageha A. Awad *et al.*, 2018; Humphrey *et al.*, 2014; Rzeznitzeck *et al.*, 2022) and the inclusion of high levels of dietary components, such as rye or protein, in the diet of broiler chickens (Qaisrani *et al.*, 2014; Tellez *et al.*, 2014; Vicuña *et al.*, 2015).

Rotavirus has been implicated as the causal organism of diarrhoea in chicken and other avian species (Day & Zsak, 2013; Pantin-Jackwood *et al.*, 2007). Most of the effects of a rotavirus infection on intestinal ion transport to a viral enterotoxin which is a non-structural protein named NSP4 which activates a protein kinase C (PKC)/inositol triphosphate (IP3) pathway, resulting in increased intracellular calcium that induces electrogenic secretion of chloride and fluid (Dong *et al.*, 1997). Also NSP4 directly inhibits glucose-mediated sodium absorption and glucose increased NSP4-stimulated chloride secretion (Yin *et al.*, 2017). Chicken astrovirus has been associated with diarrhea, runtingstunting syndrome, and other diseases and syndromes in young broiler chicks (Day & Zsak, 2013). Astroviruses cause morbidity and enteric infections in a number of mammalian hosts (Awe *et al.*, n.d.; Koci *et al.*, 2003; Raji & Omar, 2022; Zhao *et al.*, 2020) and are known to be widely prevalent in commercial chicken flocks (Baxendale & Mebatsion, 2004). Astroviruses studies in birds have demonstrated an increase in the permeability of intestinal epithelial cells, leading to the disruption of tight junctions between the cells (Meliopoulos *et al.*, 2016; Nighot *et al.*, 2010) increasing water, solute, and ion movements through compartments, limiting the capability of the reabsorption by the small intestine and thus resulting in diarrhea. Also, evidence shows that astrovirus infection causes malabsorption of sodium salts, presumably as a result of redistribution of sodium transporters, thus resulting in osmotic diarrhea (Nighot *et al.*, 2010).

Although many of the most common pathogenic bacterial species in poultry chickens, as *Clostridium perfringens, Escherichia coli, or Salmonella spp* might influence intestinal water homeostasis in mammals, it is comparatively different in chickens.

C. pefringens is a pathogen of poultry chicken and other animals but also is part of the normal microbiota in apparently healthy animals (Cooper & Songer, 2009). In chickens, different toxins of *C. perfringens* are considered important virulence factors in the pathogenesis of necrotic enteritis (NE), including the observed diarrhea and the intestinal lesions. Alpha toxin and NetB toxin are considered important for NE pathogenesis although *C. perfringens* produces more than 20 toxins and enzymes (Cooper & Songer, 2010; Sayeed *et al.*, 2010). However, in addition to the typical NE lesions, *C. perfringens* can induce changes in the intestinal homeostasis independently of evident morphological damage. It is likely that different clostridial toxins produced by *C. perfringens* disrupt intestinal ion transport and tight junction barrier which may lead to the loss of intestinal epithelial homeostasis (Laohachai *et al.*, 2003).

In view of the complexity of toxins, enzymes and virulence factors eventually produced by different strains of *C. perfringens* that can inhabit the gut of chickens, the scope of the potential changes produced by *C. perfringens* in the intestinal water transport can be exemplified by analyzing one of its most widespread toxins, alpha. All *C. perfringens*, including NE and commensal strains, codify for alpha toxin, an enzyme with phospholipase C activity which induce a secretory response in the small intestine but only when applied to the serosal, partly produced by electrogenic secretion of chloride anions (Rehman *et al.*, 2006a, 2009). In in vitro experiments, mucosal application of α toxin in jejunal sheets mounted in Ussing chamber alone failed to affect short-circuit current (Isc), which represent the sum of electrogenic ion movement produced only by active transport under iso-osmotic and electroneutral experimental conditions (Rehman *et al.*, 2006a, 2009). However, alpha toxin caused a decrease in sodium-dependent glucose absorption from the gut mucosa of chickens (Rehman *et al.*, 2006b). Considering that in birds, glucose and other nutrients are absorbed from the intestinal lumen by active transport coupled with Na^+ (W.A. Awad *et al.*, 2005) concurrently with water absorption, inhibition of this process might trigger important effects the intestinal fluid homeostasis. Therefore, this reduced response to electrogenic glucose absorption show that alpha toxin would be able to induce water imbalance or diarrhea not only by a secretory chloride pathway but also by sodium malabsorption or impairment of a substrate-dependent Na^+-absorption from the jejunum (Rehman *et al.*, 2009). The access of alpha toxin to the basolateral side of epithelial cells is facilitated when the gut integrity is weakened by hydrolytic enzymes including collagenases, sialidases, and hyaluronidases normally secreted by *C. perfringens* (Timbermont *et al.*, 2011). Intraduodenal infusion or oral inoculation of a crude alpha toxin in chickens induced necrosis of the small intestinal epithelium (Al-Sheikhly & Truscott, 1977).

A pathogenic strain of *C. perfringens* was observed to increase paracellular permeability of small intestinal tissues from infected chickens (Collier *et al.*, 2003). The increased epithelial permeability was positively correlated with the occurrence of necrotic lesions (Collier *et al.*, 2003).

E. coli infections in chickens produce important economic losses associated with avian pathogenic *Escherichia coli* (APEC) infections producing extra-intestinal disease (Antão *et al.*, 2008) although most of the resident intestinal E. coli strains of poultry are non-pathogenic and enteritis caused by them is rather uncommon compared to extra-intestinal diseases caused by the avian pathogenic variety. However, Enteropathogenic *E. coli* (EPEC), which affect the intestinal barrier function in humans, has been also isolated from clinically healthy chickens (Dhama *et al.*, 2013). In contrast to the observed effects in mammals, a decreased paracellular conductance (a tighter paracellular barrier) was observed in chickens exposed to these E. coli (Wageha A. Awad *et al.*, 2014). Similarly, asymptomatic salmonella infection in chicken produces an acute decrease in the intestinal ion permeability of chicken exposed to the presence of *Salmonella* or their LPS (W.A. Awad *et al.*, 2012), clarifying why chickens do not develop evident diarrhea after Salmonella infection. Between the complex process of the interaction of Salmonella spp. and the intestinal mucosa, the effects on intestinal ion transport. including the transepithelial migration of neutrophils through Tight junctions into the paracellular spaces causing fluid loss (Wallis & Galyov, 2000). In contrast, the inflammatory response to salmonellosis trigger important alterations in the ion transport (Quach *et al.*, 2022) and it has been difficult to deduce just the effect on ion transport.

6 The immune system and the epithelial water transport

The multilayered intestinal barrier of the gastrointestinal tract prevents the entry of various factors, like physical, chemical, and biological stimuli from the lumen while regulating the physiological absorption of water, nutrients, and electrolytes from the lumen as well as water secretion into the lumen (Farré *et al.*, 2020; Kelly *et al.*, 2015; Tabler *et al.*, 2020). This multiple barrier system consists of an outer mucus layer, a physical epithelial cell layer, and an inner functional immune barrier Thus, such an intestinal barrier, consisting of a surface epithelium with intercellular tight junctions, not only permits the selective movement of ions, nutrients and water, but also restricts the translocation of microbes and toxins from the lumen (Podolsky, 1997). Remarkably, tight junctions are under a continuous threat from proinflammatory mediators such as pathogenic microorganisms and toxins, promoting inflammation and the development of disease.

Intestinal inflammation and epithelial barrier function: The epithelial cells forming the intestinal barrier are in constant communication with the underlying immune cells in the lamina propria; thus, guaranteeing the immune maintenance of the protective barrier (Okumura & Takeda, 2017). During homeostasis, mucosal T cells have a minimal proliferative role and reduced cytokine production enabling them to regulate the balance between immune tolerance to the gut microbiota, effective immunity to environmental insults, and intestinal permeability (Hoytema Van Konijnenburg *et al.*, 2017). The transport mechanisms that underlie this physiological process are also subject to differential epithelium-T cell communication that mediate differential effects on intestinal permeability of water and ionic flux during *acute* and *chronic* inflammation (Kiela & Ghishan, 2016b; Le *et al.*, 2021).

Using an *in vitro* 3- and 2-dimensional co-culture model of acutely or chronically activated T cells, Le and colleagues (Le *et al.*, 2021) identified a biphasic epithelial permeability response. Interacting with acutely activated T cells, the epithelial barrier increased it barrier function with a reduction in ion and water flux across the barrier. However, the interaction with chronically activated T cells disrupted the epithelial barrier by inducing alterations in epithelial cell morphology, increasing cell death, induction of a efflux of ion and water, and increased expression of pro-inflammatory cytokines.

Cytokines and intestinal permeability: Although the mechanisms underlying the intestinal barrier dysfunction are not understood, the central role of pro-inflammatory cytokines in the underlying mechanisms leading to impairment of the intestinal barrier have been characterized. A number of pro-inflammatory cytokines and mediators including interleukin-1beta (IL-1β), tumor necrosis factor-alpha (TNF-α), interferon-gamma (IFN-γ), IL-6, and insulin-like growth factor (IGF-1), up-regulated during inflammation, have all been shown to stimulate increased intestinal tight junction permeability of both the pore and leak pathways (R. Al-Sadi *et al.*, 2009; R.M. Al-Sadi & Ma, 2007; Capaldo & Nusrat, 2009; Ma *et al.*, 2004, 2018).

Further, pro-inflammatory cytokines production leads to down-regulation of tight junction protein expression, promote cytoskeletal reorganization, decrease barrier function, and induction of epithelial cell apoptosis due to increased oxidative stress resulting in increased intestinal permeability through the unrestricted pathway which allows free passage of water, ions, macromolecules, and pathogens out of the lumen and into the serosal side of the epithelial barrier (De Barboza *et al.*, 2017; Edogawa *et al.*, 2020; Funderburg *et al.*, 2013; Liu *et al.*, 2019). Alternatively, anti-inflammatory cytokines including transforming growth factor-beta (TGF-β), IL-4, and IL-10, increase tight junction function; thereby, improving epithelial cell permeability (Ceponis *et al.*, 2000; Howe *et al.*, 2002; Roche *et al.*, 2000; Smith *et al.*, 2021).

7 Aquaporins and immunity

Aquaporins are highly conserved transmembrane proteins that facilitate bidirectional water across the cell membrane under osmotic gradients (Borgnia *et al.*, 1999). Aquaporins have been identified in the intestine, kidney, brain, liver, and reproductive organs of all production animals including poultry (Nowak *et al.*, 2017; Orlowski *et al.*, 2017; Sugiura *et al.*, 2008; Zaniboni & Bakst, 2004). Functionally, aquaporins in the intestine regulate water transport secretion of intestinal fluids that facilitate absorption and digestion nutrients and sustain the balance of intestinal fluid movement and the homoeostasis of the serosal and lumenal environments (S. Li *et al.*, n.d.; Liao *et al.*, 2021).

Aquaporins are found in the cell membranes of cells that mediate both the innate and acquired immunity. Functionally, aquaporins expressed on these immune cells interact with the cytoskeleton and signaling pathways of the cells. These immune cells (granulocytes, monocytes, lymphocytes) must respond to a stimulus by undergoing a rapid series of morphological alterations. These alterations are dependent on the cells' ability to alter their cytoskeleton via modulating water and solute permeability across the cell membrane by activating the aquaporin functions (Maidhof *et al.*, 2014; Meli *et al.*, 2018). Therefore, aquaporins have been found to be associated with the development of a number of cellular inflammatory processes such as apoptosis, immune cell priming, phagocytosis, cellular migration, and inflammasome activation (Boyle *et al.*, 2013; Compan *et al.*, 2012; da Silva *et al.*, 2021; Holm *et al.*, 2016; Jablonski *et al.*, 2004; Schorn *et al.*, 2011; Zhu *et al.*, 2011).

8 Diet, feed additives and epithelial water transport

In the small intestine, water transport is controlled by ion flux across the epithelial barrier. Under normal physiological conditions, many ion channels and exchangers are present in the intestinal epithelium, maintaining the balance of intestinal absorption and secretion (Kagnoff, 2014). Nutrient-coupled Na^+ absorption, electroneutral NaCl absorption, electrogenic Na^+ absorption and Cl− secretion are the main mechanisms connected with ion fluxes in the gut (Kato & Romero, 2011). Active sodium transport is one of the main drivers of water absorption and in the upper intestine both sodium and water absorption are largely dependent on the presence of glucose, and secondarily in other carbohydrates and amino acids. Hence, intestinal epithelial ion transport and concomitant water transport, are highly influenced by the diet as it is known that animal feed and water consumption are very closely related (LOTT *et al.*, 2003; MANNING *et al.*, 2007).

During animal feeding, either birds or mammals, digestive capacity needs to be physiologically adjusted to nutrient load, so a positive relationship is often observed between dietary intake of macronutrients and intestinal activity of the enzymes and transporters of their respective constituents. The ability of the intestine to adapt to these variable conditions both in type and quantity of nutrients will also influence the effectiveness of maintaining the efficiency of water absorption in the gastrointestinal tract. In all animals, as mentioned in previous sections, the small intestine's surface area is magnified by villi, cell-lined finger-like projections inward into the lumen, which are separated at their base by villus crypts that contain undifferentiated stem cells (Starck, 2003). Considering allometry, the small intestine's nominal surface area (calculated from intestinal length and perimeter) increases with a body mass exponent like that of metabolism. However, birds in general have a smaller nominal surface area compared to mammals, therefore a larger surface enlargement factor of a greater villous surface in birds may largely compensate for their smaller surface area. According to Karasov and Caviedes-Vidal (Karasov & Caviedes-Vidal, 2021), the bird increases paracellular absorption due to this enlargement, leading to more (per cm^2 NSA) tight junctions and greater clearance of passively absorbed compounds. As Brun *et al.* (Brun *et al.*, 2019) shown for bats (flying mammals) comparing to small non-flying mammals, it would be expected that birds also exceed mammals in number of enterocytes/cm^2 NSA and thus more cell-cell junctions per cm^2 NSA as the diameter of enterocytes in the epithelial layer of villi do not seem to differ between the taxa or be influenced by body mass (Brun *et al.*, 2019). Therefore, in contrast to mammals, this larger number of enterocytes in birds would influence not only the capacity for paracellular absorption of ions, water and nutrients but also would produce a potential higher influence on gut health if it is dysregulated under pathophysiological conditions.

At cellular level, the microvilli or the brush border membrane, cylindrical membrane projections of the apical membrane of enterocytes, hold a diversity of enzymes, transporters and channels that perform the final stages of intestinal digestion and absorption of carbohydrates and protein as well as ions . In the small intestine, in the proximal caecum and in the rectum of birds, glucose is absorbed by Na^+-dependent D-glucose cotransporters (SGLT1) present in the brush-border membrane (Amat *et al.*, 1999; Bindslev *et al.*, 1997; Garriga *et al.*, 1999, 2001). In contrast to the large intestine of mammals, birds can transport amino acids and glucose with Na^+ by mechanisms equivalent to those present in the small intestine (Bindslev *et al.*, 1997; Elbrønd *et al.*, 1998; Ferrer *et al.*, 1986; Holtug & Skadhauge, 1982; Laverty *et al.*, 2006; Planas *et al.*, 1986). Then, GLUT2 on the basolateral surface of enterocytes facilitates the release of hexoses into the circulatory system for reuptake by other cells (Wright *et al.*, 2004). When monosaccharide levels are high, GLUT2 may also facilitate hexose uptake from the gut lumen (Kellett & Brot-Laroche, 2005) but the knowledge regarding GLUT members and functions in birds is not completely known. Glucose transport mediated by SGLT1 is also driven by the sodium gradient generated by the sodium potassium ATPase (Na^+/K^+-ATPase) found on the basolateral membrane (Braun & Sweazea, 2008; H. Li *et al.*, 2020; Qiagen, 2014; Röhe *et al.*, 2017; Thorsen *et al.*, 2014). In chickens, expression levels of glucose-absorption-related genes remain unchanged between the age of 1- and 5 wk old, although the active transport of glucose across the jejunal epithelium declines (Shibata *et al.*, 2019). The rate-limiting factor in glucose absorption from starch containing diets is the digestion process (Riesenfeld *et al.*, 1980). So, it seems that this lower absorption of carbohydrates mediated by transporters can be compensated by a higher passive paracellular absorptive path between adjacent enterocytes (Karasov & Caviedes-Vidal, 2021). However, despite the fact that carbohydrates and sugars are one of the main sources of energy in the diet and are one of the major nutritional components, their functions in the regulation of the intestinal barrier have not been clearly defined in animals in general. Regarding dietary amino acids, these nutritional components are also transported across the brush border membrane of the intestine via mainly Na^+-dependent and Na^+-independent transporters in response to acidic, neutral, and/or basic amino acids (Bröer, 2008). An age-dependent increase in mRNA expression levels of Na^+-dependent neutral amino acid transporter (B0AT1) and anionic amino acid (aspartate and glutamate) transporter (EAAT3) are observed in the ileum, but not in the duodenum and jejunum, of broilers (Gilbert *et al.*, 2007).

On the other hand, the sodium transport does not seem to be entirely coupled to the process of nutrient digestion in chickens, but intestinal epithelial ion transport adapts extraordinarily to fluctuations in the diet. For example, it has been observed that an increase in microvillous surface area and increase in transepithelial sodium transport with relatively greater numbers of open sodium channels in the cell membranes of the lower intestine were observed following transfer of hens from a high to a low NaCl diet on which they were kept for at least 3 wk (Sodring Elbrond *et al.*, 1991). When chickens are switched to diets containing low levels of NaCl, a significant increase in the short-circuit current in the ileum, caecum and rectum is observed (Amat *et al.*, 1988; Grubb *et al.*, 1987; Skadhauge *et al.*, 1985) and higher expression of the Na^+-H^+ exchanger isoform 2(NHE2) (Donowitz *et al.*, 1998) and the epithelial Na^+channel (ENaC) (O. Goldstein *et al.*, 1997). Adaptation to low sodium diets also reduces the capacity of the large intestine of chickens to transport glucose and amino acids (Lind *et al.*, 1980), also the capacity to accumulate hexoses by jejunal enterocytes is partly reduced (Jaso *et al.*, 1995), whereas this capacity is completely abolished in enterocytes from the rectum (Jaso *et al.*, 1995; Laverty *et al.*, 2006; Lind *et al.*, 1980). Also, a low sodium diet increases Na^+/H^+ exchange in the chicken ileum and colon (Donowitz *et al.*, 1998).

9 Litter moisture, microbiota modulation, and pathogens

Typically, in intensive commercial broiler production, birds are reared in floors using different types of bedding material that include wood shavings, sawdust, rice or other grains hulls, bagasse, chopped straw, several different by-products, sand, corn cobs, and dried leaves (Monira *et al.*, 2003; Škrbić *et al.*, 2012; Swain & Sundaram, 2000). Although bedding material source varies, litter itself is comprised by the bedding material plus the excreta, feed, feathers, water, and an important and dynamic microbial community. Litter management and disposal is an important cost and challenge to the poultry industry, not only because it affects the flock health and productivity (Bjedov *et al.*, 2013; Garcês *et al.*, 2013; Macklin *et al.*, 2006), but it may act as potential reservoir and transmission vehicle for pathogens and potential pathogens. In this context, water is a determinant component of the litter that strongly influence the dynamic of the microorganisms in the environment and therefore, in the animals. The proliferation of pathogenic bacteria that increase bird diseases is more likely to be promoted in wet litter (Swiatkiewicz *et al.*, 2017) affecting gut health, respiratory infections, skin infections and lameness.

Litter becomes wet when the rate of water addition exceeds the rate of removal (evaporation) (Collett, 2007) and moisture content is usually expressed as the mass of water divided by mass of moist litter. However, the amount the water in the litter is not the best parameter reflexing freedom and availability of water in the litter. Water activity (Aw) is a thermodynamic property of the relative availability or freedom of water in a determined material. The Aw of the litter will modulate microbiota and may also affect the immune system of birds (K.W. Lee *et al.*, 2011). Labuza and Altunakar (Labuza & Altunakar, 2020) reported that different materials might show the same Aw having different moisture content. Aw is generally accepted a better measure of litter quality than total moisture content since it is a closer parameter that reflect microbial, chemical, and physical properties of natural products (Barbosa-Cánovas *et al.*, 2020). Aw is the parameter determining the process of water transfer from the excreta into the litter and then from litter into the air. Aw is defined in a scale of 0–1.

It has been shown that the growth of bacteria and fungi can be controlled by keeping the litter Aw below the minimum limit for microbial growth (Fontana, 2008; Taoukis & Richardson, 2008). However, Aw increases non-linearly with litter moisture content. Fresh excreta had high moisture content (up to 83%) with correspondingly high-water activity 0.96–0.99 (van der Hoeven-Hangoor, 2014) but according to available data (Carr *et al.*, 1995) (Van Der Hoeven-Hangoor *et al.*, 2014) when litter moisture content reached 38–55%, Aw increased to 0.98–0.99. This is important to consider how increasing levels of water in the excreta and later transfer to the litter can modulate microorganism growth and limit or support determined species to increase their presence. Based on testing of food materials, nominal Aw values for selected microorganisms with relevance in poultry production have been determined at 0.75–0.85 for Aspergillus spp.; 0.86–0.90 for Staphylococcus spp.; 0.92–0.95 for Salmonella spp.; 0.95 for *E. coli*; 0.93–0.97 for *Clostridium spp.*; and 0.98 for *Campylobacter spp.* (Fontana, 2008; Taoukis & Richardson, 2008). These growth limiting Aw values also depend on other factors including pH, temperature, oxygen, nutrient availability and presence of inhibitors (Barbosa-Cánovas *et al.*, 2020). However, Aw is crucial and it has been stablished that the multiplication of *Salmonella* can be reduced limiting Aw in the litter (Carr *et al.*, 1995), which create a cleaner environment for poultry production quality (Van Der Hoeven-Hangoor *et al.*, 2014).

Evidently, any disease that triggers diarrhea is likely to contribute to wet litter and later microbiota development. Coccidiosis is one of the most frequently disease in commercial farms, where *Eimeria maxima, Eimeria acervulina and Eimeria necatrix* are the most relevant species. *C. perfringens* Necrotic enteritis (NE) is other disease entities that is often interrelated with coccidiosis to produce gut health challenges. Dysbacteriosis, is another non-specific bacterial enteritis, which is relevant in the normal clinic analysis of animals in commercial farms. Besides non-infectious and infectious agents are thought to be the basis of dysbacteriosis, including soluble nonstarch polysaccharides, coccidia and *C. perfringens*, it is apparently produced by a malabsorption of nutrients and an imbalance in intestinal microbiota in poultry chickens (Bailey *et al.*, 2010). Several other diseases can be associated to wet litter, including Avian Infectious Bronchitis, Infectious Bursal Disease, Transmissible Viral Proventriculitis, Biogenic Amine Toxicity and the runting/stunting syndrome (Dunlop *et al.*, 2016).

Foot problems such as hock burn or pododermatitis are one of the more obvious effects of wet droppings, and the associated wet litter conditions. All the conditions related to wet droppings directly impact the bottom line. Feed conversion is a particularly important production parameter, and wet droppings are an early indication that digestion and absorption of the feed is not as good as it could be (Collett, 2012).

10 Final considerations

Modern poultry production has reached very high levels of efficiency and competitiveness that largely depend on more demanding management. The digestive system, and particularly the intestine, is key to reaching optimal efficiency levels. Water, the main nutrient of animals, and its homeostasis, has direct and indirect influences on animal health since it affects the establishment of the intestinal and environmental microbiota, the development of diseases, immunity and the digestive and nutrient absorption processes, and strongly influencing the productivity of the animals. The gastrointestinal tract absorbs a large volume of fluids daily and this transport of water through the intestine is closely related to the movement of solutes, primarily associated with the transport of ions. Any disturbance in the gastrointestinal tract creates the potential for a large loss of total body water and electrolytes. In birds, although many physiological processes regarding water and ion transport in the gut are comparable to those in mammals, there are particular differences that can be important when considering optimization of the production system. However, the information available regarding birds is often scarce or inadequate, which does not allow for a complete understanding of certain processes. The intestine of birds, as in that of mammals, is covered by a simple epithelium with numerous ion transporters, exchangers, and channels that are present on both the apical and basolateral membranes of enterocytes, and their differential expression along of the crypt-villus axis within the various intestinal segments ensures efficient intestinal absorption and effective barrier function. In birds, the available information suggests that intestinal ion transport adapts efficiently to alterations in nutrition and the environment, but for example there is little information regarding the pathophysiological processes in birds of many infections that produce intestinal imbalances, including common pathogens such as C. perfringens or *E. coli*. While it is widely accepted that the gut microbiome influences the host (tight junctions, cytokines/immune responses, diarrhea, etc.), the reciprocal interaction of the host on the microbiome is less explored. The most recent information suggests that microbial exposure and nutritional management significantly affect intestinal physiology and intestinal ion transport, which are inducers of acute and chronic inflammation processes. In turn, inflammation deregulate ion absorption/secretion balance, closing the loop. Finally, dysregulation of normal ion transport in the bird intestine manifests as diarrhea, malabsorption, and intestinal inflammation, resulting in poor production efficiency. The role of the tight junction is important in all these processes, although there are discrepancies regarding its role in the initial step of development of inflammatory processes that can lead to deregulation of intestinal homeostasis, as well as its importance in the processes of nutrient absorption in birds compared to mammals. A better understanding of the normal and pathological interaction of water and ion balance in the intestine, together with the regulation of the epithelial barrier during digestive, infectious, and inflammatory processes, will help us to optimize the management of poultry in intensive farming to maximize gut health, well-being and production.

References

Al-Sadi, R., Boivin, M., & Ma, T. (2009). Mechanism of cytokine modulation of epithelial tight junction barrier. *Frontiers in Bioscience (Landmark Edition)*, *14*(7), 2765–2778. https://doi.org/10.2741/3413.

Al-Sadi, R.M., & Ma, T.Y. (2007). IL-1beta causes an increase in intestinal epithelial tight junction permeability. *Journal of Immunology (Baltimore, Md.: 1950)*, *178*(7), 4641–4649. https://doi.org/10.4049/JIMMUNOL.178.7.4641.

Al-Sheikhly, F., & Truscott, R.B. (1977). The pathology of necrotic enteritis of chickens following infusion of crude toxins of Clostridium perfringens into the duodenum. *Undefined*, *21*(2), 241–255. https://doi.org/10.2307/1589344.

Amasheh, S., Meiri, N., Gitter, A.H., Schöneberg, T., Mankertz, J., Schulzke, J.D., & Fromm, M. (2002). Claudin-2 expression induces cation-selective channels in tight junctions of epithelial cells. *Journal of Cell Science*, *115*(Pt 24), 4969–4976. https://doi.org/10.1242/JCS.00165.

Amat, C., M. Planas, J., Díez, A., & Moretó, M. (1988). Does chicken rectal adaptation to a low nacı diet involve changes in paracellular permeability? *Comparative Biochemistry and Physiology Part A: Physiology*, *91*(2), 367–370. https://doi.org/10.1016/0300-9629(88)90432-X.

Amat, C., Piqueras, J.A., Planas, J.M., & Moretó, M. (1999). Electrical properties of the intestinal mucosa of the chicken and the effects of luminal glucose. *Poultry Science*, *78*(8), 1126–1131. https://doi.org/10.1093/ps/78.8.1126.

Antão, E.M., Glodde, S., Li, G., Sharifi, R., Homeier, T., Laturnus, C., Diehl, I., Bethe, A., Philipp, H.C., Preisinger, R., Wieler, L.H., & Ewers, C. (2008). The chicken as a natural model for extraintestinal infections caused by avian pathogenic Escherichia coli (APEC). *Microbial Pathogenesis*, *45*(5–6), 361–369. https://doi.org/10.1016/J.MICPATH.2008.08.005.

Awad, W.A., Aschenbach, J.R., Khayal, B., Hess, C., & Hess, M. (2012). Intestinal epithelial responses to Salmonella enterica serovar Enteritidis: effects on intestinal permeability and ion transport. *Poultry Science*, *91*(11), 2949–2957. https://doi.org/10.3382/PS.2012-02448.

Awad, W.A., Rehman, H., Böhm, J., Razzazi-Fazeli, E., & Zentek, J. (2005). Effects of luminal deoxynivalenol and L-proline on electrophysiological parameters in the jejunums of laying hens. *Poultry Science*, *84*(6), 928–932. https://doi.org/10.1093/PS/84.6.928.

Awad, Wageha A., Hess, C., & Hess, M. (2017). Enteric Pathogens and Their Toxin-Induced Disruption of the Intestinal Barrier through Alteration of Tight Junctions in Chickens. *Toxins 2017, Vol. 9, Page 60*, *9*(2), 60. https://doi.org/10.3390/TOXINS9020060.

Awad, Wageha A., Hess, C., & Hess, M. (2018). Re-thinking the chicken-Campylobacter jejuni interaction: a review. *Avian Pathology: Journal of the W.V.P.A.*, *47*(4), 352–363. https://doi.org/10.1080/03079457.2018.1475724.

Awad, Wageha A., Hess, C., Khayal, B., Aschenbach, J.R., & Hess, M. (2014). In vitro exposure to Escherichia coli decreases ion conductance in the jejunal epithelium of broiler chickens. *PloS One*, *9*(3). https://doi.org/10.1371/JOURNAL.PONE.0092156.

Awe, O., Kang, K., Ibrahim, M., Ali, A., … M.E.-A., & 2015, undefined. (n.d.). Age-related susceptibility of turkeys to enteric viruses. *Meridian.Allenpress.Com*. Retrieved October 4, 2022, from https://meridian.allenpress.com/avian-diseases/article-abstract/59/2/207/209877.

Bailey, M.T., Dowd, S.E., Parry, N.M.A., Galley, J.D., Schauer, D.B., & Lyte, M. (2010). Stressor exposure disrupts commensal microbial populations in the intestines and leads to increased colonization by Citrobacter rodentium. *Infection and Immunity*, *78*(4), 1509–1519. https://doi.org/10.1128/IAI.00862-09.

Barbosa-Cánovas, G.V., Fontana, A.J., Schmidt, S.J., & Labuza, T.P. (2020). Water Activity in Foods: Fundamentals and Applications. *Undefined*, 1–435. https://doi.org/10.1002/9780470376454.

Barmeyer, C., Fromm, M., & Schulzke, J.D. (2016). Active and passive involvement of claudins in the pathophysiology of intestinal inflammatory diseases. *Pflügers Archiv – European Journal of Physiology 2016 469:1*, *469*(1), 15–26. https://doi.org/10.1007/S00424-016-1914-6.

Baxendale, W., & Mebatsion, T. (2004). The isolation and characterisation of astroviruses from chickens. *Avian Pathology: Journal of the W.V.P.A.*, *33*(3), 364–370. https://doi.org/10.1080/0307945042000220426.

Baxter, M.F.A., Merino-Guzman, R., Latorre, J.D., Mahaffey, B.D., Yang, Y., Teague, K.D., Graham, L.E., Wolfenden, A.D., Hernandez-Velasco, X., Bielke, L.R., Hargis, B.M., & Tellez, G. (2017). Optimizing fluorescein isothiocyanate dextran measurement as a biomarker in a 24-h feed restriction model to induce gut permeability in broiler chickens. *Frontiers in Veterinary Science*, *4*(APR), 56. https://doi.org/10.3389/FVETS.2017.00056/BIBTEX.

Bindslev, N., Hirayama, B.A., & Wright, E.M. (1997). Na/D-glucose cotransport and SGLT1 expression in hen colon correlates with dietary Na+. *Comparative Biochemistry and Physiology. Part A, Physiology*, *118*(2), 219–227. https://doi.org/10.1016/S0300-9629(97)00071-6.

Bjedov, S., Žikić, D., Perić, L., Stojčić, M.Đ., & Milošević, N. (2013). Effect of different litter treatments on production performance of broiler chickens. *Biotechnology in Animal Husbandry*, *29*(4), 625–630. https://doi.org/10.2298/BAH1304625B.

Bonis, V., Rossell, C., & Gehart, H. (2021). The Intestinal Epithelium – Fluid Fate and Rigid Structure From Crypt Bottom to Villus Tip. *Frontiers in Cell and Developmental Biology*, *9*. https://doi.org/10.3389/FCELL.2021.661931.

Borgnia, M., Nielsen, S., Engel, A., & Agre, P. (1999). Cellular and Molecular Biology of the Aquaporin Water Channels. Annual Review Biochemestry, 68, 425–458. http://doi.org/10.1146/annurev.biochem.68.1.425.

Boyle, J.P., Bryant, C.E., & Monie, T.P. (2013). Cell Swelling and the NLRP3 Inflammasome. *Immunity*, *38*(3), 399. https://doi.org/10.1016/J.IMMUNI.2013.02.006.

Braun, E.J. (2003). Regulation of renal and lower gastrointestinal function: role in fluid and electrolyte balance. *Comparative Biochemistry and Physiology Part A: Molecular & Integrative Physiology, 136*(3), 499–505. https://doi.org/10.1016/S1095-6433(03)00170-3.

Braun, E.J., & Sweazea, K.L. (2008). Glucose regulation in birds. *Comparative Biochemistry and Physiology Part B: Biochemistry and Molecular Biology, 151*(1), 1–9. https://doi.org/10.1016/J.CBPB.2008.05.007.

Bröer, S. (2008). Amino acid transport across mammalian intestinal and renal epithelia. *Physiological Reviews, 88*(1), 249–286. https://doi.org/10.1152/PHYSREV.00018.2006.

Brugaletta, G., Teyssier, J.R., Rochell, S.J., Dridi, S., & Sirri, F. (2022). A review of heat stress in chickens. Part I: Insights into physiology and gut health. *Frontiers in Physiology, 13*, 1535. https://doi.org/10.3389/FPHYS.2022.934381/BIBTEX.

Brun, A., Fernández Marinone, G., Price, E.R., Nell, L.A., Simões, B.M.V., Castellar, A., Gontero-Fourcade, M., Cruz-Neto, A.P., Karasov, W.H., & Caviedes-Vidal, E. (2019). Morphological bases for intestinal paracellular absorption in bats and rodents. *Journal of Morphology, 280*(9), 1359–1369. https://doi.org/10.1002/JMOR.21037.

Camilleri, M. (2019). Leaky gut: mechanisms, measurement and clinical implications in humans. *Gut, 68*(8), 1516–1526. https://doi.org/10.1136/GUTJNL-2019-318427.

Campbell, M.T., Hile, K.L., Zhang, H., Asanuma, H., Vanderbrink, B.A., Rink, R.R., & Meldrum, K.K. (2011). Toll like receptor 4: A novel signaling pathway during renal fibrogenesis. *The Journal of Surgical Research, 168*(1), e61. https://doi.org/10.1016/J.JSS.2009.09.053.

Capaldo, C.T., & Nusrat, A. (2009). Cytokine regulation of tight junctions. *Biochimica et Biophysica Acta – Biomembranes, 1788*(4), 864–871. https://doi.org/10.1016/J.BBAMEM.2008.08.027.

Carr, L.E., Mallinson, E.T., Tate, C.R., Miller, R.G., Russek-Cohen, E., Stewart, L.E., Opara, O.O., & Joseph, S.W. (1995). Prevalence of Salmonella in broiler flocks: Effect of litter water activity, house construction, and watering devices. *Avian Diseases, 39*(1), 39–44. https://doi.org/10.2307/1591980.

Ceponis, P.J.M., Botelho, F., Richards, C.D., & Mckay, D.M. (2000). *Interleukins 4 and 13 Increase Intestinal Epithelial Permeability by a Phosphatidylinositol 3-Kinase Pathway LACK OF EVIDENCE FOR STAT 6 INVOLVEMENT**. https://doi.org/10.1074/jbc.M003516200.

Cheeke, P.R. (2000). Actual and potential applications of and saponins in human and animal nutrition. *Journal of Animal Science, 77*(E-Suppl), 1. https://doi.org/10.2527/jas2000.00218812007700es0009x.

Collett, S.R. (2012). Nutrition and wet litter problems in poultry. *Animal Feed Science and Technology, 173*(1–2), 65–75. https://doi.org/10.1016/J.ANIFEEDSCI.2011.12.013.

Collier, C.T., Van der Klis, J.D., Deplancke, B., Anderson, D.B., & Gaskins, H.R. (2003). Effects of tylosin on bacterial mucolysis, Clostridium perfringens colonization, and intestinal barrier function in a chick model of necrotic enteritis. *Antimicrobial Agents and Chemotherapy, 47*(10), 3311–3317. https://doi.org/10.1128/AAC.47.10.3311-3317.2003.

Compan, V., Baroja-Mazo, A., López-Castejón, G., Gomez, A.I., Martínez, C.M., Angosto, D., Montero, M.T., Herranz, A.S., Bazán, E., Reimers, D., Mulero, V., & Pelegrín, P. (2012).

Cell volume regulation modulates NLRP3 inflammasome activation. *Immunity, 37*(3), 487–500. https://doi.org/10.1016/J.IMMUNI.2012.06.013.

Cooper, K.K., & Songer, J.G. (2009). Necrotic enteritis in chickens: a paradigm of enteric infection by Clostridium perfringens type A. *Anaerobe, 15*(1–2), 55–60. https://doi.org/10.1016/J.ANAEROBE.2009.01.006.

Cooper, K.K., & Songer, J.G. (2010). Virulence of Clostridium perfringens in an experimental model of poultry necrotic enteritis. *Veterinary Microbiology, 142*(3–4), 323–328. https://doi.org/10.1016/j.vetmic.2009.09.065.

da Silva, I.V., Cardoso, C., Martínez-Banaclocha, H., Casini, A., Pelegrín, P., & Soveral, G. (2021). Aquaporin-3 is involved in NLRP3-inflammasome activation contributing to the setting of inflammatory response. *Cellular and Molecular Life Sciences: CMLS, 78*(6), 3073–3085. https://doi.org/10.1007/S00018-020-03708-3.

Day, J.M., & Zsak, L. (2013). Recent progress in the characterization of avian enteric viruses. *Avian Diseases, 57*(3), 573–580. https://doi.org/10.1637/10390-092712-REVIEW.1.

De Barboza, G.D., Guizzardi, S., Moine, L., & Tolosa de Talamoni, N. (2017). Oxidative stress, antioxidants and intestinal calcium absorption. *World Journal of Gastroenterology, 23*(16), 2841–2853. https://doi.org/10.3748/WJG.V23.I16.2841.

Dhama, K., Rajagunalan, S., Chakraborty, S., Verma, A.K., Kumar, A., Tiwari, R., & Kapoor, S. (2013). Food-borne pathogens of animal origin-diagnosis, prevention, control and their zoonotic significance: a review. *Pakistan Journal of Biological Sciences: PJBS, 16*(20), 1076–1085. https://doi.org/10.3923/PJBS.2013.1076.1085.

Diao, H., Jiao, A.R., Yu, B., Mao, X.B., & Chen, D.W. (2019). Gastric infusion of short-chain fatty acids can improve intestinal barrier function in weaned piglets. *Genes & Nutrition, 14*(1). https://doi.org/10.1186/S12263-019-0626-X.

Dong, Y., Zeng, C.Q.Y., Ball, J.M., Estes, M.K., & Morris, A.P. (1997). The rotavirus enterotoxin NSP4 mobilizes intracellular calcium in human intestinal cells by stimulating phospholipase C-mediated inositol 1,4,5-trisphosphate production. *Proceedings of the National Academy of Sciences, 94*(8), 3960–3965. https://doi.org/10.1073/PNAS.94.8.3960.

Donowitz, M., De La Horra, C., Calonge, M.L., Wood, I.S., Dyer, J., Gribble, S.M., De Sanchez Medina, F., Tse, C.M., Shirazi-Beechey, S.P., & Ilundain, A.A. (1998). In birds, NHE2 is major brush-border Na+/H+ exchanger in colon and is increased by a low-NaCl diet. *The American Journal of Physiology, 274*(6). https://doi.org/10.1152/AJPREGU.1998.274.6.R1659.

Dunlop, M.W., Moss, A.F., Groves, P.J., Wilkinson, S.J., Stuetz, R.M., & Selle, P.H. (2016). The multidimensional causal factors of 'wet litter' in chicken-meat production. *Science of The Total Environment, 562*, 766–776. https://doi.org/10.1016/J.SCITOTENV.2016.03.147.

Eckmann, L., Kagnoff, M.F., & Fierer, J. (1993). Epithelial cells secrete the chemokine interleukin-8 in response to bacterial entry. *Infection and Immunity, 61*(11), 4569–4574. https://doi.org/10.1128/IAI.61.11.4569-4574.1993.

Edogawa, S., Edwinson, A.L., Peters, S.A., Chikkamenahalli, L.L., Sundt, W., Graves, S., Gurunathan, S.V., Breen-Lyles, M., Johnson, S., Dyer, R., Graham, R., Chen, J., Kashyap, P., Farrugia, G., & Grover, M. (2020). Serine proteases as luminal mediators of intestinal barrier dysfunction and symptom severity in IBS. *Gut, 69*(1), 62–73. https://doi.org/10.1136/GUTJNL-2018-317416.

Elbrønd, V.S., Skadhauge, E., Thomsen, L., & Dantzer, V. (1998). Morphological adaptations to induced changes in transepithelial sodium transport in chicken lower intestine (coprodeum): A study of resalination, aldosterone stimulation, and epithelial turnover. *Cell and Tissue Research, 292*(3), 543–552. https://doi.org/10.1007/s004410051083.

Emami, N.K., Greene, E.S., Kogut, M.H., & Dridi, S. (2021). Heat Stress and Feed Restriction Distinctly Affect Performance, Carcass and Meat Yield, Intestinal Integrity, and Inflammatory (Chemo)Cytokines in Broiler Chickens. *Frontiers in Physiology, 12*, 1148. https://doi.org/10.3389/FPHYS.2021.707757/BIBTEX.

Farré, R., Fiorani, M., Rahiman, S.A., & Matteoli, G. (2020). Intestinal Permeability, Inflammation and the Role of Nutrients. *Nutrients, 12*(4). https://doi.org/10.3390/NU12041185.

Feng, Y., Wang, Y., Wang, P., Huang, Y., & Wang, F. (2018). Short-Chain Fatty Acids Manifest Stimulative and Protective Effects on Intestinal Barrier Function Through the Inhibition of NLRP3 Inflammasome and Autophagy. *Cellular Physiology and Biochemistry: International Journal of Experimental Cellular Physiology, Biochemistry, and Pharmacology, 49*(1), 190–205. https://doi.org/10.1159/000492853.

Ferrer, R., Planas, J.M., & Moretó, M. (1986). Characteristics of the chicken proximal cecum hexose transport system. *Pflugers Archiv: European Journal of Physiology, 407*(1), 100–104. https://doi.org/10.1007/BF00580728.

Fontana, A.J. (2008). Appendix D: Minimum Water Activity Limits for Growth of Microorganisms. *Water Activity in Foods*, 405–405. https://doi.org/10.1002/9780470376454.APP4.

Frömter, E., & Diamond, J. (1972). Route of passive ion permeation in epithelia. *Nature: New Biology, 235*(53), 9–13. https://doi.org/10.1038/NEWBIO235009A0.

Funderburg, N.T., Stubblefield Park, S.R., Sung, H.C., Hardy, G., Clagett, B., Ignatz-Hoover, J., Harding, C.V., Fu, P., Katz, J.A., Lederman, M.M., & Levine, A.D. (2013). Circulating CD4+ and CD8+ T cells are activated in inflammatory bowel disease and are associated with plasma markers of inflammation. *Immunology, 140*(1), 87–97. https://doi.org/10.1111/IMM.12114.

Gao, Y., Meng, L., Liu, H., Wang, J., & Zheng, N. (2020). The Compromised Intestinal Barrier Induced by Mycotoxins. *Toxins, 12*(10). https://doi.org/10.3390/TOXINS12100619.

Garcês, A., Afonso, S.M.S., Chilundo, A., & Jairoce, C.T.S. (2013). Evaluation of different litter materials for broiler production in a hot and humid environment: 1. Litter characteristics and quality. *Journal of Applied Poultry Research, 22*(2), 168–176. https://doi.org/10.3382/JAPR.2012-00547.

Garriga, C., Planas, J.M., & Moretó, M. (2001). Aldosterone mediates the changes in hexose transport induced by low sodium intake in chicken distal intestine. *The Journal of Physiology, 535*(Pt 1), 197. https://doi.org/10.1111/J.1469-7793.2001.00197.X.

Garriga, C., Rovira, N., Moretó, M., & Planas, J.M. (1999). Expression of Na+-D-glucose cotransporter in brush-border membrane of the chicken intestine. *The American Journal of Physiology, 276*(2). https://doi.org/10.1152/AJPREGU.1999.276.2.R627.

Ghosh, S., Whitley, C.S., Haribabu, B., & Jala, V.R. (2021). Regulation of Intestinal Barrier Function by Microbial Metabolites. *Cellular and Molecular Gastroenterology and Hepatology*, *11*(5), 1463–1482. https://doi.org/10.1016/J.JCMGH.2021.02.007.

Gilbert, E.R., Li, H., Emmerson, D.A., Webb, K.E., & Wong, E.A. (2007). Developmental regulation of nutrient transporter and enzyme mRNA abundance in the small intestine of broilers. *Poultry Science*, *86*(8), 1739–1753. https://doi.org/10.1093/PS/86.8.1739.

Glenn, G.M., Francis, D.H., & Danielsen, E.M. (2009). Toxin-mediated effects on the innate mucosal defenses: Implications for enteric vaccines. *Infection and Immunity*, *77*(12), 5206–5215. https://doi.org/10.1128/IAI.00712-09/ASSET/B170308E-B838-4FAD-87C3-5BF3BFBCE2BC/ASSETS/GRAPHIC/ZII0120983240003.JPEG.

Goldstein, D.L. (1989). Absorption by the cecum of wild birds: Is there interspecific variation? *Journal of Experimental Zoology*, *252*(3 S), 103–110. https://doi.org/10.1002/JEZ.1402520517.

Goldstein, D.L., & Skadhauge, E. (2000). Renal and Extrarenal Regulation of Body Fluid Composition. *Sturkie's Avian Physiology*, 265–297. https://doi.org/10.1016/B978-012747605-6/50012-2.

Goldstein, O., Asher, C., & Garty, H. (1997). Cloning and induction by low NaCl intake of avian intestine Na+ channel subunits. *The American Journal of Physiology*, *272*(1 Pt 1). https://doi.org/10.1152/AJPCELL.1997.272.1.C270.

Grubb, B.R., Driscoll, S.M., & Bentley, P.J. (1987). Electrical PD, short-circuit current and fluxes of Na and Cl across avian intestine. *Journal of Comparative Physiology. B, Biochemical, Systemic, and Environmental Physiology*, *157*(2), 181–186. https://doi.org/10.1007/BF00692362.

Guttman, J.A., & Finlay, B.B. (2009). Tight junctions as targets of infectious agents. *Biochimica et Biophysica Acta*, *1788*(4), 832–841. https://doi.org/10.1016/J.BBAMEM.2008.10.028.

Hallstrom, K., & McCormick, B.A. (2011). Salmonella interaction with and passage through the intestinal mucosa: Through the lens of the organism. *Frontiers in Microbiology*, *2*(APR). https://doi.org/10.3389/FMICB.2011.00088.

Holm, A., Magnusson, K.E., & Vikström, E. (2016). Pseudomonas aeruginosa N-3-oxo-dodecanoyl-homoserine Lactone Elicits Changes in Cell Volume, Morphology, and AQP9 Characteristics in Macrophages. *Frontiers in Cellular and Infection Microbiology*, *6*(MAR), 32. https://doi.org/10.3389/FCIMB.2016.00032.

Holtug, K., & Skadhauge, E. (1982). NaCl transport across hen colon. *Pflügers Archiv 1982 394:3*, *394*(3), 222–225. https://doi.org/10.1007/BF00589095.

Howe, K., Gauldie, J., & McKay, D.M. (2002). TGF-β effects on epithelial ion transport and barrier: Reduced Cl– secretion blocked by a p38 MAPK inhibitor. *American Journal of Physiology – Cell Physiology*, *283*(6 52–6), 1667–1674. https://doi.org/10.1152/AJPCELL.00414.2001/ASSET/IMAGES/LARGE/H01221255004.JPEG.

Hoytema Van Konijnenburg, D.P., Reis, B.S., Pedicord, V.A., Farache, J., Victora, G.D., & Mucida, D. (2017). Intestinal Epithelial and Intraepithelial T Cell Crosstalk Mediates a Dynamic Response to Infection. *Cell*, *171*, 783–794. https://doi.org/10.1016/j.cell.2017.08.046.

Humphrey, S., Chaloner, G., Kemmett, K., Davidson, N., Williams, N., Kipar, A., Humphrey, T., & Wigley, P. (2014). Campylobacter jejuni is not merely a commensal in commercial broiler chickens and affects bird welfare. *MBio*, *5*(4). https://doi.org/10.1128/MBIO.01364-14.

Jablonski, E.M., Webb, A.N., McConnell, N.A., Riley, M.C., & Hughes, F.M. (2004). Plasma membrane aquaporin activity can affect the rate of apoptosis but is inhibited after apoptotic volume decrease. *American Journal of Physiology. Cell Physiology*, *286*(4). https://doi.org/10.1152/AJPCELL.00180.2003.

Jaso, M.J., Vial, M., & Moretó, M. (1995). Hexose accumulation by enterocytes from the jejunum and rectum of chickens adapted to high and low NaCl intake. *Pflugers Archiv: European Journal of Physiology*, *429*(4), 511–516. https://doi.org/10.1007/BF00704156.

John, L.J., Fromm, M., & Schulzke, J.D. (2011). Epithelial Barriers in Intestinal Inflammation. Https://Home.Liebertpub.Com/Ars, *15*(5), 1255–1270. https://doi.org/10.1089/ARS.2011.3892.

Kagnoff, M.F. (2014). The intestinal epithelium is an integral component of a communications network. *The Journal of Clinical Investigation*, *124*(7), 2841. https://doi.org/10.1172/JCI75225.

Karasov, W.H., & Caviedes-Vidal, E. (2021). Adaptation of intestinal epithelial hydrolysis and absorption of dietary carbohydrate and protein in mammals and birds. *Comparative Biochemistry and Physiology Part A: Molecular & Integrative Physiology*, *253*, 110860. https://doi.org/10.1016/J.CBPA.2020.110860.

Kato, A., & Romero, M.F. (2011). Regulation of electroneutral NaCl absorption by the small intestine. *Annual Review of Physiology*, *73*, 261–281. https://doi.org/10.1146/ANNUREV-PHYSIOL-012110-142244.

Kellett, G.L., & Brot-Laroche, E. (2005). Apical GLUT2: a major pathway of intestinal sugar absorption. *Diabetes*, *54*(10), 3056–3062. https://doi.org/10.2337/DIABETES.54.10.3056.

Kelly, J.R., Kennedy, P.J., Cryan, J.F., Dinan, T.G., Clarke, G., & Hyland, N.P. (2015). Breaking down the barriers: the gut microbiome, intestinal permeability and stress-related psychiatric disorders. *Frontiers in Cellular Neuroscience*, *9*(OCT). https://doi.org/10.3389/FNCEL.2015.00392.

Kiela, P.R., & Ghishan, F.K. (2016a). Physiology of Intestinal Absorption and Secretion. *Best Practice & Research Clinical Gastroenterology*, *30*(2), 145–159. https://doi.org/10.1016/J.BPG.2016.02.007.

Kiela, P.R., & Ghishan, F.K. (2016b). Physiology of Intestinal Absorption and Secretion. *Best Practice & Research. Clinical Gastroenterology*, *30*(2), 145–159. https://doi.org/10.1016/J.BPG.2016.02.007.

Koci, M.D., Moser, L.A., Kelley, L.A., Larsen, D., Brown, C.C., & Schultz-Cherry, S. (2003). Astrovirus Induces Diarrhea in the Absence of Inflammation and Cell Death. *Journal of Virology*, *77*(21), 11798. https://doi.org/10.1128/JVI.77.21.11798-11808.2003.

Labuza, T.P., & Altunakar, B. (2020). Water Activity Prediction and Moisture Sorption Isotherms. *Undefined*, 109–154. https://doi.org/10.1002/9780470376454.CH5.

Laohachai, K.N., Bahadi, R., Hardo, M.B., Hardo, P.G., & Kourie, J.I. (2003). The role of bacterial and non-bacterial toxins in the induction of changes in membrane transport: Implications for diarrhea. *Toxicon*, *42*(7), 687–707. https://doi.org/10.1016/j.toxicon.2003.08.010.

Laverty, G., Elbrønd, V.S., Árnason, S.S., & Skadhauge, E. (2006). Endocrine regulation of ion transport in the avian lower intestine. *General and Comparative Endocrinology, 147*(1), 70–77. https://doi.org/10.1016/J.YGCEN.2006.01.008.

Le, N., Mazahery, C., Nguyen, K., & Levine, A.D. (2021). ORIGINAL RESEARCH Regulation of Intestinal Epithelial Barrier and Immune Function by Activated T Cells. *Cellular and Molecular Gastroenterology and Hepatology, 11*, 55–76. https://doi.org/10.1016/j.jcmgh.2020.07.004.

Lee, K.W., Lillehoj, H.S., Jeong, W., Jeoung, H.Y., & An, D.J. (2011). Avian necrotic enteritis: Experimental models, host immunity, pathogenesis, risk factors, and vaccine development. *Poultry Science, 90*(7), 1381–1390. https://doi.org/10.3382/ps.2010-01319.

Lee, S.H. (2015). Intestinal permeability regulation by tight junction: implication on inflammatory bowel diseases. *Intestinal Research, 13*(1), 11. https://doi.org/10.5217/IR.2015.13.1.11.

Leeson, S., Namkung, H., Antongiovanni, M., & Lee, E.H. (2005). Effect of butyric acid on the performance and carcass yield of broiler chickens. *Poultry Science, 84*(9), 1418–1422. https://doi.org/10.1093/PS/84.9.1418.

Lehmann, K., Bell, T., Bowes, M.J., Amos, G.C.A., Gaze, W.H., Wellington, E.M.H., & Singer, A.C. (2016). Trace levels of sewage effluent are sufficient to increase class 1 integron prevalence in freshwater biofilms without changing the core community. *Water Research, 106*, 163–170. https://doi.org/10.1016/j.watres.2016.09.035.

Li, H., Cheng, J., Yuan, Y., Luo, R., & Zhu, Z. (2020). Age-related intestinal monosaccharides transporters expression and villus surface area increase in broiler and layer chickens. *Journal of Animal Physiology and Animal Nutrition, 104*(1), 144–155. https://doi.org/10.1111/JPN.13211.

Li, S., Li, C., Hormones, W.W.-V. and, & 2020, undefined. (n.d.). Molecular aspects of aquaporins. *Elsevier*. Retrieved October 4, 2022, from https://www.sciencedirect.com/science/article/pii/S0083672919300792.

Liao, S., Gan, L., Lv, L., & Mei, Z. (2021). The regulatory roles of aquaporins in the digestive system. *Genes & Diseases, 8*(3), 250. https://doi.org/10.1016/J.GENDIS.2019.12.011.

Lind, J., Munck, B.G., Olsen, O., & Skadhauge, E. (1980). Effects of sugars, amino acids and inhibitors on electrolyte transport across hen colon at different sodium chloride intakes. *The Journal of Physiology, 305*(1), 315–325. https://doi.org/10.1113/JPHYSIOL.1980.SP013365.

Liu, M., Sun, T., Li, N., Peng, J., Fu, D., Li, W., Li, L., & Gao, W.Q. (2019). BRG1 attenuates colonic inflammation and tumorigenesis through autophagy-dependent oxidative stress sequestration. *Nature Communications 2019 10:1, 10*(1), 1–15. https://doi.org/10.1038/s41467-019-12573-z.

Lott, B.D., Dozier, W.A., Simmons, J.D., & Roush, W.B. (2003). Water flow rates in commercial broiler houses. *Poultry Science, 82*, 102.

Ma, T.Y., Iwamoto, G.K., Hoa, N.T., Akotia, V., Pedram, A., Boivin, M.A., & Said, H.M. (2004). TNF-alpha-induced increase in intestinal epithelial tight junction permeability requires NF-kappa B activation. *American Journal of Physiology. Gastrointestinal and Liver Physiology, 286*(3). https://doi.org/10.1152/AJPGI.00173.2003.

Ma, T.Y., Nighot, P., & Al-Sadi, R. (2018). Tight Junctions and the Intestinal Barrier. *Physiology of the Gastrointestinal Tract: Sixth Edition, 1–2*, 587–639. https://doi.org/10.1016/B978-0-12-809954-4.00025-6.

Macklin, K.S., Hess, J.B., Bilgili, S.F., & Norton, R.A. (2006). Effects of In-House Composting of Litter on Bacterial Levels. *Journal of Applied Poultry Research, 15*(4), 531–537. https://doi.org/10.1093/JAPR/15.4.531.

Maidhof, R., Jacobsen, T., Papatheodorou, A., & Chahine, N.O. (2014). Inflammation Induces Irreversible Biophysical Changes in Isolated Nucleus Pulposus Cells. *PLoS ONE, 9*(6). https://doi.org/10.1371/JOURNAL.PONE.0099621.

Manning, L., Chadd, S.A., & Baines, R.N. (2007). Water consumption in broiler chicken: a welfare indicator. *World's Poultry Science Journal, 63*(1), 63–71. https://doi.org/10.1017/S0043933907001274.

Marques, R., & Boneca, I.G. (2011). Expression and functional importance of innate immune receptors by intestinal epithelial cells. *Cellular and Molecular Life Sciences: CMLS, 68*(22), 3661–3673. https://doi.org/10.1007/S00018-011-0829-9.

Martel, J., Chang, S.H., Ko, Y.F., Hwang, T.L., Young, J.D., & Ojcius, D.M. (2022). Gut barrier disruption and chronic disease. *Trends in Endocrinology & Metabolism, 33*(4), 247–265. https://doi.org/10.1016/J.TEM.2022.01.002.

Meli, R., Pirozzi, C., & Pelagalli, A. (2018). New perspectives on the potential role of aquaporins (AQPs) in the physiology of inflammation. *Frontiers in Physiology, 9*(FEB), 101. https://doi.org/10.3389/FPHYS.2018.00101/BIBTEX.

Meliopoulos, V.A., Marvin, S.A., Freiden, P., Moser, L.A., Nighot, P., Ali, R., Blikslager, A., Reddivari, M., Heath, R.J., Koci, M.D., & Schultz-Cherry, S. (2016). Oral administration of astrovirus capsid protein is sufficient to induce acute diarrhea in vivo. *MBio, 7*(6). https://doi.org/10.1128/mBio.01494-16.

Monaco, A., Ovryn, B., Axis, J., & Amsler, K. (2021). The Epithelial Cell Leak Pathway. *International Journal of Molecular Sciences 2021, Vol. 22, Page 7677, 22*(14), 7677. https://doi.org/10.3390/IJMS22147677.

Monira, K.N., Islam, M.A., Alam, M.J., & Wahid, M.A. (2003). Effect of litter materials on broiler performance and evaluation of manureal value of used litter in late autumn. *Asian-Australasian Journal of Animal Sciences, 16*(4), 555–557. https://doi.org/10.5713/AJAS.2003.555.

Nasrin, M., Siddiqi, M.N.H., Masum, M.A., & Wares, M.A. (2012). Gross and histological studies of digestive tract of broilers during postnatal growth and development. *Journal of the Bangladesh Agricultural University, 10*(1), 69–77. https://doi.org/10.3329/JBAU.V10I1.12096.

Nighot, P.K., Moeser, A., Ali, R.A., Blikslager, A.T., & Koci, M.D. (2010). Astrovirus infection induces sodium malabsorption and redistributes sodium hydrogen exchanger expression. *Virology, 401*(2), 146–154. https://doi.org/10.1016/J.VIROL.2010.02.004.

Nowak, M., Grzesiak, M., Saito, N., Kwaśniewska, M., Sechman, A., & Hrabia, A. (2017). Expression of aquaporin 4 in the chicken ovary in relation to follicle development. *Reproduction in Domestic Animals = Zuchthygiene, 52*(5), 857–864. https://doi.org/10.1111/RDA.12990.

Nusrat, A., Turner, J.R., & Madara, J.L. (2000). Molecular physiology and pathophysiology of tight junctions. IV. Regulation of tight junctions by extracellular stimuli: nutrients, cytokines, and immune cells. *American Journal of Physiology. Gastrointestinal and Liver Physiology, 279*(5). https://doi.org/10.1152/AJPGI.2000.279.5.G851.

O'Hara, J.R., & Buret, A.G. (2008). Mechanisms of intestinal tight junctional disruption during infection. *Frontiers in Bioscience: A Journal and Virtual Library, 13*(18), 7008–7021. https://doi.org/10.2741/3206.

Okumura, R., & Takeda, K. (2017). Roles of intestinal epithelial cells in the maintenance of gut homeostasis. *Experimental & Molecular Medicine, 49*(5). https://doi.org/10.1038/EMM.2017.20.

Orlowski, S., Flees, J., Anthony, N., & Dridi, S. (2017). Differential expression of water channel- and noncoding RNA biogenesis-related genes in three lines of chickens under a short-term water restriction. *Poultry Science, 96*(12), 4172–4181. https://doi.org/10.3382/PS/PEX263.

Pan, D., & Yu, Z. (2014). Intestinal microbiome of poultry and its interaction with host and diet. *Gut Microbes, 5*(1), 108–119. https://doi.org/10.4161/GMIC.26945.

Pantin-Jackwood, M.J., Spackman, E., Day, J.M., & Rives, D. (2007). Periodic Monitoring of Commercial Turkeys for Enteric Viruses Indicates Continuous Presence of Astrovirus and Rotavirus on the Farms. Avian Diseases, 51(3), 674–680. https://doi.org/10.1637/0005-2086(2007)51.

Paradis, T., Bègue, H., Basmaciyan, L., Dalle, F., & Bon, F. (2021). Tight Junctions as a Key for Pathogens Invasion in Intestinal Epithelial Cells. *International Journal of Molecular Sciences 2021, Vol. 22, Page 2506, 22*(5), 2506. https://doi.org/10.3390/IJMS22052506.

Pavlidis, P., Powell, N., Vincent, R.P., Ehrlich, D., Bjarnason, I., & Hayee, B. (2015). Systematic review: bile acids and intestinal inflammation-luminal aggressors or regulators of mucosal defence? *Alimentary Pharmacology & Therapeutics, 42*(7), 802–817. https://doi.org/10.1111/APT.13333.

Planas, J.M., Carme Villá, M., Ferrer, R., & Moretó, M. (1986). Hexose transport by chicken cecum during development. *Pflugers Archiv: European Journal of Physiology, 407*(2), 216–220. https://doi.org/10.1007/BF00580679.

Podolsky, D.K. (1997). Healing the epithelium: Solving the problem from two sides. *Journal of Gastroenterology, 32*(1), 122–126. https://doi.org/10.1007/BF01213309.

Qaisrani, S.N., Moquet, P.C.A., Van Krimpen, M.M., Kwakkel, R.P., Verstegen, M.W.A., & Hendriks, W.H. (2014). Protein source and dietary structure influence growth performance, gut morphology, and Hindgut fermentation characteristics in broilers. *Poultry Science, 93*(12), 3053–3064. https://doi.org/10.3382/PS.2014-04091.

Qiagen. (2014). QIAamp ® Fast DNA Stool Mini Handbook For fast purification of genomic DNA from Sample & Assay Technologies QIAGEN Sample and Assay Technologies. *Tutoriais e Manuais, March.* https://doi.org/10.4103/0974-7753.153462.

Quach, A., Jayaratne, R.R., Lee, B.J., Ibeawuchi, S.R., Lim, E., Das, S., & Barrett, K.E. (2022). Diarrhoeal pathogenesis in Salmonella infection may result from an imbalance in intestinal epithelial differentiation through reduced Notch signalling. *The Journal of Physiology, 600*(8), 1851–1865. https://doi.org/10.1113/JP282585.

Raji, A.A., & Omar, A.R. (2022). Pathogenesis of Chicken Astrovirus Related Illnesses. *Frontiers in Veterinary Science*, 9, 899901. https://doi.org/10.3389/fvets.2022.899901.

Rehman, H., Awad, W.A., Lindner, I., Hess, M., & Zentek, J. (2006a). Clostridium perfringens alpha toxin affects electrophysiological properties of isolated jejunal mucosa of laying hens. *Poultry Science*, 85(7), 1298–1302. https://doi.org/10.1093/PS/85.7.1298.

Rehman, H., Awad, W.A., Lindner, I., Hess, M., & Zentek, J. (2006b). Clostridium perfringens alpha toxin affects electrophysiological properties of isolated jejunal mucosa of laying hens. *Poultry Science*, 85(7), 1298–1302. https://doi.org/10.1093/ps/85.7.1298.

Rehman, H., Ijaz, A., Specht, A., Dill, D., Hellweg, P., Männer, K., & Zentek, J. (2009). In vitro effects of alpha toxin from Clostridium perfringens on the electrophysiological parameters of jejunal tissues from laying hens preincubated with inulin and N-acetyl-L-cysteine. *Poultry Science*, 88(1), 199–204. https://doi.org/10.3382/PS.2008-00054.

Reicher, N., Melkman-Zehavi, T., Dayan, J., & Uni, Z. (2020). It's All About Timing: Early Feeding Promotes Intestinal Maturation by Shifting the Ratios of Specialized Epithelial Cells in Chicks. *Frontiers in Physiology*, 11, 1677. https://doi.org/10.3389/FPHYS.2020.596457/BIBTEX.

Riesenfeld, G., Sklan, D., Bar, A., Eisner, U., & Hurwitz, A. (1980). Glucose Absorption and Starch Digestion in the Intestine of the Chicken1. *J. Nutr*, 110, 117–121.

Roche, J.K., Martins, C.A.P., Cosme, R., Fayer, R., & Guerrant, R.L. (2000). Transforming growth factor beta1 ameliorates intestinal epithelial barrier disruption by Cryptosporidium parvum in vitro in the absence of mucosal T lymphocytes. *Infection and Immunity*, 68(10), 5635–5644. https://doi.org/10.1128/IAI.68.10.5635-5644.2000.

Röhe, I., Boroojeni, F.G., & Zentek, J. (2017). Effect of feeding soybean meal and differently processed peas on intestinal morphology and functional glucose transport in the small intestine of broilers. *Poultry Science*, 96(11), 4075–4084. https://doi.org/10.3382/PS/PEX199.

Rosenthal, R., Günzel, D., Piontek, J., Krug, S.M., Ayala-Torres, C., Hempel, C., Theune, D., & Fromm, M. (2020). Claudin-15 forms a water channel through the tight junction with distinct function compared to claudin-2. *Acta Physiologica*, 228(1), e13334. https://doi.org/10.1111/APHA.13334.

Rosenthal, R., Günzel, D., Theune, D., Czichos, C., Schulzke, J.D., & Fromm, M. (2017). Water channels and barriers formed by claudins. *Annals of the New York Academy of Sciences*, 1397(1), 100–109. https://doi.org/10.1111/NYAS.13383.

Rzeznitzeck, J., Breves, G., Rychlik, I., Hoerr, F.J., von Altrock, A., Rath, A., & Rautenschlein, S. (2022). The effect of Campylobacter jejuni and Campylobacter coli colonization on the gut morphology, functional integrity, and microbiota composition of female turkeys. *Gut Pathogens*, 14(1). https://doi.org/10.1186/S13099-022-00508-X.

Sarathy, J., Detloff, S.J., Ao, M., Khan, N., French, S., Sirajuddin, H., Nair, T., & Rao, M.C. (2017). The Yin and Yang of bile acid action on tight junctions in a model colonic epithelium. *Physiological Reports*, 5(10). https://doi.org/10.14814/PHY2.13294.

Sayeed, S., Li, J., & McClane, B.A. (2010). Characterization of virulence plasmid diversity among Clostridium perfringens type B isolates. *Infection and Immunity*, 78(1), 495–504. https://doi.org/10.1128/IAI.00838-09.

Schorn, C., Frey, B., Lauber, K., Janko, C., Strysio, M., Keppeler, H., Gaipl, U.S., Voll, R.E., Springer, E., Munoz, L.E., Schett, G., & Herrmann, M. (2011). Sodium overload and water influx activate the NALP3 inflammasome. *The Journal of Biological Chemistry, 286*(1), 35–41. https://doi.org/10.1074/JBC.M110.139048.

Shen, L., Weber, C.R., Raleigh, D.R., Yu, D., & Turner, J.R. (2011). Tight junction pore and leak pathways: a dynamic duo. *Annual Review of Physiology, 73*, 283–309. https://doi.org/10.1146/ANNUREV-PHYSIOL-012110-142150.

Shibata, M., Takahashi, T., Kozakai, T., Kakudo, M., Kasuga, S., Azuma, Y., & Kurose, Y. (2019). Active transport of glucose across the jejunal epithelium decreases with age in broiler chickens. *Poultry Science, 98*(6), 2570–2576. https://doi.org/10.3382/PS/PEZ002.

Shimada, Y., Kinoshita, M., Harada, K., Mizutani, M., Masahata, K., Kayama, H., & Takeda, K. (2013). Commensal Bacteria-Dependent Indole Production Enhances Epithelial Barrier Function in the Colon. *PLOS ONE, 8*(11), e80604. https://doi.org/10.1371/JOURNAL.PONE.0080604.

Shira, E.B., & Friedman, A. (2018). Innate immune functions of avian intestinal epithelial cells: Response to bacterial stimuli and localization of responding cells in the developing avian digestive tract. *PloS One, 13*(7). https://doi.org/10.1371/JOURNAL.PONE.0200393.

Sivaprakasam, S., Bhutia, Y.D., Yang, S., & Ganapathy, V. (2017). Short-Chain Fatty Acid Transporters: Role in Colonic Homeostasis. *Comprehensive Physiology, 8*(1), 299–314. https://doi.org/10.1002/CPHY.C170014.

Skadhauge, E., Clauss, W., Arnason, S.S., & Thomas, D.H. (1985). *Mineralocorticoid Regulation of Lower Intestinal Ion Transport*. 118–133. https://doi.org/10.1007/978-3-642-70613-4_12.

Škrbić, Z., Pavlovski, Z., Lukić, M., Petričević, V., & Milić, D. (2012). The effect of lighting program and type of litter on production and carcass performance of two broiler genotypes. *Biotechnology in Animal Husbandry, 28*(4), 807–816. https://doi.org/10.2298/BAH1204807S.

Smith, P.M., Choksi, Y.A., Markham, N.O., Hanna, D.N., Zi, J., Weaver, C.J., Hamaamen, J.A., Lewis, K.B., Yang, J., Liu, Q., Kaji, I., Means, A.L., & Daniel Beauchamp, R. (2021). Colon epithelial cell TGFβ signaling modulates the expression of tight junction proteins and barrier function in mice. *American Journal of Physiology. Gastrointestinal and Liver Physiology, 320*(6), G936–G957. https://doi.org/10.1152/AJPGI.00053.2021.

Sodring Elbrond, V., Dantzer, V., Mayhew, T., & Skadhauge, E. (1991). Avian lower intestine adapts to dietary salt (NaCl) depletion by increasing transepithelial sodium transport and microvillous membrane surface area. *Experimental Physiology, 76*(5), 733–744. https://doi.org/10.1113/EXPPHYSIOL.1991.SP003540.

Starck, J.M. (2003). Shaping up: how vertebrates adjust their digestive system to changing environmental conditions. *Animal Biology, 53*(3), 245–257. https://doi.org/10.1163/157075603322539444.

Starck, J.M., & Rahmaan, G.H.A. (2003). Phenotypic flexibility of structure and function of the digestive system of Japanese quail. *The Journal of Experimental Biology, 206*(Pt 11), 1887–1897. https://doi.org/10.1242/JEB.00372.

Sugiura, K., Aste, N., Fujii, M., Shimada, K., & Saito, N. (2008). Effect of hyperosmotic stimulation on aquaporins gene expression in chick kidney. *Comparative Biochemistry and Physiology Part A: Molecular & Integrative Physiology*, *151*(2), 173–179. https://doi.org/10.1016/J.CBPA.2008.06.014.

Suzuki, T. (2013). Regulation of intestinal epithelial permeability by tight junctions. *Cellular and Molecular Life Sciences: CMLS*, *70*(4), 631–659. https://doi.org/10.1007/S00018-012-1070-X.

Swain, B.K., & Sundaram, R.N.S. (2000). Effect of different types of litter material for rearing broilers. *Undefined*, *41*(3), 261–262. https://doi.org/10.1080/713654931.

Swiatkiewicz, S., Arczewska-Wlosek, A., & Jozefiak, D. (2017). The nutrition of poultry as a factor affecting litter quality and foot pad dermatitis – an updated review. *Journal of Animal Physiology and Animal Nutrition*, *101*(5), e14–e20. https://doi.org/10.1111/JPN.12630.

Tabler, T.W., Greene, E.S., Orlowski, S.K., Hiltz, J.Z., Anthony, N.B., & Dridi, S. (2020). Intestinal Barrier Integrity in Heat-Stressed Modern Broilers and Their Ancestor Wild Jungle Fowl. *Frontiers in Veterinary Science*, *7*, 249. https://doi.org/10.3389/FVETS.2020.00249/BIBTEX.

Taoukis, P.S., & Richardson, M. (2008). Principles of Intermediate-Moisture Foods and Related Technology. *Water Activity in Foods: Fundamentals and Applications*, 273–312. https://doi.org/10.1002/9780470376454.CH11.

Tellez, G., Latorre, J.D., Kuttappan, V.A., Kogut, M.H., Wolfenden, A., Hernandez-Velasco, X., Hargis, B.M., Bottje, W.G., Bielke, L.R., & Faulkner, O.B. (2014). Utilization of rye as energy source affects bacterial translocation, intestinal viscosity, microbiota composition, and bone mineralization in broiler chickens. *Frontiers in Genetics*, *5*(SEP), 339. https://doi.org/10.3389/FGENE.2014.00339/BIBTEX.

Thorsen, K., Drengstig, T., & Ruoff, P. (2014). Articles in PresS. *Am J Physiol Cell Physiol*. https://doi.org/10.1152/ajpcell.00068.2013.

Ticho, A.L., Malhotra, P., Dudeja, P.K., Gill, R.K., & Alrefai, W.A. (2019). Bile acid receptors and gastrointestinal functions. *Liver Research*, *3*(1), 31–39. https://doi.org/10.1016/J.LIVRES.2019.01.001.

Timbermont, L., Haesebrouck, F., Ducatelle, R., & Van Immerseel, F. (2011). Necrotic enteritis in broilers: An updated review on the pathogenesis. *Avian Pathology*, *40*(4), 341–347. https://doi.org/10.1080/03079457.2011.590967.

Van Der Hoeven-Hangoor, E., Rademaker, C.J., Paton, N.D., Verstegen, M.W.A., & Hendriks, W.H. (2014). Evaluation of free water and water activity measurements as functional alternatives to total moisture content in broiler excreta and litter samples. *Poultry Science*, *93*(7), 1782–1792. https://doi.org/10.3382/PS.2013-03776.

Vicuña, E.A., Kuttappan, V.A., Tellez, G., Hernandez-Velasco, X., Seeber-Galarza, R., Latorre, J.D., Faulkner, O.B., Wolfenden, A.D., Hargis, B.M., & Bielke, L.R. (2015). Dose titration of FITC-D for optimal measurement of enteric inflammation in broiler chicks. *Poultry Science*, *94*(6), 1353–1359. https://doi.org/10.3382/PS/PEV111.

von Buchholz, J.S., Ruhnau, D., Hess, C., Aschenbach, J.R., Hess, M., & Awad, W.A. (2022). Paracellular intestinal permeability of chickens induced by DON and/or C. jejuni is associated with alterations in tight junction mRNA expression. *Microbial Pathogenesis*, *168*, 105509. https://doi.org/10.1016/J.MICPATH.2022.105509.

Wallis, T.S., & Galyov, E.E. (2000). Molecular basis of Salmonella-induced enteritis. *Molecular Microbiology, 36*(5), 997–1005. https://doi.org/10.1046/J.1365-2958.2000.01892.X.

Wang, H.B., Wang, P.Y., Wang, X., Wan, Y.L., & Liu, Y.C. (2012). Butyrate Enhances Intestinal Epithelial Barrier Function via Up-Regulation of Tight Junction Protein Claudin-1 Transcription. *Digestive Diseases and Sciences 2012 57:12, 57*(12), 3126–3135. https://doi.org/10.1007/S10620-012-2259-4.

Wells, J.M., Rossia, O., Meijerink, M., & Van Baarlen, P. (2011). Epithelial crosstalk at the microbiota-mucosal interface. *Proceedings of the National Academy of Sciences of the United States of America, 108 Suppl 1*(Suppl 1), 4607–4614. https://doi.org/10.1073/PNAS.1000092107.

Wright, E.M., Loo, D.D.F., Hirayama, B.A., & Turk, E. (2004). Surprising versatility of Na+-glucose cotransporters: SLC5. *Physiology (Bethesda, Md.), 19*(6), 370–376. https://doi.org/10.1152/PHYSIOL.00026.2004.

Yegani, M., & Korver, D.R. (2008). Factors affecting intestinal health in poultry. *Poultry Science, 87*(10), 2052–2063. https://doi.org/10.3382/PS.2008-00091.

Yin, L., Menon, R., Gupta, R., Vaught, L., Okunieff, P., & Vidyasagar, S. (2017). Glucose enhances rotavirus enterotoxin-induced intestinal chloride secretion. *Pflugers Archiv: European Journal of Physiology, 469*(9), 1093–1105. https://doi.org/10.1007/S00424-017-1987-X.

Yu, M., Wang, Q., Ma, Y., Li, L., Yu, K., Zhang, Z., Chen, G., Li, X., Xiao, W., Xu, P., & Yang, H. (2018). Aryl Hydrocarbon Receptor Activation Modulates Intestinal Epithelial Barrier Function by Maintaining Tight Junction Integrity. *International Journal of Biological Sciences, 14*(1), 69–77. https://doi.org/10.7150/IJBS.22259.

Zaniboni, L., & Bakst, M.R. (2004). *Localization of Aquaporins in the Sperm Storage Tubules in the Turkey Oviduct. Poultry Science*, 83(7), 1209-1212. https://doi.org/10.1093/ps/83.7.1209.

Zhang, H., & Wong, E.A. (2017). Spatial transcriptional profile of PepT1 mRNA in the yolk sac and small intestine in broiler chickens. *Poultry Science, 96*(8), 2871–2876. https://doi.org/10.3382/PS/PEX056.

Zhang, H., & Wong, E.A. (2018). Identification of cells expressing OLFM4 and LGR5 mRNA by in situ hybridization in the yolk sac and small intestine of embryonic and early post-hatch chicks. *Poultry Science, 97*(2), 628–633. https://doi.org/10.3382/PS/PEX328.

Zhao, W., Wu, Z., Yao, Y., Qin, A., & Qian, K. (2020). The Isolation and Molecular Characterization of an Astrovirus From "Yellow" Chickens, China. *Frontiers in Veterinary Science, 7*, 822. https://doi.org/10.3389/FVETS.2020.581862/BIBTEX.

Zhu, N., Feng, X., He, C., Gao, H., Yang, L., Ma, Q., Guo, L., Qiao, Y., Yang, H., & Ma, T. (2011). Defective macrophage function in aquaporin-3 deficiency. *The FASEB Journal, 25*(12), 4233–4239. https://doi.org/10.1096/FJ.11-182808.

Zuo, L., Kuo, W.T., & Turner, J.R. (2020). Tight Junctions as Targets and Effectors of Mucosal Immune Homeostasis. *Cellular and Molecular Gastroenterology and Hepatology, 10*(2), 327–340. https://doi.org/10.1016/J.JCMGH.2020.04.001.

CHAPTER 14

Leveraging neurochemistry in the design of 21st century feed and microbiota strategies to improve food animal gut health

J.M. Lyte[1,*] *and V. Caputi*[1,*]

[1]Poultry Production and Product Safety Research Unit, Agricultural Research Service, United States Department of Agriculture, Fayetteville, AR 72701, USA; *Correspondence: joshua.lyte@usda.gov; vcaputi@uark.edu

Abstract

Environmental pressures due to climate change represent an omnipresent threat across the diverse production phases of food animals in the 21st century. Research directed at mitigating the deleterious effects of environmental stress, including heat stress, on the gastrointestinal tract, necessitates the identification of mechanistic pathways amenable to modulation by the producer. The research approach we have employed has its basis in common mechanistic pathways that are shared among diverse forms of environmental stressors. Those pathways principally involve the host's neurophysiological response to stress as well as the response of the gut microbiota whose composition and function are critical to gut homeostasis and animal well-being. Microbial endocrinology, which is the study of the intersection of two seemingly disparate fields, neurophysiology and microbiology, and the bi-directional communication between both, that is at the core of our approach to devising new strategies to combat environmental stressors in food production animals. The remarkable integration of an animal's neurophysiology and its microbiota is due to the evolutionary-based fact that each shares a range of neurochemicals, principally those related to the stress response, that permit bi-directional communication between the two that ultimately influences host gut function. This bi-directional communication is facilitated, in part, by the enteric nervous system which is composed of neurons that innervate the intestinal wall producing neurochemicals and sensing the microbiota. This chapter will focus on the gut of the animal as the anatomical region where the host and microbiota interact and represents the area that can most easily be leveraged to deal with environmental stressors. As will be discussed, the design of new feed formulations that seek to manage the intersection of the host neurophysiology and microbiota is one of the most promising avenues for dealing with the negative effects of environmental stress on food production animal health.

© J.M. LYTE AND V. CAPUTI, 2024 | DOI:10.3920/9789004695467_015

Keywords

neurophysiology – enteric nervous system – microbiome – stress – husbandry

1 Introduction

Stress management is a ubiquitous challenge across the diverse production environments of food animals in the 21st century (Rojas-Downing *et al.*, 2017). That monogastric and ruminant producers often share concern regarding the effect of the same environmental stressor on animal gut health highlights conserved pathways of stress adaptation across both mammalian and avian food animal species (Renaudeau *et al.*, 2012). The cross-communication between the animal's physiology and its microbiota, which are the collection of bacteria in the animal's gut, produce and respond to key stress-related neuroendocrine hormones once thought exclusive to the host, presents the novel understanding that host stress response relies on interaction between microbe and host. As a result, stress reduction strategies used in poultry and livestock production systems are increasingly designed to target the animal's microbiota (Chen *et al.*, 2021; Kraimi *et al.*, 2019). Likewise, feeding and nutrition are now seen as important not only to animal gut health but also to cultivating and maintaining a beneficial microbiota (Welch *et al.*, 2022; Yadav and Jha, 2019). Yet, much remains to be understood regarding the mechanisms of neuroendocrine bi-directional communication that govern, in part, the food animal microbiota-gut-brain and other physiological axes that intimately link the microbiota and animal physiology. Neurophysiological mechanisms of host-microbe interaction that are immediately relevant to aspects of the host stress response hold promise for the improvement of gut health in food animal species. Likewise, the extent to which the tight association between host and microbe mediates stress response to alter stress-induced changes in host behavior may hold implications for food animal welfare.

Cross communication between host and microbiota is affected by several environmental stressors that are commonly a concern across food animal production systems. The microbiota plays critical roles in maintaining host gut health under normative and stressful conditions in poultry (Aruwa *et al.*, 2021), swine (Luo *et al.*, 2022), and cattle (Cholewinska *et al.*, 2021). As studies within the field of microbial endocrinology have demonstrated host neurophysiological axes to mediate the bidirectional impact of neuroendocrine signaling molecules on both host and microbe, it is necessary to appreciate that environmental effects including stress are likewise impactful on both host gut health and the gut microbiota. Animal feed represents a tailorable input that has already largely been evidenced capable of targeting host-microbe interaction in the gut. Feed can contain a variety of both neuroendocrine and non-neuroendocrine molecules that when consumed by the host can affect both host enteric physiology and microbial function. Common feed additives in poultry and livestock such as butyrate, a short chain fatty acid (SCFA), serve as a prime example of the complex ability of feed to both promote host gut health, target specific bacterial species (*e.g.* reduce enteric colonization by bacterial foodborne pathogens), and mediate host-microbe interaction (Ahsan *et al.*, 2016). Throughout this chapter, SCFAs will serve as a prototypical example of the potential of animal feed to leverage neurochemistry to improve food animal gut health. An overwhelming reason why SCFAs, such as butyrate, is commonly referred to in this chapter is that SCFAs are already widely used in poultry and livestock feed, thereby emphasizing the design of animal feed to target neurochemical-based host-microbe interactions for the benefit of host gut health is not a future prospect but one that is already commercially available and economically practical.

In this section, we discuss the role of neurochemical-based host-microbe cross communication in mediating the effects of environmental stress on food animal gut health, with special attention towards poultry, and present suggestions how this may inform the development of next-generation feed and microbiota strategies in food animal production.

2 What is microbial endocrinology?

The recognition that microorganisms produce and respond to the same stress-related neurochemicals that constitute the neurophysiological systems of members of the animal kingdom has resulted in the discovery of new mechanisms of host-microbe interaction that affect host health and behavior. This field of study, which represents the intersection of two seemingly disparate fields, microbiology and neurobiology, is termed microbial endocrinology. It is important to highlight that this chapter, while focusing on literature that specifically utilized poultry and livestock species, does occasionally reference literature from rodent as well as human clinical studies. The reason for the decision to include rodent and human literature is two-fold: First, many of the findings regarding neurochemical-based host-microbe interaction in rodents and humans are relevant to birds and livestock; and Second, that much less research exploring microbial endocrinology has been performed in food animals compared to rodents and humans, hence rodent and human findings can provide potential insight into current gaps in food animal literature (Lyte *et al.*, 2018; Villageliu and Lyte, 2017). Although the evolutionary-based concept of microbial endocrinology as a common "language" by which host and microbiota can communicate and influence each other was founded just over two decades ago, published evidence over the preceding six decades going back to 1930 demonstrate the validity of uniting the fields of microbiology and neurobiology as a conceptual framework to understand interactions between the microbiota and host neurophysiology.

2.1 Escherichia coli – *a model bacterial organism for microbial endocrinology in food production animals*

E. coli, which is commonly found in virtually all food animal intestinal tracts, has been reported to uptake, respond to, as well as affect the biosynthesis of various stress-related neurochemicals (Burton *et al.*, 2002; Lopes and Sourjik, 2018; Nzakizwanayo *et al.*, 2015). While it is not yet known why *E. coli* can have the capacity to produce the exact same fight-or-flight neurochemicals as animals, it has been shown that exposure of *E. coli* to physiologically-relevant concentrations of stress-related neurochemicals results in orders of magnitude increase in the growth and production of virulence-related factors, such as pilus adhesion, as well as production of autoinducer metabolites that regulate growth and facilitate interkingdom signaling (Lyte, 2016). The ability of stress neurochemicals (*e.g.* norepinephrine and others) to initiate changes in gene expression has also been demonstrated (Bansal *et al.*, 2007; Xu *et al.*, 2015). Further, stress-related neurochemicals have been shown to facilitate bacterial attachment to mucosal surfaces and to increase proliferation within the intestinal tract (Beata *et al.*, 2021; Chen *et al.*, 2006). And while many reports understandably target the ability of stress-related neurochemicals to influence known *E. coli* pathogens such as *E. coli* O157:H7, it has also been shown that commensal *E. coli* respond in a similar fashion as known pathogens (Freestone *et al.*, 2002; Vlisidou *et al.*, 2004). While *E. coli* is often a commensal in the intestinal tract, different strains of *E. coli* represent significant concern in cattle production (*e.g. E. coli* O157:H7) (Callaway *et al.*, 2009), can act as the cause of colibacillosis diarrhea in pig production (Luppi, 2017), and avian pathogenic *Escherichia coli* (APEC) (Guabiraba and Schouler, 2015) is a major threat for the poultry industry. Thus, the use of a microbial endocrinology-based approach in the design of animal feed holds the potential to utilize neurochemistry in targeting microbial strains that are commonly found in the intestinal tract of almost all food production animals. Nevertheless, it is important to highlight that the overwhelming majority of reports demonstrating the ability of stress-related neurochemicals to influence *E. coli* physiology as it relates to disease pathogenesis have been carried out in non-avian animal species ranging from cattle to humans. As such, the utilization of microbial endocrinology in the design of animal feed represents a unique approach to leverage neurochemical pathways of host-microbe interaction by which environmental effects such as heat stress may influence, for example, the ability of APEC to infect chickens.

2.2 *Stress-related neurochemicals in the intestinal tract*

The presence of catecholamine neurochemicals (*i.e.* norepinephrine, epinephrine, and dopamine) in the intestinal tract of avian and mammalian species have received attention dating back several decades. Detectable concentrations of catecholamines in the gut at concentrations that are now known to impact the function of *E. coli* and other microorganisms is not unexpected as these neurochemicals play critical roles in normal intestinal physiology. Similar to the comparatively larger amount of literature describing the impact of neurochemicals on bacterial pathogens in mammals, a larger body of work also exists describing the physiological source of mammalian as compared to avian intestinal neurochemicals. For example, catecholamines that are found in the mammalian intestinal tract can derive from a variety of sources such as from synthesis by local intestinal epithelial and immune cells, spillover from the bloodstream as well as from enteric catecholaminergic innervation, and from the microbiota. Nevertheless, a growing number of studies have been performed in chickens, quail, and other food production birds that have reported concentrations of catecholamine and monoamine neurochemicals throughout the intestinal tract in both tissue and in luminal content (Fu *et al.*, 2022; Gross and Sturkie, 1975; Lyte *et al.*, 2022). Moreover, several reports exist describing the impact of the microbiota in avian species, including chickens, on intestinal neurochemical concentrations (Beaver and Wostmann, 1962; Phillips *et al.*, 1961).

3 Effects of environmental stressors on food animal gut health and microbiome in the context of climate change

Climate change is a major concern across virtually all food animal production systems. Poultry, swine, and cattle are susceptible to heat and cold stressors, and the unseasonably hot or cold temperatures that are arising more frequently as a result of climate change, are recognized to impact gut health in avian and mammalian food animals (Barrett *et al.*, 2019; Liu *et al.*, 2022; Rostagno, 2020). Likewise, the negative impact of extreme temperature stress on the gut microbiomes of poultry and livestock has been well described (Lara and Rostagno, 2013). For example, cold stress is one of the most significant environmental stressors in the post-hatch period of chickens (Tsiouris *et al.*, 2015). A major concern for both producers and consumers is that the environmental effects of climate change on food animal production, which in this chapter is specifically referring to heat or cold stress, may lead to increased susceptibility to infectious disease or carriage of foodborne pathogens (Burkholder *et al.*, 2008; Soerjadi *et al.*, 1979). Indeed, work dating back to the mid-19th century was the first to demonstrate that in birds, thermoregulation was essential in preventing bacterial infection (Thaxton *et al.*, 1974). Louis Pasteur's cold stress work in fowl showed that avian body temperature was directly related to infection by *Bacillus anthracis*. Thermoregulation has since been repeatedly shown to influence avian susceptibility to intestinal infection.

The role of thermoregulation in mediating poultry and livestock susceptibility to infection has also been described (Troxell *et al.*, 2015). Heat stress has been shown to impact bovine and swine immune function, susceptibility to infection, and pathogen shedding (Dahl *et al.*, 2020; Venegas-Vargas *et al.*, 2016). The impact of heat stress on livestock intestinal inflammation as well as microbiota composition and function has been well described. While neurophysiological axes such as the hypothalamic-pituitary-adrenal (HPA) axis, which in mammals have been shown to be in part regulated by the microbiota (Sudo *et al.*, 2004), are recognized to play a critical role in mediating animal response to environmental stressors, how environmental stressors such as heat or cold stress may modulate the stress-related neurochemical response in the intestinal tract and other peripheral sites of poultry and livestock animals has received sparse attention. Nevertheless, heat stress has been shown to impact avian brain monoamine neurochemical concentrations (Calefi *et al.*, 2019), as well as affect intestinal susceptibility to infection (Burkholder *et al.*, 2008). Likewise, heat stress was previously demonstrated to cause persistent elevation in plasma concentrations of catecholamines in cattle (Alvarez and Johnson, 1973). Similarly, plasma catecholamine concentrations were increased in piglets exposed to cold environmental temperatures (Mayfield *et al.*, 1989). As in mammals the bloodstream is a known source for spillover of neurochemicals into the intestinal tract, elevated plasma concentrations of catecholamines in cattle and swine may be relevant for study of intestinal stress and gut health.

The microbiome of poultry and livestock has also been extensively described under conditions of heat stress, and to a much lesser extent, cold stress (Cholewinska *et al.*, 2021; Patra and Kar, 2021; Zhang *et al.*, 2022b). A principal focus of the majority of studies is compositional changes in bacterial diversity utilizing 16S rRNA marker gene. Typically, temperature-based stressors have been described across avian (Emami *et al.*, 2022) and mammalian species to decrease microbial diversity in the gut – a measure of microbiota that may correlate with host health. Greater attention is needed assessing functional changes in the microbiome following host exposure to temperature-based stressors, especially as host stress response is well appreciated to alter the gut metabolome, including potentially, stress-related neurochemicals that are known to affect bacterial interaction with the host.

4 Neuronal control of gastrointestinal function: the enteric nervous system

The gastrointestinal (GI) tract is the body's system responsible for breaking down food particles, extracting nutrients, and expelling waste products. In the gut wall, the contractile activity of the muscular apparatus allows the mixing and propulsion of the intestinal content, while the local blood flow surrounding the intestinal epithelium ensures the optimal absorption of nutrients and fluids. In order to sustain these physiological functions, both intestinal mechanical (muscularis externa) and chemical (mucosal epithelium) units require a finely tuned coordination that relies upon their intrinsic innervation or enteric nervous system (ENS) (Costa *et al.*, 2000). The ENS is a complex network of more than 100 million neurons and interconnecting nerve fibers embedded in the entire GI tract from the esophagus to the anal sphincter, and also innervating gut-associated glands (*i.e.*, salivary glands and pancreas) (Furness, 2012).

Originally, the ENS was thought to be part of the autonomous peripheral nervous system (PNS), and its neurons were considered to be parasympathetic post-ganglionic neurons (Goyal and Hirano, 1996). The hypothesis of "little brain in the gut" was raised in 1899, when the physiologists W.M. Bayliss and E.H. Starling proposed the existence of enteric reflex pathways consisting of intestinal contractions coordinated by short ascending excitatory pathways and longer descending inhibitory pathways in absence of any interactions with the peripheral and central nervous systems (CNS). These findings were classified under the "law of the intestine", known as the peristaltic reflex (Bayliss and Starling, 1899).

Anatomically, the ENS includes groups of sensory and motor neurons inter-linked by interneurons, with distinct shape, neurochemical coding, and functional roles. In particular, the primary afferent neurons (IPANS) are sensitive to chemical and mechanical stimuli. The IPANS project through the intestinal mucosa up to the villi's tip and their cellular bodies are organized in small groups or ganglia located within the submucosa (*e.g.*, submucous or Meissner's plexus). The intestinal motor neurons regulate the contraction and relaxation of the intestinal smooth muscle (or intestinal motility), and are grouped in large ganglia positioned in the muscularis externa (*e.g.*, myenteric or Auerbach's plexus) (Furness, 2012; Hansen, 2003). Although capable of functioning independently from the CNS, the ENS is bi-directionally connected to the brain through the major extrinsic motor and sensory pathways of the spinal cord and vagus nerve (Spencer and Hu, 2020). Hence, the brain controls food intake (Díaz-Güemes *et al.*, 2007), nutrient sensing (Niknafs and Roura, 2018), and intestinal satiety, and the gut responds with appropriate GI reflexes, thus ensuring the organism's physiological and nutritional state.

In general, enteric neurons contain more than 30 different messenger molecules including the classical neurotransmitters such as acetylcholine (Schemann *et al.*, 1993), glutamate (Liu *et al.*, 1997), GABA (Jessen *et al.*, 1986), the monoamines noradrenaline, serotonin, dopamine and histamine (Li *et al.*, 2011; Li *et al.*, 2004) together with an array of peptide-based neuromodulators such as corticotrophin releasing hormone (Liu *et al.*, 1997), Calcitonin gene-related peptide (CGRP) (Martinez and TachÉ, 2006), vasoactive intestinal peptide (VIP) (Alves *et al.*, 2010), tachykinins (*i.e.*, substance P) (Costa *et al.*, 1985) and with the gaseous signaling molecule nitric oxide (NO) (Furness *et al.*, 1994). Individual neurons can contain more than one neurotransmitter in addition to other neuron-specific proteins or enzymes (Qu *et al.*, 2008). Therefore, the specific and dynamic distribution profile of different neurotransmitters and mediators represents and characterizes the neurochemical coding of each neuron, determining its function (Schemann and Neunlist, 2004).

Within the enteric ganglia, neuronal cell bodies are physically connected with enteric glial cells (ECGs) in a relationship highly reminiscent of the one between astrocytes and neurons in the CNS. ECGs and neurons originate from a shared pool of neural-crest-derived progenitors (Conner *et al.*, 2003; Lake and Heuckeroth, 2013), however, ECGs were discovered no later than the 19th-century (Coelho-Aguiar Jde *et al.*, 2015). These studies display how ECGs functions extend beyond the physical support for the neurons in the GI tract. ECGs can regulate synaptic transmission, supplying neurons with neurotransmitter precursors (Nagahama *et al.*, 2001) and producing neuroactive substances (Gulbransen and Sharkey, 2012). Thus, enteric neurons and glia work together to maintain intestinal epithelial barrier function and facilitate the communication between the nervous and immune systems (Gulbransen and Sharkey, 2012).

Overall, the ENS projects to various effector tissues, including non-neuronal pacemaker cells (*e.g.*, interstitial cells of Cajal), blood vessels, intestinal epithelial, immune and endocrine cells. In addition to regulating intestinal motility and visceral sensitivity, the ENS assesses intestinal luminal contents and elicits appropriate neural, endocrine, and immune responses. Given its proximity to the gut microbiota, the ENS represents the primary neuronal interface of communication between gut microbes and the host intestinal system. Studies in rodents have shown that gut microbes play a critical role in modulating ENS development and function (Caputi *et al.*, 2017) (Obata *et al.*, 2020), and in turn, the ENS maintains microbiota homeostasis through mechanisms involving different microbe-neurotransmitter interactions (Lei *et al.*, 2022; Zhang *et al.*, 2022a).

Significant progress has been achieved over the past 40 years in understanding the role of the ENS in human physiological and pathological conditions, and for this purpose, extensive pre-clinical research has been conducted in animal models (mice, rats, guinea pigs) of human diseases. However, very little is known about the ENS in food production animals. Current research in poultry and livestock is highly oriented to finding new strategies to reduce stress-induced perturbations of animals' gut health and foodborne pathogen colonization during the pre-harvest stages of production. In the context of the microbial endocrinology, the study of the ENS in poultry and livestock could provide a novel framework to understand the mechanisms that regulate host-microbe neuroendocrine communication and help design new feed alternatives that can significantly improve nutrient intake and meat quality and reduce foodborne pathogen carriage in food production animals. This approach will benefit poultry and livestock production industries, ultimately preserving consumers' health.

Chicken and quail have been extensively used in pre-clinical research as model for neurodevelopmental studies, as they are easy to manipulate, especially in the embryonic stage (Chevalier *et al.*, 2019). A large body of literature describes the origin of the avian ENS from the neural crest (Peters-van der Sanden *et al.*, 1993b) (Peters-van der Sanden *et al.*, 1993a) (Chevalier *et al.*, 2016) (Chevalier *et al.*, 2020). Nitrergic, peptidergic and monoaminergic neurotransmissions were also characterized in the avian ENS, both in early life (Boros *et al.*, 1994) (Rothman *et al.*,

1986) (Fontaine-Perus *et al.*, 1981) (Bagyánszki *et al.*, 2000) (Balaskas *et al.*, 1995) and adulthood (Jadhao *et al.*, 2017) (Yang *et al.*, 2013). Pigs have been used as animal models to study human gut neuropathies (*i.e.*, diabetes) or inflammatory GI diseases. Changes in VIP-, substance P-, CGRP- like, and NO immunoreactivity in enteric neurons occur within the pig stomach following supplementation with acrylamide (Palus *et al.*, 2018) or Bisphenol A (Szymanska *et al.*, 2018). Other studies have shown how streptozotocin-induced diabetes or inflammation upon nonsteroidal anti-inflammatory drugs administration affect the sensory neurotransmission of the ENS in the porcine GI tract (Bulc *et al.*, 2018). (Brzozowska *et al.*, 2021). The study of Rytel *et al.* assessed the effect of *Salmonella* Enteritidis infection on the ENS structure of porcine duodenum (Rytel *et al.*, 2021). These studies highlight the possibility to apply this knowledge and technical ability to study the ENS in poultry and livestock, and to further the research on microbial endocrinology-based nutritional alternatives for the food animal production industry.

5 A neurophysiological framework in approaching nutrition and feed

Improvements in the design of animal feed and understanding of nutritional requirements have accelerated the productivity of the poultry and livestock industries. The introduction of antibiotics in food animal production during the mid 20th century (Yang *et al.*, 2019) early on suggested a role for bacteria in affecting host health and growth. While the route by which antibiotics act as growth promotors (AGP) likely involve both effects on the microbiota and host (Brown *et al.*, 2017), consumer preferences, concern for the development of antimicrobial resistance, and other factors have driven food animal producers away from AGP and towards antibiotic alternatives. Despite the turn away from AGPs, interest in the role of the microbiota in affecting host health has skyrocketed. Indeed, information continues to be gleaned on the role of the microbiota in affecting host growth based on insight into the effect of different AGP's on individual microbial taxa (Costa *et al.*, 2017). Strategies seeking to harness the gut microbiota that do not rely on antibiotics include both nutritive and non-nutritive approaches that collectively center on the recognition that host-microbe crosstalk plays a role in determining poultry (Kogut, 2022) and livestock health (Lyte and Lyte, 2019; Newbold and Ramos-Morales, 2020).

Microbial endocrinology provides for an evidence-based framework in the design of next-generation animal feeds based on neuroendocrine aspects of host-microbe communication. Nutrition and feed composition must take into consideration both the requirements of the animal as well as those of its microbiota. The value of an interkingdom approach to nutrition and the design of feed is increasingly evident, especially considering common feed components such as corn, wheat, soy, animal by-products, and other plant and animal ingredients are abundant sources of neuroactive compounds including neurochemicals and their precursors (Lyte, 2018). It is important to emphasize that in human trials it was shown that neurochemicals found in ingredients that may comprise an animal's feed are not simply broken down in the upper gastrointestinal tract but do reach the intestinal tract structurally intact or are absorbed into the bloodstream (Li *et al.*, 2015b; Tohmola *et al.*, 2015). This presents considerations for the producer beyond traditional animal nutrition. For example, feeds can contain histamine and other biogenic amines which have recognized roles in affecting inflammatory processes in the gut (Bermudez and Firman, 1998). If poultry or livestock are expected to encounter stressful environmental conditions, such as a heat stress episode as discussed earlier in this chapter, feed compositions may be tailored to concentrations of neuroactive molecules, such as histamine, that could potentially affect inflammation and animal gut health.

It should also be appreciated that lateral gene transfer from prokaryotes presents an evolutionary basis for the similarity between bacteria and host in respective genetic pathways that are involved in neurochemical production (Iyer *et al.*, 2004). The routes by which nutrition and other environmental stimuli may impact host gut health through neuroendocrine physiology should therefore be appreciated through the synergy of host-microbe interaction (Lyte, 2013). This is perhaps most readily apparent in that both host and microbial pathways involved in the production of certain neurochemicals utilize the same amino acids that are supplied in feed. Surprisingly, during the initial stages of infection in patients some bacterial pathogens have even been shown to use host stores of amino acids for the synthesis of L-DOPA and 5-hydroxytryptophan (Abromaitis *et al.*, 2009) which are the precursors to the neurochemicals dopamine and serotonin, respectively. As dopamine and serotonin both affect pro- and anti-inflammatory processes in the intestinal tract of food production animals as well as patients, it would be warranted to investigate if bacterial pathogens may utilize similar strategies in the production of stress-related neurochemicals during enteric infection of poultry or livestock. Alterations in neurochemical concentrations in the intestinal tract occur in food production animals, as well as patients, following stress and do contribute to the upregulation of microbial virulence. Hence, tailoring of feed composition to drive or suppress neuroendocrine pathways in the gut, whether host or microbial, may have an effect on gut health through the modulation of inflammation and microbial interaction with the host.

Animal nutrition represents another important factor in governing gut health through microbial competition. Dietary metals, such as zinc, manganese, and iron are important in bacterial growth and are therefore a source of dynamic host-microbe competition in intestinal homeostasis (Skaar, 2010). Zinc, for example, has been shown to affect histamine concentrations in chicken intestinal tissue (Reimann *et al.*, 1971). Although it is accepted that some dietary iron regularly escapes absorption in the small intestine to reach the colonic lumen, it is not immediately useable by the microbiota as it is either largely bound to food or host iron-sequestering proteins. Interestingly, norepinephrine can release iron from host proteins making it available to bacteria, thereby enhancing bacterial growth (Freestone *et al.*, 2000). Likewise, the iron-limited environment of the colon creates conditions of microbe-microbe competition and cooperation whereby bacteria and fungi secrete iron-chelating siderophores. Interestingly, some bacterial species, including certain foodborne pathogens in the food animal gut, evolved the capability to steal siderophores secreted by other species (Raines *et al.*, 2016), thereby highlighting host nutrition as a source of enteric interkingdom stress relevant to host gut health.

Bacterial fermentation of components of the host diet is well appreciated, and the production of SCFA are a prominent example of the relevance of how bacterial metabolic potential of the host diet can affect poultry and livestock gut health. Less explored aspects of SCFA production in the gut are how these microbial metabolites may directly or through epigenetic modifications, affect host neuroendocrine-immune gene expression. Neuroendocrine framing of feed composition could also therefore present novel tailorable epigenetic strategies for improving food animal gut health. Butyrate is one such SCFA that is widely utilized as an additive in poultry and livestock feeds and has been demonstrated to effect epigenetic changes in animal physiology. Although little attention has been directed towards butyrate's neuroendocrine modulatory potential in food animals, rodent models have demonstrated butyrate's role in affecting stress-related production of glucocorticoids and monoamine neurochemicals (Neuman *et al.*, 2015).

Serotonin is widely appreciated to affect gastrointestinal inflammation through regulation of pro- and anti-inflammatory processes. Tryptophan hydroxylase (TPH)-1, which is found in the avian and mammalian intestinal epithelium, is the major rate-limiting enzyme in the conversion of serotonin from the amino acid tryptophan. Although not yet demonstrated using poultry or livestock models, in rodents it has been shown that butyrate can influence TPH-1 expression to modulate serotonin concentrations in the intestinal mucosa (Reigstad *et al.*, 2015). It would therefore be warranted to explore the epigenetic potential of butyrate as a feed additive capable of modulating gut mucosal inflammation through the serotonergic system in poultry or livestock.

Nevertheless, any potential epigenetic effect of a feed additive, such as butyrate, should be considered in light of other potentially competitive epigenetic effects from other ingredients or additives. For example, while tryptophan in poultry feed can affect physiological serotonin concentrations (Rosebrough, 1996), other ingredients such as micronutrients such as zinc are likely to also be included in the feed. Zinc has been shown to affect intestinal inflammation through modulation of DNA methylation and histone acetylation in chick mucosal tissue (Li *et al.*, 2015a). Furthermore, any single feed component, again taking butyrate as an example, should be assumed to exert potentially additive or even competitive epigenetic effects as it may act on multiple but distinct physiological pathways. For example, butyrate has also been reported to affect gut barrier integrity though epigenetic regulation of the anti-inflammatory cytokine interleukin (IL)-10 (Zheng *et al.*, 2017) which is found in both avian and mammalian intestinal tracts. Likewise, in mammals, butyrate inhibited histone deacetylase activity to affect gut macrophage function (Chang *et al.*, 2014). Together, this paints a complex epigenetic picture of the neuroendocrine potential of animal feed additives. When considering targeting neurochemical-based pathways, such as the serotonergic system, through additives like butyrate that effect epigenetic modification the producer should keep in mind such competitive actions at other physiological sites. Moreover, a microbial endocrinology framework can help producers appreciate that the effects described briefly here, while potentially affective of gut health, may affect gut host-microbe interaction with consequences for the central nervous system via the microbiome-gut-brain axis that could alter feeding behavior.

6 Microbial endocrinology as a theme in designing feed strategies to counter the impact of climate change on food animal stress, disease, and foodborne pathogen carriage

Physiological stress has been recognized for decades to influence an animal's susceptibility to enteric infection (Rostagno, 2009). Neuroendocrine axes exhibit plasticity to stress and constitute key pathways by which host stress can impact the gut microbiota, as well provide a mechanistic basis by which bacteria may affect host stress response through the utilization of feed components. Catecholamine neurochemicals that are involved in both avian and mammalian stress responses are found throughout the intestinal tract in both tissue and luminal content. It should not be surprising then, that in food animals stress has also been demonstrated to affect foodborne pathogen carriage. Indeed, catecholamines have been reported to affect foodborne pathogen dynamics in poultry (Aroori *et al.*, 2014; Xu *et al.*, 2015), swine (Toscano *et al.*, 2007), and cattle (Pullinger *et al.*, 2010). Furthermore, catecholamines in the gut exert distinct actions that affect host gut health (Dennis, 2016), including anti-inflammatory roles of dopamine (Yan *et al.*, 2015). Already in animal models of irritable bowel disease, dopaminergic agonists have shown to exert an anti-inflammatory effect (Oehlers *et al.*, 2017), however, little work in this regard has been performed in food production animals. Hence, microbial endocrinology-based frameworks for the design of animal feed could target norepinephrine in seeking to reduce foodborne pathogen carriage and dopamine if desiring to control enteric inflammation.

One plausible route by which animal feed may affect enteric concentrations of norepinephrine and dopamine is through epigenetic modulation of tyrosine hydroxylase, the rate-limiting enzyme in catecholamine synthesis (Daubner *et al.*,

2011). Butyrate, a feed additive commonly used in poultry and livestock, has been shown to influence tyrosine hydroxylase gene expression partially through epigenetic regulation (Kim *et al.*, 2003; Lenartowski and Goc, 2011). As such, butyrate may present a route by which a feed additive could leverage neurochemical biosynthesis to improve host gut health through modulation of inflammation and stress response. Considering that butyrate was reported to influence the hypothalamic-pituitary-adrenal (HPA)-axis (Gagliano *et al.*, 2014), a principal route of host-microbe neuroendocrine bi-directional communication (Sudo, 2014), it may be proposed that an effect of SCFA on host stress physiology could impact the microbiome. The adrenal gland is important in avian and mammalian physiology for glucocorticoid synthesis response to stress (Martinez-Miro *et al.*, 2016; Scanes, 2016), glucocorticoid synthesis in response to stress is affected by the microbiota. In mammals it was reported that corticosterone production in the adrenal gland can increase catecholamine synthesis (Sharara-Chami *et al.*, 2010). As such, it would be warranted to investigate in food animals if butyrate as a feed additive could work in concert with the microbiome to regulate the HPA-axis with consequences for catecholamine neurochemical production, such as dopamine, that affect host gut inflammation and health.

Aside from nutritional routes of targeting host-microbe neurochemical-based interaction, it is also important to consider how feed components can impact microbe-microbe interactions that may have consequence for host gut health. Host biosynthesis of L-3,4 dihydroxyphenylalanine (L-dopa) from the amino acid tyrosine is catalyzed by tyrosine hydroxylase. It is important to recognize that certain bacterial taxa, including some used as probiotic strains in poultry (Cao *et al.*, 2013) and livestock production (Pollmann *et al.*, 2005; Qadis *et al.*, 2014), can utilize L-dopa to produce dopamine. For example, *Enterococcus faecium* was recently demonstrated to synthesize dopamine from L-dopa (Villageliu and Lyte, 2017). As butyrate can be provided directly in feed, as well as can be produced *in vivo* via microbial fermentation of host feed components, it would be interesting to consider if feed strategies could utilize butyrate to up- or down-regulate host tyrosine hydroxylase to influence enteric concentrations of L-dopa to directly modulate the amount of L-dopa available as a substrate for *E. faecium* use in the synthesis of dopamine to ultimately affect host intestinal inflammation to improve gut health.

7 Future directions

Food animal gut health is increasingly viewed as a product of host-microbe interaction. The design of next-generation feed strategies now widely considers the impact on both host and microbiota as determinative in host health outcomes. Through the lens of microbial endocrinology, animal feed and nutrition can be understood in an evidence-based framework to leverage neurochemical-based host-microbe interactions for the benefit of host gut health. Several means are already available to poultry and livestock producers, including widely utilized feed additives such as butyrate or probiotic species, that can likely be strategically applied to target animal neuroendocrine physiological axes to counter environmental effects of stress and reduce disease as well as foodborne pathogen carriage.

References

Abromaitis, S., Hefty, P.S. and Stephens, R.S., 2009. Chlamydia pneumoniae encodes a functional aromatic amino acid hydroxylase. FEMS Immunol Med Microbiol 55: 196–205. 10.1111/j.1574-695X.2008.00511.x.

Ahsan, U., CengİZ, Ö., Raza, I., Kuter, E., Chacher, M.F.A., Iqbal, Z., Umar, S. and ÇAkir, S., 2016. Sodium butyrate in chicken nutrition: the dynamics of performance, gut microbiota, gut morphology, and immunity. World's Poultry Science Journal 72: 265–275. 10.1017/S0043933916000210.

Alvarez, M.B. and Johnson, H.D., 1973. Environmental heat exposure on cattle plasma catecholamine and glucocorticoids. J Dairy Sci 56: 189–194. 10.3168/jds.S0022-0302(73)85145-8.

Alves, E.P., Alves, A.M., Pereira, R.V., de Miranda Neto, M.H. and Zanoni, J.N., 2010. Immunohistochemical study of vasoactive intestinal peptide (VIP) enteric neurons in diabetic rats supplemented with L-glutamine. Nutr Neurosci 13: 43–51. 10.1179/147683010x12611460763841.

Aroori, S.V., Cogan, T.A. and Humphrey, T.J., 2014. Effect of noradrenaline on the virulence properties of campylobacter species. Int J Microbiol 2014: 279075. 10.1155/2014/279075.

Aruwa, C.E., Pillay, C., Nyaga, M.M. and Sabiu, S., 2021. Poultry gut health – microbiome functions, environmental impacts, microbiome engineering and advancements in characterization technologies. Journal of Animal Science and Biotechnology 12: 119. 10.1186/s40104-021-00640-9.

Bagyánszki, M., Román, V. and Fekete, E., 2000. Quantitative distribution of NADPH-diaphorase-positive myenteric neurons in different segments of the developing chicken small intestine and colon. Histochem J 32: 679–684. 10.1023/a:1004167416731.

Balaskas, C., Saffrey, M.J. and Burnstock, G., 1995. Distribution and colocalization of NADPH-diaphorase activity, nitric oxide synthase immunoreactivity, and VIP immunoreactivity in the newly hatched chicken gut. Anat Rec 243: 10–18. 10.1002/ar.1092430103.

Bansal, T., Englert, D., Lee, J., Hegde, M., Wood, T.K. and Jayaraman, A., 2007. Differential effects of epinephrine, norepinephrine, and indole on Escherichia coli O157:H7 chemotaxis, colonization, and gene expression. Infect Immun 75: 4597–4607. 10.1128/IAI.00630-07.

Barrett, N.W., Rowland, K., Schmidt, C.J., Lamont, S.J., Rothschild, M.F., Ashwell, C.M. and Persia, M.E., 2019. Effects of acute and chronic heat stress on the performance, egg quality, body temperature, and blood gas parameters of laying hens. Poult Sci 98: 6684–6692. 10.3382/ps/pez541.

Bayliss, W.M. and Starling, E.H., 1899. The movements and innervation of the small intestine. J Physiol 24: 99–143. 10.1113/jphysiol.1899.sp000752.

Beata, S., Michał, T., Mateusz, O., Urszula, W., Choroszy, M., Andrzej, T. and Piotr, D., 2021. Norepinephrine affects the interaction of adherent-invasive Escherichia coli with intestinal epithelial cells. Virulence 12: 630–637. 10.1080/21505594.2021.1882780.

Beaver, M.H. and Wostmann, B.S., 1962. Histamine and 5-hydroxytryptamine in the intestinal tract of germ-free animals, animals harbouring one microbial species and conventional animals. Br J Pharmacol Chemother 19: 385–393. 10.1111/j.1476-5381.1962.tb01443.x.

Bermudez, A.J. and Firman, J.D., 1998. Effects of biogenic amines in broiler chickens. Avian Dis 42: 199–203.

Boros, A., Timmermans, J.P., Fekete, E., Adriaensen, D. and Scheuermann, D.W., 1994. Appearance and some neurochemical features of nitrergic neurons in the developing quail digestive tract. Histochemistry 101: 365–374. 10.1007/bf00268998.

Brown, K., Uwiera, R.R.E., Kalmokoff, M.L., Brooks, S.P.J. and Inglis, G.D., 2017. Antimicrobial growth promoter use in livestock: a requirement to understand their modes of action to develop effective alternatives. Int J Antimicrob Agents 49: 12–24. 10.1016/j.ijantimicag.2016.08.006.

Brzozowska, M., Jana, B. and Całka, J., 2021. Effect of NSAIDs Supplementation on the PACAP-, SP- and GAL-Immunoreactive Neurons in the Porcine Jejunum. Int J Mol Sci 22. 10.3390/ijms222111689.

Bulc, M., Palus, K., Całka, J. and Zielonka, Ł., 2018. Changes in Immunoreactivity of Sensory Substances within the Enteric Nervous System of the Porcine Stomach during Experimentally Induced Diabetes. J Diabetes Res 2018: 4735659. 10.1155/2018/4735659.

Burkholder, K.M., Thompson, K.L., Einstein, M.E., Applegate, T.J. and Patterson, J.A., 2008. Influence of stressors on normal intestinal microbiota, intestinal morphology, and susceptibility to Salmonella enteritidis colonization in broilers. Poult Sci 87: 1734–1741. 10.3382/ps.2008-00107.

Burton, C.L., Chhabra, S.R., Swift, S., Baldwin, T.J., Withers, H., Hill, S.J. and Williams, P., 2002. The growth response of Escherichia coli to neurotransmitters and related catecholamine drugs requires a functional enterobactin biosynthesis and uptake system. Infect Immun 70: 5913–5923. 10.1128/IAI.70.11.5913-5923.2002.

Calefi, A.S., Fonseca, J., Nunes, C.A.Q., Lima, A.P.N., Quinteiro-Filho, W.M., Florio, J.C., Zager, A., Ferreira, A.J.P. and Palermo-Neto, J., 2019. Heat Stress Modulates Brain Monoamines and Their Metabolites Production in Broiler Chickens Co-Infected with Clostridium perfringens Type A and Eimeria spp. Vet Sci 6. 10.3390/vetsci6010004.

Callaway, T.R., Carr, M.A., Edrington, T.S., Anderson, R.C. and Nisbet, D.J., 2009. Diet, Escherichia coli O157:H7, and cattle: a review after 10 years. Curr Issues Mol Biol 11: 67–79.

Cao, G.T., Zeng, X.F., Chen, A.G., Zhou, L., Zhang, L., Xiao, Y.P. and Yang, C.M., 2013. Effects of a probiotic, Enterococcus faecium, on growth performance, intestinal morphology, immune response, and cecal microflora in broiler chickens challenged with Escherichia coli K88. Poult Sci 92: 2949–2955. 10.3382/ps.2013-03366.

Caputi, V., Marsilio, I., Filpa, V., Cerantola, S., Orso, G., Bistoletti, M., Paccagnella, N., De Martin, S., Montopoli, M., Dall'Acqua, S., Crema, F., Di Gangi, I.M., Galuppini, F., Lante, I., Bogialli, S., Rugge, M., Debetto, P., Giaroni, C. and Giron, M.C., 2017. Antibiotic-induced dysbiosis of the microbiota impairs gut neuromuscular function in juvenile mice. Br J Pharmacol 174: 3623–3639. 10.1111/bph.13965.

Chang, P.V., Hao, L., Offermanns, S. and Medzhitov, R., 2014. The microbial metabolite butyrate regulates intestinal macrophage function via histone deacetylase inhibition. Proc Natl Acad Sci USA 111: 2247–2252. 10.1073/pnas.1322269111.

Chen, C., Lyte, M., Stevens, M.P., Vulchanova, L. and Brown, D.R., 2006. Mucosally-directed adrenergic nerves and sympathomimetic drugs enhance non-intimate adherence of Escherichia coli O157:H7 to porcine cecum and colon. Eur J Pharmacol 539: 116–124. 10.1016/j.ejphar.2006.03.081.

Chen, S., Luo, S. and Yan, C., 2021. Gut Microbiota Implications for Health and Welfare in Farm Animals: A Review. Animals (Basel) 12. 10.3390/ani12010093.

Chevalier, N.R., Ammouche, Y., Gomis, A., Teyssaire, C., de Santa Barbara, P. and Faure, S., 2020. Shifting into high gear: how interstitial cells of Cajal change the motility pattern of the developing intestine. Am J Physiol Gastrointest Liver Physiol 319: G519–g528. 10.1152/ajpgi.00112.2020.

Chevalier, N.R., Dacher, N., Jacques, C., Langlois, L., Guedj, C. and Faklaris, O., 2019. Embryogenesis of the peristaltic reflex. J Physiol 597: 2785–2801. 10.1113/jp277746.

Chevalier, N.R., Gazguez, E., Bidault, L., Guilbert, T., Vias, C., Vian, E., Watanabe, Y., Muller, L., Germain, S., Bondurand, N., Dufour, S. and Fleury, V., 2016. How Tissue Mechanical Properties Affect Enteric Neural Crest Cell Migration. Sci Rep 6: 20927. 10.1038/srep20927.

Cholewinska, P., Gorniak, W. and Wojnarowski, K., 2021. Impact of selected environmental factors on microbiome of the digestive tract of ruminants. BMC Vet Res 17: 25. 10.1186/s12917-021-02742-y.

Coelho-Aguiar Jde, M., Bon-Frauches, A.C., Gomes, A.L., Veríssimo, C.P., Aguiar, D.P., Matias, D., Thomasi, B.B., Gomes, A.S., Brito, G.A. and Moura-Neto, V., 2015. The enteric glia: identity and functions. Glia 63: 921–935. 10.1002/glia.22795.

Conner, P.J., Focke, P.J., Noden, D.M. and Epstein, M.L., 2003. Appearance of neurons and glia with respect to the wavefront during colonization of the avian gut by neural crest cells. Dev Dyn 226: 91–98. 10.1002/dvdy.10219.

Costa, M., Brookes, S.J. and Hennig, G.W., 2000. Anatomy and physiology of the enteric nervous system. Gut 47 Suppl 4: iv15–19; discussion iv26. 10.1136/gut.47.suppl_4.iv15.

Costa, M., Furness, J.B., Pullin, C.O. and Bornstein, J., 1985. Substance P enteric neurons mediate non-cholinergic transmission to the circular muscle of the guinea-pig intestine. Naunyn Schmiedebergs Arch Pharmacol 328: 446–453. 10.1007/bf00692914.

Costa, M.C., Bessegatto, J.A., Alfieri, A.A., Weese, J.S., Filho, J.A. and Oba, A., 2017. Different antibiotic growth promoters induce specific changes in the cecal microbiota membership of broiler chicken. PLoS One 12: e0171642. 10.1371/journal.pone.0171642.

Dahl, G.E., Tao, S. and Laporta, J., 2020. Heat Stress Impacts Immune Status in Cows Across the Life Cycle. Front Vet Sci 7: 116. 10.3389/fvets.2020.00116.

Daubner, S.C., Le, T. and Wang, S., 2011. Tyrosine hydroxylase and regulation of dopamine synthesis. Arch Biochem Biophys 508: 1–12. 10.1016/j.abb.2010.12.017.

Dennis, R.L., 2016. Adrenergic and noradrenergic regulation of poultry behavior and production. Domest Anim Endocrinol 56 Suppl: S94–S100. 10.1016/j.domaniend.2016.02.007.

Díaz-Güemes, I., Sánchez, F.M., Luis, L., Sun, F., Pascual, S. and Usón, J., 2007. Continuous vagus nerve stimulation effects on the gut-brain axis in Swine. Neuromodulation 10: 52–58. 10.1111/j.1525-1403.2007.00087.x.

Emami, N.K., Schreier, L.L., Greene, E., Tabler, T., Orlowski, S.K., Anthony, N.B., Proszkowiec-Weglarz, M. and Dridi, S., 2022. Ileal microbial composition in genetically distinct chicken lines reared under normal or high ambient temperatures. Anim Microbiome 4: 28. 10.1186/s42523-022-00183-y.

Fontaine-Perus, J., Chanconie, M., Polak, J.M. and Le Douarin, N.M., 1981. Origin and development of VIP and substance P containing neurons in the embryonic avian gut. Histochemistry 71: 313–323. 10.1007/bf00495878.

Freestone, P.P., Lyte, M., Neal, C.P., Maggs, A.F., Haigh, R.D. and Williams, P.H., 2000. The mammalian neuroendocrine hormone norepinephrine supplies iron for bacterial growth in the presence of transferrin or lactoferrin. J Bacteriol 182: 6091–6098. 10.1128/JB.182.21.6091-6098.2000.

Freestone, P.P., Williams, P.H., Haigh, R.D., Maggs, A.F., Neal, C.P. and Lyte, M., 2002. Growth stimulation of intestinal commensal Escherichia coli by catecholamines: a possible contributory factor in trauma-induced sepsis. Shock 18: 465–470. 10.1097/00024382-200211000-00014.

Fu, Y., Hu, J., Erasmus, M.A., Johnson, T.A. and Cheng, H.W., 2022. Effects of early-life cecal microbiota transplantation from divergently selected inbred chicken lines on growth, gut serotonin, and immune parameters in recipient chickens. Poult Sci 101: 101925. 10.1016/j.psj.2022.101925.

Furness, J.B., 2012. The enteric nervous system and neurogastroenterology. Nat Rev Gastroenterol Hepatol 9: 286–294. 10.1038/nrgastro.2012.32.

Furness, J.B., Li, Z.S., Young, H.M. and Förstermann, U., 1994. Nitric oxide synthase in the enteric nervous system of the guinea-pig: a quantitative description. Cell Tissue Res 277: 139–149. 10.1007/bf00303090.

Gagliano, H., Delgado-Morales, R., Sanz-Garcia, A. and Armario, A., 2014. High doses of the histone deacetylase inhibitor sodium butyrate trigger a stress-like response. Neuropharmacology 79: 75–82. 10.1016/j.neuropharm.2013.10.031.

Goyal, R.K. and Hirano, I., 1996. The enteric nervous system. N Engl J Med 334: 1106–1115. 10.1056/nejm199604253341707.

Gross, K.B. and Sturkie, P.A., 1975. Concentration of serotonin in intestine and factors affecting its release. Proc Soc Exp Biol Med 148: 1261–1264. 10.3181/00379727-148-38729.

Guabiraba, R. and Schouler, C., 2015. Avian colibacillosis: still many black holes. FEMS Microbiol Lett 362: fnv118. 10.1093/femsle/fnv118.

Gulbransen, B.D. and Sharkey, K.A., 2012. Novel functional roles for enteric glia in the gastrointestinal tract. Nat Rev Gastroenterol Hepatol 9: 625–632. 10.1038/nrgastro.2012.138.

Hansen, M.B., 2003. The enteric nervous system I: organisation and classification. Pharmacol Toxicol 92: 105–113. 10.1034/j.1600-0773.2003.t01-1-920301.x.

Iyer, L.M., Aravind, L., Coon, S.L., Klein, D.C. and Koonin, E.V., 2004. Evolution of cell-cell signaling in animals: did late horizontal gene transfer from bacteria have a role? Trends Genet 20: 292–299. 10.1016/j.tig.2004.05.007.

Jadhao, A.G., Biswas, S.P., Bhoyar, R.C. and Pinelli, C., 2017. The distribution of nicotinamide adenine dinucleotide phosphate-diaphorase (NADPH-d) in the medulla oblongata, spinal cord, cranial and spinal nerves of frog, Microhyla ornata. J Chem Neuroanat 81: 76–86. 10.1016/j.jchemneu.2017.02.009.

Jessen, K.R., Hills, J.M. and Saffrey, M.J., 1986. Immunohistochemical demonstration of GABAergic neurons in the enteric nervous system. J Neurosci 6: 1628–1634. 10.1523/jneurosci.06-06-01628.1986.

Kim, H.S., Park, J.S., Hong, S.J., Woo, M.S., Kim, S.Y. and Kim, K.S., 2003. Regulation of the tyrosine hydroxylase gene promoter by histone deacetylase inhibitors. Biochem Biophys Res Commun 312: 950–957. 10.1016/j.bbrc.2003.11.012.

Kogut, M.H., 2022. Role of diet-microbiota interactions in precision nutrition of the chicken: facts, gaps, and new concepts. Poult Sci 101: 101673. 10.1016/j.psj.2021.101673.

Kraimi, N., Dawkins, M., Gebhardt-Henrich, S.G., Velge, P., Rychlik, I., Volf, J., Creach, P., Smith, A., Colles, F. and Leterrier, C., 2019. Influence of the microbiota-gut-brain axis on behavior and welfare in farm animals: A review. Physiology & Behavior 210: 112658. https://doi.org/10.1016/j.physbeh.2019.112658.

Lake, J.I. and Heuckeroth, R.O., 2013. Enteric nervous system development: migration, differentiation, and disease. Am J Physiol Gastrointest Liver Physiol 305: G1–24. 10.1152/ajpgi.00452.2012.

Lara, L.J. and Rostagno, M.H., 2013. Impact of Heat Stress on Poultry Production. Animals (Basel) 3: 356–369. 10.3390/ani3020356.

Lei, C., Sun, R., Xu, G., Tan, Y., Feng, W., McClain, C.J. and Deng, Z., 2022. Enteric VIP-producing neurons maintain gut microbiota homeostasis through regulating epithelium fucosylation. Cell Host Microbe 30: 1417–1434.e1418. 10.1016/j.chom.2022.09.001.

Lenartowski, R. and Goc, A., 2011. Epigenetic, transcriptional and posttranscriptional regulation of the tyrosine hydroxylase gene. Int J Dev Neurosci 29: 873–883. 10.1016/j.ijdevneu.2011.07.006.

Li, C., Guo, S., Gao, J., Guo, Y., Du, E., Lv, Z. and Zhang, B., 2015a. Maternal high-zinc diet attenuates intestinal inflammation by reducing DNA methylation and elevating H3K9 acetylation in the A20 promoter of offspring chicks. J Nutr Biochem 26: 173–183. 10.1016/j.jnutbio.2014.10.005.

Li, J., Zhang, Z., Liu, X., Wang, Y., Mao, F., Mao, J., Lu, X., Jiang, D., Wan, Y., Lv, J.Y., Cao, G., Zhang, J., Zhao, N., Atkinson, M., Greiner, D.L., Prud'homme, G.J., Jiao, Z., Li, Y. and Wang, Q., 2015b. Study of GABA in Healthy Volunteers: Pharmacokinetics and Pharmacodynamics. Front Pharmacol 6: 260. 10.3389/fphar.2015.00260.

Li, Z., Chalazonitis, A., Huang, Y.Y., Mann, J.J., Margolis, K.G., Yang, Q.M., Kim, D.O., Côté, F., Mallet, J. and Gershon, M.D., 2011. Essential roles of enteric neuronal serotonin in gastrointestinal motility and the development/survival of enteric dopaminergic neurons. J Neurosci 31: 8998–9009. 10.1523/jneurosci.6684-10.2011.

Li, Z.S., Pham, T.D., Tamir, H., Chen, J.J. and Gershon, M.D., 2004. Enteric dopaminergic neurons: definition, developmental lineage, and effects of extrinsic denervation. J Neurosci 24: 1330–1339. 10.1523/jneurosci.3982-03.2004.

Liu, F., Zhao, W., Le, H.H., Cottrell, J.J., Green, M.P., Leury, B.J., Dunshea, F.R. and Bell, A.W., 2022. Review: What have we learned about the effects of heat stress on the pig industry? Animal 16: 100349. https://doi.org/10.1016/j.animal.2021.100349.

Liu, M.T., Rothstein, J.D., Gershon, M.D. and Kirchgessner, A.L., 1997. Glutamatergic enteric neurons. J Neurosci 17: 4764–4784. 10.1523/jneurosci.17-12-04764.1997.

Lopes, J.G. and Sourjik, V., 2018. Chemotaxis of Escherichia coli to major hormones and polyamines present in human gut. The ISME Journal 12: 2736–2747. 10.1038/s41396-018-0227-5.

Luo, Y., Ren, W., Smidt, H., Wright André-Denis, G., Yu, B., Schyns, G., McCormack Ursula, M., Cowieson Aaron, J., Yu, J., He, J., Yan, H., Wu, J., Mackie Roderick, I. and Chen, D., 2022. Dynamic Distribution of Gut Microbiota in Pigs at Different Growth Stages: Composition and Contribution. Microbiology Spectrum 10: e00688–00621. 10.1128/spectrum.00688-21.

Luppi, A., 2017. Swine enteric colibacillosis: diagnosis, therapy and antimicrobial resistance. Porcine Health Manag 3: 16. 10.1186/s40813-017-0063-4.

Lyte, J.M., 2018. Eating for 3.8 × 10(13): Examining the Impact of Diet and Nutrition on the Microbiota-Gut-Brain Axis Through the Lens of Microbial Endocrinology. Front Endocrinol (Lausanne) 9: 796. 10.3389/fendo.2018.00796.

Lyte, J.M. and Lyte, M., 2019. Review: Microbial endocrinology: intersection of microbiology and neurobiology matters to swine health from infection to behavior. Animal 13: 2689–2698. 10.1017/S1751731119000284.

Lyte, J.M., Martinez, D.A., Robinson, K., Donoghue, A.M., Daniels, K.M. and Lyte, M., 2022. A neurochemical biogeography of the broiler chicken intestinal tract. Poult Sci 101: 101671. 10.1016/j.psj.2021.101671.

Lyte, M., 2013. Microbial endocrinology and nutrition: A perspective on new mechanisms by which diet can influence gut-to-brain communication. PharmaNutrition 1: 35–39. https://doi.org/10.1016/j.phanu.2012.11.002.

Lyte, M., 2016. Microbial Endocrinology in the Pathogenesis of Infectious Disease. Microbiology Spectrum 4: 4.2.09. 10.1128/microbiolspec.VMBF-0021-2015.

Lyte, M., Villageliu, D.N., Crooker, B.A. and Brown, D.R., 2018. Symposium review: Microbial endocrinology-Why the integration of microbes, epithelial cells, and neurochemical signals in the digestive tract matters to ruminant health. J Dairy Sci 101: 5619–5628. 10.3168/jds.2017-13589.

Martinez-Miro, S., Tecles, F., Ramon, M., Escribano, D., Hernandez, F., Madrid, J., Orengo, J., Martinez-Subiela, S., Manteca, X. and Ceron, J.J., 2016. Causes, consequences and biomarkers of stress in swine: an update. BMC Vet Res 12: 171. 10.1186/s12917-016-0791-8.

Martinez, V. and TachÉ, Y., 2006. CHAPTER 138 – Calcitonin Gene-Related Peptide and Gastrointestinal Function. In: A.J. Kastin (Ed.), Handbook of Biologically Active Peptides. Academic Press, Burlington, pp. 1005–1011. https://doi.org/10.1016/B978-012369442-3/50141-0.

Mayfield, S.R., Stonestreet, B.S., Shaul, P.W., Brubakk, A.M., Susa, J. and Oh, W., 1989. Plasma catecholamine concentrations of newborn piglets in thermoneutral and cold environments. J Dev Physiol 11: 331–334.

Nagahama, M., Semba, R., Tsuzuki, M. and Aoki, E., 2001. L-arginine immunoreactive enteric glial cells in the enteric nervous system of rat ileum. Biol Signals Recept 10: 336–340. 10.1159/000046901.

Neuman, H., Debelius, J.W., Knight, R. and Koren, O., 2015. Microbial endocrinology: the interplay between the microbiota and the endocrine system. FEMS Microbiol Rev 39: 509–521. 10.1093/femsre/fuu010.

Newbold, C.J. and Ramos-Morales, E., 2020. Review: Ruminal microbiome and microbial metabolome: effects of diet and ruminant host. Animal 14: s78–s86. 10.1017/S1751731119003252.

Niknafs, S. and Roura, E., 2018. Nutrient sensing, taste and feed intake in avian species. Nutr Res Rev 31: 256–266. 10.1017/s0954422418000100.

Nzakizwanayo, J., Dedi, C., Standen, G., Macfarlane, W.M., Patel, B.A. and Jones, B.V., 2015. Escherichia coli Nissle 1917 enhances bioavailability of serotonin in gut tissues through modulation of synthesis and clearance. Sci Rep 5: 17324. 10.1038/srep17324.

Obata, Y., Castaño, Á., Boeing, S., Bon-Frauches, A.C., Fung, C., Fallesen, T., de Agüero, M.G., Yilmaz, B., Lopes, R., Huseynova, A., Horswell, S., Maradana, M.R., Boesmans, W., Vanden Berghe, P., Murray, A.J., Stockinger, B., Macpherson, A.J. and Pachnis, V., 2020. Neuronal programming by microbiota regulates intestinal physiology. Nature 578: 284–289. 10.1038/s41586-020-1975-8.

Oehlers, S.H., Flores, M.V., Hall, C.J., Wang, L., Ko, D.C., Crosier, K.E. and Crosier, P.S., 2017. A whole animal chemical screen approach to identify modifiers of intestinal neutrophilic inflammation. FEBS J 284: 402–413. 10.1111/febs.13976.

Palus, K., Bulc, M. and Całka, J., 2018. Changes in VIP-, SP- and CGRP-like immunoreactivity in intramural neurons within the pig stomach following supplementation with low and high doses of acrylamide. Neurotoxicology 69: 47–59. 10.1016/j.neuro.2018.09.002.

Patra, A.K. and Kar, I., 2021. Heat stress on microbiota composition, barrier integrity, and nutrient transport in gut, production performance, and its amelioration in farm animals. J Anim Sci Technol 63: 211–247. 10.5187/jast.2021.e48.

Peters-van der Sanden, M.J., Kirby, M.L., Gittenberger-de Groot, A., Tibboel, D., Mulder, M.P. and Meijers, C., 1993a. Ablation of various regions within the avian vagal neural crest has differential effects on ganglion formation in the fore-, mid- and hindgut. Dev Dyn 196: 183–194. 10.1002/aja.1001960305.

Peters-van der Sanden, M.J., Luider, T.M., van der Kamp, A.W., Tibboel, D. and Meijers, C., 1993b. Regional differences between various axial segments of the avian neural crest regarding the formation of enteric ganglia. Differentiation 53: 17–24. 10.1111/j.1432-0436.1993.tb00641.x.

Phillips, A.W., Newcomb, H.R., Smith, J.E. and Lachapeller, 1961. Serotonin in the small intestine of conventional and germfree chicks. Nature 192: 380. 10.1038/192380a0.

Pollmann, M., Nordhoff, M., Pospischil, A., Tedin, K. and Wieler, L.H., 2005. Effects of a probiotic strain of Enterococcus faecium on the rate of natural chlamydia infection in swine. Infect Immun 73: 4346–4353. 10.1128/IAI.73.7.4346-4353.2005.

Pullinger, G.D., Carnell, S.C., Sharaff, F.F., van Diemen, P.M., Dziva, F., Morgan, E., Lyte, M., Freestone, P.P. and Stevens, M.P., 2010. Norepinephrine augments Salmonella enterica-induced enteritis in a manner associated with increased net replication but independent of the putative adrenergic sensor kinases QseC and QseE. Infect Immun 78: 372–380. 10.1128/IAI.01203-09.

Qadis, A.Q., Goya, S., Ikuta, K., Yatsu, M., Kimura, A., Nakanishi, S. and Sato, S., 2014. Effects of a bacteria-based probiotic on ruminal pH, volatile fatty acids and bacterial flora of Holstein calves. J Vet Med Sci 76: 877–885. 10.1292/jvms.14-0028.

Qu, Z.D., Thacker, M., Castelucci, P., Bagyánszki, M., Epstein, M.L. and Furness, J.B., 2008. Immunohistochemical analysis of neuron types in the mouse small intestine. Cell Tissue Res 334: 147–161. 10.1007/s00441-008-0684-7.

Raines, D.J., Moroz, O.V., Blagova, E.V., Turkenburg, J.P., Wilson, K.S. and Duhme-Klair, A.K., 2016. Bacteria in an intense competition for iron: Key component of the Campylobacter jejuni iron uptake system scavenges enterobactin hydrolysis product. Proc Natl Acad Sci USA 113: 5850–5855. 10.1073/pnas.1520829113.

Reigstad, C.S., Salmonson, C.E., Rainey, J.F., 3rd, Szurszewski, J.H., Linden, D.R., Sonnenburg, J.L., Farrugia, G. and Kashyap, P.C., 2015. Gut microbes promote colonic serotonin production through an effect of short-chain fatty acids on enterochromaffin cells. FASEB J 29: 1395–1403. 10.1096/fj.14-259598.

Reimann, E.M., Sunde, M.L. and Hoekstra, W.G., 1971. Effect of zinc, certain related dietary factors and age on the histamine content of chick tissues. J Nutr 101: 1623–1629. 10.1093/jn/101.12.1623.

Renaudeau, D., Collin, A., Yahav, S., de Basilio, V., Gourdine, J.L. and Collier, R.J., 2012. Adaptation to hot climate and strategies to alleviate heat stress in livestock production. Animal 6: 707–728. 10.1017/S1751731111002448.

Rojas-Downing, M.M., Nejadhashemi, A.P., Harrigan, T. and Woznicki, S.A., 2017. Climate change and livestock: Impacts, adaptation, and mitigation. Climate Risk Management 16: 145–163. https://doi.org/10.1016/j.crm.2017.02.001.

Rosebrough, R.W., 1996. Crude protein and supplemental dietary tryptophan effects on growth and tissue neurotransmitter levels in the broiler chicken. Br J Nutr 76: 87–96. 10.1079/bjn19960011.

Rostagno, M.H., 2009. Can stress in farm animals increase food safety risk? Foodborne Pathog Dis 6: 767–776. 10.1089/fpd.2009.0315.

Rostagno, M.H., 2020. Effects of heat stress on the gut health of poultry. J Anim Sci 98. 10.1093/jas/skaa090.

Rothman, T.P., Sherman, D., Cochard, P. and Gershon, M.D., 1986. Development of the monoaminergic innervation of the avian gut: transient and permanent expression of phenotypic markers. Dev Biol 116: 357–380. 10.1016/0012-1606(86)90139-9.

Rytel, L., Wojtkiewicz, J., Snarska, A. and Mikołajczyk, A., 2021. Changes in the Neurochemical Characterization of Enteric Neurons in the Porcine Duodenum After Administration of Low-Dose Salmonella Enteritidis Lipopolysaccharides. J Mol Neurosci 71: 1556–1566. 10.1007/s12031-019-01473-y.

Scanes, C.G., 2016. Biology of stress in poultry with emphasis on glucocorticoids and the heterophil to lymphocyte ratio. Poult Sci 95: 2208–2215. 10.3382/ps/pew137.

Schemann, M. and Neunlist, M., 2004. The human enteric nervous system. Neurogastroenterol Motil 16 Suppl 1: 55–59. 10.1111/j.1743-3150.2004.00476.x.

Schemann, M., Sann, H., Schaaf, C. and Mäder, M., 1993. Identification of cholinergic neurons in enteric nervous system by antibodies against choline acetyltransferase. Am J Physiol 265: G1005–1009. 10.1152/ajpgi.1993.265.5.G1005.

Sharara-Chami, R.I., Joachim, M., Pacak, K. and Majzoub, J.A., 2010. Glucocorticoid treatment-effect on adrenal medullary catecholamine production. Shock 33: 213–217. 10.1097/SHK.0b013e3181af0633.

Skaar, E.P., 2010. The battle for iron between bacterial pathogens and their vertebrate hosts. PLoS Pathog 6: e1000949. 10.1371/journal.ppat.1000949.

Soerjadi, A.S., Druitt, J.H., Lloyd, A.B. and Cumming, R.B., 1979. Effect of environmental temperature on susceptibility of young chickens to Salmonella typhimurium. Aust Vet J 55: 413–417. 10.1111/j.1751-0813.1979.tb05591.x.

Spencer, N.J. and Hu, H., 2020. Enteric nervous system: sensory transduction, neural circuits and gastrointestinal motility. Nat Rev Gastroenterol Hepatol 17: 338–351. 10.1038/s41575-020-0271-2.

Sudo, N., 2014. Microbiome, HPA axis and production of endocrine hormones in the gut. Adv Exp Med Biol 817: 177–194. 10.1007/978-1-4939-0897-4_8.

Sudo, N., Chida, Y., Aiba, Y., Sonoda, J., Oyama, N., Yu, X.N., Kubo, C. and Koga, Y., 2004. Postnatal microbial colonization programs the hypothalamic-pituitary-adrenal system for stress response in mice. J Physiol 558: 263–275. 10.1113/jphysiol.2004.063388.

Szymanska, K., Calka, J. and Gonkowski, S., 2018. Nitric oxide as an active substance in the enteric neurons of the porcine digestive tract in physiological conditions and under intoxication with bisphenol A (BPA). Nitric Oxide 80: 1–11. 10.1016/j.niox.2018.08.001.

Thaxton, P., Wyatt, R.D. and Hamilton, P.B., 1974. The effect of environmental temperature on parathyphoid infection in the neonatal chicken. Poult Sci 53: 88–94. 10.3382/ps.0530088.

Tohmola, N., Johansson, A., Sane, T., Renkonen, R., Hamalainen, E. and Itkonen, O., 2015. Transient elevation of serum 5-HIAA by dietary serotonin and distribution of 5-HIAA in serum protein fractions. Ann Clin Biochem 52: 428–433. 10.1177/0004563214554842.

Toscano, M.J., Stabel, T.J., Bearson, S.M.D., Bearson, B.L. and Lay, D.C., 2007. Cultivation of Salmonella enterica serovar Typhimurium in a norepinephrine-containing medium alters in vivo tissue prevalence in swine. Journal of Experimental Animal Science 43: 329–338. https://doi.org/10.1016/j.jeas.2006.09.007.

Troxell, B., Petri, N., Daron, C., Pereira, R., Mendoza, M., Hassan, H.M. and Koci, M.D., 2015. Poultry body temperature contributes to invasion control through reduced expression of Salmonella pathogenicity island 1 genes in Salmonella enterica serovars Typhimurium and Enteritidis. Appl Environ Microbiol 81: 8192–8201. 10.1128/AEM.02622-15.

Tsiouris, V., Georgopoulou, I., Batzios, C., Pappaioannou, N., Ducatelle, R. and Fortomaris, P., 2015. The effect of cold stress on the pathogenesis of necrotic enteritis in broiler chicks. Avian Pathol 44: 430–435. 10.1080/03079457.2015.1083094.

Venegas-Vargas, C., Henderson, S., Khare, A., Mosci, R.E., Lehnert, J.D., Singh, P., Ouellette, L.M., Norby, B., Funk, J.A., Rust, S., Bartlett, P.C., Grooms, D. and Manning, S.D., 2016. Factors Associated with Shiga Toxin-Producing Escherichia coli Shedding by Dairy and Beef Cattle. Appl Environ Microbiol 82: 5049–5056. 10.1128/AEM.00829-16.

Villageliu, D.N. and Lyte, M., 2017. Microbial endocrinology: Why the intersection of microbiology and neurobiology matters to poultry health. Poult Sci 96: 2501–2508. 10.3382/ps/pex148.

Vlisidou, I., Lyte, M., van Diemen, P.M., Hawes, P., Monaghan, P., Wallis, T.S. and Stevens, M.P., 2004. The neuroendocrine stress hormone norepinephrine augments Escherichia coli O157:H7-induced enteritis and adherence in a bovine ligated ileal loop model of infection. Infect Immun 72: 5446–5451. 10.1128/IAI.72.9.5446-5451.2004.

Welch, C.B., Ryman, V.E., Pringle, T.D. and Lourenco, J.M., 2022. Utilizing the Gastrointestinal Microbiota to Modulate Cattle Health through the Microbiome-Gut-Organ Axes. Microorganisms 10. 10.3390/microorganisms10071391.

Xu, F., Wu, C., Guo, F., Cui, G., Zeng, X., Yang, B. and Lin, J., 2015. Transcriptomic analysis of Campylobacter jejuni NCTC 11168 in response to epinephrine and norepinephrine. Front Microbiol 6: 452. 10.3389/fmicb.2015.00452.

Yadav, S. and Jha, R., 2019. Strategies to modulate the intestinal microbiota and their effects on nutrient utilization, performance, and health of poultry. Journal of Animal Science and Biotechnology 10: 2. 10.1186/s40104-018-0310-9.

Yan, Y., Jiang, W., Liu, L., Wang, X., Ding, C., Tian, Z. and Zhou, R., 2015. Dopamine controls systemic inflammation through inhibition of NLRP3 inflammasome. Cell 160: 62–73. 10.1016/j.cell.2014.11.047.

Yang, P., Gandahi, J.A., Zhang, Q., Zhang, L.L., Bian, X.G., Wu, L., Liu, Y. and Chen, Q.S., 2013. Quantitative changes of nitrergic neurons during postnatal development of chicken myenteric plexus. J Zhejiang Univ Sci B 14: 886–895. 10.1631/jzus.B1300005.

Yang, Y., Ashworth, A.J., Willett, C., Cook, K., Upadhyay, A., Owens, P.R., Ricke, S.C., DeBruyn, J.M. and Moore, P.A., Jr., 2019. Review of Antibiotic Resistance, Ecology, Dissemination, and Mitigation in U.S. Broiler Poultry Systems. Front Microbiol 10: 2639. 10.3389/fmicb.2019.02639.

Zhang, W., Lyu, M., Bessman, N.J., Xie, Z., Arifuzzaman, M., Yano, H., Parkhurst, C.N., Chu, C., Zhou, L., Putzel, G.G., Li, T.T., Jin, W.B., Zhou, J., Hu, H., Tsou, A.M., Guo, C.J. and Artis, D., 2022a. Gut-innervating nociceptors regulate the intestinal microbiota to promote tissue protection. Cell 185: 4170–4189.e4120. 10.1016/j.cell.2022.09.008.

Zhang, Y., Sun, L., Zhu, R., Zhang, S., Liu, S., Wang, Y., Wu, Y., Xing, S., Liao, X. and Mi, J., 2022b. Porcine gut microbiota in mediating host metabolic adaptation to cold stress. NPJ Biofilms Microbiomes 8: 18. 10.1038/s41522-022-00283-2.

Zheng, L., Kelly, C.J., Battista, K.D., Schaefer, R., Lanis, J.M., Alexeev, E.E., Wang, R.X., Onyiah, J.C., Kominsky, D.J. and Colgan, S.P., 2017. Microbial-Derived Butyrate Promotes Epithelial Barrier Function through IL-10 Receptor-Dependent Repression of Claudin-2. J Immunol 199: 2976–2984. 10.4049/jimmunol.1700105.

CHAPTER 15

Swine nutrition, brain, and behaviour

S. Joung[1], *L. Sutkus*[1], *R.K. Golden*[1], *S.M. Donovan*[2] *and R.N. Dilger*[1,3,*]

[1]Neuroscience Program, University of Illinois at Urbana-Champaign, Champaign, IL, USA; [2]Department of Food Science and Human Nutrition, University of Illinois at Urbana-Champaign, Champaign, IL, USA; [3]Department of Animal Sciences, University of Illinois at Urbana-Champaign, Champaign, IL, USA; *Correspondence: rdilger2@illinois.edu

Abstract

The impact of nutrition on pig growth and development has been investigated in various contexts, yet much remains unknown about specific mechanisms by which manipulation of the intestinal microbiota may influence brain structure and function. The hypothalamic-pituitary-adrenal (HPA) axis has emerged as one of the main modulators of the microbiota-gut-brain axis, and may serve as one of the most important mechanisms driving stress responses. As pigs experience various environmental stressors in commercial production settings, regulating HPA axis activation and accurately measuring stress levels in pigs has become crucial to successful pork production. Detailed in this chapter is organization of the HPA axis, a description of physiological and behavioural changes following exposure to environmental stressors, and potential beneficial effects of nutritional supplementation in relation to the microbiota-gut-brain axis. Moreover, diet can profoundly influence the timing and rate of myelination as part of cognitive development and this chapter further discusses the importance of myelin, and the effects of maternal environment and nutrition on brain structural and cognitive development in pigs.

Keywords

HPA axis – stress – diet – myelin – learning and memory

1 Introduction

Emerging evidence focused on the interplay between microorganisms and host physiology highlights the important role of diet in shaping the microbiota-gut-brain axis (Cryan and Dinan 2012). Within commercial pig production systems, implications of this dynamic relationship include the timing and rate of brain

© S. JOUNG ET AL., 2024 | DOI:10.3920/9789004695467_016

development and function, which interacts with aspects of behaviour and welfare, and the pigs' ability to respond to external stressors. Interestingly, components of the microbiota-gut-brain axis follow a developmental trajectory similar to those occurring in other organ systems in the pig (Jena *et al.* 2020). Hence, across the developmental timeline, the gastrointestinal tract and brain elicit reciprocal influences. As microbial diversity increases after birth, the enteric nervous system matures, and various brain systems undergo development and refinement (Jena *et al.* 2020). Specifically in the brain, synaptogenesis, myelination, and synaptic refinement are occurring which impact cognitive abilities and behaviour (Jena *et al.* 2020). Immediately following the birth of a pig, exposure to various microbial profiles from the mother and surrounding environment begin colonizing in gut to ultimately form the commensal microbiota (Foster 2016). This persistent microbiota remains in the gut throughout life and provides the basis for interactions with host metabolism, as well as serving a critical role in the development of immune, endocrine, and nervous systems (Jena *et al.* 2020). Furthermore, each of these systems have mediators that enable signals to pass between the gut and the brain to influence overall health and behaviour of the pig (Jena *et al.* 2020).

The most thoroughly studied behaviours influenced by early-life nutrition and microbial exposure in pigs include anxiety, stress, learning, and memory (Alemohammad *et al.* 2022; Simpson *et al.* 2021). A wide array of studies conducted in germ-free mice have concluded that disruptions in intestinal microbial colonization can cause myriad diseases (Farzi *et al.* 2018; Liu *et al.* 2015; Uzbay 2019). Additionally, abnormal microbiota colonization has been linked to changes in behaviour, specifically those involving anxiety, issues with stress regulation, and changes in sociability (Liu *et al.* 2015; Luczynski *et al.* 2016). This has led to an increased interest in identifying appropriate nutritional profiles to support beneficial bacteria, gut health, and brain function. Therefore, it is important to thoroughly understand the complicated network of the microbiota-gut-brain axis and identify various sources of environmental influencers.

One of the main physiological routes of communication between microorganisms and the host is via the hypothalamic-pituitary-adrenal (HPA) axis. The HPA axis plays a particularly important role in a commercial production setting as pigs are exposed to stressful environmental stimuli on a daily basis, not unlike that of humans. As such, the ability of a pig to mount effective and adaptive stress responses is directly related to welfare and productivity in a commercial operation. In this chapter, we will discuss not only the various brain systems, hormones, and peptides involved in the HPA axis, but also physiological and behavioural changes from HPA axis activation and exposure to environmental stressors. Examples of stressors present in the swine industry and effects of nutritional supplementations are also discussed. Additionally, we will discuss how nutrition impacts a key component of brain development, that being myelination events that are central to cognitive processing. We will further expand on how this key aspect of development plays a part in learning and memory during early life, thereby highlighting the vital influence of nutrition in the pig.

2 HPA axis

The HPA axis is an essential neuroendocrine mechanism that prepares an organism to efficiently handle environmental challenges through an intricate network of hormones and regulatory pathways. This interconnected system involves various brain and endocrine system structures, most importantly the hypothalamus, pituitary gland, and adrenal gland (Herman *et al.* 2020). Additionally, various other brain regions such as the amygdala, hippocampus, and prefrontal cortex play a role in regulating the HPA axis. These structures have an intricate system of communication and circuitry that involves various hormones, peptides, and receptors. In this section, we will introduce key players of the HPA axis and later discuss the intricate network of communication that leads to activation.

2.1 *Hypothalamus*

A critical brain region of the HPA axis is the hypothalamus, located at the base of the brain and divided into four main areas: preoptic, anterior, tuberal, and mammillary (Xie and Dorsky 2017). These areas are further subdivided into lateral, medial, and paraventricular zones each with particular functions (Xie and Dorsky 2017). Generally, these areas are involved with functions such as thermoregulation, reproduction, feed intake regulation, homeostatic regulation, stress response, and also memory. Across these vital functions, the zone that is most pivotal for the HPA axis is located in the more anterior area of the hypothalamus called the paraventricular nucleus (PVN). The PVN plays a pivotal role in neuroendocrine and autonomic nervous system regulation (Ferguson, Latchford, and Sanon 2008), with regulatory mechanisms involving the synthesis and secretion of three important neuroendocrine peptides: corticotropin-releasing hormone (CRH), arginine vasopressin (AVP), and oxytocin (Sawchenko *et al.* 1996).

Both CRH and AVP are involved in the body's response to stressful situations in the environment (Kreek *et al.* 2011). CRH is recognised as a pivotal stress hormone that is released with activation of the sympathetic nervous system (Irwin *et al.* 1992), whereas AVP is a hormone involved in managing fluid balance, regulating blood pressure and kidney function, and maintaining sodium/water balance in the body (Cuzzo *et al.* 2021). Oxytocin is a hormone that plays a more important role in various reproductive functions, though this hormone is also directly involved in stress resiliency (Cuzzo *et al.* 2021; Neumann 2002). Cell groups within the PVN have long projections to cell groups associated with the autonomic nervous system that also express CRH, AVP, and oxytocin (Sawchenko *et al.* 1996). Therefore, the neurons located in this nucleus have been identified as key regulators of integrated systems such as the thyroid axis, reproductive axis, growth and development, body fluid balance, gastrointestinal function, cardiovascular function, and, of course, the HPA axis (Ferguson *et al.* 2008).

2.2 *Pituitary gland*

During sympathetic nervous system activation, CRH and AVP are released into the portal veins and trigger the release of adrenocorticotropic hormone (ACTH) from the pituitary gland (Herman *et al.* 2020). The pituitary gland is another pivotal brain region of the HPA axis and is often referred to as the "master endocrine gland", due to its role in regulating hormone production in other tissues throughout the body. It is divided into three separate lobes: anterior, intermediate, and posterior – each displaying various mechanisms of regulation and separate cell populations (Amar and Weiss 2003). The anterior lobe is highly vascularised in mammalian tissues, receiving a large proportion of blood in the body from the portal system (Amar and Weiss 2003). Through this venous pathway, this lobe communicates with other glands by producing several hormones including thyroid stimulating hormone, corticotropin, follicle stimulating hormone, growth hormone, and prolactin (Amar and Weiss 2003). Corticotroph cells in this lobe are also responsible for producing proopiomelanocortin (POMC), a precursor protein responsible for production of neuroendocrine peptides, specifically ACTH (Day 2009). Within the pituitary gland, CRH is important for increasing POMC transcription, synthesizing peptides and eliciting intracellular calcium release to cause ACTH secretion (Day 2009). CRH has been identified to synthesize a group of peptides termed urocortins I, II, and III (Majzoub 2006; Papadimitriou and Priftis 2009). Specifically, urocortin I have high affinity for CRH receptors and appears to mediate several stress-related behaviours (Papadimitriou and Priftis 2009). On the contrary, urocortin II and III are involved in gastric functions such as appetite suppression (Majzoub 2006). Urocortins overall have been linked to decreasing an animal's stress response and are believed to be important in returning the body to homeostasis (Papadimitriou and Priftis 2009). The discovery of these peptides has determined that CRH is not solely responsible for mediating ACTH secretion from the pituitary gland (Majzoub 2006).

2.3 *Adrenal gland*

Once ACTH is secreted into the bloodstream, it travels to the adrenal glands, or adrenal cortex, which is a set of glands located on top of the kidneys. At this site, ACTH will bind to the melanocortin-2 receptor, which triggers the adrenal glands to produce glucocorticoid hormones, most notably cortisol (Soto-Rivera and Majzoub 2017). The action of ACTH is mediated by calcium levels and messengers both inside and outside of the cell (Papadimitriou and Priftis 2009). Along with production of cortisol, ACTH binding will also stimulate steroidogenesis and growth in the adrenal gland (Papadimitriou and Priftis 2009). This will stimulate the synthesis of various receptors, such as low- and high-density lipoproteins, which are important carriers of cholesterol throughout the body (Papadimitriou and Priftis 2009). Once glucocorticoid hormones are produced within the adrenal gland, they are rapidly released into circulation by diffusion (Papadimitriou and Priftis 2009). A multitude of cells across the body express glucocorticoid receptors, meaning that once glucocorticoids are released into circulation there are pleiotropic effects (Papadimitriou and Priftis 2009). In this way, the effects of cortisol span neuroendocrine, cardiovascular, immune, and nervous systems. Most notably in production animals, the effects of cortisol on growth performance, metabolism and animal physiology have been intensively studied (Mormède *et al.* 2011).

2.4 *Other players in HPA axis*

Several additional brain regions have input in regulating HPA axis activity, specifically the hippocampus, amygdala, and prefrontal cortex. The hippocampus is located in the temporal lobes of the brain and is most notable for its roles in learning and memory (Douglas 1967; Sauleau *et al.* 2009). Hippocampal neurons have been found to richly express both glucocorticoid and mineralocorticoid receptors (Herman *et al.* 2020). Mineralocorticoid receptors are known to have a high affinity for cortisol, whereas glucocorticoid receptors have low affinity (Russell and Shipston 2015). Furthermore, the hippocampus has been identified as an inhibitor of the HPA axis through cortical neuron activity that is sensitive to stress and projects to the paraventricular nucleus (Herman *et al.* 2020). Lesion and stimulation studies specifically helped identify the inhibitory role of the hippocampus, although the exact communication pathway is still not fully understood (Herman *et al.* 2020).

Whereas the hippocampus has an inhibitory effect on the HPA axis, the amygdala has an excitatory role. As part of the limbic system, the amygdala is a brain region that registers the emotional significance of events, specifically involved in an animal's fear response (Cockrem 2007; Davis 1992). The amygdala is located adjacent to the hippocampus and is responsible for influencing emotional learning, stress response, olfactory, pheromonal, and reproductive functions (Równiak and Bogus-Nowakowska 2020). In association with fear, the amygdala is involved in recognizing potential threats to prepare for a fight-or-flight response (Davis 1992). The amygdala reacts to a physical stressor and communicates with the HPA axis to activate the system and elicit downstream effects to produce CRF, ACTH, and cortisol (Russell and Shipston 2015). Higher levels of cortisol activate glucocorticoid receptors throughout the body, which in turn inhibit the amygdala and return the system to baseline (Russell and Shipston 2015). Through this inhibitory/excitatory feedback system of the hippocampus and the amygdala, the body is able to respond appropriately to stress and maintain homeostasis.

The prefrontal cortex is another brain region involved in HPA axis regulation. This region is found at the very front of the frontal lobe of the brain and is involved in a multitude of cognitive processes, such as attention and executive function

(Arnsten and Li 2005; Sauleau *et al.* 2009). The prefrontal cortex (PFC) is subdivided into 2 main regions: the ventromedial PFC and lateral PFC. The ventromedial PFC is further divided into ventral and medial. Similar to the hippocampus, the role of the medial areas of the PFC is to inhibit HPA axis functions, although the mechanism behind regulation is not well understood (Crane *et al.* 2003; Kern *et al.* 2008). Furthermore, there are no direct projections from the medial PFC to any structures of the HPA axis, but evidence suggests this structure interacts with the amygdala during stressful circumstances (Crane *et al.* 2003). It is through this interaction that the medial PFC is believed to influence HPA axis activity, but further research is warranted.

3 Stress responses in pigs

The HPA axis is an essential component of growth and development, as it allows pigs to successfully adapt to various environmental stressors and return to biological trajectories as needed. Specifically, homeostasis refers to an animal's capability to self-regulate bodily systems to adapt to stressful stimuli and return to a stable equilibrium (Fulford and Harbuz 2005; Meizlish *et al.* 2021). Dysregulation of homeostatic functions can lead to various immune inflammatory conditions in tissues across the body (Fulford and Harbuz 2005). This can occur with HPA axis activation since this system releases stress hormones that prepare an animal for action. Prolonged secretion of glucocorticoids, the main by-products of the HPA axis, can lead to pathological conditions such as immunosuppression and hypercortisolaemia that impact an animal's susceptibility to infection and disease (Fulford and Harbuz 2005). Therefore, appropriate activation and regulation of the HPA axis is necessary for general health and welfare of pigs, so physiological and behavioural stress responses are discussed.

3.1 *Physiological responses*

The HPA axis involves multiple brain regions (discussed above), which are activated in response to different types of stressors. As a result, the well-established circuitry of the HPA axis and corresponding hormones are recruited. In summary, the PVN is considered an initial activation site of the HPA axis, and releases CRH and vasopressin. CRH further stimulates the anterior pituitary gland, which leads to synthesis and release of ACTH. Upon reaching the adrenal cortex via systemic circulation, ACTH causes the release of adrenocortical hormones, including cortisol. Cortisol is considered the end-product of the HPA axis activity and plays a crucial role in the negative feedback loop within the HPA axis. Cortisol acts on various brain regions such as the PVN and anterior pituitary gland. The negative feedback loop provides an essential inhibition to continuous release of CRH and ACTH from the hypothalamus and anterior pituitary gland, respectively. A schematic of the HPA axis and involved hormones are provided in Figure 15.1. In situations of chronic exposure to stressors, the functionality of the negative feedback loop may become dysfunctional, which may result in a hyper-activated HPA axis lacking proper regulatory mechanisms and heightened release of HPA axis products such as CRH, ACTH and cortisol. Thus, proper regulation of the HPA axis plays a crucial role in successful adaptation to internal and external stressors to animals.

When an animal experiences stress, corticotropin-releasing hormone (CRH) gets released from the hypothalamic paraventricular nucleus (PVN) and delivered to the anterior pituitary gland. Adrenocorticotropic hormone (ACTH) is synthesized and released from the anterior pituitary gland and reaches the adrenal cortex via systemic circulation. Cortisol, an end-product of the HPA axis activation, is released from the adrenal cortex, and provides inhibition of continued release of CRH and ACTH via acting on negative feedback loop to various regions such as PVN and anterior pituitary gland. Abbreviation: ACTH, adrenocorticotropic hormone; CRH, corticotropin-releasing hormone; PVN, hypothalamic paraventricular nucleus.

Because of the importance of the HPA axis regulation in stress responses, measuring secretory hormones from the HPA axis, such as cortisol and ACTH, can help assess the physiological level of stress. Generally, measuring cortisol levels in blood

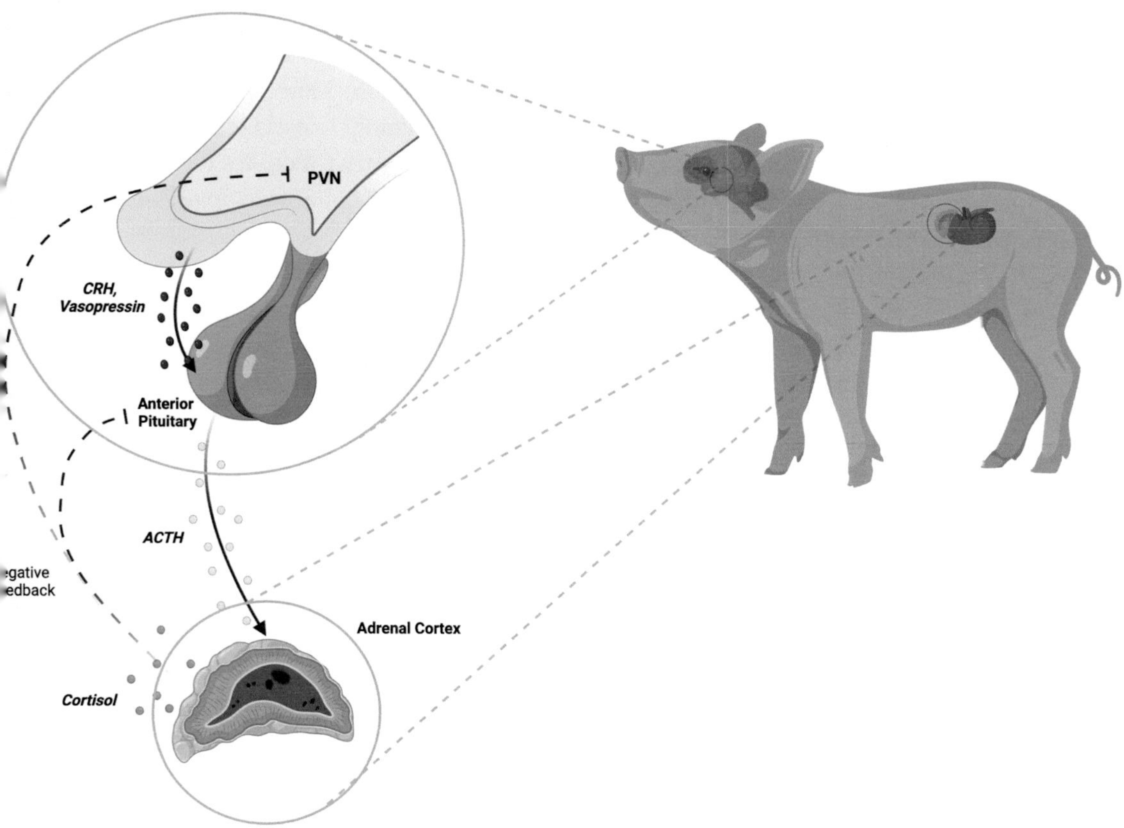

FIGURE 15.1 The schematics of activation of HPA axis in response to stress

plasma, serum, saliva, urine, and feces is one of the most abundantly used indicators of the HPA axis activity in pigs. Also, measuring hair cortisol concentrations in pigs has recently gained popularity as a valid assessment of the HPA axis activity because of its repeatability over a longer period (Heimbürge *et al.* 2020). Mormede and colleagues provide a thorough review on various methodologies utilised to evaluate animal stress levels and welfare by assessing the HPA axis activity in farm animals, including pigs (Mormède *et al.* 2007).

3.2 *Behavioural responses*

Stressful stimuli can elicit not only physiological changes, but also behavioural responses. Behavioural observations in a group-housing context can provide insights on the innate behaviours of pigs. Defecating, urinating, escape attempts, freezing, and high-pitched vocalizations such as screams are commonly measured behavioural responses to fearful stimuli (Davis 1997), and are widely used as indicators of negative emotions especially in a stressful context (Diana *et al.* 2019; Manteuffel *et al.* 2004; Norscia *et al.* 2021; Reimert *et al.* 2013). Another behavioural indicator of stress in pigs are displacement behaviours, which are often referred to as normative behaviours that may appear inappropriate in certain conflict and stressful situations (Tinbergen 1952). Displacement behaviours are present in various species, including humans, primates, and pigs (Maestripieri *et al.* 1992; Marcet-Rius *et al.* 2018, 2019). Specifically, displacement behaviours in pigs may be expressed as continuous biting, licking of pen walls, excessive salivation (Marcet-Rius *et al.* 2019), body scratching, vacuum-chewing, shaking, and yawning (Norscia *et al.* 2021). Thus, establishing and implementing an ethogram for pigs that includes displacement behaviours can enable the incidence of stress-related behaviours expressed in a pig's rearing environments.

Fear/anxiety related behaviours in pigs can also be measured by various well-established behavioural tasks. Generally, these behavioural tests present innately fearful or stressful stimuli directly or indirectly. Behavioural responses or "copying styles" measured from fear/anxiety tasks can represent how pigs express behavioural adaptations to these challenging situations. Novelty is believed to elicit fear/anxiety in pigs, and some behavioural paradigms were designed based on novelty-induced fearful/anxious behaviours. The human approach test is one of the commonly used fear assessing tests, which involves exposing pigs to an unfamiliar human (Hulbert *et al.* 2019; Janczak *et al.* 2003; Kooij *et al.* 2002; Thodberg *et al.* 1999; Wegner *et al.* 2020). The novel object test is another behavioural test that utilises an element of novelty as a source to create a challenging environment for pigs (Dalmau *et al.* 2009; Hemsworth *et al.* 1996; Kooij *et al.* 2002). In these contexts, either a human observer or novel object are introduced to a home pen or a dedicated behavioural testing environment. Changes in the frequency, duration, and latency to touch the human observer or the novel object are believed to represent novelty-induced fearful/anxious behaviours in pigs (Brown *et al.* 2009; Dalmau *et al.* 2017; Zupan *et al.* 2012).

Another common behavioural test of stress involves the back test, which is where individual pigs are physically restrained on their backs as an innately stressful context. Typically, numbers of escape attempts are measured as primary indicator of individual stress responses. These coping responses differ individually from low to high resistance, demonstrating individual variability in pig's adaptive behaviours to stress (Geverink *et al.* 2002; Hessing *et al.* 1993; Ruis *et al.* 2000). Other fear/anxiety behavioural tasks, such as the open field test and use of an elevated plus maze, have been adapted for pigs from the rodent literature. The open field test, originally designed to capture anxiety behaviours from rodents when exposed to an unfamiliar open area, has been adapted to pigs. It is common to conduct the open field test in conjunction with the novel object test, typically completed prior to the introduction of the novel object (Haigh *et al.* 2020; Jensen *et al.* 1994; Schulz *et al.* 2022; Tatemoto *et al.* 2019). The elevated plus maze, a well-established rodent anxiety paradigm by using their innate fear of heights and open areas, has also been implemented in pigs (Andersen *et al.* 2000a; Andersen *et al.* 2000b; Janczak *et al.* 2000).

It is important to note that these behavioural tests are not a complete list of fear assessing tests developed in pigs, and Forkman and colleagues provided a more in-depth review on various fear assessing behavioural tests (Forkman *et al.* 2007). Also, having an appropriate ethological basis specifically for pigs is crucial when considering what type of behavioural testing to complete for measuring stress reactivities in response to fear/anxiety-evoking challenges. Thus, different components of these behavioural tasks such as the type of stress-eliciting stimuli, apparatus design, age of the pig, and behavioural outcomes of interest play critical roles in quantifying the stress responses in pigs.

4 Environmental stressors in pork production

Pigs maintained in a commercial production environment encounter various stressors in their daily lives. In this section, the types of different stressors, their impacts on pig's growth, physiology, and behaviour, and the effects on nutritional interventions will be reviewed.

4.1 *Heat stress*

Heat stress is a well-known stressor for pigs in the production setting, having negative impacts on the efficiency and economics of the pork production. A plethora of research during the last few decades has discovered that heat stress, especially during a hot summer season, leads to impaired growth performance and altered carcass composition (Baumgard and Rhoads 2013; Christon 1988; Heath 1983; Lewis *et al.* 2011). These effects of heat stress are directly and indirectly mediated by several mechanism, including reduction in feed intake and impairment in intestinal barrier function (Pearce *et al.* 2013a; Pearce *et al.* 2013b). Physiological mechanisms of heat stress in production animals, including pigs, have been thoroughly reviewed (Liu *et al.* 2022; Ross *et al.* 2015, 2017). Specifically, *in utero* heat stress (IUHS) and its long-lasting effects on offspring have been intensively investigated (Johnson *et al.* 2020b; Johnson and Baumgard 2019). Interestingly, physiological and behavioural stress responses of pigs subjected to IUHS are known to be significantly impacted. As such, young IUHS pigs express altered HPA axis function, demonstrated by elevated cortisol and ACTH concentrations at birth (Chapel *et al.* 2017; Guo *et al.* 2020; Machado-Neto *et al.* 1987). More aggressive behaviours have also been observed when IUHS pigs experienced stress related to weaning and transportation events (Byrd *et al.* 2019). Additionally, IUHS offspring have different intestinal permeability and functionality, microbial colonization, and immune responses (Guo *et al.* 2020; He *et al.* 2020; Johnson *et al.* 2020a), potentially mediating stress responses of the young pigs.

4.2 *Early weaning stress*

Weaning is a natural stressor for any mammal, but in a commercial setting, this event is particularly stressful for pigs due to abrupt social and environmental changes. Multiple stressors such as mixing with unfamiliar pigs, separation from sows, inaccessibility to food and water during transport, and exposure to novel environments and feed sources post-weaning, are inevitably layered into a short period of time for the weaned pig. Thus, it comes as no surprise that the process of weaning results in physiological and behavioural stress responses in young pigs. Weaning that occurs relatively early in a pig's life can magnify various stress responses and thereby lead to greater disruptions in physiological mechanisms that may result in morbidity or mortality. As occurs in all mammals, a reduction in growth rate during the initial post-weaning period is typical in the swine industry, and if mismanaged, early weaning can elicit a greater reduction in feeding behaviours, feed intake and growth efficiency (Colson *et al.* 2006; Gonyou *et al.* 2011; Orgeur *et al.* 2001; Wamnes *et al.* 2011). Early weaning also interrupts intestinal health by impairing epithelial barrier structure and function, thereby compromising overall gut permeability, including significant disruptions to immune function and development of a healthy microbiota, which often manifests as diarrhoea (Brown *et al.* 2006; Pluske *et al.* 1997; Pohl *et al.* 2017; Smith *et al.* 2010).

As expected, increased cortisol concentrations have been reported after pigs have experienced weaning and transportation events (Hay *et al.* 2001; Parrott *et al.* 1998; Sutherland *et al.* 2009, 2010). Additionally, more aggressive behaviours, escape attempts, and vocalizations have been observed in weaned pigs, especially when mixed with unfamiliar conspecifics (Colson *et al.* 2006; Friend *et al.* 1983; Mason *et al.* 2003). In particular, weaning within the first two weeks postpartum has been shown to further increase behavioural patterns associated with stress such as belly-nosing, distressed vocalizations, and escape attempts in home pens (Orgeur *et al.* 2001; Worobec *et al.* 1999; Yuan *et al.* 2004). Overall, these physiological and behavioural changes represent heightened stress of pigs in response to various environmental stressors.

4.3 *Tail biting stress*

Maladaptive biting behaviours directed toward other pig's body parts, usually ears or tails, is one of the major welfare issues that pork production farms encounter (as reviewed by Valros and Heinonen 2015). Various risk factors such as poor health status and rearing environment (as reviewed by Schrøder-Petersen and Simonsen 2001), come together to express tail biting behaviours in pigs. Tail biting has shown to reduce growth rate and feed intake (Camerlink *et al.* 2012; Sinisalo *et al.* 2012; Wallgren and Lindahl 1996) and increase risk for infections (Heinonen *et al.* 2010; Munsterhjelm *et al.* 2013b; Schrøder-Petersen and Simonsen 2001), which affect overall health of pigs and serve as known risk factors for morbidity and mortality at this vulnerable stage of production. Physiological and behavioural changes in pigs participating in tail biting behaviours manifest as heightened stress responses to subsequent stressors. For example, chronic victim pigs of tail biting exhibited hypocortisolism as a response to transport stress (Valros *et al.* 2013), which suggests a dysfunctional HPA axis. Heightened cortisol concentration has also been observed in both the victims and biters (Munsterhjelm *et al.* 2013a). Moreover, both biters and the victims of tail biting have demonstrated different fear responses during fear assessment tests, such as the open field test, novel object test, and human approach test (Haigh *et al.* 2020; Ursinus *et al.* 2014; Zupan *et al.* 2012). These results collectively suggest higher anxiety-like behaviours from the victims compared with the biters, meaning that the victims of tail biting experience not only physical and immunological stresses, but also psychological stress.

4.4 *Effects of nutritional supplementation*

In an attempt to mitigate environmental stressors to which pigs in a commercial setting are exposed, many strategies have been investigated, including various dietary changes (Cottrell *et al.*, 2015). Dietary supplementation of selenium, an antioxidant has been suggested to have protective effects on heat stress-induced intestinal barrier dysfunction (He *et al.*, 2022). Heat stressed piglets receiving selenium-enriched probiotics have also demonstrated improvements in growth performance and gut microbiota ecosystem (Gan *et al.* 2014; Lv *et al.* 2015), and greater reduction in heat shock protein mRNA expression compared with heat stressed piglets receiving probiotics or sodium selenite only (Gan *et al.* 2013). Interestingly, beneficial effects of supplementing various probiotic strains have been tested, and reductions in heat shock protein expression was also observed in broilers receiving selenium-enriched probiotics (Khan *et al.* 2016). Heat stressed chickens given *Lactobacillus* strains expressed improvements in intestinal barrier functionality and gut microbiota environment as well (Lan *et al.* 2004; Zhang *et al.* 2016). Numerous studies supporting the beneficial effects of nutritional interventions suggest that microbiota manipulation throughout the various growth phases of pigs may help alleviate the negative effects of heat stress.

Probiotic supplementation has also been suggested as a strategy to alleviate various stress responses by benefitting the intestinal environment, such as maintaining a healthy microbiota, reducing gut permeability, and reducing the incidence and growth of harmful bacteria (as reviewed by Barba-Vidal *et al.* 2018; Patil *et al.* 2015). Microorganisms within the gastrointestinal tract play a key role in mediating the gut-brain-axis, including the modulation of anxiety-related behaviours (Clarke *et al.* 2013; Kobek-Kjeldager *et al.* 2022), and probiotic supplementation has been suggested to have anxiolytic effects in humans and rodents (Foster and McVey Neufeld 2013; Luna and Foster 2015; Messaoudi *et al.* 2011). Collectively, providing probiotics as a feed additive to pigs may be beneficial for mitigating environmental stress responses and anxious behaviours. However, it is important to note that findings on probiotics as a feed additive in are highly variable and inconsistent responses, likely depending on the dosing regimen (*i.e.*, amount and timing) of specific probiotic strains being supplemented. Thus, further investigations into the beneficial effects of probiotic supplementation in post-weaning pigs are warranted with regard to minimizing undesirable physiological and behavioural responses. It is important that studies report strain level identification of the administered probiotics to enable comparison across studies.

Although the exact mechanisms of interaction between the HPA axis neuroendocrine system and the microbiota are poorly understood, it has been made clear through a wide range of research that these systems communicate with each other (Farzi *et al.* 2018; Parker *et al.* 2019). Many research studies have focused on administering prebiotics and probiotics specifically to germ-free mice (Farzi *et al.* 2018). Prebiotics and probiotic administration is associated with increased volatile metabolic end-products, including short-chained fatty acids (SCFA) (Farzi *et al.* 2018; van de Wouw *et al.* 2018). Although only a minor fraction remain in circulation after colonocyte absorption, SCFA, including acetic acid, butyrate, and propionic acids, have the ability to cross the blood-brain barrier through specific transporters (Farzi *et al.* 2018; Parker *et al.* 2019; Silva *et al.* 2020). It is still unclear as to what effects, if any, SCFAs elicit in the brain since it is uncertain at what concentrations they may cross the blood-brain barrier. Additionally, it is believed that SCFA play a direct role in maintaining blood-brain barrier integrity as studies have shown greater permeability in germ-free mice without an established microbiota (Braniste *et al.* 2014; Silva *et al.* 2020). Furthermore, it has been found that supplementation of SCFA down-regulated stress signalling during various psychosocial stress tests performed on mice (van de Wouw *et al.* 2018). As such, SCFA affected gene expression, specifically leading to decreased CRF receptor and mineralocorticoid receptor expression in the hippocampus of mice (van de Wouw *et al.* 2018). Moreover, SCFA production and microbiota profiles in general have been identified as important contributors to immune system development, specifically affecting microglia cells that are essential for maturation and refinement of connections within the brain (Silva *et al.* 2020). Lastly, there is evidence for microbiota derived SCFA to directly alter levels of neurotransmitters and neurotrophic factors throughout various brain regions, which are essential for cell signalling throughout the body (Silva *et al.* 2020). It is through these various mechanisms that the gut microbiota is believed to influence the HPA axis, although future research must delve further into the exact communication method between these two systems.

5 Learning and memory

Another large area of focus in microbiota-gut-brain axis research is on learning and memory and how these aspects of behavioural development are impacted by nutrition. Breaking down learning and memory in mammals to its core, the vital structural component responsible is myelin. The maturation and development rate of myelination can have implications for both cognitive abilities and the development of cognition. In the following section we will briefly introduce what myelin is, when myelin develops, and detail how nutrition plays a role in developmental processes. We will then end by discussing how the environment influences learning and memory starting from the prenatal period and progressing to postnatal interventions.

5.1 *What is myelin?*

For the nerve cells of an animal's body to both rapidly and accurately communicate with one another, an essential structural component, known as the myelin sheath, is necessary. The myelin sheath is a layer of fatty tissue that surrounds the transport bridge, or axon, of a neuron. This fatty tissue layer covers and insulates the axon, allowing for electrical messages to travel between neurons both efficiently and precisely (Morell 2013). Myelin is found in the nerve fibers of both the central and peripheral nervous systems and a large part of the brain is populated by myelinated neurons (Knaap *et al.* 2005). Specific glial cells, called oligodendrocytes, are responsible for the creation of myelin. Oligodendrocytes accumulate plasma membrane and create a multi-layered stack of membrane that is tightly adhered (Aggarwal *et al.* 2011). This tight stack of membrane, which is referred to as myelin, is composed of specific lipids and proteins (Wiggins 1982). Additionally, across the axon, the insulating membrane has small gaps, called nodes of Ranvier, which are necessary for efficient transfer of electrical energy (Aggarwal *et al.*, 2011). Approximately 70% of myelin is lipid, while the rest is comprised of proteins, such as proteolipid, myelin basic protein, and other highly specialized proteins (Aggarwal *et al.* 2011; Wiggins 1982). Through a complicated network of gene expression and regulatory factors, oligodendrocytes synthesize, formulate, and maintain myelin to varying degrees throughout the life of an animal (Aggarwal *et al.* 2011).

5.2 *Myelination and development*

Without proper myelination, a variety of debilitating motor and neurodegenerative disorders can result, such as multiple sclerosis and other demyelination disorders (Knaap *et al.* 2005). Therefore, it is well understood that myelination, or the formation and maintenance of myelin sheaths, is a critical aspect of development. During development, both prenatally and postnatally, myelination is occurring in parallel with other processes such as neurogenesis, brain growth, and synaptogenesis (Jena *et al.* 2020).

Myelination begins relatively late in prenatal development, around the second trimester, with a large proportion occurring after birth (Jakovcevski *et al.* 2009; Williamson and Lyons 2018). Prenatally, myelination develops due to the expression of various transcription factors and growth factors that influence oligodendrocyte formation (Jakovcevski *et al.* 2009). After birth, lipid concentrations in myelin undergo remodelling with slight alterations in the composition of phospholipids (Poduslo and Jang 1984). As an organism is exposed to various environmental stimuli, myelination remodelling, such as changes in membrane thickness and length, will occur well into adulthood (Snaidero and Simons 2014). This remodelling has been observed to be due to oligodendrocytes that are generated later into adulthood and are responsible for forming myelin in the maturing adult brain (Snaidero and Simons 2014). In humans, myelination seems to peak in the first year of life whereas myelination was observed to peak around 8 weeks of age in pigs (Fil *et al.* 2021b; Snaidero and Simons 2014).

5.3 *Myelination and nutrition*

One of the main environmental influencers during early-life that affects brain development is early-life nutrition (Georgieff 2007). It is well understood that the brain and gut develop in parallel (Jena *et al.* 2020). Brain developmental processes such as myelination, synaptogenesis, and brain growth occur simultaneously with gut developmental processes such as enterocyte maturation and microbial colonization (Jena *et al.* 2020). Therefore, it is not surprising that a large amount of literature is dedicated to exploring how nutrition can influence brain development (Georgieff 2007; Jena *et al.* 2020). During early life, several of the main nutritional components found to influence myelination have been narrowed down to iron, long-chain polyunsaturated fatty acids, and choline (Georgieff 2007). An abundance of research has determined how iron deficiency in early-life interrupts myelination and oligodendrocyte functioning throughout the central nervous system (Beard *et al.* 2003; Todorich *et al.* 2009; Yu *et al.* 1986). Iron deficiency for as little as 10 days during postweaning of rats specifically affected fatty acid profiles of myelin phospholipids and decreased protein concentrations necessary for myelin development (Beard *et al.* 2003). Furthermore, long-chain polyunsaturated fatty acids, like docosahexaenoic acid (DHA) and arachidonic acid (ARA), are an important part of development due to their incorporation in membrane phospholipids (Lauritzen *et al.* 2016). As such, DHA and ARA constitute approximately 20% of the brain's fatty acid content and have been associated with increased levels of myelin development when supplemented in infant formulas (Deoni *et al.* 2018). Additionally, in the same study, choline increased myelin development (Deoni *et al.* 2018).

Choline is a nutrient, closely associated with the B-vitamin family that is important in the biosynthesis of phosphatidylcholine and sphingomyelin (Derbyshire and Obeid 2020; Zeisel *et al.* 1986). Phosphatidylcholine is essential for cellular membrane structure and function, cell division and growth, and myelin development (Derbyshire and Obeid 2020). Sphingomyelin is a phospholipid derived from phosphatidylcholine and is critical for the development of the myelin sheath in the nerve fibers of both central and peripheral nervous systems (Deoni *et al.* 2018; Derbyshire and Obeid 2020). Choline is essential for proper hippocampal development in rats (Li *et al.* 2004), and has been associated with increases in total brain volume and white matter maturation in pigs (Mudd *et al.* 2016; Mudd *et al.* 2018c). It has also been identified as a key nutrient for neurogenesis, neuronal differentiation, myelination, and synaptogenesis both prenatally and postnatally (Derbyshire and Obeid 2020).

Only a few nutritional components were highlighted in this section, but it is made clear through these examples that early-life nutrition has impacts on the development and synthesis of myelin. These nutritional impacts on structural components of the brain are most clearly observed through an animal's behaviour and their ability to learn and remember.

5.4 *Maternal environment*

The impact of inadequate nutrition on a pig begins *in utero*, which means the nutritional status of the sow can affect the development of the foetus, leading to long-term alterations in growth and performance. This phenomenon continues to be intensively studied and is best captured in the Barker Hypothesis (Barker and Osmond 1986), which can be summarized as the body becoming 'thrifty' at nutrient partitioning during times of nutrient scarcity. This thriftiness is important when resources are scarce but becomes problematic when resources are abundant. During a period of malnutrition, the body prioritises nutrients to specific tissues, including the brain and heart, and while pregnant, the foetus. When a malnourished animal is returned to a nutrient-adequate diet, it typically experiences compensatory growth in an attempt to return to typical growth patterns (Osborne *et al.* 1915). Depending on the duration and timing of the malnutrition, it may not be possible for an animal to re-establish typical rates or efficiency of body weight gain. Moreover, the composition of body weight gain is altered under such conditions, and animals experiencing compensatory growth typically gain weight as fat, as opposed to the typical muscle-to-fat ratios observed in non-affected animals.

Many studies have looked at offspring performance from sows that were placed on a protein-restricted diet. These studies concluded that offspring of protein-restricted sows were smaller, having lower birth weights (Pond *et al.* 1969), exhibited lower daily body weight gain (Atinmo *et al.* 1974), and may exhibit irreversible growth retardation (Pond *et al.* 1969). Because colostrum is enriched in protein, the protein restriction of a sow can carry over and continue impacting offspring performance even after birth, potentially leaving the offspring more susceptible to infection and potentially also influencing intestinal microbiota development (Xu and Cranwell 2003).

Brain growth and development in the foetus can be especially susceptible to sow nutrition due to the perinatal nature of brain development in pigs, with a major brain growth spurt around the time of birth (Davison and Dobbing 1966). As expected, offspring of protein-restricted sows born with lower birth weights also have lower brain weights and express reduced RNA concentration in the whole brain. However, proportional to body weight, the brain weight is significantly greater in pigs born to protein-restricted sows as compared with pigs from protein-adequate sows (Atinmo *et al.* 1974), supporting the idea that the body prioritises the foetus when partitioning nutrients within the sow's body. A greater brain weight-to-body weight ratio has been corroborated numerous times in literature looking at undernourished sows and offspring (Atinmo *et al.* 1974).

Protein restriction is not the only factor shown to impact the offspring of a sow. In a similar fashion, sows exposed to heat stress also produce litters with lower birth weights. The offspring of heat-stressed sows also have a higher body temperature set point than their non-heat-stressed counterparts (Johnson *et al.* 2015; Lucy and Safranski 2017). The offspring of heat-stressed sows express increased fat deposition at birth, but by weaning age, weigh the same as their non-heat-stressed counterparts (Lucy and Safranski 2017). The weight gain in the offspring of the heat-stressed sows suggests these animals are experiencing compensatory growth to overcome the *in-utero* restrictions they faced.

Maternal stress has been shown to have long lasting effects on the offspring, ranging from altered/decreased cognitive abilities, increased anxiety-like behaviours, and adjusted immune function and cardiovascular responses to stress (Igosheva *et al.* 2007). These points are critically important with regard to commercial pig production, especially during periods of increased stress and vulnerability, such as immediately after pigs have been weaned.

5.5 *Postnatal nutrition and brain development*

While maternal health and diet can influence the development of its offspring, postnatal intervention can also change the developmental trajectory for offspring pigs. Although there is a major brain growth spurt just before birth, there are also sensitive developmental periods postnatally, during which time, foetal brain development can be dramatically impacted (Dickerson *et al.* 1967; Pond *et al.* 2000).

Similar to how the condition of a sow can influence the growth and development of a piglet, postnatal nutritional intervention can also modify these trajectories.

Studies have shown that pigs born from a well-nourished sow, but placed on an inadequate diet postnatally can have similar developmental delays as when the sow is undernourished during pregnancy. Despite weighing more relative to body weight, the brain of a longitudinally undernourished pig is far less developed and grows at a slower rate than a pig with adequate nutrition, supporting the idea that there is significant brain maturation happening during the first postnatal year. However, during this growth period, the ratio of brain weight-to-body weight continually increased as brain development is prioritised over fat and muscle development (Dickerson *et al.* 1967). Despite the greater brain weight-to-body weight, these small-for-gestational-age pigs tend to show abnormalities in white matter development compared to average-for-gestational-age pigs (Radlowski *et al.* 2014). Similarly, young pigs placed on an iron-deficient diet postnatally for a short period of time before returning to a nutritionally adequate diet also showed abnormalities in grey and white matter concentrations, altered brain tissue composition, and smaller brain regions by volume when compared to pigs fed an adequate diet from birth as measured by fractional anisotropy, which is the degree of unidirectional diffusivity in the brain (Mudd *et al.* 2018a; Mudd *et al.* 2018b). The iron-deficient pigs, like the undersized pigs, also tend to have a larger brain-to-body weight ratio despite exhibiting significantly lower body weight (Knight and Dilger 2018), once again, suggesting that certain tissues, like the brain, are prioritised when nutrients are scarce.

Early-life dietary supplementation can also change brain growth trajectory. Piglets born to a healthy/well-nourished sow and subsequently provided additional nutritional supplementation exhibit altered rates of brain maturation. Different nutrients tend to impact the development of specific brain regions, and may or may not alter whole brain mass (Table 15.1). For example, bovine milk oligosaccharides (BMO) alone can have an impact on the size of the lateral ventricle, but when paired with human milk oligosaccharides (HMO), can increase the size of the thalamus, caudate, and hypothalamus, as well as, increase the total amount of grey matter in the pig brain (Fleming *et al.* 2020b). Similar trends of altered brain maturation have been observed with supplementation of DHA, ARA, prebiotics, bovine-derived milk fat globule membrane (MFGM), lactoferrin, bovine milk osteopontin, a phospholipid/ganglioside supplement, and a polar lipid supplement (Fil *et al.* 2021a; Hahn *et al.* 2020; Joung *et al.* 2020; Liu *et al.* 2014; Mudd *et al.* 2016). Importantly, it is not yet understood whether dietary effects are merely additive or whether synergy may exist when attempting to influence the trajectory of brain development in pigs. As such, supplementation of 2′-FL or Bi-26 separately positively impacted brain development, but when combined together lost their impact (Sutkus *et al.* 2022). The combination of BMO + HMO mentioned above was positively correlated with brain total grey matter, whereas other supplemented compounds elicited

TABLE 15.1 Nutritional interventions and their impacts on brain development

Intervention	Timing	Dosage[a]	Results	Reference
Early-life iron deficiency with follow-up repletion	Deficiency: PND 2–32 Repletion: PND 33–61	Deficiency: 1.56 mg Fe/L Repletion: standard complex diet	– Smaller absolute whole brain volume at PND 32, but not PND 61 – Decreased relative volumes in certain brain regions – Decreased fractional anisotropy in certain brain regions	Mudd *et al.* 2018a
Early-life iron deficiency with follow-up repletion	Deficiency: PND 2–32 Repletion: PND 33–61	Deficiency: 1.56 mg Fe/L Repletion: standard complex diet	– Reduced iron content in specific brain regions – Increased iron content in olfactory bulbs – Decreased grey/white matter concentrations at PND 61 – Decreased fractional anisotropy at PND 61	Mudd *et al.* 2018b
Early-life iron deficiency with follow-up repletion	Deficiency: PND 2–32 Repletion: PND 33–61	Deficiency: 1.56 mg Fe/L Repletion: standard complex diet	– Larger brain weight-to-body weight ratio – Reduced body weight gain – Reduced hematocrit and hemoglobin at PND 32 but not PND 61	Knight and Dilger 2018

a Dosages are expressed per litre of reconstituted milk replacer, when applicable..

Abbreviations: 2′-FL, 2′-fucosyllactose; ARA, arachidonic acid; Bi-26, *Bifidobacterium longum* subsp. *Infantis*; BMO, bovine milk oligosaccharides; CFU, colony forming units; DHA, docosahexaenoic acid; FA, fatty acid; HMO, human milk oligosaccharides; LNnT, Lacto-N-neotetraose; PND, postnatal day; PNW, postnatal week; SGA, small for gestational age.

TABLE 15.1 Nutritional interventions and their impacts on brain development (*cont.*)

Intervention	Timing	Dosage	Results	Reference
Bovine milk oligosaccharides and/or Human milk oligosaccharides	PND 2–33	BMO: 12.4 g/L HMO: 1.5 g/L BMO + HMO: 12.4 g/L of BMOS + 1.0 g/L of 2′-FL + 0.5 g/L of LNnT	– Larger absolute volumes for BMO + HMO – Smaller absolute volumes for BMO only – Relative volume interaction effects for specific brain regions – Relative volume main effects for specific brain regions for BMO or HMO, but not BMO + HMO	Fleming *et al.* 2020b
Early-life polar lipid supplementation	Supplementation: PND 2–PND 28 Standard diet: PNW 4–24	Alpha-lactalbumin containing high levels of sphingomyelin (950 ± 118 g/L) + phosphatidylcholine (864 ± 94 g/L) + phosphatidylethanolamine (1128 ± 286 g/L) + phosphatidylinositol (380 ± 74 g/L)	– Minor differences in hippocampal absolute brain volumes – Minor differences in fractional anisotropy	Fil *et al.* 2021
ARA and/or DHA	PND 2–30	ARA: 0.80% of total FA DHA: 0.80% of total FA ARA + DHA: 0.80% ARA + 0.80% DHA of total FA	– Supplementation with ARA or DHA increased respective concentrations in the brain – ARA + DHA increased DHA concentrations but not ARA concentrations in the brain	Hahn *et al.* 2020

TABLE 15.1 Nutritional interventions and their impacts on brain development (*cont.*)

Intervention	Timing	Dosage	Results	Reference
Early-life bovine milk osteopontin supplementation	PND 2–34	Standard diet + 250 mg of osteopontin/L	– Increased relative brain volumes in certain brain regions – Higher radial diffusivity in the corpus callosum – Lower fractional anisotropy	Joung *et al.* 2020
Early-life phospholipid/ ganglioside supplementation	PND 2–28	Standard diet with 0%, 0.8%, or 2.5% phospholipid/ ganglioside-enriched supplement	– Increased mean brain weight – Greater volume in certain brain regions – Increased grey and white matter	Liu *et al.* 2014
Early-life choline deficiency	PND 2–30	483 mg analyzed total choline/kg	– Decreased concentrations of glycerophosphocholine-phosphocholine – Smaller total brain volumes – Larger region-specific volumes proportional to total brain size	Mudd *et al.* 2016
Early-life dietary 2′-FL and Bi-26 supplementation	PND 2–35	1.0 g of 2′-FL/L + 10^9 CFU of Bi-26	– Smaller absolute brain volumes for certain regions of interest for Bi-26 – Smaller relative volumes for certain regions for Bi-26 – Interactive effects negated effects of solo test articles	Sutkus *et al.* 2022

lower accumulation of brain total grey matter. It is important to note that as the brain develops, it first generates an overabundance of grey matter before beginning a highly regulated pruning process to specialise and hone the synapses. That said, the distinction between these two studies may be that the BMOS + HMO combination caused grey matter to develop faster, but at the time of measurement, the pruning process had not yet started, which is why both timing (*i.e.*, age) and magnitude must be considered when assessing nutritional influences on tissue development rates. In the study looking at DHA, ARA, prebiotics, MFGM, and lactoferrin, it is possible that at the time of measurement, the brain had already over-developed the grey matter and was beginning the pruning process. In both studies, the volume of grey matter as measured by magnetic resonance imaging (MRI) was interpreted as indicating a faster rate of brain maturation in the pig.

5.6 *Effects of postnatal nutrition on myelination and cognition*

Learning, memory, and cognition rely heavily on when and how well the brain develops. As mentioned above, there are a number of factors that influence developmental timelines and trajectories. It is important to keep in mind that structure-function relationships within the brain must be considered for proper interpretation of nutritional effects, which necessitates a dual-focus on both structural brain development and functional cognitive development. The maternal and nutritional interventions mentioned above can also modify cognitive development. For example, low birth-weight pigs exhibit impairments on spatial and recognition memory tasks, though the severity of cognitive impairment may be associated with the level of intrauterine growth restriction experienced by the offspring (Radlowski *et al.* 2014; Schmitt *et al.* 2019). Similar trends were observed for pigs born preterm; however, persistence of the impairment may be more severe with preterm pigs as opposed to low birth-weight pigs. As such, preterm pigs were observed to have long-term learning impairments and neurodevelopmental deficits as compared to their term counterparts (Andersen *et al.* 2016). Weaning age has also been correlated with cognitive development. As mentioned previously, weaning at younger ages has been associated with those pigs expressing more aggressive and atypical behaviours than pigs weaned at an older age, and early-weaned pigs also tend to express decreased levels of cognitive abilities (Poletto *et al.* 2006).

Along with brain and body growth deficits, iron deficiency has also been associated with decreased learning and memory abilities. Some studies have demonstrated that the severity of iron deficiency is important in determining the level of cognitive deficit the pig experiences. Specifically, spatial learning and memory deficits were shown to be associated with the level of iron deficiency, with speculation that it is the amount of iron in the hippocampus causing the observed deficits (Rytych *et al.* 2012). Pigs placed on an iron-deficient diet and later returned to an iron repleted diet were also found to exhibit memory deficits, specifically reference memory, even despite the supplementation (Antonides *et al.* 2015; Mudd *et al.*

2018a). Pigs injected with scopolamine (a drug known to impact spatial memory network structures, specifically the hippocampus) at 2 weeks of age were slower to perform a spatial T-maze, suggesting that this may be a critical period for cognitive development in young pigs (Elmore *et al.* 2012).

While early-life nutrient deficiencies can hinder cognitive abilities, supplementation may help to support cognitive function in pigs and other animals. Recent evidence suggests that ingestion of a synbiotic supplement (composed of probiotics, prebiotics, a dietary fibre, and vitamin C) started on postnatal day one elicited improved performance on subsequent cognitive tasks and memory abilities, and these cognitive benefits correlated were correlated with microbiota composition of young pigs (Parois *et al.* 2021). Altering brain metabolite profiles is another mechanism by which dietary supplementation can influence cognitive function. For example, bovine milk lactoferrin supplementation has been shown to increase hippocampal protein concentrations of molecules (brain derived neurotropic factor [BDNF] and downstream cAMP response element-binding protein [CREB]) associated with learning events and memory, as well messenger RNA levels (Chen *et al.* 2015). Similarly, the ingestion of galactooligosaccharides and fructooligosaccharides were also observed to alter the molecular expression of proteins in the hippocampus related to learning and memory (Fleming *et al.* 2019). While some forms of dietary supplementation appear to elicit effects in specific brain regions, others may work on a broader scale to influence general learning and memory. Supplementation of a combination of polydextrose and galactooligosaccharides has been observed to modify behaviour of young pigs (Fleming *et al.* 2019). Similarly, pigs supplemented with polar lipids or bovine milk osteopontin also showed differences in exploratory behaviour (Fil *et al.* 2021a; Joung *et al.* 2020). Supplementation of a protein-bound source of sialic acid has been shown to enhance learning when supplemented at an early age (Wang *et al.* 2007). However, the benefits of supplemented sialic acid are still debated as other studies have found null results from the addition of it to the diet (Fleming *et al.* 2018). Memory systems supported by the hippocampus, such as long-term recognition memory, may benefit when young pigs receive diets supplemented with oligofructose or a combination of oligofructose and 2′-fucosyllactose (Fleming *et al.* 2020a). Spatial memory, such as performance on a T-maze, may also be enhanced with a phospholipid/ganglioside supplement leading to reduced deduction time and fewer errors (Liu *et al.* 2014).

The brain and cognitive development of young pigs depend on a number of factors. Environmental stressors faced by young pigs, including those associated with weaning and improper dietary nutrient intake can have long-lasting negative effects on brain structure/function and cognitive abilities, while supplementation and adequate nutrition can enhance cognitive development. The developmental trajectory and function of a young pig's brain are most vulnerable perinatally, including *in utero* and within the first 8 weeks after birth, which means that even the health, diet, and microbiota of the dam must be considered when considering modern pork production systems. Postnatally, nutrition is of utmost importance for healthy brain development of young pigs, and it is possible to develop nutritional interventions that target key developmental milestones for supporting specific cognitive and behavioural outcomes of interest to commercial pig production systems. Ensuring proper brain development is fundamental to learning and memory abilities, which have implications for the pig's overall welfare, health, and productivity.

6 Conclusion

This chapter summarized the various brain structures and systems, and environmental factors that can impact the development of pigs in commercial production settings. Much of the progress in this area has been informed by studies involving the use of domesticated pigs in biomedical contexts to study the influence of early-life nutrition on brain structure/function relationships and associations with cognitive and behavioural outcomes. Additionally, we highlighted research where the lack of, or supplementation with, specific nutrients altered behaviours in the pig, with implications to the commercial production context. As various environmental stressors are inevitable in pork production, incorporating nutritional additives may have a beneficial role not only in promoting overall health and growth performance, but also in supporting welfare of the pig. Cognitive development is another area of interest, and myelination of the brain is critical in understanding how nutrition impacts learning and memory of pigs. However, it is important to emphasise that the beneficial effects of specific nutritional supplementations, such as probiotics, depend on various factors (*e.g.*, dosing, specific strains of probiotics). Overall, there exist many exciting opportunities to leverage nutritional and management strategies to manipulate the microbiota-gut-brain axis in the pig, thereby improving sustainability of the pork industry.

References

Aggarwal, Shweta, Larisa Yurlova, and Mikael Simons. 2011. "Central Nervous System Myelin: Structure, Synthesis and Assembly." *Trends in Cell Biology* 21(10):585–93. doi: 10.1016/J.TCB.2011.06.004.

Alemohammad, Seyyed Mohammad Amin, Seyed Mohammad Reza Noori, Ehsan Samarbafzadeh, and Seyyed Mohammad Ali Noori. 2022. "The Role of the Gut Microbiota and Nutrition on Spatial Learning and Spatial Memory: A Mini Review Based on Animal Studies." *Molecular Biology Reports* 49(2):1551–63. doi: 10.1007/S11033-021-07078-2/FIGURES/4.

Amar, Arun Paul, and Martin H. Weiss. 2003. "Pituitary Anatomy and Physiology." *Neurosurgery Clinics of North America* 13:11–23.

Andersen, Anders D., Per T. Sangild, Sara L. Munch, Eline M. van der Beek, Ingrid B. Renes, Chris van Ginneken, Gorm O. Greisen, and Thomas Thymann. 2016. "Delayed Growth, Motor Function and Learning in Preterm Pigs during Early Postnatal Life." *American Journal of Physiology – Regulatory Integrative and Comparative Physiology* 310(6):R481–92. doi: 10.1152/AJPREGU.00349.2015/ASSET/IMAGES/LARGE/ZH60051689420009.JPEG.

Andersen, Inger Lise, Knut Egil Bøe, Gry Fœrevik, Andrew M. Janczak, and Morten Bakken. 2000a. "Behavioural Evaluation of Methods for Assessing Fear Responses in Weaned Pigs." *Applied Animal Behaviour Science* 69(3):227–40. doi: 10.1016/S0168-1591(00)00133-7.

Andersen, Inger Lise, Gry Færevik, Knut Egil Bøe, Andrew M. Janczak, and Morten Bakken. 2000b. "Effects of Diazepam on the Behaviour of Weaned Pigs in Three Putative Models of Anxiety." *Applied Animal Behaviour Science* 68(2):121–30. doi: 10.1016/S0168-1591(00)00098-8.

Antonides, Alexandra, Anne C. Schoonderwoerd, Gabi Scholz, Brian M. Berg, Rebecca E. Nordquist, and Franz Josef van der Staay. 2015. "Pre-Weaning Dietary Iron Deficiency Impairs Spatial Learning and Memory in the Cognitive Holeboard Task in Piglets." *Frontiers in Behavioral Neuroscience* 9:291. doi: 10.3389/FNBEH.2015.00291/BIBTEX.

Arnsten, Amy F.T., and Bao Ming Li. 2005. "Neurobiology of Executive Functions: Catecholamine Influences on Prefrontal Cortical Functions." *Biological Psychiatry* 57(11):1377–84. doi: 10.1016/J.BIOPSYCH.2004.08.019.

Atinmo, T., W.G. Pond, and R.H. Barnes. 1974. "Effect of Maternal Energy vs. Protein Restriction on Growth and Development of Progeny in Swine." *Journal of Animal Science* 39(4): 703–11. doi: 10.2527/JAS1974.394703X.

Barba-Vidal, E., S.M. Martín-Orué, and L. Castillejos. 2018. "Review: Are We Using Probiotics Correctly in Post-Weaning Piglets?" *Animal* 12(12):2489–98. doi: 10.1017/S1751731118000873.

Barker, D.J.P., and C. Osmond. 1986. "Infant Mortality, Childhood Nutrition, and Ischaemic Heart Disease in England and Wales." *The Lancet* 327(8489):1077–81. doi: 10.1016/S0140-6736(86)91340-1.

Baumgard, Lance H., and Robert P. Jr. Rhoads. 2013. "Effects of Heat Stress on Postabsorptive Metabolism and Energetics." *Annual Review of Animal Bioscience* 1:311–37. doi: 10.1146/ANNUREV-ANIMAL-031412-103644.

Beard, John L., Jason A. Wiesinger, and James R. Connor. 2003. "Pre- and Postweaning Iron Deficiency Alters Myelination in Sprague-Dawley Rats." *Developmental Neuroscience* 25(5): 308–15. doi: 10.1159/000073507.

Braniste, Viorica, Maha Al-Asmakh, Czeslawa Kowal, Farhana Anuar, Afrouz Abbaspour, Miklós Tóth, Agata Korecka, Nadja Bakocevic, Ng Lai Guan, Parag Kundu, Balázs Gulyás, Christer Halldin, Kjell Hultenby, Harriet Nilsson, Hans Hebert, Bruce T. Volpe, Betty Diamond, and Sven Pettersson. 2014. "The Gut Microbiota Influences Blood-Brain Barrier Permeability in Mice." *Science Translational Medicine* 6(263):263ra158. doi: 10.1126/SCITRANSLMED.3009759.

Brown, D.C., C.v. Maxwell, G.F. Erf, M.E. Davis, S. Singh, and Z.B. Johnson. 2006. "The Influence of Different Management Systems and Age on Intestinal Morphology, Immune

Cell Numbers and Mucin Production from Goblet Cells in Post-Weaning Pigs." *Veterinary Immunology and Immunopathology* 111(3–4):187–98. doi: 10.1016/J.VETIMM.2005.12.006.

Brown, Jennifer A., Cate Dewey, Cornelius F.M. Delange, Ira B. Mandell, Peter P. Purslow, J. Andrew Robinson, E. James Squires, and Tina M. Widowski. 2009. "Reliability of Temperament Tests on Finishing Pigs in Group-Housing and Comparison to Social Tests." *Applied Animal Behaviour Science* 118(1–2):28–35. doi: 10.1016/J.APPLANIM.2009.02.005.

Byrd, Christopher J., Nichole Chapel Anderson, Drew W. Lugar, Timothy J. Safranski, Matthew C. Lucy, and Jay S. Johnson. 2019. "Evaluating the Effects of In Utero Heat Stress on Piglet Physiology and Behavior Following Weaning and Transport." *Animals 2019, Vol. 9, Page 191* 9(4):191. doi: 10.3390/ANI9040191.

Camerlink, Irene, Piter Bijma, Bas Kemp, and J. Elizabeth Bolhuis. 2012. "Relationship between Growth Rate and Oral Manipulation, Social Nosing, and Aggression in Finishing Pigs." *Applied Animal Behaviour Science* 142(1–2):11–17. doi: 10.1016/J.APPLANIM.2012.09.004.

Chapel, N.M., C.J. Byrd, D.W. Lugar, G.M. Morello, L.H. Baumgard, J.W. Ross, T.J. Safranski, M.C. Lucy, and J.S. Johnson. 2017. "Determining the Effects of Early Gestation in Utero Heat Stress on Postnatal Fasting Heat Production and Circulating Biomarkers Associated with Metabolism in Growing Pigs." *Journal of Animal Science* 95(9):3914–21. doi: 10.2527/JAS.2017.1730.

Chen, Yue, Zhiqiang Zheng, Xi Zhu, Yujie Shi, Dandan Tian, Fengjuan Zhao, Ni Liu, Petra S. Hüppi, Frederic A. Troy, and Bing Wang. 2015. "Lactoferrin Promotes Early Neurodevelopment and Cognition in Postnatal Piglets by Upregulating the BDNF Signaling Pathway and Polysialylation." *Molecular Neurobiology* 52:256–69. doi: 10.1007/S12035-014-8856-9/FIGURES/5.

Christon, R. 1988. "The Effect of Tropical Ambient Temperature on Growth and Metabolism in Pigs." *Journal of Animal Science* 66(12):3112–23. doi: 10.2527/JAS1988.66123112X.

Clarke, G., S. Grenham, P. Scully, P. Fitzgerald, R.D. Moloney, F. Shanahan, T.G. Dinan, and J.F. Cryan. 2013. "The Microbiome-Gut-Brain Axis during Early Life Regulates the Hippocampal Serotonergic System in a Sex-Dependent Manner." *Molecular Psychiatry 2013 18:6* 18(6):666–73. doi: 10.1038/mp.2012.77.

Cockrem, John F. 2007. "Stress, Corticosterone Responses and Avian Personalities." *Journal of Ornithology* 148:169–78. doi: 10.1007/S10336-007-0175-8/FIGURES/6.

Colson, Violaine, Pierre Orgeur, Aline Foury, and Pierre Mormède. 2006a. "Consequences of Weaning Piglets at 21 and 28 Days on Growth, Behaviour and Hormonal Responses." *Applied Animal Behaviour Science* 98(1–2):70–88. doi: 10.1016/J.APPLANIM.2005.08.014.

Cottrell A, J.J., F.A. Liu, A.T. Hung A, K.A. DiGiacomo, S.S. Chauhan A, B.J. Leury A, J.B. Furness C, P.D. Celi, and F.R. Dunshea A. 2015. "Nutritional Strategies to Alleviate Heat Stress in Pigs." *Animal Production Science* 55:1391–1402. doi: 10.1071/AN15255.

Crane, J.W., K. Ebner, and T.A. Day. 2003. "Medial Prefrontal Cortex Suppression of the Hypothalamic-Pituitary-Adrenal Axis Response to a Physical Stressor, Systemic Delivery of Interleukin-1beta." *The European Journal of Neuroscience* 17(7):1473–81. doi: 10.1046/J.1460-9568.2003.02568.X.

Cryan, John F., and Timothy G. Dinan. 2012. "Mind-Altering Microorganisms: The Impact of the Gut Microbiota on Brain and Behaviour." *Nature Reviews Neuroscience* 13(10):701–12. doi: 10.1038/nrn3346.

Cuzzo, Brian, Sandeep A. Padala, and Sarah L. Lappin. 2021. "Physiology, Vasopressin." *StatPearls*.

Dalmau, Antoni, Emma Fabrega, and Antonio Velarde. 2009. "Fear Assessment in Pigs Exposed to a Novel Object Test." *Applied Animal Behaviour Science* 117(3–4):173–80. doi: 10.1016/J.APPLANIM.2008.12.014.

Dalmau, Antoni, Eva Mainau, and Antonio Velarde. 2017. "Reliability of Fear Assessment in Growing Pigs Exposed to a Novel Object Test in Commercial Conditions." *Journal of Applied Animal Welfare Science* 20(3):280–88. doi: 10.1080/10888705.2017.1310043.

Davis, Michael. 1992. "The Role of the Amygdala in Fear and Anxiety." *Annual Review of Neuroscience* 15:353–75.

Davis, Michael. 1997. "Neurobiology of Fear Responses: The Role of the Amygdala." *Journal of Neuropsychiatry and Clinical Neurosciences* 9(3):382–402. doi: 10.1176/JNP.9.3.382.

Davison, A.N., and J. Dobbing. 1966. "Myelination as a Vulnerable Period in Brain Development." *British Medical Bulletin* 22(1):40–44. doi: 10.1093/OXFORDJOURNALS.BMB.A070434.

Day, R. 2009. "Proopiomelanocortin." *Encyclopedia of Neuroscience* 1139–41. doi: 10.1016/B978-008045046-9.01197-9.

Deoni, Sean, Douglas Dean, Sarah Joelson, Jonathan O'Regan, and Nora Schneider. 2018. "Early Nutrition Influences Developmental Myelination and Cognition in Infants and Young Children." *NeuroImage* 178:649–59. doi: 10.1016/J.NEUROIMAGE.2017.12.056.

Derbyshire, Emma, and Rima Obeid. 2020. "Choline, Neurological Development and Brain Function: A Systematic Review Focusing on the First 1000 Days." *Nutrients 2020, Vol. 12, Page 1731* 12(6):1731. doi: 10.3390/NU12061731.

Diana, Alessia, Lenn Carpentier, Deborah Piette, Laura Ann Boyle, Daniel Berckmans, and Tomas Norton. 2019. "An Ethogram of Biter and Bitten Pigs during an Ear Biting Event: First Step in the Development of a Precision Livestock Farming Tool." *Applied Animal Behaviour Science* 215:26–36. doi: 10.1016/J.APPLANIM.2019.03.011.

Dickerson, J.W., J. Dobbin, and R.A. McCance. 1967. "The Effect of Undernutrition on the Postnatal Development of the Brain and Cord in Pigs." *Proceedings of the Royal Society of London. Series B. Biological Sciences* 166(1005):396–408. doi: 10.1098/RSPB.1967.0003.

Douglas, Robert J. 1967. "The Hippocampus and Behavior." *Psychological Bulletin* 67(6): 416–42. doi: 10.1037/H0024599.

Elmore, Monica R.P., Ryan N. Dilger, and Rodney W. Johnson. 2012. "Place and Direction Learning in a Spatial T-Maze Task by Neonatal Piglets." *Animal Cognition* 15(4):667–76. doi: 10.1007/S10071-012-0495-9.

Farzi, Aitak, Esther E. Fröhlich, and Peter Holzer. 2018. "Gut Microbiota and the Neuroendocrine System." *Neurotherapeutics 2018 15:1* 15(1):5–22. doi: 10.1007/S13311-017-0600-5.

Ferguson, Alastair v., Kevin J. Latchford, and Willis K. Sanon. 2008. "The Paraventricular Nucleus of the Hypothalamus A Potential Target for Integrative Treatment of

Autonomic Dysfunction." *Expert Opinion on Therapeutic Targets* 12(6):717–27. doi: 10.1517/14728222.12.6.717.

Fil, Joanne E., Sangyun Joung, Jonas Hauser, Andreas Rytz, Courtney A. Hayes, and Ryan N. Dilger. 2021a. "Influence of Dietary Polar Lipid Supplementation on Memory and Longitudinal Brain Development." *Nutrients* 13(8):2486. doi: 10.3390/NU13082486.

Fil, Joanne E., Sangyun Joung, Courtney A. Hayes, and Ryan N. Dilger. 2021b. "Influence of Rearing Environment on Longitudinal Brain Development, Object Recognition Memory, and Exploratory Behaviors in the Domestic Pig (Sus Scrofa)." *Frontiers in Neuroscience* 15: 318. doi: 10.3389/FNINS.2021.649536/BIBTEX.

Fleming, Stephen A., Maciej Chichlowski, Brian M. Berg, Sharon M. Donovan, and Ryan N. Dilger. 2018. "Dietary Sialyllactose Does Not Influence Measures of Recognition Memory or Diurnal Activity in the Young Pig." *Nutrients* 10(4):395. doi: 10.3390/NU10040395.

Fleming, Stephen A., Supida Monaikul, Alexander J. Patsavas, Rosaline v. Waworuntu, Brian M. Berg, and Ryan N. Dilger. 2019. "Dietary Polydextrose and Galactooligosaccharide Increase Exploratory Behavior, Improve Recognition Memory, and Alter Neurochemistry in the Young Pig." *Nutritional Neuroscience* 22(7):499–512. doi: 10.1080/1028415X.2017.1415280.

Fleming, Stephen A., Austin T. Mudd, Jonas Hauser, Jian Yan, Sylviane Metairon, Pascal Steiner, Sharon M. Donovan, and Ryan N. Dilger. 2020a. "Dietary Oligofructose Alone or in Combination with 2′-Fucosyllactose Differentially Improves Recognition Memory and Hippocampal MRNA Expression." *Nutrients 2020, Vol. 12, Page 2131* 12(7):2131. doi: 10.3390/NU12072131.

Fleming, Stephen A., Austin T. Mudd, Jonas Hauser, Jian Yan, Sylviane Metairon, Pascal Steiner, Sharon M. Donovan, and Ryan N. Dilger. 2020b. "Human and Bovine Milk Oligosaccharides Elicit Improved Recognition Memory Concurrent With Alterations in Regional Brain Volumes and Hippocampal MRNA Expression." *Frontiers in Neuroscience* 14:770. doi: 10.3389/FNINS.2020.00770/BIBTEX.

Forkman, B., A. Boissy, M.C. Meunier-Salaün, E. Canali, and R.B. Jones. 2007. "A Critical Review of Fear Tests Used on Cattle, Pigs, Sheep, Poultry and Horses." *Physiology & Behavior* 92(3):340–74. doi: 10.1016/J.PHYSBEH.2007.03.016.

Foster, J.A. 2016. "Gut Microbiome and Behavior: Focus on Neuroimmune Interactions." *International Review of Neurobiology* 131:49–65. doi: 10.1016/BS.IRN.2016.07.005.

Foster, Jane A., and Karen Anne McVey Neufeld. 2013. "Gut–Brain Axis: How the Microbiome Influences Anxiety and Depression." *Trends in Neurosciences* 36(5):305–12. doi: 10.1016/J.TINS.2013.01.005.

Friend, T.H., D.A. Knabe, and T.D. Tanksley. 1983. "Behavior and Performance of Pigs Grouped by Three Different Methods at Weaning." *Journal of Animal Science* 57(6):1406–11. doi: 10.2527/JAS1983.5761406X.

Fulford, Allison J., and Michael S. Harbuz. 2005. "An Introduction to the HPA Axis." *Techniques in the Behavioral and Neural Sciences* 15(PART 1):43–65. doi: 10.1016/S0921-0709(05)80006-9.

Gan, Fang, Xingxiang Chen, Shengfa F. Liao, Chenhui Lv, Fei Ren, Gengping Ye, Cuiling Pan, Da Huang, Jun Shi, Xiuli Shi, Hong Zhou, and Kehe Huang. 2014. "Selenium-Enriched Probiotics Improve Antioxidant Status, Immune Function, and Selenoprotein Gene Expression of Piglets Raised under High Ambient Temperature." *Journal of Agricultural and Food Chemistry* 62(20):4502–8. doi: 10.1021/JF501065D/SUPPL_FILE/JF501065D_SI_001.PDF.

Gan, Fang, Fei Ren, Xingxiang Chen, Chenhui Lv, Cuiling Pan, Gengping Ye, Jun Shi, Xiuli Shi, Hong Zhou, Shituleni Andreas Shituleni, and Kehe Huang. 2013. "Effects of Selenium-Enriched Probiotics on Heat Shock Protein MRNA Levels in Piglet under Heat Stress Conditions." *Journal of Agricultural and Food Chemistry* 61(10):2385–91. doi: 10.1021/JF300249J/ASSET/IMAGES/LARGE/JF-2012-00249J_0006.JPEG.

Georgieff, Michael K. 2007. "Nutrition and the Developing Brain: Nutrient Priorities and Measurement." *The American Journal of Clinical Nutrition* 85(2):614S–620S. doi: 10.1093/AJCN/85.2.614S.

Geverink, N.A., W.G.P. Schouten, G. Gort, and V.M. Wiegant. 2002. "Individual Differences in Behavioral and Physiological Responses to Restraint Stress in Pigs." *Physiology & Behavior* 77(2–3):451–57. doi: 10.1016/S0031-9384(02)00877-6.

Gonyou, H.W., E. Beltranena, D.L. Whittington, and J.F. Patience. 2011. "The Behaviour of Pigs Weaned at 12 and 21 Days of Age from Weaning to Market." *Canadian Journal of Animal Science* 78(4):517–23. doi: 10.4141/A98-023.

Guo, Huiduo, Jianwen He, Xiaojing Yang, Weijiang Zheng, and Wen Yao. 2020. "Responses of Intestinal Morphology and Function in Offspring to Heat Stress in Primiparous Sows during Late Gestation." *Journal of Thermal Biology* 89:102539. doi: 10.1016/J.JTHERBIO.2020.102539.

Hahn, Kaylee E., Irina Dahms, Christopher M. Butt, Norman Salem, Vivian Grimshaw, Eileen Bailey, Stephen A. Fleming, Brooke N. Smith, and Ryan N. Dilger. 2020. "Impact of Arachidonic and Docosahexaenoic Acid Supplementation on Neural and Immune Development in the Young Pig." *Frontiers in Nutrition* 7:214. doi: 10.3389/FNUT.2020.592364/BIBTEX.

Haigh, Amy, Jen Yun Chou, and Keelin O'Driscoll. 2020. "Variations in the Behavior of Pigs During an Open Field and Novel Object Test." *Frontiers in Veterinary Science* 7:607. doi: 10.3389/FVETS.2020.00607/BIBTEX.

Hay, Magali, Pierre Orgeur, Frédéric Lévy, Jean le Dividich, Didier Concordet, Raymond Nowak, and Pierre Mormède. 2001. "Neuroendocrine Consequences of Very Early Weaning in Swine." *Physiology & Behavior* 72(1–2):263–69. doi: 10.1016/S0031-9384(00)00404-2.

He, Jianwen, Weijiang Zheng, Chengyuan Tao, Huiduo Guo, Yongqiang Xue, Ruqian Zhao, and Wen Yao. 2020. "Heat Stress during Late Gestation Disrupts Maternal Microbial Transmission with Altered Offspring's Gut Microbial Colonization and Serum Metabolites in a Pig Model." *Environmental Pollution* 266:115111. doi: 10.1016/J.ENVPOL.2020.115111.

He, Ying, Yan Liu, Jiayong Tang, Gang Jia, Guangmang Liu, Gang Tian, Xiaoling Chen, Jingyi Cai, Bo Kang, and Hua Zhao. 2022. "Selenium Exerts Protective Effects against Heat Stress-Induced Barrier Disruption and Inflammation Response in Jejunum of

Growing Pigs." *Journal of the Science of Food and Agriculture* 102(2):496–504. doi: 10.1002/JSFA.11377.

Heath, Martha E. 1983. "The Effects of Rearing-Temperature on Body Composition in Young Pigs." *Comparative Biochemistry and Physiology. A, Comparative Physiology* 76(2):363–66. doi: 10.1016/0300-9629(83)90338-9.

Heimbürge, Susen, Ellen Kanitz, Armin Tuchscherer, and Winfried Otten. 2020. "Is It Getting in the Hair? – Cortisol Concentrations in Native, Regrown and Segmented Hairs of Cattle and Pigs after Repeated ACTH Administrations." *General and Comparative Endocrinology* 295:113534. doi: 10.1016/J.YGCEN.2020.113534.

Heinonen, Mari, Toomas Orro, Teija Kokkonen, Camilla Munsterhjelm, Olli Peltoniemi, and Anna Valros. 2010. "Tail Biting Induces a Strong Acute Phase Response and Tail-End Inflammation in Finishing Pigs." *The Veterinary Journal* 184(3):303–7. doi: 10.1016/J.TVJL.2009.02.021.

Hemsworth, P.H., E.O. Price, and R. Borgwardt. 1996. "Behavioural Responses of Domestic Pigs and Cattle to Humans and Novel Stimuli." *Applied Animal Behaviour Science* 50(1): 43–56. doi: 10.1016/0168-1591(96)01067-2.

Herman, James P., Nawshaba Nawreen, Marissa A. Smail, and Evelin M. Cotella. 2020. "Brain Mechanisms of HPA Axis Regulation: Neurocircuitry and Feedback in Context Richard Kvetnansky Lecture." *The International Journal on the Biology of Stress* 23(6):617–32. doi: 10.1080/10253890.2020.1859475.

Hessing, M.J.C., A.M. Hagelsø, J.A.M. van Beek, R.P. Wiepkema, W.G.P. Schouten, and R. Krukow. 1993. "Individual Behavioural Characteristics in Pigs." *Applied Animal Behaviour Science* 37(4):285–95. doi: 10.1016/0168-1591(93)90118-9.

Hulbert, Lindsey E., Eduarda M. Bortoluzzi, Yunzhi Luo, Jared M. Mumm, Morgan J. Coffin, Gina Y. Becker, Pamela J. Vandevord, Elizabeth M. McNeil, Tim Walilko, Zin Z. Khaing, and Laila Zai. 2019. "Noninvasive, In-Pen Approach Test for Laboratory-Housed Pigs." *JoVE (Journal of Visualized Experiments)* 148:e58597. doi: 10.3791/58597.

Igosheva, Natalia, Paul D. Taylor, Lucilla Poston, and Vivette Glover. 2007. "Prenatal Stress in the Rat Results in Increased Blood Pressure Responsiveness to Stress and Enhanced Arterial Reactivity to Neuropeptide Y in Adulthood." *Journal of Physiology* 582(2):665–74. doi: 10.1113/JPHYSIOL.2007.130252.

Irwin, Michael, Richard Hauger, and Marvin Brown. 1992. "Central Corticotropin-Releasing Hormone Activates the Sympathetic Nervous System and Reduces Immune Function: Increased Responsivity of the Aged Rat." *Endocrinology* 131(3):1047–53. doi: 10.1210/ENDO.131.3.1505449.

Jakovcevski, Igor, Radmila Filipovic, Zhicheng Mo, Sonja Rakic, and Nada Zecevic. 2009. "Oligodendrocyte Development and the Onset of Myelination in the Human Fetal Brain." *Frontiers in Neuroanatomy* 3:5. doi: 10.3389/NEURO.05.005.2009/BIBTEX.

Janczak, Andrew M., Lene J. Pedersen, Karin H. Jensen, Inger Lise Andersen, Knut Egil Bøe, and Morten Bakken. 2000. "No Effect of Variation in Handling on Behaviour in a Porcine Elevated Plus-Maze – a Brief Report." *Applied Animal Behaviour Science* 69(2):169–73. doi: 10.1016/S0168-1591(00)00111-8.

Janczak, Andrew M., Lene J. Pedersen, Lotta Rydhmer, and Morten Bakken. 2003. "Relation between Early Fear- and Anxiety-Related Behaviour and Maternal Ability in Sows." *Applied Animal Behaviour Science* 82(2):121–35. doi: 10.1016/S0168-1591(03)00055-8.

Jena, Ankita, Carlos A. Montoya, Jane A. Mullaney, Ryan N. Dilger, Wayne Young, Warren C. McNabb, and Nicole C. Roy. 2020. "Gut-Brain Axis in the Early Postnatal Years of Life: A Developmental Perspective." *Frontiers in Integrative Neuroscience* 14:44. doi: 10.3389/FNINT.2020.00044/BIBTEX.

Jensen, Karin Hjelholt, Niels Oksbjerg, and Erik Jørgensen. 1994. "Dietary Salbutamol and Level of Protein: Effects on the Acute Stress Response in Pigs." *Physiology & Behavior* 55(2):375–79. doi: 10.1016/0031-9384(94)90149-X.

Johnson, J.S., M.v. Sanz Fernandez, J.T. Seibert, J.W. Ross, M.C. Lucy, T.J. Safranski, T.H. Elsasser, S. Kahl, R.P. Rhoads, and L.H. Baumgard. 2015. "In Utero Heat Stress Increases Postnatal Core Body Temperature in Pigs." *Journal of Animal Science* 93(9):4312–22. doi: 10.2527/JAS.2015-9112.

Johnson, Jay S., and Lance H. Baumgard. 2019. "PHYSIOLOGY SYMPOSIUM: Postnatal Consequences of in Utero Heat Stress in Pigs." *Journal of Animal Science* 97(2):962–71. doi: 10.1093/JAS/SKY472.

Johnson, Jay S., Jacob M. Maskal, Alan W. Duttlinger, Kouassi R. Kpodo, Betty R. McConn, Christopher J. Byrd, Brian T. Richert, Jeremy N. Marchant-Forde, Donald C. Lay, Shelbi D. Perry, Matthew C. Lucy, and Tim J. Safranski. 2020a. "In Utero Heat Stress Alters the Postnatal Innate Immune Response of Pigs." *Journal of Animal Science* 98(12):1–13. doi: 10.1093/JAS/SKAA356.

Johnson, Jay S., Kara R. Stewart, Tim J. Safranski, Jason W. Ross, and Lance H. Baumgard. 2020b. "In Utero Heat Stress Alters Postnatal Phenotypes in Swine." *Theriogenology* 154: 110–19. doi: 10.1016/J.THERIOGENOLOGY.2020.05.013.

Joung, Sangyun, Joanne E. Fil, Anne B. Heckmann, Anne S. Kvistgaard, and Ryan N. Dilger. 2020. "Early-Life Supplementation of Bovine Milk Osteopontin Supports Neurodevelopment and Influences Exploratory Behavior." *Nutrients* 12(8):1–20. doi: 10.3390/NU12082206.

Kern, Simone, Terrence R. Oakes, Charles K. Stone, Emelia M. McAuliff, Clemens Kirschbaum, and Richard J. Davidson. 2008. "Glucose Metabolic Changes in the Prefrontal Cortex Are Associated with HPA Axis Response to a Psychosocial Stressor." *Psychoneuroendocrinology* 33(4):517–29. doi: 10.1016/J.PSYNEUEN.2008.01.010.

Khan, Alam Zeb, Shahnawaz Kumbhar, Muhammad Hamid, Samreen Afzal, Fahmida Parveen, Yunhuan Liu, Hao Shu, Berhe Mekonnen Mengistu, and Kehe Huang. 2016. "Effects of Selenium-Enriched Probiotics on Heart Lesions by Influencing the MRNA Expressions of Selenoproteins and Heat Shock Proteins in Heat Stressed Broiler Chickens." *Pakistan Veterinary Journal* 36(4):460–64.

Knaap, Marjo S. van der., J. Valk, Frederik Barkhof, and Marjo S. van der Knaap. 2005. *Magnetic Resonance of Myelination and Myelin Disorders*. 3rd Edition. Springer.

Knight, Laura C., and Ryan N. Dilger. 2018. "Longitudinal Effects of Iron Deficiency Anemia and Subsequent Repletion on Blood Parameters and the Rate and Composition of Growth in Pigs." *Nutrients* 10(5):632. doi: 10.3390/NU10050632.

Kobek-Kjeldager, Cecilie, Anna A. Schönherz, Nuria Canibe, and Lene Juul Pedersen. 2022. "Diet and Microbiota-Gut-Brain Axis in Relation to Tail Biting in Pigs: A Review." *Applied Animal Behaviour Science* 246:105514. doi: 10.1016/J.APPLANIM.2021.105514.

Kooij, E.v. Erp v.d., A.H. Kuijpers, J.W. Schrama, F.J.C.M. van Eerdenburg, W.G.P. Schouten, and M.J.M. Tielen. 2002. "Can We Predict Behaviour in Pigs?: Searching for Consistency in Behaviour over Time and across Situations." *Applied Animal Behaviour Science* 75(4): 293–305. doi: 10.1016/S0168-1591(01)00203-9.

Kreek, Mary Jeanne, Yan Zhou, and Orna Levran. 2011. "Functions of Arginine Vasopressin and Its Receptors: Importance of Human Molecular Genetics Studies in Bidirectional Translational Research." *Biological Psychiatry* 70(6):502–3. doi: 10.1016/J.BIOPSYCH.2011.07.013.

Lan, Pham Thi Ngoc, Mitsuo Sakamoto, and Yoshimi Benno. 2004. "Effects of Two Probiotic Lactobacillus Strains on Jejunal and Cecal Microbiota of Broiler Chicken under Acute Heat Stress Condition as Revealed by Molecular Analysis of 16S RRNA Genes." *Microbiology and Immunology* 48(12):917–29. doi: 10.1111/J.1348-0421.2004.TB03620.X.

Lauritzen, Lotte, Paolo Brambilla, Alessandra Mazzocchi, Laurine B.S. Harsløf, Valentina Ciappolino, and Carlo Agostoni. 2016. "DHA Effects in Brain Development and Function." *Nutrients* 8(1):6. doi: 10.3390/NU8010006.

Lewis, Craig R.G., Kim L. Bunter, Craig R.G. Lewis, and Kim L. Bunter. 2011. "Effects of Seasonality and Ambient Temperature on Genetic Parameters for Production and Reproductive Traits in Pigs." *Animal Production Science* 51(7):615–26. doi: 10.1071/AN10265.

Li, Qiang, Shirley Guo-Ross, Darrell v. Lewis, Dennis Turner, Aaron M. White, Wilkie A. Wilson, and H. Scott Swartzwelder. 2004. "Dietary Prenatal Choline Supplementation Alters Postnatal Hippocampal Structure and Function." *Journal of Neurophysiology* 91(4):1545–55. doi: 10.1152/JN.00785.2003/ASSET/IMAGES/LARGE/Z9K0040437470007.JPEG.

Liu, F., W. Zhao, H.H. Le, J.J. Cottrell, M.P. Green, B.J. Leury, F.R. Dunshea, and A.W. Bell. 2022. "Review: What Have We Learned about the Effects of Heat Stress on the Pig Industry?" *Animal* 16:100349. doi: 10.1016/J.ANIMAL.2021.100349.

Liu, Hongnan, Emily C. Radlowski, Matthew S. Conrad, Yao Li, Ryan N. Dilger, and Rodney W. Johnson. 2014. "Early Supplementation of Phospholipids and Gangliosides Affects Brain and Cognitive Development in Neonatal Piglets." *The Journal of Nutrition* 144(12): 1903–9. doi: 10.3945/JN.114.199828.

Liu, Xiaofei, Shangqing Cao, and Xuewu Zhang. 2015. "Modulation of Gut Microbiota-Brain Axis by Probiotics, Prebiotics, and Diet." *Journal of Agricultural and Food Chemistry* 63(36): 7885–95. doi: 10.1021/ACS.JAFC.5B02404/ASSET/IMAGES/LARGE/JF-2015-02404Z_0004.JPEG.

Lucy, Matthew C., and Timothy J. Safranski. 2017. "Heat Stress in Pregnant Sows: Thermal Responses and Subsequent Performance of Sows and Their Offspring." *Molecular Reproduction and Development* 84(9):946–56. doi: 10.1002/MRD.22844.

Luczynski, Pauline, Karen Anne Mc Vey Neufeld, Clara Seira Oriach, Gerard Clarke, Timothy G. Dinan, and John F. Cryan. 2016. "Growing up in a Bubble: Using Germ-Free Animals to Assess the Influence of the Gut Microbiota on Brain and Behavior." *International Journal of Neuropsychopharmacology* 19(8):1–17. doi: 10.1093/IJNP/PYW020.

Luna, Ruth Ann, and Jane A. Foster. 2015. "Gut Brain Axis: Diet Microbiota Interactions and Implications for Modulation of Anxiety and Depression." *Current Opinion in Biotechnology* 32:35–41. doi: 10.1016/J.COPBIO.2014.10.007.

Lv, C.H., T. Wang, N. Regmi, X. Chen, K. Huang, and S.F. Liao. 2015. "Effects of Dietary Supplementation of Selenium-Enriched Probiotics on Production Performance and Intestinal Microbiota of Weanling Piglets Raised under High Ambient Temperature." *Journal of Animal Physiology and Animal Nutrition* 99(6):1161–71. doi: 10.1111/JPN.12326.

Machado-Neto, R., C.N. Graves, and S.E. Curtis. 1987. "Immunoglobulins in Piglets from Sows Heat-Stressed Prepartum." *Journal of Animal Science* 65(2):445–55. doi: 10.2527/JAS1987.652445X.

Maestripieri, Dario, Gabriele Schino, Filippo Aureli, and Alfonso Troisi. 1992. "A Modest Proposal: Displacement Activities as an Indicator of Emotions in Primates." *Animal Behaviour* 44(5):967–79. doi: 10.1016/S0003-3472(05)80592-5.

Majzoub, Joseph A. 2006. "Corticotropin-Releasing Hormone Physiology." *European Journal of Endocrinology* 155(suppl_1):S71–76. doi: 10.1530/EJE.1.02247.

Manteuffel, Gerhard, Birger Puppe, and Peter C. Schön. 2004. "Vocalization of Farm Animals as a Measure of Welfare." *Applied Animal Behaviour Science* 88(1–2):163–82. doi: 10.1016/J.APPLANIM.2004.02.012.

Marcet-Rius, Míriam, Galice Kalonji, Alessandro Cozzi, Cécile Bienboire-Frosini, Philippe Monneret, Izabela Kowalczyk, Eva Teruel, Elisa Codecasa, and Patrick Pageat. 2019. "Effects of Straw Provision, as Environmental Enrichment, on Behavioural Indicators of Welfare and Emotions in Pigs Reared in an Experimental System." *Livestock Science* 221:89–94. doi: 10.1016/J.LIVSCI.2019.01.020.

Marcet-Rius, Míriam, Patrick Pageat, Cécile Bienboire-Frosini, Eva Teruel, Philippe Monneret, Julien Leclercq, Céline Lafont-Lecuelle, and Alessandro Cozzi. 2018. "Tail and Ear Movements as Possible Indicators of Emotions in Pigs." *Applied Animal Behaviour Science* 205:14–18. doi: 10.1016/J.APPLANIM.2018.05.012.

Mason, Sharon P., Susan Jarvis, and Alistair B. Lawrence. 2003. "Individual Differences in Responses of Piglets to Weaning at Different Ages." *Applied Animal Behaviour Science* 80(2):117–32. doi: 10.1016/S0168-1591(02)00209-5.

Meizlish, Matthew L., Ruth A. Franklin, Xu Zhou, and Ruslan Medzhitov. 2021. "Tissue Homeostasis and Inflammation." *Annual Review of Immunology* 39:557–81. doi: 10.1146/ANNUREV-IMMUNOL-061020-053734.

Messaoudi, Michaël, Robert Lalonde, Nicolas Violle, Hervé Javelot, Didier Desor, Amine Nejdi, Jean François Bisson, Catherine Rougeot, Matthieu Pichelin, Murielle Cazaubiel, and Jean Marc Cazaubiel. 2011. "Assessment of Psychotropic-like Properties of a Probiotic Formulation (Lactobacillus Helveticus R0052 and Bifidobacterium Longum R0175) in Rats and Human Subjects." *British Journal of Nutrition* 105(5):755–64. doi: 10.1017/S0007114510004319.

Morell, Pierre. 2013. *Myelin*. 2nd Edition. edited by P. Morell. Chaple Hill, North Carolina: Springer US.

Mormède, P., A. Foury, E. Terenina, and P.W. Knap. 2011. "Breeding for Robustness: The Role of Cortisol." *Animal* 5(5):651–57. doi: 10.1017/S1751731110002168.

Mormède, Pierre, Stéphane Andanson, Benoit Aupérin, Bonne Beerda, Daniel Guémené, Jens Malmkvist, Xavier Manteca, Gerhard Manteuffel, Patrick Prunet, Cornelis G. van Reenen, Sabine Richard, and Isabelle Veissier. 2007. "Exploration of the Hypothalamic–Pituitary–Adrenal Function as a Tool to Evaluate Animal Welfare." *Physiology & Behavior* 92(3):317–39. doi: 10.1016/J.PHYSBEH.2006.12.003.

Mudd, Austin T., Joanne E. Fil, Laura C. Knight, and Ryan N. Dilger. 2018a. "Dietary Iron Repletion Following Early-Life Dietary Iron Deficiency Does Not Correct Regional Volumetric or Diffusion Tensor Changes in the Developing Pig Brain." *Frontiers in Neurology* 8: 735. doi: 10.3389/FNEUR.2017.00735.

Mudd, Austin T., Joanne E. Fil, Laura C. Knight, Fan Lam, Zhi Pei Liang, and Ryan N. Dilger. 2018b. "Early-Life Iron Deficiency Reduces Brain Iron Content and Alters Brain Tissue Composition Despite Iron Repletion: A Neuroimaging Assessment." *Nutrients* 10(2):135. doi: 10.3390/NU10020135.

Mudd, Austin T., Caitlyn M. Getty, and Ryan N. Dilger. 2018c. "Maternal Dietary Choline Status Influences Brain Gray and White Matter Development in Young Pigs." *Current Developments in Nutrition* 2(6). doi: 10.1093/CDN/NZY015.

Mudd, Austin T., Caitlyn M. Getty, Brad P. Sutton, and Ryan N. Dilger. 2016. "Perinatal Choline Deficiency Delays Brain Development and Alters Metabolite Concentrations in the Young Pig." *Nutritional Neuroscience* 19(10):425–33. doi: 10.1179/1476830515Y.0000000031.

Munsterhjelm, C, E. Brunberg, M. Heinonen, L. Keeling, and A. Valros. 2013a. "Stress Measures in Tail Biters and Bitten Pigs in a Matched Case-Control Study." *Animal Welfare* 22:331–38. doi: 10.7120/09627286.22.3.331.

Munsterhjelm, C., O. Simola, L. Keeling, A. Valros, and M. Heinonen. 2013b. "Health Parameters in Tail Biters and Bitten Pigs in a Case-Control Study." *Animal* 7(5):814–21. doi: 10.1017/S1751731112002194.

Neumann, Inga D. 2002. "Chapter 12 Involvement of the Brain Oxytocin System in Stress Coping: Interactions with the Hypothalamo-Pituitary-Adrenal Axis." *Progress in Brain Research* 139:147–62. doi: 10.1016/S0079-6123(02)39014-9.

Norscia, Ivan, Edoardo Collarini, and Giada Cordoni. 2021. "Anxiety Behavior in Pigs (Sus Scrofa) Decreases Through Affiliation and May Anticipate Threat." *Frontiers in Veterinary Science* 8:630164. doi: 10.3389/FVETS.2021.630164/BIBTEX.

Orgeur, P., M. Hay, P. Mormède, H. Salmon, J. le Dividich, Raymond Nowak, B. Schaal, F. Lévy, J. le Dividich, and R. Nowak. 2001. "Behavioural, Growth and Immune Consequences of Early Weaning in One-Week-Old Large-White Piglets." *Reproduction Nutrition Development, EDP Sciences* 41(4):321–32. doi: 10.1051/rnd:2001134ï.

Osborne, Thomas B., Lafayette B. Mendel, Edna L. Ferry, and Alfred J. Wakeman. 1915. "The Resumption of Growth after Long Continued Failure to Grow." *Journal of Biological Chemistry* 23(2):439–54. doi: 10.1016/S0021-9258(18)87585-8.

Papadimitriou, Anastasios, and Kostas N. Priftis. 2009. "Regulation of the Hypothalamic-Pituitary-Adrenal Axis." *Neuroimmunomodulation* 16:265–71. doi: 10.1159/000216184.

Parker, Aimée, Sonia Fonseca, and Simon R. Carding. 2019. "Gut Microbes and Metabolites as Modulators of Blood-Brain Barrier Integrity and Brain Health." *Gut Microbes* 11(2):135–57. doi: 10.1080/19490976.2019.1638722.

Parois, Severine P., Susan D. Eicher, Stephen R. Lindemann, and Jeremy N. Marchant. 2021. "Potential Improvements of the Cognition of Piglets through a Synbiotic Supplementation from 1 to 28 Days via the Gut Microbiota." *Scientific Reports 2021 11:1* 11(1):1–15. doi: 10.1038/s41598-021-03565-5.

Parrott, R.F., R.H. Bradshaw, D.M. Lloyd, and J.A. Goode. 1998. "Effects of Transport and Indomethacin on Telemetered Body Temperature and Release of Cortisol and Prolactin in Pre-Pubertal Pigs." *Research in Veterinary Science* 64(1):51–55. doi: 10.1016/S0034-5288(98)90115-1.

Patil, A.K., Sachin Kumar, A.K. Verma, and R.P.S. Baghel. 2015. "Probiotics as Feed Additives in Weaned Pigs: A Review Manipulation of Rumen Microbes Using Medicinal Plant Extracts to Enhance the Nutraceutical Value of Ruminant Milk. View Project Enzyme Supplementation View Project Probiotics as Feed Additives in Weaned Pigs: A Review." *Livestock Research International* 3(2):31–39.

Pearce, S.C., V. Mani, T.E. Weber, R.P. Rhoads, J.F. Patience, L.H. Baumgard, and N.K. Gabler. 2013a. "Heat Stress and Reduced Plane of Nutrition Decreases Intestinal Integrity and Function in Pigs." *Journal of Animal Science* 91(11):5183–93. doi: 10.2527/JAS.2013-6759.

Pearce, Sarah C., Venkatesh Mani, Rebecca L. Boddicker, Jay S. Johnson, Thomas E. Weber, Jason W. Ross, Robert P. Rhoads, Lance H. Baumgard, and Nicholas K. Gabler. 2013b. "Heat Stress Reduces Intestinal Barrier Integrity and Favors Intestinal Glucose Transport in Growing Pigs." *PLOS ONE* 8(8):e70215. doi: 10.1371/JOURNAL.PONE.0070215.

Pluske, John R., David J. Hampson, and Ian H. Williams. 1997. "Factors Influencing the Structure and Function of the Small Intestine in the Weaned Pig: A Review." *Livestock Production Science* 51(1–3):215–36. doi: 10.1016/S0301-6226(97)00057-2.

Poduslo, Shirley E., and Yong Jang. 1984. "Myelin Development in Infant Brain." *Neurochemical Research 1984 9:11* 9:1615–26. doi: 10.1007/BF00964595.

Pohl, C.S., J.E. Medland, E. Mackey, L.L. Edwards, K.D. Bagley, M.P. DeWilde, K.J. Williams, and A.J. Moeser. 2017. "Early Weaning Stress Induces Chronic Functional Diarrhea, Intestinal Barrier Defects, and Increased Mast Cell Activity in a Porcine Model of Early Life Adversity." *Neurogastroenterology & Motility* 29(11):e13118. doi: 10.1111/NMO.13118.

Poletto, R., J.P. Steibel, J.M. Siegford, and A.J. Zanella. 2006. "Effects of Early Weaning and Social Isolation on the Expression of Glucocorticoid and Mineralocorticoid Receptor and 11beta-Hydroxysteroid Dehydrogenase 1 and 2 MRNAs in the Frontal Cortex and Hippocampus of Piglets." *Brain Research* 1067(1):36–42. doi: 10.1016/J.BRAINRES.2005.10.001.

Pond, W.G., S.L. Boleman, M.L. Fiorotto, H. Ho, D.A. Knabe, H.J. Mersmann, J.W. Savell, and D.R. Su. 2000. "Perinatal Ontogeny of Brain Growth in the Domestic Pig." *Proceedings of the Society for Experimental Biology and Medicine. Society for Experimental Biology and Medicine* 223(1):102–8. doi: 10.1046/J.1525-1373.2000.22314.X.

Pond, W.G., D.N. Strachan, Y.N. Sinha, E.F. Walker, J.A. Dunn, and R.H. Barnes. 1969. "Effect of Protein Deprivation of Swine During All or Part of Gestation on Birth Weight, Postnatal Growth Rate and Nucleic Acid Content of Brain and Muscle of Progeny." *The Journal of Nutrition* 99(1):61–67. doi: 10.1093/JN/99.1.61.

Radlowski, Emily C., Matthew S. Conrad, Stephane Lezmi, Ryan N. Dilger, Brad Sutton, Ryan Larsen, and Rodney W. Johnson. 2014. "A Neonatal Piglet Model for Investigating Brain and Cognitive Development in Small for Gestational Age Human Infants." *PloS One* 9(3):e91951. doi: 10.1371/JOURNAL.PONE.0091951.

Reimert, Inonge, J. Elizabeth Bolhuis, Bas Kemp, and T. Bas Rodenburg. 2013. "Indicators of Positive and Negative Emotions and Emotional Contagion in Pigs." *Physiology & Behavior* 109(1):42–50. doi: 10.1016/J.PHYSBEH.2012.11.002.

Ross, J.W., B.J. Hale, N.K. Gabler, R.P. Rhoads, A.F. Keating, and L.H. Baumgard. 2015. "Physiological Consequences of Heat Stress in Pigs." *Animal Production Science* 55:1381–90. doi: 10.1071/AN15267.

Ross, Jason W., Benjamin J. Hale, Jacob T. Seibert, Matthew R. Romoser, Malavika K. Adur, Aileen F. Keating, and Lance H. Baumgard. 2017. "Physiological Mechanisms through Which Heat Stress Compromises Reproduction in Pigs." *Molecular Reproduction and Development* 84(9):934–45. doi: 10.1002/MRD.22859.

Równiak, Maciej, and Krystyna Bogus-Nowakowska. 2020. "The Amygdala of the Common Shrew, Guinea Pig, Rabbit, Fox and Pig: Five Flavours of the Mammalian Amygdala as a Consequence of Clade-Specific Mosaic-like Evolution." *Journal of Anatomy* 236(5): 891–905. doi: 10.1111/JOA.13148.

Ruis, Marko A.W., Joop H.A. te Brake, Johan A. van de Burgwal, Ingrid C. de Jong, Harry J. Blokhuis, and Jaap M. Koolhaas. 2000. "Personalities in Female Domesticated Pigs: Behavioural and Physiological Indications." *Applied Animal Behaviour Science* 66(1–2):31–47. doi: 10.1016/S0168-1591(99)00070-2.

Russell, John A., and Michael Shipston. 2015. *Neuroendocrinology of Stress*. Wiley Blackwell.

Rytych, Jennifer L., Monica R.P. Elmore, Michael D. Burton, Matthew S. Conrad, Sharon M. Donovan, Ryan N. Dilger, and Rodney W. Johnson. 2012. "Early Life Iron Deficiency Impairs Spatial Cognition in Neonatal Piglets." *The Journal of Nutrition* 142(11):2050–56. doi: 10.3945/JN.112.165522.

Sauleau, P., E. Lapouble, D. Val-Laillet, and C.H. Malbert. 2009. "The Pig Model in Brain Imaging and Neurosurgery." *Animal* 3(8):1138–51. doi: 10.1017/S1751731109004649.

Sawchenko, P.E., E.R. Brown, R.K.W. Chan, A. Ericsson, H.Y. Li, B.L. Roland, and K.J. Kovacs. 1996. "Chapter 12 The Paraventricular Nucleus of the Hypothalamus and the Functional Neuroanatomy of Visceromotor Responses to Stress." Pp. 201–22 in *Progress in Brain Research*. Vol. 107. Elsevier.

Schmitt, Océane, Keelin O'Driscoll, and Emma M. Baxter. 2019. "Exploratory Study of the Effects of Intra-Uterine Growth Retardation and Neonatal Energy Supplementation of Low Birth-Weight Piglets on Their Post-Weaning Cognitive Abilities." *Animal Cognition* 22(3):373–85. doi: 10.1007/S10071-019-01251-8.

Schrøder-Petersen, D.L., and H.B. Simonsen. 2001. "Tail Biting in Pigs." *The Veterinary Journal* 162(3):196–210. doi: 10.1053/TVJL.2001.0605.

Schulz, Mareike, Leonie Zieglowski, Marcin Kopaczka, and René H. Tolba. 2022. "Systematic Review The Open Field Test as a Tool for Behaviour Analysis in Pigs: Recommendations for Set-Up Standardization – A Systematic Review." *European Surgical Research*. doi: 10.1159/000525680.

Silva, Ygor Parladore, Andressa Bernardi, and Rudimar Luiz Frozza. 2020. "The Role of Short-Chain Fatty Acids From Gut Microbiota in Gut-Brain Communication." *Frontiers in Endocrinology* 11:25. doi: 10.3389/FENDO.2020.00025/BIBTEX.

Simpson, Carra A., Carmela Diaz-Arteche, Djamila Eliby, Orli S. Schwartz, Julian G. Simmons, and Caitlin S.M. Cowan. 2021. "The Gut Microbiota in Anxiety and Depression – A Systematic Review." *Clinical Psychology Review* 83:101943. doi: 10.1016/J.CPR.2020.101943.

Sinisalo, Alina, Jarkko K. Niemi, Mari Heinonen, and Anna Valros. 2012. "Tail Biting and Production Performance in Fattening Pigs." *Livestock Science* 143(2–3):220–25. doi: 10.1016/J.LIVSCI.2011.09.019.

Smith, Feli, Jessica E. Clark, Beth L. Overman, Christena C. Tozel, Jennifer H. Huang, Jean E.F. Rivier, Anthony T. Blisklager, and Adam J. Moeser. 2010. "Early Weaning Stress Impairs Development of Mucosal Barrier Function in the Porcine Intestine." *American Journal of Physiology – Gastrointestinal and Liver Physiology* 298(3):352–63. doi: 10.1152/AJPGI.00081.2009/ASSET/IMAGES/LARGE/ZH30021055840012.JPEG.

Snaidero, Nicolas, and Mikael Simons. 2014. "Myelination at a Glance." *Journal of Cell Science* 127(14):2999–3004. doi: 10.1242/JCS.151043.

Soto-Rivera, Carmen L., and Joseph A. Majzoub. 2017. "Adrenocorticotrophin." *The Pituitary: Fourth Edition* 47–83. doi: 10.1016/B978-0-12-804169-7.00003-9.

Sutherland, M.A., P.J. Bryer, B.L. Davis, and J.J. McGlone. 2009. "Space Requirements of Weaned Pigs during a Sixty-Minute Transport in Summer." *Journal of Animal Science* 87(1):363–70. doi: 10.2527/JAS.2008-1078.

Sutherland, Mhairi A., Pamela J. Bryer, Brittany L. Davis, and John J. McGlone. 2010. "A Multidisciplinary Approach to Assess the Welfare of Weaned Pigs During Transport at Three Space Allowances." *Journal of Applied Animal Welfare Science* 13(3):237–49. doi: 10.1080/10888705.2010.483879.

Sutkus, Loretta T., Sangyun Joung, Johanna Hirvonen, Henrik Max Jensen, Arthur C. Ouwehand, Ratna Mukherjea, Sharon M. Donovan, and Ryan N. Dilger. 2022. "Influence of 2′-Fucosyllactose and Bifidobacterium Longum Subspecies Infantis Supplementation on Cognitive and Structural Brain Development in Young Pigs." *Frontiers in Neuroscience* 16:860368. doi: 10.3389/FNINS.2022.860368/FULL.

Tatemoto, Patricia, Thiago Bernardino, Luana Alves, Anna Cristina de Oliveira Souza, Rupert Palme, and Adroaldo José Zanella. 2019. "Environmental Enrichment for Pregnant Sows Modulates HPA-Axis and Behavior in the Offspring." *Applied Animal Behaviour Science* 220:104854. doi: 10.1016/J.APPLANIM.2019.104854.

Thodberg, Karen, Karin H. Jensen, and Mette S. Herskin. 1999. "A General Reaction Pattern across Situations in Prepubertal Gilts." *Applied Animal Behaviour Science* 63(2):103–19. doi: 10.1016/S0168-1591(99)00009-X.

Tinbergen, N. 1952. "'Derived' Activities; Their Causation, Biological Significance, Origin, and Emancipation During Evolution." *Source: The Quarterly Review of Biology* 27(1):1–32.

Todorich, Bozho, Juana M. Pasquini, Corina I. Garcia, Pablo M. Paez, and James R. Connor. 2009. "Oligodendrocytes and Myelination: The Role of Iron." *Glia* 57(5):467–78. doi: 10.1002/GLIA.20784.

Ursinus, Winanda W., Cornelis G. van Reenen, Inonge Reimert, and J. Elizabeth Bolhuis. 2014. "Tail Biting in Pigs: Blood Serotonin and Fearfulness as Pieces of the Puzzle?" *PLOS ONE* 9(9):e107040. doi: 10.1371/JOURNAL.PONE.0107040.

Uzbay, Tayfun. 2019. "Germ-Free Animal Experiments in the Gut Microbiota Studies." *Current Opinion in Pharmacology* 49:6–10. doi: 10.1016/J.COPH.2019.03.016.

Valros, Anna, and Mari Heinonen. 2015. "Save the Pig Tail." *Porcine Health Management* 1(1):1–7. doi: 10.1186/2055-5660-1-2/FIGURES/1.

Valros, Anna, Camilla Munsterhjelm, Eero Puolanne, Marita Ruusunen, Mari Heinonen, Olli A.T. Peltoniemi, and A. Reeta Pösö. 2013. "Physiological Indicators of Stress and Meat and Carcass Characteristics in Tail Bitten Slaughter Pigs." *Acta Veterinaria Scandinavica* 55(1):1–8. doi: 10.1186/1751-0147-55-75/TABLES/2.

Wallgren, P., and E. Lindahl. 1996. "The Influence of Tail Biting on Performance of Fattening Pigs." *Acta Veterinaria Scandinavica 1996 37:4* 37(4):453–60. doi: 10.1186/BF03548085.

Wamnes, S., N.J. Lewis, and R.J. Berry. 2011. "The Performance of Early-Weaned Piglets Following Transport: Effect of Season and Weaning Weight." *Canadian Journal of Animal Science* 86(3):337–43. doi: 10.4141/A06-005.

Wang, Bing, Bing Yu, Muhsin Karim, Honghua Hu, Yun Sun, Paul McGreevy, Peter Petocz, Suzanne Held, and Jennie Brand-Miller. 2007. "Dietary Sialic Acid Supplementation Improves Learning and Memory in Piglets." *The American Journal of Clinical Nutrition* 85(2):561–69. doi: 10.1093/AJCN/85.2.561.

Wegner, Birte, Ines Spiekermeier, Hendrik Nienhoff, Julia Große-Kleimann, Karl Rohn, Henning Meyer, Heiko Plate, Hubert Gerhardy, Lothar Kreienbrock, Elisabeth Grosse Beilage, and Nicole Kemper. 2020. "Application of the Voluntary Human Approach Test on Commercial Pig Fattening Farms: A Meaningful Tool?" *Porcine Health Management* 6:19. doi: 10.1186/S40813-020-00158-Y/TABLES/5.

Wiggins, Richard C. 1982. "Myelin Development and Nutritional Insufficiency." *Brain Research Reviews* 4(2):151–75. doi: 10.1016/0165-0173(82)90016-9.

Williamson, Jill M., and David A. Lyons. 2018. "Myelin Dynamics throughout Life: An Ever-Changing Landscape?" *Frontiers in Cellular Neuroscience* 12:424. doi: 10.3389/FNCEL.2018.00424/BIBTEX.

Worobec, E.K., I.J.H. Duncan, and T.M. Widowski. 1999. "The Effects of Weaning at 7, 14 and 28 Days on Piglet Behaviour." *Applied Animal Behaviour Science* 62(2–3):173–82. doi: 10.1016/S0168-1591(98)00225-1.

van de Wouw, Marcel, Marcus Boehme, Joshua M. Lyte, Niamh Wiley, Conall Strain, Orla O'Sullivan, Gerard Clarke, Catherine Stanton, Timothy G. Dinan, and John F. Cryan. 2018. "Short-Chain Fatty Acids: Microbial Metabolites That Alleviate Stress-Induced Brain–Gut Axis Alterations." *The Journal of Physiology* 596(20):4923–44. doi: 10.1113/JP276431.

Xie, Yuanyuan, and Richard I. Dorsky. 2017. "Development of the Hypothalamus: Conservation, Modification and Innovation." *Development* 144(9):1588–99. doi: 10.1242/DEV.139055.

Xu, R.J., and P.D. Cranwell. 2003. *The Neonatal Pig: Gastrointestinal Physiology and Nutrition.* edited by R.J. Xu and P.D. Cranwell. Nottingham University Press.

Yu, G.S.M., T.M. Steinkirchner, G.A. Rao, and E.C. Larkin. 1986. "Effect of Prenatal Iron Deficiency on Myelination in Rat Pups." *The American Journal of Pathology* 125(3):620–24.

Yuan, Yan, Jarno Jansen, Debby Charles, and Adroaldo J. Zanella. 2004. "The Influence of Weaning Age on Post-Mixing Agonistic Interactions in Growing Pigs." *Applied Animal Behaviour Science* 88(1–2):39–46. doi: 10.1016/J.APPLANIM.2004.01.012.

Zeisel, S.H., D. Char, and N.F. Sheard. 1986. "Choline, Phosphatidylcholine and Sphingomyelin in Human and Bovine Milk and Infant Formulas." *The Journal of Nutrition* 116(1): 50–58. doi: 10.1093/JN/116.1.50.

Zhang, Panwang, Tao Yan, Xiliang Wang, Shichang Kuang, Yuncai Xiao, Weiwei Lu, and Dingren Bi. 2016. "Probiotic Mixture Ameliorates Heat Stress of Laying Hens by Enhancing Intestinal Barrier Function and Improving Gut Microbiota." *Italian Journal of Animal Science* 16(2):292–300. doi: 10.1080/1828051X.2016.1264261.

Zupan, Manja, Andrew M. Janczak, Tore Framstad, and Adroaldo J. Zanella. 2012. "The Effect of Biting Tails and Having Tails Bitten in Pigs." *Physiology & Behavior* 106(5):638–44. doi: 10.1016/J.PHYSBEH.2012.04.025.

CHAPTER 16

Insects and gut health in food-producing animals

I. Biasato[1,#]*, E. Colombino*[2,#]*, A. Luna*[3] *and M.T. Capucchio*[2,4,*]

[1]Department of Agricultural, Forest and Food Sciences, University of Torino, Largo Paolo Braccini 2, 10095 Grugliasco (TO), Italy; [2]Department of Veterinary Sciences, University of Torino, Largo Paolo Braccini 2, 10095 Grugliasco (TO), Italy; [3]Instituto de Investigaciones Biológicas y Tecnológicas (IIByT, CONICET-UNC), Universidad Nacional de Córdoba, Av.Velez Sarsfield 1611, Córdoba, Argentina; [4]Institute of Science of Food Production, National Research Council, Largo Paolo Braccini 2, 10095 Grugliasco (TO), Italy; *Correspondence: mariateresa.capucchio@unito.it; #Equal contribution

Abstract

Insect meals, oils and live larvae can represent valuable innovative dietary ingredients for animal nutrition. In particular, *Hermetia illucens* and *Tenebrio molitor* are the most promising insect species for animal's feed. This chapter will focus on the effects of insect-derived feed ingredients on gut health of food-producing animals. In poultry low dietary inclusion levels of insect meal (< 15%) are recommended, as they maintain diet digestibility, gut health, and, adequate growth performance. In fish and pigs, both low and high inclusion levels of insect-based products (fish: from 3 to 60%; pigs: from 1.5 to 33%), can efficiently be used. Insect oils rich in lauric acid with antibacterial activities, seem to be a promising alternative fat source in rabbit nutrition as they do not impair diet digestibility and growth performance of the animals, also improving their intestinal microbiota. Less information is available on ruminants due to the European ban on the use of processed animal proteins in this species to avoid the spreading of the bovine spongiform encephalopathy. The available preliminary studies have been mainly performed in vitro using 7.5 to 25% of insects in substitution of soybean meal, suggesting that insects may represent a suitable protein and fat source also for ruminants, especially considering the ability of chitosan to reduce ruminal biohydrogenation and methane production without impairing diet digestibility. However, further studies are needed to confirm the potential effects of insect meals and oils on gut health of rabbit and ruminants and to better know the prebiotic/probiotic activity in different food producing animals.

Keywords

insects – gut health – poultry – fish – pig – ruminants

© I. BIASATO ET AL., 2024 | DOI:10.3920/9789004695467_017

1 Introduction

Considering trends towards 2050 predict a steady population increase to 9 billion people, forcing an increased food and feed output from available agro-ecosystems leads to an even greater pressure on the environment. This context placed insects in the centre of the scene. Although edible insects contain high quality protein, vitamins and amino acids not only for animal feed but also for humans, considering the objectives of the present chapter, information will focus on insects as novel alternatives to animal feed. Compared to any other animal protein source, insects have a very high food conversion rate, emit much less greenhouse gasses and ammonia, has a much shorter life cycle, can be grown on organic wastes and its production generates secondary products that can contribute to improvements in efficiency and emissions reductions on other productive systems (Halloran *et al.*, 2016). This need of increasing food production with an inexorable demand of reduced greenhouse gas emissions, and insects emerging as a great alternative, puts on regulatory agencies a huge responsibility and requires a fast response.

The European Union (EU) present an important deficiency in the supply of high protein animal feed ingredients, which led to a relative high dependency on imported animal feed (Precup *et al.*, 2022). In this context, the use of insects as feed could represent a valid and sustainable solution of animal farming in the EU. EU law regulates the conditions for food and feed business operators, such as insect producers, to produce and commercialize their products in the European Union. The general principles and standards that need to be adopted by insect producers are included in the Reg. No 178/2002, 852/2004 and 183/2005 considering just general food law, and hygiene for foodstuffs and feed. In the context of insects for animal feed, one of the most critical recent changes was in 2017, when Regulation (EU) No 2017/893 was introduced. This act amended Regulations (EC) No 999/2001 and (EU) No 142/2011, allowing the feeding of seven insect species to aquaculture animals for black soldier fly (*H. illucens*), common housefly (Musca domestica), yellow mealworm (*T. molitor*), lesser mealworm (*A. diaperinus*), house cricket (*A. domesticus*), banded cricket (*G. sigillatus*) and field cricket (*Gryllus assimilis*). This reg. 2017/893 remove requirements for producers that were not applicable for insect's producers. Furthermore, Regulation (EU) No 2019/1981 introduced a list of third countries that were authorized to export insect products complying with the mentioned Regulation (EU) No 2017/893 (Lähteenmäki-Uutela *et al.*, 2021).

In a general view about the EU legal situation, insects are subject to the rule which regulate the feeding of livestock, including the general principle enshrined in Article 4, para 1, lett. a) of Reg. (EC) 767/2009, whereby animals can be reared on substrates of vegetable origin or specifically allowed materials of animal origin such as fishmeal and hydrolysed proteins from non-ruminants (Precup *et al.*, 2022). Developments to ease some of these restrictions are ongoing, but legislative changes are not expected to be implemented in the short-term. Aquaculture animals have been allowed to be fed with insect meal since 2017 and feeding insects to pets was already permitted (Lähteenmäki-Uutela *et al.*, 2021). In August 2021 the Comm. Reg. (EU) 2021/1372 authorized the use of processed animal proteins derived from insects (insect PAPs) in poultry and pig feed, hoping to help create a "more sustainable feed chain", in line with its flagship food policy. This proposal represents a relevant milestone for the European insect sector, as it marks one of the key steps in the authorisation process. In line with the EU procedures, the entry into force of this proposal took place on the 7th of September 2021 (IPIFF, 2020). Currently, the aquafeed sector is the most relevant market for insect producers in the EU. However, as a consequence of the last regulation in 2021, the quantities of insect meal sold for the poultry and pig markets will experience a strong increase, and in particular, the black soldier fly is currently the most farmed insect species in the EU (Precup *et al.*, 2022).

By their side, in North America, insects have been used as a part of food culture in many regions, and insect's industries started to expand after 2012 (Lähteenmäki-Uutela *et al.*, 2021). Many states in the US base their feed regulations on recommendations of the Association of American Feed Control Officials (AAFCO), who has established an ingredient definition for only one insect species for livestock feed. The black soldier fly (*H. illucens*, BSF) larvae, including dried whole larvae (since 2016) and BSF meal (since 2018) can be used in aquaculture for salmonids such as salmon, trout and char (also approved by the FDA). In this case, the BSF larvae can be reared on approved feed-grade materials, including pre-consumer food waste as a substrate, as well as other food manufacturing by-products such as spent brewery grains (Lähteenmäki-Uutela *et al.*, 2018). Several states also allow insect-based pet foods, while other states wait for AATCO and FDA decisions.

Many parts of the world are responding to regulations for the insect industry to meet the incipient market. Canada do not consider crickets as novel food, and have a large number of cricket breeders in the country, however each new proposed insect species, rearing condition, and livestock species needs separate authorisation (Lähteenmäki-Uutela *et al.*, 2021). Currently, Canada, USA, the EU, Australia and New Zealand allow the trade and manufacture of BSF larvae as feed under specific conditions (Alagappan *et al.*, 2021). By their side, in Central and South America insects as feed regulations are not completely developed yet. At the moment, insects have no regulatory laws that allow them to be commercialized as animal feed ingredients in Mexico, Argentina and Brazil (Lähteenmäki-Uutela *et al.*, 2021). Recently, Argentina includes "Insect breed plant production" as a new category in their national regulations, the first step on the way of the commercial law adequation (https://www.argentina.gob.ar/noticias/apoyo-innovador-metodo-de-produccion-de-harina-de-insectos-para-consumo-animal). Chile is a vanguard country in this line, since insects are considered as feed ingredient according to RES. 6612 from the Servicio Agrícola y Ganadero del Ministerio de Agricultura de Chile (Avendaño *et al.*, 2020). In particular, in Chile BSF larvae is already being commercialized not only for aquaculture but also for pets, poultry, piglets, ruminants and reptiles (https://f4f.cl/el-futuro-de-los-alimentos).

In particular, BSF larvae and yellow mealworm (*Tenebrio molitor*, YM) are the most important insect species farmed by private companies, since they have been proven to have a good nutritional profile for feed animals. Moreover, they have been also proposed as probiotic to reduce the antibiotic usage in food-producing animals thanks to their content in chitin, a polysaccharide that show immunostimulant and antimicrobial properties (Thrastardottir *et al.*, 2021). The present chapter will focus on the effects of insect-derived feed ingredients on gut health of food-producing animals (poultry, fish, pigs, rabbit, and ruminants).

2 Main reared insect species: *Hermetia illucens* and *Tenebrio Molitor*

Hermetia illucens (Diptera: Stratiomyidae) is distributed worldwide, prominently in the equatorial tropics (Liu *et al.*, 2017). Its life cycle is characterized by four phases: eggs (4 days), larva (18 days), pre-pupa (14 days) and adults (9 days). Thus, in optimal environmental conditions (29–31 °C and 30% of relative humidity) the larvae become mature in 45 days (Ferrarezi and Bailey, 2016). The larvae can measure up to 27 mm in length and 6 mm in width and its weight can reach 220 mg, having a dull, whitish colour (Diclaro and Kaufman, 2012). The larvae feed intake is comprised between 25 and 500 mg of fresh matter/larva/day and the rearing substrate can contain a wide range of organic products (Diener *et al.*, 2011; Hardouin and Mahoux, 2004; van Huis, 2013). At the end of the larval stage, the prepupae migrates to find a dry and suitable pupation site. The adult flies are black, wasp-like and they measure around 15–20 mm in length. They are neither pest nor disease vector as they do not feed themselves and they survive on the fat stored in their larval stage (Ferrarezi and Bailey, 2016). This has converted BSF in an ideal candidate for mass rearing since the 1990s as it is able to adapt to a wide range of challenging environmental conditions and it is not linked with microbiological risks for human and animal's health (Diener *et al.*, 2011). Black soldier fly larvae are a high-value protein and fat source (Makkar *et al.*, 2014). Fresh larvae showed a quite high dry matter (DM) content, between 35.0 and 45.0%, which makes them easier and less costly to dehydrate than other fresh by-products (Newton *et al.*, 2008). They contain about 40.0–44.0% crude protein (CP) and 11.0–28.0% DM of ash (Veldkamp and van Niekerk, 2019). The larvae are also rich in calcium (5.0–8.0% DM) and phosphorus (0.6–1.5% DM) (Arango Gutiérrez *et al.*, 2004; Newton *et al.*, 2008; St-Hilaire *et al.*, 2007). Their fat content and their fatty acid composition is extremely variable, depending on the diet: 15.0–25.0% for larvae fed on poultry manure (Arango Gutiérrez *et al.*, 2004), 28.0% on swine manure (Newton *et al.*, 2005), 35.0% on cattle manure (Newton *et al.*, 1977) and 42.0–49.0% on oil-rich food waste (Makkar *et al.*, 2014). Regarding chitin content, it is similar in larvae, pre-pupae and pupae stages being around 9–10% of DM. However, it decreases in adult flies (5.0–8.0%) (Soetemans *et al.*, 2020).

Tenebrio molitor (Coleoptera: Tenebrionidae) is indigenous to Europe and is now distributed worldwide (Makkar *et al.*, 2014). Its life cycle can last from 280 to 630 days, depending on the environmental conditions and it can be divided into four stages as previously described for BSF: eggs (10–12 days at 25–27 °C), larvae (3–4 months at 18 °C), pupae (7–9 days) and adults (2–3 months) (Adámková *et al.*, 2020). The mature larva is light yellow-brown coloured, it can measure 20–32 mm and weights 130–160 mg. In the adult stage, the mealworm is a black beetle of 12–15 mm (Adámková *et al.*, 2020). Mealworms are omnivorous and they can be feed on vegetables or animal products (*e.g.*, meat and feathers) (Wang *et al.*, 2012). Their rearing substrate are typically composed by cereal bran or flour and supplemented with protein sources such as soybean flour, skimmed milk powder or yeast. Fresh fruits and vegetables can be also use (Aguilar-Miranda *et al.*, 2002; Makkar *et al.*, 2014). The rearing substrate must be balanced to contain about 20% protein on DM basis (Ramos-Elorduy *et al.*, 2002). Yellow mealworms require breeding system because a stable protein content can be obtained regardless of the rearing substrate (van Broekhoven *et al.*, 2015). For this reason, they have been chosen for industrial manufacturing of pets, zoo animals, fish and poultry feed (Bovera *et al.*, 2016; Shafique *et al.*, 2021). Fresh larvae contain about 60% of water, which makes them difficult to preserve. The CP content of YM larvae is 52.4% on average and it ranges from 47.0% to 60.2%, being greater than that of conventional soybean meal (49.4%) and lower than that of fishmeal (67.5%) (Selaledi *et al.*, 2020). They are also relatively low in ash (< 5.0 % DM) and they are low in Ca and Ca:P ratio (Ravzanaadii *et al.*, 2012). Fat content can vary between 31.0 and 43.0% and they are richer in essential polyunsaturated fatty acids such as omega-3 (47.3 g/100g) and omega-6 (31.6 g/100 g) compared to BSF (Hong *et al.*, 2020). As already described for BSF, chitin content is similar in YM larvae and pupae but lower in adults. In fact, YM larvae and pupae contained 6.41 % of chitin on average ranging from 4.9 % and 13.0% on DM (13.0 g/100.0 g of chitin) while adults contained 0.01 g/100 g (Adámková *et al.*, 2017; Hong *et al.*, 2020). Also, YM larvae contain 11.6 mg/g of chitosan, which allow them to be used as probiotics (Hong *et al.*, 2020).

3 Gut health: definition and main parameters

In recent decades, the term "gut health" has become increasingly popular and frequently used in human and veterinary medicine (Colombino *et al.*, 2021b; Kogut and Arsenault, 2016). Gut health is generally defined as "the absence, prevention or avoidance of intestinal disease so that the animal is able to perform its physiological functions in order to withstand exogenous and endogenous stressors" (Kogut and Arsenault, 2016). However, a broader definition of gut health should cover multiple aspects of the gastrointestinal tract, such as effective digestion and absorption of feed, a proper structure of the gastrointestinal barrier, the absence of gastrointestinal illness, a normal and stable intestinal microbiota, and an effective immune status (Bischoff, 2011).

The key factors for guaranteeing an optimal gut health appear to be gut morphology, gut mucin expression, local immune response, and gut microbiota, which together contribute to the "intestinal immunity" (Kogut *et al.*, 2020).

Regarding gut morphology, villus height (Vh), villus width (Vw) and crypt depth (Cd) have become the gold standard in the evaluation of the intestinal health status in animals (Ducatelle *et al.*, 2018). In fact, the physiological gut development is characterized by long, slender villi and shallow crypts. Indeed, longer villi are generally associated with increased total luminal absorptive area, representing useful indicators of gut proliferative and absorptive compartments. Moreover, shallower crypts reflect the prolonged survival of villi without the need for renewal, reducing energy expenditure for this process and maximizing nutrient utilization for growth (Laudadio *et al.*, 2012). On the contrary, lower and thickened villi and greater Cd are normally associated to poor digestion and absorption of nutrients and, as a consequence, poor growth performances of the animals. For this reason, mucosa thickness and the total area of nutrient absorption can also be evaluated as an index of the absorptive capacity of the gut. It is also important to consider the muscular layer thickness, as it controls the motility of the gut and it can affect the absorption processes (Ducatelle *et al.*, 2018).

Moreover, tight junctions (TJ) are multi-protein complexes that seal epithelial cells together, resulting in epithelial surfaces of different tightness, dynamically regulating paracellular transport of molecules (*e.g.*, water, nutrients, electrolytes) from the intestinal lumen through the epithelium into the bloodstream (Shen *et al.*, 2008). They are required for the maintenance of the mucosal barrier in order to prevent the entry of potential pathogens and result in proper nutrient absorption and utilization, leading to optimal health and performance. They can be divided into two categories: transmembrane proteins (occludins and claudins) and membrane-associated proteins (zona-occludens) that are thought to assemble into stable complexes to determine function. Within this network of proteins, pores are present to facilitate transport of water, ions, and small nutrients (Shen *et al.*, 2008). The integrity of tight junction can be assessed through immunofluorescence, immunohistochemistry or genes biomolecular expression.

Mucous layer completed the physical mucosal barrier of the gut, representing the first line of defence against pathogens. The top layer of this mucus gel surface is composed by mucins, large glycoproteins that compete with bacteria for adherence via heterogenous oligosaccharide chains, thereby preventing noxious agents from coming into contact with the underlying epithelial cells (Belley *et al.*, 1999). However, simultaneously, mucin provides a desirable environment for proliferation of specific microflora due to their high carbohydrate content (Deplancke and Gaskins, 2001). Mucins are encoded by mucin genes (MUC) and they are secreted by goblet cells located in the crypts of the small and large intestine. Histochemically, mucins can be separated into 2 broad categories: neutral and acidic (Elia *et al.*, 2018). In swine and poultry, acidic mucins can be furtherly divided and classified into sialomucins or sulfomucins based on the presence of terminating sialic acid or sulfate groups (Biasato *et al.*, 2020a; Biasato *et al.*, 2018). For these reasons, the integrity of the mucous layer can be evaluated through multiple techniques: counting of goblet cells, histochemical staining of mucins and biomolecular expression of MUC genes.

The gastrointestinal tract is also constantly exposed to infectious and non-infectious stressors which makes it an active immune organ containing more resident immune cells than any other organ in the host. The mucosal immune system can be separated into two branches: the innate immune response mediated by epithelial cells in mucosal surfaces and phagocytic cells recruited from the blood and the acquired/adaptive immune response depending on the recognition of antigens by B and T lymphocytes (Honda and Littman, 2016). The local immune response can be mainly assessed through the histological evaluation of inflammation or the biomolecular quantification of cytokines, receptors and immunoglobulins involved in the immune response (Colombino *et al.*, 2021a; Jin *et al.*, 2021; Yu *et al.*, 2019).

Finally, the gut is inhabited by a diverse bacterial community known as “microbiota”. The genome of microorganisms, symbiotic and pathogenic, living in the gut is called “microbiome”. The digestive process is strongly linked to gut microbiota as nutrient absorption, feed digestibility, energy harvest and, therefore, productivity are influenced by its composition and diversity, which also changes among species (Kogut and Arsenault, 2016). To date, most intestinal bacteria are capable of hydrolysing polysaccharides, oligosaccharides, and disaccharides into primary sugars and then ferment them, producing short-chain fatty acids, which are an important source of energy for the gut wall. Intestinal bacteria can also contribute to the metabolism of nitrogen, permitting its absorption and use for amino acids synthesis (Tellez *et al.*, 2006). Moreover, the same intestinal bacteria-particularly lactic acid bacteria- can themselves be a source of vitamins (LeBlanc *et al.*, 2013). Also, the microbiota is directly engaged in maintaining the functional innate immunity of the host, which senses the intestinal microenvironment to determine the metabolic state and colonization status. Lastly, intestinal microbiota competes with the colonizing pathogenic bacteria and is able to reduce the adhesion and colonization of pathogens in the gut (Clavijo and Flórez, 2018). Different factors can influence the composition and function of the microbiota, among which diet is the predominant one (Bailey, 2019). For this reason, it is important to evaluate how dietary ingredients can affect the composition of gut microbiota.

4 Monogastric

4.1 *Poultry*

In poultry, insects have been proposed as valuable feed ingredients for different purposes. Firstly, they have mainly been administered as BSF or YM larvae or pre-pupae meals to replace the traditional protein sources (*i.e.*, soybean meal) (Biasato *et al.*, 2016; Dabbou *et al.*, 2018; Elahi *et al.*, 2022; Gariglio *et al.*, 2019). Secondly, a few studies have also included *BSF* fats as a dietary replacement of soybean oil thanks to their advantageous fatty acid profile, rich in medium-chain fatty acids (Schiavone *et al.*, 2018). Thirdly, the potential use of insect live larvae has been proposed as an environmental enrichment in intensive turkey (Veldkamp and van Niekerk, 2019), broiler chicken (Bellezza-Oddon *et al.*, 2021) and laying hen (Star *et al.*, 2020) farming systems to reduce aggressive behaviours and increase bird mobility. Finally, both BSF and YM seem to have a prebiotic activity due to their chitin content, which seems to have immunostimulant and antimicrobial properties against some of the most important enteric pathogens in poultry production (*i.e.*, *Escherichia coli* and *Salmonella enteriditis*), thus increasing the bird gut health (Veldkamp *et al.*, 2021). To date, not only broiler chickens and laying hens have been used as target species for insect utilization, but also ducks, quails and turkeys have been taken into consideration (Biasato *et al.*, 2016; Colombino *et al.*, 2021a; Cutrignelli *et al.*, 2018; Gariglio *et al.*, 2019; Hartinger *et al.*, 2022; Sypniewski *et al.*, 2020; Wallace *et al.*, 2018). Dietary BSF meal and fat inclusion ranges from 1 to 15% (Biasato *et al.*, 2020b; He *et al.*, 2021) and from 2.5 to 6.8%, respectively (Schiavone *et al.*, 2018; Sypniewski *et al.*, 2020). Differently, YM meal has been included in rates ranging from 2.5 to 29.6% (Bovera *et al.*, 2016; Sedgh-Gooya *et al.*, 2022). Regarding insect live larvae, they were administered based on 5% (Bellezza-Oddon *et al.*, 2021) and 10% (Veldkamp and van Niekerk, 2019) of the expected daily feed intake.

Different parameters have been evaluated so far to verify the feasibility of using insects in poultry nutrition, including diet digestibility and gut health-related parameters such as gut morphology, mucin composition, intestinal immune response and microbiome. It is important to highlight that in poultry industry a diet should guarantee adequate growth performance along with an optimal animal health in order to make it economically advantageous. This explains why when a new dietary ingredient-including insect meal- is tested, digestibility must be determined. Digestibility refers to the amount of nutrient absorbed by the individual and it is generally calculated as the amount of nutrient consumed minus the amount of nutrient retained in the faeces. Overall, the insect meals have been reported to display good to high nutrient digestibility in poultry, with apparent digestibility coefficients (ADCs) for dry matter (DM) ranging from 53.0 to 77.6, for crude protein (CP) from 51.0 to 84.8, and for ether extract (EE) from 88.0 to 99.0 (Lee *et al.*, 2022).

As far as gut barrier function is concerned, it seems that dietary inclusion levels of BSF and YM meal, oil or live larvae up to 10% is able to maintain or improve gut morphology, mucin composition, and local immune response. Particularly, non-significant alterations of Vh, Cd, and mucosa, submucosa or tunica muscularis thickness have been recorded in the duodenum, jejunum or ileum of broiler chickens, free range chickens, ducks and laying hens fed YM (0.75, 2.5 ,5, 7.5, 10%) or BSF (3, 5, 9 and 10%) meals when compared to control groups fed a soybean meal-based diet (Biasato *et al.*, 2016; I. Biasato *et al.*, 2018; Chu *et al.*, 2020; Dabbou *et al.*, 2018; Gariglio *et al.*, 2019; Kawasaki *et al.*, 2019; Moniello *et al.*, 2019; Schiavone *et al.*, 2018; Sedgh-Gooya *et al.*, 2022; Sypniewski *et al.*, 2020). Moreover, all the morphometric parameters showed a proximo-distal decreasing gradient from the duodenum to the ileum. On the contrary, in some cases, 5% inclusion of BSF meal increased villus area and villus width in the duodenum and jejunum of broiler chickens (Hartinger *et al.*, 2022, 2021). The above-mentioned findings indicate a low turnover rate of the intestinal mucosa, thus suggesting that insect meal-containing diet can maintain the physiological development of the gut. Furthermore, the maintenance of the proximo-distal decreasing gradient form duodenum to ileum is in accordance with the different intensity of nutrient absorption processes that physiologically take place in each intestinal segment (Laudadio *et al.*, 2012).

Similarly, inclusion rates up to 5% do not modify the mucin staining intensity both in the villi and the crypts, as well as they do not influence the number of goblet cells along the gut (Biasato *et al.*, 2020c, 2020b; Colombino *et al.*, 2021a; Hartinger *et al.*, 2022). However, mucin staining intensity show a proximo-distal increasing gradient from the duodenum to the ileum. The preservation of the mucus layer is fundamental for an optimal gut health, as it protects the mucosa from pathogens. Indeed, mucins are the primary component of the mucous layer, and they are secreted by goblet cells, whose increase in density from the duodenum to the ileum is in accordance with the digestive physiology (Jiang *et al.*, 2013). In fact, the proximal part of the small intestine may prioritize the proliferation of absorptive cells over secreting cells, which can be associated with a lower goblet cells density, lower mucus secretion and larger goblet cells size in the duodenum when compared to the jejunum and ileum. Furthermore, the number and activity of bacteria, along with the digesta retention time, increase distally in the small intestine, which can explain the higher goblet cells density in the ileum compared to the other small gut segments (Duangnumsawang *et al.*, 2021).

Considering the gut local immune response, all the studies confirmed that histologically insect-based diets do not increase intestinal inflammation. On the contrary, evidence of the effects of insect meal in poultry immune-related gene expression is still scarce and extremely heterogeneous, demonstrating that the interpretation of the immune response is difficult, as there is an active, homeostatic balance between pro-inflammatory and anti-inflammatory responses continuously occurring in the gut (Johnson *et al.*, 2020). The use of YM live larvae in broiler chickens (5% of the expected daily feed intake) seems to reduce the expression of the pro-inflammatory cytokine IL-2 in the jejunum, but it does not affect the expression of IL-4, IL-6,

INF-γ and TNF-α (Colombino *et al.*, 2021a). Differently, 3, 6 and 9% inclusion levels of BSF meal increase the expression of sIgA in the ileum, while IL-2, IL-6 and TNF-α remain unchanged (Chu *et al.*, 2020). Further studies are needed to clarify the mode of action of insects on the local immune response gene regulation, even though the lack of inflammation, the reduction of IL-2, and the increase in sIgA can be considered positive outcomes attributable to chitin prebiotic action, which is able to modulate immune response when degraded in chitosan in the gut by the activation of the Th-2 helper pathway (Veldkamp *et al.*, 2021).

On the contrary, insect meal dosages higher than 10% (from 15 to 29%) have been reported to reduce diet digestibility, Vh, Vh/Cd and mucin staining intensity, and increase Cd and the overall length and weight of the small intestine of broiler chickens (Biasato *et al.*, 2018; Biasato *et al.*, 2018; Bovera *et al.*, 2016; Dabbou *et al.*, 2018; Moniello *et al.*, 2019). Similarly, laying hens fed 17% of BSF meal showed a reduction in Vh in the jejunum, but an increase in Vh in the ileum (Cutrignelli *et al.*, 2018). These negative effects can be attributed to the higher chitin content in the diet. In fact, it has been demonstrated that dietary chitin inclusion levels higher than 5.0 g/kg is able to decrease the efficiency of nitrogen utilization, increasing the viscosity of the ingesta, decreasing the motility of the gastrointestinal tract and, as a consequence, depressing gut morphology and growth performance (Lokman *et al.*, 2019). This is due to the low activity of chitinase enzymes in the chicken gut, where they are responsible for chitin degradation (Tabata *et al.*, 2018). In broilers, it is well documented that feeding a poorly digestible diet results in an increase in intestinal length, relative volume and weight (van der Klis and Jansman, 2002). Thus, the lower digestibility of the diet reduced the nutrient uptake for the enterocyte turn-over, the villi height decreased, and this loss was partially compensated by increased proliferation of crypt stem cells, thus resulting in increased Cd as the inclusion level of the insect meal increased (Ducatelle *et al.*, 2018). Another important consideration is that the ileal villi in chickens are smaller and lower than those of the previous tracts of the small intestine, as in poultry fed corn-soybean based diets very little amount of nutrients are available beyond the jejunum (Moniello *et al.*, 2019). However, an increased load of nutrients directly deriving from the duodenum to the ileum as a consequence of a reduction in diet digestibility might stimulate the ileal absorptive function, resulting in a compensatory development of the villi and, in turn, explaining the increase in Vh in chickens fed high inclusion rates of insect meal (Yamauchi, 2007).

The use of different types of insect-based products has also been reported to significantly modulate the intestinal microbiome of birds, with the distinction between the impact of the low and the high inclusion levels being, however, less evident when compared to the other gut health parameters. Including different levels of YM (from 5 to 15%) and BSF (from 5 to 25%) meals in diets for free-range and broiler chickens, as well as laying hens, led to unaffected (Biasato *et al.*, 2020c, 2019a; Ndotono *et al.*, 2022) or increased α-diversity (in terms of higher Shannon index [Biasato *et al.*, 2018; Borrelli *et al.*, 2017], Chao1 index [Kawasaki *et al.*, 2019]) and number of observed species (Borrelli *et al.*, 2017) than that of animals fed the control diet. Even if limited microbial diversity may be desirable as not all microbes are beneficial, high α-diversity of gut microbiota is generally favorable for the overall health and productivity of livestock animals (Gong *et al.*, 2019). Significant changes in β-diversity and relative abundances of phyla have also been highlighted after administering insect-based diets, as higher Firmicutes (Biasato *et al.*, 2018; Ndotono *et al.*, 2022), as well as lower Bacteroidetes and higher Firmicutes to Bacteroidetes ratios (Biasato *et al.*, 2018), are characteristics of insect-fed animals when compared to the control groups. This phylum profile is, in turn, mainly driven by the selection of specific genera – such as *Clostridium* (Biasato *et al.*, 2020b, 2019a; Biasato *et al.*, 2018), *Ruminococcus* (Biasato *et al.*, 2020b; Biasato *et al.*, 2018), *Lactobacillus* (Biasato *et al.*, 2020b; Ndotono *et al.*, 2022), *Oscillospira* and *Coprococcus* (Biasato *et al.*, 2018), *Alistipes* (Biasato *et al.*, 2019a), and *Faecalibacterium*, *Blautia* and *Roseburia* (Biasato *et al.*, 2020b) – that are able to produce short-chain fatty acids (SCFAs). The production of SCFAs – in particular butyrate – can enhance gut epithelial cell barrier function by acting as energy source for enterocytes and stimulating goblet cell differentiation and mucus production (Jha *et al.*, 2019), as well as reducing the enteric pathogens as a consequence of their antimicrobial properties (Bedford and Gong, 2018). Such genera signature seems to be related to the ability of SCFAs-producing bacteria to degrade the chitin, as selected species – such as *Alkaliphilus transvaalensis*, *Christensenella minuta*, and *Flavonifractor plautii* – have previously been identified in gut microbiome of BSF-fed laying hens in association with high levels of β-N-acetylhexosaminidases (K01207) and N-acetylglucosamine 6-phosphate deacetylase (K01443) (the key enzymes that cleave oligomers produced by chitinase) and increased butyrate and propionate production (Borrelli *et al.*, 2017). However, high inclusion levels of BSF and YM meals (*i.e.*, 15%) in diets for broiler chickens may negatively modulate their gut microbiota, in terms of reduced α-diversity, increased Proteobacteria and selection of *Helicobacter* (Biasato *et al.*, 2020b), and decreased Firmicutes and Firmicutes to Bacteroidetes ratios (Biasato *et al.*, 2020c), respectively. Similarly, independently of the inclusion levels, the use of full-fat YM and BSF meals has been reported to decrease the relative abundance of SCFAs-producing bacteria (*i.e.*, *Clostridium*, *Coprococcus*, and *Ruminococcus*) in broiler chickens (Biasato *et al.*, 2020c, 2019a), as well as in laying hens (*i.e.*, *Lactobacillus* and *Bifidobacterium*) (Kawasaki *et al.*, 2019). These outcomes can reasonably be attributed to the chitin-related reduction in CP digestibility, with the increase in nondigested protein at ileal level leading to hindgut protein fermentation and, in turn, formation of toxic compounds potentially capable of creating a non-healthy gut environment (Biasato *et al.*, 2020b). Finally, the use of insect fats (BSF; Dabbou *et al.*, 2021), live larvae (BSF and YM; Colombino *et al.*, 2021a) and insect meals as feed additives (YM and *Zophobas morio*; Józefiak *et al.*, 2020) seem to exert positive effects, in terms of reduction in potential pathogens (such as *Corynebacterium* Dabbou *et al.*, 2021), increase in SCFAs-producing bacteria (such as *Clostridium*, *Eubacterium* and Victivillaceae [Colombino *et al.*, 2021a], and *Lactobacillus* and *Bifidobacterium* [Józefiak *et al.*, 2020]).

4.2 *Fish*

In fish, insect meals and oils obtained from BSF and YM have been tested in both the marine and the freshwater fish including the rainbow trout (*Oncorhynchus mykiss*), the Siberian sturgeon (*Acipenser baerii*), the Atlantic salmon (*Salmo salar*), the pearl gentian grouper (*Epinephelus fuscoguttatus*), the barramundi (*Lates calcarifer*), the European sea bass (*Dicentrarchus labrax*), *the* Jian (*Cyprinus carpio* var. Jian) and the grass carp (*Ctenopharyngodon idellus*) (Basto *et al.*, 2021; Caimi *et al.*, 2021; Hender *et al.*, 2021; Huang *et al.*, 2022; Li *et al.*, 2017; Lu *et al.*, 2020; Weththasinghe *et al.*, 2021). Variable dosages of BSF and YM meal has been included in the feed as a replacement of fishmeal (FM), ranging from 3 to 50% and from 15 to 60%, respectively (Basto *et al.*, 2021; Caimi *et al.*, 2021; Józefiak *et al.*, 2019; Lu *et al.*, 2020; Zarantoniello *et al.*, 2021). Furthermore, BSF fats have been tested in aquaculture as a substitute of fish oil, with an inclusion rate varying from 1.5 to 10%. It can be noticed that higher inclusion levels of insect meals have been tested in aquaculture when compared to poultry. Apart from the different nutritional requirements, this is also potentially due to the species-specific ability to digest chitin. In fact, fish ingest chitinous substances in their natural diets, such as crabs, shrimp, and zooplankton, and they have higher chitinase activities in either the stomach or the intestine, which allow them to disrupt the exoskeleton of crustaceans and insects and give to other digestive enzymes access to nutrient-rich inner tissues (Gutowska *et al.*, 2004). Therefore, in fish the enzymatic hydrolysis of chitin has the potential to result in additional energy gain from a meal (Ikeda *et al.*, 2017). However, chitinase activity changes among different fish species, being influenced by their dietary habits (Ikeda *et al.*, 2017). This is probably the reason why insect meal has mainly been tested in carnivorous and omnivorous fish, which normally eat chitin-containing organisms in their diet, and for which vegetable protein sources as replacement of FM have the most detrimental effects on growth performance and intestinal function (Caimi *et al.*, 2020). Overall, the insect meals have been reported to display good to high nutrient digestibility in fish, with apparent digestibility coefficients (ADCs) for dry matter (DM) ranging from 61.7 to 95.8, for crude protein (CP) from 58.3 to 92.4, and for ether extract (EE) from 82.4 to 98.8 (Lee *et al.*, 2022).

As already mentioned for poultry, also in fish the effects of insect meals and oils on gut health parameters have been evaluated, mainly focusing on gut morphology, mucin, local immune response and microbiome. In fact, intestinal integrity is fundamental for nutrient digestion and absorption and, thus, for growth performance, which remains a priority for the aquaculture sector (Caimi *et al.*, 2021). In contrast to what has been observed in poultry, in aquaculture it is still not possible to determine a cut-off for the inclusion level of insect meals in the feed as the results are extremely heterogenous, probably due to the different type of insect meal (full-fat vs. defatted), the feeding regime to which fish were subjected (restricted vs. apparent satiation), the manufacturing of the feeds (pelleting vs. extrusion) and, of course, the fish species investigated, as well as the stage of development (juvenile vs. adult) (Cardinaletti *et al.*, 2019).

Regarding intestinal morphology, the main parameters that have been evaluated are Vh, villus width, and mucosa, lamina propria, submucosa and muscular thickness of anterior, middle and posterior gut. It is important to highlight that in fish the distal intestine is the main site of digestion and absorption of nutrients, playing a key role in guarantying an adequate growth of the animals (Huang *et al.*, 2022). Considering this aspect, the identification of non-significant variations in the distal intestine morphology of Atlantic salmon (*Salmo salar*) (Weththasinghe *et al.*, 2021) and rainbow trout (*Oncorhynchus mykiss*) fed BSF meal or oil can be considered a positive outcome (Caimi *et al.*, 2021; Cardinaletti *et al.*, 2019; Dumas *et al.*, 2018; Melenchón *et al.*, 2022; Palomba *et al.*, 2022; Renna *et al.*, 2017). On the contrary, contrasting results have been obtained in the proximal intestine. In some studies, no changes were recorded when trout were fed up to 40% BSF meal (Caimi *et al.*, 2021; Palomba *et al.*, 2022; Renna *et al.*, 2017). However, reduction in Vh was observed in rainbow trout receiving 26.4% of BSF meal (Dumas *et al.*, 2018). The same reduction in Vh was also recorded in the middle gut of trout fed 25 and 50% of BSF meal (Cardinaletti *et al.*, 2019). In contrast with the heterogenous results obtained in salmonids, in adult Siberian sturgeon (*Acipenser baerii*) dietary highly defatted BSF meal inclusion up to 37.5% did not affect Vh in the proximal and distal intestine while the inclusion of 15% of full-fat BSF meal reduced the thickness of mucosa and increase the muscle layer thickness, changes that have been associated with an enhancement in the digestion and absorption process (Caimi *et al.*, 2020; Józefiak *et al.*, 2019). Indeed, Siberian sturgeon juveniles fed 50% of BSF meal showed a significant reduction in Vh in the small intestine (Zarantoniello *et al.*, 2021). These results seem to suggest a dose-dependent effect of insect meal in this fish species, probably due to the chitin content of the diet. Moreover, juveniles can have a lower chitinase activity in their immature gastrointestinal system, which can provoke this worsening of Vh when a high-chitin containing diet is administered. These differences can also be attributed to the highly defatted vs full-feat BSF meal used in the different studies, as the lipid fraction can furtherly impair the digestibility of the diet. Less studies are available on other fish species. In particular, one study is available in the Barramundi (*Later calcarifer*), in which distal intestine morphology was not affected by the use of BSF meal, oil or a combination of them in the diet (Hender *et al.*, 2021). Similarly, the same results have been obtained in the only study conducted in the European sea bass (*Dicentrarchus labrax*) (Basto *et al.*, 2021). Conversely, Vh and muscular thickness were reduced in BSF meal-based diets in the pearl gentian grouper (*Epinephelus fuscoguttatus*) and in the grass carp (*Ctenopharyngodon idellus*) (Huang *et al.*, 2022; Lu *et al.*, 2020). Overall, the morphometrical results herein reported suggest that insect meal does not alter the nutrient digestion and absorption processes in fish. In fact, despite the reduction in Vh recorded in some studies, fish growth performance remains comparable to those of control group, thus suggesting that this reduction did not negatively affect gut digestive functions. However, further studies are necessary to clarify the potential mechanism behind these villi shortening.

More consistent results have been obtained regarding intestinal mucins. Regardless of the fish species, diets containing up to 40% of BSF meal maintain the mucin staining intensity along the gut, without altering the number of goblet cells (Basto *et al.*, 2021; Caimi *et al.*, 2020; Elia *et al.*, 2018; Huang *et al.*, 2022). On the contrary, dietary BSF meal inclusion level of 50% reduce the number of goblet cells in the distal gut or induce their hyperplasia (Cardinaletti *et al.*, 2019; Zarantoniello *et al.*, 2021). These findings seem to suggest a dose-dependent effect of insect meal due to its chitin content. In fact, when the chitin contained in the diet overcomes the gastrointestinal digestive capacity, it can increase the mucous shedding in the lumen, thus reducing goblet cells number and, in turn, potentially inducing dysbiosis and inflammation. As a consequence, the remaining mucous cells undergo compensatory hyperplasia in order to provide sufficient mucous to protect the gut mucosa.

Moreover, BSF meal does not seem to increase the intestinal inflammation from a histological perspective (Caimi *et al.*, 2021, 2020). However, from a biomolecular point of view, it increases the expression of myd88 (key junction molecule in the toll-like receptor signalling pathway, which induce the expression of inflammatory cytokines), tlr22 (an indicator in activating the immune system, required for downstream signalling), IL-10 and IL-1β in both the middle and distal gut of rainbow trout, Barramundi and the pearl gentian grouper. However, it does not affect TNF-α transcription levels (Cardinaletti *et al.*, 2019; Hender *et al.*, 2021; Huang *et al.*, 2022; Melenchón *et al.*, 2022). It has been hypothesized that BSF meal can activate the TLR22/MYD88/NF-κB signalling pathway, inducing intestinal inflammation. As a consequence, the increased intestinal inflammation promotes the expression of IL-10, an anti-inflammatory factor. This hypothesis provides interesting insight on the underlying immune-modulatory properties of chitin through a self-limiting mechanism (Huang *et al.*, 2022).

As already described for poultry species, the use of different insect meals and oils has also been reported to significantly modulate the intestinal microbiome of fish. Within this scenario, it is interesting to notice that, despite the wide heterogeneity of fish species being considered (rainbow trout [Biasato *et al.*, 2022; Gaudioso *et al.*, 2021; Rimoldi *et al.*, 2021, 2019; Terova *et al.*, 2021, 2019], sea bass [Panteli *et al.*, 2021; Pérez-Pascual *et al.*, 2020; Pleić *et al.*, 2022; Rangel *et al.*, 2022], Atlantic salmon [Li *et al.*, 2021; Weththasinghe *et al.*, 2022], Siberian sturgeon [Józefiak *et al.*, 2019; Zarantoniello *et al.*, 2021], sea bream [Panteli *et al.*, 2021] and olive flounder [Seo *et al.*, 2022]), the way of action of insect-based products seems to be quite constant.

Including different levels of YM (from 15 to 25%), BSF (from 3.50 to 60%) and *Musca domestica* (MD; 19.5%) meals – as well as BSF oil (6.24%) – in diets for fish led to unaffected (Gaudioso *et al.*, 2021; Li *et al.*, 2022; Pérez-Pascual *et al.*, 2020; Rangel *et al.*, 2022; Rimoldi *et al.*, 2021; Seo *et al.*, 2022; Zarantoniello *et al.*, 2021) or increased α-diversity (in terms of higher Shannon [Li *et al.*, 2021; Rimoldi *et al.*, 2019; Terova *et al.*, 2019], Simpson [Rimoldi *et al.*, 2019; Terova *et al.*, 2019] and Chao1 [Biasato *et al.*, 2022] indices, and observed species richness [Pleić *et al.*, 2022; Terova *et al.*, 2019]) than that of animals fed the control diet. Significant changes in β-diversity and relative abundances of phyla have also been highlighted after administering diets containing insect meals, as increased Firmicutes (Panteli *et al.*, 2021; Pleić *et al.*, 2022; Rangel *et al.*, 2022; Rimoldi *et al.*, 2021; Weththasinghe *et al.*, 2022) and decreased Proteobacteria (Rimoldi *et al.*, 2021, 2019; Terova *et al.*, 2021, 2019; Weththasinghe *et al.*, 2022) are constantly identified. Differently, more heterogenous outcomes can be outlined for Actinobacteria (Biasato *et al.*, 2022; Panteli *et al.*, 2021; Terova *et al.*, 2019) and Bacteroidetes phyla (Biasato *et al.*, 2022; Panteli *et al.*, 2021; Rangel *et al.*, 2022). These phyla changes may sometimes logically result in higher Firmicutes to Bacteroidetes and lower Proteobacteria to Bacteroidetes ratios in insect-fed fish when compared to those fed the control diet (Panteli *et al.*, 2021). Analogously to what was observed in poultry, this phylum profile is mainly driven by the increase in SCFAs-producing bacteria, such as *Actinomyces* (Biasato *et al.*, 2022; Gaudioso *et al.*, 2021; Li *et al.*, 2021; Rangel *et al.*, 2022; Rimoldi *et al.*, 2021; Weththasinghe *et al.*, 2022), *Bacillus* (Biasato *et al.*, 2022; Gaudioso *et al.*, 2021; Li *et al.*, 2022, 2021; Pleić *et al.*, 2022; Rangel *et al.*, 2022; Rimoldi *et al.*, 2021; Weththasinghe *et al.*, 2022), *Enterococcus* (Biasato *et al.*, 2022; Gaudioso *et al.*, 2021; Li *et al.*, 2022, 2021; Rangel *et al.*, 2022; Terova *et al.*, 2019), *Lactobacillus* (Józefiak *et al.*, 2019; Pleić *et al.*, 2022; Rimoldi *et al.*, 2021; Weththasinghe *et al.*, 2022), *Staphylococcus* (Biasato *et al.*, 2022; Panteli *et al.*, 2021), *Mycoplasma* (Rimoldi *et al.*, 2019), *Pediococcus* (Terova *et al.*, 2019), and *Clostridium coccoides – Eubacterium rectale* cluster and *Carnobacterium* (Józefiak *et al.*, 2019). However, the concomitant identification of increased and decrease SCFAs-producing bacteria in BSF-fed rainbow trout has recently suggested the hypothesis of an BSF ability to select specific genera at the expense of other ones (Biasato *et al.*, 2022). In parallel, the use of insect meals is also able to increase the relative abundance of bacteria that are characteristics of the fish gut (*i.e.*, Enterobacteriaceae, *Anaerococcus*, *Cutibacterium*, *Pseudomonas*, *Hafnia*, and *Aeromonas*), thus suggesting their ability to preserve the physiological gut microbiota in fish (Józefiak *et al.*, 2019; Panteli *et al.*, 2021). Another positive aspect – which is not clearly observed in poultry – is represented by the insect-related decrease in several pathogens, for both the fish (such as *Deefgea* [Rimoldi *et al.*, 2019; Terova *et al.*, 2021], *Streptococcus* [Fabrikov *et al.*, 2021], and *Vagococcus* and *Lactococcus* [Biasato *et al.*, 2022]) and the final consumer (such as *Listeria* and *Campylobacter* [Biasato *et al.*, 2022]). This outcome was mainly observed in BSF-fed fish, as a consequence of the synergic influence of BSF lauric acid, chitin, and antimicrobial peptides (AMPs), which both concur to its antimicrobial properties (Biasato *et al.*, 2022). The overall improvement in fish gut microbiota has also a functional relevance, as metagenomics has recently allowed highlighting an up-regulation in carbohydrate metabolism (Panteli *et al.*, 2021; Rimoldi *et al.*, 2021; Weththasinghe *et al.*, 2022), AA metabolism (Panteli *et al.*, 2021; Weththasinghe *et al.*, 2022) and FA synthesis (Weththasinghe *et al.*, 2022), and a down-regulation of lipopolysaccharide biosynthesis (Weththasinghe *et al.*, 2022), as well as a *Paenibacillus*-related increase in chitinase ChiA-encoding genes (Rangel *et al.*, 2022). Very few minor negative effects can also be identified, in terms of a reduction in α-diversity (Chao1 index in MD-fed sea bream and sea bass [Panteli *et al.*, 2021], as well as Shannon index in YM- and BSF-fed rainbow trout [Biasato *et al.*, 2022; Terova *et al.*, 2021]) decrease in characteristics bacteria (*i.e.*, *Shewanella*, *Citrobacter*, *Kluyera*, and *Aeromonas* [Rimoldi *et al.*, 2021, 2019]), and a down-regulation in vitamin metabolism (Weththasinghe *et al.*, 2022). However, these negative findings have always been observed along with predominant positive outcomes, thus reasonably counteracting them. Finally, no-insect related effects on gut microbiota may also be outlined in sea bass (Pérez-Pascual *et al.*, 2020), Siberian sturgeon (Zarantoniello *et al.*, 2021), and olive flounder (Seo *et al.*, 2022).

4.3 *Pig*

In pigs, the use of insect meal as an alternative protein source has been less extensively studied than in poultry and aquaculture. As a consequence, only a few studies are available on the use of BSF and YM meal in weaned pigs, finishing pigs and barrow pigs, mainly focusing on digestibility and growth performance to evaluate the feasibility of these new ingredients (Biasato *et al.*, 2019; Jin *et al.*, 2016; Newton *et al.*, 1977; Yu *et al.*, 2019). The inclusion rate for BSF and YM in swine ranges from 4 to 33% and from 1.5 to 10%, respectively. The available studies confirm that, overall, the inclusion of insect meals into swine diets is able to maintain or improve growth performance thanks to a higher or unaffected diet digestibility when compared to soybean meal-based diet (Biasato *et al.*, 2019; Jin *et al.*, 2016; Spranghers *et al.*, 2018). On the contrary, higher inclusion levels (*i.e.*, 33%) may decrease the nutrient digestibility, but only when full-fat meals are used (Newton *et al.*, 1977). Furthermore, when focusing on the impact of insect-based products on pig gut health, no clear negative effects have been highlighted even with the highest inclusion levels (*i.e.*, 19%).

In recent years, also in pigs some studies have evaluated the effects of feeding insect meal on gut health. The evaluated parameters were the same already mentioned above for poultry and aquaculture (gut morphology, mucin, local immune response and microbiota). To date, 5 and 10% BSF meal inclusion do not alter gut morphology of weaned pigs in terms of Vh, Cd and Vh/Cd (Biasato *et al.*, 2019; Spranghers *et al.*, 2018). Therefore, it is reasonable to hypothesize that insect meal utilization does not negatively influence gut development or nutrient absorption, maintaining the integrity of the intestinal barrier. Moreover, the integrity of the intestinal barrier has been also studied through the biomolecular expression of the tight junctions. In finishing pigs, 4 and 8% dietary inclusion of BSF meal increased the expression of Zona-occludens 1 and occludin- two of the main intestinal tight junctions- both in basal conditions and during entherotoxigenic *Escheria coli* infection, thus suggesting that BSF meal may enhance the integrity of the gut mucosal barrier by increasing its ability to protect the gut against pathogens (Jin *et al.*, 2021; Yu *et al.*, 2019). On the contrary, 5 and 10% dietary inclusion of YM meal did not influence tight junction expression in the ileum (Meyer *et al.*, 2020).

The evaluation of mucin staining intensity and the biomolecular expression of the mucin-producing related gene (MUC-1) has been also performed in BSF-fed pigs. Particularly, 5 and 10% BSF meal inclusion increased the neutral mucin staining percentage when compared to control group, with intestinal neutral mucins of the BSF animals being also higher than the sialomucins and the sulfomucins found in the gut of the control group (Biasato *et al.*, 2020a). The physiologic relevance of the distinct mucin subtypes has not yet been well understood in pigs (Lallès *et al.*, 2004). However, increase in neutral mucins during post-weaning has been suggested to be related to the physiological goblet cell differentiation and normal maturation (Turck *et al.*, 1993). Therefore, a predominance of neutral mucins is considered a positive finding, indicative of an increased intestinal maturity that facilitates the breakdown of complex nutrients (Biasato *et al.*, 2020a). Moreover, pigs fed 4% BSF meal also showed an increased MUC-1 transcription level, thus confirming that BSF inclusion in diet can enhance the mucosal barrier (Yu *et al.*, 2019).

Regarding the intestinal local immune response, a few studied investigated the effects of BSF and YM meal on small intestine inflammation and colonic mucosal response. In particular, it has been shown that BSF larvae do not enhance small intestine inflammation and they down-regulated mRNA expression of TLR4 and pro-inflammatory cytokine IFN-γ, also up-regulating the expression of anti-inflammatory cytokine IL-10 in the colon (especially supplementation with 4%) in basal conditions (Biasato *et al.*, 2019; Yu *et al.*, 2019). Moreover, 4 and 8% of BSF meal dietary inclusion increased ileal IL-10 and decreased TNF-α in weaned pigs experimentally infected with enterotoxigenic *Escherichia coli* K88 (Jin *et al.*, 2021). Similarly, YM meal down-regulated pro-inflammatory TNF in the ileal mucosa (Meyer *et al.*, 2020). It is well known that Toll-like receptors can mediate innate immune responses, recognize microbiota and their products, and then initiate pro-inflammatory pathways (Yu *et al.*, 2019). This is accordance with previous findings in fish (Huang *et al.*, 2022), and it confirms the potential immune-stimulant self-limiting mode of action of chitin.

As already described for both the poultry and fish species, the use of different insect meals has also been reported to significantly modulate the intestinal microbiota of pigs. Including different levels of YM (from 5 to 10%) and BSF (from 5 to 30%) meals in diets for weaned, growing and finishing pigs led to unaffected

(Biasato *et al.*, 2020a; Håkenåsen *et al.*, 2021; Jin *et al.*, 2021; Meyer *et al.*, 2020; Yu *et al.*, 2019) or increased α-diversity (in terms of higher Shannon index [Kar *et al.*, 2021]) than that of animals fed the control diet. Significant changes in β-diversity and relative abundances of phyla have also been highlighted after administering diets containing insect meals, as higher Actinobacteria (Kar *et al.*, 2021; Meyer *et al.*, 2020) and lower Firmicutes (Kar *et al.*, 2021), as well as heterogeneous Bacteroidetes variations (Håkenåsen *et al.*, 2021; Meyer *et al.*, 2020), are characteristics of insect-fed animals when compared to the control groups. Similarly to what was observed for poultry and fish, this phylum profile is, in turn, mainly driven by the selection of SCFAs-producing bacteria, such as Ruminococcaceae (Biasato *et al.*, 2020a; Meyer *et al.*, 2020; Yu *et al.*, 2019), *Bifidobacterium* (Kar *et al.*, 2021; Meyer *et al.*, 2020), *Roseburia* (Biasato *et al.*, 2020a; Yu *et al.*, 2019), *Lactobacillus* (Jin *et al.*, 2021; Yu *et al.*, 2019), *Pseudobutyrivibrio*, *Clostridium* and *Faecalibacterium* (Yu *et al.*, 2019), and *Blautia*, *Coprococcus*, *Eubacterium*, *Prevotella*, *Ruminococcus* and *Staphylococcus* (Biasato *et al.*, 2020a). Despite no metagenomics having been performed yet in pigs fed insect-based diets, the concomitant identification of increased SCFAs production (mainly butyrate, isobutyrate, valeric and isovaleric [Meyer *et al.*, 2020; Yu *et al.*, 2019]) suggests the same utilization of chitin by the gut microbiota already described for poultry and fish. In parallel, dietary BSF meal inclusion is also able to reduce the proliferation of potential enteric pathogens – such as *Streptococcus* (Jin *et al.*, 2021; Kar *et al.*, 2021; Yu *et al.*, 2019) and *Sarcina* (Kar *et al.*, 2021) –, as already reported in fish species. Very few minor negative effects related to the administration of BSF meal (in terms of decreased *Lactobacillus* [Håkenåsen *et al.*, 2021] and selection of potential pathogens such as *Chlamydia* [Biasato *et al.*, 2020a]) were highlighted in weaned piglets, but with no impact on health status and growth performance of animals.

4.4 *Rabbit*

In rabbit nutrition, YM and BSF have mainly been employed as an alternative fat source to replace soybean oil. In fact, intensive rabbit production requires high density of metabolizable energy in the diets to achieve high growth rates, and, simultaneously, diets with high fibre content to prevent the disruption of the sensible rabbit digestive function and health (Gidenne, 2003). Therefore, the inclusion of fat sources in rabbit feeds allows to overcome such constraints (Gidenne *et al.*, 2017). The BSF and YM oils seems to be appropriate fat sources in rabbits, as they are very rich in saturated fatty acids such as the lauric acid, a medium chain fatty acid highly abundant in BSF (Spranghers *et al.*, 2017). Lauric acid is expected to be efficiently absorbed in tissues and it presents antibacterial activities, particularly against Gram-positive bacteria (Dayrit, 2014). Therefore, it may be hypothesized that BSF fat could be efficiently used as an energy source, having also inhibitory effects on abnormal intestinal fermentation in rabbit. For the above-mentioned reasons, different inclusion levels of BSF and YM oils have been included in rabbit feed, ranging from 0.75 to 6% and from 0.75 to 1.50%, respectively (Dalle Zotte *et al.*, 2018; Gasco *et al.*, 2019; Martins *et al.*, 2018). Only one study is available on the use of 4% YM meal in the diet of meat rabbit (Strychalski *et al.*, 2021).

The majority of the studies focused on diet digestibility, growth performance and meat quality parameters (Gasco *et al.*, 2019; Martins *et al.*, 2018; Strychalski *et al.*, 2021). A few studies evaluated gut morphology and microbiota (Dabbou *et al.*, 2020; Gasco *et al.*, 2019). All the studies reported unaltered growth performance after the administration of BSF or YM oils (Gasco *et al.*, 2019; Martins *et al.*, 2018). Conversely, heterogenous results have been obtained regarding digestibility. Particularly, Gasco *et al.* (2019) reported unchanged digestibility in BSF or YM diets, while Martins *et al.*, (2018) observed a reduced digestibility of the insect diets. However, Martins *et al.* (2018) also reported that insect oils do not impair growth performance, suggesting that even though insect oil-containing diets showed a slightly lower digestibility, they still guarantee an adequate nutrient absorption.

Moreover, only Gasco *et al.* (2019) investigated the effect of insect oils on gut morphology, showing non-significant changes when compared to a commercial feed. According to what reported above on unaffected growth performance in insect-fed groups, BSF and YM lipids utilization did not negatively influence gut development and functionality.

As far as the gut microbiota is concerned, apart from the identification of unaffected α-diversity, a specific microbial signature was observed for both the BSF and YM fats (Dabbou *et al.*, 2020). In particular, dietary BSF and YM fat inclusion increased the relative abundance of SCFAs-producing bacteria (such as *Akkermansia* and *Ruminococcus*) – along with total volatile fatty acids – and decreased that of opportunistic pathogens (such), as well as decreasing opportunistic pathogens (such as *Citrobacter*). (In parallel, YM fat utilization and *Klebsiella*). All these positive outcomes were attributed to the antibacterial activities of both the insect species (Dabbou *et al.*, 2020).

4.5 *Ruminant*

Similarly to what has been described above for poultry, fish, pigs and rabbit, also in ruminants insects are gaining attention as alternative feed ingredients. Particularly, they have been proposed as a replacement for conventional protein sources (soybean meal) or oil/energy source thanks to their unsaturated fatty acids profile- rich in medium-chain fatty acids- and the ability of chitosan to modulate ruminal biohydrogenation, decrease the methane production and potentially improve milk fatty acid profile (Rumpold and Schlüter, 2013; Sánchez-Muros *et al.*, 2014).

Preliminary studies have been performed *in-vitro* to evaluate the feasibility of using insect meals or oils in cattle (Ahmed *et al.*, 2021; Jayanegara *et al.*, 2020, 2017), using rumen fermentation models. In particular, Ahmed *et al.*, (2021) evaluated the effects of four different insect species (*Acheta domesticus, Brachytrupes portentosus, Gryllus bimaculatus* and *Bombyx mori*) as an alternative protein source (substitution of 25% of the soybean meal) on diet digestibility, rumen fermentation, and methane production. These inclusion rates did not impair the nutrient digestibility or the production of volatile fatty acids even though they increased the ammonia-nitrogen production. Particularly, the use of *Gryllus bimaculatus* and *Bombyx mori* decrease the methane production. These results are confirmed by the study by Haryati *et al.* (2019) in which a significant decrease of methane emission and rumen fermentation was recorded after the administration of 2% of insect-derived chitosan extract. This can be attributable to the chitosan capacity to reduce methanogenic bacteria in the forestomaches. However, this can potentially have a detrimental effect on dry matter digestibility or acetate production, as already reported for mollusk's chitosan (Belanche *et al.*, 2016; Goiri *et al.*, 2009b, 2009a). Little information is available on the effects of insect oils on ruminal fermentation/biohydrogenation. Indeed, Jayanegara *et al.* (2020) reported a slightly negative impact of insect oils (BSF, *Oecophylla smaragdina*, *Zophobas morio*, YM *and Gryllus bimaculatus*) on diet digestibility. On the contrary, Hervás *et al.*, (2022) observed that BSF, cricket or silkworm oils (0 to 20 g/kg dry matter) and BSF chitosan (0 or 30 g/kg dry matter) did not impair ruminal fermentation, being the cricket oil the most interesting soybean oil substitute as it promoted the *trans*- 11 18:1 fatty acid production without altering *trans*-10 18:1 concentration. Moreover, BSF oils did not seem to influence the unsaturated fatty acids profile and BSF chitosan did not influence the ruminal activity, suggesting that its dietary inclusion seems to be less favorable in ruminants.

To the author's knowledge, only a few *in-vivo* studies have been performed so far to confirm the methane mitigating effects of insect and identify the best inclusion dosages in ruminant diets. The first *in-vivo* study was performed by Astuti *et al.* (2016) using 7.5 and 15% cricket meal to substitute soybean meal in lamb's diet. In particular, both dosages seem to be suitable for lamb nutrition as they did not affect diet palatability, productive performance, feed efficiency and blood metabolites profile. A similar study was performed by the same author to evaluate the potential use of a milk replacer containing 15 or 30% cricket meal for goat kids (Astuti *et al.*, 2019). Similarly to what has been previously reported for lamb, both experimental diets showed a good palatability without any negative effect on goat kid's health, rumen fermentation profile and growth performance. Also, Astuti and Wiryawan (2022) evaluated the potential use of 15 and 30% BSF larvae inclusion for creep feeding in goats, recording unaffected hematological parameter and growth performance.

Apart from small ruminants, Fukuda *et al.*, (2022) assessed the feasibility of using a BSF-based protein supplement to stimulate feed intake and digestion in steers consuming low-quality forage, confirming that it can be used for this purpose and it can maintain feed digestibility and ruminal fermentation. However, further studies are needed to better characterize the chitin digestion site, its impact on ruminal microbial composition and on diet digestibility in cattle. In fact, only Kopečný *et al.*

(1996) identified a chitinolytic bacterium in the rumen, potentially involved in chitin digestion.

Finally, Astuti and Wiryawan (2022) recently proposed the use of insects as probiotics in young ruminants. In fact, BSF live larvae reared on chicken manure harbor in their microbiota different lactic acid bacteria that can inhibit the growth of *Escherichia coli* thanks to a competitive exclusion mechanism. This could be an interesting use for insect in tropical countries (*e.g.*, Indonesia) to prevent gastrointestinal diseases.

5 Conclusion

In conclusion, insect meals and oils can represent valuable innovative dietary ingredients for animal nutrition. In particular, low dietary inclusion levels of insect meal (< 15%) are recommended in poultry, as they maintain diet digestibility, gut health, and, as a consequence, they guarantee adequate growth performance. In fish and pigs, both low and high inclusion levels of insect-based products (especially meals) can efficiently be used. Moreover, insect fats seem to be a promising alternative fat source for rabbit nutrition as they do not impair diet digestibility and growth performance of the animals, also improving their intestinal microbiota. Less information is available on ruminants due to the European ban on the use of processed animal proteins to avoid the spreading of the bovine spongiform encephalopathy. The available preliminary results suggest that insects may represent a suitable protein and fat source also for ruminants, especially considering the ability of chitosan to reduce ruminal biohydrogenation and methane production without impairing diet digestibility. However, further studies are still needed to confirm their potential beneficial results on gut health of rabbit and ruminants.

References

Adámková, A., Mlček, J., Adámek, M., Borkovcová, M., Bednářová, M., Hlobilová, V. Knížková, I., Juríková, T., 2020. Tenebrio molitor (coleoptera: Tenebrionidae)-optimization of rearing conditions to obtain desired nutritional values. Journal of Insect Science 20, 1–10. https://doi.org/10.1093/jisesa/ieaa100.

Adámková, A., Mlček, J., Kouřimská, L., Borkovcová, M., Bušina, T., Adámek, M., Bednářová, M., Krajsa, J., 2017. Nutritional potential of selected insect species reared on the island of Sumatra. Int J Environ Res Public Health 14. https://doi.org/10.3390/ijerph14050521.

Aguilar-Miranda, E.D., Lopez, M.G., Escamilla-Santana, C., Barba de la Rosa, A.P., 2002. Characteristics of maize flour tortilla supplemented with ground Tenebrio molitor larvae. J Agric Food Chem 50, 192–195. https://doi.org/10.1021/jf010691y.

Ahmed, E., Fukuma, N., Hanada, M., Nishida, T., 2021. Insects as Novel Ruminant Feed and a Potential Mitigation Strategy for Methane Emissions. Animals (Basel) 11. https://doi.org/10.3390/ANI11092648.

Alagappan, S., Rowland, D., Barwell, R., Mantilla, S.M.O., Mikkelsen, D., James, P., Yarger, O., Hoffman, L.C., 2021. Legislative landscape of black soldier fly (Hermetia illucens) as feed. https://doi.org/10.3920/JIFF2021.0111 8, 343–355. https://doi.org/10.3920/JIFF2021.0111.

Arango Gutiérrez, G.P., Vergara Ruiz, R.A., Mejía Vélez, H., 2004. Compositional, microbiological and protein digestibility analysis of larval meal of Hermetia illucens (Diptera: Stratiomyiidae) at Angelópolis-Antioquia, Colombia. Rev Fac Nac Agron Medellin 57, 2491–2499.

Astuti, D.A., Anggraeny, A., Khotijah, L., Suharti, S., Jayanegara, A., 2019. Performance, physiological status, and rumen fermentation profiles of pre- and post-weaning goat kids fed cricket meal as a protein source. Tropical Animal Science Journal 42, 145–151. https://doi.org/10.5398/tasj.2019.42.2.145.

Astuti, D.A., Khotijah, L., Rismarianty A., 2016. Utilization and Evaluation of Cricket Meal as Protein Source in Lamb Ration. The Indian Journal of Veterinary Sciences and Biotechnology 11, 49–53.

Astuti, D.A., Wiryawan, K.G., 2022. Black soldier fly as feed ingredient for ruminants. Anim Biosci 35, 356–363. https://doi.org/10.5713/AB.21.0460.

Avendaño, C., Sánchez, M., Valenzuela, C., Avendaño, C., Sánchez, M., Valenzuela, C., 2020. Insectos: son realmente una alternativa para la alimentación de animales y humanos. Revista chilena de nutrición 47, 1029–1037. https://doi.org/10.4067/S0717-75182020000601029.

Bailey, R., 2019. Gut Health in Poultry- The World Within. Aviagen, Ross Tech Note 2–10.

Basto, A., Calduch-Giner, J., Oliveira, B., Petit, L., Sá, T., Maia, M.R.G., Fonseca, S.C., Matos, E., Pérez-Sánchez, J., Valente, L.M.P., 2021. The Use of Defatted Tenebrio molitor Larvae Meal as a Main Protein Source Is Supported in European Sea Bass (Dicentrarchus labrax) by Data on Growth Performance, Lipid Metabolism, and Flesh Quality. Front Physiol 12, 473. https://doi.org/10.3389/FPHYS.2021.659567/BIBTEX.

Bedford, A., Gong, J., 2018. Implications of butyrate and its derivatives for gut health and animal production. Animal Nutrition 4, 151–159. https://doi.org/10.1016/J.ANINU.2017.08.010.

Belanche, A., Pinloche, E., Preskett, D., Newbold, C.J., 2016. Effects and mode of action of chitosan and ivy fruit saponins on the microbiome, fermentation and methanogenesis in the rumen simulation technique. FEMS Microbiol Ecol 92, 160. https://doi.org/10.1093/FEMSEC/FIV160.

Belley, A., Keller, K., Göettke, M., Chadee, K., 1999. Intestinal mucins in colonization and host defense against pathogens, in: American Journal of Tropical Medicine and Hygiene. American Society of Tropical Medicine and Hygiene, pp. 10–15. https://doi.org/10.4269/ajtmh.1999.60.10.

Bellezza-Oddon, S., Biasato, I., Imarisio, A., Pipan, M., Dekleva, D., Capucchio, M.T., Meneguz, M., Stefania, B., Barbero, R., Gariglio, M., Dabbou, S., Fiorilla, E., Gasco, L.,

Schiavone, A., 2021. Black Soldier Fly and Yellow Mealworm live larvae for broiler chickens: effects on bird performance and health status. J Anim Physiol Anim Nutr (Berl) 0, 1–9. https://doi.org/doi.org/10.1111/jpn.13567.

Biasato, I., Chemello, G., Oddon, S.B., Ferrocino, I., Corvaglia, M.R., Caimi, C., Resconi, A., Paul, A., van Spankeren, M., Capucchio, M.T., Colombino, E., Cocolin, L., Gai, F., Schiavone, A., Gasco, L., 2022. Hermetia illucens meal inclusion in low-fishmeal diets for rainbow trout (Oncorhynchus mykiss): Effects on the growth performance, nutrient digestibility coefficients, selected gut health traits, and health status indices. Anim Feed Sci Technol 290, 115341. https://doi.org/10.1016/J.ANIFEEDSCI.2022.115341.

Biasato, I., de Marco, M., Rotolo, L., Renna, M., Lussiana, C., Dabbou, S., Capucchio, M.T., Biasibetti, E., Costa, P., Gai, F., Pozzo, L., Dezzutto, D., Bergagna, S., Martínez, S., Tarantola, M., Gasco, L., Schiavone, A., 2016. Effects of dietary *Tenebrio molitor* meal inclusion in free-range chickens. J Anim Physiol Anim Nutr (Berl) 100, 1104–1112. https://doi.org/10.1111/jpn.12487.

Biasato, Ilaria, Ferrocino, I., Biasibetti, E., Grego, E., Dabbou, S., Sereno, A., Gai, F., Gasco, L., Schiavone, A., Cocolin, L., Capucchio, M.T., 2018. Modulation of intestinal microbiota, morphology and mucin composition by dietary insect meal inclusion in free-range chickens. BMC Vet Res 14. https://doi.org/10.1186/s12917-018-1690-y.

Biasato, I., Ferrocino, I., Colombino, E., Gai, F., Schiavone, A., Cocolin, L., Vincenti, V, Capucchio, M.T., Gasco, L., 2020a. Effects of dietary Hermetia illucens meal inclusion on cecal microbiota and small intestinal mucin dynamics and infiltration with immune cells of weaned piglets. J Anim Sci Biotechnol 11, 1–11. https://doi.org/10.1186/s40104-020-00466-x.

Biasato, I., Ferrocino, I., Dabbou, S., Evangelista, R., Gai, F., Gasco, L., Cocolin, L., Capucchio, M.T., Schiavone, A., 2020b. Black soldier fly and gut health in broiler chickens: insights into the relationship between cecal microbiota and intestinal mucin composition. J Anim Sci Biotechnol 11, 1–12. https://doi.org/10.1186/s40104-019-0413-y.

Biasato, I., Ferrocino, I., Grego, E., Dabbou, S., Gai, F., Gasco, L., Cocolin, L., Capucchio, M.T., Schiavone, A., 2020c. Yellow mealworm inclusion in diets for heavy-size broiler chickens: Implications for intestinal microbiota and mucin dynamics. Animals 10, 1–13. https://doi.org/10.3390/ani10101909.

Biasato, I., Ferrocino, I., Grego, E., Dabbou, S., Gai, F., Gasco, L., Cocolin, L., Capucchio, M.T., Schiavone, A., 2019a. Gut microbiota and mucin composition in female broiler chickens fed diets including yellow mealworm (Tenebrio molitor, L.). Animals 9. https://doi.org/10.3390/ani9050213.

Biasato, I., Gasco, L., de Marco, M., Renna, M., Rotolo, L., Dabbou, S., Capucchio, M.T., Biasibetti, E., Tarantola, M., Sterpone, L., Cavallarin, L., Gai, F., Pozzo, L., Bergagna, S., Dezzutto, D., Zoccarato, I., Schiavone, A., 2018. Yellow mealworm larvae (Tenebrio molitor) inclusion in diets for male broiler chickens: Effects on growth performance, gut morphology, and histological findings. Poult Sci 97, 540–548. https://doi.org/10.3382/ps/pex308.

Biasato, I., Renna, M., Gai, F., Dabbou, S., Meneguz, M., Perona, G., Martinez, S., Lajusticia, A.C.B., Bergagna, S., Sardi, L., Capucchio, M.T., Bressan, E., Dama, A., Schiavone, A., Gasco, L., 2019b. Partially defatted black soldier fly larva meal inclusion in piglet diets: Effects on the growth performance, nutrient digestibility, blood profile, gut morphology and histological features. J Anim Sci Biotechnol 10, 1–11. https://doi.org/10.1186/s40104-019-0325-x.

Bischoff, S.C., 2011. "Gut health": A new objective in medicine? BMC Med 9, 24. https://doi.org/10.1186/1741-7015-9-24.

Borrelli, L., Coretti, L., Dipineto, L., Bovera, F., Menna, F., Chiariotti, L., Nizza, A., Lembo, F., Fioretti, A., 2017. Insect-based diet, a promising nutritional source, modulates gut microbiota composition and SCFAs production in laying hens. Scientific Reports 2017 7:1 7, 1–11. https://doi.org/10.1038/s41598-017-16560-6.

Bovera, F., Loponte, R., Marono, S., Piccolo, G., Parisi, G., Iaconisi, V., Gasco, L., Nizza, A., 2016. Use of Tenebrio molitor larvae meal as protein source in broiler diet: Effect on growth performance, nutrient digestibility, and carcass and meat traits. J Anim Sci 94, 639–647. https://doi.org/10.2527/jas.2015-9201.

Caimi, C., Biasato, I., Chemello, G., Oddon, S.B., Lussiana, C., Malfatto, V.M., Capucchio, M.T., Colombino, E., Schiavone, A., Gai, F., Trocino, A., Brugiapaglia, A., Renna, M., Gasco, L., 2021. Dietary inclusion of a partially defatted black soldier fly (Hermetia illucens) larva meal in low fishmeal-based diets for rainbow trout (Oncorhynchus mykiss). J Anim Sci Biotechnol 12, 1–15. https://doi.org/10.1186/s40104-021-00575-1.

Caimi, C., Gasco, L., Biasato, I., Malfatto, V., Varello, K., Prearo, M., Pastorino, P., Bona, M.C., Francese, D.R., Schiavone, A., Elia, A.C., Dörr, A.J.M., Gai, F., 2020. Could Dietary Black Soldier Fly Meal Inclusion Affect the Liver and Intestinal Histological Traits and the Oxidative Stress Biomarkers of Siberian Sturgeon (Acipenser baerii) Juveniles? Animals (Basel) 10. https://doi.org/10.3390/ANI10010155.

Cardinaletti, G., Randazzo, B., Messina, M., Zarantoniello, M., Giorgini, E., Zimbelli, A., Bruni, L., Parisi, G., Olivotto, I., Tulli, F., 2019. Effects of Graded Dietary Inclusion Level of Full-Fat Hermetia illucens Prepupae Meal in Practical Diets for Rainbow Trout (Oncorhynchus mykiss). Animals (Basel) 9. https://doi.org/10.3390/ANI9050251.

Chu, X., Li, M., Wang, G., Wang, K., Shang, R., Wang, Z., Li, L., 2020. Evaluation of the Low Inclusion of Full-Fatted Hermetia illucens Larvae Meal for Layer Chickens: Growth Performance, Nutrient Digestibility, and Gut Health. Front Vet Sci 7. https://doi.org/10.3389/FVETS.2020.585843.

Clavijo, V., Flórez, M.J.V., 2018. The gastrointestinal microbiome and its association with the control of pathogens in broiler chicken production: A review. Poult Sci 97, 1006–1021. https://doi.org/10.3382/ps/pex359.

Colombino, E., Biasato, I., Ferrocino, I., Oddon, S.B., Caimi, C., Gariglio, M., Dabbou, S., Caramori, M., Battisti, E., Zanet, S., Ferroglio, E., Cocolin, L., Gasco, L., Schiavone, A., Capucchio, M.T., 2021a. Effect of insect live larvae as environmental enrichment on poultry gut health: Gut mucin composition, microbiota and local immune response evaluation. Animals 11. https://doi.org/10.3390/ani11102819.

Colombino, E., Prieto-Botella, D., Capucchio, M.T., 2021b. Gut health in veterinary medicine: A bibliometric analysis of the literature. Animals 11, 1–13. https://doi.org/10.3390/ani11071997.

Cutrignelli, M.I., Messina, M., Tulli, F., Randazzi, B., Olivotto, I., Gasco, L., Loponte, R., Bovera, F., 2018. Evaluation of an insect meal of the Black Soldier Fly (Hermetia illucens) as soybean substitute: Intestinal morphometry, enzymatic and microbial activity in laying hens. Res Vet Sci 117, 209–215. https://doi.org/10.1016/J.RVSC.2017.12.020.

Dabbou, S., Ferrocino, I., Gasco, L., Schiavone, A., Trocino, A., Xiccato, G., Barroeta, A.C., Maione, S., Soglia, D., Biasato, I., Cocolin, L., Gai, F., Nucera, D.M., 2020. Antimicrobial effects of black soldier fly and yellow mealworm fats and their impact on gut microbiota of growing rabbits. Animals 10, 1–19. https://doi.org/10.3390/ani10081292.

Dabbou, S., Gai, F., Biasato, I., Capucchio, M.T., Biasibetti, E., Dezzutto, D., Meneguz, M., Plachà, I., Gasco, L., Schiavone, A., 2018. Black soldier fly defatted meal as a dietary protein source for broiler chickens: Effects on growth performance, blood traits, gut morphology and histological features. J Anim Sci Biotechnol 9, 1–10. https://doi.org/10.1186/s40104-018-0266-9.

Dabbou, S., Lauwaerts, A., Ferrocino, I., Biasato, I., Sirri, F., Zampiga, M., Bergagna, S., Pagliasso, G., Gariglio, M., Colombino, E., Narro, C.G., Gai, F., Capucchio, M.T., Gasco, L., Cocolin, L., Schiavone, A., 2021. Modified Black Soldier Fly Larva Fat in Broiler Diet: Effects on Performance, Carcass Traits, Blood Parameters, Histomorphological Features and Gut Microbiota. Animals 2021, Vol. 11, Page 1837 11, 1837. https://doi.org/10.3390/ANI11061837.

Dalle Zotte, A., Cullere, M., Martins, C., Alves, S.P., Freire, J.P.B., Falcão-e-Cunha, L., Bessa, R.J.B., 2018. Incorporation of Black Soldier Fly (Hermetia illucens L.) larvae fat or extruded linseed in diets of growing rabbits and their effects on meat quality traits including detailed fatty acid composition. Meat Sci 146, 50–58. https://doi.org/10.1016/j.meatsci.2018.08.002.

Dayrit, F.M., 2014. The Properties of Lauric Acid and Their Significance in Coconut Oil. Journal of the American Oil Chemists' Society 2014 92:1 92, 1–15. https://doi.org/10.1007/S11746-014-2562-7.

Deplancke, B., Gaskins, H.R., 2001. Microbial modulation of innate defense: Goblet cells and the intestinal mucus layer, in: American Journal of Clinical Nutrition. American Society for Nutrition. https://doi.org/10.1093/ajcn/73.6.1131s.

Diclaro, J.W., Kaufman, P.E., 2012. Black soldier fly Hermetia illucens Linnaeus (Insecta : Diptera : Stratiomyidae). IFAS Extension 5.

Diener, S., Zurbrugg, C., Roa Gutiérrez, F., Nguyen, H.D., Morel, A., Koottatep, T., Tockner, K., 2011. Black soldier fly larvae for organic waste treatment – prospects and constraints. WasteSafe 2011 2nd International Conference on Solid Waste Management in Developing Countries 1315 February 2011 Khulna Bangladesh.

Duangnumsawang, Y., Zentek, J., Goodarzi Boroojeni, F., 2021. Development and Functional Properties of Intestinal Mucus Layer in Poultry. Front Immunol. https://doi.org/10.3389/fimmu.2021.745849.

Ducatelle, R., Goossens, E., de Meyer, F., Eeckhaut, V., Antonissen, G., Haesebrouck, F., van Immerseel, F., 2018. Biomarkers for monitoring intestinal health in poultry: Present status and future perspectives. Vet Res 49, 1–9. https://doi.org/10.1186/s13567-018-0538-6.

Dumas, A., Raggi, T., Barkhouse, J., Lewis, E., Weltzien, E., 2018. The oil fraction and partially defatted meal of black soldier fly larvae (Hermetia illucens) affect differently growth performance, feed efficiency, nutrient deposition, blood glucose and lipid digestibility of rainbow trout (Oncorhynchus mykiss). Aquaculture 492, 24–34. https://doi.org/10.1016/J.AQUACULTURE.2018.03.038.

El Futuro de los Alimentos – F4F [WWW Document], n.d. URL https://f4f.cl/el-futuro-de-los-alimentos/ (accessed 10.18.22).

Elahi, U., Xu, C.C., Wang, J., Lin, J., Wu, S.G., Zhang, H.J., Qi, G.H., 2022. Insect meal as a feed ingredient for poultry. Anim Biosci 35, 332. https://doi.org/10.5713/AB.21.0435.

Elia, A.C., Capucchio, M.T., Caldaroni, B., Magara, G., Dörr, A.J.M., Biasato, I., Biasibetti, E., Righetti, M., Pastorino, P., Prearo, M., Gai, F., Schiavone, A., Gasco, L., 2018. Influence of Hermetia illucens meal dietary inclusion on the histological traits, gut mucin composition and the oxidative stress biomarkers in rainbow trout (Oncorhynchus mykiss). Aquaculture 496, 50–57. https://doi.org/10.1016/j.aquaculture.2018.07.009.

Fabrikov, D., Vargas-García, M.D.C., Barroso, F.G., Sánchez-Muros, M.J., Cacua Ortíz, S.M., Morales, A.E., Cardenete, G., Tomás-Almenar, C., Melenchón, F., 2021. Effect on Intermediary Metabolism and Digestive Parameters of the High Substitution of Fishmeal with Insect Meal in Sparus aurata Feed. Insects 2021, Vol. 12, Page 965 12, 965. https://doi.org/10.3390/INSECTS12110965.

Ferrarezi, R., Bailey, D.S., 2016. UVI/AES Annual Report 2016 – Alternative Sources of Food for Aquaponics in the U.S. Virgin Islands: A Case Study with Black Soldier Flies 11.

Fukuda, E.P., Cox, J.R., Wickersham, T.A., Drewery, M.L., 2022. Evaluation of Black Soldier Fly larvae (Hermetia illucens) as a protein supplement for beef steers consuming low-quality forage. Transl Anim Sci 6, 1–6. https://doi.org/10.1093/tas/txac018.

Gariglio, M., Dabbou, S., Biasato, I., Capucchio, M.T., Colombino, E., Hernández, F., Madrid, J., Martínez, S., Gai, F., Caimi, C., Oddon, S.B., Meneguz, M., Trocino, A., Vincenzi, R., Gasco, L., Schiavone, A., 2019. Nutritional effects of the dietary inclusion of partially defatted Hermetia illucens larva meal in Muscovy duck. J Anim Sci Biotechnol 10, 1–10. https://doi.org/10.1186/s40104-019-0344-7.

Gasco, L., Dabbou, S., Trocino, A., Xiccato, G., Capucchio, M.T., Biasato, I., Dezzutto, D., Birolo, M., Meneguz, M., Schiavone, A., Gai, F., 2019. Effect of dietary supplementation with insect fats on growth performance, digestive efficiency and health of rabbits. J Anim Sci Biotechnol 10. https://doi.org/10.1186/S40104-018-0309-2.

Gaudioso, G., Marzorati, G., Faccenda, F., Weil, T., Lunelli, F., Cardinaletti, G., Marino, G., Olivotto, I., Parisi, G., Tibaldi, E., Tuohy, K.M., Fava, F., 2021. Processed Animal Proteins from Insect and Poultry By-Products in a Fish Meal-Free Diet for Rainbow Trout: Impact on Intestinal Microbiota and Inflammatory Markers. International Journal of Molecular Sciences 2021, Vol. 22, Page 5454 22, 5454. https://doi.org/10.3390/IJMS22115454.

Gidenne, T., 2003. Fibres in rabbit feeding for digestive troubles prevention: Respective role of low-digested and digestible fibre. Livest Prod Sci 81, 105–117. https://doi.org/10.1016/S0301-6226(02)00301-9.

Gidenne, T., Garreau, H., Drouilhet, L., Aubert, C., Maertens, L., 2017. Improving feed efficiency in rabbit production, a review on nutritional, technico-economical, genetic and environmental aspects. Anim Feed Sci Technol 225, 109–122. https://doi.org/10.1016/J.ANIFEEDSCI.2017.01.016.

Goiri, I., Garcia-Rodriguez, A., Oregui, L.M., 2009a. Effect of chitosans on in vitro rumen digestion and fermentation of maize silage. Anim Feed Sci Technol 148, 276–287. https://doi.org/10.1016/J.ANIFEEDSCI.2008.04.007.

Goiri, I., Oregui, L.M., Garcia-Rodriguez, A., 2009b. Dose-response effects of chitosans on in vitro rumen digestion and fermentation of mixtures differing in forage-to-concentrate ratios. Anim Feed Sci Technol 151, 215–227. https://doi.org/10.1016/J.ANIFEEDSCI.2009.01.016.

Gong, Y., Yang, H., Wang, X., Xia, W., Lv, W., Xiao, Y., Zou, X., 2019. Early Intervention With Cecal Fermentation Broth Regulates the Colonization and Development of Gut Microbiota in Broiler Chickens. Front Microbiol 10. https://doi.org/10.3389/FMICB.2019.01422.

Gutowska, M.A., Drazen, J.C., Robison, B.H., 2004. Digestive chitinolytic activity in marine fishes of Monterey Bay, California. Comp Biochem Physiol A Mol Integr Physiol 139, 351–358. https://doi.org/10.1016/J.CBPB.2004.09.020.

Håkenåsen, I.M., Grepperud, G.H., Hansen, J.Ø., Øverland, M., Ånestad, R.M., Mydland, L.T., 2021. Full-fat insect meal in pelleted diets for weaned piglets: Effects on growth performance, nutrient digestibility, gastrointestinal function, and microbiota. Anim Feed Sci Technol 281, 115086. https://doi.org/10.1016/J.ANIFEEDSCI.2021.115086.

Halloran, A., Roos, N., Eilenberg, J., Cerutti, A., Bruun, S., 2016. Life cycle assessment of edible insects for food protein: a review. Agron Sustain Dev 36. https://doi.org/10.1007/S13593-016-0392-8.

Hardouin, J., Mahoux, G., 2004. Zootechnie d'insectes Elevage et utilisation au bénéfice de l'homme et de certains animaux. Tropucultura 3–164.

Hartinger, K., Fröschl, K., Ebbing, M.A., Bruschek-Pfleger, B., Schedle, K., Schwarz, C., Gierus, M., 2022. Suitability of Hermetia illucens larvae meal and fat in broiler diets: effects on animal performance, apparent ileal digestibility, gut histology, and microbial metabolites. J Anim Sci Biotechnol 13. https://doi.org/10.1186/S40104-022-00701-7.

Hartinger, K., Greinix, J., Thaler, N., Ebbing, M.A., Yacoubi, N., Schedle, K., Gierus, M., 2021. Effect of graded substitution of soybean meal by hermetia illucens larvae meal on animal performance, apparent ileal digestibility, gut histology and microbial metabolites of broilers. Animals 11, 1–16. https://doi.org/10.3390/ani11061628.

Haryati, R.P., Jayanegara, A., Laconi, E.B., Ridla, M., Suptijah, P., 2019. Evaluation of chitin and chitosan from insect as feed additives to mitigate ruminal methane emission. AIP Conf Proc 2120, 1–5. https://doi.org/10.1063/1.5115646.

He, C., Lei, J., Yao, Y., Qu, X., Chen, J., Xie, K., Wang, X., Yi, Q., Xiao, B., Guo, S., Zou, X., 2021. Black Soldier Fly (Hermetia illucens) Larvae Meal Modulates Intestinal Morphology and Microbiota in Xuefeng Black-Bone Chickens. Front Microbiol 12. https://doi.org/10.3389/FMICB.2021.706424.

Hender, A., Siddik, M.A.B., Howieson, J., Fotedar, R., 2021. Black Soldier Fly, Hermetia illucens as an Alternative to Fishmeal Protein and Fish Oil: Impact on Growth, Immune Response, Mucosal Barrier Status, and Flesh Quality of Juvenile Barramundi, Lates calcarifer (Bloch, 1790). Biology (Basel) 10. https://doi.org/10.3390/BIOLOGY10060505.

Hervás, G., Boussalia, Y., Labbouz, Y., della Badia, A., Toral, P.G., Frutos, P., 2022. Insect oils and chitosan in sheep feeding: Effects on in vitro ruminal biohydrogenation and fermentation. Anim Feed Sci Technol 285, 115222. https://doi.org/10.1016/J.ANIFEEDSCI.2022.115222.

Honda, K., Littman, D.R., 2016. The microbiota in adaptive immune homeostasis and disease. Nature. https://doi.org/10.1038/nature18848.

Hong, J., Han, T., Kim, Y.Y., 2020. Mealworm (Tenebrio molitor larvae) as an alternative protein source for monogastric animal: A review. Animals. https://doi.org/10.3390/ani10112068.

Huang, B., Zhang, S., Dong, X., Chi, S., Yang, Q., Liu, H., Tan, B., Xie, S., 2022. Effects of fishmeal replacement by black soldier fly on growth performance, digestive enzyme activity, intestine morphology, intestinal flora and immune response of pearl gentian grouper (Epinephelus fuscoguttatus ♀ × Epinephelus lanceolatus ♂). Fish Shellfish Immunol 120, 497–506. https://doi.org/10.1016/J.FSI.2021.12.027.

Ikeda, M., Kakizaki, H., Matsumiya, M., 2017. Biochemistry of fish stomach chitinase. Int J Biol Macromol 104, 1672–1681. https://doi.org/10.1016/J.IJBIOMAC.2017.03.118.

IPIFF, 2020. The insect sector milestones towards sustainable food supply chains. The International Platform of Insects for Food and Feed 1–24.

Jayanegara, A., Haryati, R.P., Nafisah, A., Suptijah, P., Ridla, M., Laconi, E.B., 2020. Derivatization of chitin and chitosan from black soldier fly (Hermetia illucens) and their use as feed additives: An in vitro study. Adv Anim Vet Sci 8, 472–477. https://doi.org/10.17582/JOURNAL.AAVS/2020/8.5.472.477.

Jayanegara, A., Novandri, B., Yantina, N., Ridla, M., 2017. Use of black soldier fly larvae (Hermetia illucens) to substitute soybean meal in ruminant diet: An in vitro rumen fermentation study. Vet World 10, 1439–1446. https://doi.org/10.14202/VETWORLD.2017.1439-1446.

Jha, R., Fouhse, J.M., Tiwari, U.P., Li, L., Willing, B.P., 2019. Dietary Fiber and Intestinal Health of Monogastric Animals. Front Vet Sci 6, 48. https://doi.org/10.3389/FVETS.2019.00048.

Jiang, Z., Applegate, T.J., Lossie, A.C., 2013. Cloning, Annotation and Developmental Expression of the Chicken Intestinal MUC2 Gene. PLoS One 8. https://doi.org/10.1371/journal.pone.0053781.

Jin, X., Yuan, B., Liu, M., Zhu, M., Zhang, X., Xie, G., Wu, W., Wang, Z., Xu, H., Lv, Y., Huang, Y., Wang, W., 2021. Dietary Hermetia illucens Larvae Replacement Alleviates Diarrhea and Improves Intestinal Barrier Function in Weaned Piglets Challenged With Enterotoxigenic Escherichia coli K88. Front Vet Sci 8, 1392. https://doi.org/10.3389/FVETS.2021.746224/BIBTEX.

Jin, X.H., Heo, P.S., Hong, J.S., Kim, N.J., Kim, Y.Y., 2016. Supplementation of dried mealworm (Tenebrio molitor larva) on growth performance, nutrient digestibility and blood profiles in weaning pigs. Asian-Australas J Anim Sci 29, 979–986. https://doi.org/10.5713/ajas.15.0535.

Johnson, C.N., Hashim, M.M., Bailey, C.A., Byrd, J.A., Kogut, M.H., Arsenault, R.J., 2020. Feeding of yeast cell wall extracts during a necrotic enteritis challenge enhances cell growth, survival and immune signaling in the jejunum of broiler chickens. Poult Sci 99, 2955–2966. https://doi.org/10.1016/j.psj.2020.03.012.

Józefiak, A., Benzertiha, A., Kierończyk, B., Łukomska, A., Wesołowska, I., Rawski, M., 2020. Improvement of Cecal Commensal Microbiome Following the Insect Additive into Chicken Diet. Animals 2020, Vol. 10, Page 577 10, 577. https://doi.org/10.3390/ANI10040577.

Józefiak, A., Nogales-Mérida, S., Rawski, M., Kierończyk, B., Mazurkiewicz, J., 2019. Effects of insect diets on the gastrointestinal tract health and growth performance of Siberian sturgeon (Acipenser baerii Brandt, 1869). BMC Vet Res 15. https://doi.org/10.1186/S12917-019-2070-Y.

Kar, S.K., Schokker, D., Harms, A.C., Kruijt, L., Smits, M.A., Jansman, A.J.M., 2021. Local intestinal microbiota response and systemic effects of feeding black soldier fly larvae to replace soybean meal in growing pigs. Scientific Reports 2021 11:1 11, 1–16. https://doi.org/10.1038/s41598-021-94604-8.

Kawasaki, K., Hashimoto, Y., Hori, A., Kawasaki, T., Hirayasu, H., Iwase, S.I., Hashizume, A., Ido, A., Miura, C., Miura, T., Nakamura, S., Seyama, T., Matsumoto, Y., Kasai, K., Fujitani, Y., 2019. Evaluation of black soldier fly (Hermetia illucens) larvae and pre-pupae raised on household organic waste, as potential ingredients for poultry feed. Animals 9. https://doi.org/10.3390/ani9030098.

Kogut, M.H., Arsenault, R.J., 2016. Editorial: Gut health: The new paradigm in food animal production. Front Vet Sci 3, 10–13. https://doi.org/10.3389/fvets.2016.00071.

Kogut, M.H., Lee, A., Santin, E., 2020. Microbiome and pathogen interaction with the immune system. Poult Sci 99, 1906–1913. https://doi.org/10.1016/j.psj.2019.12.011.

Kopečný, J., Hodrová, B., Stewart, C.S., 1996. The isolation and characterization of a rumen chitinolytic bacterium. Lett Appl Microbiol 23, 195–198. https://doi.org/10.1111/J.1472-765X.1996.TB00063.X.

Lähteenmäki-Uutela, A., Hénault-Ethier, L., Marimuthu, S.B., Talibov, S., Allen, R.N. Nemane, V., Vandenberg, G.W., Józefiak, D., 2018. The impact of the insect regulatory system on the insect marketing system. https://doi.org/10.3920/JIFF2017.0073 4, 187–198. https://doi.org/10.3920/JIFF2017.0073.

Lähteenmäki-Uutela, A., Marimuthu, S.B., Meijer, N., 2021. Regulations on insects as food and feed: a global comparison. https://doi.org/10.3920/JIFF2020.0066 7, 849–856. https://doi.org/10.3920/JIFF2020.0066.

Lallès, J.P., Boudry, G., Favier, C., le Floc'h, N., Luron, I., Montagne, L., Oswald, I.P., Pié, S., Piel, C., Sève, B., 2004. Gut function and dysfunction in young pigs: physiology. Animal Research 53, 301–316. https://doi.org/10.1051/ANIMRES:2004018.

Laudadio, V., Passantino, L., Perillo, A., Lopresti, G., Passantino, A., Khan, R.U., Tufarelli, V., 2012. Productive performance and histological features of intestinal mucosa of broiler chickens fed different dietary protein levels. Poult Sci 91, 265–270. https://doi.org/10.3382/ps.2011-01675.

LeBlanc, J.G., Milani, C., de Giori, G.S., Sesma, F., van Sinderen, D., Ventura, M., 2013. Bacteria as vitamin suppliers to their host: A gut microbiota perspective. Curr Opin Biotechnol. https://doi.org/10.1016/j.copbio.2012.08.005.

Lee, J.H., Kim, T.K., Cha, J.Y., Jang, H.W., Yong, H.I., Choi, Y.S., 2022. How to develop strategies to use insects as animal feed: digestibility, functionality, safety, and regulation. J Anim Sci Technol 64, 409. https://doi.org/10.5187/JAST.2022.E27.

Li, S., Ji, H., Zhang, B., Zhou, J., Yu, H., 2017. Defatted black soldier fly (Hermetia illucens) larvae meal in diets for juvenile Jian carp (Cyprinus carpio var. Jian): Growth performance, antioxidant enzyme activities, digestive enzyme activities, intestine and hepatopancreas histological structure. Aquaculture 477, 62–70. https://doi.org/10.1016/J.AQUACULTURE.2017.04.015.

Li, Y., Bruni, L., Jaramillo-Torres, A., Gajardo, K., Kortner, T.M., Krogdahl, Å., 2021. Differential response of digesta- and mucosa-associated intestinal microbiota to dietary insect meal during the seawater phase of Atlantic salmon. Animal Microbiome 2021 3:1 3, 1–18. https://doi.org/10.1186/S42523-020-00071-3.

Li, Y., Gajardo, K., Jaramillo-Torres, A., Kortner, T.M., Krogdahl, Å., 2022. Consistent changes in the intestinal microbiota of Atlantic salmon fed insect meal diets. Animal Microbiome 2022 4:1 4, 1–15. https://doi.org/10.1186/S42523-021-00159-4.

Liu, X., Chen, X., Wang, H., Yang, Q., Ur Rehman, K., Li, W., Cai, M., Li, Q., Mazza, L., Zhang, J., Yu, Z., Zheng, L., 2017. Dynamic changes of nutrient composition throughout the entire life cycle of black soldier fly. PLoS One 12. https://doi.org/10.1371/journal.pone.0182601.

Lokman, I.H., Ibitoye, E.B., Hezmee, M.N.M., Goh, Y.M., Zuki, A.B.Z., Jimoh, A.A., 2019. Effects of chitin and chitosan from cricket and shrimp on growth and carcass performance of broiler chickens. Trop Anim Health Prod 51, 2219–2225. https://doi.org/10.1007/s11250-019-01936-9.

Lu, R., Chen, Y., Yu, W., Lin, M., Yang, G., Qin, C., Meng, X., Zhang, Y., Ji, H., Nie, G., 2020. Defatted black soldier fly (Hermetia illucens) larvae meal can replace soybean meal in juvenile grass carp (Ctenopharyngodon idellus) diets. Aquac Rep 18, 100520. https://doi.org/10.1016/J.AQREP.2020.100520.

Makkar, H.P.S., Tran, G., Heuzé, V., Ankers, P., 2014. State-of-the-art on use of insects as animal feed. Anim Feed Sci Technol 197, 1–33. https://doi.org/10.1016/j.anifeedsci.2014.07.008.

Martins, C., Cullere, M., Zotte, A.D., Cardoso, C., Alves, S.P., de Bessa, R.J.B., Freire, J.P.B., Falcaõ-E-Cunha, L., 2018. Incorporation of two levels of black soldier fly (Hermetia illucens L.) larvae fat or extruded linseed in diets of growing rabbits: Effects on growth performance and diet digestibility. Czech Journal of Animal Science 63, 356–362. https://doi.org/10.17221/22/2018-CJAS.

Melenchón, F., de Mercado, E., Pula, H.J., Cardenete, G., Barroso, F.G., Fabrikov, D., Lourenço, H.M., Pessoa, M.F., Lagos, L., Weththasinghe, P., Cortés, M., Tomás-almenar,

C., 2022. Fishmeal Dietary Replacement Up to 50%: A Comparative Study of Two Insect Meals for Rainbow Trout (Oncorhynchus mykiss). Animals (Basel) 12. https://doi.org/10.3390/ANI12020179.

Meyer, S., Gessner, D.K., Maheshwari, G., Röhrig, J., Friedhoff, T., Most, E., Zorn, H., Ringseis, R., Eder, K., 2020. Tenebrio molitor Larvae Meal Affects the Cecal Microbiota of Growing Pigs. Animals (Basel) 10, 1–17. https://doi.org/10.3390/ANI10071151.

Ministerio de Economia, 2022. Apoyo a innovador método de producción de harina de insectos para consumo animal. Available at: https://www.argentina.gob.ar/noticias/apoyo-innovador-metodo-de-produccion-de-harina-de-insectos-para-consumo-animal.

Moniello, G., Ariano, A., Panettieri, V., Tulli, F., Olivotto, I., Messina, M., Randazzo, B., Severino, L., Piccolo, G., Musco, N., Addeo, N.F., Hassoun, G., Bovera, F., 2019. Intestinal Morphometry, Enzymatic and Microbial Activity in Laying Hens Fed Different Levels of a Hermetia illucens Larvae Meal and Toxic Elements Content of the Insect Meal and Diets. Animals 2019, Vol. 9, Page 86 9, 86. https://doi.org/10.3390/ANI9030086.

Ndotono, E.W., Khamis, F.M., Bargul, J.L., Tanga, C.M., 2022. Insights into the Gut Microbial Communities of Broiler Chicken Fed Black Soldier Fly Larvae- Desmodium-Based Meal as a Dietary Protein Source. Microorganisms 10. https://doi.org/10.3390/MICROORGANISMS10071351.

Newton, G.L., Booram, C. v., Barker, R.W., Hale, O.M., 1977. Dried Hermetia Illucens Larvae Meal as a Supplement for Swine. J Anim Sci 44, 395–400. https://doi.org/10.2527/JAS1977.443395X.

Newton, G.L., Sheppard, D.C., Burtle, G., 2008. Black Soldier Fly Prepupae A Compelling Alternative to Fish Meal and Fish Oil.

Newton, L., Craig, S., Wes D, W., Gary, B., Robert, D., 2005. Using the black soldier fly, Hermetia illucens, as a value-added tool for the management of swine manure. Journal Korean Entomology and Applied Science 36, 17 pp.

Palomba, A., Melis, R., Biosa, G., Braca, A., Pisanu, S., Ghisaura, S., Caimi, C., Biasato, I., Oddon, S.B., Gasco, L., Terova, G., Moroni, F., Antonini, M., Pagnozzi, D., Anedda, R., 2022. On the Compatibility of Fish Meal Replacements in Aquafeeds for Rainbow Trout. A Combined Metabolomic, Proteomic and Histological Study. Front Physiol 13. https://doi.org/10.3389/FPHYS.2022.920289.

Panteli, N., Mastoraki, M., Lazarina, M., Chatzifotis, S., Mente, E., Kormas, K.A., Antonopoulou, E. 2021. Configuration of Gut Microbiota Structure and Potential Functionality in Two Teleosts under the Influence of Dietary Insect Meals. Microorganisms 2021, Vol. 9, Page 699 9, 699. https://doi.org/10.3390/MICROORGANISMS9040699.

Pérez-Pascual, D., Estellé, J., Dutto, G., Rodde, C., Bernardet, J.F., Marchand, Y., Duchaud, E., Przybyla, C., Ghigo, J.M., 2020. Growth Performance and Adaptability of European Sea Bass (Dicentrarchus labrax) Gut Microbiota to Alternative Diets Free of Fish Products. Microorganisms 2020, Vol. 8, Page 1346 8, 1346. https://doi.org/10.3390/MICROORGANISMS8091346.

Pleić, I.L., Bušelić, I., Messina, M., Hrabar, J., Žuvić, L., Talijančić, I., Žužul, I., Pavelin, T., Anđelić, I., Pleadin, J., Puizina, J., Grubišić, L., Tibaldi, E., Šegvić-Bubić, T., 2022. A plant-based diet supplemented with Hermetia illucens alone or in combination with poultry by-product meal: one step closer to sustainable aquafeeds for European seabass. Journal of Animal Science and Biotechnology 2022 13:1 13, 1–22. https://doi.org/10.1186/S40104-022-00725-Z.

Precup, G., Ververis, E., Azzollini, D., Rivero-Pino, F., Zakidou, P., Germini, A., 2022. The Safety Assessment of Insects and Products Thereof As Novel Foods in the European Union. Novel Foods and Edible Insects in the European Union 123–146. https://doi.org/10.1007/978-3-031-13494-4_7.

Ramos-Elorduy, J., González, E.A., Hernández, A.R., Pino, J.M., 2002. Use of Tenebrio molitor (Coleoptera: Tenebrionidae) to recycle organic wastes and as feed for broiler chickens. J Econ Entomol 95, 214–220. https://doi.org/10.1603/0022-0493-95.1.214.

Rangel, F., Enes, P., Gasco, L., Gai, F., Hausmann, B., Berry, D., Oliva-Teles, A., Serra, C.R., Pereira, F.C., 2022. Differential Modulation of the European Sea Bass Gut Microbiota by Distinct Insect Meals. Front Microbiol 13, 1249. https://doi.org/10.3389/FMICB.2022.831034/BIBTEX.

Ravzanaadii, N., Kim, S.-H., Choi, W.-H., Hong, S.-J., Kim, N.-J., 2012. Nutritional Value of Mealworm, Tenebrio molitor as Food Source. Int J Indust Entomol 25, 93–98. https://doi.org/10.7852/ijie.2012.25.1.093.

Renna, M., Schiavone, A., Gai, F., Dabbou, S., Lussiana, C., Malfatto, V., Prearo, M., Capucchio, M.T., Biasato, I., Biasibetti, E., de Marco, M., Brugiapaglia, A., Zoccarato, I., Gasco, L., 2017. Evaluation of the suitability of a partially defatted black soldier fly (Hermetia illucens L.) larvae meal as ingredient for rainbow trout (Oncorhynchus mykiss Walbaum) diets. J Anim Sci Biotechnol 8, 1–13. https://doi.org/10.1186/s40104-017-0191-3.

Rimoldi, S., Antonini, M., Gasco, L., Moroni, F., Terova, G., 2021. Intestinal microbial communities of rainbow trout (Oncorhynchus mykiss) may be improved by feeding a Hermetia illucens meal/low-fishmeal diet. Fish Physiol Biochem 47, 365–380. https://doi.org/10.1007/S10695-020-00918-1.

Rimoldi, S., Gini, E., Iannini, F., Gasco, L., Terova, G., 2019. The Effects of Dietary Insect Meal from Hermetia illucens Prepupae on Autochthonous Gut Microbiota of Rainbow Trout (Oncorhynchus mykiss). Animals (Basel) 9. https://doi.org/10.3390/ANI9040143.

Rumpold, B.A., Schlüter, O.K., 2013. Nutritional composition and safety aspects of edible insects. Mol Nutr Food Res 57, 802–823. https://doi.org/10.1002/MNFR.201200735.

Sánchez-Muros, M.J., Barroso, F.G., Manzano-Agugliaro, F., 2014. Insect meal as renewable source of food for animal feeding: a review. J Clean Prod 65, 16–27. https://doi.org/10.1016/J.JCLEPRO.2013.11.068.

Schiavone, A., Dabbou, S., de Marco, M., Cullere, M., Biasato, I., Biasibetti, E., Capucchio, M.T., Bergagna, S., Dezzutto, D., Meneguz, M., Gai, F., Dalle Zotte, A., Gasco, L., 2018. Black soldier fly larva fat inclusion in finisher broiler chicken diet as an alternative fat source. Animal 12, 2032–2039. https://doi.org/10.1017/S1751731117003743.

Sedgh-Gooya, S., Torki, M., Darbemamieh, M., Khamisabadi, H., Abdolmohamadi, A., 2022. Growth performance and intestinal morphometric features of broiler chickens fed on dietary inclusion of yellow mealworm (Tenebrio molitor) larvae powder. Vet Med Sci. https://doi.org/10.1002/VMS3.881.

Selaledi, L., Mbajiorgu, C.A., Mabelebele, M., 2020. The use of yellow mealworm (T. molitor) as alternative source of protein in poultry diets: a review. Trop Anim Health Prod 52, 7–16. https://doi.org/10.1007/S11250-019-02033-7.

Seo, B.-S., Park, S.-J., Hwang, S.-Y., Lee, Y.-I., Lee, S.-H., Hur, S.-W., Lee, K.-J., Nam, T.-J., Song, J.-W., Kim, J.-S., Jang, W.-J., Choi, Y.-H., 2022. Effects of Decreasing Fishmeal as Main Source of Protein on Growth, Digestive Physiology, and Gut Microbiota of Olive Flounder (Paralichthys olivaceus). Animals (Basel) 12, 2043. https://doi.org/10.3390/ANI12162043.

Shafique, L., Abdel-Latif, H.M.R., Hassan, F.U., Alagawany, M., Naiel, M.A.E., Dawood, M.A.O., Yilmaz, S., Liu, Q., 2021. The feasibility of using yellow mealworms (Tenebrio molitor): Towards a sustainable aquafeed industry. Animals. https://doi.org/10.3390/ani11030811.

Shen, L., Weber, C.R., Turner, J.R., 2008. The tight junction protein complex undergoes rapid and continuous molecular remodeling at steady state. Journal of Cell Biology 181, 683–695. https://doi.org/10.1083/jcb.200711165.

Soetemans, L., Uyttebroek, M., Bastiaens, L., 2020. Characteristics of chitin extracted from black soldier fly in different life stages. International Journal ofBiological Macromolecules 165, 3206–3214. https://doi.org/10.1016/j.ijbiomac.2020.11.041.

Spranghers, T., Michiels, J., Vrancx, J., Ovyn, A., Eeckhout, M., de Clercq, P., de Smet, S., 2018. Gut antimicrobial effects and nutritional value of black soldier fly (Hermetia illucens L.) prepupae for weaned piglets. Anim Feed Sci Technol 235, 33–42. https://doi.org/10.1016/J.ANIFEEDSCI.2017.08.012.

Spranghers, T., Ottoboni, M., Klootwijk, C., Ovyn, A., Deboosere, S., de Meulenaer, B., Michiels, J., Eeckhout, M., de Clercq, P., de Smet, S., 2017. Nutritional composition of black soldier fly (Hermetia illucens) prepupae reared on different organic waste substrates. J Sci Food Agric 97, 2594–2600. https://doi.org/10.1002/JSFA.8081.

Star, L., Arsiwalla, T., Molist, F., Leushuis, R., Dalim, M., Paul, A., 2020. Gradual Provision of Live Black Soldier Fly (Hermetia illucens) Larvae to Older Laying Hens: Effect on Production Performance, Egg Quality, Feather Condition and Behavior. Animals 2020, Vol. 10, Page 216 10, 216. https://doi.org/10.3390/ANI10020216.

St-Hilaire, S., Cranfill, K., McGuire, M.A., Mosley, E.E., Tomberlin, J.K., Newton, L., Sealey, W., Sheppard, C., Irving, S., 2007. Fish Offal Recycling by the Black Soldier Fly Produces a Foodstuff High in Omega-3 Fatty Acids. J World Aquac Soc 38, 309–313. https://doi.org/10.1111/j.1749-7345.2007.00101.x.

Strychalski, J., Juśkiewicz, J., Kowalska, D., Gugołek, A., 2021. Performance indicators and gastrointestinal response of rabbits to dietary soybean meal replacement with silkworm pupae and mealworm larvae meals. https://doi.org/10.1080/1745039X.2021.1962171 75, 294–310. https://doi.org/10.1080/1745039X.2021.1962171.

Sypniewski, J., Kierończyk, B., Benzertiha, A., Mikołajczak, Z., Pruszyńska-Oszmałek, E., Kołodziejski, P., Sassek, M., Rawski, M., Czekała, W., Józefiak, D., 2020. Replacement of soybean oil by Hermetia illucens fat in turkey nutrition: effect on performance, digestibility, microbial community, immune and physiological status and final product quality. Br Poult Sci 61, 294–302. https://doi.org/10.1080/00071668.2020.1716302.

Tabata, E., Kashimura, A., Kikuchi, A., Masuda, H., Miyahara, R., Hiruma, Y., Wakita, S., Ohno, M., Sakaguchi, M., Sugahara, Y., Matoska, V., Bauer, P.O., Oyama, F., 2018. Chitin digestibility is dependent on feeding behaviors, which determine acidic chitinase mRNA levels in mammalian and poultry stomachs. Sci Rep 8, 1–11. https://doi.org/10.1038/s41598-018-19940-8.

Tellez, G., Higgins, S.E., Donoghue, A.M., Hargis, B.M., 2006. Digestive physiology and the role of microorganisms, in: Journal of Applied Poultry Research. Poultry Science Association, pp. 136–144. https://doi.org/10.1093/japr/15.1.136.

Terova, G., Gini, E., Gasco, L., Moroni, F., Antonini, M., Rimoldi, S., 2021. Effects of full replacement of dietary fishmeal with insect meal from Tenebrio molitor on rainbow trout gut and skin microbiota. J Anim Sci Biotechnol 12, 1–14. https://doi.org/10.1186/S40104-021-00551-9/TABLES/6.

Terova, G., Rimoldi, S., Ascione, C., Gini, E., Ceccotti, C., Gasco, L., 2019. Rainbow trout (Oncorhynchus mykiss) gut microbiota is modulated by insect meal from Hermetia illucens prepupae in the diet. Rev Fish Biol Fish 29, 465–486. https://doi.org/10.1007/S11160-019-09558-Y.

Thrastardottir, R., Olafsdottir, H.T., Thorarinsdottir, R.I., 2021. Yellow Mealworm and Black Soldier Fly Larvae for Feed and Food Production in Europe, with Emphasis on Iceland. Foods 2021, Vol. 10, Page 2744 10, 2744. https://doi.org/10.3390/FOODS10112744.

Turck, D., Feste, A.S., Lifschitz, C.H., 1993. Age and Diet Affect the Composition of Porcine Colonic Mucins. Pediatric Research 1993 33:6 33, 564–567. https://doi.org/10.1203/00006450-199306000-00005.

van Broekhoven, S., Oonincx, D.G.A.B., van Huis, A., van Loon, J.J.A., 2015. Growth performance and feed conversion efficiency of three edible mealworm species (Coleoptera: Tenebrionidae) on diets composed of organic by-products. J Insect Physiol 73, 1–10. https://doi.org/10.1016/j.jinsphys.2014.12.005.

van der Klis, J.D., Jansman, A.J.M., 2002. Optimising nutrient digestion, absorption and gut barrier function in monogastrics: reality or illusion? | Groenekennis, in: Blok, M.C. (Ed.), Nutrition and Health of the Gastrointestinal Tract. Wageningen Academic Publishers.

van Huis, A., 2013. Potential of insects as food and feed in assuring food security. Annu Rev Entomol 58, 563–583. https://doi.org/10.1146/annurev-ento-120811-153704.

Veldkamp, T., Dong, L., Paul, A., Govers, C., 2021. Bioactive properties of insect products for monogastric animals – a review. J Insects Food Feed 8, 1–14. https://doi.org/10.3920/jiff2021.0031.

Veldkamp, T., van Niekerk, T.G.C.M., 2019. Live black soldier fly larvae (Hermetia illucens) for turkey poults. https://doi.org/10.3920/JIFF2018.0031 5, 301–311. https://doi.org/10.3920/JIFF2018.0031.

Wallace, P.A., Nyameasem, J.K., Aboagye, G.A., Affedzie-Obresi, S., Nkegbe, K., Murray, F., Botchway, V., Karbo, N., Leschen, W., Maquart, P.O., Clottey, V., 2018. Effects of replacing fishmeal with black soldier fly larval meal in the diets of grower-finishing guinea fowls reared under tropical conditions. Trop Anim Health Prod 50, 1499–1507. https://doi.org/10.1007/s11250-018-1588-5.

Wang, H., Liao, H., Chen, H., 2012. Tenebrio small-scale ecological farming feasibility study, in: Advanced Materials Research. pp. 267–270. https://doi.org/10.4028/www.scientific.net/AMR.356-360.267.

Weththasinghe, P., Lagos, L., Cortés, M., Hansen, J.Ø., Øverland, M., 2021. Dietary Inclusion of Black Soldier Fly (Hermetia Illucens) Larvae Meal and Paste Improved Gut Health but Had Minor Effects on Skin Mucus Proteome and Immune Response in Atlantic Salmon (Salmo Salar). Front Immunol 12. https://doi.org/10.3389/FIMMU.2021.599530.

Weththasinghe, P., Rocha, S.D.C., Øyås, O., Lagos, L., Hansen, J.Ø., Mydland, L.T., Øverland, M., 2022. Modulation of Atlantic salmon (Salmo salar) gut microbiota composition and predicted metabolic capacity by feeding diets with processed black soldier fly (Hermetia illucens) larvae meals and fractions. Anim Microbiome 4. https://doi.org/10.1186/S42523-021-00161-W.

Yamauchi, K.E., 2007. Review of a histological intestinal approach to assessing the intestinal function in chickens and pigs. Animal Science Journal 78, 356–370. https://doi.org/10.1111/J.1740-0929.2007.00448.X.

Yu, M., Li, Z., Chen, W., Rong, T., Wang, G., Ma, X., 2019. Hermetia illucens larvae as a potential dietary protein source altered the microbiota and modulated mucosal immune status in the colon of finishing pigs. J Anim Sci Biotechnol 10, 1–16. https://doi.org/10.1186/s40104-019-0358-1.

Zarantoniello, M., Randazzo, B., Nozzi, V., Truzzi, C., Giorgini, E., Cardinaletti, G., Freddi, L., Ratti, S., Girolametti, F., Osimani, A., Notarstefano, V., Milanović, V., Riolo, P., Isidoro, N., Tulli, F., Gioacchini, G., Olivotto, I., 2021. Physiological responses of Siberian sturgeon (Acipenser baerii) juveniles fed on full-fat insect-based diet in an aquaponic system. Scientific Reports 2021 11:1 11, 1–13. https://doi.org/10.1038/s41598-020-80379-x.

CHAPTER 17

Calcium sensing receptor (CaSR) and gut health in pigs

Xiaoya Zhao[1,2], *Lingyun Wei*[1] *and Chengbo Yang*[1,*]

[1]Department of Animal Science, University of Manitoba, Winnipeg, MB, Canada, R3T 2N2; [2]College of Animal Science, South China Agricultural University, Tianhe District, Guangzhou, China, 510642; *Correspondence: chengbo.yang@umanitoba.ca

Abstract

The calcium-sensing receptor (CaSR), a critical molecular component in the maintenance of mineral ion homeostasis, extends its functional repertoire to pivotal roles in nutrient sensing and the overall physiology, growth, and development of animals, particularly in swine. This chapter delineates the extensive distribution of the CaSR throughout the mammalian gastrointestinal tract and its accessory digestive organs, underscoring its multifaceted presence and utility. We delve into the intricate identification and structural elucidation of the CaSR, examining its localization and physiological roles across varied tissue types. The binding of specific ligands to the CaSR and the consequent initiation of intricate signalling pathways are scrutinized to reveal a comprehensive understanding of the CaSR's mechanisms of action. The receptor's profound influence on gastrointestinal integrity is manifested through its regulation of nutrient digestion and absorption, its promotion of intestinal cell proliferation and differentiation, and its support of gut barrier function. Additionally, the CaSR's role in mediating gut peptide secretion and energy metabolism is explored. Notably, the receptor's involvement in mitigating inflammatory responses and ameliorating secretory diarrhea highlights its therapeutic potential in managing gut health in porcine models. Throughout this book chapter, the indispensable role of the CaSR in ensuring gastrointestinal well-being in pigs is brought to the forefront, providing insights with significant implications for both animal husbandry and veterinary medicine.

Keywords

Calcium sensing receptors (CaSR) – Gut health – Gut chemosensing – Gut barrier function – Pigs

© XIAOYA ZHAO ET AL., 2024 | DOI:10.3920/9789004695467_018

It is well known that the use of in-feed prophylactic antibiotics in swine production is a precaution to reduce the risk of disease and optimize the overall welfare and health of the pigs, which makes the systematic prophylactic use redundant (Diana *et al.*, 2019). Those optimizations may also increase productivity and decrease expenditures on antibiotics (Pluske *et al.*, 2018). The intestine is the main absorptive organ of the digestive system with the assistance of digestive enzymes and microbial fermentation. Furthermore, the gut serves as a pivotal barrier that keeps out pathogens or toxins from the environment, so the gut also is the largest immune organ (Vighi *et al.*, 2008).

In this context, instead of just reducing antimicrobial usage, our goal should shift to building a healthy gut to support the pigs during all stages of production. Only a healthy gut can build a healthy pig and protect it from being disturbed by sickness during its lifespan. Moreover, effective digestion, absorption and utilization of dietary nutrients also relies on a healthy digestive system. A healthy gut allows the pig to focus its energy on eating and growing, leading to better tissue accretion and higher production efficiency, resulting in a higher return on investment in swine production. This is the reason why maintaining gut health is an "all-the-time deal" in livestock production (Feeds, 2014; Liao and Nyachoti, 2017).

The gut operates not only as an organ for nutrient digestion and absorption, but it also acts as a chemosensory system that, through a series of biochemical cascades, maintains the homeostasis in animals and humans (Reimann *et al.*, 2012; D. Burrin *et al.*, 2013; Ipharraguerre, 2013; Kaji *et al.*, 2013; Mace and Marshall, 2013; Yang *et al.*, 2013). Specifically, the gut chemosensing system is characterized by a network of G protein-coupled receptors (GPCR) that sense different types of nutrients, compounds, and microbes, and subsequently trigger a cascade of cellular signal transduction to modulate nutrient absorption, and gut health (Liu *et al.*, 2013). Understanding the relationship between gut chemosensing, integrity and body metabolism in animals and humans has emerged as an interesting research area. As an essential compartment of gut chemosensors, the activation of the CaSR induced by nutritional ligands could trigger a wide range of downstream signaling pathways, subsequently contributing to various physiological activities, including nutrient digestion and absorption, energy metabolism, gut barrier function, as well as immune response (Zhang and Mine, 2018). This implies that the CaSR has the potential to exert multiple roles on the regulation of gut health.

1 Identification and structural analysis of calcium-sensing receptor

1.1 *Identification of calcium-sensing receptor gene and protein*

The GPCRs constitute the largest type of membrane receptors and play a key role in cell-cell communication (George *et al.*, 2002). Based on the sequence difference of seven transmembrane domains (TMDs), the GPCRs can be categorized into five families (Class A, Class B, Class C, Class D, and Class E). In particular, the class C GPCRs contain metabotropic glutamate receptors (mGlu receptors), calcium-sensing receptors (CaSR), γ-aminobutyric acid B receptors (GABAB receptors), sweet and amino acid taste receptors, pheromone receptors, odorant receptors as well as some orphan receptors.

The class C GPCRs all have one large N-terminal extracellular domain (ECD), seven TMDs, and an intracellular domain. Plus, class C GPCRs usually form dimers with unique activation modes. As one member of GPCRs, the CaSR was identified as a master regulator for extracellular calcium homeostasis, and it also plays a pivotal role in sensing nutrients and modulating the physiology, growth, and development of the animals (Brennan *et al.*, 2014). Using the expression-cloning technique in *Xenopus laevis* oocytes, the CaSR was initially characterized in the bovine parathyroid (Brown *et al.*, 1993). Subsequently, the human CaSR equivalent was isolated from the parathyroid tumor in 1995. Bovine CaSR protein shares 93% identity with human CaSR, and their clones expressed in *Xenopus laevis* oocytes show similar pharmacological features as those confirmed in parathyroid cells (Garrett *et al.*, 1995). In addition to bovine and human, the CaSR has been detected in various types of vertebrates, and it is well conserved by the high similarity in sequences across different species (Figure 17.1) (Bai, 2004). In particular, the amino acid sequences of human CaSR and those from rat and rabbit share more than an 80% identity to that of bovine CaSR (Bai, 2004). So far, the presence of the CaSR has been found in a series of species, like bovine, human, rat, mouse, rabbit, dog, cat, pig, salamander, chicken, and fishes such as elasmobranch fish (Garrett *et al.*, 1995; Diaz *et al.*, 1997; Nearing *et al.*, 2002; Skelly and Franklin, 2007; Gal *et al.*, 2010; Zhao *et al.*, 2019).

It has been demonstrated that the mature form of the CaSR is expressed on the cell surface as a homodimer, while some research also found that there is cross-family covalent heterodimerization between the CaSR and Group I mGluRs (Bai *et al.*, 1998; Gama *et al.*, 2001). A previous study indicated that CaSR homodimer is formed in the endoplasmic reticulum, and the authors claimed that the dimerization is important while still inadequate for the release of the CaSR from the endoplasmic reticulum to the cell surface (Pidasheva *et al.*, 2006). The CaSR dimerization is formed before expressing on the cell surface, while researchers also found that the conformation of CaSR changes with the presence of cations binding, which possibly benefits the oxidation of free sulfhydryl groups and then boosts the CaSR dimer formation, implying that the binding of CaSR ligand could potentially stabilize the CaSR dimer (Ward *et al.*, 1998).

Identification work of CaSR protein expression is usually conducted using western blotting. Regarding the western blot analysis of CaSR protein expression, the sizes of CaSR main bands are around 120 kDa, 200 kDa, and 350 kDa, respectively (Ray *et al.*, 1998). Frequently, the two major species at nearly 130–140 kDa and 150–160 kDa represent two forms of the CaSR, that is, the immature form and the mature form, respectively. Moreover, the band of about 130–140 kDa is a fully

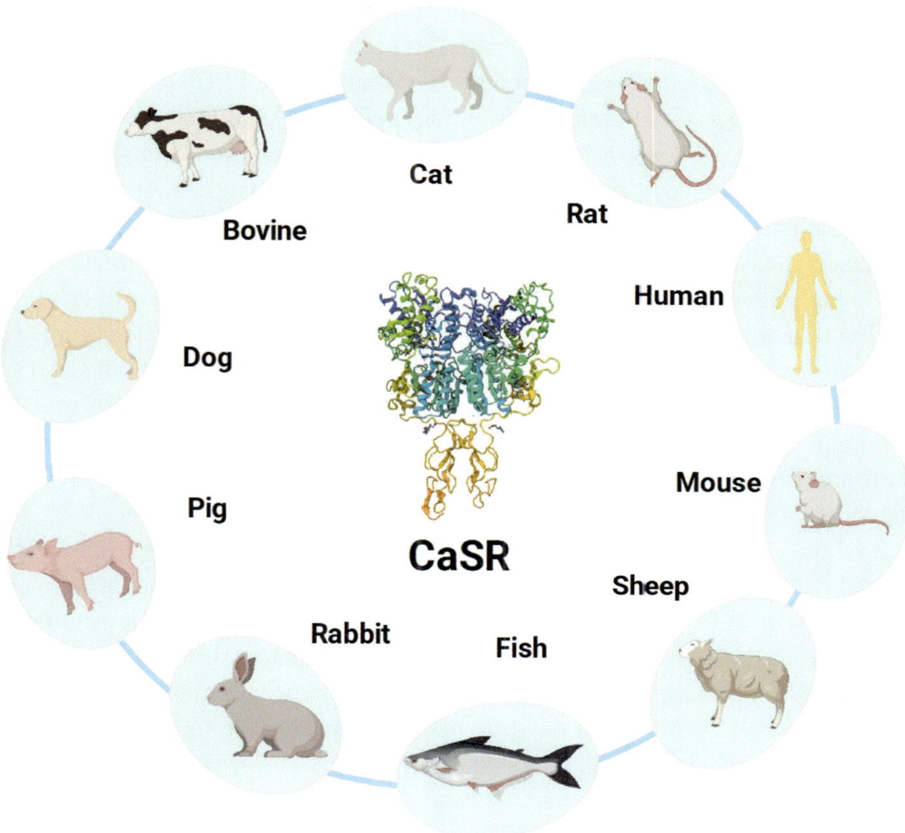

FIGURE 17.1 Identification of calcium-sensing receptor (CaSR) expression in different species

N-glycosylated monomer which is N-glycosylated with CaSR carbohydrates, while the band which is about 150–160 kDa is also a fully N-glycosylated monomer and is N-glycosylated with complex carbohydrates (Ray *et al.*, 1998). It was demonstrated that only the mature form of the CaSR was identified on the cell surface (Bai *et al.*, 1998). In addition, the minor species at 120 kDa was considered to be the non-N-glycosylated form of the CaSR. A very small fraction of the non-N-glycosylated form of the mature CaSR is expressed on the cell plasma membrane and the expression level of it is much lower compared with the other forms of the CaSR (Bai *et al.*, 1998). As for bands ranging from 200–350 kDa, they are considered the dimeric or oligomeric forms of the CaSR.

1.2 *Calcium-sensing receptor structure*

The human CaSR gene contains eight exons and spans around 100 kb which maps to 3q13.3–21 (Janicic, 1995; Yun *et al.*, 2007). Exon 1 is a 5-untranslated region (5′-UTR) and contains two parts (exon 1A and 1B) that encode alternative 5′-UTRs splicing to exon 2, suggesting the presence of two promoters driving transcription of alternative

5′-UTR exons (Hendy and Canaff, 2016). Specifically, studies have documented those two promoters found in the human CaSR: one promoter includes TATA and CAAT boxes and the second GC-rich one does not have a TATA box, which possibly contributes to the presence of multiple CaSR mRNAs through tissue-specific regulation mode (Chikatsu *et al.*, 2000). The other six exons (exons 2 to 7) have been demonstrated to encode the human CaSR protein which consists of 1078 amino acids residues (Hendy and Canaff, 2016). The translation of the CaSR gene starts in exon 2, as the AUG initiation codon is in exon 2. Plus, the 5′-UTR upstream of the ATG start codon consists of 242 nucleotides (Hendy and Canaff, 2016). Previous studies also documented that exon 5 (77 amino acids) has an alternatively spliced transcript action, as the lack of exon 5 was found in a variant human CaSR in human keratinocytes (Oda *et al.*, 1998). That change in exon 5 makes this a mutant of human CaSR with reduced sensitivity to extracellular calcium changes. In exon 7, there are two different types of 3-UTR sequences, one is short (177-nucleotide), and the other one is longer (1304-nucleotide) (Chikatsu *et al.*, 2000).

As shown in Figure 17.2, the family C GPCRs like γ-aminobutyric acid, GPRC6A, as well as CaSR are all dimers and have a large ECD which includes a biolobed "Venus flytrap" (VFT) (Reyes-Cruz *et al.*, 2001). The ECD has been demonstrated to be crucial for the regular function of the CaSR, given the fact that many spontaneous mutations located in the ECD usually result in calcium homeostatic disorders (Hu and Spiegel, 2007). Based on structural analysis of mGlu like receptors, VFT was shown to include a bilobate domain formed by two lobes (Lobe 1 and Lobe 2). Besides, the disulfide bond connecting two VFTs lobes has been shown to stabilize this dimer (Romano *et al.*, 2001). Furthermore, lobe 1 (LB1) and lobe 2 (LB2) of VFT could delineate an interlobe cavity, which allows the occurring of the ligand binding (Kunishima *et al.*, 2000). In the unliganded state, the VFT oscillates between an inactive (open form) and active (closed form) conformation. When the ligand was presented to the receptor, it could bind to LB1 in the open status of VFT and then stabilizes to become a closed state by additionally contacting with LB2 (Chun *et al.*, 2012). Furthermore, inhibitors or competitive antagonists have been shown to inhibit the receptor activation by preventing the full closure of the VFT (Bessis *et al.*, 2002; Kniazeff *et al.*, 2004). The VFT is crucial for the CaSR, as studies showed that the deletion of VFT in the CaSR caused negative responses on the CaSR in different ways which range from reducing CaSR gene expression (both the total and cell membrane expression level) resulting in lower biological activity (Reyes-Cruz *et al.*, 2001). Plus, similar to mGluRs, ligands binding sites are predominantly positioned in the VFT of the CaSR, and those ligands include cations like Ca^{2+}, Mg^{2+}, Gd^{3+} as well as L-AAs such as L-Tryptophan (L-Trp) (Reyes-Cruz *et al.*, 2001). After extracellular calcium binds in the cleft between LB1 and LB2, it results in the closing of two lobes and then leads to the rotation of VFT, which finally contributes to the activation of the CaSR. Besides ECD, the CaSR also contains seven TMDs, and a hydrophilic (C)-terminus (C-tail) (Bai, 2004). The CaSR ECD is linked to the TMD

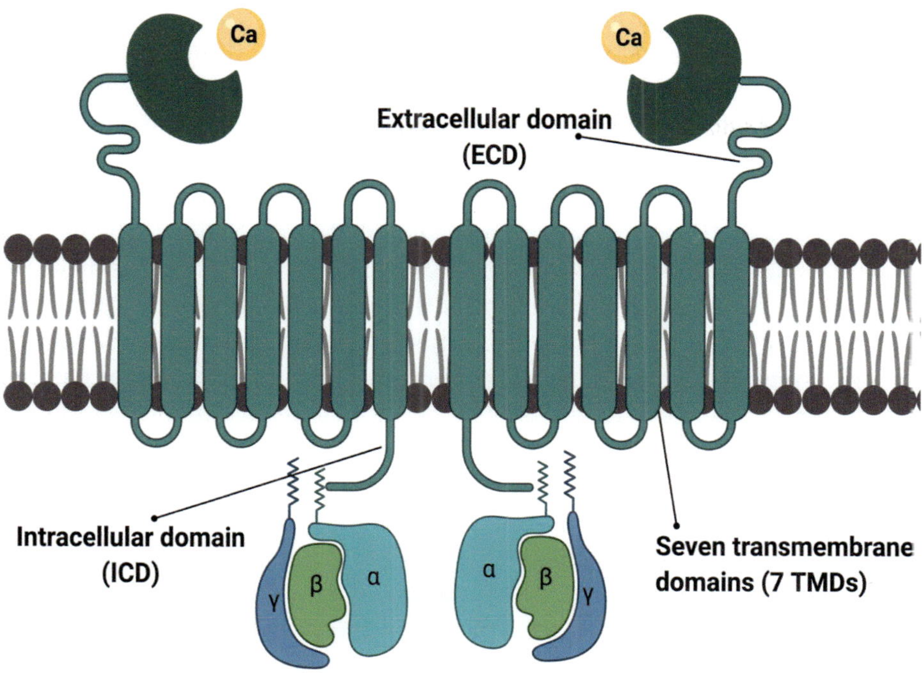

FIGURE 17.2 Schematic representation of the calcium-sensing receptor (CaSR) dimeric structure

by a cysteine-rich domain (CRD). Notably, the ECD and CRD are not linked by disulfide bonds (Hu *et al.*, 2001). Previous studies reported that there are N-linked glycosylation sites Asn-Xaa-Ser/Thr in the CaSR ECD, phosphorylation sites of protein kinase C (PKC), protein kinase A (PKA) within the intracellular loops of the CaSR (Alfadda *et al.*, 2014). Based on the site-directed mutagenesis, it was suggested that the glycosylation sites are efficiently glycosylated, and the disruption of those N-glycosylation sites could decrease the cell-surface expression of the CaSR, while N-linked glycosylation is likely not pivotal for CaSR biological action such as signal transduction (Ray *et al.*, 1998). Plus, evidence showed that a certain number of N-glycosylation sites were still needed for the cell membrane expression of the CaSR (Ray *et al.*, 1998).

The TMDs (TM1–TM7) were identified as seven hydrophobic regions that traverse the cell membrane, and those seven domains keep apart from each other in the absence of the CaSR ligand (Garrett *et al.*, 1995). Based on early studies, the important roles of TMDs are involved in the G protein-mediated cellular signaling pathways, given that the intracellular loops of other GPCRs were associated with the coupling of receptors to G proteins (Chang *et al.*, 2000). Meanwhile, it was proposed that TMDs may also affect the CaSR dimerization via noncovalent interactions (Zhang *et al.*, 2001). Moreover, an early study also proposed that the TM5 region of the CaSR may be involved in the CaSR dimerization (Hebert *et al.*, 1996). Moreover, one study found that the TMDs are also involved in the ligands binding, as additional calcium ion binding sites were confirmed in the TMDs. In this study, they found that even though they lack the ECD, the chimeric CaSR expressed at the cell surface could still respond to the extracellular calcium ion when the allosteric agonists such as NPS R-568 were subjected to the CaSR (Hauache *et al.*, 2000). According to their speculation, extracellular loops of TMDs are the only portion of TMDs exposed to the extracellular environment, and the acidic residues in extracellular loops of TMDs are likely responsible for the extracellular Ca^{2+} binding (Hauache *et al.*, 2000).

Compared with both the ECD and 7 TMDs, the residues within the signal peptide (SP), and residues in ICD are less conserved among various species. Notably, relatively conserved regions of the CaSR ICD remain between species, such as the region from positions 863 to 925 and the region spanning residues 960–984 (Ray *et al.*, 1997b). It was reported that the region spanning residues 863–925 within the ICD plays an essential role in the expression of the CaSR at the cell-surface level, and the biological activity of the CaSR, while the region spanning residues 960–984 was associated with the bonding to accessory proteins (Hjalm *et al.*, 2001). Early studies demonstrated that CaSR mutants occurring between positions 863 and 925 made the CaSR unable to respond to the extracellular calcium without impairing the CaSR expression level, ultimately contributing to the dysfunction of the CaSR such as in familial hypocalciuric hypercalcemia or autosomal dominant hypocalcemia (Bai *et al.*, 1997; Ray *et al.*, 1997b). Moreover, researchers also found that the truncation of residue 888 caused a calcium ion release from intracellular stores instead of the intracellular calcium influx induced by the CaSR activation, which ultimately blocked the PLC signaling pathway (Chang *et al.*, 2000). Interestingly, in comparation with the normal CaSR, the CaSR mutants which had a termination codon at position 892 or CaSR mutant that lost the region between positions 894 and 1075 were both shown to have a higher cell-surface expression level of the CaSR and a more active response on the release of intracellular calcium from calcium stores (Bai *et al.*, 1998; Bai, 2004). Additionally, early studies claimed that the CaSR ICD is very critical for both the characteristic positive cooperativity of the CaSR and the CaSR desensitization rate after its repeated exposure to the extracellular calcium, given the fact that CaSR mutant truncating at residue 868 showed lower sensitivity to the extracellular calcium (Bai, 2004).

2 The localization and physiological role of the calcium-sensing receptor in different tissues

As previously stated, the prominent role of the CaSR at the physiological level is to maintain the systemic calcium metabolism in the whole body (Brown, 2000). Specifically, as shown in Figure 17.3, the CaSR has been shown to strike the calcium balance by modulating PTH in parathyroid glands, modifying the calcium

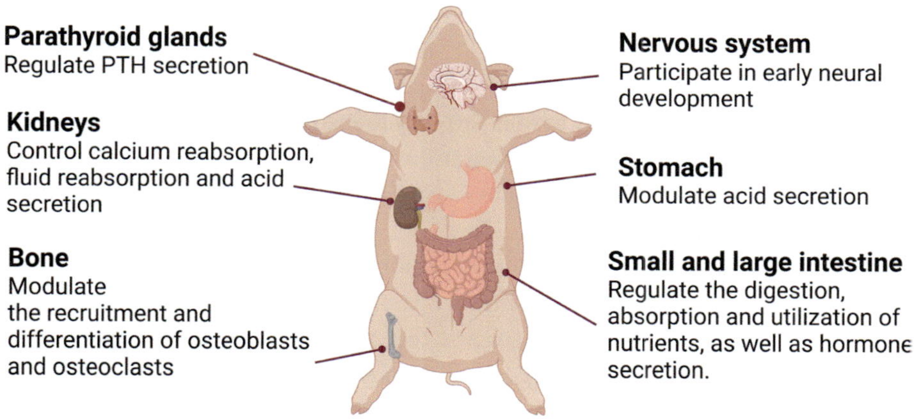

FIGURE 17.3 Summary of the distribution and localization of the calcium-sensing receptor (CaSR) in various types of tissues and relative physiological roles

reabsorption in the kidney, regulating the recruitment and differentiation of osteoblasts and osteoclasts in the bone, promoting keratinocyte differentiation in the skin, as well as controlling the ingestion and absorption of calcium in the digestive system (Brown, 2000; Bai, 2004; VanHouten *et al.*, 2004; Alfadda *et al.*, 2014). Specifically, the CaSR expressed in parathyroid and thyroid glands and those tissues mentioned above has a significant role in modulating systemic calcium homeostasis by sensing circulating concentration of extracellular calcium within the body (Chun *et al.*, 2012). For instance, when the level of extracellular calcium is low, the CaSR regulates the calcium balance by promoting the production of PTH from the parathyroid gland and reducing the calcitonin production in the thyroid gland (Hu and Spiegel, 2007).

The presence of higher levels of circulating PTH improves bone resorption, promotes calcium efflux from the skeleton, and aids the secretion of renal 1, 25-dihydroxyvitamin D3 production, finally contributing to the increased calcium absorption in the digestive tract (Quarles, 2003). Moreover, it has been reported that the CaSR-induced reduction of calcitonin production responds to a lower level of circulating calcium caused by the inhibition of bone resorption and elevating renal calcium excretion, contributing to an increased overall level of calcium (Quarles, 2003).

2.1 *Calcium-sensing receptor in the parathyroid*

The CaSR was first isolated from the parathyroid where the CaSR was found in caveolae of parathyroid cells, making parathyroid cells popular in exploring the cell-signaling role of the CaSR in most studies (Kifor *et al.*, 1998). In general, these studies found that the CaSR maintains calcium homeostasis by regulating the PTH secretion in response to the changes of extracellular Ca^{2+} levels (Conigrave, 2016). Early research showed that the CaSR maintains blood calcium levels, especially the blood ionized calcium concentration within a narrow range (around 1.2 mM) by regulating the accurate release of PTH into circulation (Chen and Goodman, 2004). The small changes in blood ionized calcium concentrations could be monitored by chief cells in the parathyroid through the response of the CaSR (Egbuna and Brown, 2008). Furthermore, the mutations of the CaSR have been reported to cause clinical disorders of calcium metabolism (Pollak *et al.*, 1993; Pollak *et al.*, 1994; Chou *et al.*, 1995).

Based on previous studies, vitamin D response elements are involved in promoter elements controlling the alternative transcripts of the CaSR (Canaff and Hendy, 2002). Plus, some findings suggest that the CaSR cell-surface expression level in parathyroid could be affected by calcitriol or 1, 25-dihydroxy-vitamin D, so vitamin D was involved in regulating the CaSR expression (Chen and Goodman, 2004). Interestingly, current evidence also found that the rising level of extracellular calcium does not affect the CaSR expression level in rat parathyroid, although some studies documented that the increasing extracellular calcium concentration could upregulate CaSR expression in the chicken parathyroid gland (Cetani *et al.*, 2000; Chang *et al.*, 2000; Chang *et al.*, 2001). In this context, the upregulation of CaSR expression and the subsequent activation of the CaSR, in turn, posed a positive response on the CaSR-mediated calcium homeostasis in parathyroid glands (Chang *et al.*, 2001). Notably, the extracellular calcium also modulates the expression of the vitamin D receptor, which could promote the effect of vitamin D on the expression level of the CaSR and boost the biological effect of the CaSR (Chang *et al.*, 1999a). Given that the CaSR plays a crucial role in the calcium metabolism under normal and pathophysiological conditions in parathyroid cells, further studies about CaSR's function in parathyroid glands are needed.

2.2 *Calcium-sensing receptor in the kidney*

Following the characterization of the CaSR in parathyroid glands, the CaSR was identified in the kidney, where the CaSR also serves an essential function in calcium homeostasis (Riccardi *et al.*, 1995; Toshimasa., 1995). Interestingly, the identification results of the CaSR protein expression in the kidney differ among various studies. For example, by employing the immunohistochemistry technique, some researchers found that the CaSR was detected throughout the main parts of the nephron, including the proximal convoluted tubule, thick ascending limb (TAL) (Ye *et al.*, 1997), distal convoluted tubule, and cortical collecting ducts (Riccardi *et al.*, 1998). However, other researchers disagree with the previously mentioned findings on the basis that they did not find detectable CaSR protein expression in the proximal convoluted tubule, distal convoluted tubule, as well as collecting duct, even though other studies found a large amount of CaSR mRNA in the distal convoluted tubule, cortical collecting ducts, and TAL in rat kidneys (Riccardi *et al.*, 1996; Yang *et al.*, 1997; Loupy *et al.*, 2012; Toka *et al.*, 2012). Besides, early studies localized the CaSR protein to the apical brush-border membrane in the proximal tubule, basolateral membranes in thick ascending limbs, and the apical plasma membrane of inner medullary collecting duct cell (Butters *et al.*, 1997; Riccardi *et al.*, 1998). Since each part of the nephron serves a specialized function, we propose that the role of the CaSR in each segment mentioned above is also different and unique.

Early research reported a high abundance of CaSR protein expression on the basolateral membranes of cells in the TAL where reabsorption of NaCl takes place (Riccardi *et al.*, 1998; Riccardi, 2012; Toka *et al.*, 2012). That resulted in a lumen-positive transepithelial difference that drives calcium and magnesium transportation (Bourdeau and Burg, 1979; Shareghi and Agus, 1982). Previous experiments documented that calcium and magnesium absorption in the TAL decreased with the presence of increasing levels of peritubular calcium or magnesium (Wong *et al.*, 1983; Quamme, 1989). When agonists like calcium and neomycin stimulated the CaSR, the subsequent activation was found to impair the activity of 70-pS apical membrane potassium channel via the cytochrome P-450-dependent metabolite of arachidonic acid-involved regulation in the TAL (Wang *et al.*, 1996). In addition, some researchers reported that the CaSR modulates the intracellular pH in a chloride-dependent manner in the medullary TAL (Aslanova *et al.*, 2006).

2.3 *Calcium-sensing receptor in the bone*

As it does in both the parathyroid glands and the kidney, the CaSR has been found to serve important roles in the bone. In 1997, CaSR expression was initially confirmed in the bone, as researchers reported the expression of the CaSR in the mouse osteoblastic cell line (Yamaguchi *et al.*, 1998). After confirming the CaSR in the mouse osteoblastic cell line MC3T3-E1, subsequent research identified the CaSR expression in the osteoblasts of mouse and bovine (Chang *et al.*, 1999b). These studies documented the CaSR-induced increase in proliferation of osteoblasts, although the related signaling pathway is still unclear (Dvorak *et al.*, 2004). Additionally, evidence shows that when the CaSR was activated, it suppressed the bone resorption within the osteoclasts (Kameda *et al.*, 1998). The stimulation of the CaSR has been demonstrated to enhance differentiation and apoptosis in osteoclasts, while the CaSR-induced differentiation promotion might result from the parathyroid hormone-related protein (PTHrP) as the stimulation of the CaSR has been shown to raise the expression level of PTHrP (Mentaverri *et al.*, 2006; Ahlstrom *et al.*, 2008).

2.4 *Calcium-sensing receptor in the nervous system*

The presence of the CaSR has also been confirmed in the enteric nervous system, like the nerve plexus. Employing immunocytochemistry and in situ hybridization technique to confirm the CaSR in almost all areas of the brain, it was shown that the expression of the CaSR was broadly found across the central nervous system, and the difference between expression levels of the CaSR is huge (Yano *et al.*, 2004). Interestingly, the expression profile of the CaSR in the nervous system resembles that of GPCRs-mGluR and GABAB receptors, and recent studies also found that the CaSR could heterodimerize with mGluR and GABAB receptors (Gama *et al.*, 2001; Chang *et al.*, 2007).

Researchers found that the heterodimerization of the CaSR with the GABAB receptor formed signaling and membrane-anchored receptor complexes, and the complexes appear to enhance CaSR expression and signaling responses to the increasing level of the extracellular calcium in the brain (Chang *et al.*, 2007; Cheng *et al.*, 2007). Even though high level of CaSR expression was found in the subfornical organ and olfactory bulbs, as well as hippocampus, there was no positive signal of the CaSR detected in mitral cell layer and external plexiform layer. In the piriform cortex, the CaSR mRNA abundance was moderate, while the expression of this receptor in the cerebral cortex was deficient (Rogers *et al.*, 1997). Plus, pronounced expression of the CaSR was found in the cornu ammonis 3 region of the hippocampus, while mild to moderate expression levels of the CaSR was observed in other regions such as the basal ganglia and ventrolateral nucleus of the thalamus (Rogers *et al.*, 1997).

Like the various expression patterns of the CaSR in many regions of the brain and cells expressing the CaSR, CaSR's function in the nervous system is also quite diverse. In neurons, the CaSR has been shown to participate in neuronal growth and migration, synaptic plasticity, and neurotransmission (Ruat and Traiffort, 2013). Plus, there is emerging evidence that the CaSR is involved in the myelinogenesis and the modulation of the central nervous system ionic microenvironment in oligodendroglial cells (Ruat and Traiffort, 2013). Moreover, researchers also proposed that the CaSR may still serve an essential role in the extracellular calcium homeostasis in the brain. As stated previously, the high densities of the CaSR mRNA in the brain were detected in the subfornical organ, implying that the CaSR may serve a significant role in region-specific neuronal functions (Ferry *et al.*, 2000). As a pivotal hypothalamic thirst center, the SFO is exposed to systemic fluid because it is located outside the blood-brain barrier (Ruat and Traiffort, 2013). There is experimental evidence indicating that the role for the CaSR in the subfornical organ is likely to sense the fluctuation of extracellular calcium concentration when the ionic strength changes and then the CaSR is involved in the central regulation of systemic fluid and electrolyte balance (McKinley *et al.*, 2004). According to the experiment conducted in hippocampal pyramidal neurons, it was shown that the CaSR participated in the regulation of the calcium-activated potassium channels and may block neuronal excitability and neurotransmission through membrane repolarization (Schneggenburger and Neher, 2000). Also, the abundant expression of the CaSR in hippocampal neurons implies that the CaSR also regulates the activation of calcium-permeable, nonselective cation channels such as Na^+-leak channel with the decrease of extracellular calcium levels (Vassilev *et al.*, 1997; Yano *et al.*, 2004). Interestingly, those stimulations of calcium-permeable, nonselective cation channels through the CaSR are activated by a CaSR agonist named amyloid-β peptide, which shows high expression levels in patients with Alzheimer's disease (Ye *et al.*, 1997). Hence, some researchers have proposed that the CaSR might play a supporting role in Alzheimer's disease, given that the other type of CaSR agonist called apolipoprotein E also has a close association with this disease (Conley *et al.*, 2009).

2.5 *Calcium-sensing receptor in the stomach*

After the first identification of the CaSR in Necturus gastric mucosa within the stomach by using RT-PCR, the CaSR was detected in the basolateral membrane of surface epithelial cells in the antrum by employing immunohistochemistry, following the confirmation of the CaSR in human gastric epithelial and glandular cells as well as cells of submucosal and myenteric plexuses (Cima *et al.*, 1997; Ray *et al.*, 1997a; Cheng *et al.*, 1999). It is well known that gastric acid is closely associated with nutrient digestion and absorption in the stomach, and the expression of the CaSR was detected in gastric cells of the basement membrane (Cheng *et al.*, 1999). Regarding acid secretion, the classical regulation pathways contain the paracrine, endocrine, and neuroendocrine systems as well as the H^+-K^+-ATPase activation, which mediates the production of HCL and the modulation of gastric acid secretion (Prinz *et al.*, 1992; Aditi and Graham, 2012; Xie *et al.*, 2014). Previous research demonstrated that Ca^{2+} is the second messenger involved in these pathways. Both in vivo and in vitro studies have shown that the CaSR serves pivotal roles in gastric acid secretion. In human studies, it was suggested that the stipulation of the CaSR results in the elevation of serum gastrin concentration and basal gastric acid secretion (Ceglia *et al.*, 2009). The complicated mechanism of this process is as follows: the activation of the CaSR located on the basal aspects of parietal cells contributes to an elevation in intracellular Ca^{2+} levels, which promotes parietal cell H^+-K^+-ATPase activity and the following gastric acid release, while inhibition of the CaSR causes a reduced acid secretion even when glands are subjected to classic secretagogues (Geibel *et al.*, 2001; Ceglia *et al.*, 2009). Experiments have suggested that factors like L-AAs, pH, and exposure to a high level of extracellular calcium could stimulate the CaSR in G cells, contributing to the stimulation of gastrin secretion (Hebert *et al.*, 2004). Specifically, the activation of the CaSR with lower extracellular Ca^{2+} levels resulted in a proliferative response in human gastric mucous epithelial cells (Rutten *et al.*, 1999; Feng *et al.*, 2010). Moreover, when the CaSR was stimulated by gadolinium (Gd^{3+}) the activation of the CaSR contributed to an elevation in intracellular Ca^{2+} levels, leading to the stimulation of gastrin secretion (Gama *et al.* 2001). Interestingly, these studies also found that when the function of the CaSR was blocked, it similarly led to a reduced acid secretion (Ruat *et al.*, 1995; Busque *et al.*, 2005). Furthermore, some researchers have found a high expression level of the CaSR on human gastrinoma cells, while the stimulation of the CaSR stimulates gastrin secretion (Feng *et al.*, 2010). Combining all of these findings, researchers proposed that the CaSR serves an essential function of acid secretion, repair of the damaged gastric mucosa, and maintaining normal G-cell numbers (Bevilacqua *et al.*, 2005; Dufner *et al.*, 2005).

2.6 *Calcium-sensing receptor in the small and large intestine*

So far, the CaSR has been confirmed along the gastrointestinal tract including the duodenum, jejunum, ileum and colon, which play very important roles in the digestion, and absorption of nutrients, as well as the secretion of hormones regulating the flow of water and ions to maintain water and electrolyte homeostasis (Chattopadhyay *et al.*, 1998). Specifically, the presence of the CaSR in the small and large intestine was identified on the apical and basolateral membranes of both colonic crypt cells and villi epithelial cells located in the small intestine (Chattopadhyay *et al.*, 1998; Cheng *et al.*, 2002).

Data obtained from early studies revealed that the CaSR was involved in the calcium homeostasis in the small intestine when the CaSR expressed in the small intestine was subjected to different concentrations of extracellular calcium as well as vitamin D (Favus *et al.*, 1981). Recent studies have proven that the CaSR could sense L-AAs such as phenylalanine, tryptophan, asparagine, arginine, and glutamine in the small intestine and stimulate gut peptide secretion with the presence of extracellular calcium in intestinal enteroendocrine cells (Mace *et al.*, 2012). According to these studies, the CaSR acts as a pivotal modulator of intestinal enteroendocrine cells like K- and L-cell activity when the CaSR was subjected to both nutrients and non-nutrients (Mace *et al.*, 2012). Specifically, these studies establish a new role for the CaSR as the modulator of gut peptide secretion that detects nutrients in the small intestine and participates in the stimulation of gut hormones including gluco-insulinotropic peptide (GIP), glucagon-like peptide-1 (GLP-1), and peptide YY (PYY) (Mace *et al.*, 2012).

As for the colon, its main role is modulating fluid absorption and secretion, while colonic surface and crypt cells have been shown to serve an important function in both nutrient absorption and secretion (Quinn *et al.*, 1997; Lin *et al.*, 1998). Previous studies focusing on colonic fluid transport and gastric acid secretion showed that the CaSR in the colon is involved in the two main roles mentioned above, making the CaSR a promising target to prevent secretory diarrhea (Geibel and Hebert, 2009). Moreover, the CaSR has been demonstrated to sense nutrient availability to epithelial cells located in the digestive system and possibly affect the rapid turnover observed in the intestinal epithelium (Geibel and Hebert, 2009).

Together with all the findings previously stated, the effect of the CaSR in the small and large intestine provides new insights for both inhibiting diarrhea and inflammatory bowel disease and reducing the risk for colon cancer, thus making the CaSR a pharmacological target. Diarrheal diseases have been a serious problem in both human health and animal production. Indeed, secretory diarrheas are caused by excessive secretion and/or impaired absorption of fluid and electrolytes across the intestinal epithelium (Thiagarajah *et al.*, 2015). Studies have demonstrated that the intestinal fluid secretion in secretory diarrhea are closely associated with a range of ion and solute transport proteins, the stimulation of cyclic nucleotides, and Ca^{2+} signaling pathways (Frizzell and Hanrahan, 2012). It was suggested that cholera toxin-mediated diarrhea is frequently accompanied by an elevation in cAMP, while the *Escherichia coli* heat stable toxin increases the production of the cGMP level (Kopic and Geibel, 2010; Foulke-Abel *et al.*, 2020). This fluid and electrolyte imbalance resulted from the secretory process because of chloride loss, and the impaired intestinal absorptive capacity (Lucas, 2001). Based on the functional assay, it was suggested that the CaSR expressed along the intestinal tract plays a negative role in the action of cAMP- or cGMP-inducing secretagogues which includes the cholera toxin and heat stable *Escherichia coli* enterotoxin STa on fluid transport in the intestine. Specifically, the CaSR performs this function through promoting the degradation of the previously stated cyclic nucleotides via phosphodiesterase, which contributes to inhibition of secretion and elevation of absorption (Geibel *et al.*, 2006). Additionally, the activation of the CaSR stimulated by small-molecule CaSR agonists was able to reverse cholera toxin- and STa endotoxin-induced fluid secretion, which provides a unique therapy for secretory diarrheas (Geibel *et al.*, 2006).

3 Calcium-sensing receptor ligands binding and signaling pathways

Current studies reveal that the CaSR is activated by a variety of ligands, including polyvalent cations (such as Ca^{2+}, Mg^{2+}), L-AAs, and drugs like cinacalcet and NPS 2143 (Leach *et al.*, 2013; Alfadda *et al.*, 2014). In addition, it has been demonstrated that charged residues in the CaSR ECD play a pivotal role in the CaSR stimulation via electrostatic interactions. Thus, researchers have proposed that pH could be a potential regulator of the CaSR activation response by its polycationic agonists (Quinn *et al.*, 2004). Besides, ionic strength is also capable of regulating the activity of the CaSR (Quinn *et al.*, 1998). The activation of the CaSR by different types of ligands may stabilize the CaSR in unique activation states, contributing to different signaling pathways (Tennakoon *et al.*, 2016). Additionally, an early study reported that a ligand could behave as an agonist for one cellular signal transduction, but it acts as a negative receptor modulator, partial ligand or even fails to alter another signaling (Thomsen *et al.*, 2012).

3.1 *Calcium-sensing receptor activation*

Even though extracellular Ca^{2+} is the principal orthostatic regulator of the CaSR, other compounds are thought to stimulate this receptor, too (Riccardi, 2002). The agonists of the CaSR could be categorized into two main types: agonists and antagonists. Agonists can also be divided into two types: Type I agonists, including direct agonists such as Ca^{2+}, other divalent such as Mg^{2+}, trivalent cations like La^{3+}, spermine, aminoglycosides, antibiotics, some dietary peptides, and Type II agonists (positive allosteric modulators such as cinacalcet and aromatic L-AAs, which can only activate the CaSR with the presence of extracellular calcium) (Table 17.1). Researchers proposed that those CaSR agonists with higher positive

TABLE 17.1 Summarized modulators of calcium-sensing receptor (CaSR)

Modulator type	Ligands	Binding region	Reference
Orthostatic agonists	Cations Ca^{2+}, Mg^{2+}, Sr^{2+}, Ba^{2+}, La^{3+}, Gd^{3+}	Extracellular domain	Navarro *et al.*, 1999; Riccardi, 2002
Allosteric agonists	L-Amino acids L-leucine, L-tryptophan, L-phenylalanine, L-arginine	Extracellular domain, Transmembrane domains	Conigrave *et al.*, 2000; Brown, 1991; Busque *et al.*, 2005; Zhao *et al.*, 2018
Allosteric agonists	Synthetic positive modulators NPS R-568, NPS R-467	Extracellular domain, Transmembrane domains	Saidak *et al.*, 2009; Sood *et al.*, 2000; Nemeth *et al.*, 1998
Orthostatic/ Allosteric agonists	Polyamines and polypeptides Spermine, γ-glutamyl peptides including GSH (γ-Glu-Cys-Gly) and γ-Glu-Val-Gly, γ-Glu-Val, γ-Glu-Leu, and γ-Glu-Val-Lys	Extracellular domain, Transmembrane domains	Canaff *et al.*, 2001; Cheng *et al.*, 2004; Ohsu *et al.*, 2010; Wang *et al.*, 2006
Allosteric antagonists	Synthetic negative modulator NPS 2143, calhex 231, compound 3	Transmembrane domains	Nemeth *et al.*, 2001; Brauner-Osborne *et al.*, 2007
Allosteric modulators	pH and ionic strength	–	Doroszewicz *et al.*, 2005; Quinn *et al.*, 2004; Saidak *et al.*, 2009

charge density tend to have higher agonistic potency (Lee *et al.*, 2007; Saidak *et al.*, 2009). As for aromatic L-AAs, they can enhance the CaSR sensitivity, thus left-shift the concentration-response curve of the CaSR activation induced by the extracellular Ca^{2+} (Conigrave *et al.*, 2000). So far, no orthostatic antagonist of the CaSR has been found, whereas negative allosteric modulators of the CaSR include NPS 2143, ronacaleret, and calhex 231 (Yanamala and Klein-Seetharaman, 2010; Gannon *et al.*, 2014). Plus, those negative allosteric modulators of the CaSR make the CaSR less sensitive to the extracellular calcium, ultimately contributing to a right-shift of the CaSR activation induced by the extracellular calcium (Conigrave *et al.*, 2000).

3.1.1 Cations

Divalent cations such as extracellular Ca^{2+} and Mg^{2+} belong to the Type I agonists of the CaSR, which could activate this receptor directly. Under normal situations, it has been demonstrated that both the extracellular Ca^{2+} and Mg^{2+} play negative regulation on the PTH secretion in a dose-dependent manner (Navarro *et al.*, 1999). Besides, even though the CaSR activity in the presence of extracellular Mg^{2+} is usually like that of extracellular Ca^{2+}, inconsistent features between these two endogenous divalent cations have also been observed. For instance, when patients with hypomagnesemia were subjected to a low level of extracellular Mg^{2+}, researchers witnessed a paradoxical decrease in PTH secretion, which was opposite to those of PTH release in the presence of a low concentration of extracellular Ca^{2+} (Anast *et al.*, 1972; Mennes *et al.*, 1978; Duran *et al.*, 1984).

Apart from Ca^{2+} and Mg^{2+}, divalent and trivalent cations including Sr^{2+}, Ba^{2-}, La^{3+}, and Gd^{3+} were also identified as agonists for the CaSR (Riccardi, 2002). Notably, when the chimeric receptor formed by both the mGluR ECD and the TMD as well as the ICD of the CaSR was subjected to Gd^{3+}, it could be activated which did not occur with the presence of the extracellular Ca^{2+}, implying that not all CaSR agonist cations bind to the same binding sites of the CaSR (Hammerland *et al.*, 1999). Moreover, the sensitivity of the CaSR to those cations might vary among different species as well as different types of cells within the body (Yamashita and Hagiwara, 1990; Garrett, 1995).

3.1.2 L-Amino acids

Family C members of GPCRs such as glutamate and GABA receptors and the CaSR all have a VFT domain, which contains several amino acid binding sites, suggesting that all of these GPCRs could sense amino acids (Wellendorph and Brauner-Osborne, 2009; Hendy and Canaff, 2016). In HEK-293 cells which shows stable expression of the CaSR, amino acids were first identified as allosteric agonists of the CaSR in the presence of extracellular Ca^{2+}, and research found that the CaSR appears to show higher affinity for large L-AAs, specifically aromatic amino acids like L-Phe, and L-Trp (Conigrave *et al.*, 2000). Besides, several aliphatic L-AAs, such as L-leucine and poly-L-arginine also affect the CaSR activity (Brown, 1991; Busque *et al.*, 2005). In general, researchers claim there is a link between the extracellular Ca^{2+} and L-AAs metabolism, given it has been known for years that L-AAs metabolism is involved in the modulation of urinary Ca^{2+} excretion and the production of PTH (Conigrave *et al.*, 2002). Meanwhile, in the gastrointestinal tract, L-AAs were also found to participate in many physiological activities through the CaSR. For example, it has been demonstrated that L-AAs participate in the CaSR-mediated control of gastric acid secretion in the stomach, and some studies also suggest that some aromatic L-AAs appear to mediate gastric luminal pH in the stomach (Conigrave *et al.*, 2002; Saidak *et al.*, 2009). It is worth noting that those physiological activities that have been shown by L-AAs were considered to be small effects, and thus researchers proposed that L-AAs might only participate in the fine-tuning of CaSR signaling (Saidak *et al.*, 2009).

3.1.3 Synthetic allosteric modulators

Except for naturally occurring CaSR agonists such as aromatic L-AAs, some synthetic pharmacological agents have recently been identified as allosteric modulators for the CaSR, and they could be categorized as calcimimetics and calcilytics (Saidak *et al.*, 2009; Alfadda *et al.*, 2014). These synthetic pharmacological compounds were designed to control the release of PTH secretion in patients with hyperparathyroidism. In addition, these ligands have been shown to share similar structural features with aromatic L-AAs, as they all have an aromatic ring and positively charged amine groups (Saidak *et al.*, 2009). In general, as previously stated, the calcimimetics and calcilytics do not activate the CaSR in the absence of Ca^{2+} but instead alter the concentration-response curves of the CaSR activation response stimulated by the extracellular calcium in a dose-dependent manner (Armato *et al.*, 2013). Plus, calcimimetics act as positive modulators or agonists for the CaSR, whereas calcilytics are considered to be CaSR antagonists that reduce the sensitivity of the CaSR to extracellular Ca^{2+} (Sood *et al.*, 2000).

The first calcimimetics were discovered in 1997 and belong to the phenylalkylamine group, as researchers found NPS R-568 and NPS R-467 could elevate the intracellular level of Ca^{2+} and thus suppress the release of PTH secretion (Nemeth *et al.*, 1998). Moreover, those small molecular drugs designed to target the CaSR have also been demonstrated to act as potent, stereoselective allosteric modulators for the CaSR (Nemeth *et al.*, 1998). As for NPS 2143, it was the first calcilytic that could inhibit the CaSR activity in the presence of extracellular calcium (Nemeth *et al.*, 2001). Since then, other calcilytics such as calhex 231 and compound 3 have also been identified as negative allosteric modulators for the CaSR (Brauner-Osborne *et al.*, 2007). In terms of orthostatic ligands, their binding sites were in the ECD of the CaSR, which is different from allosteric modulators where the seven TMD domains were their binding sites (Miedlich *et al.*, 2004; Petrel *et al.*, 2004). Except for modulating the sensitivity of the CaSR, allosteric modulators could also affect the expression level of this receptor. Several experiments have reported that the calcimimetic NPS R-568 could elevate the CaSR expression level, while the calcilytic NPS 2143 reduced the expression of this receptor (Huang and Breitwieser, 2007). Both calcimimetics and calcilytics are potential therapeutic drugs targeting diseases caused by CaSR dysfunction. Because they have specific targets, they have better performance than conventional CaSR ligands (Saidak *et al.*, 2009).

3.1.4 Polyamines and polypeptides

So far, several kinds of endogenous polyamines have been demonstrated to be modulators for CaSR activity, including spermine, spermidine, and putrescine (Quinn *et al.*, 1997). Polyamines exist in various tissues within the body and have been identified as participating in various physiological activities, including cellular metabolism, growth, differentiation, and neurotransmission (Shin *et al.*, 2005; Kurata *et al.*, 2007). Studies have also suggested that extracellular polyamines play a pivotal role in the normal postnatal development, maintenance, and function of gastrointestinal epithelia through CaSR-mediated activity (Gama *et al.*, 1997; Cheng *et al.*, 2004). Interestingly, previous studies stated that the polyamines mediated CaSR response required the presence of extracellular calcium, but later research found that spermine could result in an intracellular calcium influx without the presence of extracellular calcium (Canaff *et al.*, 2001; Cheng *et al.*, 2004). These studies suggested that parts of polyamines act as orthostatic agonists, whereas others are allosteric modulators.

Regarding polypeptides, studies document that many γ-glutamyl peptides have been identified as CaSR agonists. Recent studies have identified 46 γ-glutamyl peptides as the CaSR agonists, including GSH (γ-Glu-Cys-Gly) and γ-Glu-Val-Gly, γ-Glu-Val, γ-Glu-Leu, and γ-Glu-Val-Lys (Ohsu *et al.*, 2010). Early research documented that GSH could not activate the CaSR without the presence of the extracellular Ca^{2+} but enhanced the sensitivity of this receptor. Specifically, researchers found that the GSH-induced CaSR activity only occurred when this receptor was subjected to the 0.8 mM extracellular calcium, suggesting that those polypeptides might act as allosteric agonists for the CaSR (Wang *et al.*, 2006). These polypeptides belong to kokumi taste substances which have been applied in food production as taste enhancers, which might explain the presence of the CaSR in mammalian taste cells (Ohsu *et al.*, 2010).

3.1.5 Aminoglycoside antibiotics

Similar to the polypeptides, aminoglycoside antibiotics were initially identified to mimic the roles of extracellular Ca^{2+} on parathyroid function (Brown *et al.*, 1991) as aminoglycoside antibiotics, neomycin, gentamicin, and tobramycin are known to be agonists of the CaSR (McLarnon *et al.*, 2002). These three aminoglycoside antibiotics produced a level-dependent elevation in intracellular calcium with EC_{50} values of 43, 258, and 177 μM, respectively (McLarnon *et al.*, 2002). In addition, data from the same study showed that the minimum concentration of kanamycin necessary for the CaSR activation should be above 1 mM (McLarnon *et al.*, 2002).

Additionally, the potency of the aminoglycoside antibiotics to the CaSR is different, implying that efficacies for aminoglycoside antibiotics correlate positively with the number of their attached amino groups (McLarnon *et al.*, 2002).

3.1.6 pH and ionic strength

Both the pH and ionic strength have been shown to regulate the activity of the CaSR, which could be explained by the fact that some CaSR modulators are cations with positive charges, and those cationic compounds could bind to the CaSR via electrostatic interactions (Quinn *et al.*, 2004; Doroszewicz *et al.*, 2005). Both the increased and decreased pH were observed to affect the CaSR sensitivity to the extracellular calcium. In general, the elevated pH above the physiological range (pH 7.0–7.8) could make the CaSR more sensitive to the extracellular Ca^{2+} and Mg^{2+}, whereas a decreasing pH showed the opposite effect (Doroszewicz *et al.*, 2005). However, researchers also proposed that when the pH dropped further to below 5.5, the sensitivity of this receptor did not decrease. Instead, they found that the sensitivity of this receptor to the extracellular calcium showed some recovery (Quinn *et al.*, 2004; Saidak *et al.*, 2009). Moreover, studies also reported that the pH mediated CaSR activity might be specific to the CaSR, as the other GPCR showed little change in activity with changes in pH (Saidak *et al.*, 2009).

Apart from external pH alterations, ionic strength changes of the extracellular milieu have been shown to influence the sensitivity of the CaSR (Brown and MacLeod, 2001). Studies found that when the CaSR was subjected to increasing ionic strength, the sensitivity of this receptor to its positive modulators subsequently dropped (Quinn *et al.*, 1998). For example, an early study reported that the addition of a higher concentration of extracellular sodium chloride could right shift the EC_{50} for the CaSR activity to its agonists such as extracellular Ca^{2+} and spermine, and effects of ionic strength on the potency of this receptor have been demonstrated to be higher than that of extracellular calcium ion (Quinn *et al.*, 1998). Therefore, the CaSR was considered to be not only a pH sensor but also an ionic strength sensor (Quinn *et al.*, 1998).

3.2 *Calcium-sensing receptor regulated intracellular signaling pathways*

As described earlier, the CaSR belongs to the GPCR members, and it interacts with several types of heterotrimeric G-proteins, including Gα subunits as well as Gβγ subunits to regulate the activity of downstream signaling pathways (Conigrave and Ward, 2013). Specifically, activation of the CaSR is associated with the activation of heterotrimeric G proteins, which contributes to the dissociation of the Gα subunit from the Gβγ subunits, initiating different downstream signaling pathways (Thomsen *et al.*, 2012). The Gα subunit could be divided into several subfamilies including Gq/11, Gi/o, Gs, and G12/13. Once the ligand binds to GPCRs, it could activate Gq/11, and then elicit the activation of phospholipase Cβ (PLCβ), subsequently lead to the PLCβ cleavage be a phospholipid named diacylglycerol and inositol 1,4,5-triphosphate (IP3) (Sassmann *et al.*, 2010). This process eventually causes mobilization of intracellular Ca^{2+} and the phosphorylation of protein kinase C (PKC), subsequently causing the mitogen-activated protein kinase (MAPK) activation. Furthermore, when the CaSR is stimulated by the ligands, besides the activation of G proteins, β-arrestins could also negatively regulate the CaSR activation by regulating the phosphorylation and activation of extracellular signal-regulated kinases 1 and 2 (ERK1/2) (Rajagopal *et al.*, 2010; Hales, 2011). Furthermore, the activation of the MAPK cascade caused by the activation of Gq/1 could also lead to phosphorylation and activation of ERK1/2 (Kifor *et al.*, 2001). Besides, the activation of the G12/13 family leads to the activation of the small G protein Ras homolog family member A (RhoA), and the CaSR has been proven to be involved in the activation of RhoA (Pi *et al.*, 2002). Notably, the activation of Gs also occurs in some cell types (Hendy *et al.*, 2009). In this context, the CaSR was shown to modulate different intracellular downstream signaling pathways, as outlined in Figure 17.4.

It is well-accepted among researchers that the CaSR is closely related to intestinal development. One study found that the CaSR activation might regulate the intestinal cell proliferation and differentiation through the MAPK signaling pathway, as the upregulation of activator protein 1 (AP-1) induced by the phosphorylation and activation of ERK1/2 could mediate cell proliferation and differentiation

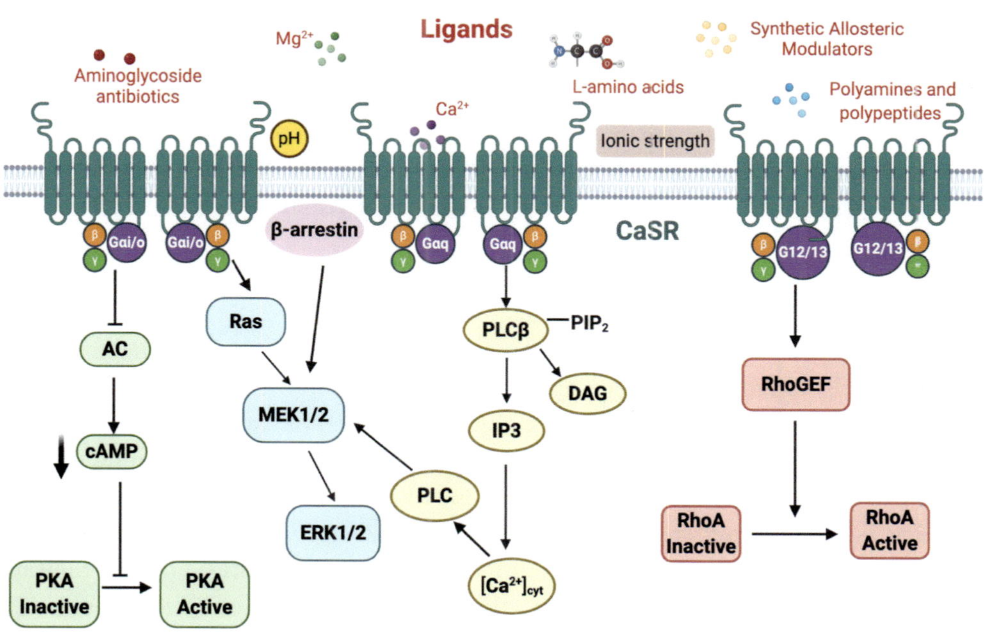

FIGURE 17.4 Generalized representations of calcium-sensing receptor (CaSR)-mediated signaling pathways

(Liu *et al.*, 2018). However, there is limited information about downstream signaling pathways triggered by the CaSR activation that improve gut health (such as nutrient absorption and gut integrity). Therefore, elucidation of the signaling pathways by the CaSR activation will provide new insights into underlying mechanisms of gut health maintained by the CaSR.

3.2.1 Gq signaling pathway

When the CaSR was initially expressed in *X. laevis* oocytes, researchers found that the extracellular calcium-induced CaSR activation could subsequently trigger the downstream intracellular PLC signaling pathway (Brown *et al.*, 1993). According to experiments conducted in HEK293 cells-expressing the CaSR, as well as in bovine parathyroid cells, it was shown that the activation of this receptor could subsequently also activate two other phospholipases, phospholipase D (PLD) and phospholipase A2 (PLA2) (Kifor *et al.*, 1997). Notably, some studies have suggested that the CaSR-mediated PLD activity might vary between different types of cell lines (Kifor *et al.*, 1997; Huang *et al.*, 2004). Research reported that the CaSR-mediated downregulation of the PKC in HEK293 cells contributed to a reduced PLD activity in the presence of extracellular Ca^{2+}, whereas the opposite results were identified in the MDCK cells stably overexpressing the CaSR (Huang *et al.*, 2004). Meanwhile, the activation of PLA2 stimulated by the CaSR activity was not fully regulated by the PKC activity (Handlogten *et al.*, 2001). As described above, the CaSR mainly modulates PLA2 activity through elevating intracellular Ca^{2+} concentrations via Gq stimulation of the PLC pathway (Handlogten *et al.*, 2001). This elevation of intracellular Ca^{2+} also had effects on calmodulin, which in turn resulted in the activation of calmodulin-dependent protein kinase. It has been demonstrated that these two proteins are needed for CaSR-regulated stimulation of PLA2 (Handlogten *et al.*, 2001). Plus, the same researchers also found that CaSR-mediated activation of ERK could also affect the phosphorylation and activation of PLA2 (Handlogten *et al.*, 2001).

3.2.2 MAPK signaling pathway

MAPKs belong to protein Ser/Thr kinases, and have been identified in all eukaryotic cells, and they participate in the regulation of apoptosis, metabolism, motility, and gene expression and differentiation (Widmann *et al.*, 1999). The conventional MAPKs signaling pathways contain ERK1/2, c-Jun amino (N)-terminal kinases (JNK1/2/3), p38 isoforms (α, β, γ, and δ), and ERK5, while the activation of the CaSR has been confirmed to be involved in those MAPK signaling cascades mentioned above (Chen *et al.*, 2001; Ogata *et al.*, 2006). A few studies investigated the regulation of ERK1/2 activation by the CaSR in both bovine parathyroid and HEK-293 cells with the CaSR stable expression, and researchers found that the elevating extracellular Ca^{2+} and CaSR agonists elicited the level-dependent phosphorylation of ERK1/2, while this up-regulated phosphorylation of ERK1/2 could also be blunted by the related kinase inhibitors (McNeil *et al.*, 1998; Kifor *et al.*, 2001). Moreover, another study confirmed that the activation of ERK by the CaSR could be inhibited with the transfection of dominant negative mutants of Ras, Raf, and MEK, suggesting that the CaSR mediated the activation of ERK via the CaSR-Src-Ras-Raf-MEK-ERK pathway (Hobson *et al.*, 2000). Meanwhile, Kifor *et al.* (2001) found that the activation of ERK could also result from the CaSR-mediated PLC pathway leading, based on the PLC pathway inhibitors experiment. In addition, another study utilizing the CaSR-transfected HEK-293 cells shared similar results, as they demonstrated that functional CaSR is necessary and sufficient for the ERK activation elicited by the calcium (Hobson *et al.*, 2003). In the same research, researchers also identified that phosphatidyl inositol 3 kinase (PI3K) was a key component of CaSR-mediated ERK activation, given that the CaSR-dependent ERK activation could be blocked by exposure to the PI3K inhibitors (Hobson *et al.*, 2003).

As the crucial members of the MAPK family, JNKs regulate various types of abiotic and biotic stress insults, as well as some key physiological processes (Zeke *et al.*, 2016). It has been demonstrated that the increasing level of extracellular calcium Ca^{2+} promoted the phosphorylation of ERK1/2 and JNK/SAPK in NIH/3T3 cells stably expressing either wild-type or mutant receptors (Hoff *et al.*, 1999). Similar to the studies employing the inhibitor experiment, this research also utilized the dominant negative kinase which is upstream of the JNK pathway to verify their results (Hoff *et al.*, 1999). Moreover, these findings were confirmed in subsequent studies. For instance, according to the western blotting results with phosphor-specific antibodies and the inhibitor targeting the JNK pathway, an investigation showed that the CaSR activation could increase the cyclooxygenase-2 expression in rat fibroblasts via the JNK signaling pathway (Ogata *et al.*, 2006).

Like other members of MAPKs mentioned above, the p38 MAPK group also serves as a signal transduction mediator that regulates a wide range of biological processes when subjected to extracellular stimuli (Zarubin and Han, 2005). The biological consequences of p38 activation include inflammation, apoptosis, cell differentiation, cardiomyocyte hypertrophy, as well as cardiomyocyte hypertrophy (Zarubin and Han, 2005). According to an early study which conducted the CaSR activation in H-500 rat Leydig cancer cells, evidence showed that high concentration of the extracellular calcium regulated PTH-related peptide secretion via the p38 signaling, while inhibition of PI3K and p38 MAPK abolished the activation of PI3K and p38 MAPK (Tfelt-Hansen *et al.*, 2003). Plus, it has been revealed that the CaSR activation promotes apoptosis via the JNK/p38 MAPK signaling pathway in focal cerebral ischemia-reperfusion in mice (Zhen *et al.*, 2016).

3.2.3 Gi/o signaling pathway

Interestingly, before the CaSR was isolated and characterized in bovine parathyroid cells, researchers found that a high level of divalent cations such as Ca^{2+} and Mg^{2+} was able to lower the dopamine-stimulated cAMP content in cultured bovine parathyroid cells, whereas the pertussis toxin could completely block the inhibitory effects of Ca^{2+} and Mg^{2+} on cAMP content (Chen *et al.*, 1989). Furthermore, in the cortical thick ascending limb of rats, it was observed that CaSR activation elicited by the CaSR agonists could suppress the production of cAMP induced by antidiuretic hormone treatment (De Jesus Ferreira and Bailly, 1998). Other research during the same period confirmed the inhibitory effects of the CaSR on the cAMP level in the CaSR stable transfected HEK-293 cells by using formalin stimulation (Chang *et al.*, 1998). In addition, it was also demonstrated that the γ-glutamyl peptides and L-AAs could inhibit the production of cAMP on HEK-293 cells that stably expressed the CaSR, implying that most of the time, the CaSR negatively regulates cAMP levels in normal cells (Broadhead *et al.*, 2011). Of note, the CaSR activation does not always suppress the cAMP content, as some studies found that the CaSR couples to Gαs subunit to stimulate cAMP production in breast cancer cells (Kim and Wysolmerski, 2016). Concretely, researchers found that the CaSR couples to Gαi and inhibits cAMP in normal mammary epithelial cells, but the CaSR couples to Gαs and stimulates cAMP production in breast cancer cells such as Comma-D cells (immortalized murine mammary cells) and MCF-7 human breast cancer cells (Mamillapalli *et al.*, 2008; Kim and Wysolmerski, 2016). These experiments suggest that there are cell-specific differences in the CaSR's role in cAMP production.

3.2.4 Akt signaling pathway

As previously described, researchers confirmed PI3K as an upstream regulator in the ERK phosphorylation signaling pathway caused by the CaSR activation (Hobson *et al.*, 2003). In the following year, another study found that CaSR-induced signaling pathways through the PI3K pathway could contribute to the phosphorylation and activation of the antiapoptotic protein kinase Akt (Dvorak *et al.*, 2004). This study reported the phosphorylation of the antiapoptotic protein kinase 1 and 2, as well as Akt, when the fetal rat calvarial cells were treated with increasing levels of extracellular Ca^{2+} and Gd^{3+} (Dvorak *et al.*, 2004). In addition, evidence indicated that the ERK and PI3K-Akt signaling was associated with cell proliferation and migration, and it was confirmed that the CaSR participates in this procedure (Ward *et al.*, 2005). Specifically, the activation of the CaSR led to the increase of intracellular Ca^{2+} via IP3-PLC signaling, and the increasing Ca^{2+} was confirmed to improve the cell proliferation via regulating ERK and PI3K-Akt signaling pathways (Lagos-Cabre *et al.*, 2018). By using an inhibition experiment, researchers demonstrated that the phosphorylation of Akt regulated by the CaSR was via the PI3K signaling pathway, while they also observed Akt phosphorylation and activation when opossum kidney cells were exposed to both extracellular Ca^{2+} and antibiotics (Ward *et al.*, 2005). Data obtained in a recent study mirrored these findings, as it revealed that the activation of CaSR induced by its agonist Gd^{3+} could promote the phosphorylation of ERK1/2, PI3K, and Akt, as well as the proliferation rate of the osteosarcoma cells MG-63 (Zhao *et al.*, 2020).

3.2.5 G12/13 signaling pathway

RhoA is a member of the Ras family, and studies have shown that ligands for GPCRs can also activate RhoA (Yu and Brown, 2015). Activation of RhoA occurs when receptors couple to G12/13 and Rho guanine nucleotide exchange factors (Palsuledesai *et al.*, 2018). Rho proteins have been confirmed to participate in the regulation of multiple physiological activities, including exocytosis, cell proliferation, and hypertrophy (Yu and Brown, 2015). By utilizing Madin-Darby canine kidney cell lines with the CaSR stable transfection, an early study showed that treating cells with a C3 exoenzyme could block the PLD activity (Huang *et al.*, 2004). PLD is a known G12/13 target, thus the CaSR is believed to be able to activate the Rho proteins by targeting G12/13 (Huang *et al.*, 2004). There is also evidence of the CaSR activation through Rho-dependent pathways, given that researchers reported the Rho is an integral part of the cellular signal transduction downstream of CaSR-regulated Gαq activation response (Pi *et al.*, 2002). Specifically, some CaSR positive modulators such as Ca^{2+} and Mg^{2+}, as well as the constitutively active Gα subunits and the constitutively active RhoA, were shown to activate the serum response element (SRE) reporter in HEK-293 cells stably transfected with the CaSR and an SRE-luciferase reporter construct, which monitors the Rho-dependent transactivation (Pi *et al.*, 2002). And the above CaSR-mediated activation of SREs could be abolished via the overexpression of the dominant negative Gαq (305–359) minigene (Pi *et al.*, 2002). Furthermore, the Rho family inhibitor C3 exoenzyme was found to block the CaSR and Gαq activation of SRE, while the C3 exoenzyme could not affect the mutant Rho, and CaSR-mediated SRE activity was partially restored (Pi *et al.*, 2002). Further evidence demonstrated that the induction of Rho by the CaSR may also be associated with the regulation of Gαq, RhoGEF, and filamin (Pi *et al.*, 2002). Another study also identified an abrogation of the CaSR activation induced by L-Phe with the treatment of Rho GTPases inhibitors toxin B as well as the C3 exoenzyme (Rey *et al.*, 2005). As with the investigation conducted by Pi *et al.* (2002), this study also found a relationship between filamin and the activation of Rho by the CaSR (Rey *et al.*, 2005).

4 Roles of the calcium-sensing receptor in gut health in pigs

The objective and overarching definition of gut health include five primary criteria: 1) effective digestion and absorption of food, 2) absence of gastrointestinal illness, 3) normal and stable intestinal microbiome, 4) adequate immune status, and 5) status of well-being (Bischoff, 2011). In general, as the pig's performance may be damaged without any clinical signs of disease, a healthy gut is not only a gut without any clinical disease but also an effective digestive organ that can resist disease effectively and deal with changes quickly. In response to infections or toxins caused by pathogens, the host's immune system can fight against certain conditions by particular actions when the gut's defense is in good working order. The complex and dynamic equilibrium composed of gut-associated lymphoid tissue, microbiota, mucus layer, and host epithelium interact is vital for making the gut work effectively and adequately (Choct, 2009). When the intestinal flora is imbalanced, dysbiosis happens between the beneficial microbiota and potentially pathogenic organisms, making it easier for disease-causing organisms to damage the gut barrier function. As mentioned previously, the activation of the CaSR induced by nutritional ligands could trigger a wide range of downstream signaling pathways, subsequently contributing to various physiological activities, including nutrient digestion and absorption, energy metabolism, gut barrier function, as well as immune response (Zhang and Mine, 2018). This implies that the CaSR has the potential to exert multiple roles in the regulation of pig gut health, and homeostasis offers great theoretical and practical significance in pig production.

4.1 *Roles of calcium sensing receptor in nutrient absorption*

The intestinal epithelial layer is highly dynamic, as intestinal epithelial cells undergo continuous turnover and intestinal cells keep migrating from the crypt to the villus, which determines gut growth and development (Cummins and Thompson, 2002). In pigs, CaSR has been detected in the digestive system, including the stomach, duodenum, jejunum, ileum, and colon, with the highest expression level of CaSR in the porcine ileum (Zhao *et al.*, 2019). Even though the absorptive cells account for 90% of the intestinal epithelial cells, the CaSR was not detected in the jejunal absorptive cells in weaned piglets (Zhao *et al.*, 2022). However, the CaSR still plays a pivotal role in the regulation of intestinal growth and development, as CaSR has been found to participate in cell proliferation and differentiation (Rey *et al.*, 2012). In this study, Rey *et al.* (2012) established an in vivo model using CaSR intestinal-specific knock-out mouse and confirmed that the activation of CaSR influenced the proliferation of colonic epithelial cells, structural changes in the colonic crypt, as well as the nuclear localization of β-catenin. Furthermore, the CaSR also has been found to regulate the intestinal development by affecting the gut morphology. For instance, Mine and Zhang (2015a) demonstrated that the activation of CaSR benefited to the recovery of damaged intestinal epithelial cells, since they observed that the poly-L-lysine activation of CaSR successfully abolished dextran sulfate sodium salt challenge induced colonic length changes and maintained intestinal morphology. Also, a similar study explored the effects of CaSR ligand spermine in suckling rats and found that the spermine administration could significantly increase specific activity of maltase and the total antioxidant capacity, as well as the jejunal villus height, villus width, surface area, and crypt depth, implying that the CaSR activation induced by dietary ligands enhance intestinal development (Cao *et al.*, 2015). Since the gut morphology and digestive enzymes activity are highly involved in the adequate nutrient utilization and absorption, the application of dietary CaSR ligands in pig diets has the potential to improve nutrient absorption and growth performance in swine production (Yang and Zhao, 2022).

Early studies documented that the presence of nutrients in the digestive system could not only activate the CaSR and trigger downstream physiological activities such as gut hormone release but also regulate the expression level of those receptors or other nutrient transporters in the EECs, which ultimately affect nutrient absorption (Cummings and Overduin, 2007). For example, in rats, researchers found that the activation of the CaSR induced by the L-Tryptophan administration contributed to the increasing level of plasma GLP-1, as they observed the attenuation of GLP-1 secretion with the CaSR antagonist NPS 2143 (Acar *et al.*, 2020). And a similar conclusion was reached by an in vitro study conducted in our lab because an increasing secretion level of GLP-1 also has been detected in STC-1 cells stimulated with pig CaSR (pCaSR) ligands (L-Tryptophan and α-casein (90–95)) (Yang and Zhao, 2022). Similarly, by employing an in vitro model (the duodenal perfusion with L-Tryptophan), Zhao *et al.* (2018) demonstrated that the activation of pCaSR by 10 mM of L-Tryptophan successfully elicited the secretion of CCK and GIP in the duodenum, implying an emerging role for the CaSR as the modulator of gut peptide secretion. Plus, the same study also observed an increased expression level of the pCaSR gene after the stimulation of L-Tryptophan (Zhao *et al.*, 2018). In the same year, Wang *et al.* (2018) demonstrated that the supplementation of L-Arginine to pCaSR resulted in the release of duodenal CCK. Through the summarized data mentioned above, we could conclude that supplementing animal feed with ligands of the CaSR stimulates the secretion of gut hormones such as CCK, PYY, and GLP-1. Given that gut hormones are known to regulate nutrient metabolism in the GIT (Dockray, 2014), these observations suggest that the CaSR plays a significant role in regulating feed digestion and nutrient absorption by affecting the gut hormone secretion in pigs.

4.2 *Roles of calcium sensing receptor in gut barrier function*

The digestive system is also an essential barrier since the digestive system has the largest interface between the host and the environment, keeping pathogens or toxins from the environment. The intestinal barrier consists of physical, chemical, immunological, and microbial barriers (Liu *et al.*, 2018), and we focus only on the roles of CaSR in physical barriers.

As a single-cell layer, the intestinal epithelium is formed by different types of epithelial cells, including enterocytes, EECs, goblet cells, and a large number of immune cells (Ali *et al.*, 2020). Plus, tight junctions connect intestinal epithelial cells lined on the surface of the intestinal epithelium, and the continuous intercellular barrier formed by tight junctions and epithelial cells is very important for the epithelial barrier function (Anderson and Van Itallie, 2009). Specifically, two major routes (transcellular and paracellular pathways) are involved in the transport of substances in the GIT, and selective permeability is the major property of the physical epithelial barrier (Odenwald and Turner, 2017). Since compromised intestinal barrier function comes with enteric disease, enhanced expression of the tight junction proteins has the potential to strengthen the gut barrier function and promote animal health.

There are several mechanisms associated with the intestinal tight junction and barrier function, namely the abundance of tight junction proteins, localization of tight junction proteins at the plasma membrane, functionality of tight junction-associated actomyosin cytoskeleton, and posttranslational regulation of tight junction proteins that might influence protein-protein interactions (González-Mariscal *et al.*, 2018). And the activation of GPCRs induced by the intraluminal nutrients has been documented to regulate intestinal tight junction and barrier function via the activation of GPCRs induced by the intraluminal nutrients (González-Mariscal *et al.*, 2008). So far, multiple studies have examined the role of the CaSR in regulating gut barrier function. Cheng *et al.* (2014) showed that CaSR deficiency in the intestinal epithelium led to decreased transepithelial resistance (TEER), increased transepithelial conductance and passive transport of FITC-conjugated dextran in the intestinal epithelium-specific CaSR-/-mice. Indeed, when intestinal inflammation occurs, it elicits the increased secretion of pro-inflammatory cytokines and then triggers the activation of the related cytokines' receptors, finally resulting in the increased transcription of myosin light-chain kinase (MLCK) (Blair *et al.*, 2006). This process further induces the phosphorylation of the regulatory myosin light chain and subsequently gives rise to more contraction of a band of actin and myosin (actin-myosin ring) located at the tight junctions, which ultimately promotes epithelial permeability (Cheng *et al.*, 2014). In this case, a leaky gut is formed with increased intestinal permeability, imbalances in the microbiota, and a greater immune response (Usuda *et al.*, 2021). Fortunately, the CaSR proves an ability to suppress those intestinal chaos limits the amplification of this cascade via the Gαq/11 signaling pathway (Geibel *et al.*, 2006). The activation of Gαq/11 downregulates the phosphorylation of NF-κB, as well as the expression of MLCK and the MLC phosphorylation inhibited, which finally ameliorates intestinal epithelial barrier impairment (Cheng *et al.*, 2014).

Besides, as stated previously, the activation of the CaSR stimulates the secretion of different types of gut hormones. Those hormones, such as GLP-2, have been demonstrated to positively regulate the intestinal tight junction, as GLP-2 treatment was shown to significantly enhance jejunal transepithelial resistance in a mice model (Hadjiyanni *et al.*, 2009). According to the above-mentioned finds, the activation of the CaSR has the potential to modulate the intestinal physical barrier. Interestingly, not all the activation of gut chemosensing benefits gut battier function, as researchers found that the activation of sweet taste receptors by artificial sweeteners increased gut permeability and decreased the expression of tight junction proteins (such as claudin 3), implying the unexpected negative effects of gut chemosensors' activation on gut barrier function (Shil *et al.*, 2020). Hence, further studies on the relationship between the CaSR and tight junctions are still required.

4.3 *Underlying the mechanisms of calcium sensing receptor-mediated in nutrient absorption and barrier function*

As stated previously, as a family C member of GPCRs, the activation of the CaSR is involved in a series of signaling pathways, including the Gq/11 signaling pathway, Gi/o signaling pathway, Gs signaling pathway, G12/13 signaling pathway (Geng *et al.*, 2016). Meanwhile, it has been documented that the activation of the CaSR induced by its dietary ligands triggers a series of downstream signaling, subsequently contributing to various physiological responses, which finally correspond to the regulation of nutrient absorption and gut barrier function (Thomsen *et al.*, 2012).

So far, several studies have explored the underlying mechanism of intestinal development and nutrient sensing via CaSR regulation. Recently, studies have documented that gut hormone secretion, such as CCK induced by the CaSR activation, was closely associated with the Gq/11 pathway in the porcine duodenum (Feng *et al.*, 2019). Also, Yang *et al.* (2019) found that the stimulation of dietary peptides to the CaSR could elicit the secretion level of CCK and GLP-1 by increasing the intracellular calcium ion mobilization, which indicates that the secretion of CCK and GLP-1 via the CaSR activation is related to the Gq/11 pathway activation. Similarly, when the pCaSR was stimulated with L-Tryptophan, it increased intracellular Ca^{2+} via the IP3-PLC signaling pathway and contributed to the secretion of CCK and GIP (Zhao *et al.*, 2018). This result ties well with a recent study since this study found that the activation of pCaSR stimulated by L-Tryptophan and dietary peptide showed biased signaling pathways, including ERK1/2 phosphorylation signaling, Gq/11 signaling, and G12/13 signaling (Zhao *et al.*, 2021). Furthermore, one recent research also identified that the activation of the CaSR induced by the dietary nutrient could trigger the secretion of GLP-1 via three signaling pathways: ERK1/2 phosphorylation signaling, Gi/o signaling, and Gq/11 signaling (Yang and Zhao, 2022).

Meanwhile, in the previous study, researchers found that when the CaSR was conditionally knocked out in mice, the colonic β-catenin phosphorylation at Ser-552 and Ser-675 was significantly decreased (Rey *et al.*, 2012). In contrast, when the CaSR stimulator was administrated to a normal human colonic epithelial cell line, the phosphorylation of β-catenin was increased, suggesting that the activation of Wnt/β-catenin signaling has been triggered (Aggarwal *et al.*, 2015). Since the Wnt/β-catenin signaling pathway is associated with cell proliferation and differentiation, CaSR activation probably mediates gut cell proliferation and differentiation via β-catenin signaling and affects intestinal development and growth (Aggarwal *et al.*, 2015). Similarly, one study found that the CaSR activation might regulate intestinal cell proliferation and differentiation through the MAPK signaling pathway, as the upregulation of activator protein one induced by the phosphorylation and activation of ERK1/2 could mediate cell proliferation and differentiation (Liu *et al.*, 2018). Meanwhile, previous studies have confirmed that the activation of the RhoA or ERK signaling pathway could subsequently regulate tight junction assembly via *in vitro* model, which suggests that the CaSR participates in the maintenance of gut barrier function via the RhoA or ERK signaling pathway (González-Mariscal *et al.*, 2008). Collectively, these findings imply that CaSR participates in nutrient absorption and gut barrier function through several cascades of intracellular signaling transduction, including the Gq/11 signaling pathway, Gi/o signaling pathway, Gs signaling pathway, and G12/13 signaling pathway.

5 Summary

The gut is not only a crucial digestive organ but also a chemosensory system that regulates the physiological activities and metabolic responses in animals through different types of cell signalling. Meanwhile, as a well-conserved and ancient GPCR, the CaSR has been demonstrated to be widely distributed along the gastrointestinal tract and its accessory organs in mammals. Furthermore, emerging evidence suggests that when the CaSR interacts with its allosteric agonists, the conformational changes of the CaSR elicit distinct cellular transduction, ultimately contributing to a broad range of physiological activities in the gut. Accordingly, the CaSR is intimately associated with intestinal homeostasis via coordinating nutrient digestion and absorption, enhancing intestinal cell proliferation and differentiation, promoting gut barrier function, mediating gut peptide secretion and energy metabolism, as well as attenuating inflammation and secretory diarrhea. Therefore, elucidation of the underlying mechanisms by which dietary CaSR ligands (like ions, L-AAs, oligopeptides, and polypeptides) preserve gut barrier function and maintain gut health will allow for the development of practical nutritional strategies to maintain intestinal homeostasis. Notably, most mechanisms were originally and mainly investigated in humans, mouses, and rats, similar mechanisms still need to be uncovered in pigs for the practical application. Except keeping screening more novel dietary ligands for the CaSR, identifying new nutrient sensors in animals, and exploring the interaction among different sensing mechanisms are also needed in further studies.

References

Acar, I., A. Cetinkaya, I. Lay, and E. Ileri-Gurel. 2020. The role of calcium sensing receptors in GLP-1 and PYY secretion after acute intraduodenal administration of L-tryptophan in rats. Nutr Neurosci 23(6):481–489. doi: 10.1080/1028415X.2018.1521906.

Aditi, A., and D.Y. Graham. 2012. Vitamin C, gastritis, and gastric disease: a historical review and update. Dig Dis Sci 57(10):2504–2515. doi: 10.1007/s10620-012-2203-7.

Aggarwal, A., J. Höbaus, S, Tennakoon, M. Prinz-Wohlgenannt, J. Graça, S.A. Price, and E. Kállay. 2016. Active vitamin D potentiates the anti-neoplastic effects of calcium in the colon: a cross talk through the calcium-sensing receptor. J Steroid Biochem Mol Biol 155, 231–238. doi: 10.1016/j.jsbmb.2015.02.006.

Aggarwal, A., Prinz-Wohlgenannt, M., Gröschel, C., Tennakoon, S., Meshcheryakova, A., Chang, W., Brown, E.M., Mechtcheriakova, D., and Kállay, E. 2015. The calcium-sensing receptor suppresses epithelial-to-mesenchymal transition and stem cell-like phenotype in the colon. Mol Cancer 14:61. doi: https://doi.org/10.1186/s12943-015-0330-4.

Ahlstrom, M., M. Pekkinen, U. Riehle, and C. Lamberg-Allardt. 2008. Extracellular calcium regulates parathyroid hormone-related peptide expression in osteoblasts and osteoblast progenitor cells. Bone 42(3):483–490. doi: 10.1016/j.bone.2007.10.025.

Al-Sadi, R., M. Boivin, and T. Ma. 2009. Mechanism of cytokine modulation of epithelial tight junction barrier. Front Biosci 14:2765–2778. doi: 10.2741/3413.

Alfadda, T.I., A.M. Saleh, P. Houillier, and J.P. Geibel. 2014. Calcium-sensing receptor 20 years later. Am J Physiol Cell Physiol 307(3):221–231. doi: 10.1152/ajpcell.00139.2014.

Ali, A., H. Tan, and G.E. Kaiko. 2020. Role of the intestinal epithelium and its interaction with the microbiota in food allergy. Front Immunol 11:604054. doi: 10.3389/fimmu.2020.604054.

Allaire, J.M., S.M. Crowley, H.T. Law, S.Y. Chang, H.J. Ko, and B.A. Vallance. 2018. The intestinal epithelium: central coordinator of mucosal immunity. Trends Immunol 39(9): 677–696. doi: 10.1016/j.it.2018.04.002.

Aluko, K., D.E. Velayudhan, E. Khafipour, L. Fang, and M. Nyachoti. 2017. Effect of chicken egg anti-F4 antibodies on performance and diarrhea incidences in enterotoxigenic *Escherichia coli* K88$^+$-challenged piglets. Anim Nutr 3(4):353–358. doi: 10.1016/j.aninu.2017.06.006.

Anast, C.S., J.M. Mohs, S.L. Kaplan, and T.W. Burns. 1972. Evidence for parathyroid failure in magnesium deficiency. Science 177(4049):606–608. doi: 10.1126/science.177.4049.606.

Anderson, J. M., and Van Itallie, C. M. 2009. Physiology and function of the tight junction. Cold Spring Harb Perspect Biol 1(2):a002584. doi: 10.1101/cshperspect.a002584.

Angus, H.C.K., A.G. Butt, M. Schultz, and R.A. Kemp. 2019. Intestinal organoids as a tool for inflammatory bowel disease research. Front Med 6:334. doi: 10.3389/fmed.2019.00334.

Armato, U., A. Chiarini, B. Chakravarthy, F. Chioffi, R. Pacchiana, E. Colarusso, J.F. Whitfield, and I. Dal Pra. 2013. Calcium-sensing receptor antagonist (calcilytic) NPS 2143 specifically blocks the increased secretion of endogenous Abeta42 prompted by exogenous fibrillary or soluble Abeta25–35 in human cortical astrocytes and neurons-therapeutic relevance to Alzheimer's disease. Biochim Biophys Acta 1832(10):1634–1652. doi: 10.1016/j.bbadis.2013.04.020.

Ashida, H., Ogawa, M., Kim, M. *et al.* 2012. Bacteria and host interactions in the gut epithelial barrier. Nat Chem Biol 8, 36–45. doi:10.1038/nchembio.741.

Aslanova, U.F., T. Morimoto, E.I. Farajov, N. Kumagai, M. Nishino, N. Sugawara, A. Ohsaga, Y. Maruyama, S. Tsuchiya, S. Takahashi, and Y. Kondo. 2006. Chloride-dependent intracellular pH regulation via extracellular calcium-sensing receptor in the medullary thick ascending limb of the mouse kidney. Tohoku J Exp Med 210(4):291–300. doi: 10.1620/tjem.210.291.

Awata, H., C. Huang, M.E. Handlogten, and R.T. Miller. 2001. Interaction of the calcium-sensing receptor and filamin, a potential scaffolding protein. J Biol Chem 276(37):34871–34879. doi: 10.1074/jbc.M100775200.

Bai, M. 2004. Structure-function relationship of the extracellular calcium-sensing receptor. Cell Calcium 35(3):197–207. doi: 10.1016/j.ceca.2003.10.018.

Bai, M., N. Janicic, S. Trivedi, S.J. Quinn, D.E. Cole, E.M. Brown, and G.N. Hendy. 1997. Markedly reduced activity of mutant calcium-sensing receptor with an inserted Alu element from a kindred with familial hypocalciuric hypercalcemia and neonatal severe hyperparathyroidism. J Clin Invest 99(8):1917–1925. doi: 10.1172/JCI119359.

Bai, M., S. Trivedi, and E.M. Brown. 1998. Dimerization of the extracellular calcium-sensing receptor (CaR) on the cell surface of CaR-transfected HEK293 cells. J Biol Chem 273(36):23605–23610. doi: 10.1074/jbc.273.36.23605.

Bandyopadhyay, S., J. Tfelt-Hansen, and N. Chattopadhyay. 2010. Diverse roles of extracellular calcium-sensing receptor in the central nervous system. J Neurosci Res 88(10):2073–2082. doi: 10.1002/jnr.22391.

Barger-Lux, M.J., R.P. Heaney, and R.R. Recker. 1989. Time course of calcium absorption in humans: evidence for a colonic component. Calcif Tissue Int 44(5):308–311. doi: 10.1007/BF02556309.

Bessis, A.S., P. Rondard, F. Gaven, I. Brabet, N. Triballeau, L. Prezeau, F. Acher, and J.P. Pin. 2002. Closure of the Venus flytrap module of mGlu8 receptor and the activation process: Insights from mutations converting antagonists into agonists. Proc Natl Acad Sci USA 99(17):11097–11102. doi: 10.1073/pnas.162138699.

Bevilacqua, M., L.J. Dominguez, V. Righini, V. Valdes, R. Toscano, O. Sangaletti, T. Vago, G. Baldi, M. Barrella, and G. Bianchi-Porro. 2005. Increased gastrin and calcitonin secretion after oral calcium or peptones administration in patients with hypercalciuria: a clue to an alteration in calcium-sensing receptor activity. J Clin Endocrinol Metab 90(3):1489–1494. doi: 10.1210/jc.2004-0045.

Bhat, A.A., S. Uppada, I.W. Achkar, S. Hashem, S.K. Yadav, M. Shanmugakonar, H.A. Al-Naemi, M. Haris, and S. Uddin. 2018. Tight junction proteins and signaling pathways in cancer and inflammation: A functional crosstalk. Front Physiol 9:1942. doi: 10.3389/fphys.2018.01942.

Bischoff S.C. 2011. "Gut health": a new objective in medicine? BMC Med 14;9:24. doi: 10.1186/1741-7015-9-24.

Blair, S. A., Kane, S. V., Clayburgh, D. R., and Turner, J. R. 2006. Epithelial myosin light chain kinase expression and activity are upregulated in inflammatory bowel disease. Lab Invest 86(2):191–201. doi: 10.1038/labinvest.3700373.

Bolger, A.M., M. Lohse, and B. Usadel. 2014. Trimmomatic: a flexible trimmer for Illumina sequence data. Bioinformatics 30(15):2114–2120. doi: 10.1093/bioinformatics/btu170.

Bourassa, P., R. Cote, S. Hutchandani, G. Samson, and H.A. Tajmir-Riahi. 2013. The effect of milk alpha-casein on the antioxidant activity of tea polyphenols. J Photochem Photobiol B 128:43–49. doi: 10.1016/j.jphotobiol.2013.07.021.

Bourdeau, J.E., and M.B. Burg. 1979. Voltage dependence of calcium transport in the thick ascending limb of Henle's loop. Am J Physiol 236(4):F357–364. doi: 10.1152/ajprenal.1979.236.4.F357.

Bourgine, J., I. Billaut-Laden, M. Happillon, J.M. Lo-Guidice, V. Maunoury, M. Imbenotte, and F. Broly. 2012. Gene expression profiling of systems involved in the metabolism and the disposition of xenobiotics: comparison between human intestinal biopsy samples and colon cell lines. Drug Metab Dispos 40(4):694–705. doi: 10.1124/dmd.111.042465.

Brauner-Osborne, H., P. Wellendorph, and A.A. Jensen. 2007. Structure, pharmacology and therapeutic prospects of family C G-protein coupled receptors. Curr Drug Targets 8(1): 169–184. doi: 10.2174/138945007779315614.

Brennan, S.C., T.S. Davies, M. Schepelmann, and D. Riccardi. 2014. Emerging roles of the extracellular calcium-sensing receptor in nutrient sensing: control of taste modulation and intestinal hormone secretion. Br J Nutr 111(1):16–22. doi: 10.1017/S0007114513002250.

Brennan, S.C., U. Thiem, S. Roth, A. Aggarwal, I. Fetahu, S. Tennakoon, A.R. Gomes, M.L. Brandi, F. Bruggeman, R. Mentaverri, D. Riccardi, and E. Kallay. 2013. Calcium sensing receptor signalling in physiology and cancer. Biochim Biophys Acta 1833(7):1732–1744. doi: 10.1016/j.bbamcr.2012.12.011.

Breves, B.S.d.a.G. 2007. Mechanisms and regulation of calcium absorption from the gastrointestinal tract in pigs and ruminants: comparative aspects with special emphasis on hypocalcemia in dairy cows. Anim Health Res Rev 7(1/2):31–41. doi: 10.1017/S1466252307001144.

Broadhead, G.K., H.C. Mun, V.A. Avlani, O. Jourdon, W.B. Church, A. Christopoulos, L. Delbridge, and A.D. Conigrave. 2011. Allosteric modulation of the calcium-sensing receptor by gamma-glutamyl peptides: inhibition of PTH secretion, suppression of intracellular cAMP levels, and a common mechanism of action with L-amino acids. J Biol Chem 286(11):8786–8797. doi: 10.1074/jbc.M110.149724.

Brosnahan, A.J., and D.R. Brown. 2012. Porcine IPEC-J2 intestinal epithelial cells in microbiological investigations. Vet Microbiol 156(3–4):229–237. doi: 10.1016/j.vetmic.2011.10.017.

Brown, E.M. 1991. Extracellular Ca^{2+} sensing, regulation of parathyroid cell function, and role of Ca^{2+} and other ions as extracellular (first) messengers. Physiol Rev 71(2):371–411. doi: 10.1152/physrev.1991.71.2.371.

Brown, E.M. 2000. The extracellular Ca^{2+}-sensing receptor: central mediator of systemic calcium homeostasis. Annu Rev Nutr 20:507–533. doi: 10.1146/annurev.nutr.20.1.507.

Brown, E.M., R. Butters, C. Katz, and O. Kifor. 1991. Neomycin mimics the effects of high extracellular calcium concentrations on parathyroid function in dispersed bovine parathyroid cells. Endocrinology 128(6):3047–3054. doi: 10.1210/endo-128-6-3047.

Brown, E.M., G. Gamba, D. Riccardi, M. Lombardi, R. Butters, O. Kifor, A. Sun, M.A. Hediger, J. Lytton, and S.C. Hebert. 1993. Cloning and characterization of an extracellular Ca^{2+}-sensing receptor from bovine parathyroid. Nature 366(6455):575–580. doi: 10.1038/366575a0.

Brown, E.M., and R.J. MacLeod. 2001. Extracellular calcium sensing and extracellular calcium signaling. Physiol Rev 81(1):239–297. doi: 10.1152/physrev.2001.81.1.239.

Bruce A. Sherf, S.L.N., Rita R. Hannah and Keith V. Wood. 1996. Dual-Luciferase reporter assay: An advanced co-reporter technology integrating firefly and renilla luciferase assays. Promega Notes Magazine:p. 02.

Bullen, T.F., S. Forrest, F. Campbell, A.R. Dodson, M.J. Hershman, D.M. Pritchard, J.R. Turner, M.H. Montrose, and A.J. Watson. 2006. Characterization of epithelial cell shedding from human small intestine. Lab Invest 86(10):1052–1063. doi: 10.1038/labinvest.3700464.

Burrin, D.G., Y. Petersen, B. Stoll, and P. Sangild. 2001. Glucagon-like peptide 2: a nutrient-responsive gut growth factor. J Nutr 131(3):709–712. doi: 10.1093/jn/131.3.709.

Burrin, D.G., B. Stoll, R. Jiang, Y. Petersen, J. Elnif, R.K. Buddington, M. Schmidt, J.J. Holst, B. Hartmann, and P.T. Sangild. 2000. GLP-2 stimulates intestinal growth in premature TPN-fed pigs by suppressing proteolysis and apoptosis. Am J Physiol Gastrointest Liver Physiol 279(6):G1249–1256. doi: 10.1152/ajpgi.2000.279.6.G1249.

Busque, S.M., J.E. Kerstetter, J.P. Geibel, and K. Insogna. 2005. L-type amino acids stimulate gastric acid secretion by activation of the calcium-sensing receptor in parietal cells. Am J Physiol Gastrointest Liver Physiol 289(4):G664–669. doi: 10.1152/ajpgi.00096.2005.

Busslinger, G.A., B.L.A. Weusten, A. Bogte, H. Begthel, L.A.A. Brosens, and H. Clevers. 2021. Human gastrointestinal epithelia of the esophagus, stomach, and duodenum resolved at single-cell resolution. Cell Rep 34(10):108819. doi: 10.1016/j.celrep.2021.108819.

Butters, R.R., Jr., N. Chattopadhyay, P. Nielsen, C.P. Smith, A. Mithal, O. Kifor, M. Bai, S. Quinn, P. Goldsmith, S. Hurwitz, K. Krapcho, J. Busby, and E.M. Brown. 1997. Cloning and characterization of a calcium-sensing receptor from the hypercalcemic New Zealand white rabbit reveals unaltered responsiveness to extracellular calcium. J Bone Miner Res 12(4):568–579. doi: 10.1359/jbmr.1997.12.4.568.

Canadian Council on Animal Care (CCAC) Guidelines 2009. CCAC guidelines on: the care and use of farm animals in research, teaching and testing. Ottawa, ON, Canada: CCAC. https://www.nfacc.ca/codes-of-practice/pig-code.

Campbell, J.M., J.D. Crenshaw, and J. Polo. 2013. The biological stress of early weaned piglets. J Anim Sci Biotechnol 4(1):19. doi: 10.1186/2049-1891-4-19.

Canaff, L., and G.N. Hendy. 2002. Human calcium-sensing receptor gene. Vitamin D response elements in promoters P1 and P2 confer transcriptional responsiveness to 1,25-dihydroxyvitamin D. J Biol Chem 277(33):30337–30350. doi: 10.1074/jbc.M201804200.

Canaff, L., J.L. Petit, M. Kisiel, P.H. Watson, M. Gascon-Barre, and G.N. Hendy. 2001. Extracellular calcium-sensing receptor is expressed in rat hepatocytes. coupling to intracellular calcium mobilization and stimulation of bile flow. J Biol Chem 276(6):4070–4079. doi: 10.1074/jbc.M009317200.

Cao, W., G. Liu, T. Fang, X. Wu, G. Jia, H. Zhao, X. Chen, C. Wu, J. Wang, and J. Cai. 2015. Effects of spermine on the morphology, digestive enzyme activities, and antioxidant status of jejunum in suckling rats. RSC Advances 5(93):76607–76614. doi: 10.1039/C5RA15793E.

Caroppo, R., A. Gerbino, L. Debellis, O. Kifor, D.I. Soybel, E.M. Brown, A.M. Hofer, and S. Curci. 2001. Asymmetrical, agonist-induced fluctuations in local extracellular Ca^{2+} in intact polarized epithelia. EMBO J 20(22):6316–6326. doi: 10.1093/emboj/20.22.6316.

Carrillo-Lopez, N., J.L. Fernandez-Martin, D. Alvarez-Hernandez, I. Gonzalez-Suarez, P. Castro-Santos, P. Roman-Garcia, J.M. Lopez-Novoa, and J.B. Cannata-Andia. 2010. Lanthanum activates calcium-sensing receptor and enhances sensitivity to calcium. Nephrol Dial Transplant 25(9):2930–2937. doi: 10.1093/ndt/gfq124.

Castano, J.P., A.J. Martinez-Fuentes, E. Gutierrez-Pascual, H. Vaudry, M. Tena-Sempere, and M.M. Malagon. 2009. Intracellular signaling pathways activated by kisspeptins through GPR54: do multiple signals underlie function diversity? Peptides 30(1):10–15. doi: 10.1016/j.peptides.2008.07.025.

Ceglia, L., S.S. Harris, H.M. Rasmussen, and B. Dawson-Hughes. 2009. Activation of the calcium-sensing receptor stimulates gastrin and gastric acid secretion in healthy participants. Osteoporos Int 20(1):71–78. doi: 10.1007/s00198-008-0637-8.

Cervenka, I., L.Z. Agudelo, and J.L. Ruas. 2017. Kynurenines: Tryptophan's metabolites in exercise, inflammation, and mental health. Science 357(6349). doi: 10.1126/science.aaf9794.

Cetani, F., A. Picone, P. Cerrai, E. Vignali, S. Borsari, E. Pardi, P. Viacava, A.G. Naccarato, P. Miccoli, O. Kifor, E.M. Brown, A. Pinchera, and C. Marcocci. 2000. Parathyroid expression of calcium-sensing receptor protein and in vivo parathyroid hormone-Ca^{2+} set-point in patients with primary hyperparathyroidism. J Clin Endocrinol Metab 85(12):4789–4794. doi: 10.1210/jcem.85.12.7028.

Chakrabarty, S., V. Radjendirane, H. Appelman, and J. Varani. 2003. Extracellular calcium and calcium sensing receptor function in human colon carcinomas: promotion of E-cadherin expression and suppression of beta-catenin/TCF activation. Cancer Res 63(1):67–71.

Chang, W., T.H. Chen, S. Pratt, and D. Shoback. 2000. Amino acids in the second and third intracellular loops of the parathyroid Ca^{2+}-sensing receptor mediate efficient coupling to phospholipase C. J Biol Chem 275(26):19955–19963. doi: 10.1074/jbc.M909613199.

Chang, W., S. Pratt, T.H. Chen, L. Bourguignon, and D. Shoback. 2001. Amino acids in the cytoplasmic C terminus of the parathyroid Ca^{2+}-sensing receptor mediate efficient cell-surface expression and phospholipase C activation. J Biol Chem 276(47):44129–44136. doi: 10.1074/jbc.M104834200.

Chang, W., S. Pratt, T.H. Chen, E. Nemeth, Z. Huang, and D. Shoback. 1998. Coupling of calcium receptors to inositol phosphate and cyclic AMP generation in mammalian cells and Xenopus laevis oocytes and immunodetection of receptor protein by region-specific antipeptide antisera. J Bone Miner Res 13(4):570–580. doi: 10.1359/jbmr.1998.13.4.570.

Chang, W., S. Pratt, T.H. Chen, and D. Shoback. 1999a. Protein kinase C activation blocks calcium receptor signaling in Xenopus laevis oocytes. Mol Cell Endocrinol 158(1–2):13–23. doi: 10.1016/s0303-7207(99)00189-6.

Chang, W., C. Tu, T.H. Chen, L. Komuves, Y. Oda, S.A. Pratt, S. Miller, and D. Shoback. 1999b. Expression and signal transduction of calcium-sensing receptors in cartilage and bone. Endocrinology 140(12):5883–5893. doi: 10.1210/endo.140.12.7190.

Chang, W., C. Tu, Z. Cheng, L. Rodriguez, T.H. Chen, M. Gassmann, B. Bettler, M. Margeta, L.Y. Jan, and D. Shoback. 2007. Complex formation with the type B gamma-aminobutyric acid receptor affects the expression and signal transduction of the extracellular calcium-sensing receptor. Studies with HEK-293 cells and neurons. J Biol Chem 282(34):25030–25040. doi: 10.1074/jbc.M700924200.

Chattopadhyay, N., G. Légrádi, M. Bai, O. Kifor, C. Ye, P.M. Vassilev, E.M. Brown, and R.M. Lechan. 1997. Calcium-sensing receptor in the rat hippocampus: a developmental study. Dev Brain Res 100(1):13–21. doi: 10.1016/s0165-3806(97)00009-6.

Chattopadhyay, N.C., I, Rogers, K., Riccardi, D., Hall, A., Diaz, R., Hebert, S.C., Soybel, D.I., Brown, Em. 1998. Identification and localization of extracellular Ca^{2+}-sensing receptor in rat intestine. Am J Physiol Gastrointest Liver Physiol 274(1):122–130. doi: 10.1152/ajpgi.1998.274.1.G122.

Chen, C.J., J.V. Barnett, D.A. Congo, and E.M. Brown. 1989. Divalent cations suppress 3′,5′-adenosine monophosphate accumulation by stimulating a pertussis toxin-sensitive guanine nucleotide-binding protein in cultured bovine parathyroid cells. Endocrinology 124(1):233–239. doi: 10.1210/endo-124-1-233.

Chen, R.A., and W.G. Goodman. 2004. Role of the calcium-sensing receptor in parathyroid gland physiology. Am J Physiol Renal Physiol 286(6):F1005–1011. doi: 10.1152/ajprenal.00013.2004.

Chen, W., J.B. Bergsman, X. Wang, G. Gilkey, C.R. Pierpoint, E.A. Daniel, E.M. Awumey, P. Dauban, R.H. Dodd, M. Ruat, and S.M. Smith. 2010. Presynaptic external calcium signaling involves the calcium-sensing receptor in neocortical nerve terminals. PLoS One 5(1): e8563. doi: 10.1371/journal.pone.0008563.

Chen, Z., T.B. Gibson, F. Robinson, L. Silvestro, G. Pearson, B. Xu, A. Wright, C. Vanderbilt, and M.H. Cobb. 2001. MAP kinases. Chem Rev 101(8):2449–2476. doi: 10.1021/cr000241p.

Cheng, H., and C.P. Leblond. 1974. Origin, differentiation and renewal of the four main epithelial cell types in the mouse small intestine. V. Unitarian Theory of the origin of the four epithelial cell types. Am J Anat 141(4):537–561. doi: 10.1002/aja.1001410407.

Cheng, I., I. Qureshi, N. Chattopadhyay, A. Qureshi, R.R. Butters, A.E. Hall, R.R. Cima, K.V. Rogers, S.C. Hebert, J.P. Geibel, E.M. Brown, and D.I. Soybel. 1999. Expression of an extracellular calcium-sensing receptor in rat stomach. Gastroenterology 116(1):118–126. doi: 10.1016/s0016-5085(99)70235-0.

Cheng, S.X. 2016. Calcium-sensing receptor: A new target for therapy of diarrhea. World J Gastroenterol 22(9):2711–2724. doi: 10.3748/wjg.v22.i9.2711.

Cheng, S. X., Lightfoot, Y. L., Yang, T., Zadeh, M., Tang, L., Sahay, B., Wang, G. P., Owen, J. L., and Mohamadzadeh, M. 2014. Epithelial CaSR deficiency alters intestinal integrity and promotes proinflammatory immune responses. FEBS Lett 588(22):4158–4166. doi: 10.1016/j.febslet.2014.05.007.

Cheng, S.X., J.P. Geibel, and S.C. Hebert. 2004. Extracellular polyamines regulate fluid secretion in rat colonic crypts via the extracellular calcium-sensing receptor. Gastroenterology 126(1):148–158. doi: 10.1053/j.gastro.2003.10.064.

Cheng, S.X., M. Okuda, A.E. Hall, J.P. Geibel, and S.C. Hebert. 2002. Expression of calcium-sensing receptor in rat colonic epithelium: evidence for modulation of fluid secretion. Am J Physiol Gastrointest Liver Physiol 283(1):G240–250. doi: 10.1152/ajpgi.00500.2001.

Cheng, Z., C. Tu, L. Rodriguez, T.H. Chen, M.M. Dvorak, M. Margeta, M. Gassmann, B. Bettler, D. Shoback, and W. Chang. 2007. Type B gamma-aminobutyric acid receptors modulate the function of the extracellular Ca^{2+}-sensing receptor and cell differentiation in murine growth plate chondrocytes. Endocrinology 148(10):4984–4992. doi: 10.1210/en.2007-0653.

Cheng, Z.B., D.F. Li, C.R. Ge, and J.J. Xing. 2006. Polyamines in sow colostrum and milk at different stages of lactation. Ani Sci 82:95–99. doi: 10.1079/Asc20051.

Chikatsu, N., S. Fukumoto, M. Suzawa, Y. Tanaka, Y. Takeuchi, S. Takeda, Y. Tamura, T. Matsumoto, and T. Fujita. 1999. An adult patient with severe hypercalcaemia and hypocalciuria due to a novel homozygous inactivating mutation of calcium-sensing receptor. Clin Endocrinol (Oxf) 50(4):537–543. doi: 10.1046/j.1365-2265.1999.00729.x.

Chikatsu, N., S. Fukumoto, Y. Takeuchi, M. Suzawa, T. Obara, T. Matsumoto, and T. Fujita. 2000. Cloning and characterization of two promoters for the human calcium-sensing receptor (CaSR) and changes of CaSR expression in parathyroid adenomas. J Biol Chem 275(11):7553–7557. doi: 10.1074/jbc.275.11.7553.

Choct M. 2009. Managing gut health through nutrition. *Bri Poult Sci* 50(1):9–15. doi: 10.1080/00071660802538632.

Choi, J., L. Wang, S. Liu, P. Lu, X. Zhao, H. Liu, L. Lahaye, E. Santin, S. Liu, M. Nyachoti, and C. Yang. 2020. Effects of a microencapsulated formula of organic acids and essential oils on nutrient absorption, immunity, gut barrier function, and abundance of enterotoxigenic Escherichia coli F4 in weaned piglets challenged with *E. coli* F4. J Anim Sci 98(9). doi: 10.1093/jas/skaa259.

Chou, Y.H., M.R. Pollak, M.L. Brandi, G. Toss, H. Arnqvist, A.A.B., S.E. Papapoulos, S. Marx, E.M. Brown, and J.G. Seidman. 1995. Mutations in the human Ca^{2+}-sensing receptor gene that cause familial hypocalciuric hypercalcemia. Am J Hum Genet 56:1075–1079. doi: 10.1016/0092-8674(93)90617-y.

Chougule, P., G. Herlenius, N.M. Hernandez, P.B. Patil, B. Xu, and S. Sumitran-Holgersson. 2012. Isolation and characterization of human primary enterocytes from small intestine using a novel method. Scand J Gastroenterol 47(11):1334–1343. doi: 10.3109/00365521.2012.708940.

Chun, L., W.H. Zhang, and J.F. Liu. 2012. Structure and ligand recognition of class C GPCRs. Acta Pharmacol Sin 33(3):312–323. doi: 10.1038/aps.2011.186.

Cima, R.R., I. Cheng, M.E. Klingensmith, N. Chattopadhyay, O. Kifor, S.C. Hebert, E.M. Brown, and D.I. Soybel. 1997. Identification and functional assay of an extracellular calcium-sensing receptor in Necturus gastric mucosa. Am J Physiol 273(5):G1051–1060. doi: 10.1152/ajpgi.1997.273.5.G1051.

Ckam, C.F. 1965. Sites of calcium absorption and the calcium concentration of gut contents in the dog. Can J Physiol Pharmacol 43(75–78). doi: 10.1139/y65-009.

Cramer, C.F.C., D H. 1959. Progress and rate of absorption of radiostrontium through intestinal tracts of rats. Proc Soc Exp Biol Med 102(2):514–517. doi: 10.3181/00379727-102-25301.

Conigrave, A.D. 2016. The calcium-sensing receptor and the parathyroid: past, present, future. Front Physiol 7:563. doi: 10.3389/fphys.2016.00563.

Conigrave, A.D., A.H. Franks, E.M. Brown, and S.J. Quinn. 2002. L-amino acid sensing by the calcium-sensing receptor: a general mechanism for coupling protein and calcium metabolism? Eur J Clin Nutr 56(11):1072–1080. doi: 10.1038/sj.ejcn.1601463.

Conigrave, A.D., S.J. Quinn, and E.M. Brown. 2000. L-amino acid sensing by the extracellular Ca^{2+}-sensing receptor. Proc Natl Acad Sci USA 97(9):4814–4819. doi: 10.1073/pnas.97.9.4814.

Conigrave, A.D., and D.T. Ward. 2013. Calcium-sensing receptor (CaSR): pharmacological properties and signaling pathways. Best Pract Res Clin Endocrinol Metab 27(3):315–331. doi: 10.1016/j.beem.2013.05.010.

Conley, Y.P., A. Mukherjee, C. Kammerer, S.T. DeKosky, M.I. Kamboh, D.N. Finegold, and R.E. Ferrell. 2009. Evidence supporting a role for the calcium-sensing receptor in Alzheimer disease. Am J Med Genet B Neuropsychiatr Genet 150B(5):703–709. doi: 10.1002/ajmg.b.30896.

Cummins, A. G.,and Thompson, F. M. 2002. Effect of breast milk and weaning on epithelial growth of the small intestine in humans. *Gut 51*(5):748–754. doi:/10.1136/gut.51.5.748.

D. Burrin, B. Stoll, and D. Moore. 2013. Digestive physiology of the pig symposium: Intestinal bile acid sensing is linked to key endocrine and metabolic signaling pathways. J Anim Sci 91:1991–2000. doi: 10.2527/jas2013-6331.

De Jesus Ferreira, M.C., and C. Bailly. 1998. Extracellular Ca^{2+} decreases chloride reabsorption in rat CTAL by inhibiting cAMP pathway. Am J Physiol 275(2):198–203. doi: 10.1152/ajprenal.1998.275.2.F198.

Desai, A.J., D.J. Roberts, G.O. Richards, and T.M. Skerry. 2014. Role of receptor activity modifying protein 1 in function of the calcium sensing receptor in the human TT thyroid carcinoma cell line. PLoS One 9(1):e85237. doi: 10.1371/journal.pone.0085237.

Diana, A., L.A. Boyle, F.C. Leonard, C. Carroll, E. Sheehan, D. Murphy, and E.G. Manzanilla. 2019. Removing prophylactic antibiotics from pig feed: how does it affect their performance and health? BMC Vet Res 15(1):67. doi: 10.1186/s12917-019-1808-x.

Diaz, R., S. Hurwitz, N. Chattopadhyay, M. Pines, Y.H. Yang, O. Kifor, M.S. Einat, R. Butters, S.C. Hebert, and E.M. Brown. 1997. Cloning, expression, and tissue localization of the calcium-sensing receptor in chicken (Gallus domesticus). Am J Physiol-Reg I 273(3): 1008–1016.

Dimke, H., P. Desai, J. Borovac, A. Lau, W. Pan, and R.T. Alexander. 2013. Activation of the Ca^{2+}-sensing receptor increases renal claudin-14 expression and urinary Ca^{2+} excretion. Am J Physiol Renal Physiol 304(6):F761–769. doi: 10.1152/ajprenal.00263.2012.

Dobin, A., C.A. Davis, F. Schlesinger, J. Drenkow, C. Zaleski, S. Jha, P. Batut, M. Chaisson, and T.R. Gingeras. 2013. STAR: ultrafast universal RNA-seq aligner. Bioinformatics 29(1): 15–21. doi: 10.1093/bioinformatics/bts635.

Dockray, G. J. 2014. Gastrointestinal hormones and the dialogue between gut and brain. J Physiol 592(14):2927–2941. doi: 10.1113/jphysiol.2014.270850.

Dong, H., J. Liu, X. Liu, Y. Yu, and S. Cao. 2017. Molecular docking and QSAR analyses of aromatic heterocycle thiosemicarbazone analogues for finding novel tyrosinase inhibitors. Bioorg Chem 75:106–117. doi: 10.1016/j.bioorg.2017.07.002.

Doroszewicz, J., P. Waldegger, N. Jeck, H. Seyberth, and S. Waldegger. 2005. pH dependence of extracellular calcium sensing receptor activity determined by a novel technique. Kidney Int 67(1):187–192. doi: 10.1111/j.1523-1755.2005.00069.x.

Drucker, D.J. 2007. The role of gut hormones in glucose homeostasis. J Clin Investig 117(1):24–32. doi: 10.1172/JCI30076.

Dufner, M.M., P. Kirchhoff, C. Remy, P. Hafner, M.K. Muller, S.X. Cheng, L.Q. Tang, S.C. Hebert, J.P. Geibel, and C.A. Wagner. 2005. The calcium-sensing receptor acts as a modulator of gastric acid secretion in freshly isolated human gastric glands. Am J Physiol Gastrointest Liver Physiol 289(6):G1084–1090. doi: 10.1152/ajpgi.00571.2004.

Duran, M.J., G.C. Borst, R.C. Osburne, and C. Eil. 1984. Concurrent renal hypomagnesemia and hypoparathyroidism with normal parathormone responsiveness. Am J Med 76(1):151–154. doi: 10.1016/0002-9343(84)90764-2.

Dvorak, M.M., A. Siddiqua, D.T. Ward, D.H. Carter, S.L. Dallas, E.F. Nemeth, and D. Riccardi. 2004. Physiological changes in extracellular calcium concentration directly control osteoblast function in the absence of calciotropic hormones. Proc Natl Acad Sci USA 101(14):5140–5145. doi: 10.1073/pnas.0306141101.

Egbuna, O.I., and E.M. Brown. 2008. Hypercalcaemic and hypocalcaemic conditions due to calcium-sensing receptor mutations. Best Pract Res Clin Rheumatol 22(1):129–148. doi: 10.1016/j.berh.2007.11.006.

Engelstoft, M.S., W.M. Park, I. Sakata, L.V. Kristensen, A.S. Husted, S. Osborne-Lawrence, P.K. Piper, A.K. Walker, M.H. Pedersen, M.K. Nohr, J. Pan, C.J. Sinz, P.E. Carrington, T.E. Akiyama, R.M. Jones, C. Tang, K. Ahmed, S. Offermanns, K.L. Egerod, J.M. Zigman, and T.W. Schwartz. 2013. Seven transmembrane G protein-coupled receptor repertoire of gastric ghrelin cells. Mol Metab 2(4):376–392. doi: 10.1016/j.molmet.2013.08.006.

Fan, M.Z., B. Stoll, R. Jiang, and D.G. Burrin. 2001. Enterocyte digestive enzyme activity along the crypt-villus and longitudinal axes in the neonatal pig small intestine. J Anim Sci 79(2):371–381. doi: 10.2527/2001.792371x.

Farre, R., M. Fiorani, S. Abdu Rahiman, and G. Matteoli. 2020. Intestinal permeability, inflammation and the role of nutrients. Nutrients 12(4):1185. doi: 10.3390/nu12041185.

Favus, M.J., S.C. Kathpalia, and F.L. Coe. 1981. Kinetic characteristics of calcium absorption and secretion by rat colon. Am J Physiol 240(5):350–354. doi: 10.1152/ajpgi.1981.240.5.G350.

Feeds, H. 2014. For more full-value pigs, managing gut health is an 'all-the-time deal'. National Hog Farmer.

Feng, J., C. Kang, C. Wang, L. Ding, W. Zhu, and S. Hang. 2019. L-phenylalanine increased gut hormone secretion through calcium-sensing receptor in the porcine duodenum. Animals (Basel) 9(8):476. doi: 10.3390/ani9080476.

Feng, J., C.D. Petersen, D.H. Coy, J.K. Jiang, C.J. Thomas, M.R. Pollak, and S.A. Wank. 2010. Calcium-sensing receptor is a physiologic multimodal chemosensor regulating gastric G-cell growth and gastrin secretion. Proc Natl Acad Sci USA 107(41):17791–17796. doi: 10.1073/pnas.1009078107.

Ferry, S., E. Traiffort, J. Stinnakre, and M. Ruat. 2000. Developmental and adult expression of rat calcium-sensing receptor transcripts in neurons and oligodendrocytes. Eur J Neurosci 12(3):872–884. doi: 10.1046/j.1460-9568.2000.00980.x.

Foulke-Abel, J., H. Yu, L. Sunuwar, R. Lin, J.M. Fleckenstein, J.B. Kaper, and M. Donowitz. 2020. Phosphodiesterase 5 (PDE5) restricts intracellular cGMP accumulation during enterotoxigenic *Escherichia coli* infection. Gut Microbes:1–12. doi: 10.1080/19490976.2020.1752125.

Frizzell, R.A., and J.W. Hanrahan. 2012. Physiology of epithelial chloride and fluid secretion. Cold Spring Harb Perspect Med 2(6):a009563. doi: 10.1101/cshperspect.a009563.

Gal, A., T.K. Ridge, and T.K. Graves. 2010. Cloning and sequencing of the calcium-sensing receptor from the feline parathyroid gland. Domest Anim Endocrinol 38(1):57–61. doi: 10.1016/j.domaniend.2009.07.004.

Galli, P., A. Brenna, P. Camilli de, and J. Meldolesi. 1976. Extracellular calcium and the organization of tight junctions in pancreatic acinar cells. Exp Cell Res 99(1):178–183. doi: 10.1016/0014-4827(76)90694-7.

Gama, L., L.M. Baxendale-Cox, and G.E. Breitwieser. 1997. Ca^{2+}-sensing receptors in intestinal epithelium. Am J Physiol 273(4):1168–1175. doi: 10.1152/ajpcell.1997.273.4.C1168.

Gama, L., S.G. Wilt, and G.E. Breitwieser. 2001. Heterodimerization of calcium sensing receptors with metabotropic glutamate receptors in neurons. J Biol Chem 276(42):39053–39059. doi: 10.1074/jbc.M105662200.

Gannon, A.W., H.M. Monk, and M.A. Levine. 2014. Cinacalcet monotherapy in neonatal severe hyperparathyroidism: a case study and review. J Clin Endocrinol Metab 99(1):7–11. doi: 10.1210/jc.2013-2834.

Garg, M.K., S. Kalra, and N. Mahalle. 2013. The intestinal calcistat: Determinant of clinical vitamin D deficiency. Indian J Endocrinol Metab 17(5):780–783. doi: 10.4103/2230-8210.117187.

Garrett, J.E. 1995. Calcitonin-secreting cells of the thyroid express an extracellular calcium receptor gene. Endocrinology 136(11):5202–5211. doi: 10.1210/endo.136.11.7588259.

Garrett, J.E., I.V. Capuano, L.G. Hammerland, B.C. Hung, E.M. Brown, S.C. Hebert, E.F. Nemeth, and F. Fuller. 1995. Molecular cloning and functional expression of human parathyroid calcium receptor cDNAs. J Biol Chem 270(21):12919–12925. doi: 10.1074/jbc.270.21.12919.

Geibel, J., K. Sritharan, R. Geibel, P. Geibel, J.S. Persing, A. Seeger, T.K. Roepke, M. Deichstetter, C. Prinz, S.X. Cheng, D. Martin, and S.C. Hebert. 2006. Calcium-sensing receptor abrogates secretagogue-induced increases in intestinal net fluid secretion by enhancing cyclic nucleotide destruction. Proc Natl Acad Sci USA 103(25):9390–9397. doi: 10.1073/pnas.0602996103.

Geibel, J.P., and S.C. Hebert. 2009. The functions and roles of the extracellular Ca^{2+}-sensing receptor along the gastrointestinal tract. Annu Rev Physiol 71:205–217. doi: 10.1146/annurev.physiol.010908.163128.

Geibel, J.P., C.A. Wagner, R. Caroppo, I. Qureshi, J. Gloeckner, L. Manuelidis, P. Kirchhoff, and K. Radebold. 2001. The stomach divalent ion-sensing receptor scar is a modulator of gastric acid secretion. J Biol Chem 276(43):39549–39552. doi: 10.1074/jbc.M107315200.

Geng, Y., L. Mosyak, I. Kurinov, H. Zuo, E. Sturchler, T.C. Cheng, P. Subramanyam, A.P. Brown, S.C. Brennan, H.C. Mun, M. Bush, Y. Chen, T.X. Nguyen, B. Cao, D.D. Chang, M. Quick, A.D. Conigrave, H.M. Colecraft, P. McDonald, and Q.R. Fan. 2016. Structural mechanism of ligand activation in human calcium-sensing receptor. Elife 5. doi: 10.7554/eLife.13662.

George, S.R., B.F. O'Dowd, and S.P. Lee. 2002. G-protein-coupled receptor oligomerization and its potential for drug discovery. Nat Rev Drug Discov 1(10):808–820. doi: 10.1038/nrd913.

Gerbino, A., and M. Colella. 2018. The different facets of extracellular calcium sensors: old and new concepts in calcium-sensing receptor signalling and pharmacology. Int J Mol Sci 19(4). doi: 10.3390/ijms19040999.

Gill, H.S., F. Doull, K.J. Rutherfurd, and M.L. Cross. 2000. Immunoregulatory peptides in bovine milk. Br J Nutr 84 Suppl 1:S111–117. doi: 10.1017/s0007114500002336.

Gonzalez, L.M., I. Williamson, J.A. Piedrahita, A.T. Blikslager, and S.T. Magness. 2013. Cell lineage identification and stem cell culture in a porcine model for the study of intestinal epithelial regeneration. PLoS One 8(6):e66465. doi: 10.1371/journal.pone.0066465.

González-Mariscal, L., Raya-Sandino, A., González-González, L., and Hernández-Guzmán, C. 2018. Relationship between G proteins coupled receptors and tight junctions. Tissue Barriers 6: e1463897. doi: 10.1080/21688370.2018.1463897.

González-Mariscal, L., Tapia, R., and Chamorro, D. 2008. Crosstalk of tight junction components with signaling pathways. Biochim Biophys Acta 1778(3):729–756. doi: 10.1016/j.bbamem.2007.08.018.

Gonzalez-Vega, J.C., C.L. Walk, Y. Liu, and H.H. Stein. 2014. The site of net absorption of Ca from the intestinal tract of growing pigs and effect of phytic acid, Ca level and Ca source on Ca digestibility. Arch Anim Nutr 68(2):126–142. doi: 10.1080/1745039X.2014.892249.

Gorvin, C.M. 2018. Insights into calcium-sensing receptor trafficking and biased signalling by studies of calcium homeostasis. J Mol Endocrinol 61(1):R1–R12. doi: 10.1530/JME-18-0049.

Gorvin, C.M., V.N. Babinsky, T. Malinauskas, P.H. Nissen, A.J. Schou, A.C. Hanyaloglu, C. Siebold, E.Y. Jones, F.M. Hannan, and R.V. Thakker. 2018. A calcium-sensing receptor mutation causing hypocalcemia disrupts a transmembrane salt bridge to activate beta-arrestin-biased signaling. Sci Signal 11(518). doi: 10.1126/scisignal.aan3714.

Grozinsky-Glasberg, S., I. Shimon, and H. Rubinfeld. 2012. The role of cell lines in the study of neuroendocrine tumors. Neuroendocrinology 96(3):173–187. doi: 10.1159/000338793.

Gu, J., S. Dai, Y. Liu, H. Liu, Y. Zhang, X. Ji, F. Yu, Y. Zhou, L. Chen, W.K.F. Tse, C.K.C. Wong, B. Chen, and H. Shi. 2018. Activation of Ca^{2+}-sensing receptor as a protective pathway to reduce Cadmium-induced cytotoxicity in renal proximal tubular cells. Sci Rep 8(1):1092. doi: 10.1038/s41598-018-19327-9.

Gunawardene, A.R., B.M. Corfe, and C.A. Staton. 2011. Classification and functions of enteroendocrine cells of the lower gastrointestinal tract. Int J Exp Pathol 92(4):219–231. doi: 10.1111/j.1365-2613.2011.00767.x.

Gundry, J., R. Glenn, P. Alagesan, and S. Rajagopal. 2017. A Practical Guide to Approaching Biased Agonism at G Protein Coupled Receptors. Front Neurosci 11:17. doi: 10.3389/fnins.2017.00017.

Hadjiyanni, I., Li, K. K., and Drucker, D. J. 2009. Glucagon-like peptide-2 reduces intestinal permeability but does not modify the onset of type 1 diabetes in the nonobese diabetic mouse. Endocrinology 150(2):592–599. doi: 10.1210/en.2008-1228.

Haid, D.C., C. Jordan-Biegger, P. Widmayer, and H. Breer. 2012. Receptors responsive to protein breakdown products in g-cells and d-cells of mouse, swine and human. Front Physiol 3:65. doi: 10.3389/fphys.2012.00065.

Hales, T.G. 2011. Arresting the development of morphine tolerance and dependence. Br J Anaesth 107(5):653–655. doi: 10.1093/bja/aer294.

Hammer, A.M., Morris, N.L., Earley, Z.M., M.A. Choudhry. 2015. The first line of defense: the effects of alcohol on post-burn intestinal barrier, immune cells, and microbiome. Alcohol research: current reviews 37(2):209–22.

Hammerland, L.G., K.J. Krapcho, J.E. Garrett, N. Alasti, B.C. Hung, R.T. Simin, C. Levinthal, E.F. Nemeth, and F.H. Fuller. 1999. Domains determining ligand specificity for Ca^{2+} receptors. Mol Pharmacol 55(4):642–648.

Hampson, D.J. 1986. Alterations in piglet small intestinal structure at weaning. Res Vet Sci 40(1):32–40. doi:10.1016/S0034-5288(18)30482-X.

Handlogten, M.E., C. Huang, N. Shiraishi, H. Awata, and R.T. Miller. 2001. The Ca^{2+}-sensing receptor activates cytosolic phospholipase A2 via a Gqalpha-dependent ERK-independent pathway. J Biol Chem 276(17):13941–13948. doi: 10.1074/jbc.M007306200.

Harnett, M.T., W. Chen, and S.M. Smith. 2009. Calcium-sensing receptor: a high-affinity presynaptic target for aminoglycoside-induced weakness. Neuropharmacology 57(5–6): 502–505. doi: 10.1016/j.neuropharm.2009.07.031.

Hassan, Y.I., L. Lahaye, M.M. Gong, J. Peng, J. Gong, S. Liu, C.G. Gay, and C. Yang. 2018. Innovative drugs, chemicals, and enzymes within the animal production chain. Vet Res 49(1): 71. doi: 10.1186/s13567-018-0559-1.

Hauache, O.M., J. Hu, K. Ray, R. Xie, K.A. Jacobson, and A.M. Spiegel. 2000. Effects of a calcimimetic compound and naturally activating mutations on the human Ca^{2+} receptor and on Ca^{2+} receptor/metabotropic glutamate chimeric receptors. Endocrinology 141(11): 4156–4163. doi: 10.1210/endo.141.11.7753.

Hebert, S.C., S. Cheng, and J. Geibel. 2004. Functions and roles of the extracellular Ca^{2+}-sensing receptor in the gastrointestinal tract. Cell Calcium 35(3):239–247. doi: 10.1016/j.ceca.2003.10.015.

Hebert, T.E., S. Moffett, J.P. Morello, T.P. Loisel, D.G. Bichet, C. Barret, and M. Bouvier. 1996. A peptide derived from a beta2-adrenergic receptor transmembrane domain inhibits both receptor dimerization and activation. J Biol Chem 271(27):16384–16392. doi: 10.1074/jbc.271.27.16384.

Hein, G.J., C. Baker, J. Hsieh, S. Farr, and K. Adeli. 2013. GLP-1 and GLP-2 as yin and yang of intestinal lipoprotein production: evidence for predominance of GLP-2-stimulated postprandial lipemia in normal and insulin-resistant states. Diabetes 62(2):373–381. doi: 10.2337/db12-0202.

Hekmati, M., Y. Ben-Shaul, and S. Polak-Charcon. 1990. A morphological study of a human adenocarcinoma cell line (HT29) differentiating in culture. Similarities to intestinal embryonic development. Cell Differ Dev 31(3):207–218. doi: 10.1016/0922-3371(90)90133-h.

Hendy, G.N., and L. Canaff. 2016. Calcium-sensing receptor gene: regulation of expression. Front Physiol 7:394. doi: 10.3389/fphys.2016.00394.

Hendy, G.N., V. Guarnieri, and L. Canaff. 2009. Chapter 3 Calcium-sensing receptor and associated diseases. Prog Mol Biol Transl Sci. p. 31–95.

Hira, T., S. Nakajima, Y. Eto, and H. Hara. 2008. Calcium-sensing receptor mediates phenylalanine-induced cholecystokinin secretion in enteroendocrine STC-1 cells. FEBS J 275(18):4620–4626. doi: 10.1111/j.1742-4658.2008.06604.x.

Hjalm, G., R.J. MacLeod, O. Kifor, N. Chattopadhyay, and E.M. Brown. 2001. Filamin-A binds to the carboxyl-terminal tail of the calcium-sensing receptor, an interaction that participates in CaR-mediated activation of mitogen-activated protein kinase. J Biol Chem 276(37):34880–34887. doi: 10.1074/jbc.M100784200.

Hobson, S.A., S.E. McNeil, F. Lee, and K.D. Rodland. 2000. Signal transduction mechanisms linking increased extracellular calcium to proliferation in ovarian surface epithelial cells. Exp Cell Res 258(1):1–11. doi: 10.1006/excr.2000.4910.

Hobson, S.A., J. Wright, F. Lee, S.E. McNeil, T. Bilderback, and K.D. Rodland. 2003. Activation of the MAP kinase cascade by exogenous calcium-sensing receptor. Mol Cell Endocrinol 200(1–2):189–198. doi: 10.1016/s0303-7207(01)00749-3.

Hoenderop, J.G., and R.J. Bindels. 2005. Epithelial Ca^{2+} and Mg^{2+} channels in health and disease. J Am Soc Nephrol 16(1):15–26. doi: 10.1681/ASN.2004070523.

Hoff, A.O., G.J. Cote, H.A. Fritsche, Jr., H. Qiu, P.N. Schultz, and R.F. Gagel. 1999. Calcium-induced activation of a mutant G-protein-coupled receptor causes *in vitro* transformation of NIH/3T3 cells. Neoplasia 1(6):485–491. doi: 10.1038/sj.neo.7900072.

Hooton, D., R. Lentle, J. Monro, M. Wickham, and R. Simpson. 2015. The ecretion and action of brush border enzymes in the mammalian small intestine. Rev Physiol Biochem Pharmacol 168:59–118. doi: 10.1007/112_2015_24.

Hu, J., J. Jiang, S. Costanzi, C. Thomas, W. Yang, J.H. Feyen, K.A. Jacobson, and A.M. Spiegel. 2006. A missense mutation in the seven-transmembrane domain of the human Ca^{2-} receptor converts a negative allosteric modulator into a positive allosteric modulator. J Biol Chem 281(30):21558–21565. doi: 10.1074/jbc.M603682200.

Hu, J., G. Reyes-Cruz, P.K. Goldsmith, and A.M. Spiegel. 2001. The Venus's-flytrap and cysteine-rich domains of the human Ca^{2+} receptor are not linked by disulfide bonds. J Biol Chem 276(10):6901–6904. doi: 10.1074/jbc.C000865200.

Hu, J., and A.M. Spiegel. 2003. Naturally occurring mutations of the extracellular Ca^{2+}-sensing receptor: implications for its structure and function. Trends Endocrinol Metab 14(6):282–288. doi: 10.1016/s1043-2760(03)00104-8.

Hu, J., and A.M. Spiegel. 2007. Structure and function of the human calcium-sensing receptor: insights from natural and engineered mutations and allosteric modulators. J Cell Mol Med 11(5):908–922. doi: 10.1111/j.1582-4934.2007.00096.x.

Huang, C., K.M. Hujer, Z. Wu, and R.T. Miller. 2004. The Ca^{2+}-sensing receptor couples to Galpha12/13 to activate phospholipase D in Madin-Darby canine kidney cells. Am J Physiol Cell Physiol 286(1):C22–30. doi: 10.1152/ajpcell.00229.2003.

Huang, Y., and G.E. Breitwieser. 2007. Rescue of calcium-sensing receptor mutants by allosteric modulators reveals a conformational checkpoint in receptor biogenesis. J Biol Chem 282(13):9517–9525. doi: 10.1074/jbc.M609045200.

Iamartino, L., T. Elajnaf, E. Kallay, and M. Schepelmann. 2018. Calcium-sensing receptor in colorectal inflammation and cancer: Current insights and future perspectives. World J Gastroenterol 24(36):4119–4131. doi: 10.3748/wjg.v24.i36.4119.

Insel, P.A., A. Snead, F. Murray, L. Zhang, H. Yokouchi, T. Katakia, O. Kwon, D. Dimucci, and A. Wilderman. 2012. GPCR expression in tissues and cells: are the optimal receptors being used as drug targets? Br J Pharmacol 165(6):1613–1616. doi: 10.1111/j.1476-5381.2011.01434.x.

Insel, P.A., A. Wilderman, A.C. Zambon, A.N. Snead, F. Murray, N. Aroonsakool, D.S. McDonald, S. Zhou, T. McCann, L. Zhang, K. Sriram, A.M. Chinn, A.V. Michkov, R.M. Lynch, A.C. Overland, and R. Corriden. 2015. G protein-coupled receptor (GPCR) expression in native cells: "novel" endoGPCRs as physiologic regulators and therapeutic targets. Mol Pharmacol 88(1):181–187. doi: 10.1124/mol.115.098129.

Ipharraguerre, I.R. 2013. Digestive physiology of the pig symposium: gut chemosensing: integrating nutrition, gut function, and metabolism in pigs. J Anim Sci 91(5):1929–1931. doi: 10.2527/jas.2013-6483.

Irwin, D.M. 2021. Variation in the evolution and sequences of proglucagon and the receptors for proglucagon-derived peptides in mammals. Front Endocrinol (Lausanne) 12:700066. doi: 10.3389/fendo.2021.700066.

Ivan Cheng, I.Q., Naibedya Chattopadhyay, Athar Qureshi, Robert R. Butters, Amy E. Hall, Robert. Cima, Kimberly V. Rogers, Steven C. Hebert, John P. Geible, Edward M. Brown, and David I. Soybel. 1999. Expression of an extracellular calcium-sensing receptor in rat stomach. Gastroenterology 116:118–126. doi: 10.1016/s0016-5085(99)70235-0.

Janicic, N.S., E., Pausova, Z., Seldin, M.F., Rivière, M., Szpirer, J., Szpirer, C., Hendy, G.N. 1995. Mapping of the calcium-sensing receptor gene (CASR) to human Chromosome3q13.3-21by fluorescencein situ hybridization, and localization to rat Chromosome 11 and mouse Chromosome 16. Mamm Genome 6:798–801. doi: 10.1007/BF00539007.

Jensen, A.A., T.A. Spalding, E.S. Burstein, P.O. Sheppard, P.J. O'Hara, M.R. Brann, P. Krogsgaard-Larsen, and H. Brauner-Osborne. 2000. Functional importance of the Ala(116)-Pro(136) region in the calcium-sensing receptor. Constitutive activity and inverse agonism in a family C G-protein-coupled receptor. J Biol Chem 275(38):29547–29555. doi: 10.1074/jbc.M910023199.

Jouret, F., J. Wu, M. Hull, V. Rajendran, B. Mayr, C. Schofl, J. Geibel, and M.J. Caplan. 2013. Activation of the Ca^{2+}-sensing receptor induces deposition of tight junction components to the epithelial cell plasma membrane. J Cell Sci 126(Pt 22):5132–5142. doi: 10.1242/jcs.127555.

Joy M Campbell, J.D.C.a.J.P. 2013. The biological stress of early weaned piglets. J Animal Sci Biotechnol 4: 19. doi: https://doi.org/10.1186/2049-1891-4-19.

K. Raun, P. v. V., L.B. Knudsen. 2007. Liraglutide, a once-daily human glucagon-like peptide-1 analog, minimizes food intake in severely obese minipigs. Obesity 15:1710–1716. doi: 10.1038/oby.2007.204.

Kaji, I., Y. Akiba, and J.D. Kaunitz. 2013. Digestive physiology of the pig symposium: involvement of gut chemosensing in the regulation of mucosal barrier function and defense mechanisms. J Anim Sci 91(5):1957–1962. doi: 10.2527/jas.2012-5941.

Kallay, E., E. Bajna, F. Wrba, S. Kriwanek, M. Peterlik, and H.S. Cross. 2000. Dietary calcium and growth modulation of human colon cancer cells: role of the extracellular calcium-sensing receptor. Cancer Detect Prev 24(2):127–136.

Kameda, T., H. Mano, Y. Yamada, H. Takai, N. Amizuka, M. Kobori, N. Izumi, H. Kawashima, H. Ozawa, K. Ikeda, A. Kameda, Y. Hakeda, and M. Kumegawa. 1998. Calcium-sensing receptor in mature osteoclasts, which are bone resorbing cells. Biochem Biophys Res Commun 245(2):419–422. doi: 10.1006/bbrc.1998.8448.

Kenakin, T. 2011. Functional selectivity and biased receptor signaling. J Pharmacol Exp Ther 336(2):296–302. doi: 10.1124/jpet.110.173948.

Kifor, O., R. Diaz, R. Butters, and E.M. Brown. 1997. The Ca^{2+}-sensing receptor (CaR) activates phospholipases C, A2, and D in bovine parathyroid and CaR-transfected, human embryonic kidney (HEK293) cells. J Bone Miner Res 12(5):715–725. doi: 10.1359/jbmr.1997.12.5.715.

Kifor, O., R. Diaz, R. Butters, I. Kifor, and E.M. Brown. 1998. The calcium-sensing receptor is localized in caveolin-rich plasma membrane domains of bovine parathyroid cells. J Biol Chem 273(34):21708–21713.

Kifor, O., R.J. MacLeod, R. Diaz, M. Bai, T. Yamaguchi, T. Yao, I. Kifor, and E.M. Brown. 2001. Regulation of MAP kinase by calcium-sensing receptor in bovine parathyroid and CaR-transfected HEK293 cells. Am J Physiol Renal Physiol 280(2):291–302. doi: 10.1152/ajprenal.2001.280.2.F291.

Kim, C.J., J.A. Kovacs-Nolan, C. Yang, T. Archbold, M.Z. Fan, and Y. Mine. 2010. l-Tryptophan exhibits therapeutic function in a porcine model of dextran sodium sulfate (DSS)-induced colitis. J Nutr Biochem 21(6):468–475. doi: 10.1016/j.jnutbio.2009.01.019.

Kim, W., and J.J. Wysolmerski. 2016. Calcium-sensing receptor in breast physiology and cancer. Front Physiol 7:440. doi: 10.3389/fphys.2016.00440.

Kleiveland, C.R. 2015. Co-cultivation of Caco-2 and HT-29MTX. In: K. Verhoeckx, P. Cotter I. Lopez-Exposito, C. Kleiveland, T. Lea, A. Mackie, T. Requena, D. Swiatecka and H. Wichers, editors, The impact of food bioactives on health: *in vitro* and *ex vivo* models, Cham (CH). pp. 135–140. doi: 10.1007/978-3-319-16104-4.

Kniazeff, J., P.P. Saintot, C. Goudet, J. Liu, A. Charnet, G. Guillon, and J.P. Pin. 2004. Locking the dimeric GABA(B) G-protein-coupled receptor in its active state. J Neurosci 24(2):370–377. doi: 10.1523/JNEUROSCI.3141-03.2004.

Kopic, S., and J.P. Geibel. 2010. Toxin mediated diarrhea in the 21 century: the pathophysiology of intestinal ion transport in the course of ETEC, V. cholerae and rotavirus infection. Toxins (Basel) 2(8):2132–2157. doi: 10.3390/toxins2082132.

Kunishima, N., Y. Shimada, Y. Tsuji, T. Sato, M. Yamamoto, T. Kumasaka, S. Nakanishi, H. Jingami, and K. Morikawa. 2000. Structural basis of glutamate recognition by a dimeric metabotropic glutamate receptor. Nature 407(6807):971–977. doi: 10.1038/35039564.

Kurata, H.T., W.W. Cheng, C. Arrabit, P.A. Slesinger, and C.G. Nichols. 2007. The role of the cytoplasmic pore in inward rectification of Kir2.1 channels. J Gen Physiol 130(2):145–155. doi: 10.1085/jgp.200709742.

Lagos-Cabre, R., M. Brenet, J. Diaz, R.D. Perez, L.A. Perez, R. Herrera-Molina, A.F.G. Quest, and L. Leyton. 2018. Intracellular Ca^{2+} increases and connexin 43 hemichannel opening are necessary but not sufficient for Thy-1-induced astrocyte migration. Int J Mol Sci 19(8): 2179. doi: 10.3390/ijms19082179.

Leach, K., A. Wen, A.E. Cook, P.M. Sexton, A.D. Conigrave, and A. Christopoulos. 2013. Impact of clinically relevant mutations on the pharmacoregulation and signaling bias of the calcium-sensing receptor by positive and negative allosteric modulators. Endocrinology 154(3): 1105–1116. doi: 10.1210/en.2012-1887.

Lee, H.J., H.C. Mun, N.C. Lewis, M.F. Crouch, E.L. Culverston, R.S. Mason, and A.D. Conigrave. 2007. Allosteric activation of the extracellular Ca^{2+}-sensing receptor by L-amino acids enhances ERK1/2 phosphorylation. Biochem J 404(1):141–149. doi: 10.1042/BJ20061826.

Li, Q., C. Yu, Y. Chen, S. Liu, P. Azevedo, J. Gong, K.O, and C. Yang. 2021. Citral alleviates peptidoglycan-induced inflammation and disruption of barrier functions in porcine intestinal epithelial cells. J Cell Physiol 237(3):1768–1779. doi: 10.1002/jcp.30640.

Liang, H., Z. Dai, N. Liu, Y. Ji, J. Chen, Y. Zhang, Y. Yang, J. Li, Z. Wu, and G. Wu. 2018. Dietary L-Tryptophan modulates the structural and functional composition of the intestinal microbiome in weaned piglets. Front Microbiol 9:1736. doi: 10.3389/fmicb.2018.01736.

Liao, S.F., and M. Nyachoti. 2017. Using probiotics to improve swine gut health and nutrient utilization. Anim Nutr 3(4):331–343. doi: 10.1016/j.aninu.2017.06.007.

Liao, Y., G.K. Smyth, and W. Shi. 2013. The Subread aligner: fast, accurate and scalable read mapping by seed-and-vote. Nucleic Acids Res 41(10):e108. doi: 10.1093/nar/gkt214.

Lin, K.I., N. Chattopadhyay, M. Bai, R. Alvarez, C.V. Dang, J.M. Baraban, E.M. Brown, and R.R. Ratan. 1998. Elevated extracellular calcium can prevent apoptosis via the calcium-sensing receptor. Biochem Biophys Res Commun 249(2):325–331. doi: 10.1006/bbrc.1998.9124.

Ling, B., and J. Alcorn. 2010. LPS-induced inflammation downregulates mammary gland glucose, fatty acid, and L-carnitine transporter expression at different lactation stages. Res Vet Sci 89(2):200–202. doi: 10.1016/j.rvsc.2010.03.004.

Liou, A.P., Y. Sei, X. Zhao, J. Feng, X. Lu, C. Thomas, S. Pechhold, H.E. Raybould, and S.A. Wank. 2011. The extracellular calcium-sensing receptor is required for cholecystokinin secretion in response to L-phenylalanine in acutely isolated intestinal I cells. Am J Physiol Gastrointest Liver Physiol 300(4):G538–546. doi: 10.1152/ajpgi.00342.2010.

Liu, G., W. Cao, G. Jia, H. Zhao, X. Chen, and J. Wang. 2018. Calcium-sensing receptor in nutrient sensing: an insight into the modulation of intestinal homoeostasis. Br J Nutr 120(8): 881–890. doi: 10.1017/S0007114518002088.

Liu, J., K. Yu, and W. Zhu. 2016. Amino acid sensing in the gut and its mediation in gut-brain signal transduction. Anim Nutr 2(2):69–73. doi: 10.1016/j.aninu.2016.03.007.

Liu, W., Y. Xie, J. Ma, X. Luo, P. Nie, Z. Zuo, U. Lahrmann, Q. Zhao, Y. Zheng, Y. Zhao, Y. Xue, and J. Ren. 2015. IBS: an illustrator for the presentation and visualization of biological sequences. Bioinformatics 31(20):3359–3361. doi: 10.1093/bioinformatics/btv362.

Liu, Y., I.R. Ipharraguerre, and J.E. Pettigrew. 2013. Digestive physiology of the pig symposium: potential applications of knowledge of gut chemosensing in pig production. J Anim Sci 91(5):1982–1990. doi: 10.2527/jas.2012-6193.

Livak, K.J., and T.D. Schmittgen. 2001. Analysis of relative gene expression data using real-time quantitative PCR and the 2(-Delta Delta C(T)) Method. Methods 25(4):402–408. doi: 10.1006/meth.2001.1262.

Loretz, C.A. 2008. Extracellular calcium-sensing receptors in fishes. Comp Biochem Physiol A Mol Integr Physiol 149(3):225–245. doi: 10.1016/j.cbpa.2008.01.037.

Loss, H., J.R. Aschenbach, F. Ebner, K. Tedin, and U. Lodemann. 2020. Inflammatory responses of porcine MoDC and intestinal epithelial cells in a direct-contact co-culture system following a bacterial challenge. Inflammation 43(2):552–567. doi: 10.1007/s10753-019-01137-4.

Loupy, A., S.K. Ramakrishnan, B. Wootla, R. Chambrey, R. de la Faille, S. Bourgeois, P. Bruneval, C. Mandet, E.I. Christensen, H. Faure, L. Cheval, K. Laghmani, C. Collet, D. Eladari, R.H. Dodd, M. Ruat, and P. Houillier. 2012. PTH-independent regulation of blood calcium concentration by the calcium-sensing receptor. J Clin Invest 122(9):3355–3367. doi: 10.1172/JCI57407.

Lu, V.B., F.M. Gribble, and F. Reimann. 2021. Nutrient-induced cellular mechanisms of gut hormone secretion. Nutrients 13(3). doi: 10.3390/nu13030883.

Lucas, M.L. 2001. A reconsideration of the evidence for *Escherichia coli* STa (heat stable) enterotoxin-driven fluid secretion: a new view of STa action and a new paradigm for fluid absorption. J Appl Microbiol 90(1):7–26. doi: 10.1046/j.1365-2672.2001.01225.x.

Lund, A., T. Vilsboll, J.I. Bagger, J.J. Holst, and F.K. Knop. 2011. The separate and combined impact of the intestinal hormones, GIP, GLP-1, and GLP-2, on glucagon secretion in type 2 diabetes. Am J Physiol Endocrinol Metab 300(6):1038–1046. doi: 10.1152/ajpendo.00665.2010.

M. Janet Barger-Lux, R.P.H., and Robert R. Recker. 1989. Time course of calcium absorption in humans: evidence for a colonic component. Calcif Tissue Int 44(5):308–11. doi: 10.1007/BF02556309.

Mace, O.J., J. Affleck, N. Patel, and G.L. Kellett. 2007. Sweet taste receptors in rat small intestine stimulate glucose absorption through apical GLUT2. J Physiol 582:379–392. doi: 10.1113/jphysiol.2007.130906.

Mace, O.J., and F. Marshall. 2013. Digestive physiology of the pig symposium: gut chemosensing and the regulation of nutrient absorption and energy supply. J Anim Sci 91(5):1932–1945. doi: 10.2527/jas.2012-5906.

Mace, O.J., M. Schindler, and S. Patel. 2012. The regulation of K- and L-cell activity by GLUT2 and the calcium-sensing receptor CaSR in rat small intestine. J Physiol 590(12):2917–2936. doi: 10.1113/jphysiol.2011.223800.

MacLeod, R.J., M. Hayes, and I. Pacheco. 2007. Wnt5a secretion stimulated by the extracellular calcium-sensing receptor inhibits defective Wnt signaling in colon cancer cells. Am J Physiol Gastrointest Liver Physiol 293(1):403–411. doi: 10.1152/ajpgi.00119.2007.

Maiellaro, I., R. Barbaro, R. Caroppo, and M. Colella. 2020. Calcium-sensing receptor signaling: it's all about multiplicity. Curr Opin Physiol 17:243–254. doi: 10.1016/j.cophys.2020.08.014.

Mamillapalli, R., J. VanHouten, W. Zawalich, and J. Wysolmerski. 2008. Switching of G-protein usage by the calcium-sensing receptor reverses its effect on parathyroid hormone-related protein secretion in normal versus malignant breast cells. J Biol Chem 283(36):24435–24447. doi: 10.1074/jbc.M801738200.

Martinez-Palomo, A., I. Meza, G. Beaty, and M. Cereijido. 1980. Experimental modulation of occluding junctions in a cultured transporting epithelium. J Cell Biol 87(3):736–745. doi: 10.1083/jcb.87.3.736.

Maruyama, Y., R. Yasuda, M. Kuroda, and Y. Eto. 2012. Kokumi substances, enhancers of basic tastes, induce responses in calcium-sensing receptor expressing taste cells. PLoS One. 7(4):e34489. doi: 10.1371/journal.pone.0034489.

McKinley, M.J., M.L. Mathai, R.M. McAllen, R.C. McClear, R.R. Miselis, G.L. Pennington, L. Vivas, J.D. Wade, and B.J. Oldfield. 2004. Vasopressin secretion: osmotic and hormonal regulation by the lamina terminalis. J Neuroendocrinol 16(4):340–347. doi: 10.1111/j.0953-8194.2004.01184.x.

McLamb, B.L., A.J. Gibson, E.L. Overman, C. Stahl, and A.J. Moeser. 2013. Early weaning stress in pigs impairs innate mucosal immune responses to enterotoxigenic *E. coli* challenge and exacerbates intestinal injury and clinical disease. PLoS One 8(4):e59838. doi: 10.1371/journal.pone.0059838.

McLarnon, S.J., D. Holden, D.T. Ward, M.N. Jones, A.C. Elliott, and D. Riccardi. 2002. Aminoglycoside antibiotics induce pH-sensitive activation of the calcium-sensing receptor. Biochem Biophys Res Commun 297(1):71–77. doi: 10.1016/s0006-291x(02)02133-2.

McNeil, S.E., S.A. Hobson, V. Nipper, and K.D. Rodland. 1998. Functional calcium-sensing receptors in rat fibroblasts are required for activation of SRC kinase and mitogen-activated protein kinase in response to extracellular calcium. J Biol Chem 273(2):1114–1120. doi: 10.1074/jbc.273.2.1114.

Mehra, A., K.H. Lee, and V. Hatzimanikatis. 2003. Insights into the relation between mRNA and protein expression patterns: I. Theoretical considerations. Biotechnol Bioeng 84(7):822–833. doi: 10.1002/bit.10860.

Mennes, P., R. Rosenbaum, K. Martin, and E. Slatopolsky. 1978. Hypomagnesemia and impaired parathyroid hormone secretion in chronic renal disease. Ann Intern Med 88(2):206–209. doi: 10.7326/0003-4819-88-2-206.

Mentaverri, R., S. Yano, N. Chattopadhyay, L. Petit, O. Kifor, S. Kamel, E.F. Terwilliger, M. Brazier, and E.M. Brown. 2006. The calcium sensing receptor is directly involved in both osteoclast differentiation and apoptosis. FASEB J 20(14):2562–2564. doi: 10.1096/fj.06-6304fje.

Metzler-Zebeli, B.U., S. Hooda, R. Mosenthin, M.G. Ganzle, and R.T. Zijlstra. 2010. Bacterial fermentation affects net mineral flux in the large intestine of pigs fed diets with viscous and fermentable nonstarch polysaccharides. J Anim Sci 88(10):3351–3362. doi: 10.2527/jas.2010-2906.

Michel, M.C., and S.J. Charlton. 2018. Biased agonism in drug discovery-is it too soon to choose a path? Mol Pharmacol 93(4):259–265. doi: 10.1124/mol.117.110890.

Miedlich, S.U., L. Gama, K. Seuwen, R.M. Wolf, and G.E. Breitwieser. 2004. Homology modeling of the transmembrane domain of the human calcium sensing receptor and localization of an allosteric binding site. J Biol Chem 279(8):7254–7263. doi: 10.1074/jbc.M307191200.

Migliore-Samour, D., and P. Jolles. 1988. Casein, a prohormone with an immunomodulating role for the newborn? Experientia 44(3):188–193. doi: 10.1007/BF01941703.

Miguel-Aliaga, I., H. Jasper, and B. Lemaitre. 2018. Anatomy and physiology of the digestive tract of drosophila melanogaster. Genetics 210(2):357–396. doi: 10.1534/genetics.118.300224.

Mine, Y., and H. Zhang. 2015a. Anti-inflammatory effects of poly-L-lysine in intestinal mucosal system mediated by calcium-sensing receptor activation. J Agric Food Chem 63(48): 10437–10447. doi: 10.1021/acs.jafc.5b03812.

Mine, Y., and H. Zhang. 2015b. Calcium-sensing receptor (CaSR)-mediated anti-inflammatory effects of L-amino acids in intestinal epithelial cells. J Agric Food Chem 63(45):9987–9995. doi: 10.1021/acs.jafc.5b03749.

Modvig, I.M., R.E. Kuhre, S.L. Jepsen, S.F.S. Xu, M.S. Engelstoft, K.L. Egerod, T.W. Schwartz, C. Orskov, M.M. Rosenkilde, and J.J. Holst. 2021. Amino acids differ in their capacity to stimulate GLP-1 release from the perfused rat small intestine and stimulate secretion by different sensing mechanisms. Am J Physiol Endocrinol Metab 320(5):874–885. doi: 10.1152/ajpendo.00026.2021.

Moeser, A.J., C.S. Pohl, and M. Rajput. 2017. Weaning stress and gastrointestinal barrier development: Implications for lifelong gut health in pigs. Animal Nutrition 3(4):313–321. doi: 10.1016/j.aninu.2017.06.003.

Moon, M.J., S. Park, D.K. Kim, E.B. Cho, J.I. Hwang, H. Vaudry, and J.Y. Seong. 2012. Structural and molecular conservation of glucagon-like Peptide-1 and its receptor confers selective ligand-receptor interaction. Front Endocrinol (Lausanne) 3:141. doi: 10.3389/fendo.2012.00141.

Moran, G.W., C. O'Neill, and J.T. McLaughlin. 2012. GLP-2 enhances barrier formation and attenuates TNFalpha-induced changes in a Caco-2 cell model of the intestinal barrier. Regul Pept 178(1–3):95–101. doi: 10.1016/j.regpep.2012.07.002.

Mun, H.-C., K.M. Leach, and A.D. Conigrave. 2019. L-amino acids promote calcitonin release via a calcium-sensing receptor: Gq/11-mediated pathway in human C-cells. Endocrinology 160(7):1590–1599. doi: 10.1210/en.2018-00860.

Mun, H.C., A.H. Franks, E.L. Culverston, K. Krapcho, E.F. Nemeth, and A.D. Conigrave. 2004. The Venus Fly Trap domain of the extracellular Ca^{2+}-sensing receptor is required for L-amino acid sensing. J Biol Chem 279(50):51739–51744. doi: 10.1074/jbc.M406164/200.

Navarro, J.F., C. Mora, A. Jiménez, A. Torres, M. Macía, and J. García. 1999. Relationship between serum magnesium and parathyroid hormone levels in hemodialysis patients. Am J Kidney Dis 34(1):43–48. doi: 10.1016/s0272-6386(99)70106-x.

Nearing, J., M. Betka, S. Quinn, H. Hentschel, M. Elger, M. Baum, M. Bai, N. Chattopadyhay, E.M. Brown, S.C. Hebert, and H.W. Harris. 2002. Polyvalent cation receptor proteins (CaRs) are salinity sensors in fish. Proc Natl Acad Sci USA 99(14):9231–9236. doi: 10.1073/pnas.152294399.

Nemeth, E.F., E.G. Delmar, W.L. Heaton, M.A. Miller, L.D. Lambert, R.L. Conklin, M. Gowen, J.G. Gleason, P.K. Bhatnagar, and J. Fox. 2001. Calcilytic compounds: Potent and selective Ca^{2+} receptor antagonists that stimulate secretion of parathyroid hormone. J Pharmacol Exp Ther 99(1):323–331.

Nemeth, E.F., and W.G. Goodman. 2016. Calcimimetic and calcilytic drugs: feats, flops, and futures. Calcif Tissue Int 98(4):341–358. doi: 10.1007/s00223-015-0052-z.

Nemeth, E.F., M.E. Steffey, L.G. Hammerland, B.C. Hung, B.C. Van Wagenen, E.G. DelMar, and M.F. Balandrin. 1998. Calcimimetics with potent and selective activity on the parathyroid calcium receptor. Proc Natl Acad Sci USA 95(7):4040–4045. doi: 10.1073/pnas.95.7.4040.

Nuovo, G. 2016. False-positive results in diagnostic immunohistochemistry are related to horseradish peroxidase conjugates in commercially available assays. Ann Diagn Pathol 25: 54–59. doi: 10.1016/j.anndiagpath.2016.09.010.

Oates, P.S., C. Thomas, and E.H. Morgan. 1997. Characterization of isolated duodenal epithelial cells along a crypt-villus axis in rats fed diets with different iron content. J Gastroenterol Hepatol 12(12):829–838. doi: 10.1111/j.1440-1746.1997.tb00380.x.

Oda, Y., C.L. Tu, S. Pillai, and D.D. Bikle. 1998. The calcium sensing receptor and its alternatively spliced form in keratinocyte differentiation. J Biol Chem 273(36):23344–23352. doi: 10.1074/jbc.273.36.23344.

Odenwald, M. A., and Turner, J. R. 2017. The intestinal epithelial barrier: a therapeutic target? Nat Rev Gastroenterol Hepatol 14(1):9–21. doi: 10.1038/nrgastro.2016.169.

Odle, J., X. Lin, S.K. Jacobi, S.W. Kim, and C.H. Stahl. 2014. The suckling piglet as an agrimedical model for the study of pediatric nutrition and metabolism. Annu Rev Anim Biosci 2(1):419–444. doi: 10.1146/annurev-animal-022513-114158.

Ogata, S., Y. Kubota, S. Satoh, S. Ito, H. Takeuchi, M. Ashizuka, and K. Shirasuna. 2006. Ca^{2+} stimulates COX-2 expression through calcium-sensing receptor in fibroblasts. Biochem Biophys Res Commun 351(4):808–814. doi: 10.1016/j.bbrc.2006.10.098.

Ohsu, T., Y. Amino, H. Nagasaki, T. Yamanaka, S. Takeshita, T. Hatanaka, Y. Maruyama, N. Miyamura, and Y. Eto. 2010. Involvement of the calcium-sensing receptor in human taste perception. J Biol Chem 285(2):1016–1022. doi: 10.1074/jbc.M109.029165.

Okumura, R., and K. Takeda. 2017. Roles of intestinal epithelial cells in the maintenance of gut homeostasis. Exp Mol Med 49(5):e338. doi: 10.1038/emm.2017.20.

Omonijo, F.A., S. Liu, Q. Hui, H. Zhang, L. Lahaye, J.C. Bodin, J. Gong, M. Nyachoti, and C. Yang. 2019. Thymol improves barrier function and attenuates inflammatory responses in porcine intestinal epithelial cells during lipopolysaccharide (LPS)-induced inflammation. J Agric Food Chem 67(2):615–624. doi: 10.1021/acs.jafc.8b05480.

Owen, J.L., S.X. Cheng, Y. Ge, B. Sahay, and M. Mohamadzadeh. 2016. The role of the calcium-sensing receptor in gastrointestinal inflammation. Semin Cell Dev Biol 49:44–51. doi: 10.1016/j.semcdb.2015.10.040.

Pais, R., F.M. Gribble, and F. Reimann. 2016. Signalling pathways involved in the detection of peptones by murine small intestinal enteroendocrine L-cells. Peptides 77:9–15. doi: 10.1016/j.peptides.2015.07.019.

Palsuledesai, C.C., Z. Surviladze, A. Waller, T.F. Miscioscia, Y. Guo, Y. Wu, J. Strouse, E. Romero, V.M. Salas, R. Curpan, S. Young, M. Carter, T. Foutz, Z. Galochkina, H. Ames, M.K. Haynes, B.S. Edwards, O. Nicolotti, L. Luo, O. Ursu, C.G. Bologa, T.I. Oprea, A. Wandinger-Ness, and L.A. Sklar. 2018. Activation of Rho family GTPases by small molecules. ACS Chem Biol 13(6):1514–1524. doi: 10.1021/acschembio.8b00038.

Partridge, I.G. 1978. Studies on digestion and absorption in the intestines of growing pigs. 3. Net movements of mineral nutrients in the digestive tract. Br J Nutr 39(3):527–537. doi: 10.1079/bjn19780068.

Petrel, C., A. Kessler, P. Dauban, R.H. Dodd, D. Rognan, and M. Ruat. 2004. Positive and negative allosteric modulators of the Ca^{2+}-sensing receptor interact within overlapping but not identical binding sites in the transmembrane domain. J Biol Chem 279(18):18990–18997. doi: 10.1074/jbc.M400724200.

Pi, M., R.F. Spurney, Q. Tu, T. Hinson, and L.D. Quarles. 2002. Calcium-sensing receptor activation of rho involves filamin and rho-guanine nucleotide exchange factor. Endocrinology 143(10):3830–3838. doi: 10.1210/en.2002-220240.

Pidasheva, S., L. Canaff, W.F. Simonds, S.J. Marx, and G.N. Hendy. 2005. Impaired cotranslational processing of the calcium-sensing receptor due to signal peptide missense mutations in familial hypocalciuric hypercalcemia. Hum Mol Genet 14(12):1679–1690. doi: 10.1093/hmg/ddi176.

Pidasheva, S., L. D'Souza-Li, L. Canaff, D.E. Cole, and G.N. Hendy. 2004. CASRdb: calcium-sensing receptor locus-specific database for mutations causing familial (benign) hypocalciuric hypercalcemia, neonatal severe hyperparathyroidism, and autosomal dominant hypocalcemia. Hum Mutat 24(2):107–111. doi: 10.1002/humu.20067.

Pidasheva, S., M. Grant, L. Canaff, O. Ercan, U. Kumar, and G.N. Hendy. 2006. Calcium-sensing receptor dimerizes in the endoplasmic reticulum: biochemical and biophysical characterization of CASR mutants retained intracellularly. Hum Mol Genet 15(14):2200–2209. doi: 10.1093/hmg/ddl145.

Pluske, J.R., D.L. Turpin, and J.C. Kim. 2018. Gastrointestinal tract (gut) health in the young pig. Anim Nutr 4(2):187–196. doi: 10.1016/j.aninu.2017.12.004.

Pollak, M.R., E.M. Brown, Y.H.W. Chou, S.C. Hebert, S.J. Marx, B. Steinman, T. Levi, C.E. Seidman, and J.G. Seidman. 1993. Mutations in the human Ca^{2+}-sensing receptor gene cause familial hypocalciuric hypercalcemia and neonatal severe hyperparathyroidism. Cell 75:1297–1303.

Pollak, M.R., E.M. Brown, H.L. Estep, P.N. McLaine, O. Kifor, J. Park, S.C. Hebert, C.E. Seidman, and J.G. Seidman. 1994. Autosomal dominant hypocalcaemia caused by a Ca^{2+}-sensing receptor gene mutation. Nat Genet 8:303–307.

Prinz, C., M. Kajimura, D. Scott, H. Helander, J. Shin, M. Besancon, K. Bamberg, S. Hersey, and G. Sachs. 1992. Acid secretion and the H,K ATPase of stomach. Yale J Biol Med 65(6):577–596.

Quamme, G.A. 1989. Control of magnesium transport in the thick ascending limb. Am J Physiol 256(2):197–210. doi: 10.1152/ajprenal.1989.256.2.F197.

Quarles, L.D. 2003. Extracellular calcium-sensing receptors in the parathyroid gland, kidney, and other tissues. Curr Opin Nephrol Hypertens 12(4):349–355. doi: 10.1097/00041552-200307000-00002.

Quinn, S.J., M. Bai, and E.M. Brown. 2004. pH Sensing by the calcium-sensing receptor. J Biol Chem 279(36):37241–37249. doi: 10.1074/jbc.M404520200.

Quinn, S.J., O. Kifor, S. Trivedi, R. Diaz, P. Vassilev, and E. Brown. 1998. Sodium and ionic strength sensing by the calcium receptor. J Biol Chem 273(31):19579–19586. doi: 10.1074/jbc.273.31.19579.

Quinn, S.J., C.P. Ye, R. Diaz, O. Kifor, M. Bai, P. Vassilev, and E. Brown. 1997. The Ca^{2+}-sensing receptor: a target for polyamines. Am J Physiol 273(4):1315–1323. doi: 10.1152/ajpcell.1997.273.4.C1315.

Rajagopal, S., K. Rajagopal, and R.J. Lefkowitz. 2010. Teaching old receptors new tricks: biasing seven-transmembrane receptors. Nat Rev Drug Discov 9(5):373–386. doi: 10.1038/nrd3024.

Ray, J.M., P.E. Squires, S.B. Curtis, M.R. Meloche, and A.M. Buchan. 1997a. Expression of the calcium-sensing receptor on human antral gastrin cells in culture. J Clin Invest 99(10):2328–2333. doi: 10.1172/JCI119413.

Ray, K., P. Clapp, P.K. Goldsmith, and A.M. Spiegel. 1998. Identification of the sites of N-linked glycosylation on the human calcium receptor and assessment of their role in cell surface expression and signal transduction. J Biol Chem 273(51):34558–34567. doi: 10.1074/jbc.273.51.34558.

Ray, K., G.F. Fan, P.K. Goldsmith, and A.M. Spiegel. 1997b. The carboxyl terminus of the human calcium receptor. Requirements for cell-surface expression and signal transduction. J Biol Chem 272(50):31355–31361. doi: 10.1074/jbc.272.50.31355.

Reimann, F., G. Tolhurst, and F.M. Gribble. 2012. G-protein-coupled receptors in intestinal chemosensation. Cell Metab 15(4):421–431. doi: 10.1016/j.cmet.2011.12.019.

Rey, O., Chang, W., Bikle, D., Rozengurt, N., Young, S.H., and Rozengurt, E. 2012. Negative crosstalk between calcium-sensing receptor and β-catenin signaling systems in colonic epithelium. J Biol Chem 287(2):1158–1167. doi: https://doi.org/10.1074/jbc.M111.274589.

Rey, O., S.H. Young, J. Yuan, L. Slice, and E. Rozengurt. 2005. Amino acid-stimulated Ca^{2+} oscillations produced by the Ca^{2+}-sensing receptor are mediated by a phospholipase C/inositol 1,4,5-trisphosphate-independent pathway that requires G12, Rho, filamin-A, and the actin cytoskeleton. J Biol Chem 280(24):22875–22882. doi: 10.1074/jbc.M503455200.

Reyes-Cruz, G., J. Hu, P.K. Goldsmith, P.J. Steinbach, and A.M. Spiegel. 2001. Human Ca^{2+} receptor extracellular domain. Analysis of function of lobe I loop deletion mutants. J Biol Chem 276(34):32145–32151. doi: 10.1074/jbc.M102977200.

Riccardi, D. 2002. Wellcome Prize Lecture. Cell surface, ion-sensing receptors. Exp Physiol 87(4):403–411. doi: 10.1111/j.1469-445x.2002.tb00053.x.

Riccardi, D. 2012. Parathyroid hormone-independent role for the calcium-sensing receptor in the control of urinary calcium excretion. J Am Soc Nephrol 23(11):1766–1768. doi: 10.1681/ASN.2012090955.

Riccardi, D., and E.M. Brown. 2010. Physiology and pathophysiology of the calcium-sensing receptor in the kidney. Am J Physiol Renal Physiol 298(3):485–499. doi: 10.1152/ajprenal.00608.2009.

Riccardi, D., A.E. Hall, N. Chattopadhyay, J.Z. Xu, E.M. Brown, and S.C. Hebert. 1998. Localization of the extracellular Ca^{2+}/polyvalent cation-sensing protein in rat kidney. Am J Physiol 274(3):611–622. doi: 10.1152/ajprenal.1998.274.3.F611.

Riccardi, D., W.S. Lee, K. Lee, G.V. Segre, E.M. Brown, and S.C. Hebert. 1996. Localization of the extracellular Ca^{2+}-sensing receptor and PTH/PTHrP receptor in rat kidney. Am J Physiol 271(4):951–956. doi: 10.1152/ajprenal.1996.271.4.F951.

Riccardi, D., J. Park, W.S. Lee, G. Gamba, E.M. Brown, and S.C. Hebert. 1995. Cloning and functional expression of a rat kidney extracellular calcium/polyvalent cation-sensing receptor. Proc Natl Acad Sci USA 92(1):131–135. doi: 10.1073/pnas.92.1.131.

Rogers, K.V., C.K. Dunn, S.C. Hebert, and E.M. Brown. 1997. Localization of calcium receptor mRNA in the adult rat central nervous system by in situ hybridization. Brain Research 744(1):47–56. doi: 10.1016/s0006-8993(96)01070-0.

Romano, C., J.K. Miller, K. Hyrc, S. Dikranian, S. Mennerick, Y. Takeuchi, M.P. Goldberg, and K.L. O'Malley. 2001. Covalent and noncovalent interactions mediate metabotropic glutamate receptor mGlu5 dimerization. Mol Pharmacol 59(1):46–53.

Roura, E., I. Depoortere, and M. Navarro. 2019. Review: Chemosensing of nutrients and non-nutrients in the human and porcine gastrointestinal tract. Animal 13(11):2714–2726. doi: 10.1017/S1751731119001794.

Ruat, M., M.E. Molliver, A.M. Snowman, and S.H. Snyder. 1995. Calcium sensing receptor: molecular cloning in rat and localization to nerve terminals. Proc Natl Acad Sci USA 92(8):3161–3165. doi: 10.1073/pnas.92.8.3161.

Ruat, M., and E. Traiffort. 2013. Roles of the calcium sensing receptor in the central nervous system. Best Pract Res Clin Endocrinol Metab 27(3):429–442. doi: 10.1016/j.beem.2013.03.001.

Ruben Diaz, S.L., Hurwitz, Naibeday Chattopadhyay, Y.Y. Mark Pines, Olga Kifor, Miriam S. Einat, and S.C.H. Robert Butters, and Edward M. Brown. 1997. Cloning, expression, and tissue localization of the calcium-sensing receptor in chicken (Gallus domesticus). Am J Physiol Cell Physiol 273(3):1008–1016. doi: 10.1152/ajpregu.1997.273.3.R1008.

Rutten, M.J., K.D. Bacon, K.L. Marlink, M. Stoney, C.L. Meichsner, F.P. Lee, S.A. Hobson, K.D. Rodland, B.C. Sheppard, D.D. Trunkey, K.E. Deveney, and C.W. Deveney. 1999. Identification of a functional Ca^{2+}-sensing receptor in normal human gastric mucous epithelial cells. Am J Physiol 277(3):662–670. doi: 10.1152/ajpgi.1999.277.3.G662.

Saidak, Z., M. Brazier, S. Kamel, and R. Mentaverri. 2009. Agonists and allosteric modulators of the calcium-sensing receptor and their therapeutic applications. Mol Pharmacol 76(6): 1131–1144. doi: 10.1124/mol.109.058784.

Santos-Hernandez, M., B. Miralles, L. Amigo, and I. Recio. 2018. Intestinal signaling of proteins and digestion-derived products relevant to satiety. J Agric Food Chem 66(39): 10123–10131. doi: 10.1021/acs.jafc.8b02355.

Saraf, M.K., B.D. Piccolo, A.K. Bowlin, K.E. Mercer, T. LeRoith, S.V. Chintapalli, K. Shankar, T.M. Badger, and L. Yeruva. 2017. Formula diet driven microbiota shifts tryptophan metabolism from serotonin to tryptamine in neonatal porcine colon. Microbiome 5(1):77. doi: 10.1186/s40168-017-0297-z.

Sassmann, A., B. Gier, H.J. Grone, G. Drews, S. Offermanns, and N. Wettschureck. 2010. The Gq/G11-mediated signaling pathway is critical for autocrine potentiation of insulin secretion in mice. J Clin Invest 120(6):2184–2193. doi: 10.1172/JCI41541.

Schneggenburger, R., and E. Neher. 2000. Intracellular calcium dependence of transmitter release rates at a fast central synapse. Nature 406(6798):889–893. doi: 10.1038/35022702.

Schucht, R., S. Lydford, L. Andzinski, J. Zauers, J. Cooper, H. Hauser, D. Wirth, and T. May. 2011. Rapid establishment of G-protein-coupled receptor-expressing cell lines by site-specific integration. J Biomol Screen 16(3):323–331. doi: 10.1177/1087057110396371.

Shahzad, A., M. Knapp, I. Lang, and G. Kohler. 2010. Interleukin 8 (IL-8) – a universal biomarker? Int Arch Med 3:11. doi: 10.1186/1755-7682-3-11.

Shareghi, G.R., and Z.S. Agus. 1982. Magnesium transport in the cortical thick ascending limb of Henle's loop of the rabbit. J Clin Invest 69(4):759–769. doi: 10.1172/jci110514.

Sheinin, Y., E. Kallay, F. Wrba, S. Kriwanek, M. Peterlik, and H.S. Cross. 2000. Immunocytochemical localization of the extracellular calcium-sensing receptor in normal and malignant human large intestinal mucosa. J Histochem Cytochem 48(5):595–602. doi: 10.1177/002215540004800503.

Shin, J., F. Shen, and J.R. Huguenard. 2005. Polyamines modulate AMPA receptor-dependent synaptic responses in immature layer v pyramidal neurons. J Neurophysiol 93(5):2634–2643. doi: 10.1152/jn.01054.2004.

Shil, A., Olusanya, O., Ghufoor, Z., Forson, B., Marks, J., and Chichger, H. 2020. Artificial sweeteners disrupt tight junctions and barrier function in the intestinal epithelium through activation of the sweet taste receptor, T1R3. Nutrients 12(6):1862. doi: 10.3390/nu12061862.

Shirazi-Beechey, S.P., A.W. Moran, D. Bravo, and M. Al-Rammahi. 2011. Nonruminant Nutrition Symposium: intestinal glucose sensing and regulation of glucose absorption: implications for swine nutrition. J Anim Sci 89(6):1854–1862. doi: 10.2527/jas.2010-3695.

Skelly, B.J., and R.J. Franklin. 2007. Mutations in genes causing human familial isolated hyperparathyroidism do not account for hyperparathyroidism in Keeshond dogs. Vet J 174(3):652–654. doi: 10.1016/j.tvjl.2006.10.018.

Song, T., J. Peng, J. Ren, H.K. Wei, and J. Peng. 2015. Cloning and characterization of spliced variants of the porcine G protein coupled receptor 120. Biomed Res Int 2015:813816. doi: 10.1155/2015/813816.

Sood, R., A.C. Porter, D.A. Olsen, D.R. Cavener, and R.C. Wek. 2000. A mammalian homologue of GCN2 protein kinase important for translational control by phosphorylation of eukaryotic initiation factor-2alpha. Genetics 154(2):787–801. doi: 10.1093/genetics/154.2.787.

Sriram, K., S.Z. Wiley, K. Moyung, M.W. Gorr, C. Salmeron, J. Marucut, R.P. French, A.M. Lowy, and P.A. Insel. 2019. Detection and quantification of GPCR mRNA: an assessment and implications of data from high-content methods. ACS Omega 4(16):17048–17059. doi: 10.1021/acsomega.9b02811.

Steinert, R.E., C. Feinle-Bisset, N. Geary, and C. Beglinger. 2013. Digestive physiology of the pig symposium: secretion of gastrointestinal hormones and eating control. J Anim Sci 91(5):1963–1973. doi: 10.2527/jas.2012-6022.

Stormo, G.D. 2009. An introduction to sequence similarity ("homology") searching. Curr Protoc Bioinformatics Chapter 3:Unit 3 1 3 1 1–7. doi: 10.1002/0471250953.bi0301s27.

Symonds, E.L., M. Peiris, A.J. Page, B. Chia, H. Dogra, A. Masding, V. Galanakis, M. Atiba, D. Bulmer, R.L. Young, and L.A. Blackshaw. 2015. Mechanisms of activation of mouse and human enteroendocrine cells by nutrients. Gut 64(4):618–626. doi: 10.1136/gutjnl-2014-306834.

Tamura, K., G. Stecher, D. Peterson, A. Filipski, and S. Kumar. 2013. MEGA6: Molecular evolutionary genetics analysis version 6.0. Mol Biol Evol 30(12):2725–2729. doi: 10.1093/molbev/mst197.

Tang, L., C.Y. Cheng, X. Sun, A.J. Pedicone, M. Mohamadzadeh, and S.X. Cheng. 2016. The extracellular calcium-sensing receptor in the intestine: evidence for regulation of colonic absorption, secretion, motility, and immunity. Front Physiol 7:245. doi: 10.3389/fphys.2016.00245.

Tedo G., Roura E., Ipharraguerre I., and M.X. 2011. Gene expression of the L-amino acid-sensing receptor T1R1/T1R3 changes in gut tissues of pigs in response to dietary protein. J Anim Sci 89:86. doi: 10.1016/j.bbrc.2011.02.016.

Tennakoon, S., A. Aggarwal, and E. Kallay. 2016. The calcium-sensing receptor and the hallmarks of cancer. Biochim Biophys Acta 1863(6):1398–1407. doi: 10.1016/j.bbamcr.2015.11.017.

Tfelt-Hansen, J., R.J. MacLeod, N. Chattopadhyay, S. Yano, S. Quinn, X. Ren, E.F. Terwilliger, P. Schwarz, and E.M. Brown. 2003. Calcium-sensing receptor stimulates PTHrP release by pathways dependent on PKC, p38 MAPK, JNK, and ERK1/2 in H-500 cells. Am J Physiol Endocrinol Metab 285(2):329–337. doi: 10.1152/ajpendo.00489.2002.

Thiagarajah, J.R., M. Donowitz, and A.S. Verkman. 2015. Secretory diarrhoea: mechanisms and emerging therapies. Nat Rev Gastroenterol Hepatol 12(8):446–457. doi: 10.1038/nrgastro.2015.111.

Thomsen, A.R., M. Hvidtfeldt, and H. Brauner-Osborne. 2012. Biased agonism of the calcium-sensing receptor. Cell Calcium 51(2):107–116. doi: 10.1016/j.ceca.2011.11.009.

Toka, H.R., K. Al-Romaih, J.M. Koshy, S. DiBartolo, 3rd, C.H. Kos, S.J. Quinn, G.C. Curhan, D.B. Mount, E.M. Brown, and M.R. Pollak. 2012. Deficiency of the calcium-sensing receptor in the kidney causes parathyroid hormone-independent hypocalciuria. J Am Soc Nephrol 23(11):1879–1890. doi: 10.1681/ASN.2012030323.

Topala, C.N., J.P. Schoeber, L.E. Searchfield, D. Riccardi, J.G. Hoenderop, and R.J. Bindels. 2009. Activation of the Ca^{2+}-sensing receptor stimulates the activity of the epithelial Ca^{2+} channel TRPV5. Cell Calcium 45(4):331–339. doi: 10.1016/j.ceca.2008.12.003.

Toshimasa., A.K.K.S.T.M.O. 1995. Molecular cloning of a putative Ca^{2+}-sensing receptor cDNA from human kidney. Biochem Biophys Res Commun 214(2):524–529. doi: 10.1006/BBRC.1995.2318.

Tyagi, A., E.B. Daliri, F. Kwami Ofosu, S.J. Yeon, and D.H. Oh. 2020. Food-derived opioid peptides in human health: a review. Int J Mol Sci 21(22)doi: 10.3390/ijms21228825.

Umar, S. 2010. Intestinal stem cells. Curr Gastroenterol Rep 12(5):340–348. doi: 10.1007/s11894-010-0130-3.

Usuda, H., Okamoto, T., and Wada, K. 2021. Leaky gut: effect of dietary fiber and fats on microbiome and intestinal barrier. Int J Mol Sci 22(14):7613. doi: 10.3390/ijms22147613.

Vahe, C., K. Benomar, S. Espiard, L. Coppin, A. Jannin, M.F. Odou, and M.C. Vantyghem. 2017. Diseases associated with calcium-sensing receptor. Orphanet J Rare Dis 12(1):19. doi: 10.1186/s13023-017-0570-z.

Van Beers, E.H., H.A. Buller, R.J. Grand, A.W. Einerhand, and J. Dekker. 1995. Intestinal brush border glycohydrolases: structure, function, and development. Crit Rev Biochem Mol Biol 30(3):197–262. doi: 10.3109/10409239509085143.

Van Boeckel, T.P., C. Brower, M. Gilbert, B.T. Grenfell, S.A. Levin, T.P. Robinson, A. Teillant, and R. Laxminarayan. 2015. Global trends in antimicrobial use in food animals. Proc Natl Acad Sci USA 112(18):5649–5654. doi: 10.1073/pnas.1503141112.

VanHouten, J., P. Dann, G. McGeoch, E.M. Brown, K. Krapcho, M. Neville, and J.J. Wysolmerski. 2004. The calcium-sensing receptor regulates mammary gland parathyroid hormone-related protein production and calcium transport. J Clin Invest 113(4): 598–608. doi: 10.1172/JCI18776.

Vassilev, P.M., C.L. Ho-Pao, M.P. Kanazirska, C. Ye, K. Hong, C.E. Seidman, J.G. Seidman, and E.M. Brown. 1997. Calcium-sensing receptor (CaR)-mediated activation of K^+ channels is blunted in CaR gene-deficient mouse neurons. Neuroreport 8(6):1411–1416. doi: 10.1097/00001756-199704140-00018.

Verrey, F., D. Singer, T. Ramadan, R.N. Vuille-dit-Bille, L. Mariotta, and S.M. Camargo. 2009. Kidney amino acid transport. Pflugers Arch 458(1):53–60. doi: 10.1007/s00424-009-0638-2.

Vighi, G., F. Marcucci, L. Sensi, G. Di Cara, and F. Frati. 2008. Allergy and the gastrointestinal system. Clin Exp Immunol 153 (1):3–6. doi: 10.1111/j.1365-2249.2008.03713.x.

Wan, J., J. Zhang, G. Wu, D. Chen, B. Yu, Z. Huang, Y. Luo, P. Zheng, J. Luo, X. Mao, J. Yu, and J. He. 2019. Amelioration of enterotoxigenic *Escherichia coli*-induced intestinal barrier disruption by low-molecular-weight chitosan in weaned pigs is related to suppressed intestinal inflammation and apoptosis. Int J Mol Sci 20(14). doi: 10.3390/ijms20143485.

Wang, C., Kang, C., Xian, Y., Zhang, M., Chen, X., Pei, M., Zhu, W., & Hang, S. 2018. Sensing of L-Arginine by Gut-Expressed Calcium Sensing Receptor Stimulates Gut Satiety Hormones Cholecystokinin and Glucose-Dependent Insulinotropic Peptide Secretion in Pig Model. J Food Sci 83(9):2394–2401. doi: 10.1111/1750-3841.14297.

Wang, M., Y. Yao, D. Kuang, and D.R. Hampson. 2006. Activation of family C G-protein-coupled receptors by the tripeptide glutathione. J Biol Chem 281(13):8864–8870. doi: 10.1074/jbc.M512865200.

Wang, W.H., M. Lu, and S.C. Hebert. 1996. Cytochrome P-450 metabolites mediate extracellular Ca^{2+}-induced inhibition of apical K^+ channels in the TAL. Am J Physiol 271(1):103–111. doi: 10.1152/ajpcell.1996.271.1.C103.

Wang, Y., R. Chandra, L.A. Samsa, B. Gooch, B.E. Fee, J.M. Cook, S.R. Vigna, A.O. Grant, and R.A. Liddle. 2011. Amino acids stimulate cholecystokinin release through the Ca^{2+}-sensing receptor. Am J Physiol Gastrointest Liver Physiol 300(4): 528–537. doi: 10.1152/ajpgi.00387.2010.

Ward, B.K., A.L. Magno, E.A. Davis, A.C. Hanyaloglu, B.G. Stuckey, M. Burrows, K.A. Eidne, A.K. Charles, and T. Ratajczak. 2004. Functional deletion of the calcium-sensing receptor in a case of neonatal severe hyperparathyroidism. J Clin Endocrinol Metab 89(8): 3721–3730. doi: 10.1210/jc.2003-031653.

Ward, B.K., A.L. Magno, J.P. Walsh, and T. Ratajczak. 2012. The role of the calcium-sensing receptor in human disease. Clin Biochem 45(12):943–953. doi: 10.1016/j.clinbiochem.2012.03.034.

Ward, D.T., E.M. Brown, and H.W. Harris. 1998. Disulfide bonds in the extracellular calcium-polyvalent cation-sensing receptor correlate with dimer formation and its response to divalent cations *in vitro*. J Biol Chem 273(23):14476–14483. doi: 10.1074/jbc.273.23.14476.

Ward, D.T., D. Maldonado-Perez, L. Hollins, and D. Riccardi. 2005. Aminoglycosides induce acute cell signaling and chronic cell death in renal cells that express the calcium-sensing receptor. J Am Soc Nephrol 16(5):1236–1244. doi: 10.1681/ASN.2004080631.

Wasserman, R.H. 2004. Vitamin D and the dual processes of intestinal calcium absorption. J Nutr 134:3137-3139. doi: 10.1093/jn/134.11.3137.

Wellendorph, P., and H. Brauner-Osborne. 2009. Molecular basis for amino acid sensing by family C G-protein-coupled receptors. Br J Pharmacol 156(6):869–884. doi: 10.1111/j.1476-5381.2008.00078.x.

Widmann, C., S. Gibson, M.B. Jarpe, and G.L. Johnson. 1999. Mitogen-activated protein kinase: conservation of a three-kinase module from yeast to human. Physiol Rev 79(1):143–180. doi: 10.1152/physrev.1999.79.1.143.

Wong, N.L., J.H. Dirks, and G.A. Quamme. 1983. Tubular reabsorptive capacity for magnesium in the dog kidney. Am J Physiol 244(1):78–83. doi: 10.1152/ajprenal.1983.244.1.F78.

Wongdee, K., M. Rodrat, J. Teerapornpuntakit, N. Krishnamra, and N. Charoenphandhu. 2019. Factors inhibiting intestinal calcium absorption: hormones and luminal factors that prevent excessive calcium uptake. J Physiol Sci 69(5):683–696. doi: 10.1007/s12576-019-00688-3.

Worthington, J.J., F. Reimann, and F.M. Gribble. 2018. Enteroendocrine cells-sensory sentinels of the intestinal environment and orchestrators of mucosal immunity. Mucosal Immunol 11(1):3–20. doi: 10.1038/mi.2017.73.

Wouters, E., L. Vasudevan, R.A.J. Crans, D.K. Saini, and C.P. Stove. 2019. Luminescence-and fluorescence-based complementation assays to screen for GPCR oligomerization: current state of the art. Int J Mol Sci 20(12): 2958. doi: 10.3390/ijms20122958.

Wu, H.X., C. Wang, K.J. Gregory, G.W. Han, H.P. Cho, Y. Xia, C.M. Niswender, V. Katritch, J. Meiler, V. Cherezov, P.J. Conn, and R.C. Stevens. 2014. Structure of a Class C GPCR metabotropic Glutamate receptor 1 bound to an allosteric modulator. Science 344(6179): 58–64. doi: 10.1126/science.1249489.

Wusigale, L. Liang, and Y. Luo. 2020. Casein and pectin: Structures, interactions, and applications. Trends Food Sci Technol 97:391–403. doi: 10.1016/j.tifs.2020.01.027.

Xie, R., B. Tang, X. Yong, G. Luo, and S.M. Yang. 2014. Roles of the calcium sensing receptor in digestive physiology and pathophysiology (review). Int J Oncol 45(4):1355–1362. doi: 10.3892/ijo.2014.2560.

Xu, J., A. Zeug, B. Riederer, S. Yeruva, O. Griesbeck, H. Daniel, B. Tuo, E. Ponimaskin, H. Dong, and U. Seidler. 2020. Calcium-sensing receptor regulates intestinal dipeptide absorption via Ca^{2+} signaling and IKCa activation. Physiol Rep 8(1):e14337. doi: 10.14814/phy2.14337.

Yamaguchi, T., N. Chattopadhyay, O. Kifor, R.R. Butters, Jr., T. Sugimoto, and E.M. Brown. 1998. Mouse osteoblastic cell line (MC3T3-E1) expresses extracellular calcium Ca^{2+}-sensing receptor and its agonists stimulate chemotaxis and proliferation of MC3T3-E1 cells. J Bone Miner Res 13(10):1530–1538. doi: 10.1359/jbmr.1998.13.10.1530.

Yamashita, N., and S. Hagiwara. 1990. Membrane depolarization and intracellular Ca^{2+} increase caused by high external Ca^{2+} in a rat calcitonin-secreting cell line. J Physiol 431: 243–267. doi: 10.1113/jphysiol.1990.sp018329.

Yanamala, N., and J. Klein-Seetharaman. 2010. Allosteric modulation of G protein coupled receptors by cytoplasmic, transmembrane and extracellular ligands. Pharmaceuticals 3(10):3324–3342. doi: 10.3390/ph3103324.

Yang, C., D.M. Albin, Z. Wang, B. Stoll, D. Lackeyram, K.C. Swanson, Y. Yin, K.A. Tappenden, Y. Mine, R.Y. Yada, D.G. Burrin, and M.Z. Fan. 2011. Apical Na^{+}-D-glucose cotransporter 1 (SGLT1) activity and protein abundance are expressed along the jejunal crypt-villus axis in the neonatal pig. Am J Physiol Gastrointest Liver Physiol 300(1):60–70. doi: 10.1152/ajpgi.00208.2010.

Yang, C., M.A. Chowdhury, Y. Huo, and J. Gong. 2015. Phytogenic compounds as alternatives to in-feed antibiotics: potentials and challenges in application. Pathogens 4(1):137–156. doi: 10.3390/pathogens4010137.

Yang, C., T. Huang, L. Wu, T. Li, G. Wu, and Y. Yin. 2013. Molecular cloning and characterization of porcine calcium-sensing receptor (CaSR). Amino Acids 45(3):598–598.

Yang, C., X. Yang, and M.Z. Fan. 2016a. Apical Na^{+}-dependent neutral amino acid exchanger ASCT2 (ATB0) and mTOR-signaling components are expressed along the entire jejunal crypt-villus axis in young pigs fed a liquid milk replacer. Can J Anim Sci 97(1): 2016–2021. doi: 10.1139/cjas-2016-0021.

Yang, C., X. Yang, D. Lackeyram, T.C. Rideout, Z. Wang, B. Stoll, Y. Yin, D.G. Burrin, and M.Z. Fan. 2016b. Expression of apical Na^{+}-L-glutamine co-transport activity, B(0)-system neutral amino acid co-transporter ($B^{0}AT1$) and angiotensin-converting enzyme 2 along the jejunal crypt-villus axis in young pigs fed a liquid formula. Amino Acids 48(6):1491–1508. doi: 10.1007/s00726-016-2210-7.

Yang, C., Zhao, X, 2022. Calcium-sensing receptor and the regulation of nutrient absorption and gut health in pigs. J Anim Sci 100(3):64–65. doi: https://doi.org/10.1093/jas/skac247.127.

Yang, J., W. Bai, X.F. Zeng, and Cui Chun. 2019. γ-$[Glu]_{(n = 1,2)}$-Phe/-Met/-Val stimulates gastrointestinal hormone (CCK and GLP-1) secretion by activating the calcium-sensing receptor. Food Funct 10(7): 4071–4080. doi: 10.1039/c9fo00313d.

Yang, T., S. Hassan, Y.G. Huang, A.M. Smart, J.P. Briggs, and J.B. Schnermann. 1997. Expression of PTHrP, PTH/PTHrP receptor, and Ca^{2+}-sensing receptor mRNAs along the rat nephron. Am J Physiol 272(6):751–758. doi: 10.1152/ajprenal.1997.272.6.F751.

Yano, S., E.M. Brown, and N. Chattopadhyay. 2004. Calcium-sensing receptor in the brain. Cell Calcium 35(3):257–264. doi: 10.1016/j.ceca.2003.10.008.

Yao, K., J. Fang, Y.L. Yin, Z.M. Feng, Z.R. Tang, and G. Wu. 2011. Tryptophan metabolism in animals: important roles in nutrition and health. Front Biosci (Schol Ed) 3:286–297. doi: 10.2741/s152.

Ye, C., C.L. Ho-Pao, M. Kanazirska, S. Quinn, K. Rogers, C.E. Seidman, J.G. Seidman, E.M. Brown, and P.M. Vassilev. 1997. Amyloid-β proteins activate Ca^{2+}-permeable channels through calcium-sensing receptors. J Neurosci Res 47(5):547–554. doi: 10.1002/(sici)1 097-4547(19970301)47:5<547::Aid-jnr10>3.0.Co;2-v.

Yu, O.M., and J.H. Brown. 2015. G protein-coupled receptor and RhoA-stimulated transcriptional responses: links to inflammation, differentiation, and cell proliferation. Mol Pharmacol 88(1):171–180. doi: 10.1124/mol.115.097857.

Yun, F.H., B.Y. Wong, M. Chase, A.Y. Shuen, L. Canaff, K. Thongthai, K. Siminovitch, G.N. Hendy, and D.E. Cole. 2007. Genetic variation at the calcium-sensing receptor (CASR) locus: implications for clinical molecular diagnostics. Clin Biochem 40(8):551–561. doi: 10.1016/j.clinbiochem.2006.12.011.

Zarubin, T., and J. Han. 2005. Activation and signaling of the p38 MAP kinase pathway. Cell Res 15(1):11–18. doi: 10.1038/sj.cr.7290257.

Zeke, A., M. Misheva, A. Remenyi, and M.A. Bogoyevitch. 2016. JNK signaling: regulation and functions based on complex protein-protein partnerships. Microbiol Mol Biol Rev 80(3):793–835. doi: 10.1128/MMBR.00043-14.

Zhang, C., Y. Huang, Y. Jiang, N. Mulpuri, L. Wei, D. Hamelberg, E.M. Brown, and J.J. Yang. 2014. Identification of an L-phenylalanine binding site enhancing the cooperative responses of the calcium-sensing receptor to calcium. J Biol Chem 289(8):5296–5309. doi: 10.1074/jbc.M113.537357.

Zhang, C., C.L. Miller, R. Gorkhali, J. Zou, K. Huang, E.M. Brown, and J.J. Yang. 2016a. Molecular basis of the extracellular ligands mediated signaling by the calcium sensing receptor. Front Physiol 7: 441. doi: 10.3389/fphys.2016.00441.

Zhang, C., T. Zhang, J. Zou, C.L. Miller, R. Gorkhali, J.Y. Yang, A. Schilmiller, S. Wang, K. Huang, E.M. Brown, K.W. Moremen, J. Hu, and J.J. Yang. 2016b. Structural basis for regulation of human calcium-sensing receptor by magnesium ions and an unexpected tryptophan derivative co-agonist. Sci Adv 2(5):e1600241. doi: 10.1126/sciadv.1600241.

Zhang, H., C.A. Hu, J. Kovacs-Nolan, and Y. Mine. 2015a. Bioactive dietary peptides and amino acids in inflammatory bowel disease. Amino Acids 47(10):2127–2141. doi: 10.1007/s00726-014-1886-9.

Zhang, H., J. Kovacs-Nolan, T. Kodera, Y. Eto, and Y. Mine. 2015b. gamma-Glutamyl cysteine and gamma-glutamyl valine inhibit TNF-alpha signaling in intestinal epithelial cells and reduce inflammation in a mouse model of colitis via allosteric activation of the calcium-sensing receptor. Biochim Biophys Acta 1852(5):792–804. doi: 10.1016/j.bbadis.2014.12.023.

Zhang, H., and Y. Mine. 2018. Is calcium-sensing receptor a new molecular target toward improving gastrointestinal health? J Agric Food Chem 66(16):3995–3997. doi: 10.1021/acs.jafc.8b01150.

Zhang, M., D. Chen, Z. Zhen, J. Ao, X. Yuan, and X. Gao. 2018a. Annexin A2 positively regulates milk synthesis and proliferation of bovine mammary epithelial cells through the mTOR signaling pathway. J Cell Physiol 233(3):2464–2475. doi: 10.1002/jcp.26123.

Zhang, Q., Ke, H., Blikslager, A., Fujita, T., and Yoo, D. 2018b. Type III interferon restriction by porcine epidemic diarrhea virus and the role of viral protein nsp1 in IRF1 signaling. J virol 92(4): e01677-17. doi: 10.1128/JVI.01677-17.

Zhang, R., and X. Xie. 2012. Tools for GPCR drug discovery. Acta Pharmacol Sin 33(3): 372–384. doi: 10.1038/aps.2011.173.

Zhang, S., F. Chen, Y. Zhang, Y. Lv, J. Heng, T. Min, L. Li, and W. Guan. 2018b. Recent progress of porcine milk components and mammary gland function. J Anim Sci Biotechnol 9:77. doi: 10.1186/s40104-018-0291-8.

Zhang, Z., S. Sun, S.J. Quinn, E.M. Brown, and M. Bai. 2001. The extracellular calcium-sensing receptor dimerizes through multiple types of intermolecular interactions. J Biol Chem 276(7):5316–5322. doi: 10.1074/jbc.M005958200.

Zhang, Z.L., Z.R. Li, J.S. Li, and S.R. Wang. 2020. Calcium-sensing receptor antagonist NPS-2143 suppresses proliferation and invasion of gastric cancer cells. Cancer Gene Ther 27(7–8):548–557. doi: 10.1038/s41417-019-0128-4.

Zhao, X., Q., Hui, P., Azevedo, C. Nyachoti, K.O, C. Yang. 2022. The calcium-sensing receptor is not expressed in the absorptive enterocytes of weaned piglets. J Anim Sci. 100 (4): skac085.

Zhao, X., Hui, Q., Azevedo, P., O., K., Yang, C. 2021. In vitro characterization of potential pig calcium-sensing receptor (CaSR) ligands and cell signaling pathways related to CaSR activation by a dual-luciferase reporter assay. J Anim Sci 199(3):82–83. doi: https://doi.org/10.1093/jas/skab235.149.

Zhao, X., Xian, Y., Wang, C., Ding, L., Meng, X., Zhu, W., and Hang, S. 2018. Calcium-sensing receptor-mediated L-tryptophan-induced secretion of cholecystokinin and glucose-dependent insulinotropic peptide in swine duodenum. J Vet Sci 19(2):179–187. doi: 10.4142/jvs.2018.19.2.179.

Zhao, W., Z. Zhang, L. Zheng, C. You, H. Chi, T. Zhang, and G. Xu. 2020. Calcium-sensing receptor activating ERK1/2 and PI3K-Akt pathways to induce the proliferation of osteosarcoma cells. Clin Exp Pharmacol Physiol 47(3):517–519. doi: 10.1111/1440-1681.13209.

Zhao, X., B. Dong, M. Friesen, S. Liu, C. Zhu, and C. Yang. 2021. Capsaicin attenuates lipopolysaccharide-nduced inflammation and barrier dysfunction in intestinal porcine epithelial cell line-J2. Front Physiol 12:715469. doi: 10.3389/fphys.2021.715469.

Zhao, X., B. Schindell, W. Li, L. Ni, S. Liu, C.U.B. Wijerathne, J. Gong, C.M. Nyachoti, K.O., and C. Yang. 2019. Distribution and localization of porcine calcium sensing receptor in different tissues of weaned piglets1. J Anim Sci 97(6):2402–2413. doi: 10.1093/jas/skz096.

Zhen, Y., C. Ding, J. Sun, Y. Wang, S. Li, and L. Dong. 2016. Activation of the calcium-sensing receptor promotes apoptosis by modulating the JNK/p38 MAPK pathway in focal cerebral ischemia-reperfusion in mice. Am J Transl Res 8(2):911–921.

Zhou, L., and L.M. Bohn. 2014. Functional selectivity of GPCR signaling in animals. Curr Opin Cell Biol 27:102–108. doi: 10.1016/j.ceb.2013.11.010.

CHAPTER 18

Manipulation of gut microbiome composition by genetic selection of the host

Francesco Tiezzi[1,*] *and Christian Maltecca*[2,*]

[1]University of Florence – DAGRI, Piazzale delle Cascine 18, Firenze, Italy; [2]Department of Animal Science, North Carolina State University, 120 W Broughton Dr, Raleigh, NC, USA; *Correspondence: francesco.tiezzi2@unifi.it; cmaltec@ncsu.edu

Abstract

This chapter explores the intricate relationship between host selection, gut microbiome dynamics, and the holo-biont concept, shedding light on the impact of artificial selection on the host and its microbial symbionts, oriented towards the maintenance of gut health in farm animals. We examine the potential evolution of the host throughout its life, highlighting dynamic changes in gut microbial colonization from birth to adulthood. We emphasize the symbiotic link between the host and microbiota, playing a pivotal role in selection and evolution. An analytical model for microbiome composition is introduced, incorporating the influence of host genes, maternal effects, and the environment. The environmental impact on the microbiome, mainly through diet and stress, is explored. Examples illustrate the host's influence on microbiome composition, emphasizing bi-directional host-microbiome interactions in the holo-biont.

The co-diversification of host and commensal communities under selective pressure is examined. The impact of host genes on microbial composition in livestock is investigated, providing evidence across ruminants, monogastric species, and poultry. Genetic progress in shaping the holo-biont is explored, considering the heritability of both host traits and microbial composition. Finally, caveats are explored, including challenges in disentangling the effect of vertical transmission from Mendelian inheritance, the intricate nature of microbial community composition, and the dynamic interaction factor contributing to genetic progress over generations.

Keywords

Selective breeding – gut microbiome – host genotype – co-diversification – holobiont

© FRANCESCO TIEZZI AND CHRISTIAN MALTECCA, 2024 | DOI:10.3920/9789004695467_019

1 The selection of the host and the microbiome of the gut

Genetic improvement through selective breeding has dramatically shaped the livestock populations used in commercial production over time. Modern selection started with the thorough and accurate recording of animal performance and pedigree information. Its efficiency then has then been increased with the use of genomic markers (in what is called 'genomic selection'), which resulted in a significant boost in genetic progress.

Selection has considerably changed the genomic make-up of all farm animals. When (host) genomic markers began to be used, the gain in efficiency of selection confirmed the relevance of the host genotype in the determination of the phenotype. No other genome variants (*e.g.* the microflora) was considered, other than the host itself.

Evidence emerging over the last decade suggests that (host) selection along with increasing phenotypic performance might have shaped the host gut microflora composition, to the point that such gut microflora has played a major role in securing the genetic gain achieved.

This chapter will explore the controversial relationship between host genotype and gut microbiome composition. Being the latter that ultimately determines gut health, it is straightforward that the study of the impact of artificial selection on microbiome composition is pivotal.

In fact, hosts that have been selected could represent a profoundly different environment for the gut microbiome, compared to their ancestral counterparts. This would suggest that the microbiota's effect on gut health, in turn, could potentially be affected by the host genomic makeup. After a brief review of the possible dynamics outlining how the host could provide a different environment for its microflora, we will see how genetic selection could potentially be used as a tool to improve gut health by modifying (or acting as a buffer to modification of) the gut microbiome composition. We will focus on the two major monogastric livestock species (swine and poultry) still using references from the study of the species, from insects (*e.g.* aphids) to humans.

2 How artificial selection works (in a tiny nutshell)

Animal (and plant) breeders make use of statistical models to partition the variability of a phenotype (*e.g.* growth, feed efficiency, fat deposition) into genetic and environmental components. Since selection can only rely on the former component (in the form of the 'breeding value' of an individual), it is important that these models work correctly and provide an estimate of the individual's breeding value that is as accurate as possible.

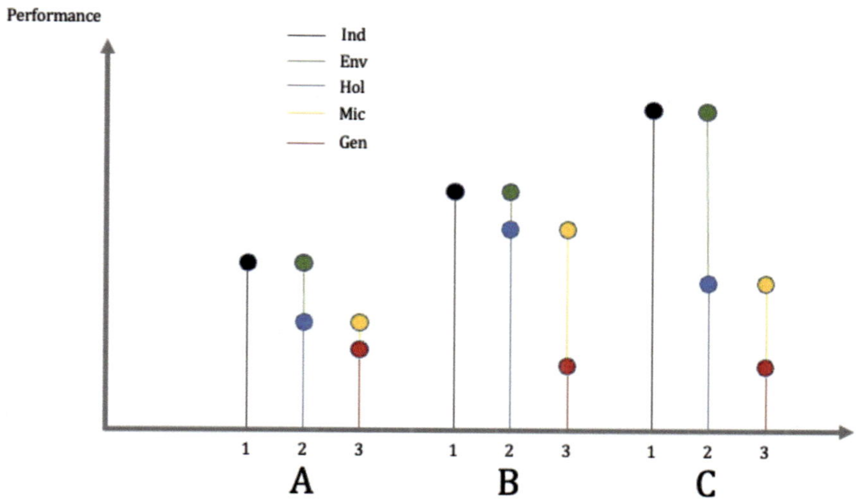

FIGURE 18.1 Schematic example of how breeders use statistical models to partition the variability of a phenotype (*e.g.* growth, feed efficiency, fat deposition) into genetic and environmental components

We will refer to figure 18.1 for a schematic example. In the figure, three individuals are depicted on the x-axis, with their performance for some hypothetical trait is measured on the y-axis.

If we were to just look at its phenotypic performance (value 1) and pick the individual with the highest value for this trait, we would probably select individual C. This is what has happened for thousands of years where selection has been based on the phenotype and still happens in many cases. However, if we could (forgoing all details on how) separate the individuals' phenotypic performance into a genetic and an environmental component (value 2), we would notice that the large phenotypic "advantage" of individual C was indeed due to environmental effects. These could be some superior diet fed to the individual by the farmer, the fact that this individual did not undergo heat stress or that was not exposed to any pathogens. Individual B, instead, shows the larger genetic potential once the (small) environmental component has been removed from its phenotypic value. Maybe individual B received little help from the farmer, was growing under stressful conditions or was exposed to a strong load of pathogens. Note that all these hypothesized environmental components are non-heritable: there could be no impact of the individual's genes on these occurrences. Being non-heritable, these components could not be transmitted to the progeny and would therefore be useless in selection.

Individual A shows an intermediate scenario between B and C. Animal breeders would formulate a statistical model to separate the phenotypic variation into genetic and environmental, using a model as follows:

$$P = E + D + G \tag{1}$$

Here, P is the phenotype, G is the component due to the genes of the individual itself, D is the contribution from the dam of the individual (the maternal effect), and E is, in a broad sense, the impact of the environment on the phenotype of the individual.

The maternal effect D is normally included in animal breeding models to account for the contribution that a mother would have on the individual's development, either during the intra-uterine life or the fostering period, or for any care provided to the growing individual (we did not include the component D in figure 18.1 for sake of simplicity). The E component, in quantitative genetics terms, encompasses all the contributions from a large number of factors, *e.g.* management, diet, climatic conditions, exposure to pathogens etc.

In quantitative genetics, the transmissibility of a trait from parents to offspring is referred to as 'heritability', which is usually expressed in its narrow-sense with the symbol h^2. The heritability is the portion of variation in the phenotype P that is due to host additive genetic effects G, *i.e.* that is heritable. The phenotypic variance is the overall variance observable in a population, such variation is meant in undistinguishable terms as the sum and interaction of all terms in formula 1. Given the phenotypic variance as the sum of all components in formulas 1 and 3, the heritability is the magnitude of the G component proportionally to the sum of all the components.

This model though might still be missing something. The model assumes that the G component is determined by the genomic make-up of the host. That is, all the genomic variants that are physically recognizable on the chromosomes of the host (we will neglect epigenetic variation, for sake of simplicity). Unfortunately, not all the 'genetic' variation can always be found on the host chromosomes. When the genetic variation in a population of the host is not fully explained by the physical variation on the genome, we talk of the "missing heritability" phenomenon. Simply put, the gene variants segregating in a population don't fully capture the genetic variation in the population itself. Several complex phenomena, which are beyond the scope of this chapter, contribute to missing heritability.

Among others, though, it has been suggested that the (host) genome should be 'extended' to include the gut microbiome and all its assembly of genes as a source of genetic variability, since these might significantly contribute to the individual's performance (Hurst *et al.*, 2017; Suzuki, 2017; Henry *et al.*, 2021). Such extension could be so well constructed that the effect of the host genes or the microbes' genes would in fact be indistinguishable from a functional standpoint. Doolittle and Booth (2017) stated that 'it's the song, not the singer' that matters, suggesting that the host performance could be determined by host or the microbiome, without any particular difference on how the host trait (*i.e.* the phenotype) appears. In figure 18.1 (value 3), the genetic component of the phenotype of the three individuals has been broken down in a part being properly determined by the host genes (G) and on part being determined by the microbiome genes (M). The revised formula would now be:

$$P = E + D + G + M \quad (2)$$

Similarly to the concept of heritability (as the impact of *G* on the overall variation in *P*), the term 'microbiability' has been coined, meant as the proportions of the overall variation of *P* which is due to *M* (Difford *et al.*, 2017; Camarinha-Silva *et al.*, 2017, Khanal *et al.*, 2020). This concept will be further considered later in this chapter.

As for value 3 reported in figure 18.1, individuals B and C appear to be substantially equivalent for what is conferred by their own (the host) genes, with individual B having a strong microbial contribution compared to the others. If the breeder were to choose based on the host genes only (*i.e.* *G*), individual A should be chosen. Would that be correct? Would it be beneficial to select individual B, which is being helped by *its* microbial genes?

3 The holo-biont: how the host and the microbiome come together and its role in evolution and selection

The answer to the questions above would be: individual B can take advantage of its microbial genes to the extent that *M* component can be transmitted to the progeny, *i.e.* by how 'faithfully' the microbiome in inherited by its progeny (Henry *et al.*, 2021). The use of '*G* + *M*' could be supported by Wright's early view on selection, which 'applies to the organism as a whole and thus to the effects of the entire gene system rather than to single genes' (Wright, 1931; Borderstein and theis, 2015). If we assume that *M* is a component that belongs, to some extent, to the host, then it is straightforward to acknowledge its relevance in host performance. The challenge would rather be to properly verify and estimate the transmissibility of *M* from parent to progeny (*i.e.* the heritability of *M*). Throughout this chapter, we will explore to what extent factors *G*, *M* and *E* are interrelated and how difficult, or impossible would be to separate and estimate each of their (marginal) effects (Brucker and Borderstein, 2016).

The question about the control of *M* by *G* is triggered once the relationship between the host and the microbiota is found to be so strong. In front of this strength of the relationship, the term 'holo-biont' has been coined, to refer to the two (*G* and *M*) as parts of a whole (Zilber-Rosemberg and Rosemberg, 2008; Doolittle and Booth, 2017; Rosenberg, 2010).

And what would be the conditions for considering the host and its microbiome as a single entity?

Precisely, the holo-biont is formed when the conditions of 1) the presence of abundant and diverse microbiota, 2) the impact of the symbiotic microbiota on the host fitness, and 3) fidelity of transmission of the microbiota from parent to offspring (*i.e.* inheritance) apply. The holo-biont can be seen as 'a multispecies consortia of cells with many genomes that can contribute to multiple functions throughout the body', with the genome being those of the host and the thousands of species that live in symbiosis (Bordenstein and Theis, 2015). The selection of species that inhabit it would not be random, since the host would provide the environment to a 'selected' set of species (Spor *et al.*, 2011). Some of these species might even be 'obligate symbionts', which can only live in such an environment provided by the host (Moran *et al.*, 2008), to the extent that some bacteria have shown a loss in genome size and functionality as a sign of adaptation to specific conditions (Shapira *et al*, 2016; Michelland *et al.*, 2010). On the other side, some hosts are obligate in acquiring a symbiont for certain functions, as it is common in invertebrates (Moran *et al.*, 2008; Henry *et al.*, 2021). Such interest (or necessity) to acquire a host or a symbiont is called 'partner fidelity' (Hurst, 2017). It should also be noted that some microbes would fall into the pool of 'symbionts' but others would not. For example, *Mycobacterium tuberculosis* is a pathogen that can live with their host for decades without conferring any form of disease so it would be considered as a symbiont, while *Vibrio cholera* would kill the host in a few days, therefore it is not to be considered as a symbiont (Roughgarden *et al.*, 2018). Therefore, there is supporting evidence about the legitimate consideration of the host and its microbiome as a whole, this consideration relies on the transmissibility of the microbiome from parent to progeny. In other words, it is legitimate to consider the microbiome as a heritable feature of the host species.

The concept of holo-biont is *per se* not new or relegated to microbes as they are: in the 19th and early 20th century it was already hypothesized that mitochondria and chloroplasts were symbiotic bacteria that are now systematically transmitted through the cytoplasm. (Portier, 1918; Baldauf, 2003).

In the context of this chapter, we will refer to the holo-biont as the assembly of genes of the host and the microbiome, which forms a dynamic entity when it comes to selection, response to selection and evolution. When all the genes are assembled, it should be reminded that the number of genes contained in the gut microbiota of mammals (*e.g.* humans) by far exceeds the number of genes contained on the chromosomes of the host. Therefore, from a host genetic standpoint, the 'extension' to holo-biont is remarkable (Alberdi *et al.*, 2016).

4 The holo-biont and its plasticity in front of environmental challenge

In addition to the three properties of the holo-biont as listed above, there should be another one listed: the variation measurable on the holo-biont phenotype can depend on the host, the microbiome or both, to different extents and depending on the conditions (Zilber-Rosemberg and Rosemberg, 2008).

Figure 18.2 shows the performance (*e.g.* growth rate, gut health, welfare) of an individual (y-axis) depending on the level of environmental stress (x-axis). The latter is meant in a broad sense here, but for example this could be represented by the

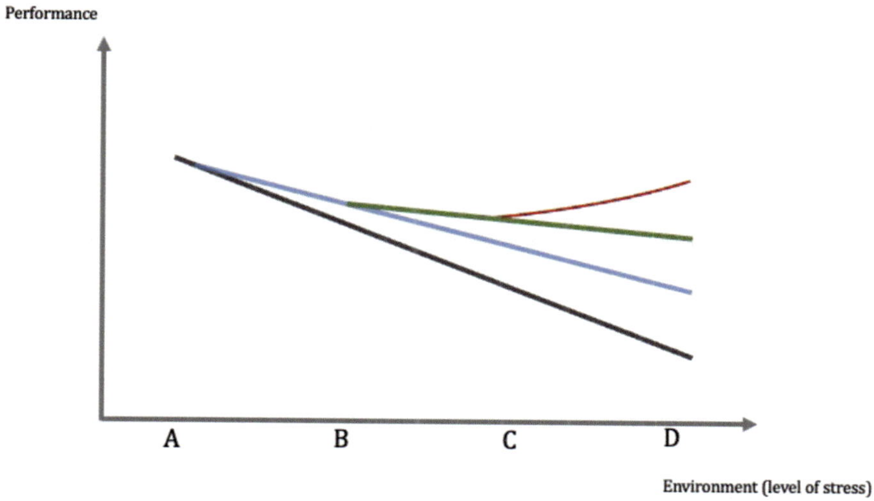

FIGURE 18.2 Performance of an individual (y-axis) depending on the level of environmental stress (x-axis)

complexity of the diet (from mother's milk to scavenged food), the presence of fiber in the diet or even the pathogenic load in an environment. For simplicity, we will consider the environmental stress as the complexity of the diet fed to a pig that is first fostered by its mother but then finished in the woodland: from the individual being nourished with the milk of the mother (reported with A on the x-axis), to the addition of feed (starch) to the diet (reported with B), to the exclusion of mother's milk but the inclusion of fiber (reported with C) to a diet rich in fiber and tannins (reported as D).

The solid black line is the fitness of the individual as determined by the genes of the host. This line shows its higher value on the y-axis under the conditions depicted as A (*i.e.* milk suppling nutrients). With that being the most complete and readily available sources of nutrients, the holo-biont can exploit its potential as coded by the host's genes. The performance shows a dramatic fall when diet becomes more complex and less readily available: the starch in the diet (condition B) requires amylase to be synthesized, fiber is not digestible by the pig itself (condition C) and tannins can have a detrimental effect on gut health (condition D). Fortunately, the gut microbiome comes in help to the host. The blue line shows the contribution of the 'core microbiota', *i.e.* that set of microbes that is species-specific and is not usually found to change across individuals (Carthey *et al.*, 2018). Here, we see a contribution to the host performance that is present across the diets. Such contribution is smaller under high-nutritional-value diets but increases when diet become of lower nutritional value. In fact, the host is in 'no need of help' when the feed is already broken down in components that are easy to absorb (*e.g.* milk peptides), rather it might need some contribution from the microbiota when molecules become more complex and some anti-nutritional compounds are to be found in the diet. The microbiome has the possibility to adapt to the diet faster than the individual can. In fact, the microbiome is clearly more 'plastic' than the host genome, with the latter being substantially fixed. An important feature of the microbiome is to change quickly in the abundance of the different species, as for the (temporary) needs of the host. In case of sudden changes in the environmental conditions, the diverse microbial symbiont community can quickly expand or shrink, conferring acclimating ability to the host and buy the time necessary for the host genome to evolve (Rosemberg, 2010; Alberdi *et al.*, 2016; Hurst, 2017).

The green line shows the contribution to host performance from those species that can change in abundance given the diet. This contribution is null under condition A (feed easy to digest) and negligible under condition B (amylase being synthesized by the host), given its lack of use. But the contribution increases with the presence fiber in the diet (and need of *Ruminococcaceae*, Walker *et al.*, 2011; Yang *et al.*, 2021) and poisonous molecules (Kohl and Dearing, 2012; Kohl *et al.*, 2014). Under challenging conditions, the abundance of the beneficial (necessary) species would increase and proportionally replace the others, without bringing them completely to absence. Some diversity is usually maintained, in case the change in diet was to be reverted. Diversity is often seen as 'insurance' against further fluctuation of the environmental conditions, and the not-so-useful microbes are maintained presence as a gene pool reservoir and would allow to thrive different, unexplored niches (Shapira, 2016; Rosemberg, 2010; Henry *et al.*, 2021). This insurance policy hypothesis could be the basis for host resilience against a number of stressors and has probably already been exploited in agriculture, without knowledge of the microbiome's contribution. As for diversity, it should be pointed out that under conditions similar to A and B it is less needed. Under non-challenging conditions and diets that make nutrients readily available, we have showed the microbial contribution to taper. This could lead to a loss in diversity because the microbes needed under condition C are no longer needed. Evidence has been found in domesticated wild animals, both for experimental reasons and livestock production (Carrier and Reitzel, 2017; Kohl and Dearing, 2014; Suzuki, 2017). In synthesis, diversity seems to increase when more challenging conditions are experienced, while seems to be lost when conditions change towards 'less challenging'.

All the contribution from the microbiome has come from species that were already present in the gut environment of the host. The further step of microbiome plasticity would be to acquire additional functions. New microbes could 'provide' new metabolic pathways allowing a shift in the use of food resources otherwise unusable (Hehemann *et al.*, 2010; Sudakaran *et al.*, 2017). This could happen by horizontal transfer of microbes, which occurs when new species enter the formation of the gut microbiota because the host simply enters in physical contact with the microbes. The red line shows the contribution of these acquired new species. It is evident how such contribution is not linear but is present only under conditions than confer some sort of challenge to the host and is much stronger under the challenging conditions. This is because the host would probably acquire the new symbiont only when strictly necessary, as we should not forget that any symbiont would be, to some extent, commensal and its presence should be limited to the cases that such symbiont is strictly beneficial. From a quantitative genetic standpoint, it should be pointed out that the host could have the ability to control the composition of the facultative microbiome (conferring the effect reported in the green-shaded area) area and the acquisition of new species (conferring what reported in the yellow-shaded area). This will be further discussed later in this chapter.

Here, it should be remarked how horizontal transfer can increase the diversity of microbiome in an individual or species. Again, the complexity of the diet is a good example since it has been shown that species with diets that are more complex and include non-readily available nutrients have a more diverse composition of the microbiome, as in the case of herbivores that have a more diverse microbiota than carnivores (Ley *et al.*, 2008).

5 Evolution of gut microbiome over lifetime

Figure 18.3 shows the potential evolution of three hosts over their lifetime. This figure is an adaptation of figure 18.3 from Carrier and Reitzel (2017). The figure shows the different points in life of an individual on the *x*-axis and the number of species that colonize the gut on the *y*-axis. While it would be difficult to pinpoint the different stages of life on the *x*-axis (*e.g.* in utero, birth, fostering, growth, adulthood), this graph is meant to be interpreted with some lack of precision on the actual evolution of the microbiome over the life of the host, still pointing out a *climax* towards a host-specific community. The most important point would be a rapid increase in microbial richness during the early stages of life, until a plateau is reached towards adulthood (Carrier and Reitzel, 2017). The adult composition of the microbiome would include a core microbiota that is conserved at the species level, a specific microbial pool, which is more dependent on the environmental exposure (Shapira, 2016) and finally a flexible microbial pool that has maintained some degree of

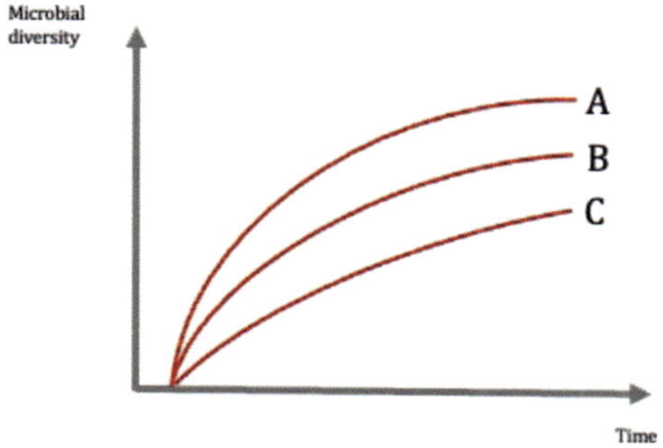

FIGURE 18.3
Potential evolution of three hosts over their lifetime

plasticity. This adult composition would be relatively resilient to changes (as compared to the young's one) but could also differ substantially from individual to individual. Remarkably, the core microbiome could be conserved the high taxonomic levels, leaving the plasticity to changes in genera and species within a *phylum* or family (Benson *et al.*, 2010). In other words, inter-individual differences could be found as differences in genera or species, while all individuals would include the same families or *phyla* with similar proportions. The life progression could also be seen as a continued exposure to microbes, under the assumption that the host would be exposed to every possible microbe if the lines were to be extrapolated to an infinite (life)time (Carrier and Reitzel, 2017).

5.1 *In-utero and in-egg colonization*

It was accepted knowledge that the fetus would be sterile until parturition (Benson *et al.*, 2010). However, this hypothesis has recently been revisited under the hypothesis that some bacteria could colonize the host's intestine before birth (Gilbert, 2014). These bacteria would pass through placental and amniotic barriers, since some have been found the umbilical cord, amniotic fluid and in the bowels of the newborn (Funkhouser and Borderstein, 2013; Gilbert, 2014). In poultry, pathogens such as *Salmonella* and *Mycoplasma* can potentially be passed to the offspring as carried in the eggshell but also in the egg yolk (MacOwan *et al.*, 1984; Bencina *et al.*, 1988; Gantois *et al.*, 2009).

This type of colonization would be the results of vertical transfer from the dam of the individual, which could be seen as a 'maternal effect' in quantitative genetics terms. Intuitively, the dam could only pass the microbes that colonize her own body, therefore there would be some environmental component to this maternal contribution, still being filtered by the mother's body and physiology.

5.2 *Colonization at birth*

The passage of bacteria to the amniotic fluid would imply their passage into the newborn's gut, since the amniotic fluid could be ingested at birth (Funkhouser and Borderstein, 2013). In addition, the passage of the newborn through the birth canal would imply the exposure to the microbes that colonize the canal. It is well known, at this point, that such exposure helps the colonization of the newborn, since birth by C-section implies a dramatically reduced colonization. In fact, when C-section is practiced there is a lack of association between the newborn's and mother's microbiomes, while the newborn would mostly acquire the microbes from the environment and any person that is helping with the delivery. For an extensive review on the colonization at birth, during nursing (mammals) or through eggs (birds), see Funkhouser and Borderstein (2013) and Gilbert (2014). While the dynamics and consequences of gut colonization go beyond the scope of this work, it is important to point out that such colonization could again be seen as a maternal/environmental component affecting the host microbial richness. While the maternal component (as vertical transmission) would be stronger in case of vaginal birth, the environmental component (as horizontal transmission) would be stronger in case of C-section. Remarkable differences have been found between individuals born either way, and this will be considered when defining the list of factors that shape the microbiome.

5.3 *Colonization during later life*

The relevance of vertical transmission over horizontal transmission decreases with weaning and progression of life of the host (either mammal or bird). Though some colonization by vertical transmission could still happen, if multiple generations share the same environment. For example, the rumen of calves can be colonized with parents' fecal or ruminal microbes when the calves graze on grass where saliva or feces of parents can be found (Dehority, 2003, Hurst, 2017). Undeniably, on such grass there could be presence of individuals that are not related to the calf, which would imply some 'social' horizontal transmission. Here, it is again the vertical transmission due to the presence of the parents' microbes to represent a particular issue when it comes to genetic improvement, due to collinearity between the parents' genomes and the calf's genome on one side and parents' microbes and calf's microbes on the other (see later sections).

Horizontal transmission would be more frequent and relevant later in life (Zilber-Rosemberg and Rosemberg, 2008; Moeller *et al.*, 2018) and would depend on the host exposure to the microbes. If, for instance, the host were provided a different feed and had to shift to a new ecological niche but the beneficial symbiont was to be lacking in the environment, then the host would never be able to reach the microbiome composition necessary for the new niche, with lower chances of acclimation and survival. The process of horizontal transfer appears to be aleatory to some extent and could have explained the lack of adaptation of certain species or breeds to certain environments, denying the possibility of a host to move towards the right side of the graph in figure 18.3.

It should also be noted what reported by Moeller *et al.* (2018), that not all symbionts will have equal ability to be acquired horizontally. For example, obligate symbionts can only be acquired vertically, obligate anaerobes will likely be acquired vertically, while obligate aerobes will likely be acquired horizontally. Which suggest that there is a limited window of time under which a given microbe can be acquired, and this would affect the curve of potential microbiome richness pictured in figure 18.3.

To complicate things, it could be the microbiome to hamper the acquisition of new microbes. The well-established species could produce antibiotics to protect their niches and, in turn, protect the host from colonization by pathogens or other symbionts (Donia *et al.*, 2014; King *et al.*, 2016; Rosenberg and Zilber-Rosenberg, 2018). It would be the early colonizers, those acquired vertically, to occupy the niche and hold the other microbes from colonizing, at least in part (Ochman *et al.*, 2010).

This would imply that the maternal component of the microbiome, acquired by vertical transmission, would leave a footprint throughout the life of the individual without fading away.

Last but not least, an important feature of the microbes to be potentially acquired would be the ability to confer novel adaptive traits to their host (Sudakaran *et al.*, 2017). In figure 18.2, we hypothesized that the symbiont(s) conferring the performance advantage pictured with the yellow shaded area would have a beneficial impact. However, several microbes would not be able to confer such advantage despite of the fact of not being pathogens. While the acquisition might still happen, this would probably come at a cost for the host and would not be considered in evolutionary or breeding terms. The conferred advantage could be host-specific: for example, host carrying and expressing the amylase genes (*AMY-1* and *AMY-2*) would not benefit from a symbiosis with starch-breaking symbionts. Conversely, individuals that lack of these genes would benefit from such symbiosis. From a symbiont's standpoint, the individuals of the first group would be more difficult to colonize because of the lack of starch arriving to the large intestine and a smaller niche to occupy, while the latter group would provide an abundant starch substrate for the microbes to thrive. Therefore, colonization of the host is not trivial. We will revise how the different modalities of symbiosis acquisition can be seen from the animal breeding and quantitative genetics standpoint.

6 Shaping the microbiome

In the previous sections of this chapter, the concept of holo-biont has been introduced, together with a snapshot of the factors that could influence its establishment, composition and dynamics. Here, we will put these concepts in analytical terms before we can proceed with hypothesizing selection of the host oriented to change the microbiome of the intestine and, ultimately, gut health.

6.1 *A statistical model for microbiome composition*

In a similar fashion as we did above for a common phenotypic trait of the host, we could extend the model to the microbiome, in order to list the main factors that determine it. The factors shaping the microbial composition (M) could be summarized in:

$$M = D + E + G \tag{3}$$

Where, again, G is the component due to the genes of the individual itself, D is the contribution from the dam of the individual (the maternal effect), E is the impact of the environment, in a broad sense, on the gut microbiome composition of the individual. All components entered the model for *P* as well: this is because the diet could alter the microbiome but also the growth of the individual, the dam could have good fostering ability but also promote the establishment of certain microbes, the genes of the host could regulate the digestive efficiency of an individual but also control the microbial composition of the intestines. Or, each factor could shape the microbiome composition, which in turn affects the phenotype (as *M* entered the model for *P*). In this chapter we are interested in understanding how the host's genes can shape the microbiome so that this one can affect a trait, indirectly. Such trait was probably considered to be affected by the host's genes directly, until microbiome information was made available and properly accounted for in the model (Maltecca *et al.*, 2020; Tiezzi *et al.*, 2021). This justifies the fact that the holo-biont effect on the phenotype ($M + G$, as explained above) has now been broken down and the two components (*G* and *M*) now appear on different sides of the equation. While in formula 2 they contributed to causing the phenotypic variation, now it's the host's genes (*G*) that are causing the gut microbiome variation (*M*). Thus, we're establishing a directional causality. This will be done just for convenience in explaining how to shape microbiome (*M*) through changes in host genetics (*G*), but we acknowledge that the two components could covary without a specified directionality of effects, or perhaps there could be reversed causality (with *M* affecting *G*).

6.2 *How the maternal environment (D) affects the microbiome*

As discussed above, the formation of the holo-biont is strongly shaped by the colonization at birth and in early life, but also *in utero* or though the egg yolk. This has been found to happen in mammals, birds and invertebrates (Spor *et al.*, 2011; Funkhouser and Bordenstein, 2013; Moran *et al.*, 2019). This allows the symbionts in partnership with the mother (or dam) of the individual to be vertically transmitted to the offspring and be the first to form the new holo-biont. Such symbionts for the new holo-biont will likely be the same taking part to the maternal holo-biont (Suzuki *et al.*, 2022). Here we should consider the privilege such symbionts have in colonizing the offspring, but also the contribution of the dam itself in regulating the establishment of the new holo-biont. In mammals, for example, the composition of the milk (fat, protein, sugars) might alter the composition of the symbionts thriving in the gut. Therefore, it is the whole maternal holo-biont to determine the formation of the offspring holo-biont.

From a host standpoint, there are examples of absolute necessity to establish a symbiosis with that specific microbe in order to show certain traits. This has been found to be more frequent in invertebrates than mammals or chickens.

From a symbiont's standpoint, the microbes that pass from mother to offspring by vertical transmission happen to be mostly obligate symbionts, needing that specific host to survive because they don't show replicative or dormant phase outside hosts (Moran *et al.*, 2019) and often they are obligate anaerobes (Moeller *et al.*,

2018). For example, *Buchnera aphidicola* in needs the aphids for its life cycle and has not been found to thrive in symbiosis with any other host (Moran *et al.*, 2008). As a sign of the obligation to form a symbiosis with the host, obligate vertically transmitted microbes have been found to have genomes of smaller size as compared to their relatives that are facultative symbionts (Moran *et al.*, 2019). A smaller genome size would imply a reduced functionality and lack of ability to thrive in different environments. At the same time, obligate symbionts show the highest level of 'fidelity' in the host-microbe partnership. Therefore, we can imply that the effect of D on M will likely be strong and will leave a footprint on the evolution of the offspring holo-biont. As in figure 18.3, the microbial community might differ between the three holo-bionts because of the early establishment of the vertically-transmitted microbes, since later-life horizontal transmission might be hampered by early-established vertically-transmitted microbes (Donia *et al.*, 2014; King *et al.*, 2016; Ochman *et al.*, 2010).

Such a strong impact of D on the formation of the holo-biont has reflection on the evolution of species and genetic selection, ultimately. In the case of evolution of population and speciation, the term phylo-symbiosis has been coined (Brucker and Bordenstein, 2013; Moeller *et al.*, 2016; Hurst, 2017). This introduces the concept of a relationship between the phylogeny of the host and the symbiosis that follows (Bordenstein and Theis, 2015). In simple terms, individuals from a population that diverged from the others will likely share the presence of an ancillary symbiosis, since this one actively drove the differentiation by supporting the adaptation to new environments. Likewise, the symbiosis might support the host in being selected in the case of genetic selection, because of the added functionality given by the symbiont.

Within the context of artificial genetic selection, however, we need to dig a bit deeper in the different scenarios that may occur. We need to differentiate between 1) both the mother and the offspring living in the same environment as their relatives, 2) the mother and the offspring living in the same environment which is different than the rest of the population and 3) the relatives and the mother living in an environment and the offspring living in a different one. We could make the example of the 'environment' being defined as the diet, on one side being a high-starch diet and on the other being a high-fiber diet. In case 1, with the mother, the offspring and the rest of the population living on the same diet, the effect D will likely be beneficial. The mother will likely pass selected microbes with beneficial effect, in most of the cases. In case 2, the mother could have been exposed to a new diet. Depending on her success in shifting her gut microbial composition towards a beneficial one (including horizontally acquiring new ones), she could be a good provider of symbionts to the offspring. With the mother being the first to be fed a high-fiber diet, her ability to shift to the beneficial bacteria be one of the many factors of her success. This crucial role of the mother as a holo-biont would determine the direction and strength of the D effect on M. Case 3 is the most likely to show a detrimental effect of *D* on *M*. With the offspring colonizing a new environment, the vertically-acquired symbionts and their 'imposed' presence could hamper their acclimation to new conditions. Again, mothers that were fed starch and passed high levels of *Prevotella* might play a negative effect to offspring that is fed mostly fiber, with the magnitude of such effect being dependent on the strength of the vertical transmission and the difference between the two diets. We should also consider the example that the vertically-transmitted microbes could play a role as pathogens rather than symbionts (Funkhouser and Bordenstein, 2013). In cases 1 and 3, it is the whole population (mother and relatives) that warrants the non-pathogenic nature of the symbiont, a large number of individuals will unlikely host a pathogen over generations. In case 2, it is the mother only that needs hold the pathogens from colonizing the offspring. We could assume that the survival of the mother guarantees the moderate pathogenicity of the microbe, although this is probably too optimistic. Here, the ability of the mother as a holo-biont to contrast the pathogen would drive the risk of vertical transfer of pathogens. Using the example of vertically-transmitted *Salmonella* in chickens (Gantois *et al.*, 2009), is there an inherent capacity of the mother to control the abundance of the microbe in the gut, even under exposure? If any, is this capacity due to the mother's genetic composition of the gut community genetic composition? This will be pivotal when the use of the holo-biont as a unit of selection for beneficial microbes is considered.

In general, we should consider the presence of vertically-transmitted microbes as something detrimental as well. Most of the effect, however, will depend on the history of the whole population (*e.g.* stable and optimal farming conditions over generations) as well as the occurrence of a rapid shift in the farming conditions from mother to offspring.

In this chapter, we included the impact of *D* on *M* for the undeniable role that this could have in any breeding scheme (either mammals or birds) but also because of its overlap with host genetic inheritance. Under most circumstances, the individual that carries the fetus or the egg (*e.g.* the sow, the hen) will also provide half of the genetic material to the offspring. This inherent collinearity between microbial vertical transmission and host genetic inheritance will have to be considered when planning to change gut microbial composition through genetic selection of the host.

6.3 *How the environment (E) affects the microbiome*

The impact of diet and any form of stress on microbiome and gut health is better explained in other chapters of this book. Here, we will just review this part with the purpose of understanding the functionality of the holo-biont.

First, we should reiterate that any change in gut microbiome composition that could occur during the life of the individual will likely just affect the set of 'flexible' microbiome, leaving unchanged the 'core' microbiome (Shapira, 2016).

Furthermore, such change might happen at the level or the lower taxonomy, altering proportions of genera and species, while leaving unchanged the proportional at the family level.

Given a change in the diet, the community composition would shift towards the microbes that are able to thrive on the substrate provided by the new diet, from a symbiont standpoint, or use the diet to provide useful nutrients, from a host standpoint (Walker *et al.*, 2011; Walter and Ley, 2011). For example, a restriction in the diet and a change in the availability of nutrients might quickly induce an increase in the *Firmicutes* to *Bacteroidetes* ratio, which is known for its metabolic efficiency. The symbiont community composition of the holo-biont is not set to be constant but rather dynamic. The changes in microbial composition will be quick, given the short generation interval of the bacteria that likely outcompetes, in responsiveness, the physiology of the host (*e.g.* fat mobilization). The host should have the ability to increase or decrease the species that appear necessary or superfluous in a given timepoint, in a way that it not overly dissimilar, from a functional standpoint, from gene expression (Zilber-Rosemberg and Rosemberg, 2008).

This responsiveness is also pictured in figure 18.2, which was previously interpreted from a holo-biont standpoint. In that case, the change in diet was seen as an increased contribution from the symbiont bacteria in providing necessary nutrients otherwise unavailable. The same graph could be interpreted from the mere perspective of the symbionts. On the left-hand side, where diet is less complex and provides readily available nutrients, the symbiont could perhaps be absent and the host would tried with no problem. Moving towards a more complex diet and nutrients less readily available does change the microbial composition. The change would be in the core microbiome (or its function), the flexible microbiome, until the inclusion of new symbionts acquired from the environment by horizontal transmission (Moeller *et al.*, 2018) as long as a substrate is provided by the host and its diet. As mentioned above, such new symbionts would be acquired if providing a beneficial contribution to the host (Sudakaran *et al.*, 2017), therefore the environmental exposure alone (*i.e.*, the environment) is not sufficient to guarantee a change in the microbiome composition.

Also figure 18.3 could be re-interpreted from the symbiont standpoint. In fact, the lifetime of the individual (x-axis) could be seen as a cumulative sum of all the environmental exposure that the host went through, as affecting the microbial community composition (y-axis). While vertical transmission would be more frequent in the early stages of life, horizontal transmission would be frequent towards the adult phase. Again, the continued exposure would alter the microbiome composition, which would continue to evolve and asymptotically reach a maximum richness in species if the host is infinitely exposed to new environmental conditions (Carrier and Reitzel, 2017).

In summary, the environment and all of its forces can contribute to temporary changes in the community composition, as long as there is exposure to new symbionts and conditions. The exposure is necessary to make a shift in any direction, although we will see that the host and the microbial community itself might hamper the change, in some cases.

6.4 *How the host genotype (G) affects the microbiome*

Quantitative genetics has studied the impact that the host genetic variants have on the determination of phenotypic traits of the host (equation 1). When the composition of the gut microbial community is considered as a phenotype, its study under quantitative genetics terms is straightforward (equation 3).

In quantitative genetics, we imply a direct causal effect of the host genetic variant on the phenotype. Therefore, we are about to make the same assumption here: whichever the host gene and its variant, this will have a direct impact on the presence and abundance of each of the microbes that can potentially colonize the gastro-intestinal tract.

Examples of this strong causal association have been suggested in humans, specifically between the variation at the *LCT* gene in the host and *Bifidobacterium* abundance in the gut microbial community and between the variation at the *AMY-1* and *AMY-2* genes and the *Ruminococcus*/*Butyrivibrio*/*Roseburia* abundance.

The LCT gene is known for its involvement in the digestion of lactose, encoding for the lactase enzyme in the mammal's small intestine enterocytes (Troelsen *et al.*, 1997; Bonder *et al.*, 2016; Suzuki *et al.*, 2020). When the lactase-encoding variant of the *LCT* gene is present, lactose is digested by the host-produced enzymes before it can arrive to the large intestines. The lack of substrate for *Bifidobacterium* limits its possibility to multiplicate. Therefore, we can infer that it is the enzymatic product of the different *LCT* variants to control *Bifidobacterium*'s abundance. On top of this, Bonder *et al.* (2016) have also suggested a 'gene by diet' interaction: it would be also the assumption of dairy products to regulate *Bifidobacterium*'s abundance, with the lactase-producing host genotype holding the *Bifidobacterium*'s abundance constant regardless of the intake in dietary products, while the lactase-deficient host genotype would see an increase in *Bifidobacterium*'s abundance when the intake of dairy products increased.

The *AMY-1* and *AMY-2* genes are known for their involvement in the digestion of starch, which has a pivotal role in livestock production. The digestion of the starch starts with the salivary amylase, which is encoded by the *AMY-1* gene in the salivary glands and the continues in the small intestines thanks to the pancreatic amylase encoded by the *AMY-2* gene (Walter and Ley, 2011; Suzuki *et al.*, 2020). Since it can potentially start in the mouth and continue before its arrival to the large intestine, this mechanism has the potential to digest most of the starch in the diet. Whatever amount of starch is undigested by the time of its arrival to the colon, an assembly of *Bacteria* carrying the amylase gene would provide to its degradation to glucose and dextrins. Again, the abundance of the *Bacteria* carrying the amylase (*e.g. Ruminococcus*, *Butyrivibrio*, *Roseburia*) would be determined by the host genes

AMY-1 and *AMY-2*, probably in interaction with the diet (the more starch ingested, the more starch to degrade).

These cases were discussed as an example of direct host control on the gut microbial community composition. Other, more complex cases are probably waiting to be described. One important component of this examples is how the host is, ultimately, controlling the abundance but also addition of microbes to the community by exerting selection on the community composition in order to maintain only the beneficial functions of the microbiome (Benson *et al.*, 2010). For example, an individual might be born carrying the *LCT* gene from a mother that carried the *LCT* gene, both of which did not need the presence of *Bifidobacterium* as a symbiont. However, this species could be present in the environment and the individual could, eventually, get in contact with it as it was pictured in figure 18.3. Without the need of *Bifidobacterium* to join the gut community, the host would not establish such symbiosis. This case would be pictured in figure 18.3 as the individual that does not reach the maximum number of species even when maximum exposure is reached, proving that the host genotype would also limit the overall diversity and complexity of the microbial community. In terms of the holo-biont, such host would show smaller variability on the microbial side but larger variability on the host side because of the *LCT* gene being carried. Which suggest that the different holo-bionts can reach the same level of overall diversity and functionality on either side of the symbiosis. Equal lactose digestibility could be reached using host or microbial genes. The Doolittle and Booth (2017) concept for which 'it's the song, not the singer' perfectly applies here.

7 Host and commensal community co-diversify in front of selective pressure

The presence of the host genes capable of digesting starch was probably dictated by a shift in diet with the advent of agriculture in the Neolithic age (Walter and Ley, 2011). Such shift can again be represented in figure 18.2, with the addition of starch being a shift towards the right side of the graph (in fact, adding to the diet something that can not be digested could be seen as a form of stress). Before the adoption of an agriculture-enabled diet we could assume little necessity of the host genes *AMY-1*, *AMY-2* and *LCT* or the starch/lactose degrading bacteria. With the addition of new molecules to the diet, the need for new enzymes for digesting such molecules is needed. Since the horizontal acquisition of symbionts is a faster way to develop new functions (Rosemberg, 2010; Alberdi *et al.*, 2016; Hurst, 2017), the microbes that are now widespread symbionts were probably acquired at the time of diet shift. Later on, the host might have developed the ability to synthesize the necessary enzymes (amylase, lactase) making unnecessary the presence of the symbionts. Actually, the presence of the symbionts could have been disadvantageous as compared to the presence and number of copies of the host gene. The digestion through the host-produces enzymes starts earlier in the gastro-intestinal tract and doesn't require microbial cell proliferation, allowing to extract more energy from the diet. This would have provided better fitness to the hosts that carried the *AMY-1*/*AMY-2*/*LCT* genes as compared to those that took advantage of the microbes, leading to an increase in the frequency of these genes in the population. The rise in frequency of the host genes would have brought to a decrease in the associated bacterial species just mentioned, creating a co-diversification within the human species. As showed by Suzuki *et al.* (2022), species from the *Firmicutes* phylum and from the *Bifidobacterium* genus showed more evidence of co-phylogeny with human diversification.

A change in environmental exposure to a diet rich in starch or lactose would have applied some selective pressure, with response happening first in the commensal bacterial community and later in the host gene frequency. It should be noted that often the evolution at the level of the microbial community brings to a reduction in its diversity (Carrier and Reitzel, 2017; Henry and Ayroles, 2021), since there is a selection for the useful symbionts and, consequently, against those that do not provide the necessary function. Overall, this could be an example of selection and evolution at the level or the holo-biont, which encompasses both host's and symbiont's genes.

8 How the host genes control microbiome: examples in livestock

The influence of the host on microbial composition, in livestock, has been first investigated in ruminants. This should not be surprising, since the rumen, the 'quintessential bioreactor', is responsible for the conversion of non-digestible plant material into nutrients the animal and, ultimately, by humans. Weimer and colleagues, (2010), after observing a restoration of the original microbiome features following an exchange of ruminant content between cows, were the first ones to postulate a potential role of the host influence on the rumen microbial composition. In the following decade, several studies have linked host genetics to rumen microbial composition in relation to a variety of performance and welfare traits. Heritable sets of rumen microbial features linked to animal production have been identified by Wallace *et al.* (2019): the authors identified a small number of host-determined, heritable microbes contributing to host phenotypes and proposed host microbiome breeding strategies as an efficient way to increase efficiency and lower emissions from ruminant livestock. Similar results were presented by Li and colleagues (2019), who identified heritable rumen microbial features associated with host feed efficiency traits and rumen metabolic measures. The authors also identified several genomic regions (as marked by single nucleotide polymorphisms variants) associated with rumen microbial taxa, and with known quantitative trait loci for feed efficiency in cattle. Microbiota composition has also been reported to be heritable among others by Saborío-Montero and colleagues (2020), Roehe *et al.* (2016), and Martínez-Álvaro *et al.* (2020, 2022). Genome-wide association studies searching for causal polymorphisms explaining ruminal microbial abundances have identified SNPs associated with muscle contraction and passage rate (Zhang *et al.*, 2020), absorption of nutrients by the rumen epithelium (Li *et al.*, 2019), immune-related signaling pathways, tight junction of epithelial cells (Fan *et al.*, 2021), mucin-encoding genes critical for gut mucosal health (Fan *et al.*, 2019), and cell division and cell cycle (Cardinale and Kadarmideen, 2022).

Similarly, monogastric species have a growing body of evidence linking the host genomic make-up to microbial composition. Camarinha-silva *et al.* (2017) were among the first to suggest a direct link between swine genetic composition and microbial abundance in swine. Lu *et al.* (2018) estimated a genetic component in the alpha-diversity of pigs with estimates in the 0.2 to 0.4 range. Bergamaschi *et al*, (2020a) identified genetic differences in microbial composition associated with swine breeds selected primarily for fast growth and breeds primarily selected for reproductive efficiency. Similarly, Aliakbari *et al.* (2021) evaluated the genetic relationship between fecal microbial composition and feed efficiency, and production traits finding significant genetic correlations between genera of the *Lachnospiraceae*, *Ruminococcaceae*, *Prevotellaceae*, *Lactobacillaceae*, *Streptococcaceae*, *Rikenellaceae* and *Desulfovibrionaceae* families and feed efficiency traits. Ramayo-Caldas *et al.* (2020) estimated alpha-diversity heritability in pigs to range from 0.15 to 0.28 and identified host genomic regions associated with microbial abundance. In the same fashion, Bergamaschi *et al.* (2020b), using fecal microbial samples from a large commercial swine population, estimated heritability values ranging from 0.05 to 0.50 for several microbial features. Furthermore, they identified several genomic regions associated with microbial abundance, these regions appeared to co-locate in proximity to known quantitative trait loci (QTL) related to growth and carcass composition feed efficiency as well as health-related traits. Tiezzi *et al*, (2021), using structural equation modeling, identified host genomic variants affecting fat deposition through microbiome-mediated effects. These results were similar to those obtained by Wang *et al*, (2022), who found that fat deposition and intramuscular fat content were significantly influenced by host genetics–gut microbiome interactions. Finally, Reverter *et al.* (2021) identified a gene co-association network regulating gut microbial communities in pigs.

While the majority of these studies have focused on the link between the gut microbiome and growth performance parameters, given the direct link of the microbiome to metabolism and absorption of fiber-rich sources of feed in the intestine, other studies have also linked microbiome variability to traits related to immune functions (Estelle *et al.*, 2014) and sow prolificacy traits (Sanglard *et al.*, 2020). All these results point to a substantial contribution of the host to microbial composition. Weishaar *et al.* (2020) and Maltecca *et al.* (2020) suggested selecting the intersection between genome and microbiome to improve energy efficiency in swine.

Similar studies have been published in poultry (Zhao *et al.*, 2013; Ji *et al.*, 2020). For all these studies, 5% to 15% of genera in the duodenum, jejunum, ileum, cecum, and feces had significant heritability estimates, with most genera being from the *Firmicutes* and *Proteobacteria phyla* (Ryu and Davenport, 2022). Haas *et al.* (2022) identified microbiota as a mediator between the host genome and phosphorus utilization and other efficiency traits in Japanese quail.

9 Applying the breeder's equation to the holo-biont

Selective breeding is based on the control of reproduction of the individuals. Specifically, the individuals with transmissible beneficial characteristics are to be selected, so that those beneficial characteristics can be found with higher frequency in the next generation of individuals.

At this point, some questions come straightforward:

1. Can we select the holo-biont as a whole? Can we leverage on both the host and symbiont's genes so that we can transfer the of the beneficial characteristics to the offspring?
2. If yes, to what extent can we expect the genome-metagenome selection to be shaped in a favorable way?

And additionally:

3. Are there any caveats?

9.1 *Selection of the holo-biont*

Considering all of what we listed in the previous parts of this chapter, we support the hypothesis that selective breeding can be extended to the holo-biont, such that the both the host genomic variants and the gut microbial composition can be used to rank and select the most desirable individuals. Consequently, the host genome and gut metagenome can be shaped by having one 'individual' being selected rather than another. Such hypothesis is not ours, but it has been well described by Roughgarden *et al.* (2018). The individual, being selected for some desirable property, appears as the 'unit of selection', meaning that its reproductive success depends on both genomes (host and microbes) altogether. With this, we're not arguing that the holo-biont has a fixed composition, but simply that it is functionally stable enough to be selected, or not.

Considering individuals in figure 18.1, individual B should be selected, since it shows the largest '$G + M$' contribution for the trait considered. Individual B has an advantage in being selected as compared to individual C. Their host genomic value (G) is the same, but the microbial contribution is different to the extent of giving individual B a strong selective advantage. Since we assume the Doolitte and Booth's

(2017) concept of 'it's the song, not the singer' to apply, we would choose individual B over individuals C and A.

By selecting individual B, variants from both genomes will be transferred to the next generation, to some extent. Following the Zilber-Rosenberg and Rosenberg (2008) claim 'the holo-genome is transmitted from one generation to the next with reasonable accuracy', we postulate that part of the advantage found in individual B will be found in its offspring. Please note that the transfer of the genomes, in this sense, is not bound to the Mendelian transmission of allelic variants but could also happen by vertical transmission or by a combination of both, like a host-controlled horizontal transmission (*e.g.* the ability of the host genotype *LCT* to control for the *Bifidobacterium* abundance).

Once the preferred individual has been selected, to what extent can we expect the performance of our selected population to change?

9.2 *The genetic progress*

We will forget about the contribution of *M*, for now.

When genetic selection is performed, we can expect some genetic progress to be realized. Genetic progress is meant as the gain in some phenotypic trait of interest *P* given the selection applied and is measured on the same scale of the phenotypic traits on a per-generation basis. The most common formula used is defined such as:

$$\Delta_{G,P} = \sigma_P * h_P^2 * i \tag{4}$$

Where $\Delta_{G,P}$ is the genetic progress measurable on trait *P*, σ_P is the standard deviation in the phenotype *P*, h_P^2 is the heritability of trait *P*, *i* is the selection intensity. The selection intensity is to be meant as a measure of selective pressure: the higher the intensity, the stricter the requirements for an individual to be selected, the smaller the number of individuals that can potentially meet the requirements. Since we're just considering h_P^2, only the host genome contributes to the determination of *P*, as reported in formula 1.

With the holo-biont encompassing the microbial contribution as well, we need *M* to be brought back for consideration. First, we can shift the focus from *P* to *M*, same as it was down moving from formula 1 to formula 3, *i.e.* considering the microbial composition as trait of the individual itself. Therefore:

$$\Delta_{G,M} = \sigma_M * h_M^2 * i \tag{5}$$

the change due to host selection in gut microbiome composition is quantified by $\Delta_{G,M}$, σ_M is the overall variation in microbial composition (the sum of the variation in all species abundance), h_M^2 is the heritability of the microbial composition and *i* is the intensity of selection applied. To the extent that the gut microbiome is controlled by the host genes, we could assume that selecting the host will shape the microbial community. Again, an example is the strong relationship between the host *LCT* gene and the *Bifidobacterium* in the large intestine (Walter and Ley, 2011): an increase of the frequency in the *LCT* gene in the population will bring to a decrease in the *Bifidobacterium* abundance in the large intestine microflora of the population.

However, in figure 18.1 we admitted individual B to be selected for the contribution from both the *G* and *M* components. So far, we've only described the impact of selection impacting the phenotype *P* and microbial composition *M* due to the change in the host genetic component *G*, but we lack an understanding on how the holo-biont will change.

We could consider the three conditions posed by Bordenstein and Theis (2015) about the formation of the holo-biont to infer the change due to host selection in microbiota composition.

By incorporating these three conditions to formulas 4 and 5:

$$\Delta_{GM,P} = (\sigma_P * h_P^2 + m_P^2 * \sigma_M * h_M^2) * i \tag{6}$$

Hence, the change in the holo-biont phenotype *P* is quantified by $\Delta_{GM,P}$, since the holo-biont includes both the *G* and *M* components; $\sigma_P * h_P^2$ accounts for the progress obtained by shaping the host genome as in formula 4, $m_P^2 * \sigma_M * h_M^2$ accounts for the progress obtained by shaping the gut microbial community, with m_P^2 measuring the relationship between the variation in the microbial community and the phenotype of the holo-biont; *i* is the selection intensity.

The first condition by Bordenstein and Theis (2015), *i.e.* the presence of abundant and diverse microbiota, relates to the variability in microbiota, here accounted for by σ_M. While abundance and diversity do not necessarily imply variability in abundance between individuals, we could assume that a complex microbial community will show some degree of variation among holo-bionts. The second condition, *i.e.* the fidelity of transmission of the microbiota from parent to offspring, is here represented by h_M^2. This inheritance is not to be meant in strict Mendelian terms but will be revisited in the next paragraph. The third condition, *i.e.* the impact of the symbiotic microbiota on the host fitness, is here represented by m_P^2. This parameter was introduced earlier in this chapter, being names as the microbiability of a trait, which is a measure of the impact of the microbial variability on the holo-biont phenotypic variability.

Please note that the part $m_P^2 * \sigma_M * h_M^2$ could be seen in two different ways. On one side, we could see it as the microbiability (m_P^2) that translates the genetic change in terms of microbial community ($\sigma_M * h_M^2$) to changes in the phenotype. Selection might alter the community in different components of it, but not all these components necessarily impact the phenotype, either meant as its performance, gut health and, ultimately, its fitness. On the other hand, we could see it as the overall impact of the microbial variation on the phenotype of the holo-biont

$(m_P^2 * \sigma_M)$, which is being 'filtered' by the inheritance of the microbial variation (h_M^2) in its contribution to the genetic progress at the holo-biont level. This dissection of the microbial contribution to the phenotype as partially regulated by the host: it is also pivotal in selection, since the contribution of the microbial community to the genetic variation in the phenotype is limited to the extent that the host genome can control the community itself. As an example, the contribution of *Bifidobacterium* to the digestion of lactose is limited to the extent that the host *LCT* variants allow it. These complex interactions have been modeled using multi-path analysis, for this we suggest the reading of Weishaar *et al.* (2020) and Tiezzi *et al.* (2021).

Finally, the selection intensity i is here included as a parameter that affect the host and microbial genomes equally: simply put, the holo-biont is to be selected as a whole and it therefore undergoing the same selective pressure.

9.3 *Caveats*

In the previous paragraph we defined h_M^2 as the degree of inheritance of the microbial composition. While such inheritance is meant in mere Mendelian terms when discussing the phenotype (*i.e.* h_P^2), this may be not so straightforward when we discuss the microbiome. Likewise, the environmental components (E) exerted on either the phenotype or the microbiome might have to be interpreted in profoundly different ways, as it was more than once mentioned in this chapter.

Looking that figures 18.2 and 18.3, we must bear in mind that vertical and horizontal transmission might be strong components of inheritance.

The host will acquire half of its genes (those located on the chromosome of the host cell) from the parents. Also, the host will likely acquire part of the microbial community from its mother by vertical transmission (the D component in formula 3). These components are strongly collinear also in controlled conditions such as those that apply to swine and poultry. Sows will vertically transmit microbes to piglets though the vaginal canal, the udder skin, and the milk. Hens will lay eggs with the likely contaminated by fecal bacteria. Even under the most thorough design of an experiment, the separation of the Mendelian contribution and the vertical transmission could be virtually impossible.

The host will be exposed to many microbes during its life (figure 18.3) and the acquisition of those as symbionts might depend on actual benefits provided (figure 18.2). The acquisition of new symbionts might be dictated by the host genes (*e.g.* the *LCT* variants), the presence of competitive symbionts already established, but also by the actual chances of encountering such microbe, which ultimately depend on exposure. The composition of the microbial community will climax to full complexity towards the advanced life of the individual (figure 18.3). First and foremost, the parameter that will be affected is σ_M, with reduced variation in case of limited exposure, remarkable vertical transmission from the mother, and presence of host genetic variants that control the colonization of the gut (*e.g. LCT*). While all of these should be properly accounted for in formula 6, that goes beyond the purpose of this chapter. In general, we should consider that the microbial composition is an overly complex trait, with some of the factors that shape it working in interaction with each other.

We expressed the genetic progress in the holo-biont as a simple measure ($\Delta_{GM,P}$), likely to stay the same over the generations. However, Selection itself will likely alter the parameters entering the formula, such that the response at the first generation of selection might differ from what is coming later. For example, when selection was applied to a population of *Drosophila*, the microbial composition was seen to response sooner and faster, with the host genomic variants to be altered later in the generation of selection (Henry and Ayroles, 2021). This suggest the contribution of the factors listed in formula 6 to be heterogeneous over the generations, with multiple forces acting to maintain the homeostasis of the holo-biont, to some extent.

References

Alberdi, A., Aizpurua, O., Bohmann, K., Zepeda-Mendoza, M.L., & Gilbert, M.T.P. (2016). Do vertebrate gut metagenomes confer rapid ecological adaptation? Trends in ecology & evolution, 31(9), 689–699.

Aliakbari, A., Zemb, O., Billon, Y., Barilly, C., Ahn, I., Riquet, J., & Gilbert, H. (2021). Genetic relationships between feed efficiency and gut microbiome in pig lines selected for residual feed intake. *Journal of animal breeding and genetics*, *138*(4), 491–507.

Baldauf, S.L. (2003). The deep roots of eukaryotes. Science, 300(5626), 1703–1706.

Bencina D., Tadina T., Dorrer D. (1988) Natural infection of ducks with Mycoplasma synoviae and Mycoplasma gallisepticum and Mycoplasma egg transmission. Avian Pathol 17: 441,449.

Benson, A.K., Kelly, S.A., Legge, R., Ma, F., Low, S.J., Kim, J., ... & Pomp, D. (2010). Individuality in gut microbiota composition is a complex polygenic trait shaped by multiple environmental and host genetic factors. Proceedings of the National Academy of Sciences, 107(44), 18933–18938.

Bergamaschi, M., Maltecca, C., Schillebeeckx, C., McNulty, N.P., Schwab, C., Shull, C., ... & Tiezzi, F. (2020). Heritability and genome-wide association of swine gut microbiome features with growth and fatness parameters. *Scientific reports*, *10*(1), 1–12.

Bergamaschi, M., Tiezzi, F., Howard, J., Huang, Y.J., Gray, K.A., Schillebeeckx, C., ... & Maltecca, C. (2020). Gut microbiome composition differences among breeds impact feed efficiency in swine. *Microbiome*, *8*(1), 1–15.

Bonder, M.J., Kurilshikov, A., Tigchelaar, E.F., Mujagic, Z., Imhann, F., Vila, A.V., ... & Zhernakova, A. (2016). The effect of host genetics on the gut microbiome. Nature genetics, 48(11), 1407–1412.

Bordenstein S.R., Theis K.R. (2015) Host Biology in Light of the Microbiome: Ten Principles of Holobionts and Hologenomes. PLoS Biol 13(8): e1002226. doi:10.1371/journal.pbio.1002226.

Brucker R.M., Bordenstein S.R. 2013 The hologenomic basis of speciation: gut bacteria cause hybrid lethality in the genus Nasonia. Science 341,667 ,Äì 669. (doi:10.1126/science.1240659).

Brucker, R.M., & Bordenstein, S.R. (2013). The capacious hologenome. Zoology, 116(5), 260–261.

Camarinha-Silva, A., Maushammer, M., Wellmann, R., Vital, M., Preuss, S., & Bennewitz, J. (2017). Host genome influence on gut microbial composition and microbial prediction of complex traits in pigs. *Genetics, 206*(3), 1637–1644.

Cardinale, S., & Kadarmideen, H.N. (2022). Host Genome–Metagenome Analyses Using Combinatorial Network Methods Reveal Key Metagenomic and Host Genetic Features for Methane Emission and Feed Efficiency in Cattle. *Frontiers in genetics, 13*.

Carrier T.J. and Reitzel A.M. (2017) The Hologenome Across Environments and the Implications of a Host-Associated Microbial Repertoire. Front. Microbiol. 8:802. doi: 10.3389/fmicb.2017.00802.

Carthey, A.J., Gillings, M.R., & Blumstein, D.T. (2018). The extended genotype: microbially mediated olfactory communication. Trends in Ecology & Evolution, 33(11), 885–894.

Dehority, B.A. (2003) Rumen microbiology. Nottingham University Press, Nottingham, UK. Environ Microbiol 72: 1027–1033.

Difford, G.F., Plichta, D.R., Løvendahl, P., Lassen, J., Noel, S.J., Højberg, O., ... & Sahana, G. (2018). Host genetics and the rumen microbiome jointly associate with methane emissions in dairy cows. PLoS genetics, 14(10), e1007580.

Donia M.S., Cimermancic P., Schulze C.J., Fischbach M.A. A systematic analysis of biosynthetic gene clusters in the human microbiome reveals a common family of antibiotics. Cell. 2014;158:1402,Äì1414.76.

Doolittle W.F., Booth A. 2017 I's the song, not the singer: an exploration of holobiosis and evolutionary theory. Biol. Philos. 32, 5 ,Äì 24. (doi:10.1007/s10539-016-9542-2).

Estellé, J., Mach, N., Ramayo-Caldas, Y., Levenez, F., Lemonnier, G., Denis, C., ... & Rogel-Gaillard, C. (2014, August). The influence of host's genetics on the gut microbiota composition in pigs and its links with immunity traits. In *10th World Congress of Genetics Applied to Livestock Production, Vancouver, BC, Canada*.

Funkhouser L.J., Bordenstein S.R. (2013) Mom Knows Best: The Universality of Maternal Microbial Transmission. PLoS Biol 11(8): e1001631. doi:10.1371/journal.pbio.1001631.

Gantois I., Ducatelle R., Pasmans F., Haeseb-rouck F., Gast R., *et al.* (2009) Mechanisms of egg contamination by Salmonella Enteritidis. FEMS Microbiol Rev 33: 718,Äì738.

Gilbert, S.F. (2014). A holobiont birth narrative: the epigenetic transmission of the human microbiome. Frontiers in genetics, 5, 282.

Haas, V., Vollmar, S., Preuß, S., Rodehutscord, M., Camarinha-Silva, A., & Bennewitz, J. (2022). Composition of the ileum microbiota is a mediator between the host genome and phosphorus utilization and other efficiency traits in Japanese quail (*Coturnix japonica*). *Genetics Selection Evolution, 54*(1), 1–13.

Hehemann, J.H., Correc, G., Barbeyron, T., Helbert, W., Czjzek, M., and Michel, G. (2010). Transfer of carbohydrate-active enzymes from marine bacteria to Japanese gut microbiota. Nature 464, 908,Äì912. doi: 10.1038/nature 08937.

Henry, L.P., & Ayroles, J.F. (2021). Meta-analysis suggests the microbiome responds to Evolve and Resequence experiments in Drosophila melanogaster. BMC microbiology, 21(1), 1–14.

Henry, L.P., Bruijning, M., Forsberg, S.K.G. *et al.* The microbiome extends host evolutionary potential. Nat Commun 12, 5141 (2021). https://doi.org/10.1038/s41467-021-25315-x.

Hurst G.D.D. 2017 Extended genomes: symbiosis and evolution. Interface Focus 7: 20170001. http://dx.doi.org/10.1098/rsfs.2017.0001.

Ji J., Xu Y., Luo C., He Y., Xu X., *et al.* 2020. Effects of the DMRT1 genotype on the body weight and gut microbiota in the broiler chicken. Poultry Sci. 99(8):4044–51.

Khanal, P., Maltecca, C., Schwab, C., Fix, J., & Tiezzi, F. (2021). Microbiability of meat quality and carcass composition traits in swine. *Journal of Animal Breeding and Genetics, 138*(2), 223–236.

King, K.C., Brockhurst, M.A., Vasieva, O., Paterson, S., Betts, A., Ford, S.A., ... & Hurst, G.D. (2016). Rapid evolution of microbe-mediated protection against pathogens in a worm host. The ISME journal, 10(8), 1915–1924.

Kohl, K.D., and Dearing, M.D. (2012). Experience matters: prior exposure to plant toxins enhances diversity of gut microbes in herbivores. Ecol. Lett. 15, 1008,Äì1015. doi: 10.1111/j.1461-0248.2012.01822.x.

Kohl, K.D., and Dearing, M.D. (2014). Wild-caught rodents retain a majority of their natural gut microbiota upon entrance into captivity. Environ. Microbiol. Rep. 6, 191,Äì195. doi: 10.1111/1758-2229.12118.

Kohl, K.D. *et al.* (2014) Unique and shared responses of the gut microbiota to prolonged fasting: a comparative study across five classes of vertebrate hosts. FEMS Microbiol. Ecol. 90, 883,894.

Lu, D., Tiezzi, F., Schillebeeckx, C., McNulty, N.P., Schwab, C., Shull, C., & Maltecca, C. (2018). Host contributes to longitudinal diversity of fecal microbiota in swine selected for lean growth. *Microbiome, 6*(1), 1–15.

MacOwan K.J., Atkinson M.J., Bell M.A., Brand T.F., Randall C.J. (1984) Egg transmission of a respiratory isolate of Mycoplasma synoviae and infection of the chicken embryo. Avian Pathol 13: 51,Äì58.

Maltecca, C., Bergamaschi, M., & Tiezzi, F. (2020). The interaction between microbiome and pig efficiency: A review. *Journal of Animal Breeding and Genetics, 137*(1), 4–13.

Maltecca, C., Dunn, R., He, Y., McNulty, N.P., Schillebeeckx, C., Schwab, C., ... & Tiezzi, F. (2021). Microbial composition differs between production systems and is associated with growth performance and carcass quality in pigs. Animal microbiome, 3(1), 1–20.

Martínez-Álvaro, M., Auffret, M.D., Stewart, R.D., Dewhurst, R.J., Duthie, C.A., Rooke, J.A., ... & Roehe, R. (2020). Identification of complex rumen microbiome interaction within diverse functional niches as mechanisms affecting the variation of methane emissions in bovine. *Frontiers in microbiology, 11*, 659.

Martínez-Álvaro, M., Mattock, J., Auffret, M., Weng, Z., Duthie, C.A., Dewhurst, R.J., ... & Roehe, R. (2022). Microbiome-driven breeding strategy potentially improves beef fatty acid profile benefiting human health and reduces methane emissions. *Microbiome, 10*(1), 1–21.

Michelland, R.J., Monteils, V., Combes, S., Cauquil, L., Gidenne, T., & Fortun-Lamothe, L. (2010). Comparison of the archaeal community in the fermentative compartment and faeces of the cow and the rabbit. Anaerobe, 16(4), 396–401.

Moeller A.H. *et al.* 2016 Cospeciation of gut microbiota with hominids. Science 353, 380 ‚Äì 382. (doi:10.1126/science.aaf3951).

Moran, N.A., McCutcheon, J.P., & Nakabachi, A. (2008). Genomics and evolution of heritable bacterial symbionts. Annual review of genetics, 42, 165–190.

Ochman H., Worobey M., Kuo C.-H., *et al.* Evolutionary relationships of wild hominids recapitulated by gut microbial communities. PLoS Biol. 2010;8: e1000546.

Portier, P. (1918). Les symbiotes. Masson.

Ramayo-Caldas, Y., Prenafeta-Boldú, F., Zingaretti, L.M., Gonzalez-Rodriguez, O., Dalmau, A., Quintanilla, R., & Ballester, M. (2020). Gut eukaryotic communities in pigs: diversity, composition and host genetics contribution. Animal microbiome, 2(1), 1–12.

Reverter, A., Ballester, M., Alexandre, P.A., Mármol-Sánchez, E., Dalmau, A., Quintanilla, R., & Ramayo-Caldas, Y. (2021). A gene co-association network regulating gut microbial communities in a Duroc pig population. *Microbiome*, *9*(1), 1–16.

Roehe R., Dewhurst R.J., Duthie C.-A., Rooke J.A., McKain N., Ross D.W., *et al.* (2016) Bovine Host Genetic Variation Influences Rumen Microbial Methane Production with Best Selection Criterion for Low Methane Emitting and Efficiently Feed Converting Hosts Based on Metagenomic Gene Abundance. PLoS Genet 12(2): e1005846.

Rosenberg, E., Sharon, G., Atad, I., & Zilber-Rosenberg, I. (2010). The evolution of animals and plants via symbiosis with microorganisms. Environmental microbiology reports, 2(4), 500–506.

Roughgarden, J., Gilbert, S.F., Rosenberg, E., Zilber-Rosenberg, I., & Lloyd, E.A. (2018). Holobionts as units of selection and a model of their population dynamics and evolution. Biological Theory, 13(1), 44–65.

Ryu, E.P., & Davenport, E.R. (2022). Host Genetic Determinants of the Microbiome Across Animals: From Caenorhabditis elegans to Cattle. *Annual Review of Animal Biosciences*, *10*, 203–226.

Saborío-Montero, A., López-García, A., Gutiérrez-Rivas, M., Atxaerandio, R., Goiri, I., García-Rodriguez, A., ... & González-Recio, O. (2021). A dimensional reduction approach to modulate the core ruminal microbiome associated with methane emissions via selective breeding. *Journal of Dairy Science*, *104*(7), 8135–8151.

Saborío-Montero A, *et al.* Structural equation models to disentangle the biological relationship between microbiota and complex traits: Methane production in dairy cattle as a case of study. J Anim Breed Genet. 2020;137:36–48.

Sanglard, L.P., Schmitz-Esser, S., Gray, K.A., Linhares, D.C.L., Yeoman, C.J., Dekkers, J.C.M., ... & Serão, N.V.L. (2020). Vaginal microbiota diverges in sows with low and high reproductive performance after porcine reproductive and respiratory syndrome vaccination. *Scientific reports*, *10*(1), 1–11.

Shapira, M. (2016). Gut microbiotas and host evolution: scaling up symbiosis. Trends in ecology & evolution, 31(7), 539–549.

Spor, A., Koren, O., & Ley, R. (2011). Unravelling the effects of the environment and host genotype on the gut microbiome. Nature Reviews Microbiology, 9(4), 279–290.

Sudakaran, S., Kost, C., & Kaltenpoth, M. (2017). Symbiont acquisition and replacement as a source of ecological innovation. Trends in Microbiology, 25(5), 375–390.

Suzuki, T.A. (2017). Links between natural variation in the microbiome and host fitness in wild mammals. Integrative and comparative biology, 57(4), 756–769.

Suzuki, T.A., & Ley, R.E. (2020). The role of the microbiota in human genetic adaptation. Science, 370(6521), eaaz6827.

Suzuki, T.A., Fitzstevens, J.L., Schmidt, V.T., Enav, H., Huus, K.E., Mbong Ngwese, M., ... & Ley, R.E. (2022). Codiversification of gut microbiota with humans. Science, 377(6612), 1328–1332.

Tiezzi, F., Fix, J., Schwab, C., Shull, C., & Maltecca, C. (2021). Gut microbiome mediates host genomic effects on phenotypes: a case study with fat deposition in pigs. Computational and structural biotechnology journal, 19, 530–544.

Troelsen, J.T., Mitchelmore, C., Spodsberg, N., Jensen, A.M., Noren, O., & Sj√∂str√∂m, H. (1997). Regulation of lactase,Äìphlorizin hydrolase gene expression by the caudal-related homoeodomain protein Cdx-2. Biochemical Journal, 322(3), 833–838.

Walker, A.W., Ince, J., Duncan, S.H., Webster, L.M., Holtrop, G., Ze, X., ... & Flint, H.J. (2011). Dominant and diet-responsive groups of bacteria within the human colonic microbiota. The ISME journal, 5(2), 220–230.

Wallace, R.J., Sasson, G., Garnsworthy, P.C., Tapio, I., Gregson, E., Bani, P., ... & Mizrahi, I. (2019). A heritable subset of the core rumen microbiome dictates dairy cow productivity and emissions. *Science advances*, *5*(7), eaav8391.

Walter, J., & Ley, R. (2011). The human gut microbiome: ecology and recent evolutionary changes.

Wang, Y., Zhou, P., Zhou, X., Fu, M., Wang, T., Liu, Z., ... & Liu, B. (2022). Effect of host genetics and gut microbiome on fat deposition traits in pigs. *Frontiers in Microbiology*, *13*.

Weishaar, R., Wellmann, R., Camarinha-Silva, A., Rodehutscord, M., & Bennewitz, J. (2020). Selecting the hologenome to breed for an improved feed efficiency in pigs – A novel selection index. *Journal of Animal Breeding and Genetics*, *137*(1), 14–22.

Wright S. Evolution in Mendelian populations. Genetics. 1931; 16(2):0097,Äì159.

Yang, L., Wang, G., Zhou, J., Yang, Y., Pan, H., Zeng, X., Qiao, S. Exploration of the Potential for Efficient Fiber Degradation by Intestinal Microorganisms in Diqing Tibetan Pigs. Fermentation 2021, 7, 275. https://doi.org/10.3390/fermentation7040275.

Zhang, Q., Difford, G., Sahana, G., Løvendahl, P., Lassen, J., Lund, M.S., ... & Janss, L. (2020). Bayesian modeling reveals host genetics associated with rumen microbiota jointly influence methane emission in dairy cows. *The ISME journal*, *14*(8), 2019–2033.

Zhao L., Wang G., Siegel P., He C., Wang H., *et al.* 2013. Quantitative genetic background of the host influences gut microbiomes in chickens. Sci. Rep. 3:1163.

Zilber-Rosenberg, I., & Rosenberg, E. (2008). Role of microorganisms in the evolution of animals and plants: the hologenome theory of evolution. FEMS microbiology reviews, 32(5), 723–735.

Printed in the United States
by Baker & Taylor Publisher Services